Kyoji Sassa
Hiroshi Fukuoka
Fawu Wang
Gonghui Wang
(Editors)

Progress in Landslide Science

Kyoji Sassa
Hiroshi Fukuoka
Fawu Wang
Gonghui Wang
(Editors)

Progress in Landslide Science

With 431 Images, 349 in Color

Springer

Editors

Sassa, Kyoji

President of the International Consortium on Landslides
Research Centre on Landslides, Disaster Prevention Research Institute,
Kyoto University, Uji, Kyoto 611-0011, Japan
Tel: +81-774-38-4110, Fax: +81-774-32-5597, E-mail: sassa@SCL.kyoto-u.ac.jp

Fukuoka, Hiroshi

Research Centre on Landslides, Disaster Prevention Research Institute,
Kyoto University, Uji, Kyoto 611-0011, Japan
Tel: +81-774-38-4111, Fax: +81-774-38-4300, E-mail: fukuoka@SCL.kyoto-u.ac.jp

Wang, Fawu

Research Centre on Landslides, Disaster Prevention Research Institute,
Kyoto University, Uji, Kyoto 611-0011, Japan
Tel: +81-774-38-4114, Fax: +81-774-38-4300, E-mail: wangfw@landslide.dpri.kyoto-u.ac.jp

Wang, Gonghui

Research Centre on Landslides, Disaster Prevention Research Institute,
Kyoto University, Uji, Kyoto 611-0011, Japan
Tel: +81-774-38-4114, Fax: +81-774-38-4300, E-mail: wanggh@landslide.dpri.kyoto-u.ac.jp

Library of Congress Control Number: 2007920433

ISBN 978-3-540-70964-0 **Springer Berlin Heidelberg New York**

Springer is a part of Springer Science+Business Media
springeronline.com

Cover Photo: Guinsaugon landslide, Southern Leyte, Philippines

Cover design: deblik, Berlin
Typesetting: Stasch · Bayreuth (stasch@stasch.com)
Production: Almas Schimmel

Printed on acid-free paper 32/3141/as – 5 4 3 2 1 0

Foreword for the Publication of "Progress in Landslide Science" from UNESCO

Natural disasters are increasing in terms of frequency, complexity, scope and destructive capacity. They have been particularly severe during the last few years when the world has experienced several large-scale natural disasters: the Indian Ocean earthquake and tsunami, Hurricane Katrina and Hurricane Rita; the Kashmir earthquake in Pakistan; floods and forest fires in Europe, India and China; and drought in Africa. Images of these events have shocked us all and will remain with us for a long time. Numerous landslides and mudflows have also occurred, causing deaths, injuries and material losses. The most recent tragic ones were the large-scale landslides which struck the Philippines in 2006, hitting the Albay province on 02 December and the Leyte Island on 17 February respectively, resulting in terrible loss of life, suffering and damage. National authorities and the international community, of course, should continue to provide the practical support needed by the affected communities. At the same time, it is important to quickly learn appropriate lessons that may help individuals, families, communities and whole societies to be better prepared for other disasters, whether caused by natural forces or otherwise.

The time has come for putting more emphasis on pre-disaster action rather than remaining content with post-disaster reaction. We must mobilize scientific knowledge and technological know-how to assess natural hazards and to strengthen disaster mitigation measures. We should promote a better understanding of natural disasters. We must promote and enforce sound scientific, engineering and construction principles. And we must promote education and public awareness about natural disaster reduction.

Landslides pose considerable risks to people's livelihoods and to the environment. They cause great disruption and economic losses by the destruction of infrastructure works such as roads and other communications and utility lines and of cultural heritage and the environment. Today there is a need more than ever before to address the problem of landslides in an integrated and internationally concerted way. These are the purposes of the International Consortium on Landslides (ICL) and the International Programme on Landslides (IPL). Both initiatives encompass research, education and capacity-building in landslide risk reduction. They both enjoy the participation and support of numerous international, governmental and non-governmental organizations and entities. They contribute to the International Strategy for Disaster Reduction (ISDR) and to the implementation of the Hyogo Framework for Action 2005–2015 which was adopted at the World Conference on Disaster Reduction held in Kobe, Japan in January 2005. The 2006 Tokyo Action Plan on Landslides, which was adopted during the Tokyo Round Table Discussion on Landslides in January 2006, provides a roadmap for strengthening international collaboration and identifying focus areas for reducing landslide risk worldwide.

UNESCO had the privilege to accompany from the very beginning the establishment of the ICL and the launching of the IPL. In so doing, the Organization enjoys partnership with a large number of stakeholders including the World Meteorological Organization (WMO), the United Nations University (UNU), the Food and Agriculture Organization (FAO) of the United Nations, the UN/ISDR Secretariat and the International Council for Science (ICSU) and its Unions. I am glad that the ICL and IPL have also been marked with the establishment of a UNITWIN Cooperation Programme on Landslide Risk Mitigation for Society and the Environment in the framework of the UNITWIN/UNESCO Chairs Programme, at Kyoto University (KU). This Programme is now hosted in the UNESCO-KU-ICL UNITWIN Headquarters at the Research Centre on Landslides of the Disaster Prevention Research Institute in Kyoto University. Furthermore UNESCO and ICL have established in August 2006 a Memorandum of Understanding for cooperation. Finally I am especially pleased that UNESCO will serve as co-organizer with ICL of the World Landslide Forum scheduled to take place in 2008. This Forum will constitute a milestone in our efforts to strengthen global risk preparedness.

The "Landslides" Journal of the International Consortium on Landslides plays a key role for the progress of landslide study as an integrated research field by putting together knowledge and technologies in many related fields of natural sciences, engineering, social sciences and culture. The present Publication "Progress in Landslide Science" comes in this context to provide an overview of the current status of this science. The diversity of subjects which are presented in this publication represents a rich collaborative work regarding landslides. I wish to commend the editors and the numerous authors involved in it. My particular greetings go to Professor Kyoji Sassa, Chairperson of ICL who continues to spare no effort in promoting ICL and IPL. It is with great pleasure that I praise the edition of this publication as a means of disseminating good knowledge in the area of landslide risk reduction.

Koïchiro Matsuura
Director-General of UNESCO

Foreword for the Publication of "Progress in Landslide Science" from UN/ISDR

A series of extremely high-profile disasters – the Indian Ocean tsunami of December in 2004, Atlantic hurricane season, the South Asian earthquake and the East African drought in 2005 underscored the importance of how better cooperation between Government authorities and the international community including scientific community would have played a critical role in helping people make life changing decisions about where and how they live before the disaster strikes, in particular high-risk urban areas.

Landslide, floods, drought, wildfire, storms, tsunami, earthquakes and other types of natural hazards are increasingly affecting the world. In the decade 1976–1985, close to billion people were affected by disasters. But by the most recent decade, 1996–2005, the decade total had more than doubled, to nearly two and a half billion people. In the last decade alone, disasters affected 3 billion people, killed over 750 000 people and cost around US$ 600 billion[1]. We cannot let this trend continue. Disaster risk concerns every person, every community, and every nation; indeed, disaster impacts are slowing down development, and their impact and actions in one region can have an impact on risks in another, and vice versa. Without taking into consideration the urgent need to reduce risk and vulnerability, the world simply cannot hope to move forward in its quest for sustainable development and reduction of poverty.

The *Hyogo Framework for Action 2005–2015: Building the Resilience of Nations and Communities to Disasters*, adopted at the World Conference on Disaster Reduction (WCDR, Kobe, Hyogo, Japan, in January 2005), represents the most comprehensive action-oriented policy guidance in universal understanding of disasters induced by vulnerability to natural hazards and reflects a solid commitment to implementation of an effective disaster reduction agenda. In order to ensure effective implementation of the Hyogo Framework at all levels, tangible activities must be carried out. For the last two years as post WCDR, we have seen many activities and initiatives developed to implement the Hyogo Framework in various areas. As a concrete activity in the area of landslide risk reduction, the International Programme on Landslides has maintained the momentum created at the WCDR and has been moving forward, led by the International Consortium on Landslides.

This publication is a valuable contribution to the implementation of the priority area 2 of the HFA – "Identify, assess and monitor disaster risks and enhance early warning", by gearing landslide risk assessment, both hazard identification, monitoring and vulnerability analysis, as well as preparedness and landslide risk management. The combination of landslide scientific knowledge and risk reduction measures are essential to reduce the impact of landslides. The Hyogo Framework calls for the international coordination and collaboration among different actors dealing with disaster risk reduction. In this sense, the Global Cooperation Platform for research and investigation for landslide risk reduction in the 2006 Tokyo Action Plan consist of very

[1] Data derived from the EM-DAT: The OFDA/CRED International Disaster Database, *www.en-dat.net*, Universite Catholiqué de Louvain, Brussels, Belgium.

important activities to promote the thematic coordination to research and reduce impacts of landslides. This initiative also contributes to the priority area 3 of the HFA, which emphasizes the importance of education and public awareness of the disaster risk reduction. Education and public awareness about the hazards, in this case landslide, are also key for people to be able to reduce risks and their vulnerabilities.

I welcome the work of the Global Cooperation Platform for research and investigation for landslide risk reduction and the International Consortium on Landslides. I look forward to collaboration with them, in particular through the International Programme on Landslides (IPL) and its Global Promotion Committee.

Sálvano Briceño
Director of United Nations Secretariat of the International Strategy for Disaster Reduction

UNESCO's Contribution to Landslide Risk Reduction

International scientific programmes provide a forum for an in-depth study of natural phenomena, their characteristics and their occurrence. This study is an essential prerequisite for a logical approach to the understanding of natural hazards. Furthermore the mitigation of risks arising from these hazards is based on the applied science and technology as well as on educational and information campaigns and programmes. The problem of risk reduction therefore requires a multidisciplinary approach, involving co-operation between specialists in several sectors of science, technology, education and culture.

The activities of the United Nations Educational, Scientific and Cultural Organization (UNESCO) in the study of natural disasters and the protection against them date from the beginning of the 1960s. Originally concerned with basic seismology, these activities were later extended to the reduction of earthquake hazards and still later to other categories of natural hazards including landslides and their socio-economic aspects. UNESCO has brought an interdisciplinary approach to the study of geohazards and the mitigation of their effects. Being at the crossroads of several sectors, UNESCO provides a unique intellectual setting, linking, within a single organization, the natural sciences with education, culture, communication and the social sciences. With its broad mandate and breadth of expertise, UNESCO is able to integrate many of the essential ingredients for disaster reduction.

The purposes of UNESCO in the field of disaster risk prevention can be described as follows:

- to promote a better understanding of the distribution in time and space of natural hazards and of their intensity;
- to help set up reliable early warning systems;
- to foster rational land use plans;
- to encourage the adoption of suitable building design;
- to provide policy advise on the protection of educational buildings and cultural monuments;
- to strengthen environmental protection for the prevention of natural disasters;
- to enhance preparedness and public awareness through education and training;
- and, when catastrophes do strike, to foster post-disaster investigation, recovery and rehabilitation.

Facing the increasing vulnerability of its Member States to natural hazards, UNESCO has constantly advocated that risk prevention policies, including warning systems related to natural hazards like landslides, must be established or improved. Hence the promotion of landslide risk reduction, environmental protection, and sustainable development has become among the objectives of UNESCO. Various studies have been conducted, and efforts supported by the Organization on the cause and prevention of landslides. In the 1970s, UNESCO books including guidelines on landslides hazards zonation have been published. As early as 1981 a joint project initiated jointly with the United Nations Environment Programme (UNEP), under the title "The Protection of the Lithosphere as a Component of the Environment", resulted in significant research being carried out on landslide mitigation. Both the International Hydrological Programme (IHP) and the International Geoscience Programme (IGCP) have taken prominent roles in promoting activities related to landslides. The International Flood Initiative, which is led by UNESCO, addresses landslide risks as an integral part of hydrological extremes.

The foundation of the International Consortium on Landslides (ICL), in itself, lies within an IGCP project titled "Landslide Hazard Assessment and Mitigation for Cultural Heritage Sites and other Locations of High Societal Value". Approved in 1998, this project served as an initial platform for the establishment of a Memorandum of Understanding concerning cooperation in research for landslide risk mitigation and protection of the cultural and natural heritage between UNESCO and the Disaster Prevention Research Institute, Kyoto University, Japan.

UNESCO has been closely associated with the process that led to the establishment of ICL in January 2002. Subsequently the UNESCO UNITWIN Programme on Landslide Risk Mitigation for Society and the Environment came into effect in 2003, followed by the construction in 2004 of the headquarters building of the UNESCO/Kyoto University/ICL UNITWIN cooperation program within the Research Centre on Landslides, which hosts the secretariat of the International Programme on Landslides (IPL).

Under the above-mentioned cooperative mechanisms, a number of undertakings have been carried out, including the landslide investigation in the Inca's World Heritage Machu Picchu in Peru. Above all, UNESCO has fully participated in the adoption of the 2006 Tokyo Action Plan on Landslides which represents a global road map for the assessment and mitigation of landslides.

Much more recently an understanding has been reached between UNESCO and ICL in order to cooperate in the following areas:

1. The promotion of landslide research for the benefit of society and the environment, learning and capacity-building in landslide risk reduction, notably in developing countries;
2. the integration of earth sciences, water sciences, geophysical and geotechnical sciences, technology and disaster management within the appropriate cultural and social contexts in order to evaluate landslide risk in urban and rural areas, including cultural and natural heritage sites, as well as to contribute to the protection of the human and natural environment, including lifelines and buildings of high societal value;
3. the promotion of some or all of the global cooperation fields that have been agreed in 2006 Tokyo Action Plan.

UNESCO expects to enhance its contribution to the above-mentioned areas in shaping its participation in the preparations and follow-up of the First World Landslide Forum to take place in Tokyo from 18 to 21 November 2008. This participation will build on interdisciplinary scientific work through and among UNESCO's intergovernmental scientific programmes, and will capitalize further on the transdisciplinary activities between the programmes of the Organization in the sciences, education, culture and communication.

Badaoui Rouhban
Chief
Section for Disaster Reduction
Natural Sciences Sector
UNESCO, Paris

Establishment of the Technical Journal "Landslides" as the Successor to "Landslide News"

Landslides pose a hazard and threat in most countries. Worldwide, they annually cause billions of dollars in damages and thousands of casualties. To counter these losses, scientists and engineers are successfully conducting research that aims to investigate and mitigate landslide hazards. To sustain this research requires continuing communication among researchers and those who apply its results.

From 1987–2003 this goal of communication between researchers and practitioners was aided by the international newsletter "Landslide News," which was edited, published, and distributed by the Japan Landslide Society. Professor Sassa served as Editor-in-Chief of this comprehensive newsletter, which was a very successful technical news publication. It reported on landslide news, landslide research, the efforts of landslide research organizations, and news of landslide-related technical meetings.

At the first session of the Board of Representatives of the International Consortium on Landslides (ICL), which was held at UNESCO Headquarters in Paris on 19–21 November 2002, the Board decided to launch the International Programme on Landslides (IPL). As the first coordinating project of this new program, the Board conceived the concept of the new scientific journal "Landslides" as the successor to "Landslide News." This new journal was planned and activated by ICL as the international organ of scientific communication on landslides. Thereafter, extensive deliberation on the necessary financial resources, editorial board, editorial secretariat, and publishing firm was conducted. As a result of this deliberation, Professor Sassa was appointed Editor-in-Chief of this new journal and Kyoto University was named as the site of the secretariat.

The first issue of "Landslides" was published by Springer-Verlag in April 2004 as the successor to "Landslide News." This new quarterly journal with color illustrations was created by and for landslide researchers and practitioners as a means of communication to promote landslide research and to contribute to landslide disaster reduction.

"Landslides" is supported by the United Nations Educational, Scientific and Cultural Organization (UNESCO), the World Meteorological Organization (WMO), the Japan Ministry of Education, Culture, Sports, Science and Technology (MEXT) and Kyoto University. In addition to these organizations, ICL-supporting organizations and landslide experts worldwide are cooperating in writing, editing, and publishing this journal.

Publication of "Landslides" represents an important landmark in promoting research on landslide hazards worldwide, and thus in the mitigation of landslide hazards. The journal contributes to the development and continuance of landslide research by providing an international forum for the exchange and coordination of expertise in landslide risk assessment and mitigation.

We thank Professor Sassa for his efforts as Editor-in-Chief of "Landslide News" and its successor, the very successful technical journal "Landslides." We wish this new journal a long life in service of the international landslide research community.

Robert L. Schuster
U.S. Geological Survey, E-mail: rschuster@usgs.gov

Preface – Aims of This Volume

Large and small landslides occur almost every year in nearly all regions of the world. However, the number of landslides is difficult to ascertain, and even the number of landslide-caused casualties is not correctly counted worldwide. Most casualties caused by rain-induced landslides are included in those tabulated for hurricane or storm disasters, and casualties caused by earthquake-induced landslides are often included in those for earthquake disasters. Thus, the casualties due to landslide disasters are often extremely underestimated. Japan has statistics of casualties by various types of landslides (small and large debris or rock slides, debris flows, rock falls, et al. since 1967), even though they occurred during typhoons, earthquakes, or volcanic activities. Figure 0.1 shows the statistics of casualties caused by landslides, earthquakes, and volcanic activities in Japan for the period of 1967–2004. Casualties by earthquake-induced landslides are included both in the landslide disasters and earthquake disasters. Landslide disasters in Japan for this period have occurred every year; the total number of deaths (3 285) due to landslides is about one half of the deaths (7 008) caused by earthquakes, including the catastrophic 1995 Kobe earthquake. Extensive landslide prevention works have been constructed in Japan. Thanks to those preventive works, the number of casualties has gradually decreased as seen in Fig. 0.1 in spite of progress of urban and mountain development all over Japan during those years. Although there are no reliable data for damages, or even for casualties due to landslides in many countries of the world, Fig. 0.1 provides a clear evidence of the strong negative impact of landslides to society. One clear difference of landslides, as compared to earthquakes and volcanic eruptions, is that humans can prevent or mitigate many landslide phenomena, while earthquakes and volcanic eruptions can not be prevented. The hazard assessment of landslides is very effective for disaster reduction because of relatively small affected area comparing to earthquakes and typhoons. Therefore, we can do much for landslide disaster mitigation. In addition to human damages, landslides often destroy cultural and natural heritage, and the natural environment, which cannot

Fig. 0.1.
Comparison of the numbers of victims in Japan from 1967–2004 due to landslide disasters, earthquake disasters including deaths by earthquake-induced-landslides, and volcanic disasters including deaths due to volcanic gas (The data on victims due to landslide disasters since 1967 were published by the Sabo Technical Center)

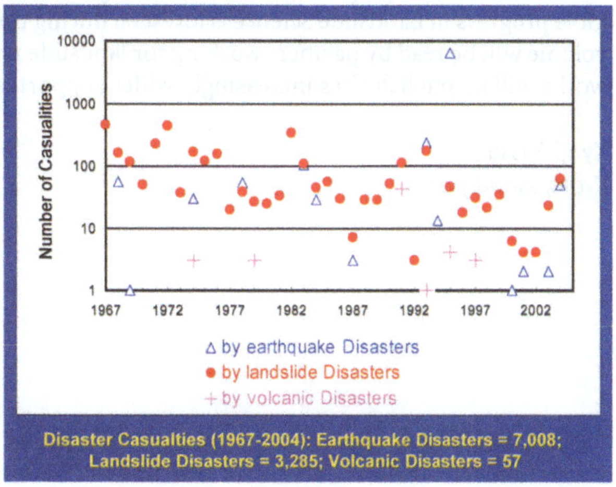

be recovered. Landslide disaster reduction, protection of cultural and natural heritage sites, and the invaluable environment are vital factors for human society. However, studies of landslides have not been conducted in an integrated manner, although the phenomena are targets of many scientific and engineering fields. Landsides are a target of application of many technical fields, but they are not the major interest of any individual field. Therefore, no national landslide society has been established, except the Japan Landslide Society (JLS), and the very recent Nepal Landslide Society. No international society on landslides has been established because there are only one or two national societies.

The Japan Landslide Society was founded in 1963, and the society organized the first International Symposium on Landslides (ISL) in Kyoto in 1972, and the second International Symposium on Landslides in Tokyo in 1977. The Japan Landslide Society also initiated the International Conference and Field Workshop on Landslides (ICFL) in Tokyo, 1985. The society started to publish the International Newsletter "Landslide News" in 1987. Five thousand copies were printed in three colors and 2 000 copies were distributed free of charge over the world. Landslide News continued to be published until 2003 with supports from the United Nations Educational, Scientific and Cultural Organization (UNESCO), the Food and Agriculture Organization of the United Nations (FAO), the United Nations Secretariat for International Strategy for Disaster Reduction (UN/ISDR). This long-term activity gradually built up an infrastructure of international landslide community. It led to the formation of the International Consortium on Landslides (ICL) in 2002 and the new journal of ICL, "Landslides", in 2004. The publication of journal is the essential and necessary platform to create an independent field of science dealing with landslides. This book "Progress in Landslide Science" aims to present an overview of the current status of research by major landslide researchers worldwide. This volume can be a part of series of publications of ICL that may contribute to the education system of landslide science and landslide disaster management in the future.

The International Consortium on Landslides (ICL), United Nations Educational, Scientific and Cultural Organization (UNESCO), World Meteorological Organization (WMO), Food and Agriculture Organization of the United Nations (FAO), United Nations International Strategy for Disaster Risk Reduction (UN/ISDR), United Nations University (UNU), United Nations Environment Programme (UNEP), United Nations Development Programme (UNDP), World Bank (IBRD), International Council for Science (ICSU), World Federation of Engineering Organizations (WFEO), Kyoto University (KU), and the Japan Landslide Society (JLS) have jointly been organizing the First World Landslide Forum for Strengthening Research and Learning on Earth System Risk Analysis and Sustainable Disaster Management within the UN-ISDR as Regards "Landslides" which will be held in November 2008 in Tokyo. This forum will be a milestone in developing research and learning on landslide disaster reduction. Tangible progress in Landslide Science is foreseen during this process. It is hoped that this volume will be read by partners working for landslide risk mitigation and that further works will be published as increasingly wider support is obtained.

Kyoji Sassa
Kyoto, January 2007

Landslides and Cultural Heritage: the Common Thread from IGCP-425 to ICL/IPL

The protection of cultural heritage from natural hazards is an issue of worldwide concern that regards advanced nations and developing countries alike. The history and identity of every people is preserved in and portrayed by the unique monuments, art, literature, archeology and artefacts that have been created during its existence. The destruction of this wealth would represent an irreparable loss for the whole of humanity. The damage caused by natural disasters is increasing steadily as both the vulnerability of rapidly developing urban areas and the consequences of climate change tend to amplify the effects caused by their occurrence.

Within this framework landslides represent a major threat both for the safety of people and for the preservation of the built environment, including many important sites of renowned international cultural or natural value. For instance, the Mediterranean area is an excellent example of this situation as the immeasurable wealth of the entire region is often exposed to landslide hazard due to its particular topographic, geologic and climatic settings. Although the concept of preservation has already taken hold in many of the European nations bordering on the Mediterranean, much still has to be accomplished as the awareness of both the cultural and economic implications of loss is only just starting to be appreciated. The situation is rather more serious in developing countries, where even minimum levels of both the awareness of the unique value represented by cultural heritage and the economic, scientific and technical means for mitigating landslide hazard are often lacking.

For these reasons, the International Geological Correlation Programme (IGCP) project 425 titled "Landslide hazard assessment and mitigation for cultural heritage sites and other locations of high societal value" was approved during the 26th Session of the IGCP Scientific Board. The main objectives of IGCP-425, proposed in 1998 by Kyoji Sassa, were to assess and mitigate the effects of natural hazards on the cultural heritage in countries from around the world and to propose reliable, cost-effective slope stabilization techniques suitable for use on a global scale, including developing countries.

Before ending in 2002, IGCP-425 strengthened previous projects for the preservation of cultural heritage sites carried out under the Safeguarding Campaigns of UNESCO, in which it played a key role in protecting 26 sites of global relevance. The IGCP-425 was set up to provide a framework for assembling the results of national projects carried out by the individual participants (over 50 national and regional institutions and universities were involved) concerning various aspects of landslide hazard that are still useful today for drawing up prevention and mitigation plans of general validity.

In particular, the specific aims of the IGCP-425 project were:

- to develop high precision and durable slope monitoring equipment for potential landslides in urban and rural areas, especially in relation to outstanding cultural heritage sites;
- to develop reliable, cost-effective techniques for assessing rapid landslide motion, delimit hazard areas and reduce damage to life, cultural heritage, infrastructures and property;

- to develop economical and effective slope stabilization works and disaster mitigation measures suitable for use on a global scale, including developing countries;
- to research reliable landslide hazard assessment and risk evaluation methods;
- to detect potential landslides and identify precursory phenomena.

To promote the project, UNESCO and the Disaster Prevention Research Institute of Kyoto University, Japan (DPRI/KU) exchanged in 1999 a Memorandum of Understanding concerning cooperation in research for landslide risk mitigation and protection of the cultural and natural heritage as a key contribution to environmental protection and sustainable development in the first quarter of the twenty-first century.

The Tokyo Declaration "Geoscientists tame landslides" was released in the 2001 UNESCO/IGCP Symposium on Landslide Risk Mitigation and Protection of Cultural and Natural Heritage to propose an initiative that led to the creation of the International Consortium on Landslides (ICL) for the worldwide promotion of landslide research.

This initiative derived from two observations: (*1*) that the specific points listed previously correspond to the main themes of landslide research in general and (*2*) that the specificity of the protection of cultural heritage from landslides must necessarily take into account not only landslide research methods but also those relative to the study of the vulnerability and worth of the cultural and natural artifact, which entails the involvement of other disciplines and professions. These factors led to a practical approach in which landslides are tackled comprehensively and cultural heritage is considered as a related, highly important field.

In 2002, during the International Symposium "Landslide Risk Mitigation and Protection of Cultural and Natural Heritage", co-organized by UNESCO and Kyoto University, international experts coming from different national, scientific, and governmental institutes, academic institutions, regional and international organizations, international non-governmental organizations and United Nations organizations unanimously agreed and declared to launch the ICL in the 2002 Kyoto Declaration. In the months following its launch, a great effort by the IGCP-425 leaders led to the definition of the objectives and the structure of the Consortium, and the results achieved during the projects were adopted as the basis for the construction of a world research program on landslides (International Programme on Landslides, IPL) as one of the main initiatives of the recently created ICL.

The IPL is today a pillar of the ICL and both are rooted in the initial endeavor: IGCP-425, the objectives of which are still the guiding principles. The IPL also gave birth to the journal Landslides, indeed one of its most important outcomes, which hosts many papers dedicated to the topic of landslides and cultural heritage.

Paolo Canuti
Firenze, January 2007

Contents

List of Contributors

Araiba, Kiminori · (Chapter 16)

National Research Institute of Fire and Disaster, Jindaiji Higashimachi 4-35-3, Chofu, Tokyo 182-8508, Japan
Corresponding author of Chapter 16:
Tel: +81-422-44-8331, Fax: +81-422-42-7719, E-mail: araiba@fri.go.jp

Bhandary, Netra P. · (Chapter 13)

Graduate School of Science and Engineering, Ehime University, Bunkyo-3, Matsuyama 790-8577, Japan
Corresponding author of Chapter 13:
Tel: +81(0)89-927-8566, Fax: +81(0)89-927-8566, E-mail: netra@dpc.ehime-u.ac.jp

Bromhead, Edward N. · (Chapter 2)

School of Engineering, Kingston University, Penrhyn Road, Kingston upon Thames, KT1 2EE, UK
Corresponding author of Chapter 2:
Tel: +44-20-8547-2000 ext 62225, E-mail: e.bromhead@kingston.ac.uk

Cai, Fei · (Chapter 23)

Graduate School of Eng., Gunma University, Tenjin-cho 1-5-1, Kiryu 376-8515, Japan

Canuti, Paolo · (Chapter 18)

Department of Earth Sciences, University of Firenze, Via Giorgio La Pira 4, 50121 Firenze, Italy

Casagli, Nicola · (Chapter 18)

Department of Earth Sciences, University of Firenze, Via Giorgio La Pira 4, 50121 Firenze, Italy
Corresponding author of Chapter 18:
Tel: +39 055 2757523, Fax: +39 055 2756296, E-mail: nicola.casagli@unifi.it

Catani, Filippo · (Chapter 18)

Earth Sciences Department, University of Firenze, Via Giorgio La Pira 4, 50121 Firenze, Italy

Davies, Tim · (Chapter 8)

Department of Geological Sciences, University of Canterbury, Christchurch, New Zealand

Falorni, Giacomo · (Chapter 18)

Earth Sciences Department, University of Firenze, Via Giorgio La Pira 4, 50121 Firenze, Italy

Farina, Paolo · (Chapter 18)

Department of Earth Sciences, University of Firenze, Via Giorgio La Pira 4, 50121 Firenze, Italy

Fukuoka, Hiroshi · (Chapters 6, 7, 14)

Research Centre on Landslides, Disaster Prevention Research Institute, Kyoto University, Gokasho, Uji, Kyoto 611-0011, Japan
Corresponding author of Chapter 7:
Tel: +81-774-38-4111, Fax: +81-774-38-4300, E-mail: fukuoka@scl.kyoto-u.ac.jp

Greif, Vladimir · (Chapter 19)

Department of Engineering Geology, Comenius University Bratislava, Slovakia
Corresponding author of Chapter 19:
Tel: +421-2-6029-6627, Fax: +421-2-6029-6702, E-mail: greif@nic.fns.uniba.sk

Hencelova, Lucia · (Chapter 19)

Department of Engineering Geology, Comenius University Bratislava, Slovakia

Holzer, Rudolf · (Chapter 19)

Department of Engineering Geology, Comenius University Bratislava, Slovakia

Hsu, Yu-Charn · (Chapter 26)

PhD Candidate, Department of Civil Engineering, National Taiwan University

Hungr, Oldrich · (Chapter 4)

Department of Earth and Ocean Sciences, University of British Columbia, 6339 Stores Rd., Vancouver, B.C. V6T 1Z4, Canada
Corresponding author of Chapter 4:
Tel: +1-604-822-8471, Fax: +1-604-822-6088, , E-mail: ohungr@eos.ubc.ca

Ibsen, Maia L. · (Chapter 2)

School of Engineering, Kingston University, Penrhyn Road, Kingston upon Thames, KT1 2EE, UK

Iwahashi, Junko · (Chapter 17)

Geography and Crustal Dynamics Research Center, Geographical Survey Institute, Ministry of Land, Infrastructure and Transport, 1 Kitasato, Tsukuba, Ibaraki 305-0811, Japan

Jezny, Michal · (Chapter 19)

Department of Engineering Geology, Faculty of Natural Sciences, Comenius University Bratislava, 84215 Bratislava, Mlynska dolina, Slovak Republic

Johansson, Jörgen · (Chapter 20)

Institute of Industrial Science, University of Tokyo, 4-6-1 Komaba, Meguro-ku, Tokyo 153-8505, Japan

Klimeš, Jan(Chapter 24)

Institute of Rock Structure and Mechanics, Academy of Sciences, V Holešovièkách 41, 182 09 Prague 8, Czech Republic

Koarai, Mamoru · (Chapter 17)

Geography and Crustal Dynamics Research Center, Geographical Survey Institute, Ministry of Land, Infrastructure and Transport, 1 Kitasato, Tsukuba, Ibaraki 305-0811, Japan

Konagai, Kazuo · (Chapter 20)

Institute of Industrial Science, University of Tokyo, 4-6-1 Komaba, Meguro-ku, Tokyo 153-8505, Japan
Corresponding author of Chapter 20:
Tel: +81-3-5452-6142, Fax: +81-3-5452-6144, E-mail: konagai@iis.u-tokyo.ac.jp

Li, Hsin-Chi · (Chapter 26)

Researcher, socio-economic system division, National Science & Technology Center for Disaster Reduction

Liu, Ko-Fei · (Chapter 26)

Department of Civil Engineering, National Taiwan University, Slope Land Disaster Reduction Division, National Science & Technology Center for Disaster Reduction
Corresponding author of Chapter 26:
Tel: +886-2-2365-5405, Fax: +886-2-2363-1558, E-mail: kfliu@ntu.edu.tw

Margottini, Claudio · (Chapter 25)

ENEA CR Casaccia, Via Anguillarese 301, 00060 S. Maria di Galeria, Rome, Italy
Corresponding author of Chapter 25:
Tel: +39-06-3048-4688, Fax: +39-06-3048-4029, E-mail: margottini@casaccia.enea.it

Marui, Hideaki · (Chapter 21)

Research Center for Natural Hazards and Disaster Recovery, Niigata University, 8050 Ikarashi-Ninocho, Niigata, 950-2181, Japan
Corresponding author of Chapter 21:
Tel: +81-25-262-7055, Fax: +81-25-262-7050, E-mail: maruihi@cc.niigata-u.ac.jp

McSaveney, Mauri · (Chapter 8)

GNS Science, Lower Hutt, New Zealand
Corresponding author of Chapter 8:
Tel: +64-4-570-4734, Fax: +64-4-570-4600, E-mail: M.McSaveney@gns.cri.nz

Numada, Muneyoshi · (Chapter 20)

Institute of Industrial Science, University of Tokyo, 4-6-1 Komaba, Meguro-ku, Tokyo 153-8505, Japan

Ochiai, Hirotaka · (Chapter 15)

Forestry and Forest Products Research Institute, 1 Matsunosato, 305-8687 Tsukuba, Ibaraki, Japan
Corresponding author of Chapter 15:
Tel: +81-29-829-8237, Fax: +81-29-874-3720, E-mail: ochi@ffpri.affrc.go.jp

Ohtsuka, Satoru · (Chapter 22)

Department of Civil & Environmental Engineering, Nagaoka University of Technology, 1603-1 kamitomioka-cho, Nagaoka, Niigata 940-2199, Japan
Corresponding author of Chapter 22:
Tel: +81-258-47-9633, Fax: +81-258-47-9600, E-mail: ohtsuka@nagaokaut.ac.jp

Okada, Yasuhiko · (Chapter 15)

Forestry and Forest Products Research Institute, 1 Matsunosato, 305-8687 Tsukuba, Ibaraki, Japan

Picarelli, Luciano · (Chapter 3)

Dipartimento di Ingegneria Civile, Seconda Università di Napoli, Aversa, Italy
Corresponding author of Chapter 3:
Tel: +39-081-5010213, Fax: +39-081-5037370, E-mail: luciano.picarelli@unina2.it

Sammori, Toshiaki · (Chapter 15)

Forestry and Forest Products Research Institute, 1 Matsunosato, 305-8687 Tsukuba, Ibaraki, Japan

Sassa, Kyoji · (Chapters 1, 6, 7, 11, 12, 14)

Research Centre on Landslides, Disaster Prevention Research Institute, Kyoto University, Uji, Kyoto 611-0011, Japan
Corresponding author of Chapters 1, 6, 14:
Tel: +81-774-38-4110, Fax: +81-774-32-5597, E-mail: sassa@scl.kyoto-u.ac.jp

Sato, Hiroshi P. · (Chapter 17)

Geography and Crustal Dynamics Research Center, Geographical Survey Institute, Ministry of Land, Infrastructure and Transport, 1 Kitasato, Tsukuba, Ibaraki 305-0811, Japan
Corresponding author of Chapter 17:
Tel: +81-29-864-5946, Fax: +81-29-864-2655, E-mail: hsato@gsi.go.jp

Sekiguchi, Tatsuo · (Chapter 17)

Geography and Crustal Dynamics Research Center, Geographical Survey Institute, Ministry of Land, Infrastructure and Transport, 1 Kitasato, Tsukuba, Ibaraki 305-0811, Japan

Suemine, Akira · (Chapter 16)

Research Centre on Landslides, Disaster Prevention Research Institute, Kyoto University, Gokasho, Uji, Kyoto 611-0011, Japan

Takahashi, Tamotsu · (Chapter 5)

Prof. Emeritus of Kyoto University, 3-6-26, Kita-Oji, Otsu, Shiga 520-0843, Japan
Corresponding author of Chapter 5:
E-mail: ta-taka@iris.eonet.ne.jp

Tiwari, Binod · (Chapter 9)

Department of Civil and Environmental Engineering, College of Engineering and Computer Sciences, California State University, Fullerton, 800 N. State College Blvd. E-419, Fullerton, CA 92834, U.S.A.
Corresponding author of Chapter 9:
Tel: +1-714-278-3968, Fax: +1-714-278-3916, E-mail: btiwari@fullerton.edu

Towhata, Ikuo · (Chapter 10)

Department of Civil Engineering, University of Tokyo, 7-3-1, Hongo, Bunkyo-Ku, Tokyo 113-8656, Japan
Corresponding author of Chapter 10:
Tel: +81-3-5841-6121, Fax: +81-3-5841-8504, E-mail: towhata@geot.t.u-tokyo.ac.jp

Ugai, Keizo · (Chapter 23)

Graduate School of Eng., Gunma University, Tenjin-cho 1-5-1, Kiryu 376-8515, Japan

Vilímek, Vít · (Chapter 24)

Department of Physical Geography and Geoecology, Faculty of Science, Charles University, Albertov 6, 128 43 Prague 2, Czech Republic
Corresponding author of Chapter 24:
Tel: +420-22-0561-361, Fax: +420-22-1951-367, E-mail: vilimek@natur.cuni.cz

Vlčko, Ján · (Chapter 19, 24)

Department of Engineering Geology, Faculty of Natural Sciences, Comenius University Bratislava, Mlynska dolina G, 842 15 Bratislava, Slovakia

Wakai, Akihiko · (Chapter 23)

Graduate School of Eng., Gunma University, Tenjin-cho 1-5-1, Kiryu 376-8515, Japan
Corresponding author of Chapter 23:
Tel: +81-277-30-1624, Fax: +81-277-30-1624, E-mail: wakai@ce.gunma-u.ac.jp

Wang, Fawu · (Chapters 6, 11, 14)

Research Centre on Landslides, Disaster Prevention Research Institute, Kyoto University, Gokasho, Uji, Kyoto 611-0011, Japan
Corresponding author of Chapter 11:
Tel: +81-774-38-4114, Fax: +81-774-38-4300, E-mail: wangfw@landslide.dpri.kyoto-u.ac.jp

Wang, Gonghui · (Chapters 6, 7, 12, 14)

Research Centre on Landslides, Disaster Prevention Research Institute, Kyoto University, Japan, Uji, Kyoto, 611-0011, Japan
Corresponding author of Chapter 12:
Tel: +81-774-38-4114, Fax: +81-774-38-4300, E-mail: wanggh@landslide.dpri.kyoto-u.ac.jp

Yagi, Hiroshi · (Chapter 17)

Yamagata University, Japan

Yatabe, Ryuichi · (Chapter 13)

Graduate School of Science and Engineering, Ehime University, Bunkyo-3, Matsuyama 790-8577, Japan

Yoshimatsu, Hiroyuki · (Chapter 21)

Sabo Technical Center, 4-8-1, Kudan-minami, Chiyoda-ku Tokyo, 102-0074, Japan

Part I Progress in Landslide Science

Chapter 1

Landslide Science as a New Scientific Discipline

Kyoji Sassa

Abstract. Landslides cause great disasters and their impact to society is very great. Thus, they are studied in many scientific and engineering fields. However, studies on landslides from various fields have not been conducted in an integrated manner. There was neither international society, nor international journal, and the meaning of landslides was not defined internationally and interdisciplinary. During the United Nations International Decade for Natural Disaster Reduction (IDNDR) in 1990–2000, landslide researchers worldwide agreed the definition of landslides as "the movement of a mass of rock, debris or earth down a slope". This is a basis for the development of the study of landslides as a scientific field. This paper describes the progress in landslide science as an integrated discipline together with the development of international landslide community and a global cooperation platform as its infrastructure.

Keywords. Classification of landslides, landslide dynamics, landslide science, ISDR, IDNDR, ICL

1.1 Definition of Landslides

Landslides are studied in many countries because they occur almost worldwide, from high mountain areas to coastal areas and even in marine geologic units, from very wet or heavy rainfall areas to very dry areas, and from seismic or volcanic areas to tectonically non-active areas. Some landslides are triggered by human activities, such as road and railway construction, mining, and development in urban and mountain areas. The landslide phenomena have been studied in may countries and many areas. Landslide studies have been conducted in many fields of science and engineering. Landslide disasters have been dealt with many governments, ministries, and they are related to many other disasters. As shown in Fig. 1.1, landslides are phenomena that involve many disciplines, ministries, and individuals. However, landslide phenomena have not been studied in an integrated way. Even the definition of landslides, what is a "landslide", has been diverse in countries and disciplines worldwide.

David Varnes presented "Classification of Landslides" by the type of material (bedrock and soils) and type of movement (falls, slides, flows) in "Landslides and Engineering Practice" (Eckel, ed.) in 1958 (Varnes 1958). This paper presented the concept of landslides in wider meaning, including debris flows, rock falls, debris avalanches, creep, etc. However, many opinions and criticism occurred whether land-SLIDE can include the phenomena of FLOW, FALL and other movements. Therefore, Varnes (1978) presented "Types of Slope Movements" in "Landslides-Analysis and Control" using the type of material (bed-

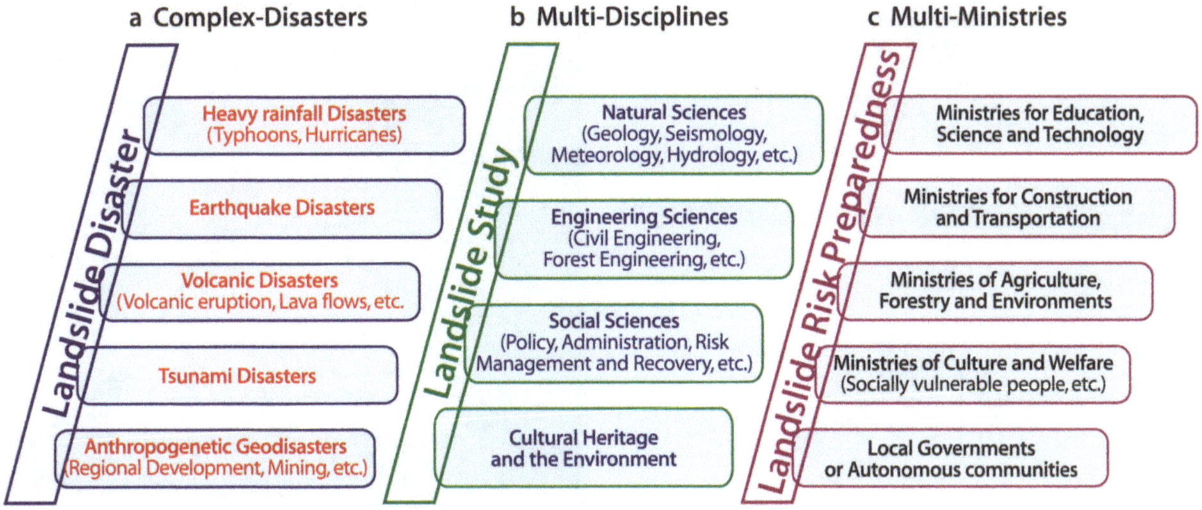

Fig. 1.1. Characteristics of landslide disasters (from 2006 Tokyo Action Plan; Sassa 2006)

rock, debris, earth) and the type of movement (falls, topples, slides, lateral spreads, flows; Fig. 1.2). The International Geotechnical Societies' UNESCO Working Party on World Landslide Inventory was established in conjunction with the United Nations International Decade for Natural Disaster Reduction (IDNDR). This Working Party was formed by the Commission on Landslides and other Mass Movements of the International Association of Engineering Geology and the Environment (IAEG), the Technical Committee on Landslides of the International Society for Soil Mechanics and Geotechnical Engineering (ISSMGE) and the International Society for Rock Mechanics (ISRM). This working party defined various terms of landslide features, landslide velocities, landslide dimensions, and state of activities (Cruden and Varnes 1996). The most important of these factors is the definition of landslides by the Working Parties as "the movement of a mass of rock, debris or earth down a slope". Namely various types of gravitational mass movements were integrated in the category of "Landslides". Even some movements are not "Slide". Cruden (1991) explained that an English word combining two words can express something different from original two words. Thus, landslide is not always necessary to be "*Slide of Land*".

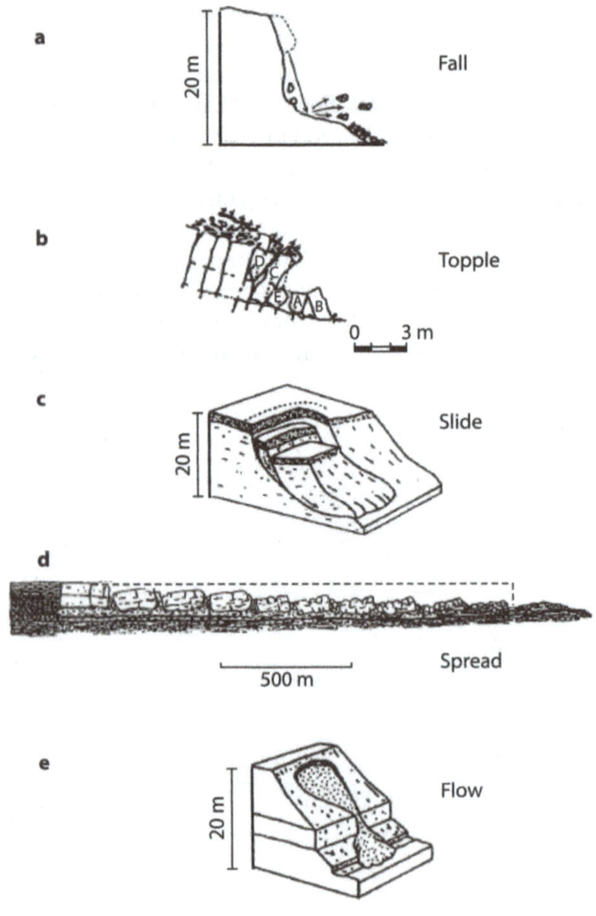

Fig. 1.2. Types of landslides (modified from Cruden and Varnes 1996)

Landslides were classified by the Working Party by the type of material (rock, debris: predominantly coarse soils, and earth: predominantly fine soils) and the type of movement (fall, topple, slide, spread, flow). Basically the idea was to return to the Varnes' Classification of Landslides (Varnes 1958). It was introduced by Cruden and Varnes (1996) in "Landslides – Investigation and Mitigation". The definition of landslides as the "the movement of a mass of rock, debris or earth down a slope" was widely accepted. It is the very important basis for the development of Landslide Science.

The International Consortium on Landslides (ICL) which was established in 2002 as the first and unique international organization dedicated to landslide research, created a new award called the "Varnes Medal", which recognizes professional excellence in landslide research and education, the basis for David Varnes definition of the area of landslide study. The first Varnes Medal was bestowed to Robert Schuster of the U.S. Geological Survey (Canuti 2004).

1.2 Landslide Science as a New Scientific Discipline and Landslide Dynamics as its Core

The International Strategy for Disaster Risk Reduction of the United Nations (UN-ISDR) was initiated in 2000 following the International Decade for Natural Disaster Reduction (IDNDR) for 1990–2000. Landslide disaster risk reduction is one of major task of this strategy. Landslide

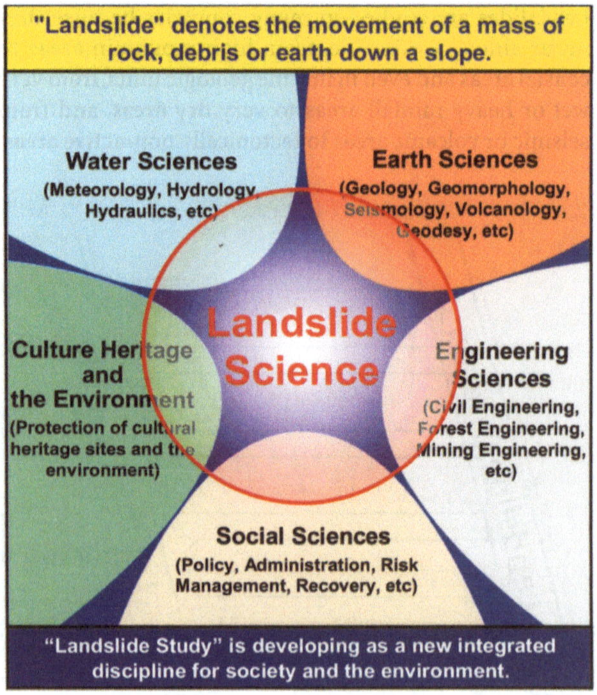

Fig. 1.3. Landslide Science as a new integrated discipline

study by integrating knowledge obtained in many fields of science and engineering related to landslides is imperative to effectively mitigate landslide risk. Figure 1.3 presents an illustration of landslide study as a possible integrated discipline. However, It necessitates at least two factors to be an independent discipline, namely its own core study in research and an international journal to bring together common scientific knowledge.

1.2.1 Core Study of Landslide Science

The core study in landslide research may develop as an integrated and independent discipline, namely landslide science. One of the initial core studies in landslide science can be landslide dynamics because landslides have been defined as "the *Movement of a Mass* of rock, debris or earth down a slope". Dynamics are studied in the field of sciences dealing fluids: air, water, and other liquids. Geology, geomorphology, and geotechnology dealing with solids on earth do not include dynamics, at least as their central interests. The stability analysis for the failure of slopes is a major task for geotechnology, but post-failure motion is not a concern because major interests are to design construction works, such as roads, dams, and others, in a state that does not allow failure. The tools to study dynamics of soil masses have not been fully developed. Sassa (1992) developed an apparatus to simulate the initiation and post-failure motion of earthquake induced landslides. This apparatus was not well functioned to monitor pore-pressure generation in the shear zone and to keep a shear box in the undrained condition during tests. The apparatus was later improved in its capability to maintain an undrained state in the soil sample and to monitor pore-pressure generation near the sliding surface. It was applied to the Nikawa landslide triggered by the 1995 Kobe earthquake, which killed 34 residents by its rapid and long travel movement (Sassa 1996). By obtaining a special budget to investigate the 1995 Kobe earthquake disaster and to prevent further similar disasters, a new advanced undrained dynamic loading ring shear apparatus (DPRI-5 and 6) was developed (Sassa 2000). The apparatus is called as a "ring-shear simulator of earthquake-induced landslides", with capability to reproduce 5 Hz real seismic wave loading, to maintain undrained condition in the soil sample during a maximum 224 cm s^{-1} velocity at the center of ring shear sample box (250 mm inside diameter, 350 mm outside diameter for DPRI-6). The latest version is the ring shear apparatus (DPRI-7) which has a transparent shear box enabling the direct observation of shearing and crushing of grains at high speed under a high normal stress at a maximum speed of 300 cm s^{-1} (Sassa et al. 2004). These apparatuses, DPRI-5,6,7, can reproduce the stress in the slope during ground-water rise (pore-pressure increase) or during earthquakes (seismic wave loading up to 5 Hz in both normal stress and shear stress). Therefore, the apparatuses can study pore-pressure generation and mobilized shear resistance during sliding-surface formation and post-failure motion in various types of landslides; such as rain-induced landslides, earthquake-induced landslides, transformation of debris slides to debris flows, and enlargement of landslide masses during the process of downslope movement. Those applications were introduced in the initial issue of the new international journal *Landslides* (Sassa et al. 2004). This study is one of core studies in landslide science as a new discipline on landslides. Within this initial issue of *Landslides,* a new monitoring technology for landslides, the ground-based Synthetic Aperture Radar system (GB-SAR), was also reported on. SAR was originally developed as a satellite monitoring method; however, this new system is installed on a short rail on the ground in front of landslides. The radar antenna is repeatedly moved on the short rail. Therefore, monitoring for a short time span is possible (Antonello et al. 2004). Application of GIS and geophysical technology for landslide mapping and investigation, and application of stability technology to protect the Bamiyan Budda niches in Afghanistan were also reported in this initial issue.

1.2.2 A New International Journal, *Landslides*

The publication of an international journal is imperative for an independent discipline. Such publication requires a stable financial background and a strong involvement of the international scientific community to contribute quality articles. It is not easy to create a new international journal. The number of individuals, organizations, and entities that are involved in landslide research are many, as is shown in Fig. 1.1. However, the total number of professionals studying landslides as their main interest is small. The target of readers of new landslide journal must be the wide variety of groups and disciplines shown in Fig. 1.1. Therefore, the articles of the journal must be enjoyed by those people who have not professionally studied and those who have no time to read in detail, but have time to glance at the articles. Thus, articles in the new journal must be something understood by a glance, and enjoyed by looking over pages of the journal. Ideally, the journal should be printed in full color. Mono-color landslide photos can neither be fully understood nor attractive. Figures must be drawn in color to present their meaning immediately in an understandable way. However, publication of a full color journal is very expensive. As far as the author knows, no full-color scientific journal has been published without advertising, or without additional charges for the color figures. Sassa proposed publication of the journal "Landslides" as the initial and the life-time project of the International Programme on Landslides (IPL) at the first session of Board of Representatives of ICL held at UNESCO Headquarters on November 19–21, 2002. Publication of "Landslides" was decided to begin as the IPL C100 project. Sassa surveyed and negotiated with major international publication companies. Agreement was reached with Springer-Verlag at Heidelberg, Germany. Dr. Wolfgang Engel, the Executive Editor of Geosciences of Springer-Verlag, was the main partner. "Landslides" began publication in April 2004 as a quarterly journal. The journal was accepted at Thompson ISI for coverage in Science Citation Index Expanded from 2005. It is extremely fast to be accepted in this index after only one year of publication. The journal is distributed through the web at most of universities, institutes, and other organizations that have contracts to purchase web journals with Springer-Verlag. The printed version of the journal is distributed to ICL member organizations and others through Springer-Verlag sales worldwide. It is said that the journal is fulfilling the initial purpose to attract groups, organizations, and individuals in many fields in both contribution and reading. The number of pages submitted and the quality of papers are constantly increasing.

1.3 Foundation of the International Consortium on Landslides

Landslide science as an integrated discipline is developing through the journal publication and progress of core studies. As its background for the development, the International Consortium on Landslide (ICL) and the recent development of the international landslide community under the initiative of ICL are introduced in this and the next section.

As a part of the Japanese contribution to the IDNDR (International Decade for Natural Disaster Reduction) in the last decade of the 20th century, the Ministry of Education, Culture, Sports, Science and Technology of the Government of Japan (MEXT) conducted international joint research projects. The projects included the Japan-China Joint Project "Assessment of Landslide Hazards in Lishan (Yang-Que-Fe Palace), Xian, China", which was proposed by Kyoji Sassa, Disaster Prevention Research Institute, Kyoto University. The Palace is an important Cultural Heritage site, attracting more than three million visitors per year. The report of the joint research clearly demonstrated evidence of the risk of large-scale rock slide, based on detailed monitoring and observation of two investigation tunnels. The Secretary-General of the Communist Party of the Shaanxi Provincial Government, the former director of the Chinese Seismological Bureau, and the honorary chairperson of the International Symposium on Landslide Hazard Assessment, Xian, China organized by this group understood the landslide risk at the Lishan Palace. An extensive landslide prevention work was conducted to stabilize the slope. This work was funded at a level of three million U.S. dollars by municipal, regional, and national governments of China. Probably this is the first case of the initiation of extensive landslide remedial measures at a cultural heritage site for mitigation of potential landslides at the precursor stage in the world. This investigation of landslides at the precursor stage was evaluated as a contribution of geoscientists to protection of Cultural Heritage.

Based on the invitation by Edward Derbyshire, Chairman of the Scientific Board of International Geological Correlation Programme (IGCP), which is a joint program funded by UNESCO and the International Union of Geological Sciences (IUGS), and encouragement by Hideo Noguchi (Programme specialist of the Division of Cultural Heritage of UNESCO), Kyoji Sassa proposed an IGCP project in 1998. Then, the UNESCO–IUGS joint project, International Geological Correlation Programme (IGCP) No. 425, "Landslide hazard assessment and mitigation for cultural heritage sites and other locations of high societal value" began. Thirty-one subprojects were proposed to join this IGCP-425 project worldwide. This project obtained about 4000 US dollars from UNESCO and IUGS. The budget was shared to 31 subproject leaders as a part of their travel fees to attend the IGCP-425 meeting. This small amount of budget was very effective in promoting subproject leaders and in raising fund in their countries. Because IGCP projects have to be terminated within 5 years, the IGCP-425 group wished to establish its own international program on landslides by creating an international organization on landslides similar to the IUGS in IGCP program.

As a first step, a cooperative agreement was proposed between UNESCO and the institute of the IGCP-425 leader, the Disaster Prevention Research Institute, Kyoto University. This proposal was made at the International Conference "Cultural Heritage at Risk", which was organized by UNESCO and the IGCP-425 group at UNESCO Headquarters, Paris on 20–24 September 1999. Then, the Memorandum of Understanding (MoU) between the United Nations Educational, Scientific and Cultural Organization (UNESCO) and the Disaster Prevention Research Institute (DPRI), Kyoto University concerning "Cooperation in Research for Landslide Risk Mitigation and Protection of the Cultural and Natural Heritage as a Key Contribution to Environmental Protection and Sustainable Development in the First Quarter of the Twenty-First Century" was signed by Koichiro Matsuura, Director General of UNESCO, on 26 November and by Shuichi Ikebuchi, Director of the Disaster Prevention Research Institute, Kyoto University, on 3 December 1999. Based on this MoU, UNESCO and Kyoto University jointly organized the international symposium "Landslide Risk Mitigation and Protection of Cultural and Natural Heritage" in Kyoto.

Fig. 1.4. Group photo commemorating the establishment of the International Consortium on Landslides on 23 January 2005 at the Kyoto Campus Plaza

Eight representatives of UNESCO, the World Meteorological Organization (WMO), and the United Nations Secretariat for the International Strategy for Disaster Reduction (UN/ISDR) participated in the symposium. The participants decided to establish the ICL. The Statutes of the ICL were adopted and the first President (Kyoji Sassa) and interim steering committee members were nominated. By releasing the 2002 Kyoto Appeal "Establishment of a New International Consortium on Landslides", the International Consortium on Landslides was inaugurated on 21 January 2002. Figure 1.4 is a group photo of the participants.

The International Consortium on Landslides (ICL), created during the Kyoto Symposium in January 2002, is an international non-governmental and non-profit scientific organization, which has been supported from the beginning by the United Nations Educational, Scientific and Cultural Organization (UNESCO), the World Meteorological Organization (WMO), the Food and Agriculture Organization of the United Nations (FAO), the United Nations International Strategy for Disaster Risk Reduction (UN/ISDR), and intergovernmental programmes such as the International Hydrological Programme of UNESCO, the International Union of Geological Sciences (IUGS), the Ministry of Education, Culture, Sports, Science and Technology (MEXT) of the Government of Japan, the U.S. Geological Survey, and other governmental bodies. The ICL was registered as a legal body under the Japanese law for non-profit organizations in August 2002 by the Government of Kyoto Prefecture, Japan.

The objectives of the consortium are to:

1. promote landslide research for the benefit of society and the environment, and capacity building, including education, notably in developing countries;
2. integrate geosciences and technology within the appropriate cultural and social contexts in order to evaluate landslide risk in urban, rural and developing areas including cultural and natural heritage sites, as well as contribute to the protection of the natural environment and sites of high societal value;
3. combine and coordinate international expertise in landslide risk assessment and mitigation studies, thereby resulting in an effective international organization, which will act as a partner in various international and national projects; and
4. promote a global, multidisciplinary program on landslides.

The central activity in the ICL is the International Programme on Landslides (IPL). The necessity of establishment of a new Research Centre on Landslides to support IPL was proposed in the inaugural meeting of the ICL on 21 January 2002. The new Research Centre on Landslides (RCL) was established on 1 April 2003 in the Disaster Prevention Research Institute, Kyoto University (DPRI/KU). The secretariat of the IPL is now hosted at the headquarters building of UNESCO's UNITWIN (university twinning and networking) Cooperation Programme "Landslide Risk Reduction for Society and the Environment", which was constructed by Kyoto University and the International Consortium on Landslides on the Uji campus of Kyoto University in 2004. The relationship of ICL, IPL and RCL is illustrated in Fig. 1.5. The ICL was briefly introduced in "Landslides" by Sassa (2004); the full history of ICL was introduced in the proceedings of the first General Assembly of the ICL held at the U.S. National Academy of Sciences in Washington D.C. on 13–14 October 2005.

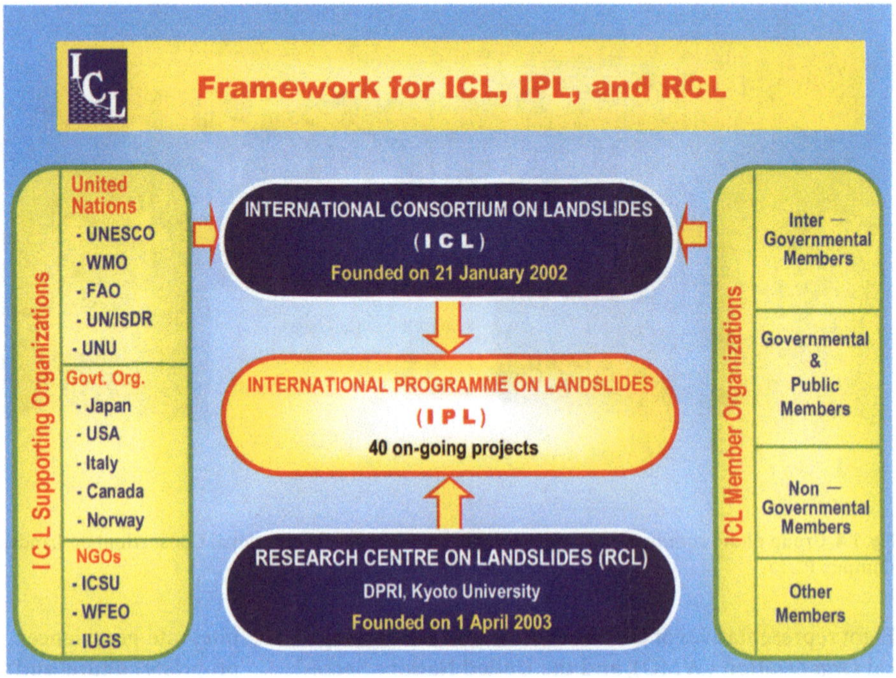

Fig. 1.5. Cooperative structure of the ICL, IPL and RCL for global landslide risk reduction

1.4 Development of the International Landslide Community: the 2005 Letter of Intent, the 2006 Tokyo Action Plan, and the 2008 First World Landslide Forum

The World Conference on Disaster Reduction (WCDR) was held on 18–22 January in Kobe, Japan. At this conference, a session titled "New International Initiatives for Research and Risk Mitigation of Floods (IFI) and Landslides (IPL)" was organized by the ICL, the United Nations Educational, Scientific and Cultural Organization (UNESCO), the World Meteorological Organization (WMO), the Food and Agriculture Organization of the United Nations (FAO), the Ministry of Education, Culture, Sports, Science and Technology of the Government of Japan (MEXT), the United Nations University (UNU), Kyoto University (KU), and others. Within this session, a Letter of Intent to promote further joint global activities in disaster reduction and risk prevention through "Strengthening research and learning on 'Earth System Risk Analysis and Sustainable Disaster Management' within the framework of the 'United Nations International Strategy for Disaster Risk Reduction' (ISDR)" by global partners: UNESCO, WMO, FAO, UNU, ICSU, WFEO. The Letter of Intent can be an umbrella for all initiatives of Earth-system risk reduction. It was approved and signed by seven global stakeholders, as shown in Fig. 1.6.

Based on this Letter of Intent, a round-table discussion was organized at the United Nations University in Tokyo on 18–20 January 2006, in order to examine the global plan promoting research and learning on landslides and developing the International Programme on Landslides as a dynamic global network. Organizers of this discussion were the International Consortium on Landslides (ICL), United Nations Educational, Scientific and Cultural Organization (UNESCO), World Meteorological Organization (WMO), Food and Agriculture Organization of the United Nations (FAO), United Nations International Strategy for Disaster Risk Reduction (UN/ISDR), United Nations Environment Programme (UNEP), United Nations University (UNU), Kyoto University (KU). It was cosponsored by Cabinet Office of Japan (CAO), Ministry of Foreign Affairs, Japan (MOFA), Ministry of Education, Culture, Sports, Science and Technology, Japan (MEXT), Ministry of Agriculture, Forestry and Fisheries of Japan (MAFF), Ministry of Land Infrastructure and Transport, Japan (MLIT), Ministry of Foreign Affairs, Italy, Italian Civil Protection Department (Presidency of the Council of Ministers), Ministry of Environment of the Slovak Republic, Ministry of Environment of the Czech Republic, National Emergency Management Agency of Korea, Science Council of Japan (SCJ), Japan International Cooperation Agency (JICA), International Union of Geological Sciences (IUGS), Academy of Forest, Wood and Environment, Japan (AFWE), and Japan Landslide Society (JLS).

The Honorary Chairpersons for the round-table discussion were Salvano BRICENO (Director of UN/ISDR), Hosny El-Lakany (Assistant Director-General of FAO), Walter ERDELEN (Assistant Director-General of UNESCO), Michel JARRAUD (Secretary-General of WMO), and Kazuo OIKE (President of KU). The Chairpersons were Hans van GINKEL (Rector of UNU), Yoshiaki KAWATA (Director of Disaster Prevention Research Institute of

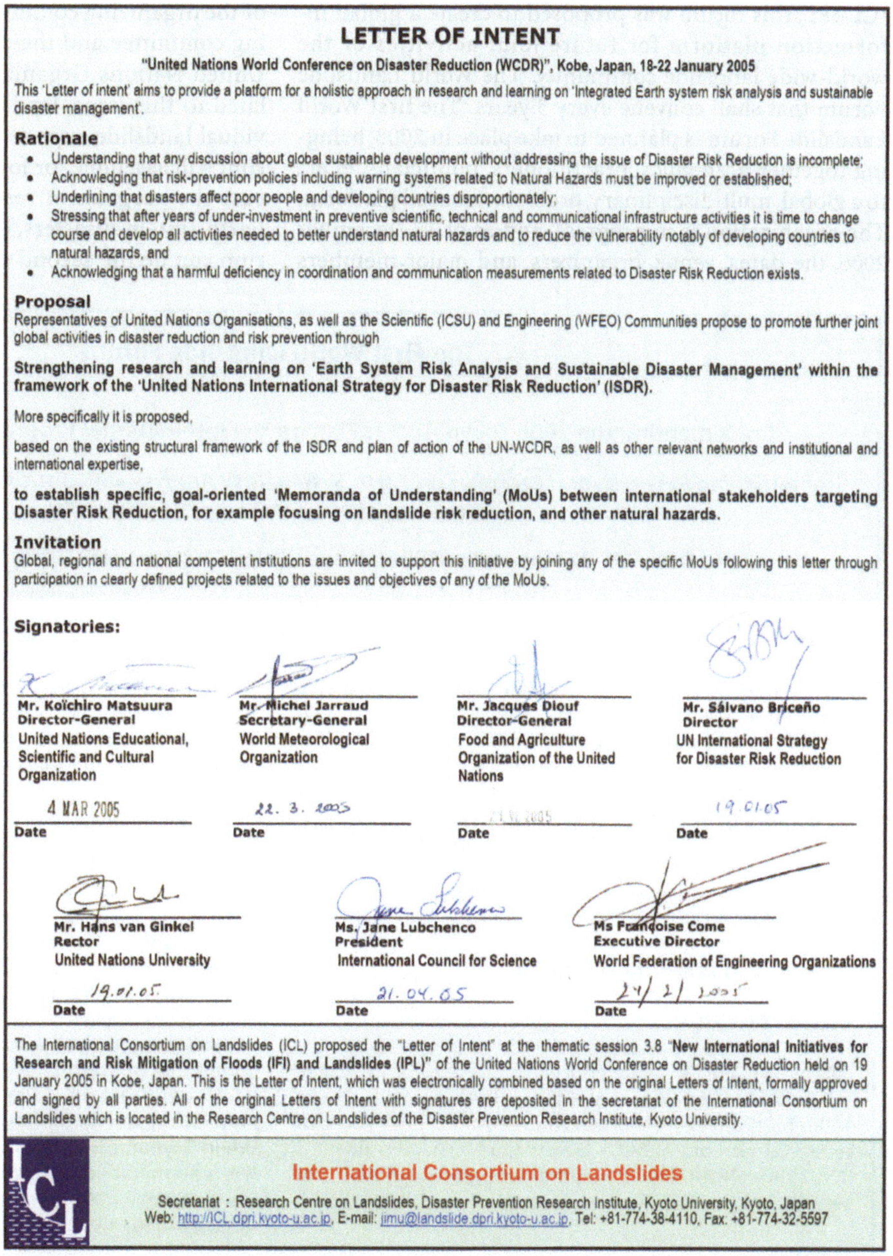

LETTER OF INTENT

"United Nations World Conference on Disaster Reduction (WCDR)", Kobe, Japan, 18-22 January 2005

This 'Letter of intent' aims to provide a platform for a holistic approach in research and learning on 'Integrated Earth system risk analysis and sustainable disaster management'.

Rationale

- Understanding that any discussion about global sustainable development without addressing the issue of Disaster Risk Reduction is incomplete;
- Acknowledging that risk-prevention policies including warning systems related to Natural Hazards must be improved or established;
- Underlining that disasters affect poor people and developing countries disproportionately;
- Stressing that after years of under-investment in preventive scientific, technical and communicational infrastructure activities it is time to change course and develop all activities needed to better understand natural hazards and to reduce the vulnerability notably of developing countries to natural hazards, and
- Acknowledging that a harmful deficiency in coordination and communication measurements related to Disaster Risk Reduction exists.

Proposal

Representatives of United Nations Organisations, as well as the Scientific (ICSU) and Engineering (WFEO) Communities propose to promote further joint global activities in disaster reduction and risk prevention through

Strengthening research and learning on 'Earth System Risk Analysis and Sustainable Disaster Management' within the framework of the 'United Nations International Strategy for Disaster Risk Reduction' (ISDR).

More specifically it is proposed,

based on the existing structural framework of the ISDR and plan of action of the UN-WCDR, as well as other relevant networks and institutional and international expertise,

to establish specific, goal-oriented 'Memoranda of Understanding' (MoUs) between international stakeholders targeting Disaster Risk Reduction, for example focusing on landslide risk reduction, and other natural hazards.

Invitation

Global, regional and national competent institutions are invited to support this initiative by joining any of the specific MoUs following this letter through participation in clearly defined projects related to the issues and objectives of any of the MoUs.

Signatories:

Mr. Koïchiro Matsuura
Director-General
United Nations Educational, Scientific and Cultural Organization
4 MAR 2005
Date

Mr. Michel Jarraud
Secretary-General
World Meteorological Organization
22. 3. 2005
Date

Mr. Jacques Diouf
Director-General
Food and Agriculture Organization of the United Nations
Date

Mr. Sálvano Briceño
Director
UN International Strategy for Disaster Risk Reduction
19.01.05
Date

Mr. Hans van Ginkel
Rector
United Nations University
19.01.05
Date

Ms. Jane Lubchenco
President
International Council for Science
21.04.05
Date

Ms Françoise Come
Executive Director
World Federation of Engineering Organizations
24/2/2005
Date

The International Consortium on Landslides (ICL) proposed the "Letter of Intent" at the thematic session 3.8 "**New International Initiatives for Research and Risk Mitigation of Floods (IFI) and Landslides (IPL)**" of the United Nations World Conference on Disaster Reduction held on 19 January 2005 in Kobe, Japan. This is the Letter of Intent, which was electronically combined based on the original Letters of Intent, formally approved and signed by all parties. All of the original Letters of Intent with signatures are deposited in the secretariat of the International Consortium on Landslides which is located in the Research Centre on Landslides of the Disaster Prevention Research Institute, Kyoto University.

ICL

International Consortium on Landslides

Secretariat : Research Centre on Landslides, Disaster Prevention Research Institute, Kyoto University, Kyoto, Japan
Web: http://ICL.dpri.kyoto-u.ac.jp, E-mail: jimu@landslide.dpri.kyoto-u.ac.jp, Tel: +81-774-38-4110, Fax: +81-774-32-5597

Fig. 1.6. Letter of Intent strengthening research and learning on "Earth System Risk Analysis and Sustainable Disaster Management" within the framework of the United Nations International Strategy for Disaster Risk Reduction (ISDR)

Kyoto University), Badaoui ROUHBAN (Chief, Section for Disaster Reduction of UNESCO), and Kyoji SASSA (President of ICL, Director of Research Centre on Landslides (RCL/DPRI-KU)). As the result of the discussion, participants adopted the 2006 Tokyo Action Plan Strengthening Research and Learning on Landslides and Related Earth System Disasters for Global Risk Preparedness. The action plan is outlined from its preface: The 2006 Tokyo Round Table Discussion "Strengthening Research and Learning on Earth System Risk Analysis and Sustainable Disaster Management within UN-ISDR as Regards Landslides"-towards a dynamic global network of the International Programme on Landslides (IPL) was held at the United Nations University, Tokyo, from 18–20 January, 2006 to formulate a framework for cooperation and to identify focus areas to reduce landslide risk worldwide. The action plan was adopted as a summary of the meeting, to be implemented within the scope of the Hyogo Framework for Action 2005–2015, "*Building the Resilience of Nations and Communities to Disasters*", declared at the United Nations World Conference on Disaster Reduction held in Kobe, Japan, in 2005. The full action plan is attached in the appendix of this volume.

The Tokyo Action Plan proposed the *World Landslide Forum.* Capitalizing on the competence, international experience, and established organizational network of

ICL-IPL, this forum was proposed to create a global information platform for future joint activities of the world-wide landslide community. The World Landslide Forum that shall convene every 3 years. The first World Landslide Forum is planned to take place in 2008, bringing together academics, practitioners, politicians, et al., to a global, multidisciplinary, problem-focused platform. The organization is in progress, and as of 15 December 2006, the dates, venue, organizers, and major members of the organizing committee were decided. The organizing committee and their members consist of all related United Nations Organizations, global stakeholders related to this issue, landslide research groups and individual landslide experts. This first World Landslide Forum will assemble for four days and will extensively discuss the progress of research and learning to mitigate Earth system disasters, focusing on landslides. This forum can be the second step for the global development

The First World Landslide Forum

Implementing the 2006 Tokyo Action Plan on the International Programme on Landslides:

Strengthening Research and Learning on Earth System Risk Analysis and Sustainable Disaster Management within UN-ISDR as Regards "Landslides"

Date: 18–21 November 2008, Venue: United Nations University, Tokyo, Japan

Organizers

International Consortium on Landslides (ICL), United Nations Educational, Scientific and Cultural Organization (UNESCO), World Meteorological Organization (WMO), Food and Agriculture Organization of the United Nations (FAO), United Nations International Strategy for Disaster Risk Reduction (UN/ISDR), United Nations University (UNU), United Nations Environment Programme (UNEP), United Nations Development Programme (UNDP), World Bank (IBRD), International Council for Science (ICSU), World Federation of Engineering Organizations (WFEO), Kyoto University (KU), and Japan Landslide Society (JLS)

International Organizing Committee

Honorary Chairpersons

Salvano BRICENO (Director of UN/ISDR), Jacques DIOUF (Director-General of FAO), Hans van GINKEL (Rector of UNU), Michel JARRAUD (Secretary-General of WMO), Koïchiro MATSUURA (Director-General of UNESCO), Goverdhan MEHTA (President of ICSU), Shuzo NISHIMURA (Executive Vice President of Kyoto University)

Chairpersons

Edward BROMHEAD (University of London, Kings College, London), Paolo CANUTI (European Centre of ICL, University of Florence), Srikantha HERATH (Senior Academic Programme Officer of UNU), Yoshiaki KAWATA (Director of Disaster Prevention Research Institute of Kyoto University), Howard MOORE (Senior Advisor of ICSU), Badaoui ROUHBAN (Chief, Section for Disaster Reduction of UNESCO), Kyoji SASSA (President of ICL, IPL World Centre), Robert SCHUSTER (U.S. Geological Survey)

Members

Robert F. ADLER (Senior Scientist, Goddard Space Flight Center of National Aeronautics and Space Administration (NASA), USA), Peter BOBROWSKY (Secretary General, International Union of Geological Sciences (IUGS), Canada), Earl BRABB (President Emeritus, International Landslide Research Group, USA), Ryu FUKUI (Manager, Tokyo Development Learning Center of the World Bank), See Sew GUE (World Federation of Engineering Organizations (WFEO), Institution of Engineers, Malaysia), Thomas HOFER (Forestry Officer, Specialist on watershed management and mountain development, Forestry Department of FAO, Italy), Alik ISMAIL-ZADEH (Chair, Commission on Geophysical Risk and Sustainability of International Union of Geodesy and Geophysics (IUGG), Germany/Russia), Saroj Kumar JHA (Team Leader, Global Facility for Disaster Reduction and Recovery the World Bank Group, USA), Suzanne LACASSE (Managing Director, Norwegian Geotechnical Institute (NGI)), Willy LACERDA (Chair, Joint Technical Committee 1- Landslides and Engineered Slopes, Brazil), Peter LYTTLE (Coordinator, Landslide Hazard Programme, U.S. Geological Survey), Hideaki MARUI (President, Japan Landslide Society), Gordon MACBEAN (Chair, ICSU Planning Committee on Natural and Human-Induced Environmental Hazards and Disasters, Canada), Norio OKADA, N. (President, Japan Society for Natural Disaster Science), Hari SRINIVAS (Chief, Urban Environmental Management, UNEP, Japan), Niria SANZ (Programme Specialist, World Heritage Centre of UNESCO, France), Yueping YIN (Director, Department of Environmental Geology, China Geological Survey)

Local Organizing Committee

Chairperson

Kaoru TAKARA (Executive Director of ICL, Deputy Director of DPRI/KU)

Deputy Chairpersons

Libor JANSKY (Senior Academic Programme Officer of UNU)
Hirotaka OCHIAI (Vice President of the Japan Landslide Society)

Secretary General

Hiroshi FUKUOKA (Treasurer of ICL, Associate Professor of DPRI/KU)

of the landslide research community and the progress of as a new scientific discipline on landslides, namely "Landslide Science". The already confirmed structure of the Forum is attached at the end of this article. Those who are willing to contribute to the development of research and learning on landslides and other Earth system disasters in the framework of the United Nations International Strategy for Disaster Risk Reduction (UN-ISDR) are requested to join this forum. Any groups or any organizations are invited to propose and organize various types of session, workshops, or symposia to promote research and learning on landslides and other related Earth-system disasters in this forum. Rooms and facilities will be organized by the organizing committee.

References

Antonello G, Casagli N, Farina P, Leva D, Nico G, Sieber AJ, Tarchi D (2004) Ground-based SAR interferometry for monitoring mass movements. Landslides 1(1):21–28

Canuti P (2004) ICL 2003 Varnes Medal. Landslides 1(2):163–164

Cruden DM (1991) Cowboys and Landslides. Landslide News 5:31–32

Cruden DM, Varnes DJ (1996) Landslide types and processes. In: Turner AK, Schuster RL (eds) Landslides: investigation and mitigation. Special Report 247, Transportation research board, US National research council, Washington, D.C., pp 36–75

Sassa K (1992) Access to the dynamics of landslides during earthquakes by a new cyclic loading high-speed ring-shear apparatus (keynote paper). In: 6th International Symposium on Landslides, "Landslides", A.A. Balkema. Christchurch, 3, pp 1919–1937

Sassa K (1996) Prediction of earthquake induced landslides. In: Proceedings of 7th International Symposium on Landslides, A.A. Balkema, Trondheim, 17–21 June, 1, 115–132

Sassa K (2000) Mechanism of flows in granular soils. In: Proceedings of the International Conference of Geotechnical and Geological Engineering, GEOENG2000, Melbourne, 1, pp 1671–1702

Sassa K (2004) The International Consortium on Landslides. Landslides 1(1):91–94

Sassa K (2005) ICL history and activities. In: Sassa K, Fukuoka H., Wang FW, Wang G (eds) Landslides –Risk analysis and sustainable disaster management. Springer-Verlag, Berlin Heidelberg, pp 3–21

Sassa K (2006) "2006 Tokyo Action Plan"-strengthening research and leaning on landslides and related earth system disasters for global risk preparedness. Landslides 3(4):361–369

Sassa K, Fukuoka H, Wang G, Ishikawa N (2004) Undrained dynamic loading ring-shear apparatus and its application to landslide dynamics. Landslides, 1 (1), pp 7–19

Varnes DJ (1958) Landslide types and processes. In: Eckel EB (ed) Landslides and engineering practice. Highway Research Board, Special Report 29, pp 20–47

Varnes DJ (1978) Slope movement types and processes. In: Schuster RL, Krizek RJ (eds) Special report 176: Landslides: analysis and control. Transportation research board, National research council, Washington, D.C., pp 11–33

An Overview of Landslide Problems in the British Isles, with Reference to Geology, Geography and Conservation

Edward N. Bromhead* · Maia L. Ibsen

Abstract. An overview is presented of the landslide problems experienced in Britain, and their primary causes. Principally, landslides occur in Britain where the strata are argillaceous and there is sufficient topographic relief. This combination occurs in swathes through the southern and central parts of Britain and along the lengthy coastline. In these parts of Britain, the main rock types are sedimentary, and they often exhibit low angles of dip. This gives rise to the occurrence of compound landslides, often with a bedding-controlled flat basal shear surface. Issues relating to this type of landslide are discussed in the article, which concludes with a discussion of the conflicts between interests of land users and the need for conservation, especially in some areas that are of significant interest in the history of the geological and other sciences, and to research in the present day.

Keywords. Bedding, slip-surface, landslide distribution

2.1 Introduction

The British Isles contains geomaterials from most of the geological record, and is one of the locations where major steps were taken in unravelling earth history in the early years of the geological sciences. This is evident in the many names which emanate from the British Isles within the geological time scale: these names range from the names of tribes resident in Britain 2 millennia ago (*Ordovician, Silurian*) to the names of particular localities (e.g., *Devonian, London Clay*), and in this respect, the country is grossly overrepresented in geological nomenclature. In general terms, the rocks are oldest in the north of Scotland, and in the west, although episodes of glaciation have emplaced a variety of soils of glacial origin to add interest to the landscape by putting an irregular veneer of weak soils over the oldest and strongest rocks. The southern half of the main island (England, Scotland and Wales) of the British Isles is composed of weak Mesozoic and Tertiary sedimentary rocks, and these usually give rise to the most interesting of the landslides.

The native inhabitants of Britain think that their islands are generally wet, and that this is the main factor governing slope instability, but the appearance of wetness comes not from the total rainfall, but from its generally all year round occurrence, and the absence of a long dry season to generate a build up of soil moisture deficit. Indeed, although changes in rainfall are a major factor in activation and reactivation of landslides, a survey of landslides in the literature revealed (Jones and Lee 1994; DOE/GSL 1987) that the majority of the recorded instances were associated with erosion along the coastline (especially where argillaceous rocks and soils formed sea cliffs of moderate height); along the outcrops of clays of Mesozoic and Tertiary age, which sweep across country in several well-defined bands (Fig. 2.1); in the Coal Measures (Carboniferous); and with a few areas where glacial erosion, and to a lesser extent, river erosion have oversteepened inland valleys – again, in the clay strata, and where glacial materials have been cut into by stream or coastal erosion.

A further factor affecting the coast is rising sea level during the past few millennia, and the consequent increase in erosion rates, so that with only a comparatively few exceptions, the most active landslides are found on the coast where the slide-prone strata are intercepted by the evolution of the coastline. Inland, several generations of transportation technologies (roads, canals, railways and roads again) have cut swathes through the landscape, and these have generated first-time failures in both cut and fill slopes (some of which merit designation as "engineered" slopes) and have reactivated ancient slides, again, where they cross the outcrops of the most slide prone strata. Failures in "engineered" slopes either involve the reactivation of fossil (usually periglacial) landslides, or are generally similar to first time failures in natural slopes, so the inclination of British practitioners is not to rigorously separate natural slopes from man made ones, and that approach is followed here.

Landslides are not confined to particular types of geomaterials, but given a steep enough and high enough slope, together with an input of water and in some cases, ground accelerations due to seismicity (uncommon, but not completely unknown in Britain), landslides of vari-

* The corresponding author of each chapter is marked with an asterisk.

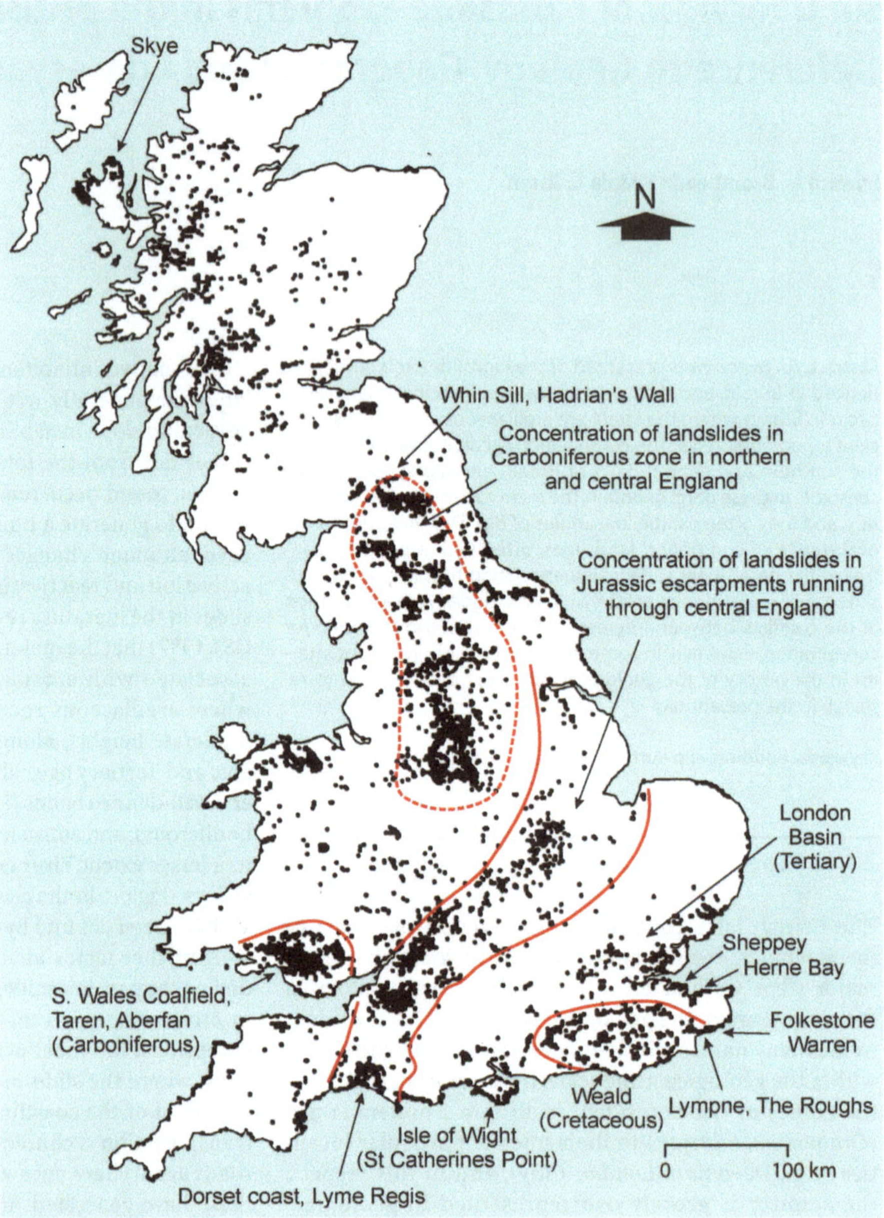

Fig. 2.1. Sketch map of Britain, showing principal geological units and the landslides found in a survey of the literature (Jones and Lee 1994; DOE/GSL 1987). Main locations mentioned in the text are also shown

ous types occur in every rock type. As a result, the usual initial classification of geomaterials on the basis of their origin (igneous, sedimentary, metamorphic) is less use than it might at first seem.

The geotechnical engineer usually finds it convenient to divide geomaterials into "rocks" and "soils" largely on the basis of the ease with which they can be excavated, and developments in the power of excavation plant have made this junction rather indistinct. The landslide specialist may well consider a classification based on whether or not stability is conditional principally on failure through pre-existing discontinuities or through the mass of material between discontinuities as respectively indicating that the material is a "rock" or a "soil".

2.2 Strong Rocks and Discontinuity-controlled Slope Instability

Vulcanism in Britain is ancient, and there are no loose pyroclastic deposits susceptible to slide or flow mechanisms, so that in Britain igneous rocks are usually seen as strong rocks. Most igneous rocks contain joints and other discontinuities resulting from cooling and subsequent regional tectonic stresses. The resulting joints have close to moderate spacing, resulting in comparatively small block sizes when falls occur. A good example of this is the Whin Sill (Fig. 2.2) of columnar-jointed eclogites intruded into limestones of Carboniferous age, and

Fig. 2.2.
The Whin Sill is formed from rocks with columnar jointing, and exhibits scree slopes, although the rock blocks have been exploited for building for two millennia, not least in the construction of a defensive wall in the 2nd century AD which still remains (albeit in ruins) at the crest of the cliff

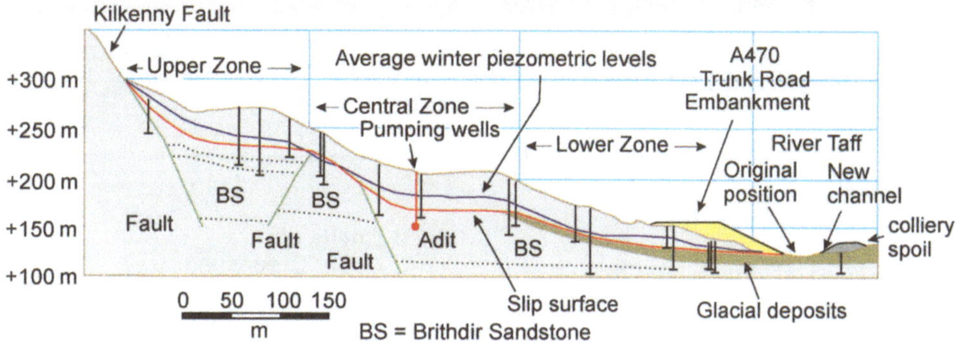

Fig. 2.3.
Bedding control of the basal shear or slip surface of the Taren landslide forms a "staircase", following a downfaulted seatearth in the Coal Measures (Kelly and Martin 1985)

forming a pronounced outcrop across northern England. This was exploited by the Roman Emperor Hadrian for part of his wall (2nd century AD) separating the barbarians to the north from the civilized south. Angular fragments of rock, falling few at a time, tend to accumulate in scree slopes, as seen in the photograph, and provided that care is taken to avoid the areas of highest impact risk, generally pose little problem.

Occasionally, Tertiary lavas have capped some deposits of Mesozoic rocks (e.g., the Jurassic clays of the Lias in the Isle of Skye forming the Trotternish landslides), so that the resulting landslides are compound block slides with strong caprocks, but these are few and far between.

What passes for a mountain even in the highest parts of Britain would be seen as a hill in many other parts of the world, and the relief is not conducive to the formation of large landslides in the stronger rock types. In consequence, large landslides in Britain are often slow moving.

2.3 Weak Rocks and Strong Soils

Glaciation appears to have removed any areas of deep weathering, and so problems of, for example, the appearance of clays in weathered granites such as plague Hong Kong are not experienced. However, materials of glacial origin are found in most of the upland, and some lowland, areas in northern and central Britain, and these prove to be susceptible to being involved in debris flows. Naturally-occurring large debris flows are unusual, although failures in mine waste tips in coal mine and china clay waste have provided examples, including the Aberfan mine waste tip landslide in October 1966 that generated by far the UK's highest toll of casualties – 144 – as fatality-inducing landslides are not unknown, but tend to cause the loss of one or two lives per incident.

Many of the mudrocks of the late Paleozoic, Mesozoic and the Tertiary are comparatively weak, and fall in that transitional group of "weak rocks or strong soils". Perhaps the oldest of these in Britain are the Coal Measures (Carboniferous), best known from its outcrop in the Welsh Coalfield, an upland syncline in south Wales cut through by numerous valleys, and therefore exhibiting landslide after landslide in essentially the same strata on both sides of, and throughout the length of, each valley (Siddle et al. 2000). The downfaulting present in many of the valleys leads to the weak beds forming a "staircase", so that landslides with a step-form of basal slip surface are found. Perhaps the best known example of this from the literature is the Taren landslide (Kelly and Martin 1985; Fig. 2.3) reanalysed in 3D by Bromhead and Martin (2004).

Similar staircase forms are found in many slopes, not generated by down-faulting of the critical beds, but where slipping drapes shallow mudslide debris over a slope with a series of more competent beds, for example, as found in the East Cliffs at Lyme Regis (Clark et al. 2000). These

Fig. 2.4. Landsliding in the coastal cliffs of Barton on Sea (Barton et al. 2006). Several weak zones in the mainly clay sequence are exploited for basal shear surfaces of the slides, most of which are "perched", i.e. daylight in the cliff. This length of coastal slope has been exempted from slope stabilization in view of its geological interest

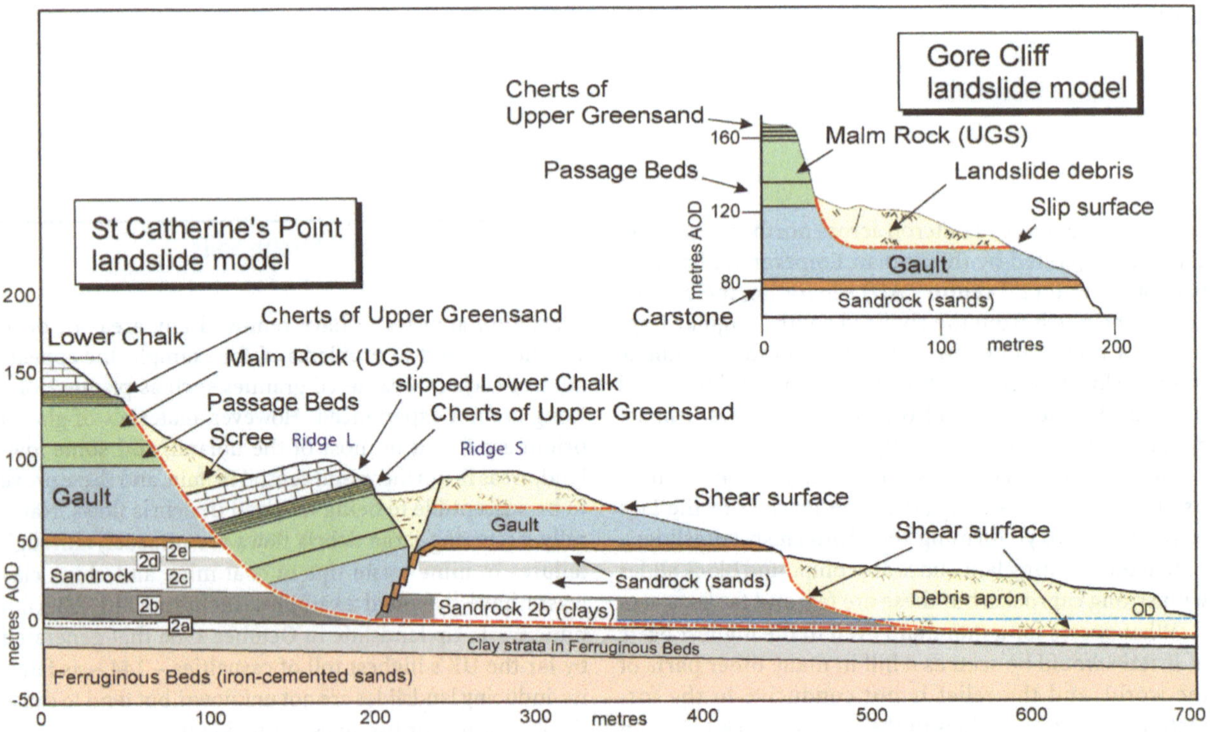

Fig. 2.5. A major bedding controlled landslide at St. Catherine's Point, Isle of Wight (Hutchinson et al. 1991). The slip surface is located in a thin clay layer, which is below sea level. Geological sequence is primarily mid to lower Cretaceous in age. Dips are seaward

staircase sequences are very common, with numerous small steps, or fewer large steps. Examples with fewer steps are seen in the Barton Beds (Barton 1977, 1984; Fig. 2.4), and in the SW coast of the Isle of Wight (Hutchinson et al. 1985), although in these cases the more competent beds are weak sandstones, often of "locked sand" and not the thin muddy limestones of the Lias.

Such landslides that owe their overall morphology to one or a few slide prone horizons in a complicated sequence of beds are commonly termed "bedding-controlled slides" (Bromhead and Ibsen 2002). In exceptional cases, the staircase landform and matching basal slip surface is involved in a massive landslide involving displacement often a deeper bedding surface, usually through a plastic clay layer, as in the case of the St. Catherine's Point landslide on the Isle of Wight (Hutchinson et al. 1991; Bromhead and Ibsen 2002 – see Fig. 2.5), although equally, there are instances of such landslides occurring not as a few blocks, but as many successive more rotational slides, as at Folkestone Warren (Hutchinson et al. 1980). However, bedding-control of the basal shear is a well-established phenomenon. Indeed, the Skye landslides are best understood as bedding-controlled and thus compound landslides with a massive caprock.

Naturally, the presence of caprocks permit the multiple block nature of many of these deep seated bedding controlled landslides to be more readily understood than where they occur in single lithologies, although the same general processes occur in the coastal London Clay cliffs of North Kent (Bromhead 1978; Dixon and Bromhead 2002).

Bedding controlled basal shear locations are not confined to natural slopes, but occur commonly in cuttings, especially for railways (which are often steeper and deeper than the corresponding road cuts). Examples about in the literature, from as long ago as the mid nineteenth century (Gregory 1844), through the "golden era" of London Clay and Imperial College slopes research (Henkel 1957) right through to modern times (Kovacevic et al. 2003).

2.4 Erosion

The dominant form of erosion experienced in the UK is coastal erosion. Sea level has been both slightly higher, and substantially lower, in the comparatively recent past. Raised beaches at Black Rock, Brighton, and on the Isle of Portland, attest to a sea level 6 or 7 m higher during a particular time in the Pleistocene than at present, and sea levels have been much lower in the historic past. For example, the maps of the Italian Coastline by Margottini and Vai (2004) show a sea level 100 m lower than present as recently as 22 ± 2 ka BP, and those few meters higher around 8 ± 1 ka BP. Hutchinson and Bromhead (2002) reproduce a sea level rise curve which shows 15 m of rise in the past 8 millennia. There is conflicting evidence of precise levels, positions of the coastline, and dates, but the overall picture is one of the sea level lowering and abandoning coastal cliffs, which were then able to degrade subaerially with extensive aprons of slide debris at the toe, before being caught up in rising sea level and the buried sea cliff being re-exhumed from underneath its covering of landslides. Such coastal slopes are far from being in equilibrium, and are thus subject to rapidly worsening stability as the slope toes are eroded.

Fig. 2.6.
The world's first cast-iron bridge across the Ironbridge Gorge (a modern name) is threatened by slope movements from both sides of the gorge. A concrete structure has already been built in the riverbed

Inland, glacial erosion has oversteepened slopes in many upland areas, and combined with stream or river erosion, in some cases by streams rejuvenated by lowered sea levels, has had further effects. In the Ironbridge Gorge, where the river Severn leaves a formerly ice-dammed lake (only the river meander across the sediments of the lake bottom attest to its former extent) to pass through a glacial lake erosion channel, the resulting cut through Coal Measures (which is still eroding at the present time) has led to landsliding on both sides of the valley. This gorge exposed all the necessary minerals for iron smelting (iron ore, limestone, low-sulfur coal and fireclay for refractories) in close juxtaposition, and the area became the focus for the industrial revolutions. Inevitably, the stability of the area was further compromised by underground workings and surface spoil heaps, as is everywhere that the Coal Measures have been mined underground. Movement of the as-yet incompletely understood landslides threatens the historic Ironbridge (Fig. 2.6), the world's first cast-iron bridge. (This bridge was assembled with wedges and dovetail joints, because it predated the regular application of bolts or rivets in structural work!). Erosion of these types and slope movements pose problems in the conservation of historic sites and monuments, although equally and paradoxically, some locations are of high cultural value due to erosion. Perhaps the best example of this is the designation of the Dorset Coast as a World Heritage site, on account of its historic role in the development of the geological sciences.

2.5 The Coastline of South Eastern Britain

The coastline of Southeast Britain is formed in sedimentary rocks of Jurassic, Cretaceous and Tertiary age, the majority of these deposits contain thick strata of mudrocks, which have very low angles of dip. Where these strata are appropriately exposed, particularly at the foot of a coastal slope, they give rise to landslides where all or part of the sliding surface follows a single bed of mudrock. Where the dip of the bedding is steep, such landslides are referred to as dip-slope failures, and they are usually translational. However, landslides where the basal sliding surface is controlled by the location and orientation of a single argillaceous bed in the sequence are better termed "bedding-controlled landslides" and they are usually compound slides. In addition, where coastal slopes contain several layers of mudrocks, geometrically similar landslides may occur with perched slide surfaces breaking out at a higher level within the slopes. Landslides with strong bedding-controlled basal shear surfaces are the predominant form of instability along the Southeast coastline of Britain.

In this part of Britain there are numerous coastal towns, which are on, or near, these active landslides. As a consequence of the various threats posed by these landslides, such as cliff retreat or the movement of built-up areas, detailed studies of many of the landslides have been undertaken. Such studies have revealed the control exerted on landslide morphology by geological structure. Often, bedding has a low angle of dip, and such geological conditions usually give rise to a graben like structure or multiple/successive rotational slips. Successive failures lead to a regular, stepped pattern in the landscape. A graben is the zone of subsidence adjacent to the detachment surface of a landslide. As well as subsidence, there may also be block rotation and disruption in this zone.

Studies of aspects of the groundwater behavior, in particular its response to weather and to erosion rate, have provided deep insights into the mechanics of failure and of post-failure movement. However, the geological structure provides the primary control on slip surface position. Slides may take place where the beds dip strongly out of the slope, i.e. coastward or valleyward, or where the dip is gently out of the slope. Paradoxically, they can also occur where the dip is into the slope, but only when the dip is comparatively gentle, since a steep dip into the slope inhibits sliding. The loose terms 'steep' and 'gentle' when applied to dip are relative to the mechanical properties of the soils and rocks involved. For the purposes of this paper, the boundary between the two occurs at a dip equal to approximately half of the residual angle of shearing resistance. In addition, a very steep dip is above about half of the peak angle of shearing resistance.

A major contribution to landslide science from Britain is related to the understanding of the interplay between geological structures and the internals of the landslide systems that result, together with an understanding of the morphology of the topographic surface. The principal factor is the existence of a low-angle clay bed in which the main basal shear surface can form. In one or two rather exceptional cases, this follows the line of an unconformity. There are therefore numerous similarities between sites that have been studied, and broad correlations may be identified. However, the precise mechanics of each landslide are different, affected by local dips, folding, faulting and lithologies. Indeed, the position within the coastal cliff at which the clay beds are found, has an important role in governing, inter alia, the response to toe erosion, itself a variable due to the differing degrees of exposure to marine action around the coastline. Some of these differences are drawn out in the following sections.

2.6 Bedding-Control of Slip Surfaces

The development of the 'slip circle' by Petterson (1955), although a great advance in the understanding of the mechanics of slope failure, locked engineers into a mindset which ignores the role of geological structure on slope failure. Petterson himself merely simplified the geometry of rotational slope failures after Collin (1856) whose elaborate spiral shape curve-fitting exercises ignore the geological control which is evidently present even in thick comparatively uniform strata of mudrocks. By contrast, Gregory (1844) clearly indicates the presence of a slip surface aligned to bedding in his failed slope section even if it is not referred to explicitly in his text.

Terzaghi and Peck (1948) classified slip surface positions as 'slope failure', 'toe failure' and 'base failure' (Fig. 2.7a) where the emergent slip surface position is respectively on the slope face, precisely at the toe of the slope, or where the slip surface is sufficiently deep-seated to emerge beyond the toe of the slope. His sketch was drawn in terms of "slip circles", and a redrawing of the figure to include bedding-controlled landslides creates the variant shown in Fig. 2.7b according to Ibsen and Bromhead (1999). Note that in this figure the dip of the beds is zero, Similar diagrams could be drawn with non-zero dips, both in a seaward sense and in a landward one (for a coastal slope).

Reasons cited in an extensive research of the literature on these landslides for the occurrence of bedding-controlled slide surfaces include:

a zones of high plasticity or liquid limit in which the slide surface forms;
b change in rigidity above a hard band, causing localized strains to occur under stress relief;
c perched water table conditions resulting from a change in permeability;
d basal erosion cutting down to a limiting horizon (hard band).

Various authors have speculated that slip surfaces develop in the unsheared ground at or about the base of a cut slope, thus providing a pre-existing slip surface to be followed by a major failure. Attempts to model or detect such a developing surface have been made on numerous occasions, with varying degrees of success. Such a feature has been found in practice by Burland et al. (1977), and by Cooper et al. (1998). Both of these slopes were in pits where overconsolidated clays were extracted for brick making, but the principle is otherwise the same. In no natural slope or coastal landslide case has the necessary field instrumentation been installed or monitored, although modeling by both limit equilibrium and finite element methods is strongly indicative of such a development.

The evidence as to whether such a basal slide surface is the result of a specific lithology in the geological sequence is mixed, with strong support only at Folkestone, where the slip surface location of the Folkestone Warren landslide complex is clearly associated with a band of high liquid limit clay. No doubt even extremely subtle lithological differences may provide the necessary conditions for slip surface formation.

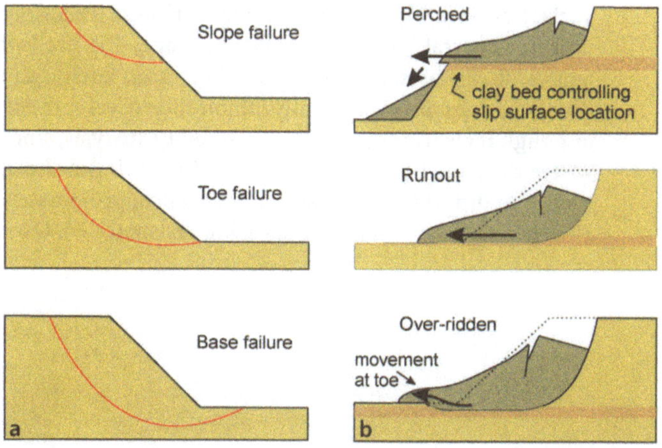

Fig. 2.7. Toe breakout position in the slope for different types of landslide. Terminology in (**a**) by Terzaghi and Peck (1948), and (**b**) by Ibsen and Bromhead (1999)

2.7 Slides in Strata with Low Angle Dips

Landslides where the bedding surfaces follow strata with low angles of dip are common in Southeast Britain, not merely on the coast. A cross section of a slope failure in a railway cutting is given by Gregory (1844) of a landslide that occurred in 1841 in a railway cutting at New Cross in south London. He describes the slip surface as "glass like", and clearly shows it following the bedding (Fig. 2.8), which here is sub-horizontal and is clearly recognisable as it is the transition between the weathered and unweathered beds of the London Clay. Much more recently, Cooper et al. (1998) created a small landslide in Gault Clay in a pit dug for brick-making at Selborne. Their failure (Figs. 2.9 and 2.10) followed the bedding a little above the toe of the slope, and the slip surface could also fairly be described as glass like if one refers to the reflectivity and polish (not transparency).

However, failures in cut slopes are likely to be cleared away during remedial works. Failures in natural slopes, especially along the coast, are rather more permanent features of the landscape.

At Red Cliff (Bromhead and Ibsen 2003), Folkestone (Hutchinson et al. 1980) and Herne Bay (Bromhead 1978), the basal slip surface locations are close to the base of the clay stratum in which they have formed. This has given rise to speculation that the slip surface has formed where the differential shear strains are concentrated. In such a case, the basal shear surface location could result from stress-relief related expansion of the slope following erosion. At other locations, notably Gore Cliff, Bromhead et al. (1991), and the Roughs, Hythe, Bromhead et al. (1998), the critical slip surface location lies, as at Selborne, in the middle of a thick stratum and is not easily correlated with any major lithological change. Chandler (1984) does implicate a change in plasticity for the location of the slip surface at Gore Cliff. The dip at Red Cliff, as in a number of other coastal localities, is into the slope. For the low angles of dip present over most of the area, such a dip direction does not significantly inhibit slide development, although it clearly has an effect in stability analysis, since where the slide behavior is dominated by its basal shear, each 1° of dip adds or subtracts the effect of approximately 1° to the mobilized angle of shearing resistance ϕ'. Over

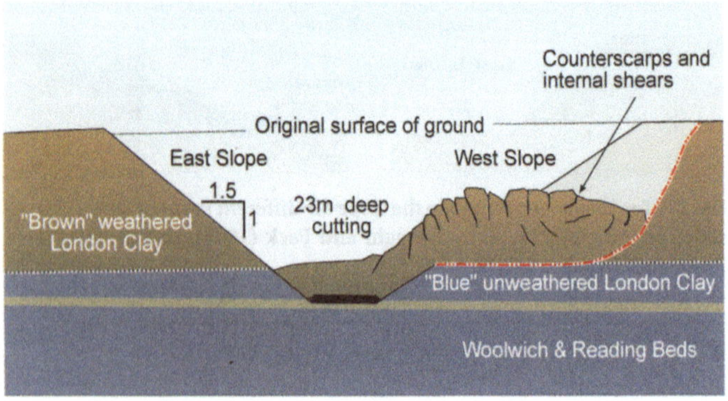

Fig. 2.8. Slide in a railway cutting at New Cross (after Gregory 1844). The basal shear surface was described as "glass-like". It follows a bedding surface at the base of the London Clay formation

a long enough period, however, coastal retreat will encounter significantly different geological structures, with a consequent change in landslide activity and the rate of coastal retreat.

For landslides primarily in debris, it has been speculated that the process of *basal incorporation*, Hutchinson (1972), may occur until numerous mudslides or debris slides scour down to a common basal level. This is inconsistent with some of the mechanisms which pre-shear a particular horizon. Basal incorporation appears to be a major factor with large and active mudslide systems, and these appear to be able to gouge down through hard bands in the sequence, and easily cut through pre-existing bedding-controlled basal shear zones. Underneath existing landslide accumulations, particularly when they are inactive for long periods, the processes of softening and weathering may continue. The slip surface is not a significant barrier to this and it is not surprising, therefore, that the basal shears of bedding-controlled landslides are rather easier to find in active landslides than in quiescent ones, or in some abandoned slopes. This may appear to contradict the observation of basal erosion or basal incorporation, where the materials underneath the slide surface of a mudslide are softened and sheared out

Fig. 2.9.
Slope failure at Selborne induced by groundwater recharge. The slip surface breaks out above the toe of the slope

Fig. 2.10.
Cross section of the failure induced in the Selborne pit showing correlation of slip surface position with the bedding, although the weak bed here (sketched in as a white band) was not detected in the investigation or instrumentation stages in the project

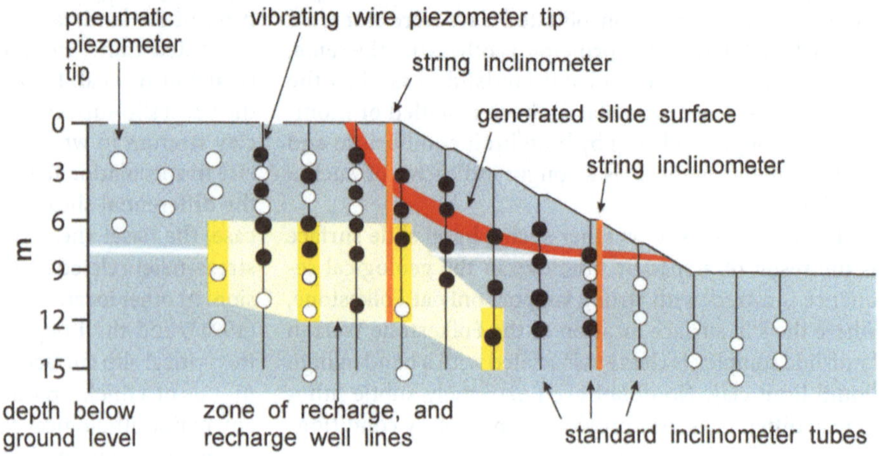

as the mudslide develops. Such basal erosion, scouring down to a common horizon in many, closely adjacent or superimposed mudslides, might be the reason for the occurrence of bedding-controlled landslides where further penetration is inhibited by a strong horizon (e.g., a limestone in the Lias), but this is improbable where any thickness of mudrock separates the strong bed from the slip surface. Indeed, mudslide cascades appear capable of cutting down through virtually any geology if they are active enough.

2.8 Grabens and Graben Geometry

Bedding-controlled slip surfaces, especially where dips are low, give rise to the development of grabens at the head of the slide. A part of the slide mass moves largely in a downwards direction and may rotate slightly, however, the bulk of the slide mass travels in a different direction, along the line of the bedding, and there has to be an accommodation displacement between these two major parts of the landslide. This often takes the form of a shear surface, and thus a counterscarp or antithetic scarp (Cruden et al. 1991) appears in the surface morphology.

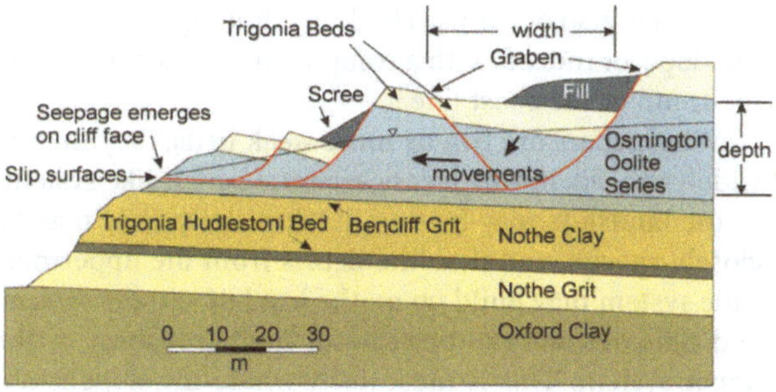

Fig. 2.11. Sketch cross section of the graben-type compound landslide at Red Cliff, Dorset

This situation is illustrated in Fig. 2.11, which is a sketch cross of the section of the Red Cliff landslide. In the upper part of the figure, both the main landslide scarp and the antithetic scarp break out in flat ground at the top of the slope. In this case, the resulting ridge in the slide mass after movement is at, or close, to the elevation of the crest of the pre-slide slope. An impression of decrease in level is gained, however, when the main slide movement is down-dip. Alternatively, the breakout position of the antithetic scarp may be in the main pre-slide slope. This leads to a somewhat different morphology, particularly where the slope has a thick, strong, caprock (Fig. 2.5 – the St. Catherine's Point landslide). The graben in such a case, if it may still be called a graben, is a separately back-tilted mass of soil and rock.

Cruden et al. (1991) suggest a correlation between graben depth and width and Ibsen and Bromhead (1999) have extended this correlation for coastal landslides in Southeast Britain. A limited amount of experience suggests that the head-to-toe distance is longer where the dip is out of the slope (seaward) than when it is into the slope (landward). Furthermore, the latter case also tends to push the counterscarp towards the toe, and therefore away from any flat ground at the head of the slope.

Grabens in landslides take their characteristic morphology (Figs. 2.12 and 2.13) from the changes in direction in the slip surface shape, and the abruptness or otherwise of those changes. The model for graben geometry developed by Cruden et al. (1991) stated that it was possible to estimate the depth of the slip surface, from the original ground surface, using the initial graben width, based on observations of ten small landslides in Alberta, Canada. They indicated that the depth to the translational slip surface was approximately 1.1 times the original graben width. Ibsen and Bromhead (1999) have extended this field relationship between graben width and depth to slip surface, with additional

Fig. 2.12.
Perched slip surfaces result from a weak bed that daylights in the face of a slope

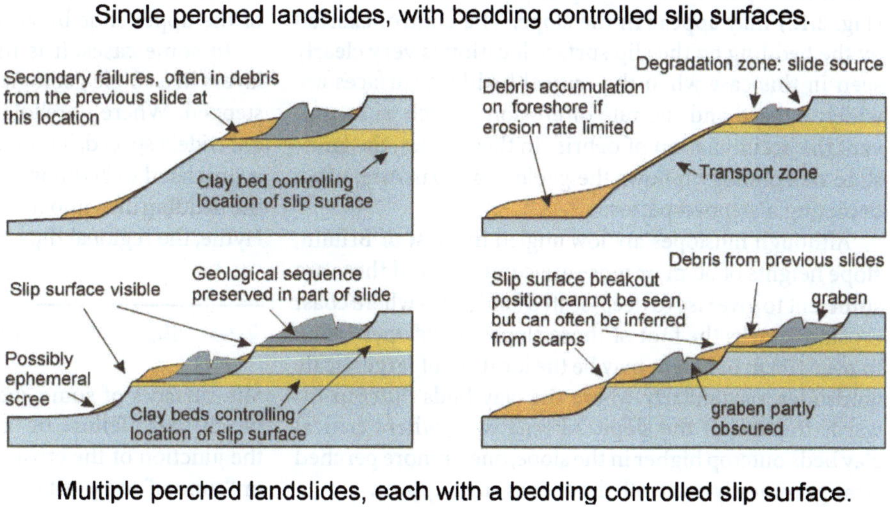

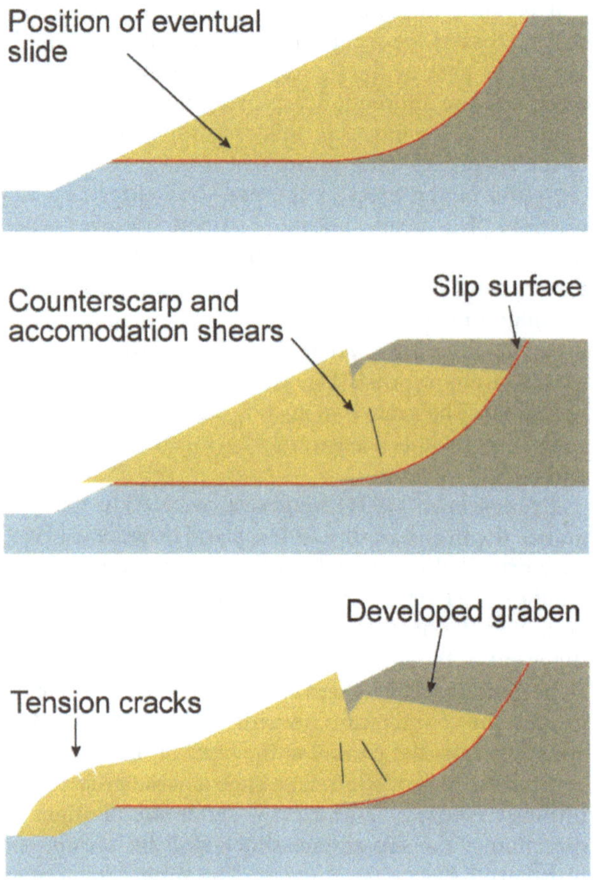

Fig. 2.13. Evolution of a graben in a compound landslide

data and refining the relationship using further regression analysis techniques.

2.9 Perched Slip Surfaces

Where clay beds outcrop above the overall toe of a slope, one or more perched stratigraphically-controlled benches (Fig. 2.12) may appear in the slope. The control exerted by the bedding on the slip surface location is very clearly seen in this case when the critical bedding surfaces are widely spaced and the rate of erosion is such as to prevent the accumulation of debris. In these cases, the landslide morphology follows the geological structure, often depicting a stepped pattern.

Although hillslopes are low angled in most of Britain, slope heights of 30 m or more are common, and these are sufficient to give rise to deep-seated landslides where coast erosion attacks the foot of those slopes. Hillslopes rising to over 100 m in height may be the location of larger-scale landslides, particularly where the clay beds outcrop towards the foot of the slope. Alternatively, where critical clay beds outcrop higher in the slope, one or more perched stratigraphically-controlled benches may appear.

The control exerted by bedding on slip surface location is very clearly seen where the slip surfaces are perched in the slope. Perched systems of landslides are related to the occurrence of a weak bed or beds (usually clays) in the geological sequence and located above beach level. In these cases, the landslide morphology follows the geological structure, often depicting a stepped pattern.

In the simplest cases, the sequence contains only one weak bed, and a single perched slide appears in the slopes. Even where the elevation of the weak bed is only a little above beach level and the sea cliff is therefore low, any debris which spills over the sea cliff from slide activity at a higher level is readily removed, thus keeping the sequence clean. Even if debris is not removed, then it takes the form of mudslides and screes (depending on the nature of materials present and their water content), which are easily recognised as different from the higher-level slide morphology.

Single perched slides may also occur in abandoned or defended cliffs. For example, the slopes at The Roughs, Hythe, Bromhead et al. (1998) or at Hadleigh Castle, Hutchinson and Gostelow (1976). In such cases, the elevation of the basal slip surface may be obscured by the build-up of slide debris and vegetation. The slopes underneath the rotational slide may be occupied by a *transport zone* where mudsliding occurs, predominantly parallel to the terrain slope, and an *accumulation zone.* The slides may also occur with bedding dipping into or out of the slope, or indeed, with a component of dip obliquely to the trend of the coast line.

Where there are two or more weak beds, they may be widely spaced, in which case separate bench-like systems of the landslide may be found. Alternately, they may be closely spaced, and then the debris from the uppermost slide system may build up on the head of a lower system, and thus obscure the position of the basal shear of the upper system. This is most likely where the slope is not subject to active erosion at its toe, because the process of undrained loading (Hutchinson and Bhandari 1971) leads to increased activity in the lower slopes and re-exposure of the upper slide break-out position.

In some cases, it is inferred that there are two weak layers in a single geological unit, so that the slip surface is stepped. Where borehole data is used, and the boreholes are widely spaced, it may appear that the basal slide surface is unrelated to bedding. This is particularly marked where the bedding dips into the slope. For example, at the Roughs, Hythe, the regional dips have a component inland.

2.10 Slip Surfaces at or close to the Base of a Slope

Slip surfaces of some coastal landslides break out in the beach (base failure or "over-ridden" toe position) or at the junction of the erosion platform and slope (toe failure or "runout" toe position). Classic examples are the major landslides at Sheppey and Herne Bay in the London Clay, Dixon and Bromhead (1991, 2002), Bromhead (1978); in the Chalk, Upper Greensand and Gault of Folkestone Warren, Hutchinson et al. (1980), Trenter and Warren (1996); and in the Ventnor Undercliff, e.g., at St. Catherine's Point, Hutchinson et al. (1991), where the Sandrock sequence of the Lower Greensand contains the basal shears.

Folkestone Warren (Hutchinson et al. 1980) occupies the entire length of coastal cliff where the Gault outcrops. At its western end, the base of the Gault emerges above the shore platform. The Warren landslides extend to the east where the Gault disappears beneath sea level. Trenter and Warren (1996) report movement patterns which are greater to the west, where there is less passive restraint to the breakout of the toe of the landslides. These slides, therefore, appear to be rotating in plan. Also, the Warren becomes generally wider towards the west possibly as a consequence of this increased activity.

The Herne Bay landslides, Bromhead (1978), show another interesting facet of behavior. The three major landslide remnants encountered by slope stabilization works are more clearly bedding-controlled where the junction with the underlying Oldhaven Beds is close to sea level (Miramar landslide), and less so where the junction lies deep below sea level (Beacon Hill landslide). The Miramar landslide was a first time failure in 1953, and contemporary photographs show a clear graben. In contrast to Folkestone Warren, the Miramar landslide was more active at its western end, where the slip surfaces were below sea level, although their depth was never sufficient to create a major passive resistance, and marine erosion was able to make a significant impact on what passive restraint there was, until the site was stabilized in 1969.

At Warden Point, Isle of Sheppey, Kent (Dixon and Bromhead 2002), the basal slip surface position lies below sea level, and the rapid coastal retreat there has left scars of "planed off" landslides in the foreshore, where they are clearly visible at low tide (Fig. 2.14).

2.11 Identification of Slip Surface Position

Hutchinson (1983) discusses the problems associated with locating slip surfaces in pits, boreholes and soil samples. However, the breakout of a bedding-controlled slip surface may be evident from the morphology of the slope alone. For example, the location of bedding-controlled sliding surfaces perched in a slope in an active landslide system can be identified visually. Clearly, for slopes covered in vegetation, or greatly degraded and covered in debris, this is not possible. Furthermore, since the outcrop of the line of the slip surface is on or near a vertical face, it may be identified most easily visually in the field, or on photographs. Oblique air photographs are best: vertical photographs are of little use in showing the detail on vertical faces.

Fig. 2.14. Scars in the foreshore from previous landslides at Warden Point, Isle of Sheppey (Dixon and Bromhead 2002). The basal slip surfaces are below sea level, and ongoing coastal recession planes off the slide masses, leaving these scars in the beach

Many of the smaller landslides are primarily slides of debris, with the original in-situ geological succession above the basal shear unrecognisable except in small blocks. This contrasts with the larger landslides, where massive blocks of landslipped material may retain its original appearance and succession. Where the slides are of debris, and are perched in the cliff, the basal shear surface outcrop may be marked by seepages, by a color change, or by a step as the debris moves forward *en masse*. The latter feature appears from time to time, and will be recognised in the field or in terrestrial photographs (Fig. 2.15 – a bedding-controlled basal shear of a compound landslide in the rapidly-eroding cliffs at Fairlight in Sussex). It is rarely if ever large enough to be identified in an air photograph (although conceivably it could be if a shadow was cast) since sufficient forward projection to be seen would result in the collapse of the toe of the slide mass which being composed of debris is weak and to a certain extent lacks cohesion.

A major problem with studies of these coastal landslides is that it is difficult to view them from a suitable angle to understand their morphology. Stereoscopic vertical air photographs can help, but often need a trained eye to interpret them. However, low-angle oblique air photographs clearly reveal the details of the overall morphology and in many cases, the location of the breakout position of the bedding-controlled slip surfaces can be identified. The depth of slip surface in the slide mass may also be estimated from the graben geometry.

Fig. 2.15. Bedding-controlled slip surface in the Fairlight Clays (Sussex)

2.12 Three Dimensions and the Plan Shape of Landslides

Bromhead et al. (2002) describe a small coastal landslide of the bedding-controlled type at Hanover Point, Isle of Wight. This landslide occurs in mudrocks of Wealden age. At the site of the landslide, they dip inland, and also have a coast-wise component of dip. The morphology of the landslide is seen to have a dip into the slope as erosion progressively uncovers the critical bed. At first, the bed daylights high in the slope face, and the only possible landslide is a small one. As the slope retreats, the landslide increases in size and volume. Eventually, at extreme stages in erosion, the critical bed disappears below the toe of the slope and landsliding related to it can no longer occur. The progression when coastal retreat occurs through a sequence which contains a slide-prone horizon (usually a plastic clay) dipping landward, follows the slide as it enlarges while cliff retreat takes place. Eventually, the slide-prone horizon plunges beneath the foot of the sea cliff, and active sliding ceases.

In contrast to the situation where the beds are horizontal, and where parallel retreat of the slope is possible, where the beds dip into the slope, the nature and morphology of the slides relate to the relative erosion of the slope.

It is possible to view this series of sections not merely as a progression in time at a particular location, but also for them to be a series of sections along a valley side or coastline. Interestingly, the resulting slide would have its least width where the slip surface daylights at a high level in the slide, broadening as the bed containing the basal shear progressively descends towards the foot of the slope, and very rapidly narrowing again as the critical bed descends further, to a position below the toe of the slope which does not permit sliding to occur. The plan shape of the landslide is therefore controlled by bedding in a related way to the cross-sectional shape.

2.13 Conservation and Conflict

In the late nineteenth and early twentieth century, it became the practice in Britain to protect vulnerable coastal sites from erosion and inundation. Such sites included both low-lying ground and some sites where coastal slopes and cliffs of weak rocks are susceptible to landsliding. In the latter case, cliff retreat rates in south-eastern England can be of the order of a meter per year. It has probably never been economic to protect agricultural land, nor rural settlements, and whatever form of cost-benefit analysis is employed (Lee and Jones 2004), coastal protection and stabilization schemes are only meaningful where the built environment has a significant value, for example, a large village or small town is the minimum that generates the benefit to counter the major cost of building the works in the first instance, and of maintaining them thereafter. However, some works have been constructed to protect land of marginal value, possibly for political rather than economic reasons, although this practice is less prevalent than some decades ago. Since it is difficult to separate the effects of coast erosion from those of the movement of coastal landslides, the costs of the former include the stabilization of the latter.

Towards the end of the twentieth century, the governmental agencies in the UK that fund major coastal defense and protection schemes decided to reconsider their policies regarding coastal defense and protection (see Clarke and Lee 2003). In part, this was driven by financial motives, but in part, it has been for scientific reasons. Successful defense of soft cliffs from erosion in some places has depleted coastal sediments, and changed the littoral hydraulics to the extent that neighbouring areas have suffered accelerated erosion. When taken in conjunction with recognition of the likely effects of (*a*) rising sea levels, (*b*) climate change, and (*c*) ongoing eustatic changes in land levels in SE England, the economic and technical arguments have swung against further new construction, or even maintaining existing schemes. Finally, the scientific and environmental lobbies have caused some coastal areas to be designated and listed as of such scientific importance that the presumption is that they will not be defended by major engineering works. Such designations include being listed as being a Site of Special Scientific Importance (SSSI), and Area of Outstanding Natural Beauty (AONB), or being protected as a wildlife habitat under the European Wildlife Habitats Directive.

Such conflicting interests, especially at a time of changing climate and rising sea levels, guarantee that the management of unstable slopes will never degenerate to a simple problem in elementary geotechnics, but will remain a challenge for the foreseeable future.

References

Barton ME (1977) Landsliding along bedding planes. Bull Int Assoc Eng Geol 16:5–7

Barton ME (1984) The preferred path of landslide shear surfaces in over-consolidated clays and soft rocks. In: Proceedings 4th International Symposium on Landslides, Toronto, 3, pp 75–79

Bromhead EN (1978) Large landslides in London Clay at Herne Bay, Kent. Q J Eng Geol 11:291–304

Bromhead EN, Ibsen M-L (2004a) Bedding-controlled coastal landslides in Southeast Britain between Axmouth and the Thames Estuary. Landslides 1(2):131–141

Bromhead EN, Martin PL (2004b) Three-dimensional limit equilibrium analysis of the Taren landslide. In: Jardine RJ, Potts DM, Higgins KG (eds) Advances in geotechnical engineering: the Skempton Conference. Thomas Telford, London, 2, pp 789–802

Bromhead EN, Chandler MP, Hutchinson JN (1991) The recent history and geotechnics of landslides at Gore Cliff, Isle of Wight. Slope Stability Engineering – Developments and Applications, Thomas Telford, pp 189–196

Bromhead EN, Hopper AC, Ibsen M-L (1998) Landslides in the Lower Greensand Escarpment of South Kent. Bull Eng Geol Environ 57(2):131–144

Bromhead EN, Ibsen M-L, Papanastassiou X, Zemichael AA (2002) Three-dimensional slope stability analysis of a coastal landslide at Hanover Point, Isle of Wight. Q J Eng Geol Hydroge 35:79–88

Brunsden D, Jones DKC (1976) The evolution of landslide slopes in Dorset. Philos Trans R Soc London, Ser A 283:605–631

Burland JB, Longworth TI, Moore JFA (1977) A study of ground movement and progressive failure caused by a deep excavation in Oxford Clay. Géotechnique 27:557–591

Chandler MP, Hutchinson JN (1984) Assessment of relative slide hazard within a large pre-existing coastal landslide at Ventnor, Isle of Wight. In: Proceedings 4th International Symposium Landslides, Toronto, pp 517–522

Clark AR, Fort DS, Davis GM (2000) The strategy, management and investigation of coastal landslides at Lyme Regis, Dorset, UK. In: Bromhead E, et al. (eds) Landslides in research, theory and practice, Vol. 1, Thomas Telford. London, pp 279–286

Collin A (1856) (trans. W. Schriever) Landslides in clay. University of Toronto Press

Cooper MR, Bromhead EN, Petley DJ, Grant DI (1998) The Selborne cutting slope stability experiment. Géotechnique 48(1):83–101

Cruden DM, Thomson S, Hoffmann BA (1991) Observation of graben geometry in landslides. Slope Stability Engineering – Developments and Applications, Thomas Telford, pp 33–35

Dixon N, Bromhead EN (1991). The mechanics of first-time slides in the London Clay Cliffs at the Isle of Sheppey, England. In: Chandler RJ (ed) Slope stability engineering – developments and applications. Thomas Telford, pp 277–282

Dixon N, Bromhead EN (2002) Landsliding in London Clay Coastal Cliffs. Q J Eng Geol Hydroge (35)4:327–343

DOE/GSL (Department of the Environment, UK/Geomorphological Services Ltd.) (1987) Review of research into landsliding in Great Britain (12 vols)

Gregory CH (1844) On railway cuttings and embankments; with an account of some slips in the London Clay, on the line of the London and Croydon Railway. Minutes Proc Inst Civ Engrs 3: 135–145

Hutchinson JN (1969) A reconsideration of the coastal landslides at Folkestone Warren, Kent. Géotechnique 19:6–38

Hutchinson JN (1970) A coastal mudflow on the London clay cliffs at Beltinge, North Kent. Géotechnique 20(4):412–438

Hutchinson JN (1983) Methods of locating slip surfaces in landslides. Bull Assoc Eng Geol XX(3):235–252

Hutchinson JN, Bhandari, RK (1971) Undrained loading: a fundamental mechanism of mudflows and other mass movements. Géotechnique 21(4):353–358

Hutchinson JN, Bromhead EN (2002) Keynote paper: Isle of Wight landslides. International Conference on Landslides and Coastal Management. Thomas Telford, London, 1, pp 3–70

Hutchinson JN, Gostelow TP (1976) The development of an abandoned cliff in London Clay at Hadleigh, Essex. Philos Trans R Soc London, Ser A 283:557–604

Hutchinson JN, Bromhead EN, Lupini JF (1980) Additional observations on the Folkestone Warren landslides. Q J Eng Geol 13:1–31

Hutchinson JN, Chandler MP, Bromhead EN (1981) Cliff recession on the Isle of Wight S-W coast. In: Proceedings 10th International Conference on Soil Mechanics and Foundation Engineering, Stockholm 3(12):429–434

Hutchinson JN, Bromhead EN, Chandler MP (1991) Investigations of the landslides at St Catherine's Point, Isle of Wight. In: Chandler RJ (ed) Slope stability engineering – developments and applications. Thomas Telford, London, pp 169–179

Ibsen M-L, Bromhead EN (1999) Head scarps and toe heaves. Z Geomorphol, special issue

Jones, DKC, Lee EM (1994) Landsliding in Great Britain. HMSO, London

Kelly JMH, Martin PL (1985) Construction works on or near Landslides. In: Morgan CS (ed) Landslides in the South Wales Coalfield. Polytechnic of Wales, pp 85–106

Lee EM, Jones DKC (2004) Landslide Risk Assessment. Thomas Telford, London

Margottini FA, Vai GB (2004) Litho-palaeoenvironmental maps of Italy during the last two climatic extremes. Museo Geologico Giovanni Capellini, Bologna

Petterson KE (1955) The early history of circular sliding surfaces. Géotechnique 5:275–296

Pitts J (1983) The temporal and spatial development of landslides in the Axmouth-Lyme Regis undercliffs, National Nature Reserve, Devon. Earth Surf Proc Land 8:589–603

Pitts J, Brunsden D (1987) A reconsideration of the Bindon landslide of 1839. Proc Geol Assoc 98:1–18

Siddle HJ, Bromhead EN, Bassett MG (2000) Landslides and landslide management in South Wales. Geological Series No 18. National Museum of Wales, Cardiff, Wales

Terzaghi K, Peck RB (1948) Soil mechanics in engineering practice. John Wiley & Son, New York

Trenter NA, and Warren CD (1996) Further investigations at the Folkestone Warren landslide. Géotechnique 46(4):589–620

Chapter 3

Considerations about the Mechanics of Slow Active Landslides in Clay

Luciano Picarelli

Abstract. Slow active landslides in clay include slides, mudslides and spreads. Movement is induced by any change of effective stress and by creep; in the very long-time, some role may be played by a change in soil properties. Looking at geological phenomena causing movement, pore pressure fluctuations and erosion have a strong influence on shallow translational slides and mudslides, while creep or erosion and other geological phenomena of stress relief govern movement of deep seated slides and spreads. In several cases, excess pore pressures generated by changes of boundary conditions may play a significant role.

Keywords. Clay, slide, mudslide, spread, active slow slope movement

3.1 Introduction

Slow landslides are widespread in geomorphological contexts where stiff clays or clay shales crop out. According to old chronicles it can be argued that there are landslides which have been active from thousands of years.

The risk posed by slow slope movements is rather low, but management of landslides interacting with urban areas, infrastructures and lifelines raises peculiar problems since, even though the evacuation of people is not a pressing need, in the long-term movements can severely damage structures or interrupt the serviceability of lifelines, while slope stabilization or reinforcement of exposed elements is too expensive or non-effective. As a consequence, land management requires a complex and delicate cost-benefit approach in which assessment of future slope behavior and comparison among the effects of different remedial measures take a crucial role.

This paper concerns natural slopes in clay, and is mainly based on the state-of-the art report presented by L. Picarelli and C. Russo at the 9th International Symposium on Landslides (Rio de Janeiro 2004), but includes further data and considerations drawn from more recent papers and reports.

3.2 General Features of Slow Active Landslides

Slow active landslides involve a number of natural slopes much larger than what would be expected. The main types of slow active landslides are slides, mudslides and spreads. Movement is a result displacement along internal discontinuities (as the slip surface) and of internal strains. The landslide body is generally subjected to a constant driving force due to self-weight, thus movement is triggered by change of the resisting force caused by change of boundary conditions, and by viscous deformations. Furthermore, in the long-term, any change of soil properties due to weathering or to other processes of soil deterioration, can play some role. Finally, change of slope morphology caused by movement itself may restrain further movement. Therefore, the displacement rate depends on geometric features of the landslide body, on the rate of effective stress change and of soil deterioration and on viscous properties of both landslide body and discontinuities.

Usually, the expression "active landslide" refers to (those) soil bodies which have experienced a general failure (in the sense discussed by Urciuoli et al. 2007) and are still moving. Their behavior is represented by the last part of the schematic diagram reported in Fig. 3.1, which depicts a typical time-slope displacement relationship from the so-called pre-failure stage, i.e. before general slope failure, to the post-failure stage (Leroueil et al. 1996). Slow slope movements have much a longer life than rapid landslides: in some cases they last centuries or even thousands of years. Even though movement can appear permanent, i.e. characterised by a constant velocity, very often it displays a continuous change of the displacement rate, as a function of pore pressure fluctuations.

According to the classification proposed by Cruden and Varnes (1996), landslides can be classified slow as far as their velocity is less than 13 m month^{-1}, but for a velocity less than 16 mm yr^{-1}, they are categorised as extremely slow. However, here, the adjective slow will be used in a flexible sense, to indicate a long-lasting movement unable to provoke, in the short-time, any significant consequence on people and goods.

Many active landslides in clay (slides and mudslides) present a slip surface, generated by shear strain localization and soil rupture, over which the soil mass slides advancing on the slope. However, just after failure, mudslides display a peculiar flow-like style attaining a rapid to moderate velocity followed by a slow decline, while the style of movement progressively turns to slide (Picarelli 2001).

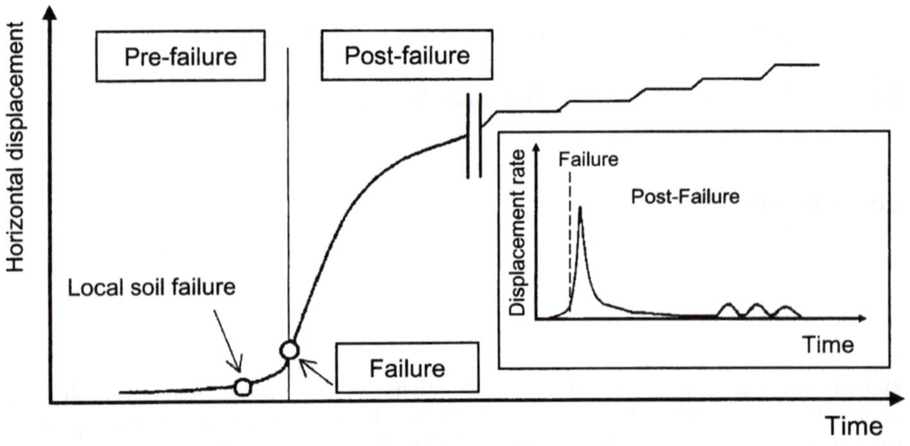

Fig. 3.1. Simplified scheme of slope movement from pre-failure to arrest (after Picarelli 2000)

In some cases people include in the category of landslides soil masses which did not experience any general slope failure, but are subjected to continuous movement due to perceivable internal strains. For instance, lateral spreads caused by valley rebound can be driven by deformation of even large soil masses, but not by slope failure. This implies that a persistent slip failure does not necessarily exist. Also movements caused by creep or by internal soil deformation provoked, for instance, by ongoing progressive failure or by internal stress redistribution, reveal a pre-failure more than a the post-failure stage. Moreover, these phenomena will not necessarily lead to general slope failure.

3.3 Considerations about the Mechanics of Active Slides and Mudslides

3.3.1 Active Slides

Slow active slides generally move over a shear zone located at the base of the landslide body. This zone is the result of a complex process of strain localisation occurring in the pre-failure stage, whose final effect is the formation of a shear surface, also called slip or sliding surface, internal to the shear zone.

Urciuoli (2002) analyses the formation of the shear zone in the case of infinite slope, showing that is accompanied by rotation of the principal stresses and formation of minor shears (Skempton 1967; Morgenstern and Tchalenko 1967). As the direction of principal stresses becomes consistent with the direction of the theoretical failure plane, i.e. parallel to the ground surface, a persistent slip surface eventually forms and slope failure takes place. This occurs simultaneously in all points of the failure plane because of uniformity of the state of stress. The thickness of the shear zone depends on the initial state of stress and on the shear strength of soil: for a critical value of the coefficient of earth pressure at rest, it is nil; for high values of OCR it is quite small; for small values, it is relatively large. In other cases formation of the shear zone is a more complicated process because of non uniformity of the state of stress. Generally it can be said that it starts forming in the most stressed part of the slope, propagating into the soil mass as a result of a mechanism of progressive failure which culminates in the general failure (Urciuoli and Picarelli 2004).

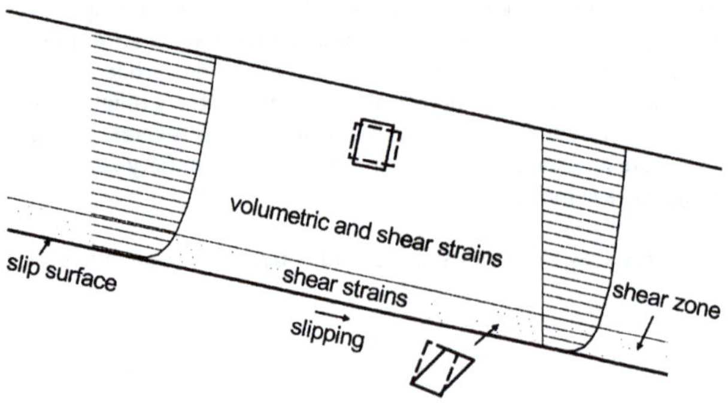

Fig. 3.2. The nature of movement of slides (from Picarelli and Russo 2004)

In the post-failure stage, the mobilized soil mass can eventually move over the slip surface. Even if immediately after failure the landslide may experience a strong acceleration because of brittle soil behavior, eventually it progressively slows down due to change of both slope morphology and Generalised Brittleness Index of soil (D'Elia et al. 1998), while the shear strength along the slip surface approaches the residual value. At residual, the resisting force along the slip surface depends only on the normal effective stress and, possibly, on viscous effects associated with the displacement rate. In the final stage, i.e. once the friction angle has attained a constant value, slope movement is governed by the cyclic balance between driving force and resisting force, being the result of sliding along the slip surface and of internal strains (Fig. 3.2). Slipping occurs every time the residual strength is mobilized along the shear surface, for instance when pore pressure trespasses a critical value. Internal strains are caused by changes of the effective stress field and by viscous phenomena, but in the very long-term additional strains can be provoked by change of stiffness due to soil deterioration. Concerning this point, Picarelli (2000) shows that the residual strength of clay shales may decrease with time because of changes of grain size due to progressive breaking of bonds linking aggregates of clay particles. In addition, accounting for data provided by Di Maio (1996) and by Di Maio and Onorati (2000), Picarelli et al. (2006) illustrate possible long-term effects of infiltration of fresh water in marine clay of high plasticity leading to a decrease of both peak (softening) and residual shear strength, but certainly, even soil stiffness experiences some decay. Some numerical analyses on the effects of soil deterioration on slope movements are reported by Yoshida (1990) through a manipulation of the Hoek and Brown (1980) strength criterion.

The behavior of translational slides (an example is reported in Fig. 3.3a) is governed by the component of the driving force in the direction of movement. Since such a component is essentially constant, the displacement rate depends on changes of the resisting force caused by changes of effective stress and on soil viscosity. Quite a similar framework held for compound slides, most of which are characterised by an almost rectilinear and subhorizontal lowermost part of the slip surface (Fig. 3.3b). However, in this last case the non uniformity of the state of stress may determine a complex interaction between the rear and the front part of the landslide body. The behavior of circular slides strongly depends on changing balance between the driving and resisting force due to modifications of the morphology caused by movement itself (D'Elia et al. 1998). As a consequence, circular slides generally run short distances and are active for quite a short time.

In clay deposits the groundwater table is located at rather a small depth from the ground surface, and fluctuates during the year as a consequence of seasonal water recharge and discharge. In uniform clay, the magnitude of pore water fluctuations decreases with depth because of the time required by pore pressure to reach equilibrium with boundary conditions, compared to the duration of the wet season. Kenney and Lau (1984) present excellent experimental data regarding a slope in Ontario, while Cavalera (1977) reports analytical solutions of this problem for sinusoidal changes with time of hydraulic conditions at the ground surface. This suggests that pore pressure fluctuation due to rainfall (or to ice melting) strongly governs the behavior of shallow translational slides, while its role is minor for deep-seated landslides. In contrast, the behavior of deep seated slides should be essentially governed by creep, i.e. by soil deformation occurring under an essentially constant state of stress, or by stress changes due to geological phenomena, as erosion. A well known slide driven by fluvial erosion is the huge La Frasse slide, which has an extension of 42 km^3. Monitoring dates back 170 years, since construction of the major cantonal road connecting the town of Aigle to the

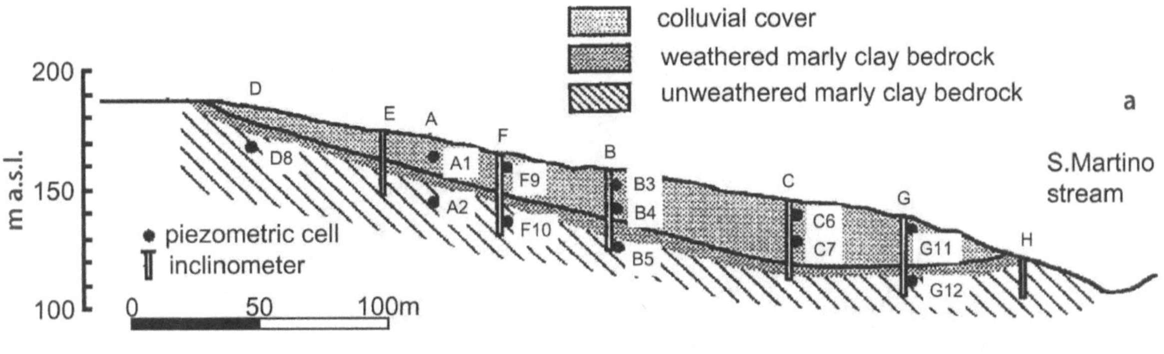

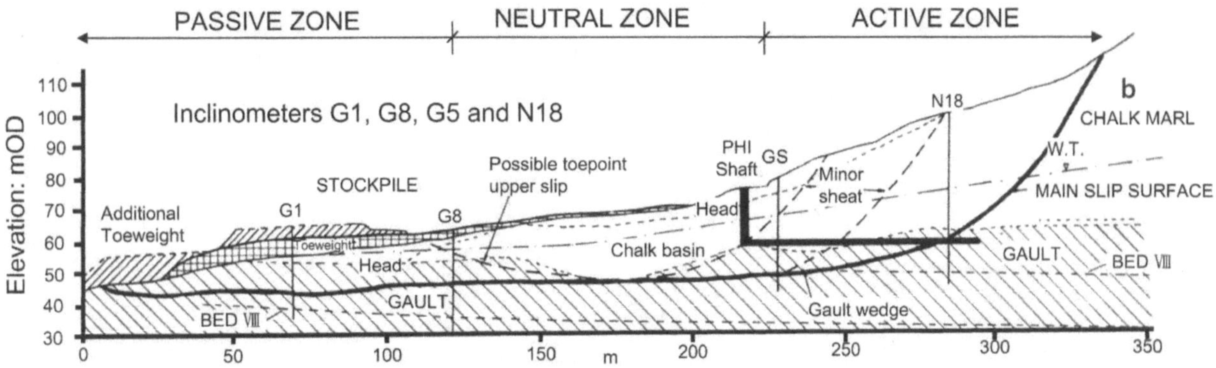

Fig. 3.3. Examples of slow active slides. **a** The essentially translational Fosso San Martino slide (from Bertini et al. 1986); **b** the compound Castle Hill slide (from Varley and Warren 1995)

Col des Mosses Pas, but the landslide is much older, as dating of wood fragments suggests (Tacher et al. 2005).

Literature reports numerous examples of slow slides. A number of these, as the La Frasse slide, are have been active for hundreds or even thousands of years, moving with an average displacement rate as low as a few centimeters per year or less. An example of long-lasting landslide is Castle Hill slide shown in Fig. 3.3b, which is believed to have formed about 10 000 years ago. Its displacement rate in the last tens of years, argued from deformation of a pipeline located within the landslide body, is about 1–2 mm yr^{-1}. The displacement rate of long-lasting slow movements is often assumed to be uniform with time, mainly because of a lack in continuous measurements, but once more data are available, some landslides show themselves to be intermittent (stick slip movements) being driven by pore pressures changes occurring over very short periods of time (Van Genuchten 1984). This naturally disproves the assumption of creep as cause of movement.

An example of creeping slide is reported by Picarelli and Simonelli (1991), who discuss deformation phenomena affecting the neighborhood of a small Italian town (Fig. 3.4), located on a gentle slope. The buildings rise on coarse calcareous debris covering a deposit of highly fissured sheared clay shales. At the time of investigations, the majority of dwellings and walls built in the last tens of years presented only some very small cracks, while only the oldest constructions (a monastery built some centuries ago, a building and a high retaining wall) displayed large cracks. Monitoring through some inclinometer purposely installed in the area revealed the presence of a slow active slide whose slip surface is located just at the top of clay shales. The displacement rate was quite constant (around 1 cm yr^{-1}) and uniform all over the area (Fig. 3.5). Readings carried out with Casagrande piezometers indicated that the groundwater table is located at the base of the landslide body, some decimeters above the top of clay, and that it experiences only very little annual fluctuations because of high permeability of the debris, preventing formation of an aquifer. Therefore the state of stress is practically constant and no significant future changes of this can be predicted. The small internal deformation of the landslide body depends on the high stiffness of debris; moreover, the absence of significant differential movements can explain the absence of severe damages on dwellings and walls built recently, while large cracks across the walls of the monastery and of the other old structures testify deformations cumulated over a much longer period of time.

Another example of slope movement presumably caused by creep is reported by Corominas et al. (2005), who describe the Vallcebre landslide in fine grained stiff deposits (Fig. 3.6a,b). Figure 3.7 reports the water levels measured

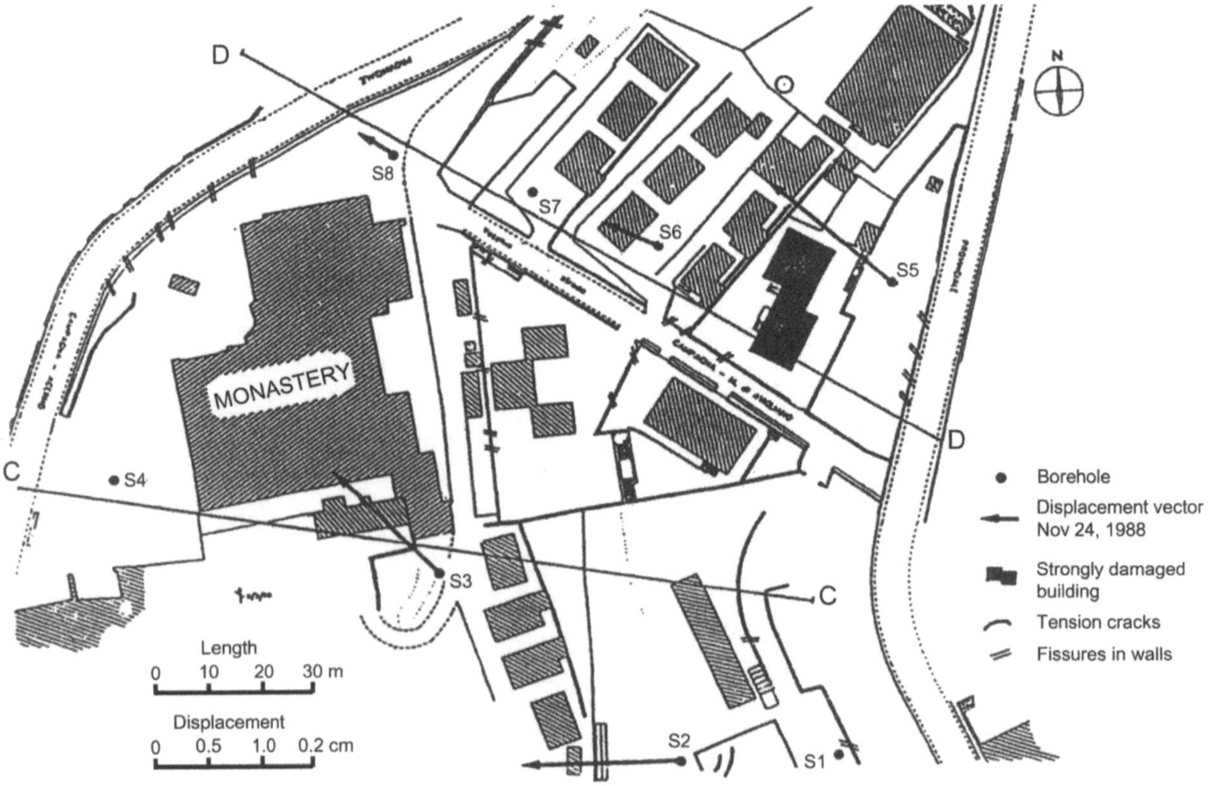

Fig. 3.4. A urban area experiencing slow movements and displacement vectors between 1987 and 1990 (after Picarelli and Simonelli 1991)

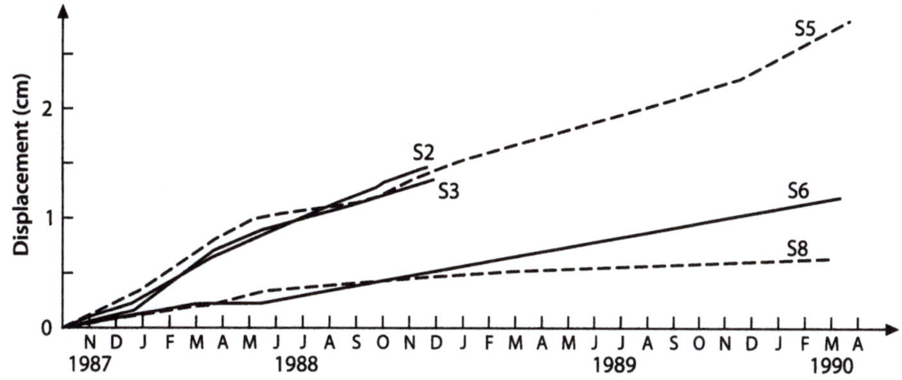

Fig. 3.5. Cumulated displacements in the urban area of Fig. 3.4 (from Picarelli and Simonelli 1991)

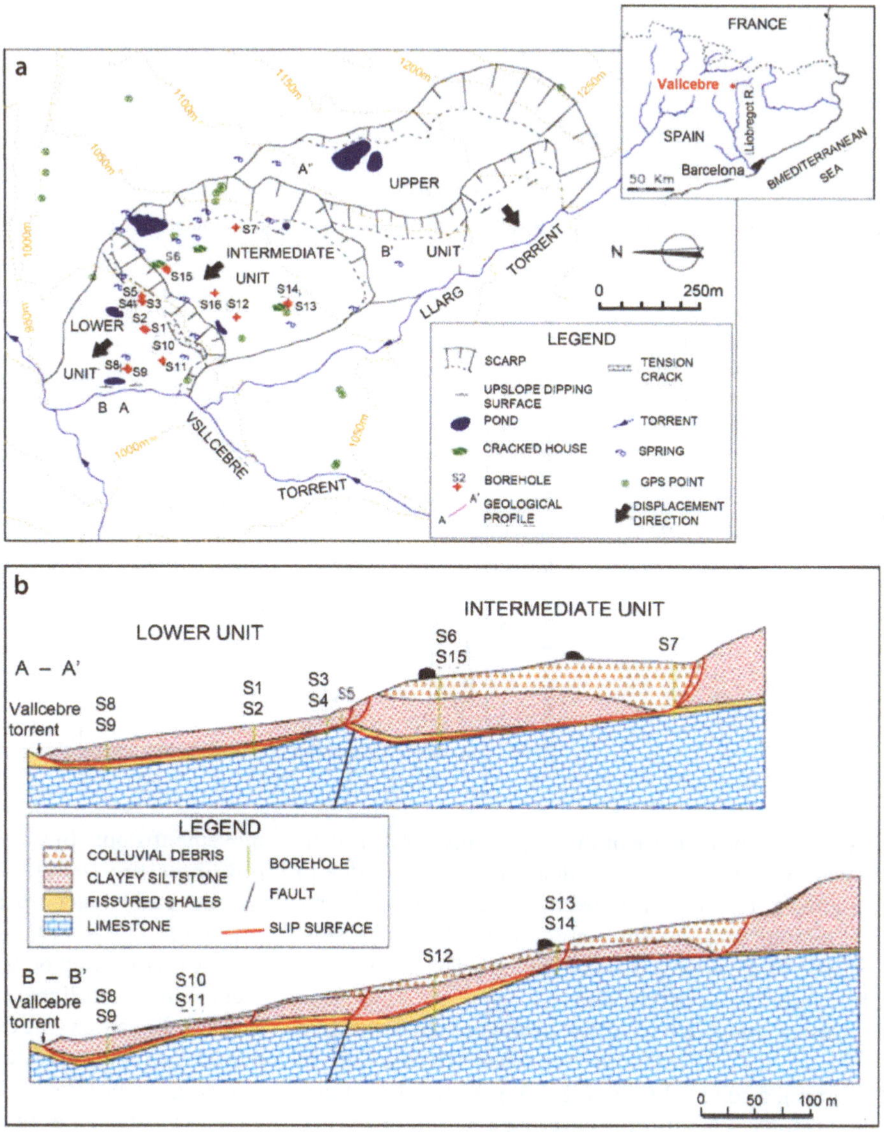

Fig. 3.6. The Vallcebre slide (from Corominas et al. 2005): (**a**) plan, (**b**) longitudinal sections

with a standpipe piezometer and displacements at vertical S2 (Fig. 3.6a). It shows that in some periods of the year pore pressures remain constant, even though the landslide moves anyway. For instance, between April and June, 1997, the displacement rate was around 15 mm month^{-1}, pore pressure being constant, but a sudden increase of this, followed by a decrease to a value slightly higher than before, determined a doubling of velocity. The Authors

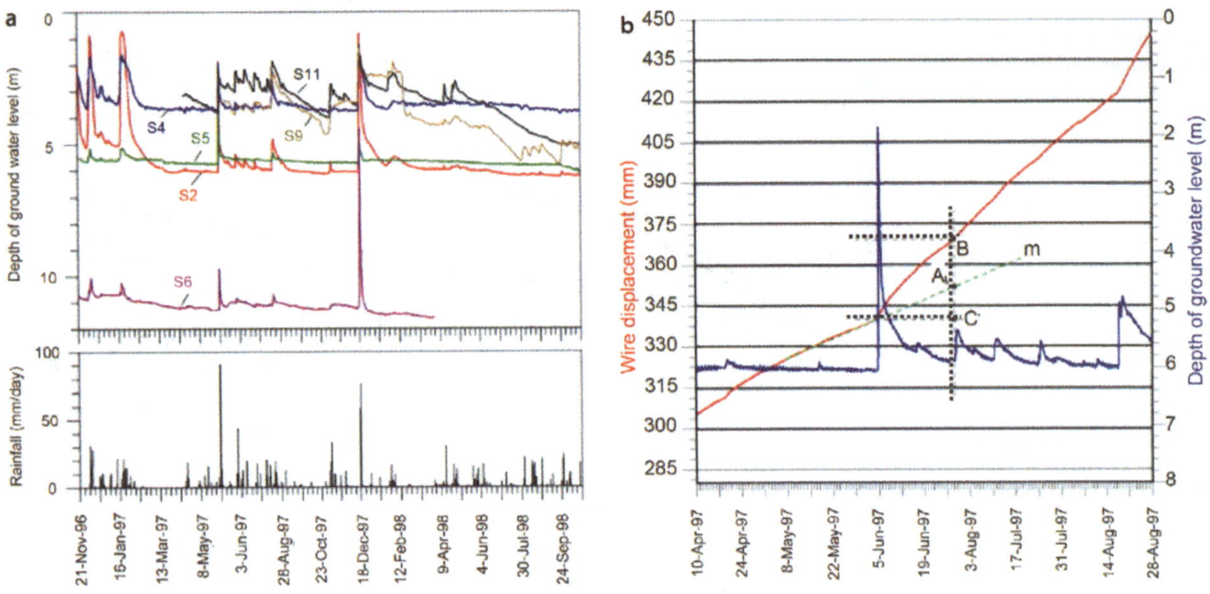

Fig. 3.7. Groundwater level, rainfall and displacements at vertical S2, Vallcebre landslide (from Corominas et al. 2005)

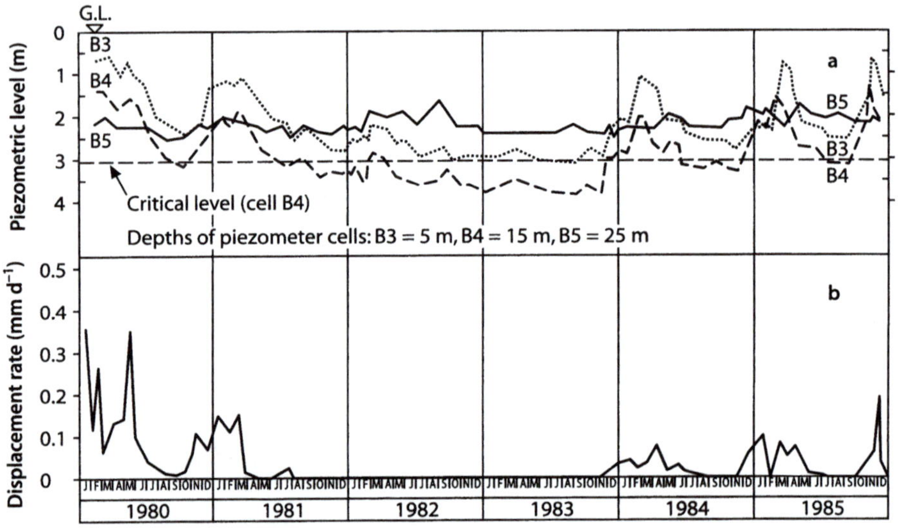

Fig. 3.8. Fosso San Martino slide. **a** Ground-water levels; **b** displacements of the ground surface (from Bertini et al. 1986)

assume that the increase of the displacement rate is contrasted by increase of the residual shear strength due to rate effects. This idea is shared by other Authors, who report similar considerations about the behavior of soils involved in slow movements (Vulliet 1986; Angeli et al. 1996; Vulliet and Hutter 1998).

A more complex viscous soil behavior is hypothesised by Bertini et al. (1986) for the Fosso San Martino slide, in fine grained colluvial soils (Fig. 3.3a). Figure 3.8 reports some results of monitoring covering a period of 6 years. Movement is directly correlated to rainfall and to consequent pore pressures changes, starting once the mobilized shear strength along the slip surface is about 95% of the residual value. When pore pressure decreases below a threshold value, as between summer, 1981, and fall, 1983, movement stops. In wetter periods, it re-starts following pore pressure rising. As remarked by other Authors (Cartier and Pouget 1988; Corominas et al. 2005), the displacement rate increases as the pore pressure increases, but the relationship between pore pressure and displacement rate is non-linear. In addition, Bertini et al. show that, for the same value of pore pressure, the velocity displayed during groundwater rising is higher than during lowering. Hence, two different relationships can be established between pore pressure and displacement rate; for the same reason, reactivation and arrest occur for different pore pressure thresholds (Fig. 3.9a). The Authors attribute such a behavior to the viscous behavior of soil along the slip surface, accounting for both primary and secondary creep. In fact, they argue that:

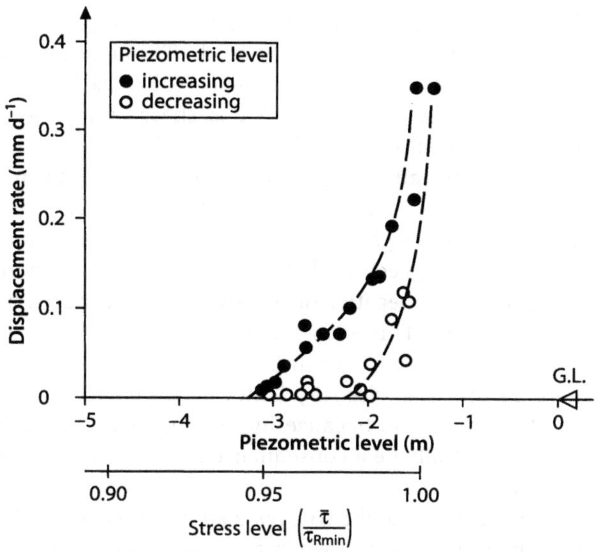

Fig. 3.9. Relationship between displacement rate and water level (from Bertini et al. 1986)

1. during the stage of pore pressure rising, the displacement rate is the result of two contrasting phenomena: increase of the rate of primary creep, which is associated with increasing stress level, and decrease of the rate of steady-state creep, which depends on time;
2. in the stage of pore pressure decrease, the two effects have the same negative sign and the resulting velocity is smaller.

The existence of different thresholds for increasing and decreasing pore pressure is stressed by other Authors too, but sometimes their observations are opposite, suggesting different interpretations. Moore and Brunsden (1996), as Bertini et al., find that the threshold pore pressure required to reactivate the Worbarrow Bay mudslide, Dorset, is lower than the one required to stop movement, but they explain this apparent inconsistency with a change in pore water chemistry during periods of rest. They also note that each reactivation requires a pore pressure larger than previous reactivations. Differently, in the case of the Alverà mudslide, Angeli et al. (1996) note the existence of two different thresholds corresponding to reactivation and arrest, but in this case the lowest one corresponds to arrest. They assume that the higher threshold required for reactivation is due to some increase of the residual strength occurring during the period of rest.

A non linear relationship between pore pressure and displacement rate has been found by Mandolini and Urciuoli too (1999) for the Miscano mudslide in softened clay shales (Fig. 3.10). According to Fig. 3.10b, movement starts when the groundwater table is located at a depth around 2 m: in this condition, the mobilized friction angle is about 13°. A full shear strength mobilization, corresponding to a high displacement rate (tending to infinite),

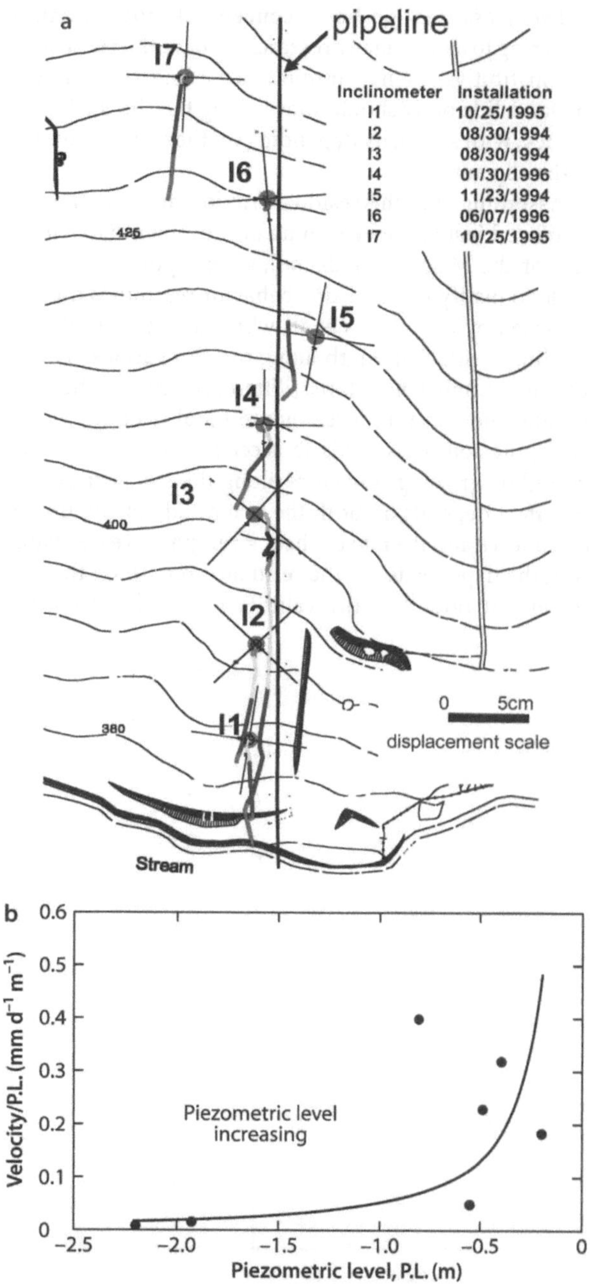

Fig. 3.10. Displacements (**a**) of the Miscano mudslide (from Picarelli et al. 1999), and relationship (**b**) between velocity divided by groundwater level and groundwater level (from Mandolini and Urciuoli 1999)

seems to occur once the groundwater table reaches the ground surface, i.e. for a mobilized friction angle of 18°. The ratio between the first stress level (creep threshold) and the second one (full shear strength mobilization) is then only 70%, which seems to be very low when compared to data provided by Bertini et al. (95%) and to laboratory experiences (Boucek and Pardo-Praga 1984). This raises some doubts about the real mechanisms of movement that will be discussed below.

Every assumption has a conceptual model behind. Referring to slopes, many researchers have the rigid-plastic constitutive law in their mind. According to this model, the landslide body should move along the slip surface as a block, with a velocity depending on the constitutive law of the slip surface.

Assuming that the residual friction angle is the one mobilized just at the beginning of movement (13° in the case of the Miscano mudslide), a conceptual model capable to justify the landslide behavior resembles the one proposed by Corominas et al., which assumes a rate-depending residual strength along the slip surface. Therefore, once a unbalanced force (difference between the driving and the resisting force due to basal friction) establishes, the soil mass tends to accelerate, but this is contrasted by growing resistance along the slip surface: the net effect depends on both the magnitude of the unbalanced force and the rate of shear strength increase. Naturally, the dependence of the residual strength on the displacement rate, thus the validity of proposed model, should be checked by laboratory tests. Unfortunately, available data do not help very much. Through tests on clay, Kenney (1967) and Skempton (1985) remark a negligible or little increase of the residual shear strength with the displacement rate. Furthermore, no influence is noticed by Hungr and Morgenstern (1983) for sand, while data collected by Picarelli and Urciuoli (1988) and by Tika et al. (1996) through laboratory tests on clay, are more problematic, suggesting that in some cases excess pore pressure (either negative or positive) can be induced by fast movement. This adds a complication in the model, which should account for drainage conditions. Furthermore, in this case viscosity is not necessary to explain the slope behavior, because increasing shear strength would essentially be a consequence of induced negative pore pressure.

If we assume that the true residual friction angle is the one which is mobilized when the displacement rate attains an "infinite" value (18° in the case of Miscano mudslide, when the piezometric level reaches the ground surface), the model to be adopted for interpretation of the landslide behavior should include pre-failure creep (as assumed by Bertini et al.), which is able to justify movements occurring for pore pressures less than the critical value required to provoke a full shear strength mobilization. Therefore, once again, viscosity of the slip surface could play a fundamental role, but in this case, prior to mobilization of the shear strength, which does not necessarily depend on the rate of movement. In this last case, a full shear strength mobilization would lead to catastrophic consequences.

The two models of slope behavior could be easily unified, but both consider the soil mass as a block moving along a viscous interface. Reality is more complex since soil is deformable and both initial and induced state of stress are not uniform in the slope: as a consequence, internal strains may play a significant role on landslide behavior. This last point is stressed by Picarelli et al. (2004) who argue that in many cases the landslide body is not entirely mobilized by pore pressure fluctuations, which cause internal strains associated with slipping along only part of the sliding surface.

Figure 3.10a shows that the displacement field of the Miscano mudslide at a given time is not uniform along the slope, with decreasing values downward. This can be explained accounting for pore pressure changes in the wet season (Fig. 3.11a), which are faster upslope than downslope, causing a temporary mobilization of the uppermost part of the landslide only, while the local shear stress mobilized at the base of the lowermost remains lower than the shear strength. This causes a compression of the landslide body induced by the overstress caused by restraint imposed by the lowermost part of it. Such a phenomenon has been simulated by the FEM. The slip surface has been characterised with elastic-perfectly plastic elements along

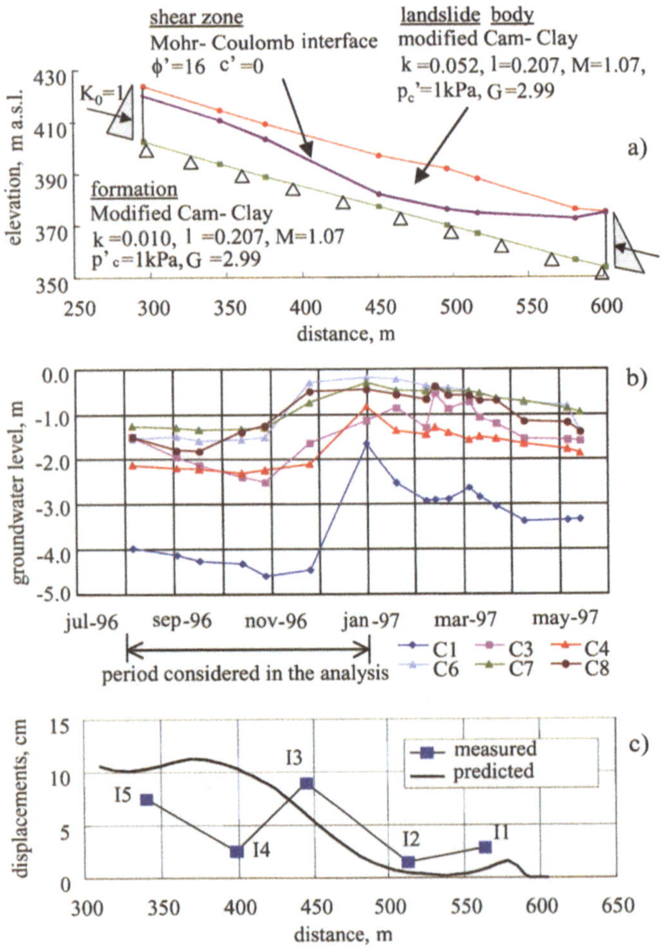

Fig. 3.11. Analysis of displacements of the Miscano mudslide. **a** Adopted model; **b** measured groundwater levels; **c** measured and calculated displacements (from Picarelli et al. 1999)

which is operative the residual shear strength, while the landslide body has been modeled with an elastic-plastic constitutive law (Cam-Clay) adopting the parameters reported in Fig. 3.11b. Slope displacements have been calculated for the same pore pressure increase which has been measured in site, as in Fig. 3.11a. The results of the analysis are comparable to reality (Fig. 3.11b): in particular, significant movement develop even though the safety factor of the lowermost part of the mudslide body remains higher than one. Therefore, this simple model confirms that movement can be essentially caused by internal strains more than by full mobilization of the landslide body.

The same mechanism can also explain the non linear increase of displacement with increasing pore pressure, discussed above (Figs. 3.9a and 3.10b): in fact, this result depends on the increase of the length of the part of the landslide body which is mobilized by pore pressure changes (Picarelli and Russo 2004). In addition, since pore pressures vary with time, movement too varies with time, so that soil viscosity is not necessary to explain the rate-depending behavior of slide.

Naturally, previous simplified interpretation of slope behavior does not exclude mobilization of a viscous component of displacement within the landslide body and along the slip surface, which can be easily incorporated in the model (Vulliet 1986).

Similar effects on slope behavior, i.e. displacement caused by internal deformation of the landslide body, can be provoked by uniform pore pressure increase, if movement is locally constrained by:

- variation of the geometry of the landslide body, because of increasing thickness and/or decreasing slope of the slip surface moving from upward to downward;
- local change of the width of the landslide (3D effect), as in mudslides which present a neck in between the alimentation zone and the main track;
- any change of the shear strength parameters along the slip surface.

These considerations offer alternative scenarios about slope behavior. In fact, at least in the case of long landslides, even very small strains, when integrated to the entire length of the landslide, could give rise to a significant component of displacement. As an extreme case, for a 100 m long landslide having a free upper boundary (the crown) and a fixed lower boundary (the toe), an average compressive strain of 10^{-2}% due to pore pressure rising, can trigger a displacement of the uppermost part of the landslide body of 1 cm, solely due to compression. It is worth noting that, in stiff clay, a strain equal to 10^{-2}% can be provoked by increase of pore pressure in the order of 1 kPa, i.e. by a rising of the water table in the order of 10 cm.

Described mechanisms of slope movement, which might be governed by constraints due to boundary conditions or non uniform geometric and mechanical parameters of soil, have been investigated by Russo (1997) and discussed by Picarelli and Russo (2004) using both an elastic-plastic and an elastic-viscous-plastic constitutive law of soil. In particular, the use of an elastic-viscous-plastic constitutive law allows to perform more sophisticated analyses. As an example, a slope model as the one used to interpret movement of the active Miscano mudslide, can account for the effects of relaxation occurring in any phase of pore pressures decrease (dry season), when the compressive state of stress stored in the landslide body in previous wet season (when pore pressure increases) progressively disappears as a consequence of stress relaxation. It is worth noting that the magnitude of deformation occurring in the wet season depends on viscous soil properties and on time span prior to the new pore pressure increase. Such a deformation history governs the magnitude of pore pressure (threshold) required to activate new movements, which decreases as the magnitude of relaxation increases.

Similar considerations could be made about opposite mechanisms inducing extensive strains in the slope.

All these remarks about the possible role of soil deformability on slope movement appear consistent with data provided by some Authors (Wilson 1969; Jappelli et al. 1977; Nakamura 1984; Pouget 1996).

3.3.2 Active Mudslides

The typical flow-like style of mudslides is revealed just at failure or immediately after failure, when the landslide body displays a high mobility, spreading along the slope and filling pre-existing tracks. In this stage the mobilized soil mass moves quite rapidly over the ground surface, eroding and incorporating the top soil and the upper part of the underlying deposit: as a consequence, it can experience some subsidence (Hutchinson 1970; Corominas 1995). The described mechanism of movement implies that the base of the mudslide body is subjected to high shear stresses which are presumably responsible for intense remoulding of quite a thick basal slice of soil. In stages following first post-failure movements, the soil mass decelerates assuming a slide style, but the occurrence of previous flow-like mechanisms can be still easily recognized from the assumed morphology of the landslide body. This typically displays an alimentation zone, a track and a fan-shaped accumulation zone. In this last stage of slope deformation prior to definitive arrest, a well defined persistent slip surface, which could not survive to movements occurring in the flow-like stage, presumably governs the mudslide behavior.

Mudslides are bounded by a main scarp that dominates a depletion area (Fig. 3.12). This alimentation zone feeds the landslide body with "fresh" soil masses (surges),

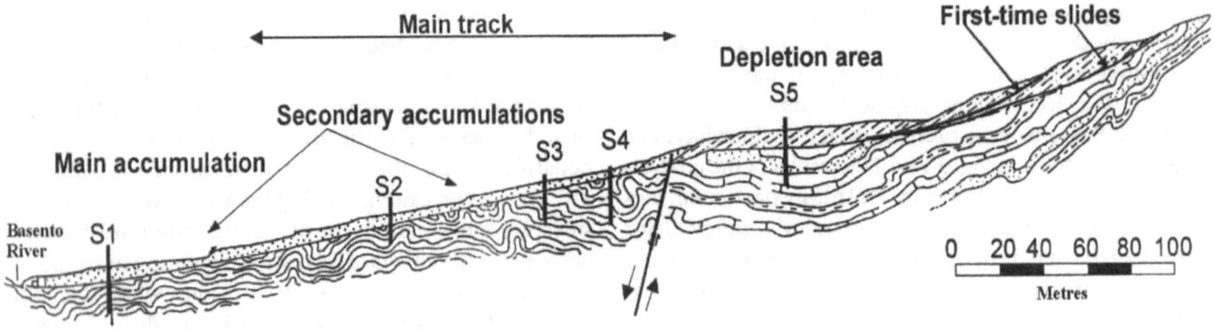

Fig. 3.12. The Brindisi di Montagna mudslide (from Cotecchia et al. 1984)

mobilized by local failures of the scarp, which are conveyed into the main track, causing periodical flow-like reactivations of movement. Further alimentation is provided by slips occurring along the flanks of the track. Surges form secondary accumulations along the track, but can reach the toe of the slope, thickening the accumulation zone. The presence of independent active or temporarily quiescent landslide bodies within the area occupied by the mudslide body is revealed by inclinometer profiles which show more shear zones along the same vertical.

To sum up, the main features of mudslides concern:

- their morphology, which reveals the mechanism of movement which characterises the initial post-failure flow-like phase: when channellized within a pre-existing track, long-term movements are restrained and governed by a complex 3D condition;
- the features of the soil mass, which is highly remoulded as a consequence of high deformations and softening experienced during first movements (Hutchinson 1988; Picarelli 1993);
- the thickness and fabric of the shear zone: while in the case of slides this is quite thin (generally a few centimeters) and displays a set of minor shears, in the case of mudslides it is rather thick (up to one meter) and fully remoulded (Picarelli et al. 2005): often the structure of the parent formation is completely obliterated.

Several Authors (Hutchinson and Bandhari 1971; Picarelli 1988;) argue that the initial flow-like style is a consequence of building-up of positive excess pore pressures. Pellegrino et al. (2004) discuss the main mechanisms which are supposed to be responsible for triggering of excess pore pressures. They include:

- rupture itself;
- seismic loading;
- accumulation of debris over a pre-existing mudslide body as a consequence of secondary failures or of rapid erosion along the main scarp or flanks of the track;
- loading caused by surges traveling over the mudslide body.

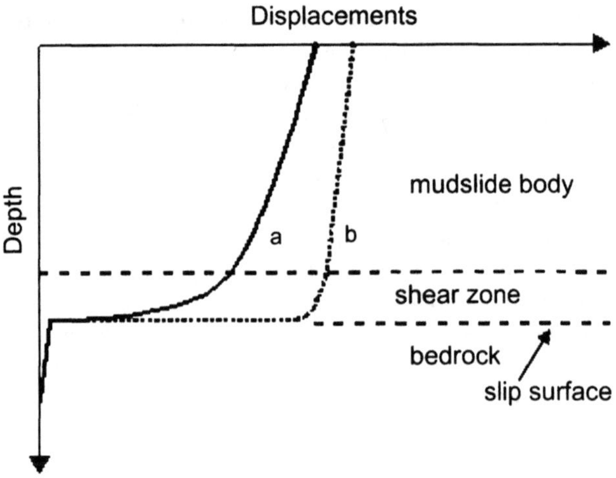

Fig. 3.13. Typical displacement profiles of a mudslide. **a** In the first stage of movement (flow-like style); **b** in the last stage of movement (slide style) (from Comegna et al. 2005)

Dissipation of excess pore pressures leads to a decrease of the rate of movement while the soil mass takes the features of a slide: in this stage movement localizes along a slip surface, while shear strains within the landslide body become smaller than in previous flow-like stage (Fig. 3.13). However, some data suggest that even this sliding phase can be characterised by development of moderate excess pore pressures because of internal stress changes.

Figure 3.14 shows pore pressures measured within the Masseria Marino mudslide at a depth of about 3 m. Figure 3.13a concerns a quiescent zone and Fig. 3.13b an active zone subjected to slow movement. While in the quiescent zone pore pressures display smooth seasonal fluctuations, quite in a good agreement with the results of numerical simulations carried out using pluviometer data and adopting the permeability of soil obtained by in situ and laboratory tests, in the most active zone pore pressures display rapid fluctuations which do not seem to follow a logic course (Comegna et al. 2004). The most reliable explanation is that any local soil deformation can trigger excess pore pressure: where deformation is by compression, induced pore pressure is positive.

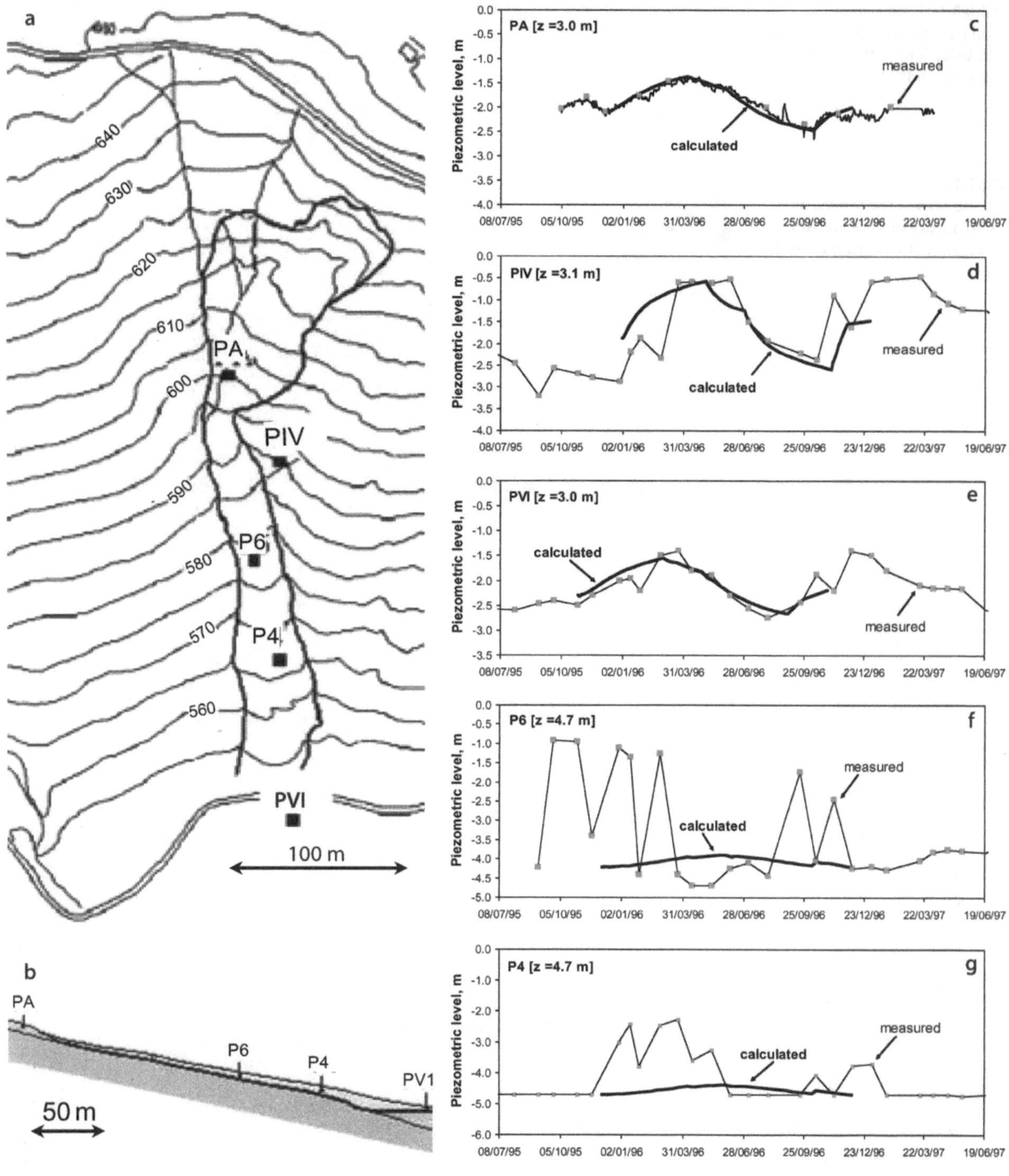

Fig. 3.14. Masseria Marino mudslide: location of the piezometers considered in the analysis (**a, b**); calculated and measured water levels in quiescent (**c, d, e**) and active zones (**f, g**) of the mudslide (after Comegna et al. 2005)

Such a mechanism has been checked through a simplified analysis. The mudslide body has been modeled with a non linear elastic constitutive law characterized by an isotropic yielding law ("Soft-Soil Model" present in the library of the code PLAXIS). In addition, has been considered a basal 1 m thick softer shear zone. The soil parameters adopted in the analysis have been obtained by the best fitting of laboratory tests (Comegna 2005): their values are reported in Table 3.1.

The sliding surface located at the base of the shear zone, has been simulated by interface elements along which is operative a residual friction angle of 13°. The analysis

Table 3.1. Soil parameters used to simulate the behaviour of the Masseria Marino mudslide

	γ_{sat} (kN m^{-3})	λ	k	e_0	ν	K_0^{NC}	O.C.R.	c' (kPa)	ϕ' (°)	K (m s^{-1})
Mudslide	20	0.057	0.026	0.59	0.35	0.58	3	8	25	10^{-9}
Shear zone	20	0.103	0.038	0.68	0.35	0.61	1	4	23	10^{-9}

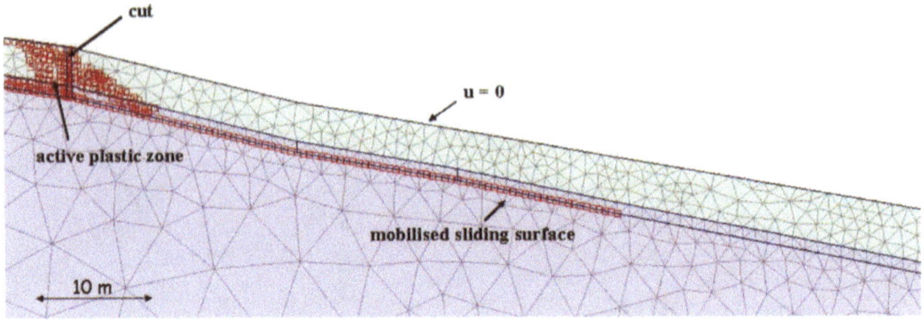

Fig. 3.15. Plastic active zone formed as a consequence of pore pressure rising (from Comegna et al. 2007)

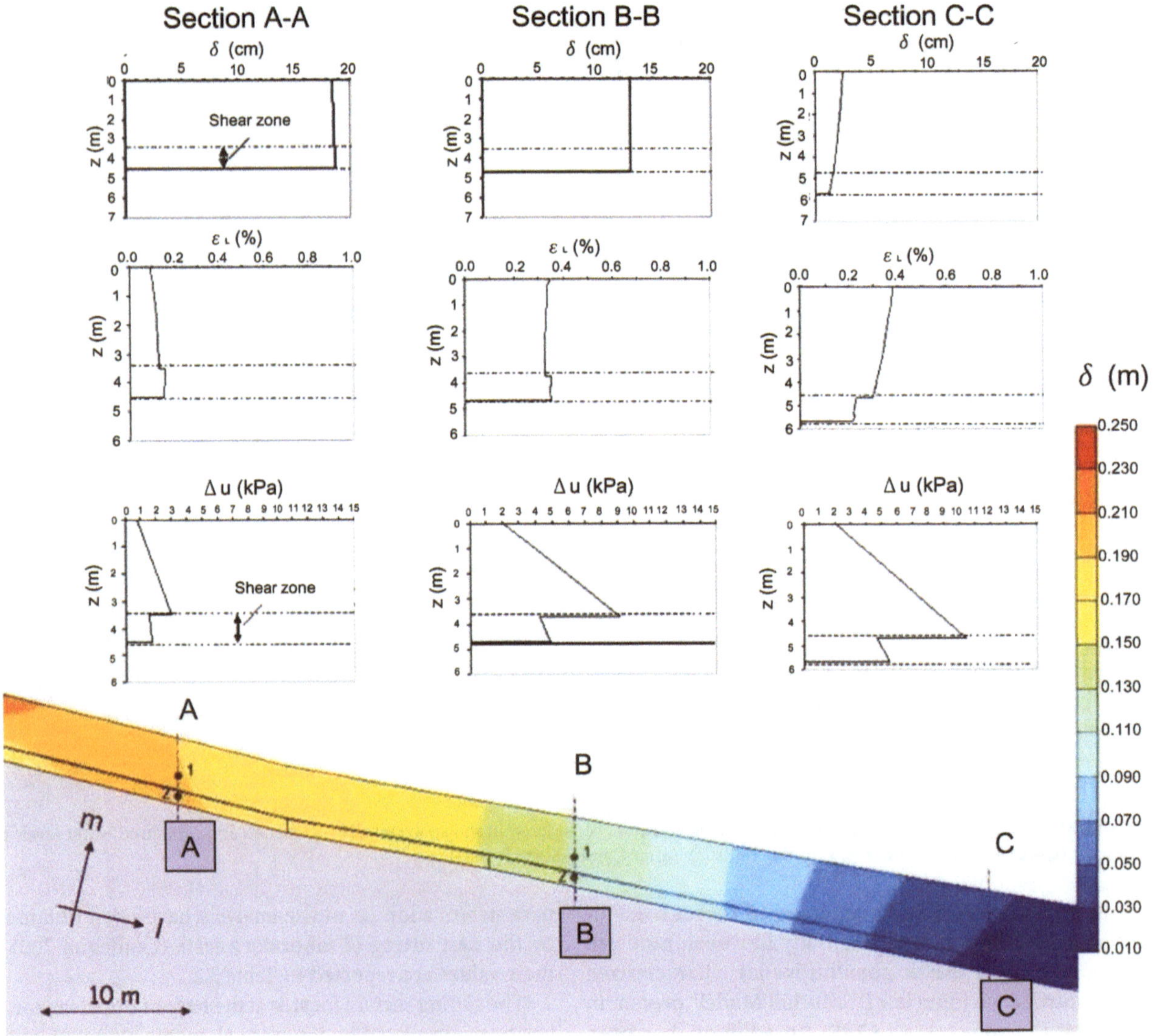

Fig. 3.16. Displacements, longitudinal strains and excess pore pressures at three sections of the mudslide (from Comegna et al. 2007)

simulates the pore pressure rising induced by rainfall, by imposing a water film along the ground surface (pore pressure equal to zero). During this phase, only part of the slip surface is mobilized, but the length of the mobilized part propagates downward as pore pressure increases. As a result, the active mudslide body slides downward, compressing the still stable part located ahead; at the same time, an active zone forms immediately upslope (Fig. 3.15). In order to simulate local soil rupture in the active zone, a vertical cut has then been imposed. Assuming that the deformation of the landslide body after cracking is fast enough to trigger excess pore pressures, the concerned stage of analysis has been performed assuming "short-term" (undrained) conditions. As a consequence of cracking, the mobilized part of the sliding surface further propagates downward while pore pressures rise due to undrained compression. However, the longitudinal strain ε_L is not uniform because of the different stiffness of mudslide body and shear zone; this difference in stiffness is also responsible for lower excess pore pressure in the shear zone. Longitudinal strains, pore pressures and displacements at three different sections of the slope are reported in Fig. 3.16.

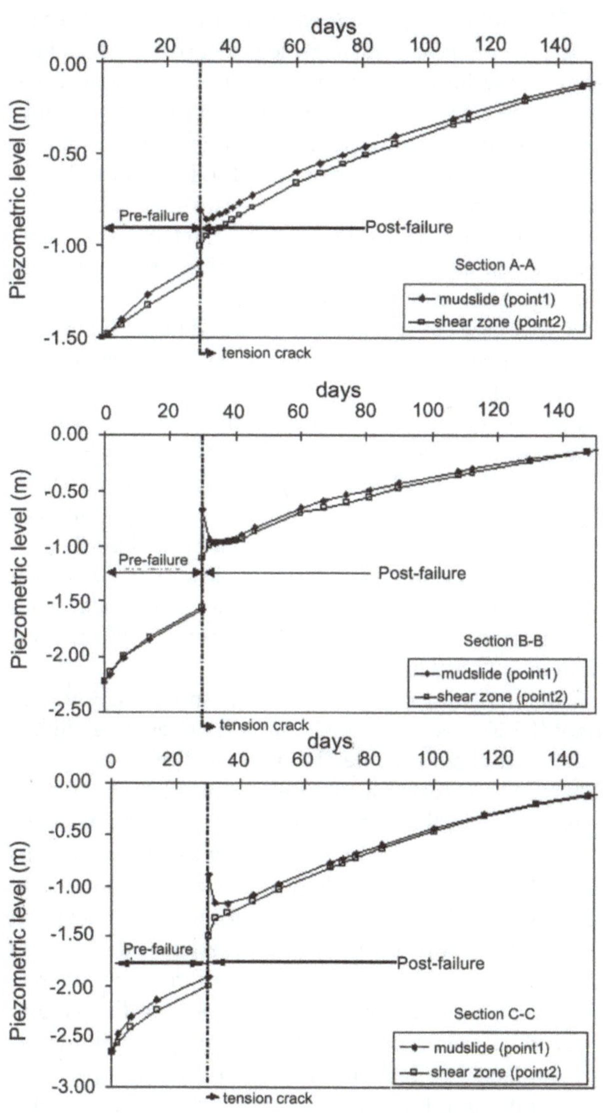

Fig. 3.17. Pore pressure evolution at two points within the mudslide body (from Comegna et al. 2007)

During the following stage of analysis, excess pore pressures are allowed to equalize. This phase is characterized by the overlap of two opposite phenomena: dissipation of excess pore pressure and continuing rising of water table due to infiltration. Figure 3.17 reports the evolution of pore pressure at two points, one located in the shear zone and the other one in the mudslide body, in three sections of the landslide. Because of the high gradient of the piezometer head closely around the two points, excess pore pressures rapidly dissipate: a decrease in the mudslide body is associated with increase in the shear zone. Such a result could explain the anomalous drops of pore pressures monitored in the Masseria Marino mudslide (Fig. 3.14b).

3.4 Consideration about the Mechanics of Active Lateral Spreads

Lateral spreads represent a special category of slow active landslides in clay. They are so slow to be recognized only through sophisticated monitoring or from continuing damage to old structures. According to literature, lateral spreads can be triggered by tectonic uplift, by removal of lateral support following glacial retreat or by river erosion (Radbruch-Hall et al. 1976).

An interesting example is movement which involves the Monte Verna hill (Fig. 3.18) where San Francis retired in the last years of his life (Canuti et al. 1990). The hill is constituted by jointed massive calcarenites and bedded sandstones which rest on a deposit of highly fissured sheared clay shales. It is subjected to lateral movements, essentially driven by squeezing out of clay shales. Movement causes opening of vertical joints in the rock slab.

A schematic description of the mechanisms which govern spreading induced by valley formation is shown in Fig. 3.19. This concerns the case of a rock slab resting on clay, as in Fig. 3.18. Erosion starts developing along fractures of the slab (stages 1 and 2). The stress relief caused by unloading provokes rising of the valley floor; as a consequence, the clay deforms forming an anticline (stages 3 and 4). Further deepening of the valley reaches the buried top of clay, which squeezes out bringing about lateral deformation of slopes. As a consequence of imposed drag forces, a shear zone forms at the contact between clay and rock bed, and the slab is subjected to fracturing as a consequence of tensile stresses imposed by shear (stage 5). Induced movement causes opening of

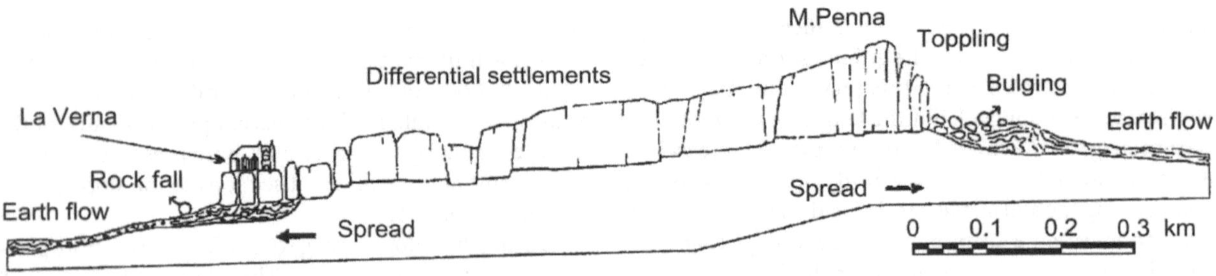

Fig. 3.18. Deformation phenomena of the Monte Verna mountain (from Canuti et al. 1990).

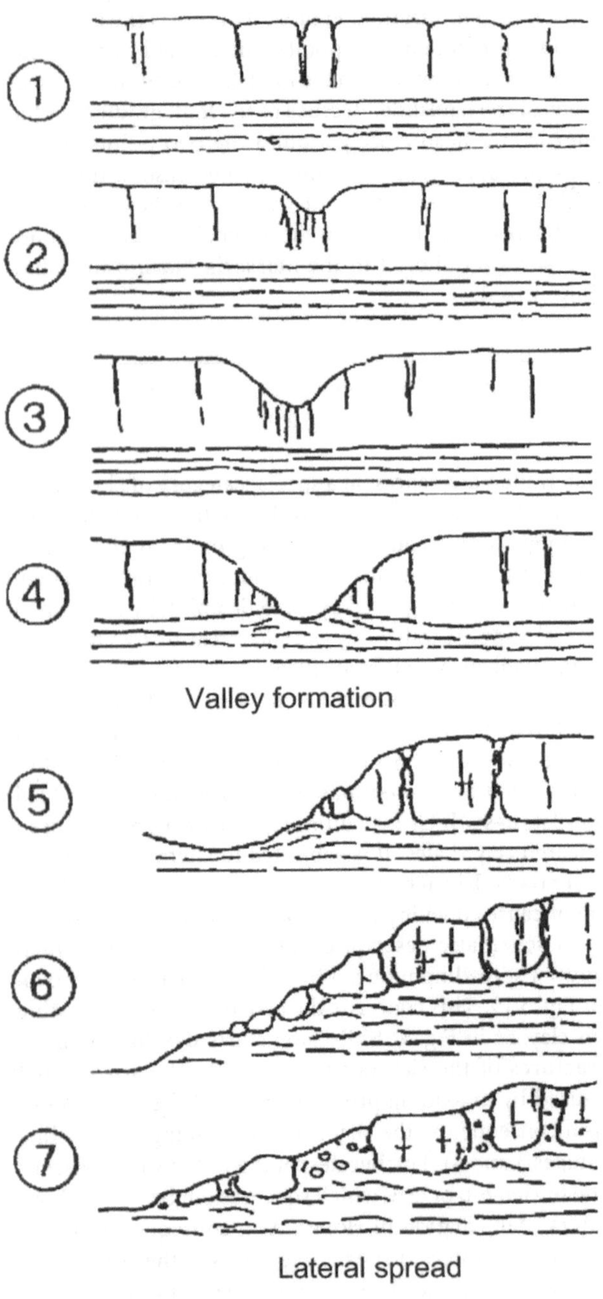

Fig. 3.19. Effects of river erosion on the deformation pattern of a rock slab resting on clay (from Pasek 1974).

vertical joints which can be filled by clay spread from the bottom. In addition, internal rock blocks subside, while lateral blocks tilt, forming a trench at the top of the slab. Further movements trigger landslides along the boundaries of the slab contributing to their complete split up (stages 6 and 7).

The geological phenomena which govern such a process, even when very fast in geological terms, are extremely slow in the human perception, thus consequent movements are extremely slow. However, in case of very thick clay deposits, unloading due to erosion can be slower than the time required for dissipation of induced negative excess pore pressures (Neuzil 1993), causing a delay of associated deformations. In seismically active areas, earthquakes can add their effects to erosion and creep, producing acceleration of deformation.

Other cases of spreads can be found in the Italian scientific literature (Crescenti et al. 1994; Fenelli et al. 1996; Chelli et al. 2006). Several of them concern towns located on the top of hills and mountains in the geologically active area of Apennines. A well known example is the Orvieto hill, in the Tiber valley (Lembo-Fazio et al. 1984; Tommasi et al. 2006), which is constituted by a tuff slab resting on stiff clay. A similar situation is described by Tommasi and Rotonda (1995) who discuss the deformations of the calcareous San Leo cliff, on top of which stands a Renaissance's Castle where the famous Cagliostro was hold in jail until his death.

An interesting case is the Bisaccia spread, in Southern Italy (Di Nocera et al. 1995). Bisaccia rises on a slightly cemented conglomerate slab with thickness exceeding 100 m, which rests on intensely fissured highly plastic clay shales. In the last 300 000 years, the area has been deeply eroded by downcutting along two parallel faults (Fig. 3.20), through a geological process very similar to the one described in Fig. 3.19. Concentrated erosion left an elongated hill in the middle. Presently, the floor of the two valleys is lower than the bed of the conglomerate slab over which stands the town. The slab is divided into large blocks separated by vertical cracks. Its boundaries are subjected to landslides involving both upper conglomerates and lower clay shales. Some movements involve both conglomerate blocks and clay at the foot (slides); others involve only clay and generally display a flow-like style (Picarelli et al. 2006). The top of the hill presents a series of morphological steps. In the past, these have been smoothed with man-made ground or bridged with stairs. In addition, buildings show fissures and cracks; old reparations reveal the occurring of a process of general deformation.

The area is shaken by earthquakes having a return time of some tens of years. In the last century, two strong earthquakes occurred in 1930 and in 1980 (the last one characterized by a Magnitude $M = 6.3$). Seismic events systematically cause cracks in the masonry of old buildings and in pavements. The main effects of the 1930 and 1980 earthquakes were a series of fractures on pavements. Most of the cracks generated by the two events coincide, and are located along natural morphological steps; in addition, buildings most severely damaged by the 1980 quake are located along such alignments (Fenelli 1986). This means that the damages were mostly caused by opening of fractures.

The area has been systematically investigated since 1981. The first campaigns included field and laboratory investigations which allowed to obtain a complete mechanical characterization of clay shales (Picarelli et al. 2002). In 1985 and 1989, thus after the Irpinia earthquake (1980), two verticals, one in the eastern valley and the other in the urban area, were instrumented with Casagrande and vibrating wire piezometers (Fenelli and Picarelli 1990; Di Nocera et al. 1995). Finally, on February, 1981, a number of benchmarks were installed in the town to monitor deformations induced by the quake. Monitoring covered more than seven years, until October, 1988.

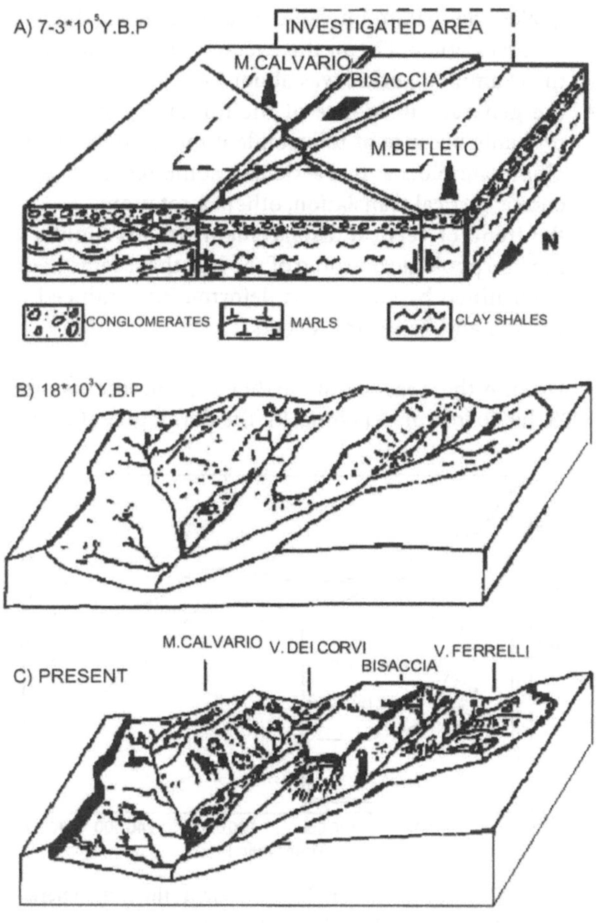

Fig. 3.20. The Bisaccia hill (from Fenelli and Picarelli 1990)

Figure 3.21 reports pore pressures measured along the two instrumented verticals. Deficient values (locally negative, i.e. below the atmospheric pressure) have been measured below the bed of the eastern valley where unloading has its major effect. In contrast, pore pressures above their theoretical value have been measured just below the slab. However, accurate readings have shown a constant decrease in time.

These results have been interpreted accounting for the geological phenomena affecting the site. In fact, accounting for the thickness and low permeability of clay shales, erosion is quite fast in geological terms and is assumed to trigger negative excess pore pressures (Fenelli and Picarelli 1990). Di Nocera et al. (1995) show that swelling of clay can explain the geomorphological shape of the two valleys and the distribution of the water content in the subsoil; other consequences of erosion are cracking of the slab, squeezing out of clays and a shear zone recognized at the conglomerate-clay shale contact.

Formation of the hill has been simulated by FEM analyses (Picarelli and Urciuoli 1993; Di Nocera et al.

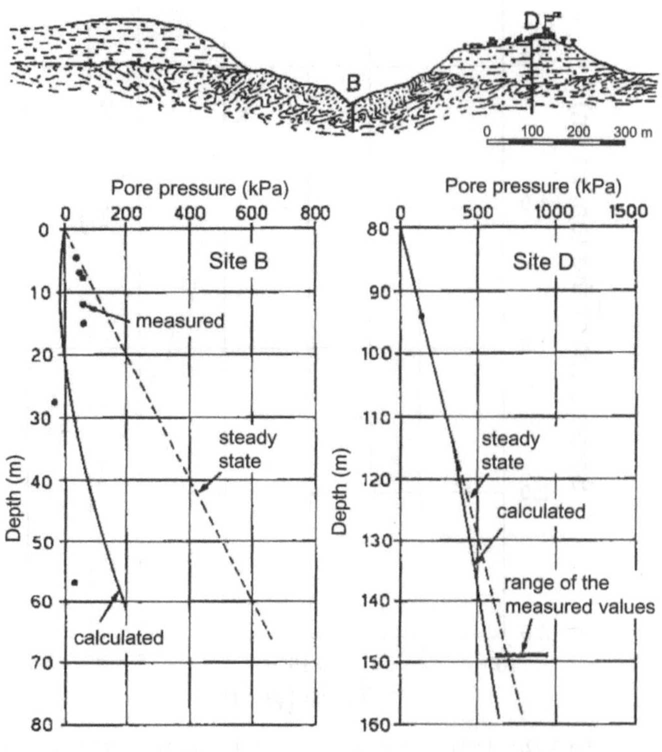

Fig. 3.21. Pore pressures measured in the Bisaccia area (from Fenelli and Picarelli 1990)

1995) assuming that erosion started 300 000 years ago, starting from an elevation corresponding to the present top of the hill. Figure 3.22 shows calculated vertical displacements of the valley floor, at the depth of its present elevation, and of the top of the hill. The figure shows that the valley floor is rising with a velocity of about 0.04 mm yr^{-1}, while the hill sinks in clay shales with a velocity of one order of magnitude less. Horizontal displacement at the contact between conglomerate and clay shale is shown in Fig. 3.23. Calculated movement is in the order of 0.2 mm per century; the differential displacement along the conglomerate-clay shale interface is responsible for formation of a shear zone. It is worth noting that the deformations calculated in the analysis not only depend on the assumed rate of erosion, but also on the rate of pore pressure equalization.

Previous data about the effects of the 1980 earthquake suggest that strong seismic events can accelerate movement of the hill. Figure 3.24 shows the settlements measured from February, 1981, to respectively February, 1982, and October, 1988, along a longitudinal section of the town (Fenelli et al. 1992). The location of major and minor fractures in the slab is represented by arrows. A profile of the hill along the same section is also shown in a magnified scale.

The figure shows that:

- until October, 1988, the entire urban area underwent a general subsidence with settlements increasing from South to North, where the hill is truncated by a steep high slope;
- settlements increased with time; eight years after the quake, the maximum value measured along this section was about 12 cm: the average settlement rate in the examined period was about 1 cm yr^{-1};
- the general subsidence of the hill is discontinuous, probably because of independent movements of conglomerate blocks; in particular, some blocks experience a vertical translation, others a rotation;
- the trend of vertical displacements is very similar to the morphological profile of the hill, that seems hence determined by cumulated deformations induced by erosion and by seismic events.

Despite the absence of further measurements after 1988 due to loss of benchmarks, the development of new

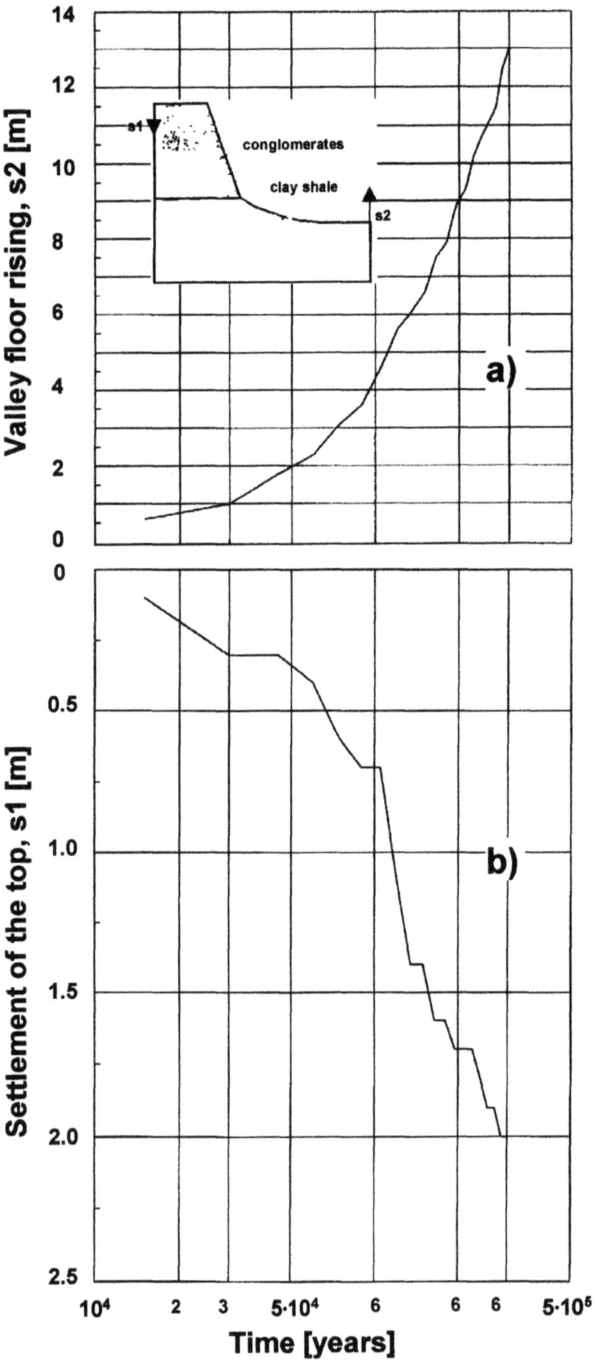

Fig. 3.22. Calculated vertical displacement of the valley floor (**a**) and (**b**) of the top of the hill (Picarelli and Russo 2004)

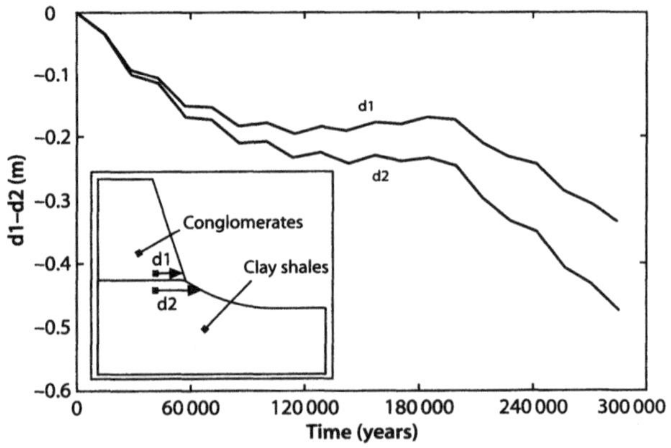

Fig. 3.23. Calculated horizontal displacement at the conglomerate-clay shale interface (Picarelli and Russo 2004)

Fig. 3.24.
Vertical displacements measured after the 1980 earthquake and longitudinal profile of the hill (from Fenelli et al. 1992)

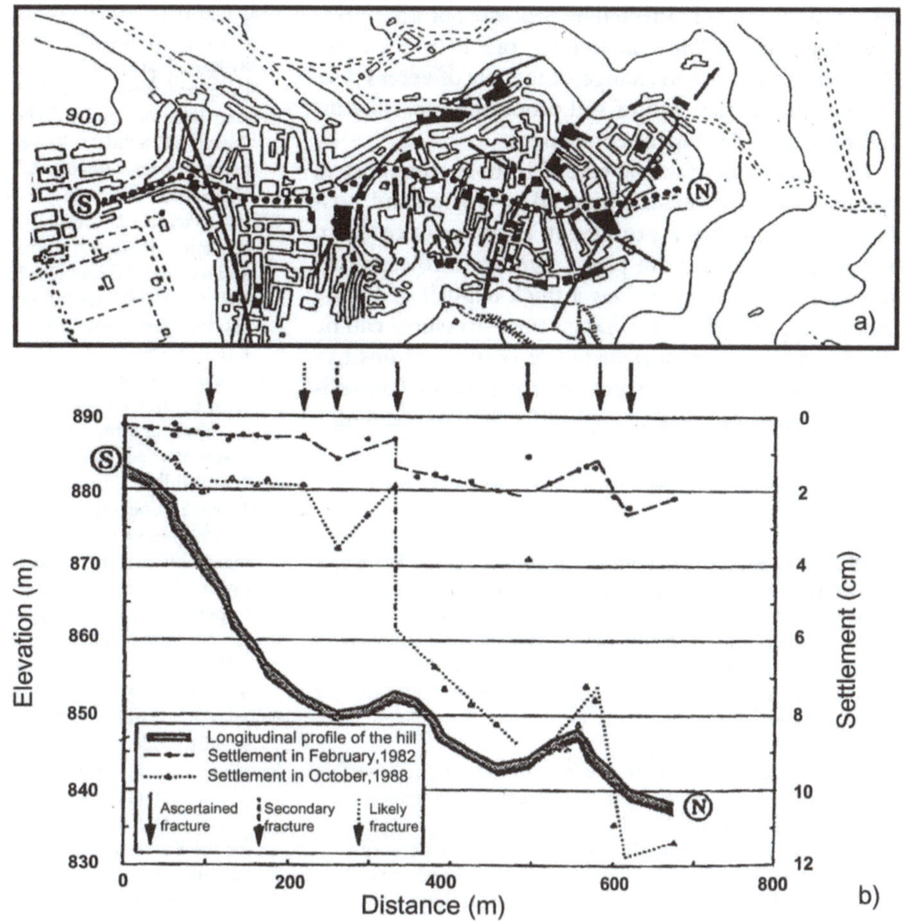

cracks in buildings demonstrate that movements were still continuing. Unfortunately, no data are available about horizontal displacements, but certainly they were not negligible.

More information are provided by numerical analyses (Olivares 1997; Lampitiello et al. 2001). The simulations have been carried out using seismogram data recorded in the town during the earthquake, deconvoluted to reproduce the seismic input motion at the top of bedrock. The analyses demonstrate that, in a large part of the clay shale deposit under the slab, the quake induced shear strains larger than the volumetric threshold, causing a local building up of positive excess pore pressures. The dissipation of these is responsible for delayed vertical displacements, as shown above.

Summing up, the hill is subjected to two different geological phenomena: erosion and earthquakes. Both contribute to slope movements. Erosion brings about a deepening of the valleys surrounding the hill. The superimposed effects of earthquakes cause a general subsidence of the slab, with sinking of the blocks into the clay shale deposit.

3.5 Conclusions

Typical slow active landslides in clay are translational slides and mudslides, but other types of movements, as spreads, also belong to the category of slow active landslides.

Active slides and mudslides advance along a pre-existing slip surface and are driven either by changes of boundary conditions, as those caused by rainfall or erosion, or by viscous deformations. The effects of pore pressure fluctuations prevail in the case of landslides of moderate thickness, while those of erosion and of creep can be significant in the case of deep-seated landslides. To understand the mechanics of slides and mudslides, the model of rigid-plastic body is not adequate. In fact, since any change of boundary condition provokes a non homogeneous variation of the stress field, internal deformation of the landslide body is a normal condition: since slow landslides present very small displacements, internal deformation is a significant component of movement. However, available data show that positive excess pore pressures self-generated by movement itself may play a not negligible role, at least in the case of mudslides; following movements is affected by processes of pore pressure equalization. The displacement rate is then a function of the rate of stress change, of the rate of excess pore pressure dissipation and of soil properties, possibly including a time-dependent component.

The mechanics of lateral spread is more complicated. In some cases, deformation induced by erosion and, possibly, by earthquakes, is prominent. However, once again, excess pore pressures can play a significant role, mainly when the movement involves a thick deposit of highly plastic clay: in such a case, excess pore pressures can be either negative, as a consequence of erosion, or positive due to seismic input. Pore pressure equalization (consolidation or swelling) may strongly affect soil movement.

References

Angeli MG, Gasparetto P, Menotti RM, Pasuto G, Silvano S (1996) A visco-plastic model for slope analysis applied to a mudslide in Cortina d'Ampezzo, Italy. Q J Eng Geol 29:233–240

Bertini T, Cugusi F, D'Elia B, Rossi-Doria M (1986) Lenti movimenti di versante nell'Abruzzo adriatico: caratteri e criteri di stabilizzazione. In: Proceedings 16th National Geotechnical Conference "La Progettazione Geotecnica per la Stabilizzazione dei Pendii", Bologna, CLEUP, Rome, 1, pp 91–100

Boucek B, Pardo-Praga D (1984) Fluage d'une argile sur la surface de glissement. In: Proceedings 4th International Symposium on "Landslides", Toronto, 2, pp 247–252

Canuti P, Casagli N, Garzonio CA, Vannocci P (1990) Lateral spreads and landslide hazards in the Northern Apennine: the example of the Mt. Fumaiolo (Emilia-Romagna) and Chiusi della Verna (Tuscany). In: Proceedings 6th Congress IAEG, Amsterdam, 3, pp 1525–1533

Cartier G, Pouget P (1988) Étude du comportement d'un remblai construit sur un versant instable: le remblai de Sallèdes (Puy-de-Dôme). Rapport de recherche LPC n°153, Paris

Cavalera L (1977) Consolidazione per variazione periodica delle pressioni interstiziali al contorno. Rivista Italiana di Geotecnica 11(4):187–205

Chelli A, Mandrone G, Trufelli G (2006) Field investigations and monitoring as tools for modelling the Rossena castle landslide (Northern Appennines, Italy). Landslides 3(3):252–259

Comegna L (2005) Proprietà e comportamento delle colate in argilla. PhD Thesis, Seconda Università degli Studi di Napoli

Comegna L, Urciuoli G, Picarelli L (2004) The role of pore pressures on the mechanics of mudslides. In: Lacerda W (ed) Proceedings 9th International Symposium on "Landslides", Rio de Janeiro, 2, pp 1141–1176

Comegna L, Picarelli L, Urciuoli G (2007) The mechanics of mudslides as a cyclic undrained-drained process. Landslides (accepted)

Corominas J (1995) Evidence of basal erosion and shearing as mechanism contributing to the development of lateral ridges in mudslides, flow-slides, and other flow-like gravitational movements. Eng Geol 39:45–70

Corominas J, Moya J, Ledesma A, Lloret A, Gili JA (2005) Prediction of ground displacements and velocities from groundwater level changes at the Vallcebre landslide (Eastern Pyrenees, Spain). Landslides 2(2):83–96

Cotecchia V, Del Prete M, Federico A, Fenelli GB, Pellegrino A, Picarelli L (1984) Some observations on a typical mudslide in a highly tectonized formation in Southern Apennines. In: Proceedings 4th International Symposium on "Landslides", Toronto, 2, pp 39–44

Crescenti U, Dramis F, Prestininzi A, Sorriso-Valvo M (1994) Deep-seated gravitational slope deformations and large-scale landslides in Italy. A regional guide. Special volume for the International Congress of the IAEG, Lisboa, pp 7–10

Cruden DM, Varnes DJ (1996) Landslide types and processes. In: Turner AK, Schuster RL (eds) Landslides: investigation and mitigation. Special Report 247, Transportation research board, US National research council, Washington, D.C., pp 36–75

D'Elia B, Picarelli L, Leroueil S, Vaunat J (1998) Geotechnical characterisation of slope movements in structurally complex clay soils and stiff jointed clays. Rivista Italiana di Geotecnica 32(3):5–47

Di Maio C (1996) The influence of pore fluid composition on the residual shear strength of some natural clayey soils. In: Senneset K (ed) Proceedings 7th International Symposium on "Landslides", Trondheim, Balkema, Rotterdam, 2, pp 1189–1194

Di Maio C, Onorati R (2000) Influence of pore liquid composition on the shear strength of an active clay. In: Bromhead E, Dixon N, Ibsen ML (eds) Proceedings 8th International Symposium on "Landslides". Cardiff, 1, pp 463–468

Di Nocera S, Fenelli GB, Iaccarino G, Pellegrino A, Picarelli L, Urciuoli G (1995) An example of the geotechnical implications of geological history. In: Proceedings 11th ECSMFE, "The interplay between Geotechnical Engineering and Engineering Geology", Copenhagen, 8, pp 39–48

Fenelli GB (1986) Alcuni peculiari aspetti relativi alla stabilità del colle di Bisaccia. In: Proceedings 16th National Geotechnical Conference, Bologna, CLEUP, Roma, 3, pp 191–196

Fenelli GB, Picarelli L (1990) The pore pressure field built up in a rapidly eroded soil mass. Can Geotech J 27:387–392

Fenelli GB, Pellegrino A, Picarelli L (1986) Stability problems of old towns built on relict plateaux resting on clay deposits. In: Viggiani C (ed) International Symposium on "Geotechnical Engineering for the Preservation of Monuments and Historic Sites". Napoli, Balkema, Rotterdam, pp 163–174

Fenelli GB, Picarelli L, Silvestri F (1992) Deformation process of a hill shaken by the Irpinia earthquake in 1980. In: Faccioli E, Pecker A (eds) Proceedings French-Italian Conference on "Slope Stability in Seismic Areas". Bordighera, Ouest Éditions pp 47–62

Hoek E, Brown CT (1980) Empirical strength criterion for rock masses. J Geotech Eng-ASCE, 106(GT9):1013–1035

Hungr O, Morgenstern NR (1983) High velocity ring shear tests on sand. Géotechnique 34(3)415–421

Hutchinson JN (1970) A coastal mudflow on the London clay cliffs at Beltinge, North Kent. Géotechnique 20(4):412–438

Hutchinson JN (1988) General report: Morphological and geotechnical parameters of landslides in relation to geology and hydrogeology. In: Bonnard C (ed) Proceedings 5th International Symposium on Landslides, Lausanne, Balkema, Rotterdam1, pp 3–35

Hutchinson JN Bandhari RK (1971) Undrained loading: a fundamental mechanism of mudflows and other mass movements. Géotechnique 21(4):353–358

Jappelli R, Musso A, Umiltà G (1977) An observational approach to a creep problem in debris due to reservoir filling. In: Proceedings International Symposium "The Geotechnics of Structurally Complex Formations", Capri, 2, pp 115–121

Kenney TC (1967) The influence of mineral composition on the residual strength of natural soils. In: Proceedings Geotechnical Conference, Oslo, 1, pp 123–129

Kenney TC, Lau KC (1984) Temporal changes of groundwater pressure in a natural slope of non fissured clay. Can Geotech J 21:138–146

Lampitiello S, Olivares L, Silvestri F (2001) Numerical simulation of seismic and post-seismic response of Bisaccia Hill. In: Proceedings TC4 Satellite Conference on "Lesson Learned on Recent Strong Earthquakes", Istanbul

Lembo-Fazio A, Manfredini G, Ribacchi R, Sciotti M (1984) Slope failure and cliff instability in the Orvieto Hill. In: Proceedings 4th International Symposium on "Landslides", Toronto, 1, pp 155–178

Leroueil S, Vaunat J, Picarelli L, Locat J, Faure R (1996) Geotechnical characterization of slope movements. In: Senneset K (ed) Proceedings 7th International Symposium on "Landslides", Trondheim, Balkema, Rotterdam,1, pp 53–74

Mandolini A, Urciuoli G (1999) Previsione dell'evoluzione cinematica dei pendii mediante un procedimento di simulazione statistica. Rivista Italiana di Geotecnica 33(1):37–44

Moore R, Brunsden D (1996) Physical-chemical effects on the behaviour of a coastal mudslide. Géotechnique 46(2):259–278

Morgenstern N, Tchalenko JS (1967) Microscopic structures in kaolin subjected to direct shear direct shear. Géotechnique 17:309–328

Nakamura H (1984) Landslides in silts and sands mainly in Japan. In: Proceedings 4th International Symposium on "Landslides", Toronto, 1, pp 155–178

Neuzil CE (1993) Low fluid pressure within the Pierre shale: a transient response to erosion. Water Resour Res 29(7):2007–2020

Olivares L (1997) Caratterizzazione dell'argilla di Bisaccia in condizioni monotone, cicliche e dinamiche e riflessi sul comportamento del Colle a seguito del terremoto del 1980. PhD Thesis, Università di Napoli Federico II

Pasek K (1974) Gravitational block-type slope movements. In: Proceedings 2nd International Congress IAEG, Sao Paulo, 2: V-PC-1

Pellegrino A, Picarelli L, Urciuoli G (2004) Experiences of mudslides in Italy. In: Picarelli L (ed) Proceedings International Work on "Occurrence and Mechanisms of Flow-Like Landslides in Earthfills and Natural Slopes", Sorrento. Patron, Bologna, pp 191–206

Picarelli L (1988) Modellazione e monitoraggio di una colata in formazioni strutturalmente complesse. In: Cascini L (ed) Proceedings Conference on "Cartografia e Monitoraggio dei Movimenti Franosi", Bologna, pp 119–130

Picarelli L (1993) Structure and properties of clay shales involved in earthflows. In: Anagnastopoulos A, Schlosser F, Kalteziotis N, Frank R (eds) Proceedings International Symposium on "The Geotechnical Engineering of Hard Soils-Soft Rocks", Athens, 3, pp 2009–2019

Picarelli L (2000) Mechanisms and rates of slope movements in fine grained soils. In: Proceedings GeoEng2000, International Conference on "Geotechnical and Geological Engineering", Melbourne, Technomic Publ. Co., 1, pp 1618–1670

Picarelli L (2001) Transition from slide to earthflow, and the reverse. In: Proceedings TC-11 Satellite Conference on "Transition from Slide to Flow-Mechanisms and Remedial Messures", Trabzon, in CD-ROM

Picarelli L, Russo C (2004) Mechanics of slow active landslides and interaction with man-made works. In: Lacerda W (ed) Proceedings 9th International Symposium on "Landslides", Rio de Janeiro, 2, pp 1141–1176

Picarelli L, Simonelli AL (1991) Monitoring of a slow movement in colluvial soils. In: M. Durand M, Fabre D (eds) Proceedings 4ème Entretiens du Centre Jacques Cartier, Colloque 4: Géotechnique et Environnement, Grénoble

Picarelli L, Urciuoli G (1988) Influenza della velocità di scorrimento sulla resistenza residua delle argille. Gruppo Naz. di Coordin. per gli Studi di Ingegneria Geotecnica, Attività di ricerca nell'anno 1987–88, pp 101–104

Picarelli L, Urciuoli G (1993) Effetti dell'erosione in argilliti di alta plasticità. Rivista Italiana di Geotecnica 17:29–47

Picarelli L, Olivares L, Di Maio C, Silvestri F, Di Nocera S, Urciuoli G (2002) Structure, properties and mechanical behaviour of the highly plastic intensely fissured Bisaccia Clay Shale. In: Tan TS, Phoon KK, Hight DW, Leroueil S (eds) Proceedings International Symposium on "Characterisation and Engineering Properties of Natural Soils", Singapore, 2, pp 947–962

Picarelli L, Urciuoli G, Russo C (2004) The role of groundwater regime on behaviour of clayey slopes. Can Geotech J 41(3):467–484

Picarelli L, Urciuoli G, Ramondini M, Comegna L (2005) Main features of mudslides in tectonised highly fissured clay shales. Landslides 2(1):15–30

Picarelli L, Urciuoli G, Mandolini A, Ramondini M (2006) Softening and instability of natural slopes in highly fissured plastic clay shales. Natural Hazards and Earth System Sciences 6(4):529–539

Radbruch-Hall D, Varnes DJ, Savage WZ (1976) Gravitational spreading of steep-sided ridges (sackung) in western United States. Bull IAEG 14:23–35

Russo C (1997) Caratteri evolutivi dei movimenti traslativi e loro interpretazione meccanica attraverso l'analisi numerica. PhD Thesis, Università di Napoli Federico II

Skempton AW (1967) Some observations on tectonic shear zones. In: Proceedings II International Conference on Rock Mech., Belgrade, 1, pp 329–335

Skempton AW (1985) Residual strength of clays in landslides, folded strata and the laboratory. Géotechnique 35(1):3–18

Tacher L, Bonnard C, Laloui L, Parriaux A (2005) Modelling the behaviour of a large landslide with respect to hydrogeological and geomechanical parameter heterogeneity. Landslides 2(1):3–14

Tika TM, Vaughan PR, Lemos LL (1996) Fast shearing of pre-existing shear zones in soils. Géotechnique 46(2):197–233

Tommasi P, Rotonda T (1995) Analysis of the deformations occurring at the edge of a rock cliff underneath a historical center. In: Fujii T (ed) Proceedings 8th International Congress on Rock Mechanics, Tokyo, pp 383–388

Tommasi P, Pellegrini P, Boldini D, Ribacchi R (2006) Influence of rainfall on hydraulic condition and movement rate in the overconsolidated clayey slope of the Orvieto Hill (Central Italy). Can Geotech J 70–86

Urciuoli G (2002) Strain preceding failure in infinite slopes. Int J Geomech 2(1):93–112

Urciuoli G, Picarelli L (2004) The shear strength mobilized in first-time slides in highly overconsolidated clays. In: Jardine RJ, Potts DM, Higgins KG (eds) Proceedings International Conference on "Advances in Geotechnical Engineering. The Skempton Conference", Thomas Telford, 2, pp 1005–1018

Urciuoli G, Picarelli L, Leroueil S (2007) Local soil failure before general slope failure. Geotech Geol Eng 25(1):103–123

Van Genuchten PMB (1984) Intermittent sliding of a landslide in varved clays. In: Bonnard C (ed) Proceedings 5th International Symposium on "Landslides", Lausanne, Balkema, Rotterdam,1, pp 471–476

Varley PM, Warren CD (1995) The Channel tunnel project. In: Proceedings 11th ECSMFE, Copenhagen, 10, pp 71–117

Vulliet L (1986) Modélisation des pentes naturelles en mouvement. PhD Thesis no 635. École Politechnique Fédérale de Lausanne

Vulliet L, Hutter K (1988) Viscous-type sliding laws for landslides. Can Geotech J 25(3):467–477

Wilson ST (1969) Observational data on ground movements related to slope instability. J Soil Mech Found Div-ASCE p 96

Yoshida N (1990) Time-dependent instability in fissured overconsolidated clays and mudstones. PhD Thesis, Univ. of Alberta, Edmonton, Canada

Chapter 4

Dynamics of Rapid Landslides

Oldrich Hungr

Abstract. Velocity is the most important parameter determining the destructive potential of landslides. "Catastrophic" velocities of the order of several meters per second are attained only by certain types of landslides. High velocities are the consequence of a range of strength loss mechanisms, reviewed in this chapter. Strength loss can occur instantly during the process of failure, through loss of cohesion, liquefaction of granular material or remoulding of sensitive clay. Further important loss of strength can occur during movement, including rock joint roughness reduction, shearing in clays, sliding surface liquefaction, frictional heating, loss of internal coherence of the sliding body, material entrainment, rapid undrained loading, and entrainment of water. Extremely rapid landslides include rock, debris and earth fall, rock block topple, rock slide, debris slide, flow slide in granular soil or clay, debris avalanche, debris flow and rock avalanche. There is a need to study the post-failure behavior of materials, in order to facilitate predictions of the behavior of extremely rapid landslides for hazard assessment.

Keywords. Velocity, rapid landslide, dynamics, liquefaction, shear strength

4.1 Introduction

The destructive potential of a landslide depends on velocity. This can be seen in published examples of landslide incidents, where both the typical velocity of movement and the corresponding human response were documented. A compilation of such cases by Hungr (1981) was later paraphrased by the IUGS Working Group on Landslides (1995), as seen in Table 4.1. In a summary, only landslides in the first and second response class, exhibiting velocities of the order of 5 m s^{-1}, stand out in terms of danger to human life. Slower movements usually cause only material damage. The limit of 5 m s^{-1} corresponds to the speed of a person running and could be called "catastrophic velocity". Using terminology suggested by Varnes (1978) and Cruden and Varnes (1996), landslides capable of reaching catastrophic velocities classify as "extremely rapid".

From the Work-Energy Theorem, the mean kinetic energy of a landslide with a mass M can be calculated as the work of the net driving force (gravity force resultant, G, minus resisting force, R, derived from material strength), moving along the landslide path, s:

$$\frac{MV^2}{2} = \int_0^s (G - R)\mathrm{d}s \tag{4.1}$$

Thus, landslide velocity is primarily the consequence of strength loss, i.e. the decrease of R, relative to G that is a function of slope only. This chapter reviews the causes of strength losses that are responsible for generating catastrophic landslide velocities. Examples of extremely rapid landslides are presented.

4.2 Mechanisms Causing Strength Loss in Landslides

Loss of strength can occur instantaneously as part of a brittle failure of the soil or rock material, or during displacement as a results of changes in friction and pore-pressure along the rupture surface.

Table 4.1. Landslide velocity scale (Cruden and Varnes 1996)

Velocity class	Description	Velocity (mm s^{-1})	Typical velocity	Human response
7	Extremely rapid			Nil
		5×10^{3}	5 m s^{-1}	
6	Very rapid			Nil
		5×10^{1}	3 m min^{-1}	
5	Rapid			Evacuation
		5×10^{-1}	1.8 m h^{-1}	
4	Moderate			Evacuation
		5×10^{-3}	13 m month^{-1}	
3	Slow			Maintenance
		5×10^{-5}	1.6 m yr^{-1}	
2	Very slow			Maintenance
		5×10^{-7}	16 mm yr^{-1}	
1	Extremely slow			Nil

4.2.1 Instantaneous Strength Loss Mechanisms

4.2.1.1 Loss of Cohesion in Rocks and Cemented or Unsaturated Soils

The true cohesion of granular soils cemented by calcium carbonate, iron oxide or similar will disappear instantaneously, as connected grains break apart under excessive shear or tensile stress. The same may occur in partially saturated soils as suction elements at particle contacts are severed by excessive stress. The brittle initial failure of a silt column shown in Fig. 4.4, later in this chapter, was caused by a combination of cement and suction cohesion loss. Root cohesion in shallow soils can also be destroyed suddenly by overstress, albeit with a few cm of displacement. Cohesion of rock masses results largely from the presence of intact rock bridges, separating non-persistent joints. (e.g., Eberhardt et al. 2004). When subjected to excessive stress, the bridges are destroyed through shear or crack propagation, often with a dramatic loss of strength combined with very small displacement.

4.2.1.2 Liquefaction of Granular Soils

Spontaneous or earthquake liquefaction refers to a collapse of a loose soil skeleton due to excessive strain, followed by an increase in pore-pressure and loss of strength under undrained conditions. The liquefaction process has been studied in the laboratory for many decades (e.g., Castro 1975), usually with the help of a triaxial testing apparatus. Numerous conditions influence the amount of strength loss, including: density of the soil, grain-size distribution, stress history, stress level and failure stress path (e.g., McRoberts and Sladen 1985). In recent years, it was realized that liquefaction affects not only sands, but a variety of well-graded materials, including those that contain significant coarse fractions (e.g., Dawson et al. 1998).

A serious disadvantage of the triaxial test is that it is limited to very small displacements. The sophisticated ring shear tests reported by Sassa and his co-workers (e.g., Sassa et al. 2004), allow the liquefaction process to be followed to large displacements. An example is shown in Fig. 4.1, presenting the time series and stress paths of an undrained cyclic ring shear test on loose sand. The sand was collected from the vicinity of the rupture surface of an extremely rapid landslide, triggered by the 2004 Mid-Niigata Prefecture earthquake (Sassa et al. 2005). The ring shear test was conducted by transferring simulated dynamic normal and shear stress signals into the annular test box (green and red traces in Fig. 4.1a). As the sample deformed under the cyclic load, pore-pressure steadily increased (blue line). Initial pore-pressure increase of a few percent of the normal stress occurred with very small displacement. This phenomenon could be compared to the "cyclic mobility" observed frequently in triaxial tests. More severe pore-pressure increase occurred with increasing displacement. After 50 mm of movement, the pore pressure ratio within the box reached a level of approximately 0.75, indicating that a high degree of liquefaction has occurred. The displacement of 50 mm is about twice as large as can be achieved in a typical triaxial test. Additional pore-pressure generation followed this initial result, as discussed below.

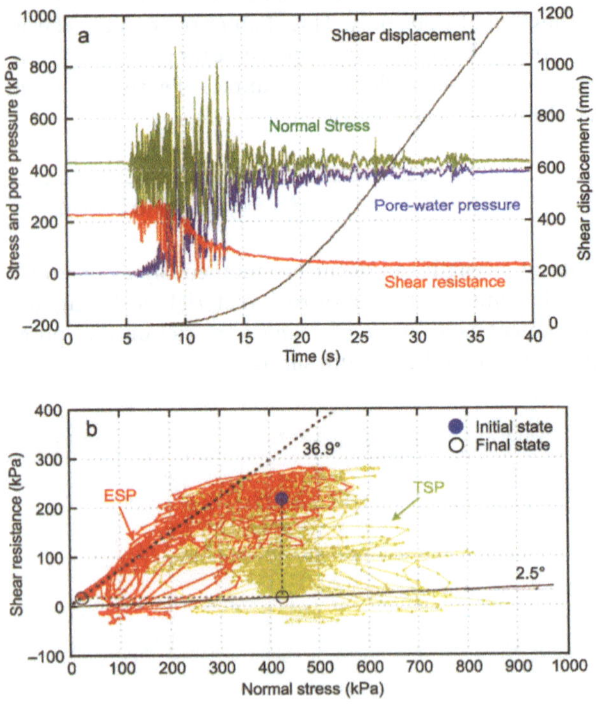

Fig. 4.1. Undrained cyclic ring shear test on sand from the Higashi-Takezawa landslide (from Sassa et al. 2005)

Recent studies of earthquake liquefaction showed that the undrained effective stress response in soils of non-uniform permeability can be very different from that which would take place in a homogeneous soil. This is an effect that cannot easily be studied in laboratory tests (Byrne et al. 2006).

The above considerations and the fact that many landslides involve heterogeneous soils, indicate that the results of laboratory testing must be interpreted with caution, if they are to provide constitutive relationships for use in quantitative analyses of flow slides. This is true especially of the traditional undrained triaxial laboratory test with limited displacement.

4.2.1.3 Remoulding of Sensitive Clays

Certain clays of marine origin, such as the "quick clays" of north-eastern North America and southern Scandinavia are characterized by extreme loss of strength on remoulding following shear failure (e.g., Crawford 1968). The rheological properties of remoulded extra-sensitive clay can be those of a Newtonian viscous fluid, with negligible shear strengths. Unfortunately, laboratory-derived viscosity values cannot be used directly to analyze full-scale flows, due to a scale-dependence of the rheology (Locat 1993).

4.2.2 Strength Loss Requiring Large Displacements

4.2.2.1 Surface Roughness Reduction in Rock Discontinuities

Large shearing displacements applied to rock discontinuities will reduce the dilatancy component of frictional strength, producing residual friction (e.g., Hoek and Bray 1974). For example, Cruden and Krahn (1978) tested the peak friction angle of joints and bedding planes in the limestone of Turtle Mountain (the Frank Slide) as ranging between 32 and 52° with small cohesion, while the ultimate friction of friction-polished joints was 14 to 32°. No laboratory tests have yet been carried out that would approximate the effects of long-displacement shearing at high normal stresses, such as must occur at the base of thick translational or compound slides in rock. Therefore, little is known about the amount of displacement needed to produce true "ultimate strength" of natural rock joints. The effects of scale will undoubtedly contribute to the interpretation of such tests. Nevertheless, for runout analyses of large rock slides, a substantial reduction of the friction angle following a few meters of displacement is to be expected.

4.2.2.2 Shearing in Clays

Frictional strength of clay is also known to decrease after long shear displacements, as a result of reorientation of the platy clay particles and the development of a sheared structure (Skempton 1985). The amount of decrease ("brittleness") depends on the type of clay mineral and the clay content and can amount to more than 50%. In the laboratory, the residual strength is fully developed after only a few hundreds of mm. The corresponding minimum displacement in the field may be much greater. Recent laboratory testing indicated that the residual friction of certain clays may decrease further by as much as 60% at high rates of shearing, increasing brittleness (Tika and Hutchinson 1999).

4.2.2.3 Sliding Surface Liquefaction

The concept of sliding surface liquefaction in granular material, proposed by Sassa (2000) is somewhat analogous to the formation of residual strength in clays. Once failure occurs and is followed by large displacement, the intense shearing within a thin shear band leads to a textural change within the material (grain crushing). The modified soil contains a greater proportion of fines and is capable of denser packing. This generates pore pressure under undrained conditions and a corresponding loss of strength. The effect can be observed to some degree in Fig. 4.1. The initial structural collapse of the soil skeleton reduced the effective stress by some 75% after 50 mm of displacement. Following this, however, a further gradual reduction of about 15% occurred, to reach a steady-state pore-pressure ratio of about 0.9 after 200 mm of displacement. Fukuoka et al. (2006) showed that measurable changes in grain size distribution in the vicinity of the sliding surface follow such tests. The phenomenon appears to be most prominent in soils with relatively weak grains, especially pumice.

4.2.2.4 Frictional Heating

Friction across a sliding surface produces heat. The coefficient of volumetric thermal expansion of water is about an order of magnitude larger than that of mineral particles. As a result, undrained heating of saturated soil will be accompanied by an increase in pore pressure, hence a reduction of shear strength. Such effects have been shown theoretically by Voight and Faust (1982). Earlier, Hendron and Patton (1985) proposed that frictional heating led to the reduction of shear strength at the base of the Vaiont Slide. This is a plausible idea for certain large rock slides although, to the Author's knowledge, no experimental or field confirmation exists to date.

4.2.2.5 Loss of Internal Coherence of the Sliding Body

The resisting forces in a slide derive not only from the strength of the rupture surface, but also to some extent from the mobilized internal strength of the sliding body. Therefore, the internal brittleness of a rigid slide may in some cases add to the brittleness of the sliding surface. Hutchinson (1988) proposed that the rapid acceleration of the Vaiont Slide was partly due to the brittle loss of cohesion within the limestone mass, sliding on a clay-coated bedding plane. Based on kinematic considerations, the mobilization of internal strength requires greater displacement than the mobilization of the boundary strength on the rupture surface.

4.2.2.6 Material Entrainment and Rapid Undrained Loading

The rapid undrained loading mechanism was documented by Hutchinson and Bhandari (1971), who observed pore pressure increases in a piezometer embedded near the basal surface of an earth flow subjected to rapid deposition of material from upslope. This mechanism appears to be the principal means of mobilizing earth flow movements on gentle slopes. Sassa (1985) extended the same principle to debris avalanches and de-

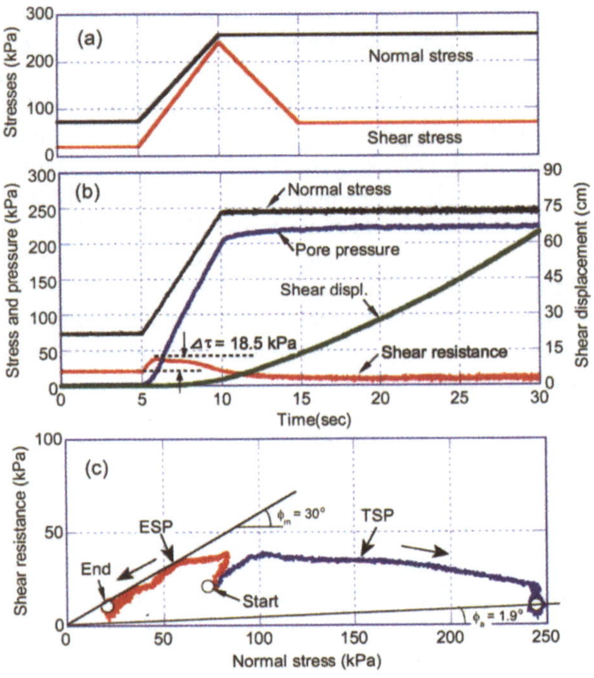

Fig. 4.2. Time series and stress path plots of rapid undrained loading of a sample from a debris flow in a weathered volcanic soil (Sassa et al. 2004)

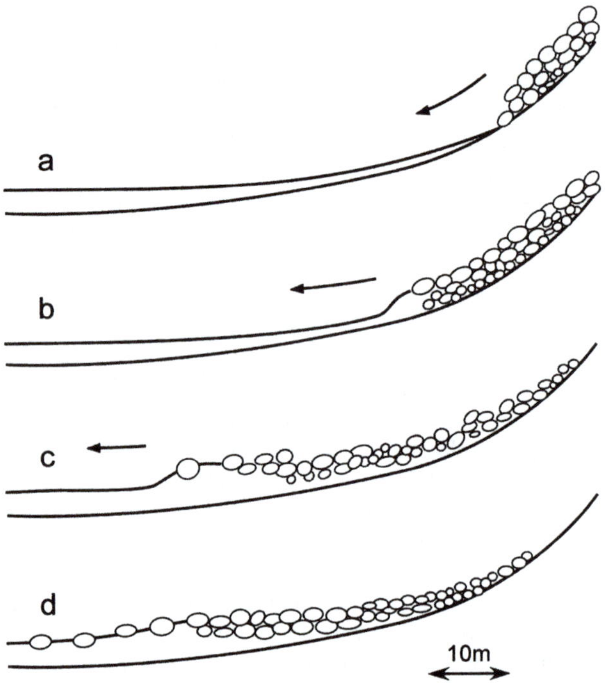

Fig. 4.3. Process of simultaneous entrainment, plowing and lubrication at the front of a rock avalanche, described by Hungr and Evans (2004)

bris flows. It has long been known that debris flows derive most of their volume by entraining loose saturated material from their path (e.g., Hungr et al. 2005). Such material is in all probability mobilized by rapid undrained loading, caused by debris surges arriving from upstream. Figure 4.2 shows the results of an undrained ring shear test under sudden loading by Sassa et al. (2005), illustrating this phenomenon.

The entrainment mechanism has also been described for rock avalanches and rock slide-debris avalanches (Hungr and Evans 2004). The process has two, often simultaneous consequences, both of which reduce strength: (*1*) It leads to lubrication of the base of the slide by liquid saturated material, and (*2*) it incorporates such material into the body of the landslide, increasing its volume and internal fluidity (Fig. 4.3).

4.2.2.7 Entrainment of Water, Dilution

Rapidly moving landslides also incorporate surface water flowing in the path. This is especially true of confined debris flows that often follow natural drainage paths or stream channels. Wallance (2005) gives a description of gradual dilution that can change a large volcanic debris flow into a flood-like hyperconcentrated flow. Dilution is often accompanied by a gradual reduction in peak discharge of debris surges and this leads to some decrease in their destructive potential. On the other hand, it can greatly increase the ultimate runout distance.

4.3 Types of Extremely Rapid Landslides

A simple form of the popular Varnes (1978) classification of landslides can be represented by 26 keywords, each representing a unique type of slope movement. The keywords are shown in Table 4.2 within Varnes' two-dimensional classification framework. A letter type code is used in the table to distinguish phenomena considered capable of producing catastrophic velocities from those that do not. Some of the keywords, related to landslides of flow-like character, were later re-defined by Hungr et al. (2001), as shown in Table 4.3.

A number of individual landslide types capable of extremely rapid motion, chosen from a combination of the two classification systems shown in Tables 4.2 and 4.3 are described as follows.

Rock Fall (fragmental rock fall, cf. Evans and Hungr 1993) is the free fall, bouncing and rolling of rock fragments down a slope. The initial acceleration results usually from loss of cohesion. Subsequent propagation mechanisms involve free fall, bouncing and rolling and the movement is invariably extremely rapid. The volumes are limited. Figure 4.4 shows a dynamic back-analysis of the fall of a 100 tonne boulder which caused two fatalities in British Columbia in 1982 (Evans and Hungr 1993).

Debris Fall involves similar free movement of solid fragments derived from coarse soil.

Table 4.2. Landslide classification (after Varnes 1978) Code: *RAPID*, SLOW (in most cases)

	Bedrock	Debris (<80% sand and finer)	Earth (>80% sand and finer)
Falls	*ROCK FALL*	*DEBRIS FALL*	*EARTH FALL*
Topples	*BLOCK TOPPLE* *FLEXURAL TOPPLE*	–	*BLOCK TOPPLE*
Slides	*ROCK SLUMP* *ROCK SLIDE*	*DEBRIS SLIDE*	EARTH SLUMP EARTH SLIDE
Spreads	*ROCK SPREAD*	–	*EARTH SPREAD*
Flows	ROCK CREEP SLOPE SAGGING	*DEBRIS FLOW* *DEBRIS AVALANCHE* SOIL CREEP SOLIFLUCTION	*WET SAND AND SILT FLOW* *RAPID EARTH FLOW* *LOESS FLOW* DRY SAND FLOW EARTH FLOW
Complex	*ROCK AVALANCHE* EARTH SLUMP-EARTHFLOW		

Table 4.3. Classification of landslides of the flow type (Hungr et al. 2001)

Material	Water content[a]	Special condition	Velocity	Name
Silt, sand, gravel, debris (talus)	Dry, moist or saturated	▪ No excess pore-pressure ▪ Limited volume	Various	*NON-LIQUEFIED SAND (SILT, GRAVEL, DEBRIS) FLOW*
Silt, sand, debris, weak rock[b]	Saturated at rupture surface	▪ Liquefiable material[c] ▪ Constant water content	Ex. rapid	*SAND (SILT, DEBRIS, ROCK) FLOW SLIDE*
Sensitive clay	At or above liquid limit	▪ Liquefaction in situ[c] ▪ Constant water content[d]	Ex. rapid	*CLAY FLOW SLIDE*
Peat	Saturated	▪ Excess pore-pressure	Slow to very rapid	*PEAT FLOW*
Clay or earth	Near plastic limit	▪ Slow movements ▪ Plug flow (sliding)	<Rapid	*EARTH FLOW*
Debris	Saturated	▪ Established channel[e] ▪ Increased water content[d]	Ex. rapid	*DEBRIS FLOW*
Mud	At or above liquid limit	▪ Fine-grained debris flow	>Very rapid	*MUD FLOW*
Debris	Free water present	▪ Flood[f]	Ex. rapid	*DEBRIS FLOOD*
Debris	Partly or fully saturated	▪ No established channel[e] ▪ Relatively shallow, steep source	Ex. rapid	*DEBRIS AVALANCHE*
Fragmented rock	Various, mainly dry	▪ Intact rock at source ▪ Large volume[g]	Ex. rapid	*ROCK AVALANCHE*

[a] Water content of material in the vicinity of the rupture surface at the time of failure.
[b] Highly porous, weak rock (examples: weak chalk, weathered tuff, pumice).
[c] The presence of full or partial in situ liquefaction of the source material of the flow slide may be observed or implied.
[d] Relative to in-situ source material.
[e] Presence or absence of a defined channel over a large part of the path, and an established deposition landform (fan). *Debris flow* is a recurrent phenomenon within its path, while *debris avalanche* is not.
[f] Peak discharge of the same order as that of a major flood or an accidental flood. Significant tractive forces of free flowing water. Presence of floating debris.
[g] Volume greater than 10 000 m^3 approximately. Mass flow, contrasting with fragmental rock fall.

Earth Fall results from localized failure of fine-grained, cohesive soil. High movement velocity again results from loss of cohesion. Figure 4.5 shows the flow of dry glacio-lacustrine silt following the failure of a part of a vertical face, supported by cohesion. The cohesion of the silt material *in situ* is a combination of suction forces in the unsaturated soil and cementing by a trace of clay and calcium carbonate cement. A column of silt, separated from the cliff face along a vertical stress-relief joint, toppled, fell and disintegrated, to produce an extremely rapid flow

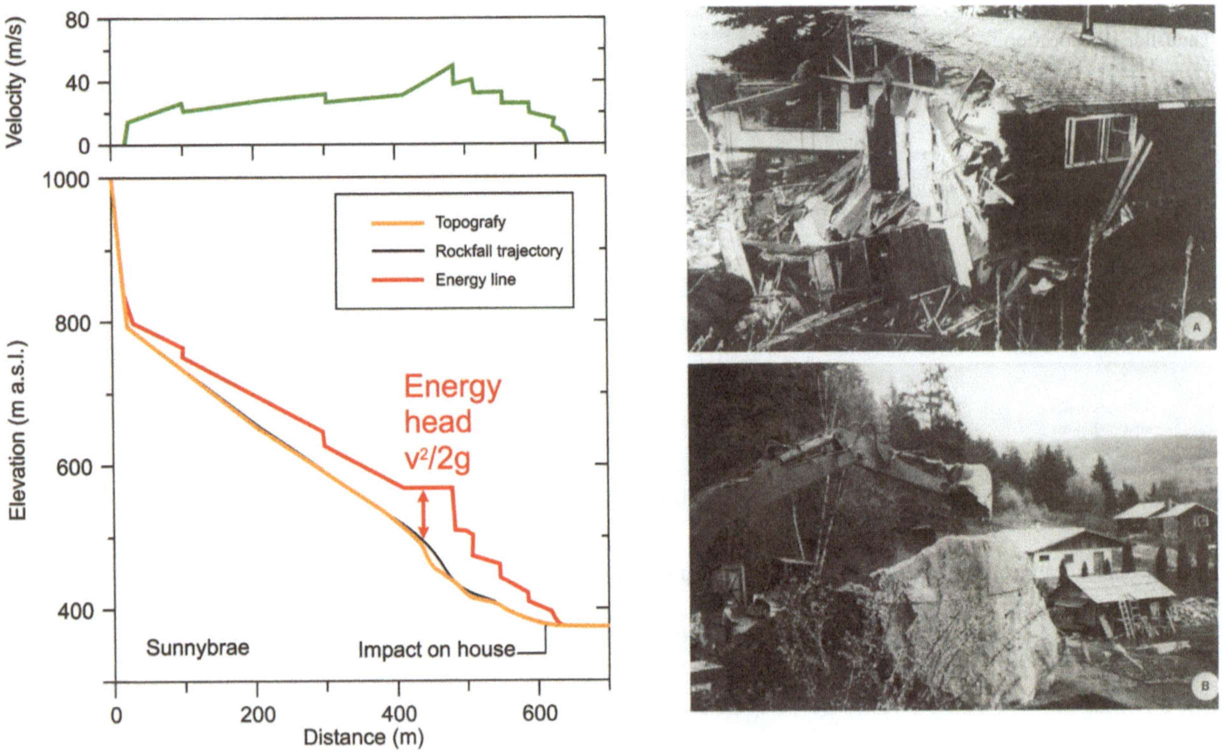

Fig. 4.4. Dynamic analysis of a 100 tonne rock fall. **a** Damage, **b** boulder (Hungr and Evans 1985)

Fig. 4.5. Dry silt flow near Peachland, British Columbia (photo courtesy of Mr. J. Valentinuzzi, British Columbia Ministry of Transportation and Highways)

that crossed a major highway and covered a distance of 100 m in about 8 seconds. A dynamic back-analysis of the failure using the program DAN (Hungr 1995), confirmed that the disintegrated silt was dry and frictional, traveling with a bulk friction angle of 30°, corresponding to the dynamic friction angle of dry silt. The high velocity in this case was thus due solely to the release of potential energy as a result of cohesion loss in course of the initial failure. No pore-pressure was involved.

Rock Block Topple. Block toppling involves relatively massive blocks separated by near-vertical joints and resting on structurally-defined basal surfaces. Each block originally derives its stability from the position of the center of gravity with respect to the base of the block. The block rotates forward gradually under the influence of some destabilizing agent (e.g., water or ice pressure in the tension crack). Once the point of overturning is reached, however, further rotation becomes irreversible and ex-

Fig. 4.6.
A block topple, Czech Republic

tremely rapid. Block toppling is the only type of slope failure where high velocity does not require strength loss, but results from the inherent instability of the failure mechanism. Block toppling mechanism affects single blocks, small groups of blocks (see Goodman and Bray 1976 and Fig. 4.6) and possibly also certain large rock masses (e.g., Chigira and Kiho 1994).

Rock Slide. Sliding failure in strong rock requires discontinuities. Where the discontinuities are persistent, failure brittleness will result from surface roughness reduction, as discussed above. It is probably for this reason that structurally-controlled translational and compound sliding in strong rocks produces highly mobile rock slides. The initial sliding block sometimes remains semi-coherent (e.g., Vaiont, Hendron and Patton 1985), but more often disintegrates to produce a rock avalanche (Fig. 4.7).

The presence of non-persistent discontinuities, separated by "bridges" of intact rock, contribute a cohesion, the loss of which usually leads to dramatic acceleration. Hungr and Evans (2004a) referred to this type of sliding mechanism as a "rock collapse".

Debris Slide. Debris slides in granular material involve translational sliding of a thin colluvial or residual veneer over strong substrate (e.g., Savage and Baum 2005). Slow-moving debris slides exist, but are common only in relatively dry loose soils (for example, slip faces of sand dunes). Usually, strength losses in the form of cohesion loss and full or partial liquefaction combine to produce

Fig. 4.7.
Translational failure of sedimentary sequence of siltstone and sandstone, Rockslide Pass, Mackenzie Mountains, Canada. The landslide began by sliding of a 600 m thick block on a bedding plane. After colliding with the valley fill to the right of the center of the photo, the block disintegrated and flowed for another 4 km as a rock avalanche (Hungr and Evans 2004)

acceleration, while rapid undrained loading or sliding surface liquefaction (Sassa 2000) lead to further increase of speed, disintegration of the granular mass and the formation of an extremely rapid debris avalanche.

Flow Slide in Granular Soil. The classical concept of flow sliding, defined by Casagrande (1976) and others, involves extremely rapid, flow-like movement following sudden strength loss due to spontaneous or earthquake liquefaction of loose, saturated granular soil. The process takes place at constant volume, with no increase in water content. Many flow slides occur in poorly consolidated sandy sediments under water. Older accounts of flow slides describe very loose sands or silty sands. More recently, however, it was realized that liquefaction can involve a wide variety of granular materials, including those that are relatively coarse and well-graded such as colluvium and mine waste (e.g., Dawson et al. 1998; Hunter and Fell 2002, see Fig. 4.8).

While saturation is key to the basic liquefaction mechanism, this does not mean that liquefaction flow slides are limited to completely saturated slopes. In many cases, the presence of spontaneous or earthquake liquefaction can only be deduced from the sudden acceleration of the initial failure and rapidity and flow-like character of the subsequent motion. The bulk of the soil or rock may be in an unsaturated condition, as long as sufficient saturation exists in the vicinity of the rupture surface.

Flow Slide in Extra-Sensitive Clay. A slide in an extra-sensitive clay, often on a river bank undermined by erosion, produces dramatic loss of strength by remoulding. The mobile debris evacuates the source area and a steep scarp at the current slide head becomes exposed to further failure. The process may take the form of multiple retrogressive rotational slides, or a spontaneous displacement of a mass of earth supported by a failing sensitive layer (cf. "flake slide", Hutchinson 1988). The liquefied clay flows out of a "slide crater" and follows the slope, often along a river channel. The flow is that of a Newtonian viscous substance, often in the turbulent regime (Locat 1993) The flow carries intact floes of broken desiccated surficial crust, that can influence the flow behavior (Hungr 1981).

Fig. 4.8. A flow slide in coal mining waste, South-east British Columbia, Canada. The 60 000 m^3 slide originated on the waste pile in the background and traveled 2 km

Debris Avalanche. As described earlier, a debris avalanche is the end result of shallow sliding of granular material on steep hillslopes (Fig. 4.9). The initial slide is metamorphosed into a rapid flow through a combination of strength loss mechanisms, that may include cohesion loss, spontaneous liquefaction, rapid loading of material in the path and sliding-surface liquefaction. In conducting dynamic analysis it is exceedingly difficult to separate the effects of these various contributing effects. The concept of "equivalent fluid", introduced by Hungr (1995) avoids this complex issue, by assuming that the solid landslide material in the source area turns instantly into a simple fluid, the characteristics of which must be determined by back analysis.

Fig. 4.9. A debris avalanche in colluvium, Jasper National Park, Canadian Rocky Mountains.

The boundary between flow slides, dominated by liquefaction, and debris avalanches that are mobilized by several simultaneous strength loss processes, is difficult to define. A well-known case of an extremely rapid landslide from Hong Kong illustrates the problem:

The Shum-Wan Road landslide took place on August 13, 1995 (Fig. 4.10). Failure began at the end of a rain storm with total precipitation of 159 mm and hourly intensity of up to 48 mm, preceded by 846 mm of rain during the preceding 13 days. As interpreted by Hong Kong Geotechnical Engineering Office (GEO 2006), the landslide began by the failure of a road fill, coinciding with the release of large amounts of surface water onto the slope below Shum Wan Road. Immediately after the initial failure, a spoon-shaped rupture surface developed in the slope made up of volcanic residual soil, utilizing relict joints and seams filled with kaolinite (Fig. 4.11). The main body of the landslide accelerated to a high speed, mobilizing a thin slab of residual soil further downslope. The whole landslide of 26 000 m^3 came to a stop within a few minutes of the initial movement, having covered some 80 m distance at the toe.

A dynamic analysis of the landslide using a "bulk friction coefficient" of 20° (including pore pressure effects) yielded good correspondence with the observed movement behavior (Ayotte et al. 1999). In contrast, static slope stability analysis used several surfaces with friction angles between 26° and 38° and small cohesion. The fall of strength responsible for the rapid movement may have resulted partly from cohesion loss, from shearing of kaolinite from peak to residual friction angle (26° to 21°) and from undrained loading. Both the fill and colluviated surficial residual soil were probably susceptible to liquefaction. These diverse sources of potential strength loss cannot

Fig. 4.10. The Shum Wan Road debris avalanche, Hong Kong (photo courtesy of Geotechnical Engineering Office, Hong Kong)

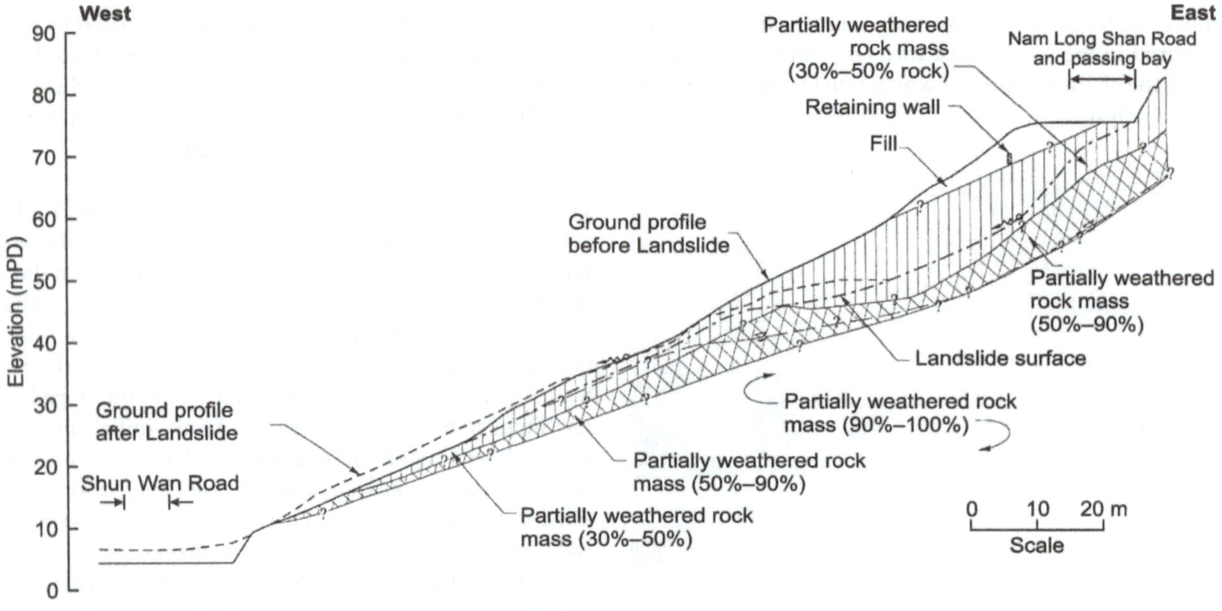

Fig. 4.11. Profile of the Shun Wan Road debris avalanche (GEO 2006)

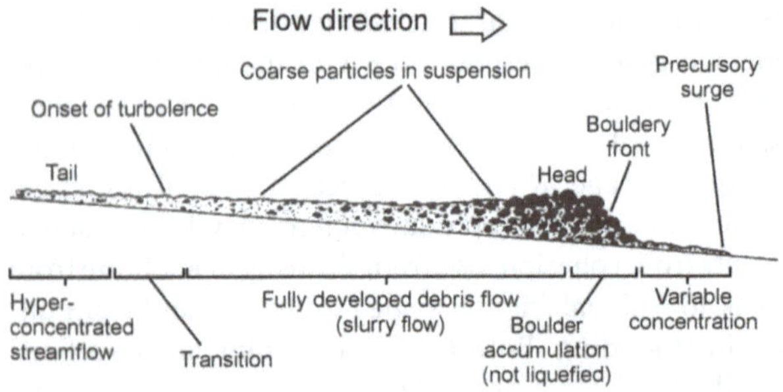

Fig. 4.12. A schematic profile of a debris flow surge (Pierson 1986)

easily be separated from each other and the author prefers to refer to this occurrence as a debris avalanche.

Debris Flow. Debris flows are extremely rapid, surging flows of unsorted, saturated debris in pre-defined channels (Hungr et al. 2001). Many debris flows form from debris avalanches starting on steep slopes and entering a channel.

The process of surge formation, conditioned by the presence of a channel, give debris flows their extraordinary mobility and destructive character. In coarse debris, the surges result from longitudinal sorting. A concentration of coarse clasts forms at the surge front, followed by increasingly more fine-grained and dilute debris (Pierson 1986, Fig. 4.12). As a result, the peak depth and discharge of the surge increase in course of movement down the channel. The physical properties of the material forming the surge are heterogeneous and velocity of the surge movement is determined by the composition and water content of material at the surge peak (Hungr 2000). Material entrainment and dilution by surface water are important processes controlling the dynamics of debris flow.

Rock Avalanche. A rock avalanche is the extremely rapid flow-like movement of fragmented rock from a large rock slide. The initial acceleration is provided by cohesion loss and joint surface roughness reduction. In case of certain planar detachments such as that shown on Fig. 4.7, frictional heating may also play a role. When the rock avalanche mass over-rides saturated surficial soils, a process of rapid loading and material entrainment will take place to mobilize long-runout movement (e.g., Sassa 1988; Hungr and Evans 2004). Several other physical explanations of rock avalanche mobility have been advanced, but none has so far been shown to be universal (e.g., Hungr 2006).

4.4 Conclusion

Formation of high velocity in landslides occurs primarily through strength loss. The physical processes leading to loss of strength in earth materials are many and often they work together. Some strength loss occurs instantaneously at the point of failure, other involves gradual change of material during motion. Existing research on slope stability and failure behavior has concentrated almost exclusively on processes leading to instability. In order to predict landslide behavior and runout for hazard assessment, it is necessary to study the mechanisms that determine post-failure loss of strength. One step in the right direction is the use of long-displacement testing methods, advanced in recent decades by the Disaster Prevention Research Institute at Kyoto University (Sassa 2000).

Acknowledgment

The author thanks the Head of the Geotechnical Engineering Office and the Director of Civil Engineering and Development, Hong Kong SAR Government, for the permission to publish the Figs. 3.10 and 3.11.

References

Ayotte D, Evans N, Hungr O (1999) Runout analysis of debris flows and avalanches in Hong Kong. In: Proceedings Slope Stability and Landslides, Vancouver, Geotechnical Society Symposium, May 1999, pp 39–46

Byrne PM, Naesgaard E, Seid-Karbasi M (2006) Analysis and design of earth structures to resist seismic soil liquefaction. Keynote Paper, 59th Canadian Geotechnical Conference, Canadian Geotechnical Society, Vancouver

Casagrande A (1976) Liquefaction and cyclic deformation of sands; a critical review. Harvard Soil Mechanics Series, No 88, p 51

Castro G (1975) Liquefaction and cyclic deformation of sands. J Geotech Eng-ASCE 101:551–569

Chigira M, Kiho K (1994) Deep-seated rockslide-avalanches preceded by mass rock creep of sedimentary rocks in the Kaishi Mountains, central Japan. Eng Geol 38:221–230

Crawford CB (1968) Quick clays of Eastern Canada. Eng Geol 2:239–265

Cruden DM, Krahn J (1978) Frank rockslide, Alberta, Canada. In: Voight B (ed) Rockslides and avalanches, Vol. 1. Elsevier, Amster-dam, pp 97–112

Cruden DM, Varnes DJ (1996) Landslide types and processes. In: Turner AK, Schuster RL (eds) Landslides: investigation and mitigation. Special Report 247, Transportation research board, US National research council, Washington, D.C., pp 36–75

Dawson RF, Morgenstern NR, Stokes A (1998) Liquefaction flow slides in Rocky Mountain coal mine waste dumps. Can Geotech J 35:328–343

Eberhardt E, Stead D, Coggan JS (2004) Numerical analysis of initiation and progressive failure in natural rock slopes – the 1991 Randa rockslide. International Journal of Rock Mech and Mining Sciences 41(1):69–87

Evans SG, Hungr O (1993) The assessment of rockfall hazards at the base of talus slopes. Can Geotech J 30:620–636

Fukuoka H, Sassa K, Wang G, Sasaki R (2006) Observation of shear development in ring-shear apparatus with a transparent shear box. Landslides 3:239–251

Geotechnical Engineering Office (2006) Report on the Shum Wan Road landslide of 13 August, 1995. GEO Report No. 178, Government of Hong Kong, 115p

Goodman RE, Bray JW (1976) Toppling of rock slopes. Rock Engineering, ASCE Geotechnical Engineering Division Conference, Boulder, 2, pp 201–237

Hendron AJ, Patton FD (1985) The Vaiont Slide. US Corps of Engineers Technical Report GL-85-8

Hoek E, Bray JW (1997) Rock slope engineerig. 2nded., Inst. Mining and Metallurgy, London

Hungr O (1981) Dynamics of rock avalanches and other types of slope movements. Ph.D. Thesis, University of Alberta, 500 p

Hungr O (1995) A model for the runout analysis of rapid flow slides, debris flows and avalanches. Can Geotech J 32(4):610–623

Hungr O (2000) Analysis of debris flow surges using the theory of uniformly progressive flow. Earth Surf Proc Land 25:1–13

Hungr O (2006) Rock avalanche occurrence, process and modelling. Springer NATO Science Series, IV. Earth Environ Sci 49:243–266

Hungr O, Evans SG (2004) The occurrence and classification of massive rock slope failure. Felsbau 22:16–23

Hungr O, Evans SG, Bovis M, Hutchinson JN (2001) Review of the classification of landslides of the flow type. Environ Eng Geosci VII:221–238

Hungr O, McDougall S, Bovis M (2005) Entrainment of material by debris flows. In: Jakob M, Hungr O (eds) Debris flow hazards and related phenomena. Springer Verlag, Heidelberg, in association with Praxis Publishing Ltd, Chapter 7, pp 135–158

Hunter GJ, Fell R (1992) Mechanics of failure of soil slopes leading to "rapid" failure. In: Picarelli L (ed) Fast slope movements prediction and prevention for risk mitigation. Patron Editore, Bologna, 1, pp 283–290

Hutchinson JN (1988) General report: Morphological and geotechnical parameters of landslides in relation to geology and hydrogeology. In: Bonnard C (ed) Proceedings 5th International Symposium on Landslides, Lausanne, Balkema, Rotterdaml, pp 3–35

Hutchinson JN, Bhandari RK (1971) Undrained loading: a fundamental mechanism of mudflows and other mass movements. Géotechnique 21(4):353–358

IUGS Working Group on Landslides (1995) A suggested method for describing the rate of movement of a landslide. IAEG Bull 52:75–78

Locat J (1993) Fra fjell til fjord: consideration on viscous flows. In: Proceedings, Pierre Beghin International Workshop on Rapid Gravitational Movements, CEMAGFEF, Grenoble University Press, pp 197–207

McRoberts ED, Sladen JA (1985) Observations on static and earthquake liquefaction methodologies. In: Proceedings 43rd Canadian Geotechnical Conference, Quebec City, Canadian Geotechnical Society, 1, pp 215–226

Pierson T (1986) Flow behaviour of channelized debris flows, Mt. St. Helens. In: Abrahams AD (ed) Hillslope processes. Allen and Unwin, Boston, pp 269–296

Sassa K (1985) The mechanism of debris flows. In: Proceedings of XI International Conference on Soil Mechanics and Foundation Engineering, San Francisco, 3, pp 1173–1176

Sassa K (1988) Geotechnical model for the motion of landslides. In: Special Lecture of 5th International Symposium on Landslides, "Landslides", 10–15 July, 1, pp 37–55

Sassa K (2000) Mechanism of flows in granular soils. In: Proceedings of the International Conference of Geotechnical and Geological Engineering, GEOENG2000, Melbourne, 1, pp 1671–1702

Sassa K, Fukuoka H, Wang GH, Ishikawa N (2004) Undrained dynamic-loading ring-shear apparatus and its application to landslide dynamics. Landslides 1:7–19

Sassa K, Fukuoka H, Wang FW, Wang G (2005) Dynamic properties of earthquake-induced large-scale rapid landslides within past landslide masses. Landslides 2(2):125–134

Savage W, Baum R (2005) Instability of steep slopes. In: Jakob M, Hungr O (eds) Debris flow hazards and related phenomena. Springer-Verlag, Berlin Heidelberg, in association with Praxis Publishing Ltd, chapt. 4, pp 53–77

Skempton AW (1985) Residual strength of clays in landslides, folded strata and the laboratory. Géotechnique 35(1):3–18

Tika TE, Hutchinson JN (1999) Ring shear tests on soil from the Vaiont landslide slip surface. Géotechnique 49(1):59–74

Vallance J (2005) Volcanic debris flows. In: Jakob M, Hungr O (eds) Debris flow hazards and related phenomena. Springer-Verlag, Berlin Heidelberg, in association with Praxis Publishing Ltd, pp 247–271

Varnes DJ (1978) Slope movement types and processes. In: Landslides, analysis and control. National Academy of Sciences, National Research Council, Washington, DC., Special Rep. 176:11–33

Voight B, Faust C (1982) Frictional heat and strength loss in some rapid landslides. Géotechnique 32:43–54

Chapter 5

Progress in Debris Flow Modeling

Tamotsu Takahashi

Abstract. The processes of the two major types of debris flow initiation; the bed erosion type and the landslide-induced type, are discussed and the methods to analyze the respective phenomena are introduced. Especially for the landslide-induced type, a new model of liquefaction of the slid earth mass while in motion is introduced. The earth mass is liquefied at the bottom without water supply from the outside and this liquefied layer gets behind the mass as a following debris flow. The reviews of previous investigations on the mechanics of developed debris flow confirms that the two-phase model is influential, in which debris flow consists of the mixture of two continuum media of fluid phase and solid phase. Using this model, developed debris flows are classified into three types from the point of the dominant stresses within flow. The characteristics of the respective types of flow such as the solids concentration distribution, the velocity distribution, the equilibrium sediment transport concentration are given.

Keywords. Debris flow, erosion type debris flow, landslide-induced debris flow, single-phase continuum model, mixture theory, constitutive relations, quasi-static debris flow, stony type, turbulent muddy type, viscous type, immature debris flow, inertial debris flow

5.1 Introduction

Debris flow is composed of a dense mixture of water and sediment, and it surges down as if it were comprised of a kind of continuous fluid. For the initiation of debris flow, as the solid particles in debris flow have originally been the components of a stable soil mass on a gully bed or on a slope, some natural forces that destroy the skeleton of the mass must operate. The condition for destruction of the skeleton of soil mass is common to that of landslides, but the condition for keeping the flow to surge down require a steep slope and an ample quantity of water to fill the enlarged void space between particles.

Three types of debris flow initiation predominate. The first type is due to the erosion of channel bed. Following a severe rainfall, surface water flow appears on a steep channel bed on which plenty of debris is accumulating, and the water flow destabilizes the bed and entrains the debris rapidly to form debris flow. The second type is due to landslide. The slid earth mass is successively destroyed while in motion, and it transforms into debris flow by mixing of debris with water that has been contained originally in the earth mass or supplied from behind. The third type is the destruction of natural dam. A landslide sometimes dams up a river, and sooner or later, the dam is destroyed by the overtopping of river water or by the collapse of the dam body itself under the effects of seepage water and/or hydraulic pressure. From the mechanical point of view, the debris flow initiation process due to overtopping of a natural dam can be considered similar to the first type, and that due to collapse of the dam body is akin to the second type.

High fluidity of debris flow is secured by sustaining the particles apart from each other. The resistance to flow must be the sum of the resistance due to deformation of interstitial fluid and/or turbulent mixing of particles and fluid and that due to consumption of energy to disperse the particles by inter-particle collision, and the larger the particle concentration, the larger the resistance to flow becomes. The necessary energy to disperse particles is supplied from the shearing of the mixture. In an equilibrium state, the resistance to flow balances with the driving force due to gravity. If debris flow plunges into a gently sloping area, the particle dispersing force becomes small because the velocity and hence the shearing rate of the mixture becomes small, and the particles tend to settle down. Then, the particles concentrate denser than before in the lower part of flow. This makes the resistance larger than in the upstream channel, thereby the particle dispersing force becomes smaller. Thus, an abrupt deposition occurs at the channel slope change.

The mechanism to disperse the particles in flow is different depending on the properties of the materials, channel slope, flow rate, particle concentration, etc., and the behavior of flow is also diverse. Theoretical discussions on the mechanism to disperse the particles come to a conclusion that debris flows can be classified into three typical types; the stony type, the turbulent muddy flow type and viscous type debris flows. The stony and turbulent muddy flow types can be grouped as the inertial debris flow.

In this paper the author attempts to systematize the mechanical understandings of debris flows and gives the methods or the perspectives for the quantitative analyses.

5.2 Processes of Initiation and Development in the Erosion Type Debris Flow

When a surface water flow abruptly enters the channel upstream, the bed will be eroded and sediment and water that have been entrapped within bed will be entrained into flow, thereby the discharge and solids concentration increase downstream. This process is similar to the fluvial process brought about by bed load transport on a gentler slope channel, and the process can be quantitatively treated by the following well established system of equations:

The one-dimensional momentum conservation equation:

$$\frac{\partial(Uh)}{\partial t}+\frac{\partial(U^2h)}{\partial x} = gh\sin\theta_o - gh\cos\theta_o\frac{\partial(z_b+h)}{\partial x}-\frac{\tau_b}{\rho_T} \tag{5.1}$$

The continuity equation of the total volume:

$$\frac{\partial h}{\partial t}+\frac{\partial(Uh)}{\partial x}=i\{C_*+(1-C_*)s_b\} \tag{5.2}$$

The continuity equation of the solids:

$$\frac{\partial(Ch)}{\partial t}+\frac{\partial(CUh)}{\partial x}=iC_* \tag{5.3}$$

The equation for bed surface elevation change:

$$\frac{\partial z_b}{\partial t}+i=0 \tag{5.4}$$

where U is the depth-wise averaged mean velocity of flow, h the depth of flow, g the acceleration due to gravity, θ_o the gradient of the original channel bed, z_b the thickness of bed measured from the original bed surface, i the erosion velocity, C the volume concentration of solids, C_* the solid concentration in static bed, s_b the degree of saturation in the bed, t the time, x the distance to downstream, τ_b the shearing stress on the bed, ρ_T $(= C(\sigma-\rho)+\rho)$ the specific density of flow in which σ is the density of particles and ρ is the density of water.

Usually in the analyses of fluvial process in a mild slope channel, the equation for bed surface elevation change is described by the difference between the upstream and downstream sediment discharges obtained by applying some appropriate equilibrium sediment discharge formula, but in very steep slope channel, especially in the erosion process, the rate of erosion is so large that the sediment discharge cannot be obtained by equilibrium sediment discharge formula. Hence, Eq. 5.4 is used.

The shearing stress on the bed in Eq. 5.1 can be described as

$$\tau_b = f\rho_T U^2 \tag{5.5}$$

where f is the resistance coefficient, and it is differently described corresponding to the different types of flow that will be explained later.

According to Bagnold (1957), τ_b in the flow of the mixture of non-cohesive large grains and fluid would be described as

$$\tau_b=\tau_c+\tau_f \tag{5.6}$$

where τ_c is shearing stress due to collision of particles and τ_f is the shearing stress in the interstitial fluid. Herein, grains are so large that they are not suspended by turbulence. Hence, the total submerged weight of transported sediment is sustained by the inter-particle collisions that produce τ_c. Bed erosion is induced by the effects of τ_f, and so, under a given value of τ_b, as long as sediment concentration is less than the equilibrium value that will be attained under the operating bed shear stress (when τ_c is smaller than that in the equilibrium), τ_f has an ability to erode bed. Thus, sediment concentration in flow increases accompanying the increment in τ_c and hence the reduction in τ_f, finally some equilibrium state is attained, in which no more erosion takes place. Takahashi (1991) applied this concept to obtain the erosion velocity of unsaturated bed as following:

$$\frac{i}{\sqrt{gh}} = K\sin^{3/2}\theta\left\{1-\frac{\sigma-\rho}{\rho}C\left(\frac{\tan\varphi}{\tan\theta}-1\right)\right\}^{1/2} \times\left(\frac{\tan\varphi}{\tan\theta}-1\right)(C_\infty - C)\frac{h}{d_p} \tag{5.7}$$

where K is a numerical coefficient, θ the energy slope, $\tan\varphi$ the internal friction coefficient in the bed material, d_p the representative diameter of the bed material, and C_∞ the equilibrium solids concentration in flow.

Giving the water discharge hydrograph at the upstream end or along the channel as the boundary conditions, one

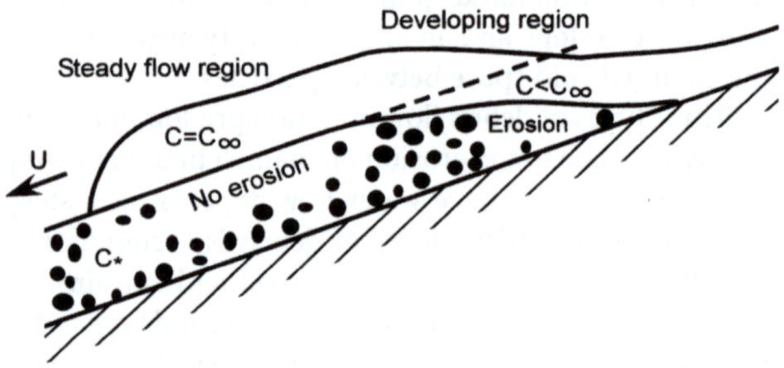

Fig. 5.1. Longitudinal profile of debris flow on uniform bed under the constant rate water supply

can obtain the downstream debris flow hydrograph under appropriate parameter values appeared in the governing equations. Figure 5.1 is the schematic illustration of debris flow on a uniform erodible bed under a constant rate water supply from the upstream end. The steady flow region in which no bed erosion takes place elongates itself downstream and the erosion proceeds to the bedrock in the developing region that follows the steady flow region as long as the steady water supply at the upstream end continues. Examples of the application of these equations such as to obtain the hydrograph due to the collapse of natural dam by overtopping and that due to flood runoff in the actual basin can be found elsewhere (Takahashi and Nakagawa 1994; Takahashi et al. 2001). Sometimes, a shallow surface landslide transforms into debris flow while it is moving on slope. Even in such a case the analysis of debris flow development in the channel can be carried out without difficulty by giving the debris flow hydrograph and sediment concentration in it as the boundary conditions. The prediction of that boundary condition is, however, difficult problem to be investigated.

5.3 Criteria of Debris Flow Occurrence for Erosion Type

Aforementioned discussion ignores how dense the equilibrium solids concentration is. The analysis is common to the cases of bed load transport and debris flow as long as the appropriate f and C_∞ values are selected to the respective cases. Therefore, the criteria for occurrence of various modes in sediment transportation must be made clear.

Imagine a thick uniform layer of loosely packed noncohesive grains, whose slope angle is θ. It is assumed that at the moment when the surface water flow of depth h_0 appears, the pore spaces among grains are saturated with water and a parallel seepage flow without any excessive pore pressure occurs. The characteristic distribution of shear stress in the bed should be like one of those shown in Fig. 5.2, in which τ is the operating tangential stress and τ_L the internal resistive stress.

Case 1 in Fig. 5.2 occurs under the condition

$$\tan\theta \geq \frac{C_*(\sigma-\rho)}{C_*(\sigma-\rho)+\rho}\tan\varphi \tag{5.8}$$

When Case 2 occurs, the following equation should be satisfied:

$$\tan\theta = \frac{C_*(\sigma-\rho)}{C_*(\sigma-\rho)+\rho(1+h_0/a_L)}\tan\varphi \tag{5.9}$$

in which a_L is the depth where τ and τ_L coincide.

The entire bed in Case 1 and the part above the depth a_L in Case 2 will begin to flow as soon as the sur-

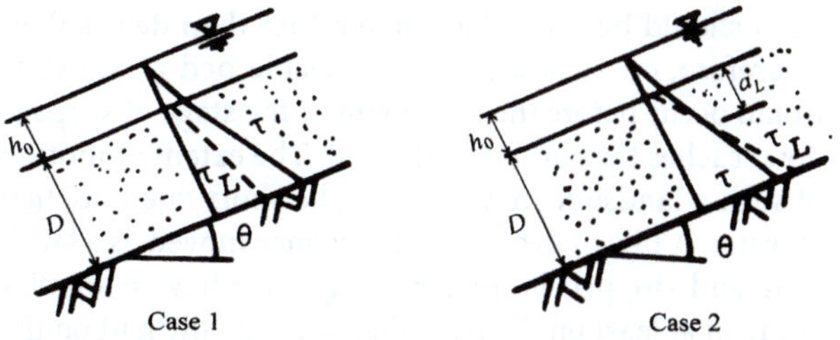

Fig. 5.2. Characteristic shear distributions within the water saturated bed under the surface water flow

face flow appears. This type of instability in bed is due not to the dynamic force of fluid flow but to static disequilibrium, so that the flow should be called sediment gravity flow.

The condition for the occurrence of sediment gravity flow is, therefore, $a_L \geq d_p$. Substitution of this condition into Eq. 5.9 obtains

$$\tan\theta \geq \frac{C_*(\sigma-\rho)}{C_*(\sigma-\rho)+\rho(1+h_0/d_p)}\tan\varphi \tag{5.10}$$

Even though the condition $a_L \geq d_p$ is satisfied, when a_L is far less than h_0 grains cannot be uniformly dispersed throughout the whole depth of flow due to rather small ability to disperse by the effect of particle collision but they concentrate in the lower layer of flow. Takahashi (1991) named this kind of sediment gravity flow as immature debris flow. Therefore, a sediment gravity flow that is appropriately called a debris flow should meet the condition $a_L \geq kh_0$, in which k is a numerical coefficient, determined from experiment to be about 0.7~1. Substitution of the condition $a_L \geq kh_0$ into Eq. 5.9 gives

$$\tan\theta \geq \frac{C_*(\sigma-\rho)}{C_*(\sigma-\rho)+\rho(1+1/k)}\tan\varphi \tag{5.11}$$

Debris flow in which grains are uniformly dispersed throughout the depth occurs when Eqs. 5.10 and 5.11 are simultaneously satisfied.

The surface water flow on the unsaturated bed would entrain grains by the effect of fluid dynamic force and develop to debris flow. Sometimes when permeability of the bed is very large, the infiltration of water from debris flow into the bed retards or even stops the motion and the accumulation of surface water flow from behind rejuvenates the debris flow motion. Thus, the motion of debris flow due to rather small surface runoff is sometimes peristaltic.

Attention must be paid to Case 1 where Eqs. 5.10 and 5.11 are satisfied. A grain bed that satisfies Eq. 5.8 might slip even though the seepage flow did not reach the surface. In such a case, in the neighborhood of the surface layer, τ_L would be larger than τ, and there the skeletal structure of the bed would be maintained. This phenomenon should be called landslide rather than debris flow. Therefore, on such a steep permeable bed a landslide would occur before the increment of the stage of seepage flow reaches the surface of the bed. The extent of motion of such a landslide may, however, be short due to deficit of water. A pile of debris in a gully once moved as a landslide and stopped in a short range may have a smaller gradient at least on the back than it originally had on the gully bed, so that it may be moved again as debris flow waiting for the appearance of a surface water flow. The time lag between the occurrence of a landslide and the removal of its deposit by the generation of debris flow would depend on the water supply (rainfall) condition, and it sometimes could be so short that one would hardly recognize the transition between the two phenomena.

To sum up the discussion above, fully developed debris flow can occur on the bed that satisfies Eq. 5.11.

As shown in Fig. 5.3, the domains of occurrence of various type sediment transportations are defined by Eqs. 5.9, 5.10, 5.11, and the equation of critical tractive force on a steep channel (Ashida et al. 1977). The broken lines labeled D and E are the experimentally obtained critical lines for the occurrence of debris flow on unsaturated bed (Tognacca et al. 2000) and on saturated bed (Armanini and Gregoretti 2000), respectively. The second abscissa axis that represents the relative depth h_0/d_p is attached for the convenience to get the general idea of necessary depth of surface flow and the scale markings on the two abscissa axes are not rigorously correspondent. Attention must be paid to this figure that the discussion is for the onset of debris flow by the supply of plain surface water flow, and once debris flow is formed it can flow on a flatter slope than that is necessary to generate.

Consider a situation as illustrated by Fig. 5.4 on a saturated bed whose slope angle satisfies the condition for Case 2. The depth a_L where τ and τ_L coincide is given as

$$a_L = \frac{\{\rho \tan\theta - C(\sigma-\rho)(\tan\varphi - \tan\theta)\}h}{C_*(\sigma-\rho)(\tan\varphi-\tan\theta) - \rho\tan\theta} \tag{5.12}$$

In this case the denominator of Eq. 5.12 is positive and if

$$C = \frac{\rho\tan\theta}{(\sigma-\rho)(\tan\varphi-\tan\theta)} \quad (\equiv C_\infty) \tag{5.13}$$

is satisfied, a_L becomes 0 meaning that the bed will not be eroded, and if $C < C_\infty$, $a_L > 0$ and the bed will be eroded and the grain concentration in flow will increase. Therefore, Eq. 5.13 indicates the equilibrium grain concentration in debris flow on a slope angle that satisfies the condition for Case 2 but milder than the slope given by Eq. 5.8.

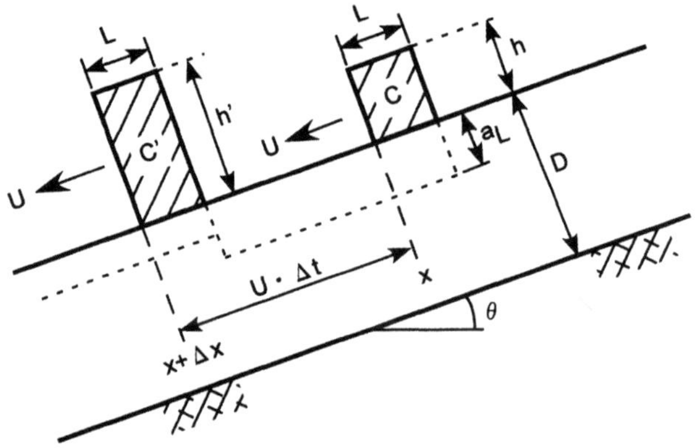

Fig. 5.4. The process of debris flow front development on a uniform erodible bed

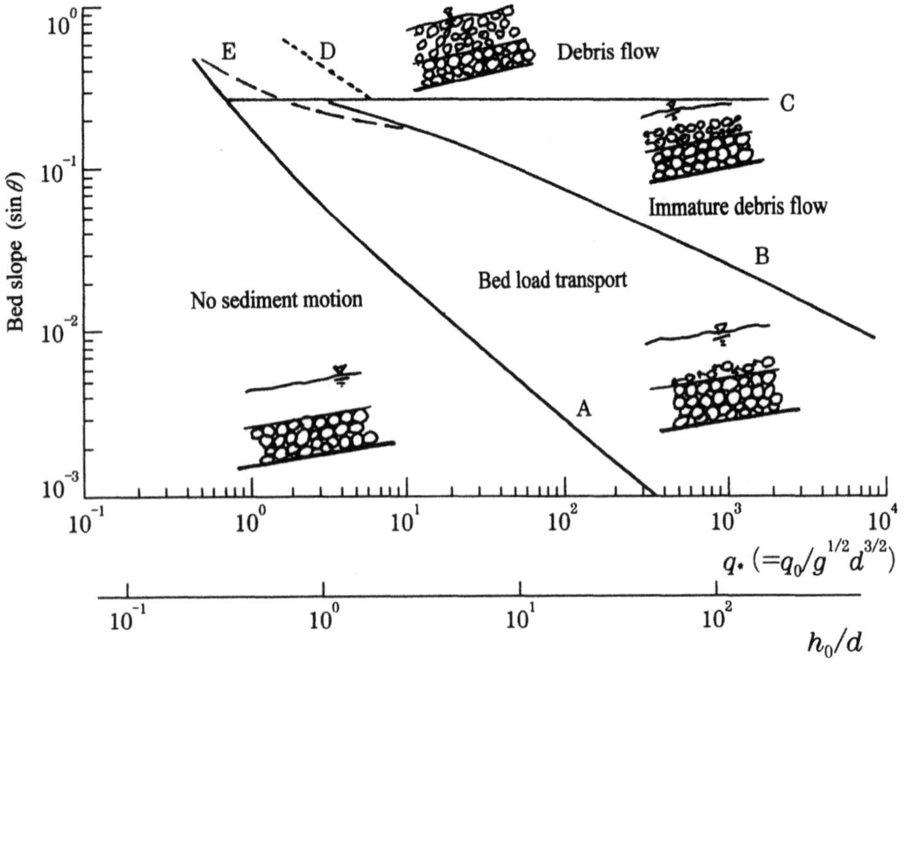

Fig. 5.3. Criteria for the occurrence of various sediment transportations. *A:* Threshold to the individual particle motion, *B:* Threshold for the sediment gravity flow, *C:* Threshold for the occurrence of debris flow, *D:* Threshold by Tognacca et al. (2000), *E:* Threshold by Arumanini et al. (2003)

5.4 Processes in Transformation and Stoppage of Landslide-induced Debris Flow

A severe rainfall from 6th to 10th July, 1997 amounting about 400 mm attacked Southern Kyushu, Japan. On 10th July, 4 hours after the stopping of rainfall, a large landslide of 80 m in the maximum width, 190 m in length, and 30 m in the maximum depth occurred in the Harihara River basin. There was a dam to check debris flows, but the volume of the landslide (130 000 m^3) was so large that some parts that transformed into debris flow overflowed the dam. The debris flow destroyed 19 houses and killed 21 people (Fig. 5.5). After the disaster a trench cut inspection of the deposit upstream of the dam was carried out. It revealed that there were two different deposit units as shown in Fig. 5.6; one was the relatively undisturbed earth block in the lower part (deposit A) and the other was the disturbed debris flow deposit (deposit B) laid on the deposit A (Sabo Technical Center 1998). This fact suggests that the considerable parts of solid earth block reached the trench cut position with little disturbance; the small trees originally on the surface of mountain slope were found on the boundary between the deposits A and B, and after the stoppage of the earth block the debris flow passed and then stopped on the already halted earth block. Some parts of debris flow overflowed the dam and caused the severe disasters. When landslide occurred, rainfall had already ceased about 4 hours before, and hence, the river discharge had to be little. It must not be enough to generate debris flow by the erosion of halted earth block. Therefore, the debris flow should have been generated from the sliding earth block without water supply from outside.

The description above suggests a model for the generation of landslide-induced debris flow as illustrated in Fig. 5.7. If a landslide is triggered under a plenty of groundwater within the slid earth block, liquefaction may take place near the slip surface; it may easily occur if the earth block has rich void space. Velocity distribution in the liquefied layer must be as shown in the upper diagram; zero at the bottom and the maximum at the boundary between the liquefied layer and the earth block, so that the earth block supported by the pressure within the liquefied layer goes faster than the liquefied layer and the liquefied layer is left behind as debris flow. The motion of earth block continues as long as the slope

Fig. 5.5.
The Harihara River debris flow

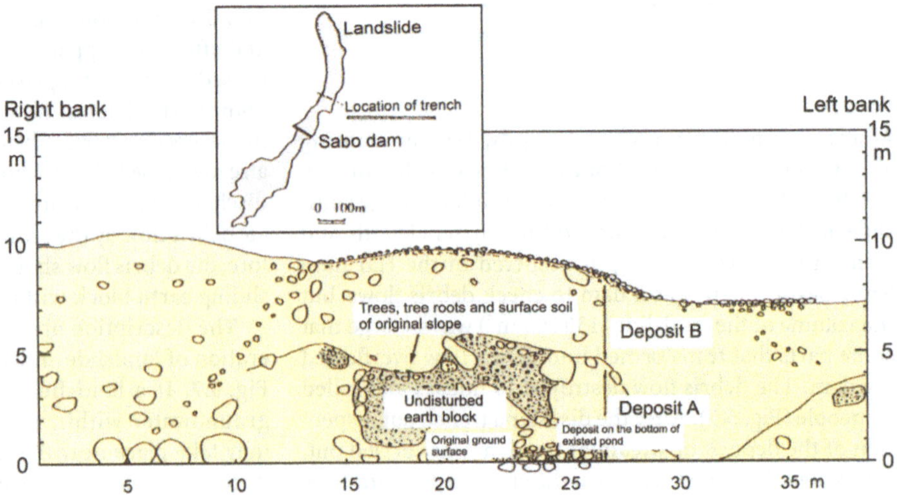

Fig. 5.6. Sketch of the inside structure of deposit at the trench wall

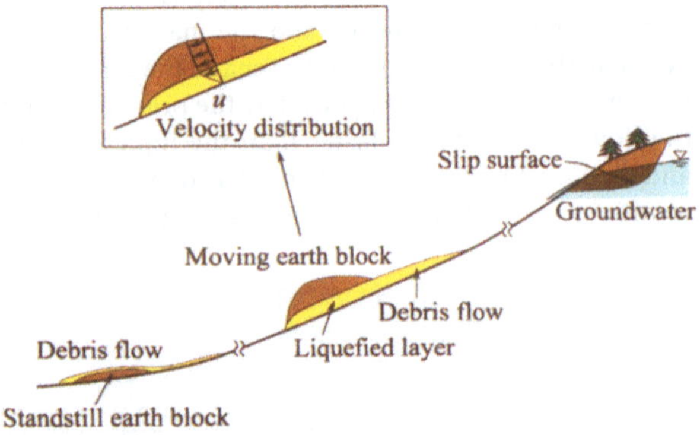

Fig. 5.7. Process model of landslide earth mass transformation into debris flow

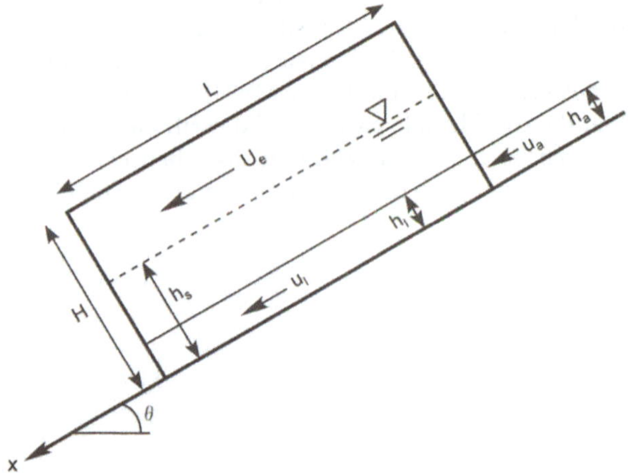

Fig. 5.8. A moving earth mass on a slope with the partial liquefaction

gradient is steep and liquefaction at the boundary between the earth block and the ground continues. Thus, the earth block becomes thinner and thinner with its advance down the slope cannibalizing itself to left debris flow behind. When the earth block arrives at a flatter area, its motion stops, and the succeeding debris flow gets over it and continues to run down. Therefore, the earth block remains underneath of the debris flow deposit. In the Harihara River debris flow, the earth block was worn out from its maximum thickness of 30 m to about 5 m during the motion of about 200 m and stopped upstream of the sabo dam.

5.5 Simplified Mathematical Model

For the sake of simplicity, a moving earth mass of rectangular parallelepiped is considered on a uniformly inclined plane as shown in Fig. 5.8. The lowest part of the earth mass is liquefied by the shearing action between the mass and the slip surface. To accomplish the liquefaction, the skeletal structure of the mass must be destroyed and the particles must become free from others. The void space between the particles must be filled with water or slurry. The condition of no water supply from the outside of the layer to be liquefied requires that the skeletal structure of the mass must be full of pores so as to be able to form the mixture of well dispersed particles and fluid. If the mass is not saturated by water, the liquefaction must occur with volume contraction.

The equation of conservation of momentum for the earth block is as following:

$$\begin{aligned}\frac{\mathrm{d}}{\mathrm{d}t}(\rho_t LHU_e) &= \rho_t gLH\sin\theta - \alpha\rho_t gLH\cos\theta\mu_k \\ &-(U_e - u_l)h_l\rho_l u_l - (1/2)\rho_l f u_l^2 L \\ &+\rho_l(u_a - U_e)^2 h_a + (1/2)\rho_a g h_a^2 - F_{side} \quad (5.14)\end{aligned}$$

where U_e is the velocity of the earth mass, h_l the thickness of liquefied layer, u_l the mean velocity of liquefied layer, H the thickness of the mass, L the length of the mass, h_s the thickness of the saturated layer in the mass, h_a the thickness of debris flow that follows the moving mass, u_a

the mean velocity of debris flow, ρ_t the apparent density of the mass, ρ_l the apparent density of the liquefied layer, ρ_a the apparent density of debris flow, μ_k the kinetic friction coefficient of the mass, f the resistance coefficient of the liquefied layer, α a coefficient, and F_{side} the friction at the side wall. In Eq. 5.14, the left-hand side term represents the change in the momentum of the mass. The first term of the right-hand side is the driving force due to gravity operating on the mass. The second term is the friction force at the bottom due to the load that is directly transmitted through the particles. The third term is the momentum that is left behind. The fourth term is the fluid resistance of the liquefied layer. The fifth term is the momentum supply from the following debris flow (this term becomes zero if $u_a \le U_e$). The sixth term is the hydrostatic pressure operating to the mass, and the seventh term is the side friction.

As an extreme case, if all the load of the mass is transmitted to the bottom as the fluid pressure (perfectly liquefied case), α becomes zero. If the mass is saturated but no liquefied layer is produced, $\alpha = (\sigma - \rho)C_* / \{(\sigma - \rho)C_* + \rho\}$, and if the mass is not saturated by water even at the bottom layer, α becomes 1. In an actual case α would have a value between one and zero; it would depend on the particle concentration and the diameter. The smaller the α, the milder the slope the mass can reach. If α is small and h_s is nearly equal to H, almost all of the mass will be transformed into debris flow.

The conservation equation of the mass volume is given as

$$\frac{\mathrm{d}(HL)}{\mathrm{d}t} = -(U_e - u_l)h_l \tag{5.15}$$

and the change of the thickness of liquefied layer is given as

$$\frac{\mathrm{d}(h_l L)}{\mathrm{d}t} = iL - (U_e - u_l)h_l \tag{5.16}$$

where i is the velocity at which the mass is abraded at the boundary between the bottom of the mass and the upper surface of the liquefied layer.

The shearing stress from the liquefied layer operates to the bottom of the mass, and this shearing stress, τ, does the work τu_l in unit time. Because the mass has a certain strength, τ_s, the amount of energy of $i\tau_s$ must be applied to the mass to be abraded with the rate i. Therefore, i would be regulated by the relation $i\tau_s = \tau u_l$. If one writes $\beta = \tau / \tau_s$, the following relation is obtained:

$$i = \beta u_l \tag{5.17}$$

If the strength of the mass is determined by the cohesion, c, and if the liquefied layer is considered as a Newtonian fluid, the following relationships will be satisfied:

$$\tau_s = c \tag{5.18}$$

$$\tau = \mu \frac{\mathrm{d}u}{\mathrm{d}z} \approx \mu \frac{u_l}{h_l} \tag{5.19}$$

where μ is the viscosity of the liquefied layer. Then,

$$\beta \approx \frac{\mu u_l}{c h_l} \tag{5.20}$$

As a field standard, if one assumes c=35×10[5] Pa, μ=100 Pas, h_l=100 cm and u_l=1000 cms^{-1}, one obtains β=0.03.

This simplified one-dimensional model was extended to analyze the motion across a three-dimensional terrain, in which we considered a rigid body is moving on the surface of a fluid flow interacting each other and the motion of fluid was analyzed by the Eulerian continuous fluid equations and the motion of the earth mass was analyzed by the Lagrangian solid mass equations (Takahashi et al. 2003). This hybrid simulation model was verified by the laboratory experiments and it was successfully applied to reproduce the actual debris flow occurred at Hogawati, Minamata City, Japan on 20th July 2003 (Takahashi 2004).

5.6 Models for Debris Flow Dynamics

Aforementioned discussions excluded the details of resistance coefficient to flow that will depend on the characteristics of flow such as the size and distribution of composing material, solids concentration, velocity, etc. Hereafter, the existing models of debris flow are reviewed categorizing into single-phase flow model and the two-phase flow model and applying a two-phase model that considers the debris flow is comprised of the mixture of two continuum media of fluid phase and solid phase debris flows are classified and made it possible to analyze respective types of debris flows.

Many simulation models for granular flow with the negligible effects of interstitial fluid like that in vacuum or heavy particles' motion in the atmosphere have been given, that is to say single-phase model. They can be categorized into the continuum media model and the discrete particle model. The former describes the granular assembly as a continuum media and its characteristics in flow are analyzed by the Eulerian forms of the continuity and the momentum conservation equations. The characteristics of discrete particles' motion are necessarily neglected. The latter one, on the other hand, traces the motion of individual particles by the Lagrangian equations.

In the continuum media model, the characteristics of flow as a granular assembly are indirectly involved in the stress terms in the Eulerian momentum equation. These terms are obtained from the relationship between the stress and the strain of the continuum that is called the constitutive equation. Although the general constitutive equation under the arbitrary conditions of particle density, particle size distribution, and particle interactions is difficult to obtain, if an appropriate form under a particular flow condition is given, the continuum media model can give an operative method to explain the flow characteristics en masse under that condition.

The discrete particle model describes the motions of individual particles by the Newton's second law equation under the effects of contact forces among the neighboring particles. Two kinds of approach for the inter-particle contact have been adopted: the hard-contact technique and the soft-contact technique. In the hard-contact technique, particles behave as rigid body and the contacts arise only between the two adjacent particles in an instantaneous moment (Campbell and Brennen 1985; Straub 2001). This technique has a shortcoming that it cannot handle the simultaneous collision of multiple bodies.

The soft-contact technique is free from the defect of hard-contact technique (Campbell 2001). At the moment of soft-contact, particles deform a little and the forces proportional to the stiffness of particles act on the contact points. The interaction between particles on a contact point is modeled by the deformation of the interface instrument that consists of a spring and a dash-pot connected in parallel, and a pair of such an instrument is mounted normally and tangentially on each contact point. To handle non-cohesive particle, a joint that is nonresistant to tension toward normal direction and a slipping joint that is effective under a tangential force more than a threshold should be serially connected to the respective instruments (Gotoh and Sakai 1997). Thus, many parameter values must be given beforehand, and if particle number becomes large, load on computer becomes enormous. However, the implementation of simulation makes possible to observe the motion of individual particles that may be difficult to observe in the physical model tests. The quantitative characteristics of overall flow of bulk mixture such as depth and mean velocity can be obtained only after the completion of simulation.

In the case of debris flow, the interstitial fluid is water or slurry, so the relative density of particles in flow is small, and therefore, the effects of interstitial fluid often play important role. Then, flow should be considered as that consists of two phases. Liquid phase is naturally described by the Eulerian equation, but as mentioned above, the solid phase may be described as a continuum media or as an assembly of discrete particles.

If solid phase is considered as a continuum media and described by the Eulerian equations, mathematically the model becomes Euler-Euler coupling model. In this model, two momentum conservation equations each for liquid and solid phases are given as following (Iverson 1997):

$$\sigma C(\partial \mathbf{v}_s / \partial t + \mathbf{v}_s \cdot \nabla \mathbf{v}_s) = \nabla \cdot \mathbf{T}_s + \sigma C \mathbf{g} + \mathbf{F} \quad (5.21)$$

$$\rho(1-C)(\partial \mathbf{v}_f / \partial t + \mathbf{v}_f \cdot \nabla \mathbf{v}_f) = \nabla \cdot \mathbf{T}_f + \rho(1-C)\mathbf{g} - \mathbf{F} \quad (5.22)$$

where $\mathbf{v}_s$ is the velocity of solids, $\mathbf{v}_f$ the velocity of fluid, $\mathbf{g}$ the acceleration due to gravity, $\mathbf{T}_s$ the stress tensor of solid phase, $\mathbf{T}_f$ the stress tensor of fluid phase, and $\mathbf{F}$ the interaction force per unit volume that results from the momentum exchange between the solid and the fluid constituents.

Addition of these two equations yields a momentum conservation equation applicable to the bulk mixture. This procedure eliminates the assessment of complicated interactions between fluid and solid phases. If nearly steady uniform flow is considered and the relative velocity between solids and fluid is neglected in Cartesian coordinate system, following two equations are finally obtained (Takahashi 1991):

$$P_{ds} + P_f = (\sigma - \rho) g \cos\theta \int_z^h C dz \quad (5.23)$$

$$T_{ds} + T_f = (\sigma - \rho) g \sin\theta \int_z^h \rho_T dz \quad (5.24)$$

where P_{ds} is the pressure in solid phase, P_f the pressure in fluid phase in excess over hydro-static one, T_{ds} the shear stress in solid phase, T_f the shear stress in fluid phase, ρ_T $\{= (\sigma - \rho)C + \rho\}$ the apparent density of debris flow material. These equations are the fundamental equations in the two-phase fluid model describing the stress balance in the flow of mixture. The essential problem in this two-phase mixture theory is how the left-hand sides of Eqs. 5.23 and 5.24 are described.

If solid phase is considered as an assembly of discrete particles, flow should be analyzed by Euler-Lagrange coupling model, in which the individual particle's motion is solved by the discrete element method mentioned earlier taking the interaction between particles and fluid into account. Some investigations on the bed load transport have been done by this method (Gotoh et al. 1994), but debris flow has not been analyzed by this method.

5.7 Single-phase Continuum Models

In the single-phase continuum models, the mixture of particles and fluid is considered as a kind of continuum fluid whose constitutive relation between stress and rate of strain is to be found empirically by the laboratory tests of samples collected from flow or by the deduction from the back analysis of the model applied to the flow in field.

5.7.1 Visco-plastic Fluid Model

Debris flow that induced by landslide initiates from the gradual or the abrupt motion of a debris' block under the effects of water, and after a motion of considerable distance, it comes out to flat area and stops. Therefore, if one considers the material has a strength intrinsically, one can explain the starting of motion attributing to the operation of a stress more than the (yielding) strength, and flow stops if the stress in flow becomes smaller than the strength. Debris flow models that consider the material is comprised of a kind of viscous fluid having strength are called 'visco-plastic fluid model', and they assume the relation between the stress and the strain rate as following:

$$\tau = \tau_y + K(du/dz)^n \quad (5.25)$$

where τ_y is the yielding strength, u the velocity at height z. K and n are the numerical coefficients.

The simplest in the visco-plastic models is the one setting $n = 1$, that is called 'Bingham fluid model'. This model was first proposed independently by Yano and Daido (1965) and Johnson (1965). Later, Johnson (1970) modified his model to the following 'Coulomb-viscous model':

$$\tau = c + \sigma_n \tan\varphi + \mu_B(du/dz) \quad (5.26)$$

where σ_n the internal normal stress and μ_B the Bingham viscosity.

Visco-plastic fluid models, including the Herschel-Bulkley fluid model that sets $n < 1$ (Coussot 1995) and the polynomial models that add the term $(du/dz)^2$ to Bingham fluid model (Julien and O'Brien 1997) or add $(du/dz)^n$ to Eq. 5.26 (Chen 1988), are still frequently used especially for the cases of viscous type debris flow.

For example, the stress balance equation for steady uniform Bingham fluid flow is given by

$$\tau_y + \mu_B(du/dz) = \rho_T g(h - z)\sin\theta \quad (5.27)$$

The integration of Eq. 5.27 under the boundary condition; $u = 0$ at $z = 0$, gives the velocity distribution as shown in Fig. 5.9, in which z' $(\tau_y/\rho_T g \sin\theta)$ is the thickness of the plug that behaves like a rigid body.

The serious shortcoming of visco-pastic models is the difficulty in determining the crucial parameter values such as τ_y, μ_B, etc. Those values obtained by the laboratory tests and by the field observations are usually much different, and they cannot account for the changes of parameter values due to addition or subtraction of water or sediment during motion. According to the author's observation, the typical viscous type debris flow in the Jiangjia ravine, China lacks the plug that should exist in a visco-plastic fluid flow (Takahashi 1999).

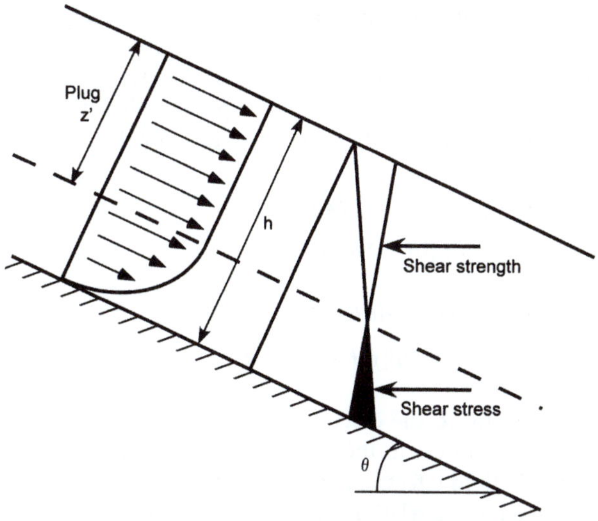

Fig. 5.9. Velocity distribution in Bingham fluid flow

5.7.2 Dilatant Fluid Model

Debris flow that consists of mainly coarse grains larger than gravel size can often reach to an area flatter than about 4° keeping its high mobility and competence to freight big boulders. Takahashi (1978) inferred such small resistance to flow was caused by the dispersion of grains separated with one another, and the inter-particle collision gave rise to the dispersion of heavier grains than the ambient fluid.

Bagnold (1954) was the first to discover the significance of inter-particle collision as the cause of particle dispersion. His experiments and a simple binary grain collision analysis gave the stress-strain relationships in inertial and viscous regimes. In the inertial regime, those relationships were given as following:

$$p = a_i \cos\alpha_i \cdot \sigma\lambda^2 d_p^2 (du/dz)^2 \quad (5.28)$$

$$\tau = p \cdot \tan\alpha_i = a_i \sin\alpha_i \cdot \sigma\lambda^2 d_p^2 (du/dz)^2 \quad (5.29)$$

where p is the dispersive pressure in flow, a_i the numerical coefficient, α_i the collision angle, λ the linear concentration of grains defined by $\lambda = \{(C_*/C)^{1/3} - 1\}^{-1}$. He directly measured the pressure within the flow suspending neutrally buoyant grains that was in excess of the pressure in plain fluid flow, so that p in Eq. 5.28 is equivalent to P_{ds} in Eq. 5.23. He could not directly measure T_{ds} corresponding to P_{ds}, but it was obtained by subtracting the periphery drug in the plain fluid flow from the periphery drug in the mixture flow. Therefore, thus deduced τ value in Eq. 5.29 is equivalent to T_{ds} in Eq. 5.24.

Bagnold in his experiment used neutrally buoyant solid spheres immersed in a Newtonian fluid, but Takahashi (1978) exploited his experimental results to the open channel flow in which heavy particles are very densely concentrated, and he also assumed the uniform distribution of grains in the entire depth. Then, the stress balance equations for the normal and the parallel directions to the bed were obtained, respectively, as following:

$$a_i \cos\alpha_i \cdot \sigma\lambda^2 d_p^2 (du/dz)^2 = C(\sigma-\rho)g(h-z)\cos\theta \tag{5.30}$$

$$a_i \sin\alpha_i \cdot \sigma\lambda^2 d_p^2 (du/dz)^2 = \{C(\sigma-\rho)+\rho\}g(h-z)\sin\theta \tag{5.31}$$

Referring to the discussion above, this model is an equivalent model to the two-phase model in which the dynamic fluid effects; P_f and T_f, are neglected. A kind of fluid having the relationship; $\tau \sim (du/dz)^n$, $n > 1$, is called 'dilatant fluid', so Takahashi's model is called the dilatant fluid model.

Equations 5.30 and 5.31 are integrated under the boundary condition; $u = 0$ at $z = 0$, to obtain the flow velocity at height z. The result of integration of Eq. 5.31 is given as

$$u = \frac{2}{3d_p}\left(\frac{g\sin\theta}{a_i \sin\alpha_i}\frac{\rho_T}{\sigma}\right)^{1/2} \frac{1}{\lambda}\left\{h^{3/2} - (h-z)^{3/2}\right\} \tag{5.32}$$

Mathematically, the velocity distribution function obtained from Eq. 5.30 has not the same expression with the one obtained from Eq. 5.31. These two functions become physically equivalent only when the following formula is satisfied:

$$C = \frac{\rho\tan\theta}{(\sigma-\rho)(\tan\alpha_i - \tan\theta)} \tag{5.33}$$

For example, if one substitutes $\sigma = 2.65$ g cm^{-3}, $\rho = 1.0$ g cm^{-3}, $\tan\alpha_i = 0.75$ and $\theta = 18°$ into Eq. 5.33, one obtains $C = 0.46$. Then, if one supplies the mixture of water and grains whose grain diameter is 4 mm and the volume concentration is 0.45 into a rigid bed flume of 18° in gradient, one observes velocity distribution that fits very well to Eq. 5.32 by giving $a_i = 0.04$ that was obtained by Bagnold for wax beads as shown in Fig. 5.10. As another example, if one supplies water with a steady rate onto an erodible bed that is saturated by water and having the slope gradient steeper than 15°, debris flow develops downstream and approaches an equilibrium state as illustrated in Fig. 5.1. As explained earlier, in the steady equilibrium flow region grain concentration is given by Eq. 5.13 and the validity of Eq. 5.13 as the equilibrium grain concentration formula has been proved by many flume experiments. This fact and the good fitness of Eq. 5.32 under Eq. 5.33 to the rigid bed experiments suggests that in the equilibrium stage $\tan\alpha_i = \tan\varphi$. Hence,

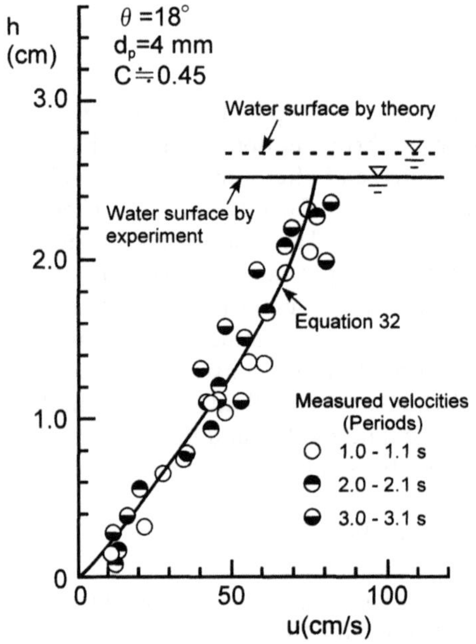

Fig. 5.10. Velocity distribution of debris flow nearly in equilibrium concentration on rigid bed

in the equilibrium stage, the dynamic collision angle is almost equal to the static internal friction angle. Therefore, one can consider Eq. 5.13 is the equilibrium grain concentration formula obtained from the static as well as the dynamic equilibrium conditions. Thus, the dilatant fluid model can well explain the dynamics of debris flow in which inter-particle collision plays the dominant role, moreover, different from visco-plastic fluid model, it can account for the change in solid concentration during motion (Takahashi 1991).

The assumption of uniform grain distribution throughout the depth is, however, at least theoretically, too much simplification, and this causes the contradiction that there exists two independent formulae 5.30 and 5.31 for one unknown velocity (Iverson and Denlinger 1987). To dissolve this contradiction, Takahashi (1991) assumed the following formula:

$$\tan\alpha_i = (C_*/C)^{1/3}\tan\varphi \tag{5.34}$$

This formula means the dynamic collision angle between particles is a little larger than the static internal friction angle when C is smaller than C_*. This would be a reasonable assumption because the wider space between particles than in the static bed will result in the deeper colliding angle.

Substitution of Eq. 5.34 into Eqs. 5.28 and 5.29 results in rather good agreement with Bagnold's experiments, and also the substitution of Eq. 5.34 into Eqs. 5.30 and 5.31 gives two independent formulae necessary and sufficient to obtain u and C at height z. The analysis using these two equations can well explain the vertical grain

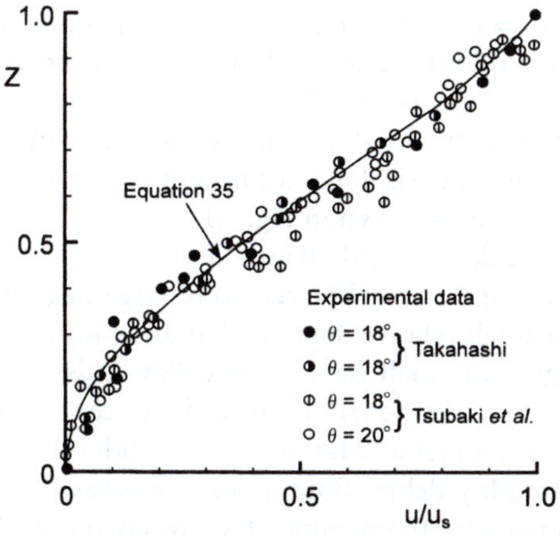

Fig. 5.11. Theoretical velocity distribution and the experimental data

concentration and velocity distributions on an erodible bed. The velocity distribution formula is given by

$$\frac{u}{u_s} = 1 - \frac{1}{2}(1 - Z)^{3/2}(2 + 3Z) \tag{5.35}$$

where u_s is the velocity at the flow surface and $Z = z/h$. Equation 5.35 is compared with the experimental data on erodible bed in Fig. 5.11. The vertical velocity distribution on erodible bed has a peculiar form with two inflection points; one near the bed and the other near the surface of flow, whereas that on rigid bed, especially when flow contains slightly thinner concentration than equilibrium, has simple concave form (Fig. 5.10) as that is explained by the first approximation under the assumption of uniform concentration distribution.

5.8 Two-phase Fluid Flow Model (Mixture Theory)

As mentioned earlier, how one considers the left-hand side terms in Eqs. 5.23 and 5.24 is the origin of diverse debris flow theories. The particle dispersive pressure due to inter-particle collision; p_c, and the static skeletal pressure due to enduring contact between particles; p_s, would be the candidates to consist P_{ds}. P_f would exist if the sediment mass contracts volumetrically or if there is a net flux of sediment toward the bed (Iverson 1997). Herein, only quasi-steady flow is considered, so $P_f = 0$. The shearing stress due to inter-particle collision; τ_c, the shearing stress due to enduring contact between particles; τ_s, and the streaming stress due to migration of particles in one layer to other layer; τ_k, would be the candidate to consist T_{ds}. The deformation stress within interstitial fluid; τ_μ, and the turbulent mixing stress; τ_t, would be the candidate to consist T_f. All such candidates cannot prevail simultaneously, but some among others prevail in a particular regime. Therefore, via the discussion of dominant stresses, debris flows can be classified.

5.8.1 Coulomb Mixture Theory (Quasi-static Debris Flow)

Iverson and Denlinger (2001) claims that the Coulomb friction term; τ_s, is almost always larger than the particle collision term; τ_c, stating that the Savage number, defined by Eq. 5.36, in various field examples and the large scale experiments by USGS is smaller than 0.1 as the evidence, and they develop a Coulomb mixture theory as the only one suitable model for debris flow.

$$N_s = \frac{\sigma(\mathrm{d}u/\mathrm{d}z)^2 d_p^2}{(\sigma - \rho)gh} \tag{5.36}$$

The numerator of Eq. 5.36 is, as is clear from Eq. 5.28, the term representing the particle collision effects, and the denominator represents the load operating on the bed due to submerged weight of particles. They say the denominator is the gravitational grain contact stress that produces Coulomb friction, but it must be noted that the static gravitational contact stress can be transferred from the surface to the bottom of flow only when a skeletal structure due to the enduring contact of grains extends throughout the depth. To furnish this necessity, the local solid fractions everywhere within flow must exceed a threshold level C_3. According to Bagnold (1966), C_3 for natural beach sand is about 0.51. It is true that denominator represents the effective pressure at the bottom of flow, but it is not necessarily transferred as the static gravitational contact stress. It is transferred as the collision stress if grain concentration is smaller than C_3. If $C\cos\theta$ is multiplied to the denominator, the term precisely represents the effective pressure at the bottom, but the term multiplied by $C\cos\theta$ to the numerator does not represent the magnitude of collision stress. To represent the collision stress not $C\cos\theta$ but $a_i\cos\alpha_i\lambda^2$ must be multiplied to the numerator as is evident in Eq. 5.28. As the consequence, Eq. 5.30 indicates the effective pressure is balanced with collision stress. Therefore, N_s that is smaller than 0.1 cannot be the proof of the predomination of Coulomb stress term. Coulomb mixture theory is considered as a model suitable for very densely grain concentrated quasi-static flow.

Herein, we do not dwell on the details of Coulomb mixture theory.

5.8.2 Classification of Dynamic Debris Flows

Dynamic debris flow herein defined is the one in which dynamic stresses predominate in governing the behaviors of flow. The total shear stress $\tau (= T_{ds} + T_f)$ in a grain and water mixture would be described as

$$\tau = \tau_c + \tau_k + \tau_s + \tau_\mu + \tau_t \tag{5.37}$$

If C is larger than about 0.2, τ_k becomes smaller than τ_c and the difference between the two tends to become drastically large as C becomes larger (Takahashi and Tsujimoto 1997), and if C is smaller than about 0.5, τ_s is almost equal to zero (Bagnold 1966). Therefore, in a dynamic debris flow, if the average grain concentration (except for the fine constituent making up the interstitial slurry) is between 0.2 and 0.5, and if τ_t/τ_c and τ_μ/τ_c are small, $\tau_c/\tau \approx 1$ is satisfied. Debris flow in which the stress due to inter-particle collision dominates among others is called 'stony debris flow'. On the other hand, if $\tau_t/\tau \approx 1$ is satisfied, it is called 'turbulent muddy debris flow', and if $\tau_\mu/\tau \approx 1$ is satisfied, it is called 'viscous debris flow'.

Equation 5.29 and the other previous theoretical considerations on the constitutive relations of inertial granular flows confirm (Campbell 1990)

$$\tau_c = \sigma d_p^2 f_1(C)(\mathrm{d}u/\mathrm{d}z)^2 \tag{5.38}$$

and fluid mechanics give the following forms:

$$\tau_\mu = \mu f_2(C)(\mathrm{d}u/\mathrm{d}z) \tag{5.39}$$

$$\tau_t = \rho_T l^2 (\mathrm{d}u/\mathrm{d}z)^2 \tag{5.40}$$

where $f_1(C)$ and $f_2(C)$ are the functions of C, μ the viscosity of interstitial fluid, and l the turbulent mixing length. From these equations one obtains the following formulae:

$$\frac{\tau_t}{\tau_c} \sim \left(\frac{\rho_T}{\sigma}\right)\left(\frac{h}{d_p}\right)^2 \frac{1}{f_1(C)} \tag{5.41}$$

$$\frac{\tau_\mu}{\tau_c} = \frac{\lambda^{1/2}}{B_a}\frac{f_2(C)}{f_1(C)} \tag{5.42}$$

$$\frac{\tau_t}{\tau_\mu} \sim R_e \frac{1}{f_2(C)} \tag{5.43}$$

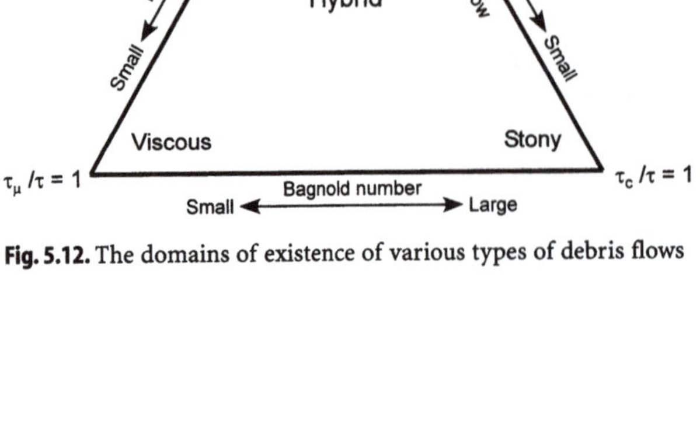

Fig. 5.12. The domains of existence of various types of debris flows

where B_a $(= \sigma\lambda^{1/2} d_p^2(\mathrm{d}u/\mathrm{d}z)/\mu)$ is called Bagnold number and R_e $(= \rho_T U h/\mu)$ is Reynolds number, in which $l \approx h$ and $l(\mathrm{d}u/\mathrm{d}z) \approx U$ are assumed.

Summarizing the above discussions, one can conclude that dynamic debris flows occur in the domain of the ternary diagram as shown in Fig. 5.12, where the three axes represent relative depth, Bagnold number and Reynolds number, respectively. The region of large Bagnold number and small relative depth is that for stony debris flow. The region of small Bagnold and Reynolds numbers is that for viscous debris flow, and the region of large Reynolds number and large relative depth is that for turbulent muddy debris flow. Thus, the areas close to the three apices are the regions of stony, viscous and turbulent muddy debris flows, respectively, and the other area in the triangle is the region of 'hybrid debris flows'.

Previous experiments reveal that if the transport concentration of large grains is smaller than about 0.2 (herein, large grains are defined as the ones not being suspended by turbulence, and the transport concentration is defined as the ratio of large grain's discharge to the total discharge of large grains plus fluid phase including suspended sediment), large grains cannot disperse in the entire depth of flow. The mixture layer of large grains and fluid appears only in the lower area of flow. This type flow is, as mentioned earlier, 'immature debris flow' in contrast to '(mature) debris flows' in which large grains are dispersed in the entire depth.

Thus, sediment transport or sediment moving phenomena can be categorized by the transport concentrations into the plain liquid flow, the individual particle motion (bed load and suspended load; $0 < C \leq 0.02$), the immature debris flow ($0.02 < C \leq 0.2$), the debris flow ($0.2 < C \leq 0.5$), the quasi-static debris flow ($0.5 < C \leq 0.6$), and the sliding of rigid earth body ($0.6 < C$). The boundary C values indicated are only for reference, they must depend on properties of particles such as cohesion, shape and size distribution.

5.8.3 Generalized Theory for Inertial Debris Flow

Debris flows that occur in the domain adjacent to the axis representing the relative depth in Fig. 5.12 can be generically called 'inertial debris flow', because the inertial terms; τ_c or τ_t, dominate in flow. The hybrid flow in the inertial debris flow consists of the lower particle collision layer and the upper turbulent suspension layer as illustrated in Fig. 5.13. The ratio of these two layers in the depth depends on the relative depth; h/d_p, and the concentration. If the relative depth is small and the particle collision layer occupies almost entire depth, it is stony debris flow, and if the relative depth is large and the turbulent suspension layer occupies almost entire depth, it is turbulent muddy flow. Herein, the theory of Takahashi and Satofuka (2002) is introduced.

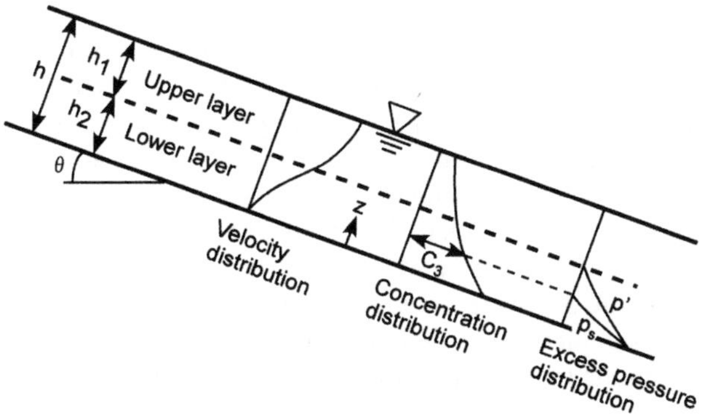

Fig. 5.13. The general structure of inertial debris flow

Consider an inertial debris flow as illustrated in Fig. 5.13. Noticing τ_μ can be neglected, the balance of force equations in the lower layer to the directions of main flow and perpendicular to it are written, respectively, as following:

$$\tau_c + \tau_k + \tau_s + \tau_t = g\sin\theta \int_z^h \{(\sigma-\rho)C + \rho\}dz \quad (5.44)$$

$$p_c + p_s = g\cos\theta \int_z^{h_2} (\sigma-\rho)C dz \equiv p' \quad (5.45)$$

where p' is the effective pressure transmitted directly on particle to particle, and h_2 the thickness of lower layer. The upper limit of integration of Eq. 5.45 is h_2, because the particle's load in the upper layer is supported by turbulence and it is transmitted to the lower layer as hydro-static pressure.

If solids concentration at the bottom of flow is C_*, the dynamic pressure; p_c, at the bottom must be zero because no flow can exist at such a high concentration. Therefore, at the bottom all the effective pressure; p', must be borne by the static pressure; p_s. If solids concentration decreases upward, at the height where C becomes equal to C_3, p_s should become zero. The vertical distribution function of p_s is not yet known, but following function is assumed:

$$p_s = \begin{cases} \dfrac{C - C_3}{C_* - C_3} p' & ; \ C > C_3 \\ 0 & ; \ C \le C_3 \end{cases} \quad (5.46)$$

The relation between p_s and τ_s is assumed as the following Coulomb's equation:

$$\tau_s = p_s \tan\varphi \quad (5.47)$$

The balance of force equation in the upper layer is given as

$$\tau_c + \tau_t = \int_z^h \rho_T g \sin\theta dz \quad (5.48)$$

Concerning with the descriptions of τ_c, τ_k, and p_c following constitutive equations for dry granular flow obtained by Takahashi and Tsujimoto (1997) are used:

$$\tau_c = \frac{4}{5}\sqrt{\frac{1}{15\pi}}\frac{1+e}{\sqrt{1-e}}C^2(1+\lambda)\sigma d_p^2 \left(\frac{du}{dz}\right)^2 \quad (5.49)$$

$$\tau_k = \frac{1}{3(1+\lambda)}\sqrt{\frac{1}{15\pi(1-e)}}\sigma d_p^2 \left(\frac{du}{dz}\right)^2 \quad (5.50)$$

$$p_c = \frac{2}{15}C^2(1+\lambda)\frac{1+e}{1-e}\sigma d_p^2 \left(\frac{du}{dz}\right)^2 \quad (5.51)$$

where e is the restitution coefficient of particles.

Turbulent mixing stress; τ_t, is given by Eq. 5.40, where the mixing length is assumed as $l = \kappa z$, in which κ is Kármán constant. Question may arise that in the densely concentrated lower layer, mixing length would have the scale of void between particles. Previously, Takahashi (2000) developed a theory adopting such a concept; $l = d_p/\lambda$ in the lower particle colliding layer and $l = \kappa z$ in the upper turbulent layer. But, interstitial fluid is widespread as a continuum fluid in the entire depth, so that the mixing length may have larger scale than the void between particles. Furthermore, no matter which mixing length is adopted in the lower layer, τ_t is far smaller than τ_c or τ_s, hence, it little affects the velocity and solids concentration there. However, considering the intensity of turbulence, ρ_T in the lower layer would be almost equal to the density of interstitial fluid, whereas in the upper layer, it would be equal to the density of the mixture.

Substitution of these constitutive relations and Eq. 5.46 into Eqs. 5.44 and 5.45 gives the velocity and the solids concentration distributions in the lower layer.

In the upper layer, the substitution of Eqs. 5.40 and 5.49 into Eq. 5.48 gives the velocity distribution. The solids concentration distribution can be obtained assuming that particles are dispersed by turbulence obeying the equation of diffusion used in the normal suspended load analysis.

The criteria whether the upper layer exists and how high the boundary between the upper and the lower layers from the bottom of flow are as following:

1. $w_s/u_* < 1$, where w_s is the settling velocity of particles and $u_* = (gh\sin\theta)^{1/2}$ is the shear velocity of flow. If this criterion is not satisfied, particles are not suspended and $h_2 = h$.
2. The minimum solids concentration in the lower layer appears at the upper boundary of the lower layer, above that height τ_c is no more larger than τ_t. The experiments for immature debris flow show that the minimum solids concentration; C_{lim}, attained by inter-particle collision is about $0.5C_*$ (Takahashi 1991). Therefore, although it is somewhat expedient, the boundary height is set where the solids concentration calculated in the lower layer, from lower to upper direction, at first becomes equal to $0.5C_*$.

If calculated solids concentration is larger than C_{lim} in the entire depth, even the criterion 1) is satisfied, there is no upper layer, and $h_2 = h$. If, at a certain height, calculated solids concentration becomes smaller than C_{lim}, but the criterion 1) is not satisfied, it is immature debris flow.

Theoretical solids concentration and velocity distributions were obtained corresponding to the conditions; $\sigma = 2.65$ g cm^{-3}, $\rho = 1.0$ g cm^{-3}, $\tan\varphi = 0.7$, $e = 0.3$, $\kappa = 0.3$, $C_* = 0.65$, and $C_3 = 0.5$. Figure 5.14 shows the examples of the calculated results, where flow depths are held constant at 2 cm throughout the cases. When particle diameter is 7 mm, even for steep channel slope of 18°, the condition for particle suspension is not satisfied but particles are dispersed in the entire depth due to the action of inter-particle collision, and the velocity is small. It is stony debris flow. For the same size particles, when channel slope is 6°, particles cannot be dispersed in the entire depth, and it is immature debris flow. When particle diameter is 1 mm, even the channel slope is as flat as 6°, the condition for particle suspension is satisfied, and there exists the upper layer in which particles are in suspension. It is hybrid type debris flow.

Figure 5.15 compares the theoretically obtained velocity distributions with the experimental data of Hirano et al. (1992), where the velocity is normalized by the surface velocity. Debris flows in the experiments were generated by supplying water from upstream onto the water saturated erodible bed flume 20 cm in width and 7 m in length. Bed slope was held constant at 14°. The relative depth was changed by changing both the supplying water discharge and the particle size. The theoretical as well as experimental velocity distribution curves in the figure have a tendency to become lower the break point on the curve; i.e., the boundary between the upper and the lower layer, and to increase the degree of concavity as the relative depth increases. Moreover, the theoretical curves well fit to the experimental results.

Figure 5.16 is a comparison of the theoretical equilibrium transport concentrations; C_{tr}, with author's experimental results. In the experiments debris flows were generated by abruptly supplying water from upstream onto the water saturated erodible bed in a flume of 9.9 cm wide and 10 m long. Bed slope was changed, and the discharge of sediment plus water; Q_T, was held constant at about 531 cm^3 s^{-1}. Four kinds of materials; 0.201, 0.066, 0.030 and 0.017 cm in median diameters with densities between 2.64 and 2.66 g cm^{-3} and internal friction angles between 38.5° and 39°, were used for the erodible bed. The debris flow samples were collected with a bucket at the outlet of the flume to measure Q_T and C_{tr}. As shown in the figure, the theory can well explain the tendency that the smaller the particle diameter, the larger the equilibrium transport concentration becomes under a given channel slope.

In the same figure, other than those corresponding to the experiments, the theoretical curves for $d_p = 1$ cm and 0.5 cm are also drawn. Under the experimental conditions herein, particles larger than 1 cm in diameter cannot be suspended by turbulence even on the steep slope of 15°. The theoretical curve for $d_p = 1$ cm almost coincides with the respective concentrations calculated by substituting 0.8 into $\tan\varphi$ in Eq. 5.13 or into $\tan\alpha_i$ in Eq. 5.33 and into $\tan\varphi$ in the following equilibrium concentration formula for immature debris flow (Takahashi 1991):

$$C_{tr} = 6.7 C_\infty^2 \tag{5.52}$$

where C_∞ is given by Eq. 5.13 and it is valid only when C_{tr} by this equation is smaller than C_∞ calculated by Eq. 5.13. The reason for branching of curves corresponding to $d_p = 0.5$ cm and 0.2 cm from the curve for $d_p = 1$ cm is due to the beginning of suspension on the slopes steeper than those corresponding to the respective branching points.

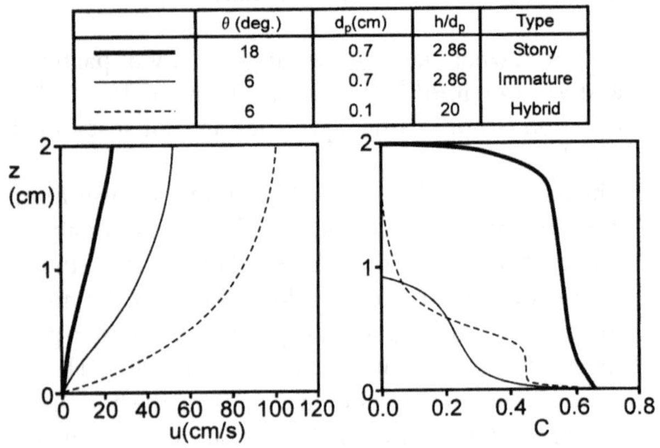

Fig. 5.14. Velocity and particle concentration distributions by the theory

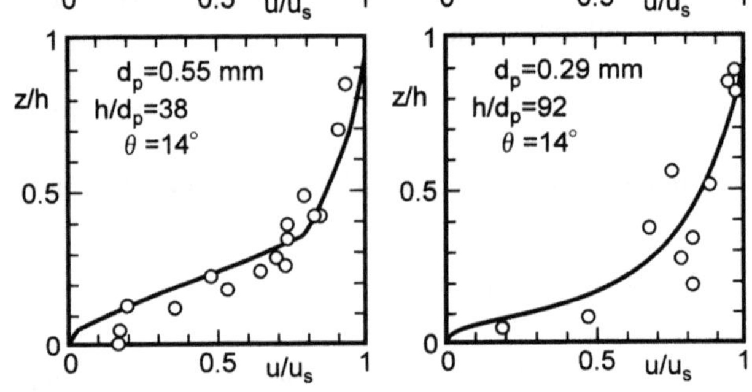

Fig. 5.15. Theoretical and experimental velocity distributions on erodible bed

It must be noted here that $\tan\varphi$ is set equal to 0.8 in the calculation of Eqs. 5.13 and 5.52, but in the calculation of theoretical curves in Fig. 5.16, $\tan\varphi$ is set equal to 0.7, and moreover, $\tan\alpha_i$ is set at 0.75 to obtain the solids concentration nearly equal to the equilibrium on rigid bed. It may seem somewhat expedient. But, why such a different value should be used will become clear by the following explanation.

If particle diameter is large ($d_p > 1$ cm), the settling velocity of particle is given by Newton's equation, and the criterion 1) for particle suspension is written as

$$\frac{h}{d_p} > \frac{\sin\theta}{3.31(\sigma/\rho - 1)} \tag{5.53}$$

If $\theta = 10°$ and $(\sigma/\rho - 1) = 1.65$ are substituted into this equation, $h/d_p > 32$ is obtained. This means that if the relative depth in an inertial debris flow is smaller than about 30, it is stony debris flow. From the Eqs. 5.41, 5.49 and 5.53, the ratio of τ_t to τ_c in stony debris flow is given as following:

$$\frac{\tau_t}{\tau_c} < \frac{\rho}{\sigma}\left\{\frac{\sin\theta}{3.31(\sigma/\rho - 1)}\right\}^2 \frac{5\sqrt{15\pi(1-e)}}{4(1+e)C^2(1+\lambda)} \tag{5.54}$$

The largest range value of the right-hand side term is obtained as 0.05 by substituting $C = 0.2$, $e = 0.3$, $\theta = 15°$, $\sigma/\rho = 2.65$, and $C_* = 0.65$. Namely, in stony debris flow, the turbulent mixing stress is at most 5% of the particle collision stress. This is the reason for a wide applicability of single-phase dilatant fluid model in which fluid stress is neglected. Being the effective fluid resistance and τ_k negligibly small, energy in stony debris flow is almost solely dissipated by the solids interfacial friction that occurs intermittently during the instant of inter-particle collision. The effective friction angle α_i (the angle of encounter between two particles) for densely solids concentrated flow ($C \approx 0.5$) must be approximately equal to φ (Bagnold 1966 and the earlier description in this paper). If solid concentration decreases as small as about 0.2 in stony debris flow, the angle of encounter between two particles becomes larger than φ due to enlarged free distance between particles. Thus, referring to Eq. 5.34, if $\tan\varphi$ is 0.7, the average value of $\tan\alpha_i$ for stony debris flow whose grain concentration is between 0.5 and 0.2 would reasonably estimated as 0.8.

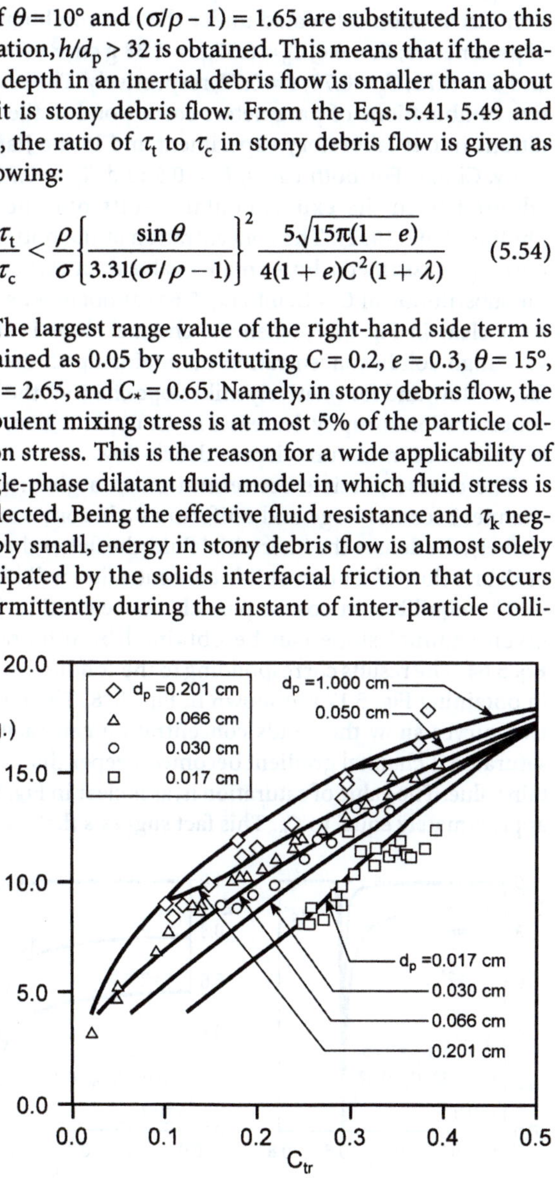

Fig. 5.16. Theoretical and experimental equilibrium transport concentrations

5.8.4 Newtonian Fluid Model for Viscous Debris Flow

Quasi-steady debris flows that occur in the domain adjacent to the axis representing Bagnold number in Fig. 5.12 flow in an orderly manner in which coarse particles move nearly parallel to the bed keeping tracks within their embedded layer.

Experiments (Bagnold 1954) revealed that the transition from inertia to viscous conditions occur in the range between 450 and 40 in terms of Bagnold number. In the transition range, the total shear stress would be given by

$$\tau = \tau_c + \tau_k + \tau_\mu \tag{5.55}$$

And as the Bagnold number increases, grain elements; $\tau_c + \tau_k$, surpass the fluid element; τ_μ, so it approaches stony debris flow. Contrary, if Bagnold number becomes small, the dynamic grain elements diminish gradually and flow becomes viscous debris flow. But, in viscous range, according to Bagnold's discussion, although the grain collision stress and grain streaming stress are negligibly small, the effect of grain existence remains to increase the apparent viscosity of mixture. Therefore, the total shear stress in viscous debris flow is a mixed one due to the viscosity of fluid itself and that due to presence of grains, and it cannot be split into grain and fluid elements. Thus, the following formula is given:

$$\tau = \tau_{\mu a} \tag{5.56}$$

where $\tau_{\mu a}$ is the viscous stress due to the deformation of mixture having the apparent viscosity of μ_a.

The interesting issues in viscous debris flow is why coarse grains heavier than the ambient fluid can disperse in laminar viscous flow without inter-particle collision effect and how the apparent viscosity of the mixture is estimated. Takahashi et al. (2000) proposed a new Newtonian fluid flow model extending an investigation of neutrally buoyant particles in viscous fluid (Philips et al. 1992).

The theory considers that particles experience encounters between adjacent shearing surface and they give rise to particle flux; N_c, perpendicular to the main flow direction as following:

$$N_c = -K_c d_p^2 (C^2 \nabla \gamma + C\gamma \nabla C) \tag{5.57}$$

where $\gamma = du/dz$, and K_c the numerical coefficient. The word encounter in this explanation does not necessarily mean the collision, but only an approach of particles can give rise to squeezing flow in the interstitial fluid that moves the particle perpendicular to the main flow.

If coarse particles distribute anisotropically, the apparent viscosity varies spatially. Because the apparent viscosity is larger in the higher concentration region than in the lower concentration region, the resistance to particle migration into the higher concentration region is larger than into the thinner concentration region. This mechanism gives rise to particle flux; $N_{\mu a}$, as following:

$$N_{\mu a} = -K_{\mu_a} \gamma C^2 \frac{d_p^2}{\mu_a} \frac{d\mu_a}{dC} \nabla C \tag{5.58}$$

where $K_{\mu a}$ is a numerical coefficient.

The coarse particles tend to deposit due to gravity, and the flux is given as following:

$$N_s = -\frac{2}{9} C \frac{d_p^2 (\sigma - \rho) g \cos\theta}{\mu} G(C) \tag{5.59}$$

where $G(C)$ is the hindrance function to account for highly concentrated group settling, and here

$$G(C) = (1 - C)/\eta \tag{5.60}$$

is assumed, where η is the specific viscosity that is given by (Krieger 1972)

$$\eta = \frac{\mu_a}{\mu} = \left(1 - \frac{C}{C_*}\right)^{-1.82} \tag{5.61}$$

In a steady state, these three fluxes must balance so as to perform zero particle flux perpendicular to the main flow direction. This condition is given by

$$K_c \left(C^2 \frac{d\gamma}{dz} + C\gamma \frac{dC}{dz} \right) + K_{\mu_a} \gamma C^2 \frac{d\mu_a}{dC} \frac{dC}{dz} + \frac{2}{9} C \frac{(\sigma - \rho) g \cos\theta}{\mu} G(C) = 0 \tag{5.62}$$

The balance of force equation in such a steady uniform flow is given by

$$\tau = \mu_a \gamma = \rho g h \sin\theta \left\{ \left(1 - \frac{z}{h}\right) + \frac{\varepsilon}{h} \int_z^h C dz \right\} \tag{5.63}$$

where $\varepsilon = (\sigma - \rho)/\rho$.

Substitution of Eq. 5.63 into Eq. 5.62 yields the particle concentration distribution function as following:

$$\frac{dC}{dZ} = \frac{F_1}{F_2 F_3} \tag{5.64}$$

where

$$\left. \begin{aligned} F_1 &= -(2/9)\varepsilon(1 - C)/(K_c \tan\theta) + C(1 + \varepsilon C) \\ F_2 &= 1 + 1.82\{(K_{\mu_a}/K_c) - 1\} C/C_* (1 - C/C_*)^{-1} \\ F_3 &= 1 - Z + \varepsilon \int_Z^1 CdZ \end{aligned} \right\} \tag{5.65}$$

Figure 5.17 shows the examples of solids concentration distribution calculated under the two combinations of parameters; $C_* = 0.6, C_b = 0.57, \rho = 1.38$ g cm^{-3} ($\varepsilon \approx 0.9$), and $C_* = 0.72, C_b = 0.70, \varepsilon = 0.77$, respectively. The respective combinations of parameters correspond to the conditions in our laboratory experiment and in the Jiangjia Gully, China. For both cases, $K_c = 0.5$ and $K_{\mu a} = 0.75$ are adopted from the experimental results obtained by Phillips et al. (1992). The concentration at the bottom of flow; C_b, is assumed a little smaller value than C_*, because the substitution of $C = C_*$ into Eq. 5.64 cannot make sense. As is clear in Fig. 5.17, when the channel slope is steep, very high solids concentration maintains up to the surface of flow, and for relatively mild slope channel, the high concentration abruptly decreases in the upper region. This is the case of immature viscous debris flow.

The coarse particle concentration in a given slope channel becomes large and uniform as ε becomes small. Namely, the denser the interstitial fluid, the flatter the area of debris flow of a given solids concentration will reach.

The equilibrium coarse particle concentration for a given channel slope can be obtained by integrating Eq. 5.64. The results corresponding to the conditions used in obtaining Fig. 5.17 are shown in Fig. 5.18. The respective curves show the solids concentration can easily be saturated if channel gradient becomes steeper than a certain value. The value of saturation is, as is clear in Fig. 5.17, approximately equal to C_b. This fact suggests that, as long

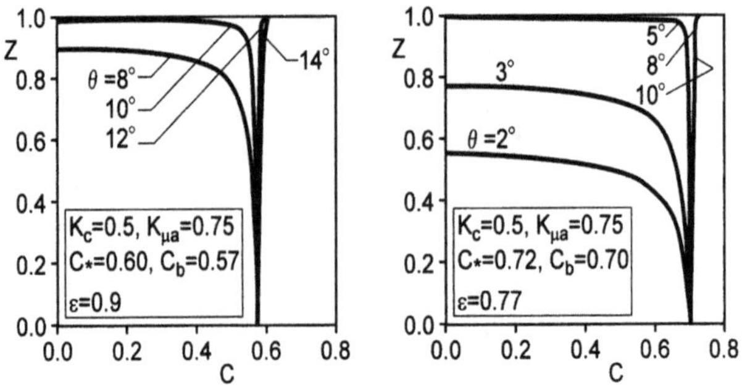

Fig. 5.17. Distributions of coarse particle concentrations on various slopes

as flow rate is sufficiently large, once a saturated flow is formed in an upstream steep channel reach, it runs down the gradually flattening channel reach keeping its original concentration until the slope becomes flatter than a critical one on which the equilibrium concentration is smaller than the saturated value. Deposition commences if the solids concentration is larger than the equilibrium value.

The velocity distribution can be obtained by solving Eq. 5.63 under the boundary condition; at $Z = 0$, $u' = 0$, where $u' = u/u_*$. The solution cannot be obtained analytically, but provided the viscosity is approximated as uniform from bottom to a height ($Z = Z_2$), it has the tendency like shown by the line in Fig. 5.19. It is given by the following equation:

$$u' = \begin{cases} \dfrac{\rho u_* h}{\mu_a}(1+\varepsilon\bar{C})\left(Z - \dfrac{1}{2}Z^2\right) & ;0 \le Z \le Z_2 \\ \dfrac{\rho u_* h}{\mu}\left\{Z - \dfrac{1}{2}Z^2 + X\left(Z_2 - \dfrac{1}{2}Z_2{}^2\right)\right\} & ;Z_2 < Z \le 1 \end{cases} \tag{5.66}$$

where $X = (1 + \varepsilon\bar{C}\mu/\mu_a) - 1$ and $\bar{C}$ is the mean particle concentration below Z_2.

Well mixed debris flow material in the Jiangjia Gully except for the component larger than 10 mm was imitated by mixing silica sand and kaolin. This material was mixed with water to make the constant solids concentration of 0.56~0.57, and it was supplied into the upstream end of experimental flume from a hopper.

The velocity distributions given by Eq. 5.66 are compared with the experimental data in Fig. 5.20, in which $Z_2 = 1$ and $\mu_a = 8.1$ Pa s obtained from the theory explained above are used. Figure 5.20a is the case the flow is produced on the deposit formed by a previous debris flow surge, and Fig. 5.20b is the case the flow is produced on rigid bed. This figure shows that not only the velocity distribution obeys the Newtonian laminar flow equation equally well on erodible and rigid beds, but also the absolute velocity values are well predicted by this theory.

Recently, Armanini et al. (2003) did the similar theoretical and experimental discussion that took the yield strength of interstitial fluid into account.

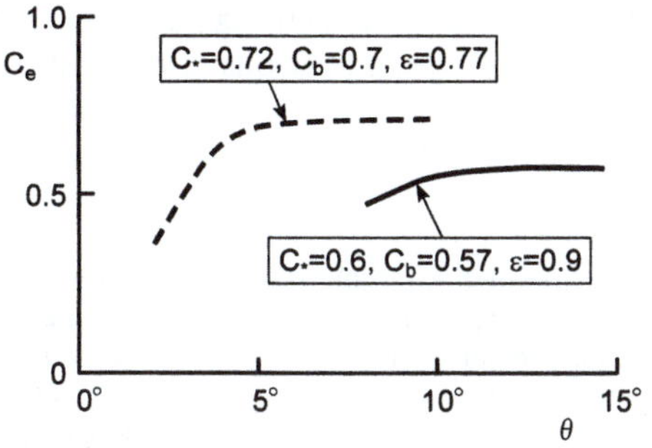

Fig. 5.18. Equilibrium coarse particle concentrations versus channel slopes

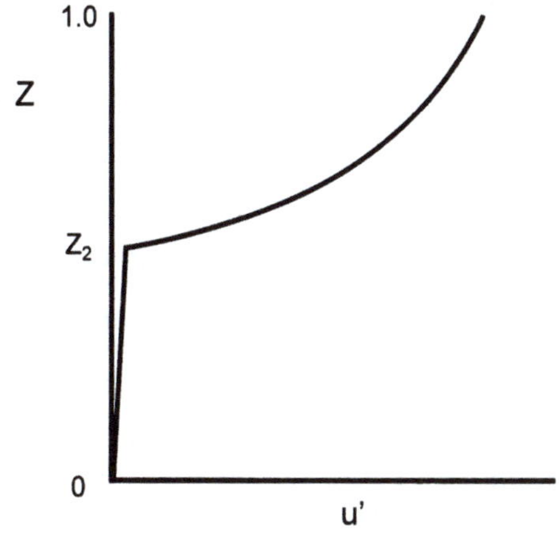

Fig. 5.19. Schematic velocity distribution in an immature viscous debris flow

5.9 Conclusion

The processes of initiation and the mechanics of flow of various types of debris flow were discussed from the point of view of systematizing the theory of debris flows.

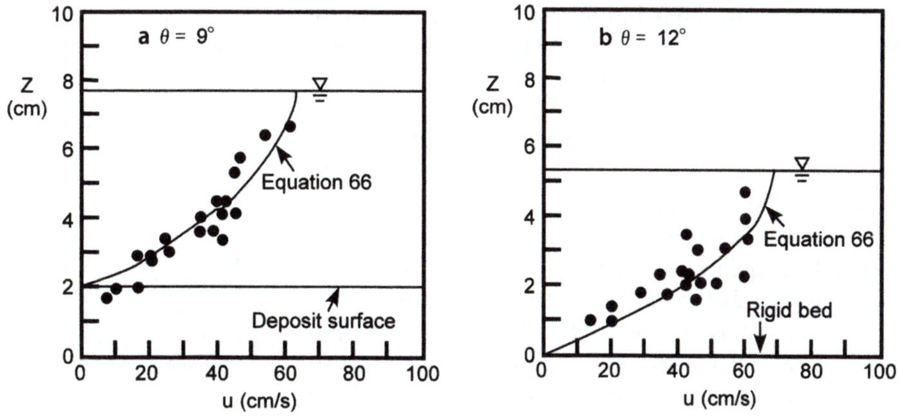

Fig. 5.20. Calculated and experimental velocity distributions

The causes of initiation of debris flows were classified into the erosion type and the landslide-induced type. The process of formation of the erosion type debris flow can be analyzed similar to the known methods for the analyses of fluvial processes by the bed load transportation. The differences between the analyses of debris flow and bed load transport appear in the flow resistance law and the erosion velocity equation.

A model of liquefaction of the landslide earth mass was introduced. At least the lower part of the slid earth mass is assumed saturated with water. The saturated part will liquefy from lower to upper by the effect of shearing between the mass and the ground. The liquefied layer not only acts as a lubricant to move the mass on top of it, but also gets behind the mass to form a following debris flow. Therefore, if the mass is completely saturated and liquefaction proceeds to the extent of disappearance of the mass, the mass completely transforms into debris flow. If the liquefaction ceases halfway, the following debris flow may overflow the stopped mass. The liquefaction without supply of water from the outside of the slid earth mass will be accomplished under the contractive destruction of the skeleton of the mass. This rationale leads to an inference that the mass of dilative nature will not transform into debris flow without supply of water from the outside.

The basic system of equations to describe the aforementioned process was given. Although the detailed comparison of the theory to the laboratory experiments were abbreviated due to limitation of pages, the validation of the theory including its expanded form capable to analyze the three-dimensional motion on a topographically complicated ground is referred by introducing authors' recent investigations.

Various concepts for modeling debris flows are reviewed. Under the present state of the art, the mixture theory that considers the debris flow material is a mixture of two continua; one is the interstitial fluid and the other is a kind of continuum media comprised of granular assembly, is influential. By the adoption of appropriate constitutive relations in the mixture theory, debris flows were classified into three; stony type, turbulent muddy type and viscous type, and their respective existing regions were defined in the ternary diagram whose three axes represent the Bagnold number, the Reynolds number, and the relative depth.

The inertial debris flow was defined as the unified concept of the stony and the turbulent muddy debris flows. The theoretical solutions of the inertial debris flow; the particle concentration distribution, the velocity distribution, and the equilibrium transport concentration, were given. The equilibrium transport concentration of the large particles that cannot be suspended by turbulence is given uniquely as a function of the channel gradient. But, for the smaller particles that are at least partly suspended by turbulence, the equilibrium transport concentration depends on both the channel slope and the particle diameter. The theoretically obtained results rather well coincide to the experimental results.

The viscous type debris flow was modeled as a laminar flow of a Newtonian fluid, in which coarse particles were dispersed in flow almost uniformly under the balance between the upward dispersing fluxes due to squeezing flow and viscosity gradient on the encounter of particles embedded in the two adjacent shearing layers and the downward flux due to gravity settling. The theory not only well explains the high fluidity of very densely particle freighting debris flow but also predicts the velocity of viscous debris flow without using the empirical parameters by contrast to many previous visco-plastic models.

References

Armanini A, Gregoretti C (2000) Triggering of debris-flow by overland flow; A comparison between theoretical and experimental results. In: Wieczorek GF, Naeser ND (eds) Debris flow hazards mitigation: mechanics, prediction, and assessment; Proceedings International Conference, Taipei, Balkema, Rotterdam, pp 117–124

Armanini A, Dalri C, Fraccarolo L, Larcher M, Zorin E (2003) Experimental analysis of the general features of uniform mud-flow. In: Rickenmann D, Chen CL (eds) Debris flow hazards mitigation: mechanics, prediction, and assessment. Proceedings International Conference, Davos, Millpress, Rotterdam, pp 423–434

Ashida K, Takahashi T, Mizuyama T (1977) Study on the initiation of motion of sand mixtures in a steep slope channel. J JSECE 29(4): 6–13 (in Japanese)

Bagnold RA (1954) Experiments on a gravity-free dispersion of large solid spheres in a Newtonian fluid under shear. Proc R Soc London, Ser A 225:49–63

Bagnold RA (1957) The flow of cohesionless grains in fluid. Philos Trans R Soc London, Ser A 249:235–298

Bagnold RA (1966) The shearing and dilatation of dry sand and 'singing' mechanism. Proc R Soc London, Ser A 295:219–232

Campbell CS (1990) Rapid granular flows. Annu Rev Fluid Mech 22:57–92

Campbell CS (2001) Granular flows in the elastic limit. Spec Publs Int Ass Sediment 31:83–89

Campbell CS, Brennen CE (1985) Computer simulation of granular shear flows. J Fluid Mech 151:167–188

Chen CL (1988) Generalized viscoplastic debris flow. J Hydraul Eng-ASCE 114(3):237–258

Coussot P (1995) Structural similarity and transition from Newtonian to non-Newtonian behavior for clay-water suspensions. Phys Rev Lett 74:3971–3974

Gotoh H, Sakai T (1997) Numerical simulation of sheetflow as granular materials. J Waterw Port C Div 123(6):329–336

Gotoh H, Tsujimoto T, Nakagawa H (1994) Numerical model of interface momentum transfer and interparticle collision in bedload layer. In: Proceedings APD-IAHR, pp 565–572

Hirano M, Hashimoto H, Fukutomi K, Taguma K, Pallu MS (1992) Nondimensional parameters governing hyperconcentrated flow in an open channel. J Hydraul Eng-ASCE 36:221–226 (in Japanese)

Iverson RM (1997) The physics of debris flows. Rev Geophys 35(3): 245–296

Iverson RM, Denlinger RP (1987) The physics of debris flows-a conceptual assessment. In: Beschesta et al. (ed.) Erosion and sedimentation in the Pacific rim. IAHS publ. 165:155–165

Iverson RM, Denlinger RP (2001) Flow of variably fluidized granular masses across three-dimensional terrain. J Geophys Res 106(B1):537–552

Johnson AM (1965) A model for debris flow. Ph.D. Dissertation Pa. State Univ., State College

Johnson AM (1970) Physical processes in geology. Freeman, New York

Julien PY, O'Brien JS (1997) On the importance of mud and debris flow rheology in structural design. In: Proceedings 1st Inter-national Conference on Debris-Flow Hazards Mitigation: Mechanics, Prediction, and Assessment, New York, ASCE, pp 350–359

Krieger JS (1972) Rheology of monodisperse lattices. *Advance of Colloid Interface Science* 3:111–136

Phillips RJ, Armstrong RC, Brown RA, Graham AL, Abbot JR (1992) A constitutive equation for concentrated suspensions that accounts for shear-induced particle migration. Phys Fluids A4(1): 30–40

Sabo Technical Center (1998) Reality of sediment disasters in 1997. pp 25–27 (in Japanese)

Straub S (2001) Bagnold revisited: implications for the rapid motion of high-concentration sediment flows. Spec Publs Int Ass Sediment 31:91–109

Takahashi T (1978) Mechanical characteristics of debris flow. J Hydr Eng Div-ASCE 104:1153–1169

Takahashi T (1991) Debris flow. Balkema, Rotterdam

Takahashi T (1999) Mechanics of viscous debris flow. In: Takahashi (ed) Japan-China joint research on the mechanism and the countermeasures for the viscous debris flow. Research report of Group C3 of special project associated with IDNDR sponsored by Ministry of Education, Science, Sports and Culture, Japan, pp 64–84

Takahashi T (2000) Initiation and flow of various types of debris-flow. In: Wieczorek GF, Naeser ND (ed) Debris flow hazards mitigation: mechanics, prediction, and assessment. Proceedings International Conference, Taipei, Balkema, Rotterdam, pp 15–25

Takahashi T (2004) Mechanics and countermeasures for debris flow. Kinmiraisha, Nagoya (in Japanese)

Takahashi T, Nakagawa, H (1994) Natural dam formation and the disaster-a possible explanation of one extreme event. In: Proceedings International Workshop on Floods and Inundations related to Large Earth Movements, A8.1–A8.12

Takahashi T, Satofuka Y (2002) Generalized theory of stony and turbulent muddy debris-flow and its practical model. J JSECE 55(3): 33–42 (in Japanese)

Takahashi T, Tsujimoto H (1997) Mechanics of granular flow in inclined chute. J Hydraul Coast Environ Eng JSCE565:57–71 (in Japanese)

Takahashi T, Nakagawa H, Satofuka Y (2000) Newtonian fluid model for viscous debris-flow. In: Wieczorek GF, Naeser ND (ed) Debris flow hazards mitigation: mechanics, prediction, and assessment. Proceedings International Conference, Taipei, Balkema, Rotterdam, pp 255–262

Takahashi T, Nakagawa H, Satofuka Y, Kawaike K (2001) Flood and sediment disasters triggered by 1999 rainfall in Venezuela; a river restoration plan for an alluvial fan. J Natural Disaster Science, JSNDS 23(2):65–82

Takahashi T, Satofuka Y, Kashimoto S (2003) Motion of landslide-induced debris flow. In: Rickenmann D, Chen CL (eds) Debris flow hazards mitigation: mechanics, prediction, and assessment; Proceedings International Conference, Davos, Millpress, Rotterdam, pp 399–410

Tognacca C, Bezzola GR, Minor HE (2000) Threshold criterion for debris-flow initiation due to channel-bed failure. In: Wieczorek GF, Naeser ND (ed) Debris flow hazards mitigation: mechanics, prediction, and assessment. Proceedings International Conference, Taipei, Balkema, Rotterdam, pp 89–97

Yano K, Daido A (1965) Fundamental study on mud-flow. Bull Disaster Prev Res Inst Kyoto Univ 14:69–83

Part II Landslide Dynamics

Undrained Stress-controlled Dynamic-loading Ring-shear Test to Simulate Initiation and Post-failure Motion of Landslides

Kyoji Sassa* · Hiroshi Fukuoka · Gonghui Wang · Fawu Wang

Abstract. Landslides are gravitational mass movements of rock, debris or earth. Shear deformation in landslides before failure conforms to the field of statics. But shear deformation during seismic loading and post-failure motion of landslides conforms to the field of dynamics. Thus, study of the initiation of earthquake-induced landslides and rapid landslide motion needs to develop "Landslide dynamics" involving dynamic loading and dynamic generation/dissipation of excess pore-water pressure during motion. New developments in science can be facilitated by new technological advances. This study aimed to develop a new testing method that can geotechnically simulate the formation of the shear zone and the following long and rapid shear displacement that occurs in high-velocity landslides. Sassa K and his colleagues at DPRI (Disaster Prevention Research Institute), Kyoto University, have worked to develop an undrained stress-controlled dynamic-loading ring-shear apparatus and its testing method for this purpose. This paper describes the development of this testing method, and its application to the study of earthquake-induced landslides and landslide triggered debris flows in Japan.

Keywords. Ring-shear test, landslide dynamics, earthquake-induced landslides, landslide-triggered debris flows, undrained loading

6.1 Introduction

The ring-shear apparatus was designed initially to investigate the residual shear resistance mobilized along the sliding surface at large shear displacements in landslides because it allows unlimited deformation of the specimen. The test configuration for the ring-shear device was introduced by Hvorslev (1939); Hvorslev's device forced the sample to begin shearing on a predefined plane located at the separation of the upper and lower confining rings. This concept was utilized and improved by Bishop et al. (1971), Bromhead (1979), Savage and Sayed (1984), Sassa (1984), Hungr and Morgenstern (1984), Tika (1989), and Garga and Sendano (2002).

The most well-known and widely adopted type of ring-shear apparatus was developed jointly by scientists and engineers at the Imperial College of Science and Technology (United Kingdom) and the Norwegian Geotechnical Institute (Bishop et al. 1971). The main advantage of the Bishop type of ring-shear apparatus compared with other previous models was provided by its ability to measure the friction between the soil sample and the sidewalls of the upper shear box. This feature is particularly important for the ring-shear apparatus as well as for the direct shear apparatus, because it ensures an accurate determination of the total normal stress acting on the soil specimen during the experiment.

The apparatus developed by Bromhead (1979, 1986), and by Savage and Sayed (1984), and the apparatus modified by Garga and Sendano (2002) do not have two (upper half and lower half) sample boxes to accommodate shear. These apparatus have one sample box loaded and sheared by the upper loading platen. This mechanism is simpler than others because it is not always easy to prevent leakage of samples through the gap between two rings. Shearing takes place between the upper loading platen, bottom of the shear box and the soil sample. The mobilized shear resistance may differ from that mobilized during shearing within the soil specimen, especially for sandy materials; however, this probably poses no problem for clays. The ring shear apparatus are also used for the study on faults under much higher pressure and much smaller specimens (Shimamoto and Tsutsumi 1994; Zhang et al. 1999).

The original high-speed ring shear apparatus (DPRI-1) developed by Sassa (1984) used a conventional shear-speed control motor; it could not provide cyclic shear-stress loading. The first stress-controlled dynamic-loading ring shear apparatus (DPRI-3) was developed to reproduce seismic loading using a torque-control motor and a servo-control system that utilized the feed-back signal from a load cell that measured torque (Sassa 1994, 1995). Following DPRI-3, a series of stress-controlled dynamic-loading ring-shear apparatus has been developed that incorporated different features (DPRI-4, DPRI-5, DPRI-6 and DPRI-7 in Table 6.1).

The general purpose of the DPRI ring shear testing program is to design an apparatus that can quantitatively simulate the entire process of failure of a soil sample, from initial static or dynamic loading, through shear failure, pore-pressure changes and possible liquefaction, to large-displacement, steady-state shear movement. No other laboratory apparatus has so far been able to provide an integrated simulation of this natural process. The latest devices in the DPRI ring shear series have succeeded in this goal.

Table 6.1. Features of previous ring shear apparatus, compared with the DPRI family of undrained dynamic loading ring shear apparatus

Shear box parameter	Author(s)								
	Bishop et al. (1971)	Hungr and Morgenstern (1984)	Tika (1989)	Garga and Sendano (2002)	Sassa (1992) DPRI-3	Sassa (1996[a]) DPRI-4	Sassa (1997) DPRI-5	Sassa (1997) DPRI-6	Sassa (this paper) DPRI-7
Inner diameter (cm)	10.16 (4.0 in)	22.0	10.16	9.2	21.0	21.0	12.0	25.0	27.0
Outer diameter (cm)	15.24 (6.0 in)	30.0	15.24	13.3	31.0	29.0	18.0	35.0	35.0
Max. height of sample (cm)	1.9	2.0	1.9	2.0	9.0	9.5	11.5	15.0	11.5
Ratio of max. height/width	0.75	0.5	0.75	0.98	1.8	2.38	3.83	3.0	2.88
Shear area (cm^2)	101.34	326.73	101.34	72.45	408.41	314.16	141.37	471.24	389.56
Max. normal stress (kPa)	980	200	980	660	500	3 000	2 000	3 000	500
Max. shear speed ($cm\ s^{-1}$)	–	100.0	9.33	–	30.0	18.0	10.0	224.0	300.0
Cyclic torque control testing (max. frequency)	No	No	No	No	Yes (0.5 Hz)	Yes (5 Hz)	Yes (5 Hz)	Yes (5 Hz)	Yes (5 Hz)
Undrained testing and pore pressure monitoring	No	No	No	No	Yes	Yes	Yes	Yes	Yes
Max. data acquisition rate (readings s^{-1})	–	–	10	–	12	200	200	200	1 000

[a] The apparatus was developed in 1995. The apparatus and the test results were reported in Vankov and Sassa (1999).

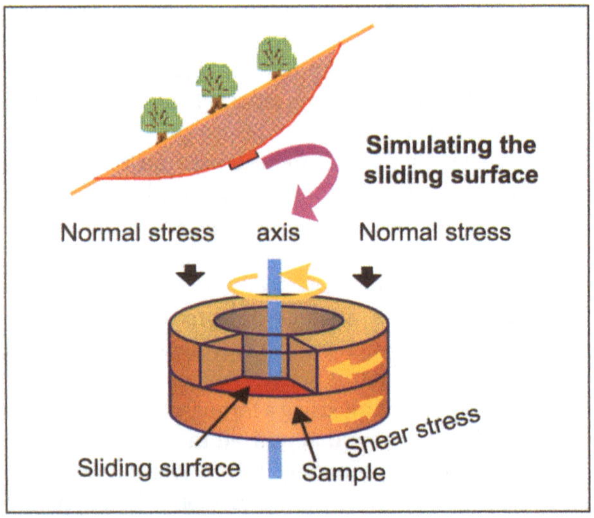

Fig. 6.1. Design concept of the ring shear apparatus to simulate initiation and post-failure motion of landslides

The mechanism of the undrained stress-controlled dynamic-loading ring shear apparatus introduced in this paper (Fig. 6.1) has basically the same principle as the ring shear device developed by Bishop et al. (1971) in shearing soils for unlimited displacement; however, the major purpose and, consequently, the design concept is rather different. The ring shear apparatus described by Bishop et al. (1971) was designed to study the post-peak interval of the shear resistance-displacement curve with emphasis on residual strength developed in slow clayey landslides. However, the current undrained ring shear apparatus geotechnically simulates the formation of the shear zone and the post-failure mobility of high-speed landslides and observes the consequence of mobilized shear resistance, as well as the post-failure shear displacement and generated pore-water pressure.

Many unknown factors exist in landslide motion. To simulate the natural sliding phenomena as clearly as possible, the following conditions were necessary, dictating the new design features summarized in Table 6.1 in comparison with some previous ring shear apparatus:

1. Shearing is provided primarily by the stress-control (more precisely by torque-control) condition because the triggering factors in natural phenomena, such as earthquake shaking, ground-water fluctuation, or change of slope profiles due to toe erosion or filling, are associated with changes in stress within the slope. However, the speed-control test, which is the same as that developed in the previous apparatus by Bishop et al. (1971), can also be conducted to obtain the residual friction angle of the sample.
2. The maximum shear speed along the mid-diameter circle of the specimen is 33 $cm\ s^{-1}$ (DPRI-3) to 300 $cm\ s^{-1}$ (DPRI-7) to simulate rapid landslide motion.
3. The undrained condition under rapid shearing can be maintained by pressing rubber edges onto the bottom of an upper pair of shear boxes at a necessary contact pressure using a servo-oil piston and gap sensor with a precision of 1/1 000 mm.

4. Pore pressure is monitored by a pressure gauge connected to the filtered gutter (4×4 mm) along the circumference of the inner wall of the upper outer ring of the shear box.
5. Rapid loading and high-speed data acquisition are available to simulate earthquake loading and other phenomena involving rapid stress changes. A seismic wave form with frequency of up to 5-Hz can be reproduced, and a data-acquisition rate from 12 readings per second (DPRI-3) to 1000 readings per second (DPRI-7) is available.
6. The shear box is deep enough to allow for the development of a well-defined shear zone within the soil specimen in order to avoid the possibility of measuring the shear resistance between soil specimen and the loading platen or the bottom of the shear box.
7. It is also possible to conduct cyclic shear-displacement controlled tests, torque-controlled tests, and shear-speed controlled tests in DPRI-4 and DPRI-7.

The undrained stress-controlled dynamic-loading ring shear apparatus were developed to accomplish the above conditions while reproducing the formation of the shear zone and the resulting motion along the shear zone. Therefore, the aim and design features are different from previous ring shear apparatus. Table 6.1 presents the major characteristics of some previous ring shear apparatus in comparison to the features of the undrained stress-controlled dynamic loading ring shear apparatus developed by Sassa and colleagues at the DPRI (Disaster Prevention Research Institute), Kyoto University.

6.2 Comparison with Triaxial Tests

Most of geotechnical tests are conducted by triaxial tests. Often question is raised on the difference. This question especially focusing on the undrained shear behavior of sands is examined in this section.

Undrained shear behavior of sands is of great importance to engineers in the anti-liquefaction designs. By now, a great number of experimental studies have been conducted for a better understanding on the undrained shear behavior of sands. Based on the results of mostly triaxial tests, a concept of steady state was proposed and widely used in the procedure of analyzing liquefaction susceptibility of a soil in a given geoenvironment (Castro 1969; Ishihara 1993; Vaid and Chern 1985; Poulos et al. 1985; Alarcon-Guzman et al. 1988; Kramer and Seed 1988). However, the determination of steady state is still a problem needed to be scrutinized, because according to Poulos (1981), steady state means the mass is continuously deforming at constant volume, constant normal effective stress and constant velocity. It is achieved only after all particle orientation has reached a steady state condition and after all particle breakage, if any, is complete so that the shear stress needed to continue deformation and the velocity of deformation remain constant. As well known, although triaxial apparatus showed best achievement in measuring the stress-strain relationship and maximum shearing resistance, it is not suitable in determining the temporary or permanent decrease of the shearing resistance after failure (Bishop et al. 1971), due to the limitation in shear displacement. Thus, the potential particle orientation and particle breakage may be incomplete due to the limitation of shear displacement, i.e., the steady state of deformation would be difficult, if not impossible, to be reached, say in triaxial tests, especially for those sands in medium or dense state. In fact, in many cases, such as some landsides, the shearing behavior may involve runout of several meters to tens of kilometers, often on slopes of only a few degrees (Voight 1978, 1979; Rouse 1984). Therefore, quite large shear displacement is needed to examine these failure phenomena where the shearing resistance may decrease along with shear displacement.

To overcome this limitation in shear displacement and measure the shear resistance after failure, ring shear apparatus had been developed and improved, and then widely used in the analysis of slope stability (Bishop et al. 1971; Bromhead 1979; Gibo 1994; Tika and Hutchinson 1999). However, almost all of the ring shear apparatuses used in these studies were not enable to perform undrained shearing tests. Quite recently, a series of undrained ring shear apparatuses was developed and improved by Sassa and his colleagues. These series newly developed undrained ring shear apparatuses offered an available method to study the undrained shear behavior of sands in virtually limitless displacement levels.

Difference from conventional ring shear apparatus is not only undrained test capability, but capability of shear-stress controlled tests. Shearing by conventional ring shear apparatuses is in the shear-speed controlled way (Bishop et al. 1971; Bromhead 1979; Hungr et al. 1984; Garga et al. 2002). The series of stress-controlled dynamic-loading ring-shear apparatus can conduct shear stress-controlled test. The shear stress in any form, including monitored real seismic waves up to 5 Hz and an impact loading in a way which a landslide mass falls on the torrent deposit, can be loaded. Pore-pressure-control test to simulate the ground water rise in the slope is also conducted. Therefore, the apparatus is not to measure shear resistance as a parameter, but to reproduce the stress and pore pressure acting on a potential sliding surface and the resulting sliding-surface formation by failure and the post-failure motion. The generated excess pore pressure in the shear zone and the shear resistance mobilized on the sliding surface are monitored. Thus, the stress-controlled dynamic-loading ring shear apparatus and its test is not a method to measure shearing parameter of an element of soil, but a test to reproduce shearing phenomenon, sliding surface formation and post-failure motion, which will occur in the natural slopes during rainfalls, earthquakes and other triggering stress changes.

Note that a comparison study on the undrained shear behavior of granular material in ring shear tests and triaxial compression tests has been performed and can be obtained from Okada et al. (2000).

6.3 Structure and Control System of Apparatus

6.3.1 Outlines

Figure 6.2 presents the overall frontal view of DPRI-6 (the largest type within the DPRI ring shear family). The total height is 5.1 m. The apparatus was installed in a pit at the DPRI Landslide laboratory with the level of the sample box approximately 1 m above the floor for easy access. The rear of the machine in Fig. 6.2 is the main apparatus, and the front instrument box is the computerized control and monitoring system. The shear box and the oil piston for normal stress loading are indicated in the figure; the servo-motor for shear-stress loading as well as speed-control shearing is hidden behind the control and monitoring unit.

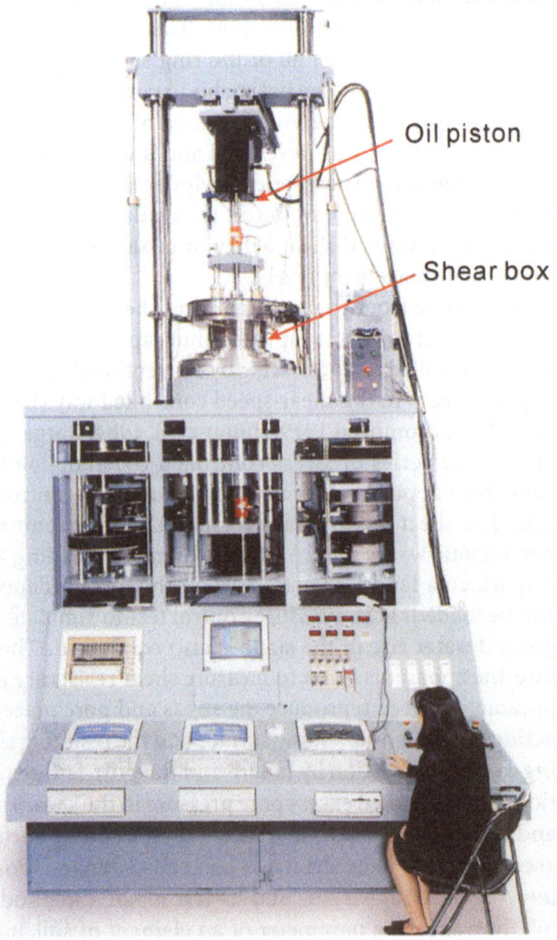

Fig. 6.2. Photo of the undrained dynamic-loading ring shear apparatus (DPRI-6)

Figure 6.3 shows a schematic diagram of the apparatus, while Fig. 6.4 presents a brief illustration of the electrical control system of DPRI-5, 6, and 7. As shown in both figures, the sample is loaded by a loading platen through an oil piston (OP1), and the loaded normal force (for normal stress) is measured by a load cell (N1). The sum of the friction between the sample and the sidewall of the upper pair of rings in addition to the self-weight of the upper pair of rings is measured by a load cell (N2). The actual normal force acting on the shear surface is obtained from the difference between N1 and N2. This value is sent to a servo-amplifier as a feed-back signal (as shown in Fig. 6.4). Then, the normal stress on the shear surface is automatically kept the same as the control signal given by the computer. Shear stress is supplied by a torque-controlled servo-motor, which can be switched to a speed-controlled mode as well. The applied torque is measured by torque transducer T1. Using the monitored value of T1 as the feed-back signal (Fig. 6.4), the applied shear stress is automatically controlled by the servo-amplifier and servo-motor and then kept the same as the pre-determined value given by the computer. The shear resistance acting on the shear surface is monitored by two load cells (S1 and S2), through which the upper shear box is restrained from rotation. The second shear-stress loading system (servo-motor 2 and torque transducer T2) in Fig. 6.3 is a spare because the manufacturing or repairing of this special motor is a lengthy procedure.

Four computers are used in DPRI-6. One computer is for controlling the test. The controlling signals could be either edited signals of monotonic/cyclic loading, or seismic records of earthquakes (as listed in Table 6.1). The second computer monitors the response of each transducer, records the data, and then draws the effective-stress path, simultaneously. Note that a multi-pen recorder is also used for recording the data on paper. The third computer is installed for monitoring the system functions and providing the safeguard of an automatic alarm system. If the apparatus is operated incorrectly or something happens to the mechanical parts, the alarm system will sound the alarm and automatically terminate the test. The fourth computer is mounted for memorizing the details of the test procedures and data analysis.

6.3.2 Prevention of Water Leakage

The most essential part of the undrained ring shear apparatus is the undrained shear box. Its design is illustrated as Fig. 6.5, an enlarged diagram of the left half of the cross section of the undrained shear box and its surroundings, including the water-pressure measurement system.

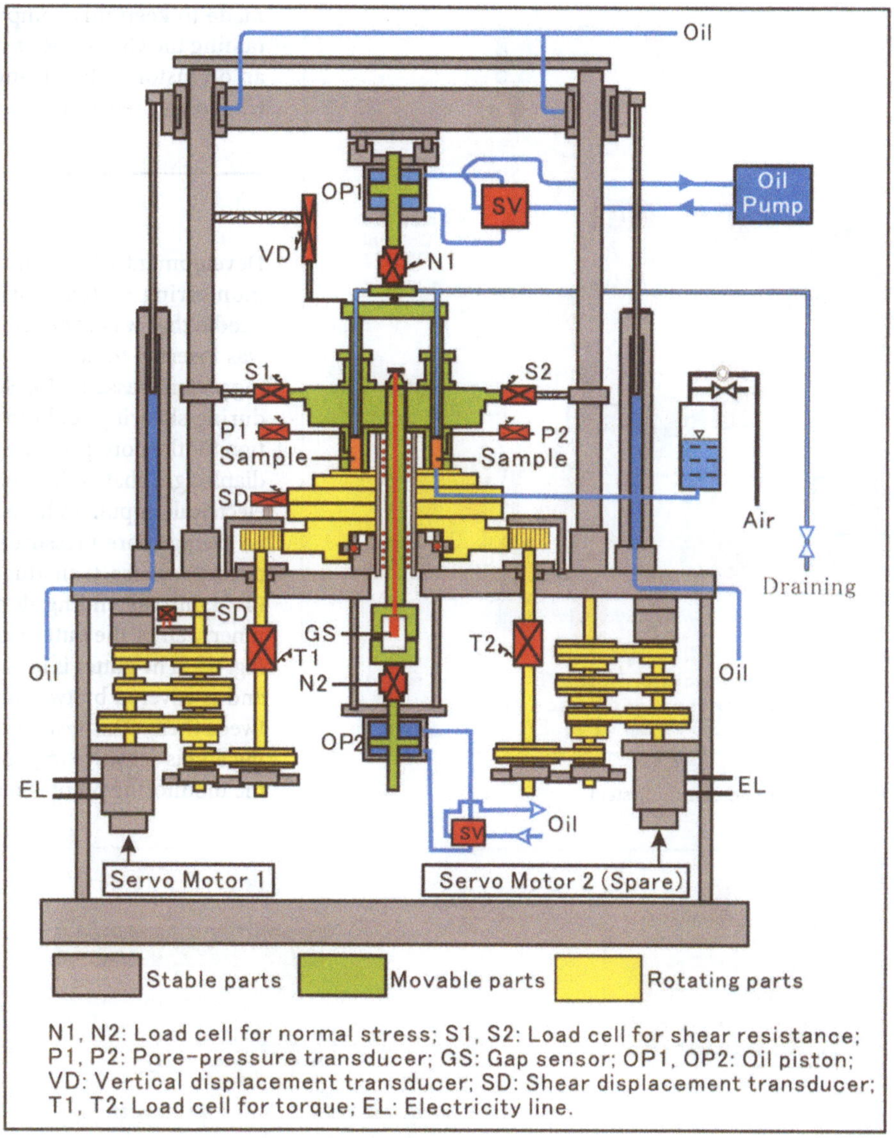

Fig. 6.3. Mechanical structure of the apparatus

The most difficult part is prevention of leakage of water through the gap between the upper shear box and the lower shear box during high-speed shearing (33–300 cm s^{-1}). This was perfected for DPRI-3 in 1992 and 1993 (Shoaei and Sassa 1994). A material with strong abrasion resistance, low friction characteristics and good performance as a sealant had to be found. Shoaei and Sassa tried various types and shapes of rubber edges, Teflon (polytetrafluorethylene) plates, and combined layers of Teflon plate underlain by rubber plate. The final selection was a stair-shaped rubber ring (rubber hardness index, 45°JIS) as shown in Fig. 6.5. High surface smoothness and equal height of both the outer and inner rings were found to be extremely important. After pasting rubber edges on the two confining rings of the lower rotary box, rubber was cut by a lathe and then polished to a smooth surface in the assembly process. The outer confining rings of the shear box for DPRI-7 were made of transparent Acrylic. Acrylic is deformable under elevated temperature, a factor taken advantage of in the assembly process. Periodically, the tops of the rubber edges were leveled in the laboratory by pressing a hand file (fixed at a certain level) to both edges and rotating the lower rotary box. Before each test, the rubber edges were covered with a friction coat of Teflon (polytetrafluorethylene), and then daubed with vacuum silicon grease.

During the test, a certain amount of compression is applied between the upper pair of rings and the rubber edges by lowering the gap control oil piston (OP2 in Figs. 6.3 and 6.4) to a specific value (hereinafter termed as Gap Value). To maintain the undrained condition, contact pressure between the rubber edges and the upper pair of rings should be kept greater than the generated pore pressure in the sample. Then during the test, an effort is

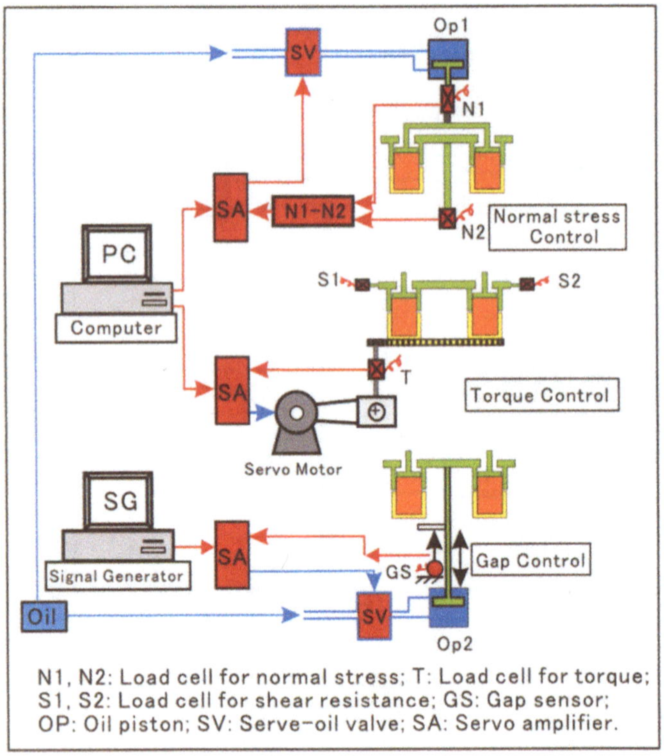

Fig. 6.4. Electronic Control System

made to keep this compression force constant by maintaining the Gap Value as constant as possible by means of an oil piston (OP2), using the feed-back signal obtained from a gap sensor (GS) with a precision of 1/1000 mm.

6.3.3 Pore-pressure Monitoring

Development of an effective and durable pore-pressure monitoring system was also a difficult task. At first a needle that was connected to a pore-pressure transducer was inserted to near the shear zone from the upper loading platen (Sassa 1995). However, the needle was deformed during shearing, and did not have large enough inlet section for the pore-pressure transducer. The transducer has a diaphragm that is deformed by water pressure to provide electrical output. To have a large inlet section and provide an average pore-pressure value throughout the soil sample, pore-pressure transducers are connected to a gutter (4×4 mm) extending along the entire circumference of the inner wall of the outer ring in the upper box, as shown in Fig. 6.5. The gutter is located 2 mm above the shear surface and is covered by two metal filters, with a filter cloth between them. This system is quite durable in regard to shearing and is sensitive to pore-pressure monitoring, although the monitoring point is not at the center of the shear zone.

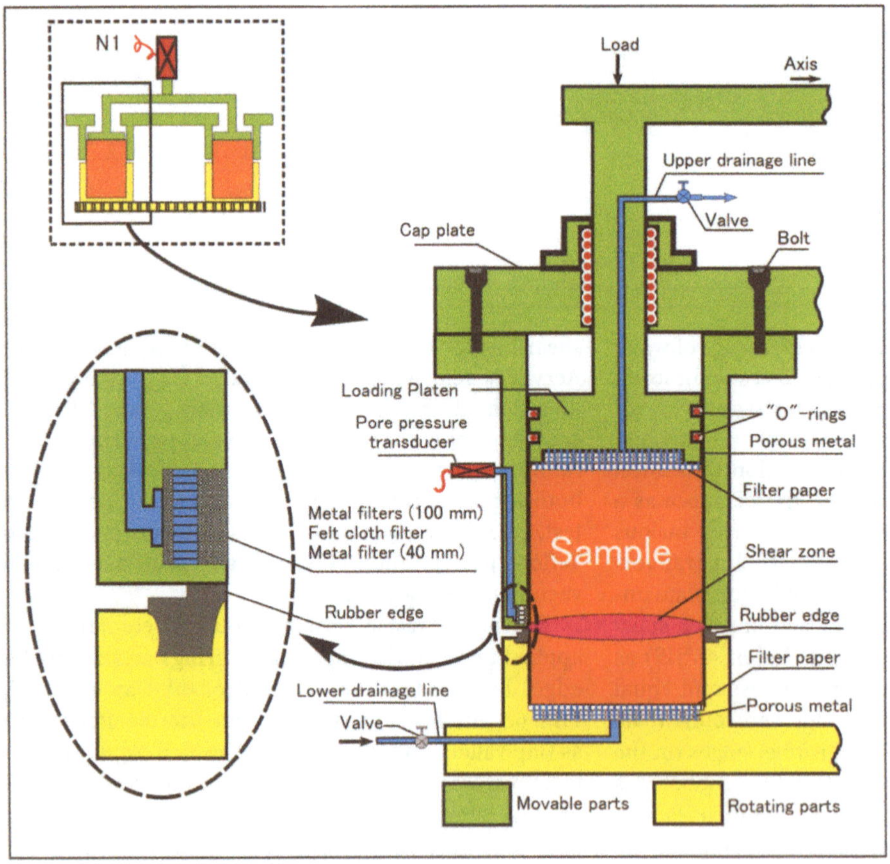

Fig. 6.5. A half section of the shear box and the close-up diagram of the edges

6.3.4 Efficiency of Servo-control System and Water Pressure Measurement System

The efficiency of servo-control system and water pressure measuring system was evaluated by conducting cyclic loading (normal stress) on water within the undrained shear box. During test, the target cyclic loading was designed and input in the controlling computer, while the output response was checked by monitoring the response of pore water pressure transducers. The time series data of loaded normal stress are shown in Fig. 6.6a, while the measured response of pore water pressure transducer is plotted in Fig. 6.6b. Figure 6.6c presents these two components against time in one cycle. It is seen that essentially the loaded normal stress behaved the same as the monitored pore water pressure, showing the compatibility of controlling system and water pressure measuring system.

As mentioned above, there are many kinds of shearing methods using DPRI-Ver.6. The following section will introduce the basic operating methods by performing undrained shearing tests under monotonic loading condition. To observe the shear behavior of soil accompanying the increasing shear stress before failure, torque-controlled method is selected.

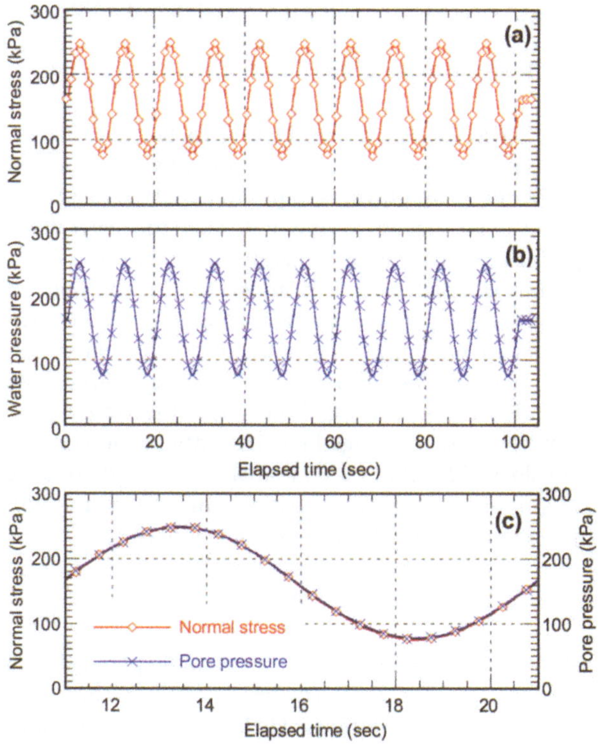

Fig. 6.6. Response of the loading system and monitoring system in the case of loading on only water in the shear box. (**a**) and (**b**) show 10 cycles of the applied normal stress and the resulting monitored water pressure; (**c**) shows one cycle of applied normal stress and the resulting water pressure (pore pressure)

6.4 Testing Procedure

The test procedures could be separated as two parts, namely preparation test and test on saturated sand. The purposes of the preparation test are adjusting the Gap Value to give an appropriate pressing force between the upper pair of rings and the rubber edges, at which time there will be no water leakage during undrained shearing tests; and then measuring the friction between the rubber edge and the upper parts of ring box corresponding to the adjusted gap value. The tests procedures are listed below.

6.4.1 Preparation Test for Adjustment and Correction

6.4.1.1 Gap Value Adjusting

An appropriate Gap Value is necessary; too small pressing force could not ensure water leakage proof, while too great pressing force would cause damage to the rubber edges and give a greater rubber edge friction. After the rubber edge was cleaned, covered with Teflon, daubed by silicon grease, the apparatus was set up without sample. And then, the upper parts (including the upper pair of rings and the cap plate shown in Fig. 6.5) were raised by increasing the Gap Value until the upper parts were separated completely from the rubber edges and the reading of the amplifier for N2 in Fig. 6.2 did not change anymore, while the loading platen was kept free. Thereafter, the amplifier for N2 was adjusted to zero, and the upper parts were pulling down by lowering the oil piston of OP2 until a certain Gap Value, at which the reading of N2 (which is the Gap Contact Force caused by the touching and pressing of the upper pair of rings on the rubber edges) reached a certain value (determined according to different test conditions, and was kept as approximately 250 kgf in the tests presented in this report). If this contact pressure between the rubber edges and the bottoms of upper pair of rings is always kept greater than the generated pore pressure inside the sample along the whole rubber edge contact surface, any leakage of water should not occur. The rubber edge contact surface is precisely processed to be smooth by machine after being pasted onto the lower parts. The Gap Value was kept constant by the servo-oil piston in the commenced undrained shear test on saturated sample, while the readings in the amplifier for N2 (250 kgf) was adjusted to zero before the loading of normal stress so to give a pure reading for the total value of friction between the sample and the sidewalls of upper rings.

6.4.1.2 Leakage Proof Checking

Water leakage proof was checked by putting water into the shear box and applying normal stress on water and

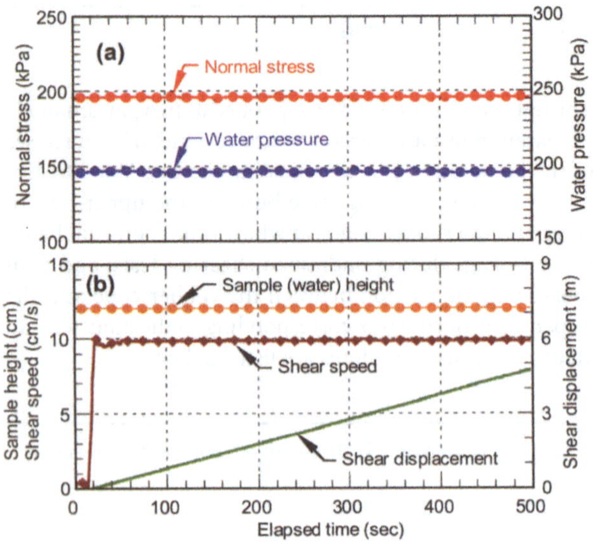

Fig. 6.7 Checking the water leakage proof at the adjusted gap value

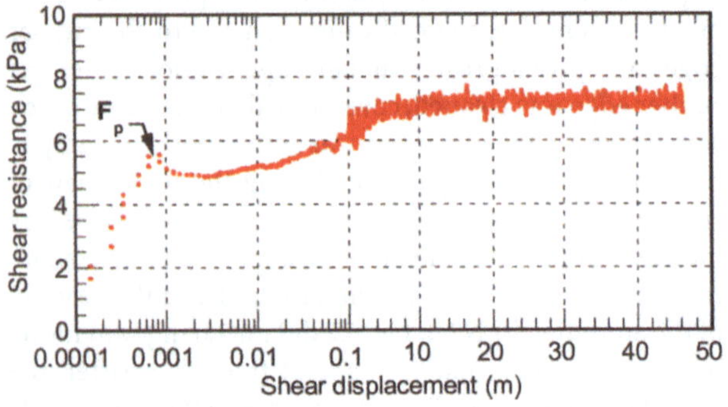

Fig. 6.8. Friction of rubber edge against shear displacement (shear speed = 10 mm s^{-1})

shearing, while keeping the Gap Value as the same as adjusted. If there is no decrease in the water volume within the shear box during rotation under the loaded normal stress, it is ensured that there is no water leakage using the adjusted Gap Value. Figure 6.7 presents the applied normal stress, monitored water pressure, and the volume change (in the form of sample height) against the time. As shown, no change occurred in them with time, showing the efficiency of leakage proof under the adjusted Gap value.

6.4.1.3 Rubber Friction Measuring

After Leakage proof checking was finished, the friction between the upper pair of rings and the rubber edges was measured. In this apparatus, there are three kinds of rotating gear with final speed of Low (10 mm s^{-1}), Medium (32.3 cm s^{-1}) and High (2.24 m s^{-1}). In this test, the Low gear was selected. Figure 6.8 presents the measured rubber friction corresponding to the adjusted Gap Value at the shear speed of 10 mm s^{-1} of the Low Gear. As seen, the friction showed a transient reduction after the peak value (mobilized maximum static friction at Point Fp) due to the fact that friction was shifted to kinetic friction after Point Fp, and thereafter increased with progress of shear displacement, probably due to the abrasion of the covered Teflon and the daubed silicon grease with progress of shear displacement; finally, it became almost a constant. This measured rubber-friction with shear displacement was subtracted from the monitored shear strength for the correction of shear strength.

6.4.2 Test on Saturated Sand

6.4.2.1 Sample Setting

The samples were made by means of moist placement or dry deposition (Ishihara 1993), according to different test purposes. For moist placement method, de-aired water was first added to the oven-dried samples to make the initial water content rise up to 5 percent, and then the sand was stirred evenly. After that, the sample was placed into the shear box. To make the sample uniform, while packing, the sample was placed in a series of layers of 3 cm thickness, and then each layer was tamped. For dry deposition method, the oven-dried sample was fallen into the shear box freely by layers, and each layer was and was not tamped differently to make the initial density different.

6.4.2.2 Sample Saturation

Sample was saturated with help of carbon dioxide and de-aired water. After the sample was packed, CO_2 was then percolated through the sample to expel the air in the sample pores out, by flowing in through the lower drainage line very slowly, and discharging from the upper drainage line. Usually, this process took 4 to 12 hours, depending on the samples (for the sample used in this research, 4 hours are enough). After hours of percolation of CO_2, de-aired water was infiltrated into the sample through the lower drainage line to expel the CO_2 in the sample pores from the upper drainage line. This infiltration was kept very slow with help of a very small water head. To expel the CO_2 as completely as possible, this water saturation process was usually kept 24 hours or longer.

6.4.2.3 Saturation Checking

Saturation degree was checked by using B_D parameter. B_D is a parameter of saturation degree in the direct shear state, which was proposed by Sassa (1988), and formulated as:

$$B_D = \Delta u / \Delta \sigma \tag{6.1}$$

where Δu and $\Delta \sigma$ are the increment of pore pressure and normal stress, respectively. During checking, the sample was firstly consolidated under normal stress of 49 kPa (0.5 kgf cm^{-2}) in drained condition. Thereafter, a normal stress increment, $\Delta\sigma = 49$ kPa, was applied in undrained condition, and the resultant increment of excess pore pressure (Δu) was measured. And then, the saturation degree was checked by the ratio (B_D) of excess pore pressure increment and normal stress increment ($\Delta u/\Delta\sigma$). Because high saturation degree is necessary for acquiring correct monitoring data, in this study, all the tests were carried out with $B_D \geq 0.95$.

6.4.2.3 Sample Consolidation

All the samples were normally consolidated in this series tests under the pre-decided normal stress and shear stress. After the checking of B_D, the normal stress was decreased in undrained condition to a value (usually smaller than 49 kPa, due to the plasticity deformation nature of saturated sand) where the pore pressure decreased to zero, and then the upper drainage valve was switched to open. Thereafter, normal and then shear stresses were loaded slowly to the decided values. In the present research, all the tests were carried out under the same initial stress state, with normal stresses being 196 kPa and shear stress being 0.

6.4.2.4 Undrained Shearing

The studies by both Castro et al. (1982) and Sladen et al. (1985) have made it clear that there is no difference between the results of either load (stress)-controlled or strain-controlled tests. The torque-controlled method was selected to take more data between the start point of shearing and the point where peak shear strength was mobilized, and then to obtain a well-defined effective stress path until a steady state after a long shear displacement of tens of meters. To avoid the phenomenon of quick motion in contractive sand once the peak shear stress has been reached, which usually appears in stress-controlled triaxial compression tests, in this test, the Low gear (as mentioned above) was used for careful monitoring, providing that shear speed has no effects on the key results

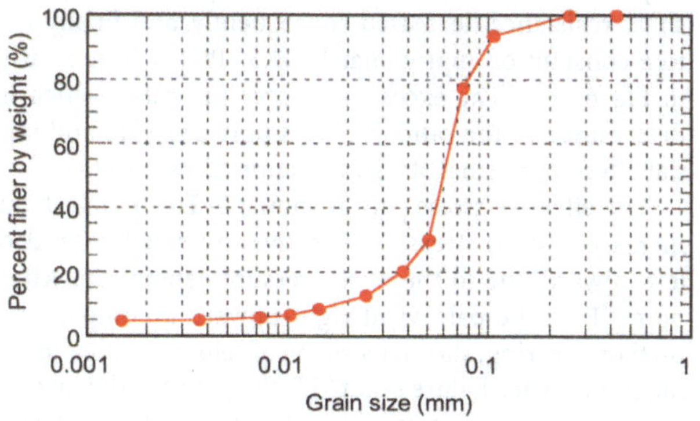

Fig. 6.9. Grain size distribution of silica sand no. 8 (S8)

relating to the undrained shear behavior (Hungr and Morgenstern 1984). After consolidation, undrained shear stress was subsequently applied at a loading rate of 0.098 kPa s^{-1} (0.001 kgf cm^{-2} s^{-1}). Transducers were scanned at an interval of 1 second before the peak shear strength; after that, the sampling rate was increased to 20 samples per second because of rapid shearing, then decreased again to 1 sample per second when the change in each parameter became not so quick (Note that a greater sampling rate on 200 samplesper second is used for seismic loading or other faster phenomena).

6.4.2.5 Sample Characteristics

In this test, silica sand no. 8 (S8) was selected as the sample. Silica sand is a kind of sand material for building use made from silica sandstone by grinding, comprised of subangular to angular quartz. The particles ranged from fine sand to silt sizes with a specific gravity of approximate 2.63. The maximum and minimum dry densities are found to be approximately 1.42 and 0.99 g cm^{-3}, respectively. The grain size distribution is presented in Fig. 6.9. As shown, the mean diameter is approximately 0.05 mm.

6.5 Undrained Shear Behavior on Sands

Since the trust of this note is on the introduction of the design of the undrained ring shear apparatus and its operating procedures, here only selected typical results from the tests on loose and dense sands are presented. More detailed discussion of the undrained behavior of sands in ring shear tests will be presented in a separate paper.

6.5.1 Undrained Shear Test on Loose Sand

The results of one test on loose silica sand S8, showing the same effective stress path as that of liquefaction behavior reported for many undrained triaxial tests on loose sands, are presented in Fig. 6.10d. The specimen for this test was set by means of moist placement and consolidated under the normal stress of 196 kPa. After consolidation, the relative density reached 63.3% (1.15 in void ratio).

Figure 6.10a shows the variation of normal stress and pore pressure in relation to shear displacement, and Fig. 6.10b superimposes the monitored shear resistance and rubber friction (shown in Fig. 6.8) as well as the corrected shear resistance of sand against shear displacement. Figure 6.10c plots the time series data of normal stress, corrected shear resistance and pore pressure. In Figs. 6.10a and 6.10b, to make a clear view on the generation of pore pressure accompanying the shear displacement in the initial shearing period, a logarithmic abscissa of shear displacement within the range of 0.1 m was taken, and thereafter, linear abscissa was used to show that the test had been sheared to a steady state (point SSP in Fig. 6.10a). It is seen that some pore pressure was built-up before the peak shear strength (point F in Fig. 6.10b), while after failure, pore pressure showed a sharp increase, and shear resistance underwent a quick reduction in terms of time (dropped from F to P in Fig. 6.10c). This period is usually known as the collapse period, mainly due to the failure of metastable structure. After point P, accompanying the further shearing, pore pressure built up gradually and then shear resistance decreased slowly as the subsequence.

Figure 6.10d presents the effective stress path and failure line. The residual failure line (R.F.L.) was measured after the undrained shearing test; after the undrained shearing was completed (reached at point SSP in Figs. 6.10a and 6.10d), the upper drainage line was switched to drained condition so that the generated pore pressure could dissipate, while the low part of ring shear apparatus was kept rotating at a constant speed. The stress shifted from SSP to RS1, when the generated pore pressure dissipated completely. Then, the loaded normal stress was reduced at a very small rate under drained condition while the shear resistance was measured. The stress moved from RS1 to RS2. The line connecting RS1 and RS2 shows the residual failure line (R.F.L.) of this sample. As shown in Fig. 6.10d, soon after the start of shearing, with increasing shear stress, stress path moved towards but did not reach the failure line until the final point, i.e., the steady state. This effective stress path showed the same changing tendency as those in undrained triaxial tests on loose sand.

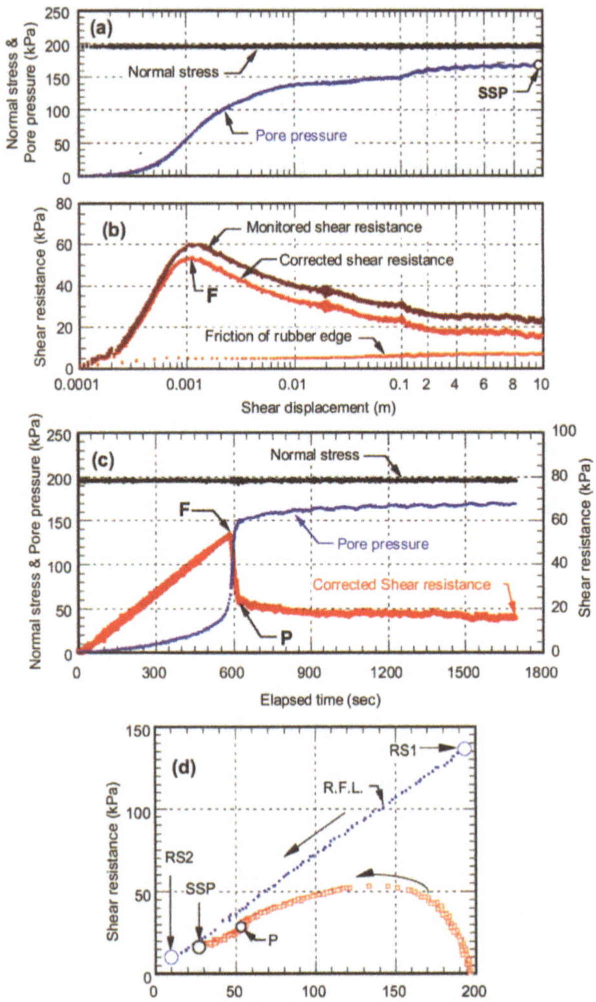

Fig. 6.10. Ring shear test on loose sand ($B_D = 0.99$, $D_r = 63.3\%$, $\sigma = 196$ kPa). **a** Variation of normal stress and pore pressure with shear displacement; **b** monitored shear resistance, rubber edge friction and corrected shear resistance against shear displacement; **c** time series data of normal stress, pore pressure and corrected shear resistance; **d** effective stress path

6.5.2 Undrained Shear Test on Dense Sand

Figure 6.11 shows the results of a test on dense sand. This sample was made by means of dry deposition with heavy tamping. After saturation and normal consolidation, the sample was sheared in undrained condition to a large displacement of approximate 73 m. Figure 6.11a plots the normal stress and pore pressure against shear displacement, while Fig. 6.11b shows the variation of the monitored shear strength, rubber friction, and the corrected shear resistance of sand in relation to shear displacement. Note that the rubber friction shown in Fig. 6.8 was 7.3 kPa at the terminal shear displacement of 46 m, and was treated as the same as 7.3 kPa until the final shear displacement of 73 m in this test. Figure 6.11c presents the time series data of these components. As shown in Fig. 6.11a, in the initial period after undrained shear stress was applied, with increasing shear displacement, pore pressure built up gradually. However, after point "PT" (Phase Transformation), pore pressure decreased due to the dilatancy of dense sand. After the peak shear strength was reached (Point "F" in Figs. 6.11b and 6.11c), sample failed, and thereafter, pore pressure built up with shear displacement, finally reached 135 kPa approximately. The shear resistance decreased consequently, and finally fell to a constant of approximately 53.5 kPa (as seen clearly in Fig. 6.11c). The excess pore pressure ratio, which is determined as the ratio of excess pore pressure and initial effective normal stress, was approximately 0.69.

The effective stress path is shown in Fig. 6.11d. Upon increase of shear stress, the effective stress path extended left-upward due to the pore pressure generation. After point "PT", the path went right-upward accompanying further shearing, showed a shape of "elbow" with a turning point. After failure point "F", the path fell downward until a small shear stress along the residual failure line.

The stress path before the point "F" in Fig. 6.11d was described as same as typical dilative behavior of the undrained shear behavior in triaxial apparatus (Castro 1969). Due to the limitation of triaxial apparatus in shear displacement, the behavior after "F" was not usually obtained. This phenomenon of large stress reduction along the failure line (from F to SSP) could not be reproduced well before the development of undrained ring shear apparatus by Sassa and his co-workers, and it is termed as "sliding surface liquefaction", due to the grain crushing with progress of shear displacement (Sassa 1996; Sassa et al. 1996).

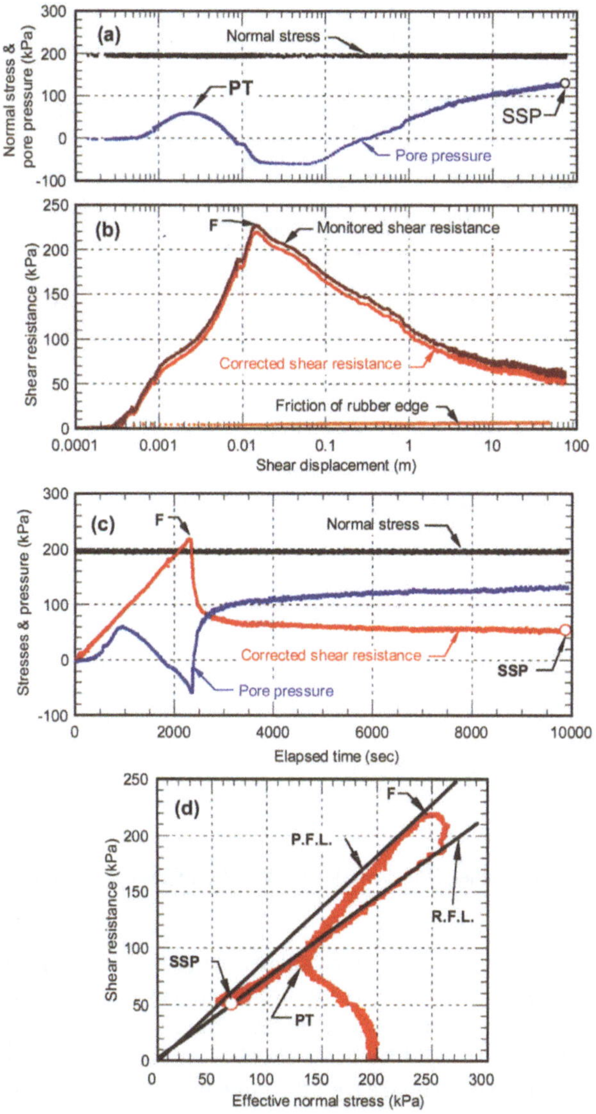

Fig. 6.11. Ring shear test on dense sand ($B_D = 0.99$, $D_r = 95.2\%$, $\sigma = 196$ kPa.). **a** Variation of normal stress and pore pressure with shear displacement; **b** monitored shear resistance, rubber edge friction and corrected shear resistance against shear displacement; **c** time series data of normal stress, pore pressure and corrected shear resistance; **d** effective stress path

6.6 Geotechnical Simulation of Earthquake-induced Landslides

Figure 6.12a shows the stress conditions in a slope, where the weight of the soil column (W) and the dynamic stress (which can be expressed by kW) are acting along a potential sliding surface. The stress path due to dynamic stress loading is expressed by TSP in terms of total-stress and ESP in terms of effective-stress in Fig. 6.12b. TSP is determined automatically, but ESP depends on excess pore-pressure generation during stress loading, as well as during motion. Generally speaking, dynamic stress on slopes is not limited to seismic stress and the direction of seismic stress is not constant in the natural condition. So the loading angle (α) and the seismic coefficient (k) are time functions in nature.

Earthquake-induced landslides often move rapidly, have long paths of travel, and cause catastrophic disasters. For these reasons, a project called APERIF (Areal Prediction of Earthquake and Rainfall Induced Rapid and Long-traveling Flow Phenomena) was launched by the Special Coordinating Fund for Science and Technology of the Ministry of Education, Culture, Sports, Science and Technology (MEXT) of Japan. In 2002, this project was approved as one of International Programme on Landslides (IPL M101-APERITIF Project of the International Consortium on Landslides).

As a part of this project, the upper slope of the Nikawa landslide site, at which 34 people were killed as a result of the 1995 Hyogo-Ken Nanbu earthquake (Sassa 1996; Sassa et al. 1996), was investigated to determine whether or not this slope could undergo retrogressive rapid landslide activity if a possible similar earthquake were to occur. Figure 6.13 is a geological cross section through the slope as estimated by geological drilling. Within the slope, we found a geological contact that can be a potential sliding surface. The angle and shape of the slope are very similar to those of the 1995 Nikawa landslide. There is a granitic sand layer above and a gravel layer below the potential sliding surface. The gravel layer includes silt with low permeability. Samples were taken from both layers and laboratory seismic simulation tests were performed using the ring shear apparatus. The details can be found in Sassa (2002) and Sassa et al. (2003). During the tests, the applied seismic loadings acting on the potential sliding surface during the earthquake were synthesized by using the seismic waves that were recorded at the Japan Railway Takarazuka station (7 km from the earthquake fault) during the Hyogo-Ken Nanbu earthquake. This synthesis took into account attenuation laws for horizontal peak acceleration proposed by Fukushima and Tanaka (1990) and vertical peak acceleration proposed by Ambraseys and Bommer (1991), as well as the amplification of seis-

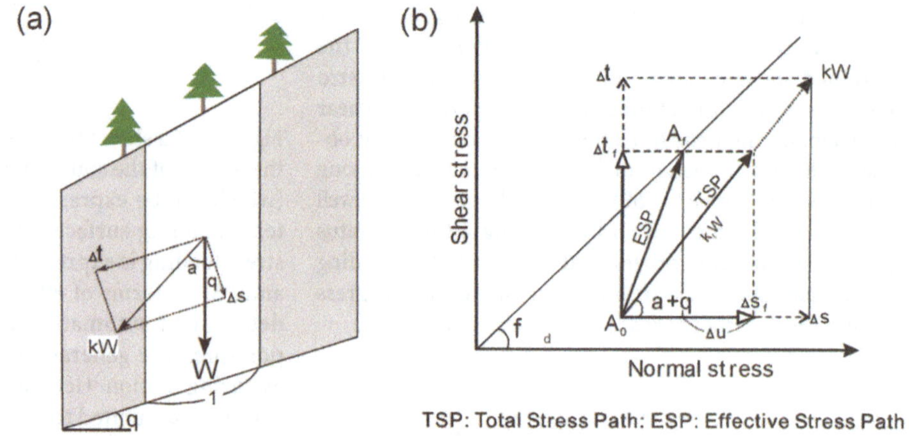

Fig. 6.12. Stress conditions in a slope and in the ring shear apparatus

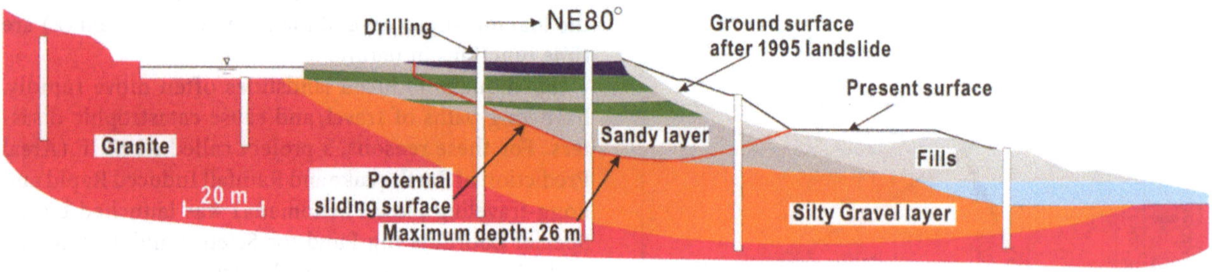

Fig. 6.13. Cross section through the Nikawa slope

mic waves (usually 1.4 times) when these waves transfer from hard rock to a weak soil layer (Fukushima and Tanaka 1990). The synthesized seismic loadings are presented in Fig. 6.14.

The seismic loadings shown in Fig. 6.14 were applied to samples taken from the granitic sand layer and from the silty gravel layer. These samples were placed and normally consolidated in the ring shear apparatus. The results of tests on the sample of the granitic sand layer under undrained conditions are presented in Fig. 6.15. The monitored test results show that the normal stress expected in Fig. 6.14 was approximately reproduced inside the apparatus, while the mobilized shear resistance was less than the given shear stress because the shear stress reached the failure line and failed there. In other words, failure occurred. Shear displacement began during the main shock and increased its speed during the post-failure shear process. The total stress path (TSP) is shown in blue, its initial stress corresponding to the stress acting on the potential sliding surface at a depth of 26 m with an inclination of 20 degrees. The effective-stress path (ESP) is shown in red. The initial difference between TSP and ESP corresponds to the ground-water level that is 16 m above the potential sliding surface. During seismic loading, the stress path deviated considerably because pore-pressure monitoring did not always follow the rapid loading. However, it was found that the ESP reached the failure line, and then reduced along the failure line until it reached a certain low steady-state stress. The mobilized

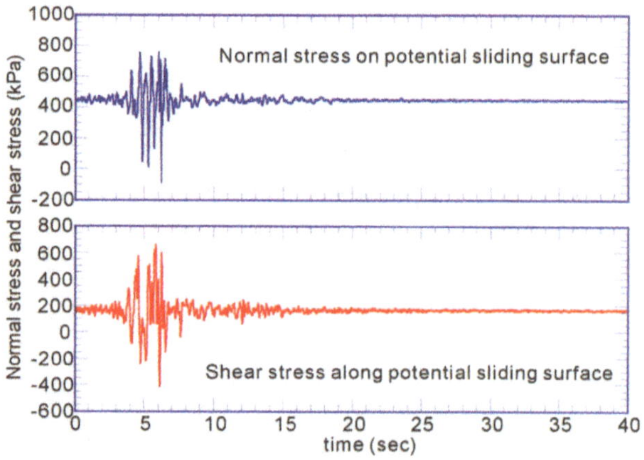

Fig. 6.14. Synthesized seismic loadings on the inferred potential sliding surface

apparent friction angle calculated from the arc-tangent of the mobilized shear resistance divided by the total normal stress was about 3.5 degrees. Therefore, it was estimated that a rapid landslide could be initiated retrogressively if an earthquake with the same magnitude and seismic waveform as the 1995 Hyogo-Ken Nanbu earthquake were to occur in the future under the undrained condition.

A key assumption in this test is that a sliding surface can be formed passing through a normally consolidated part, or a well-softened part similar to a normally consolidated state. Observation of cores of five drillings in this site showed that the sand and silty gravel layers were

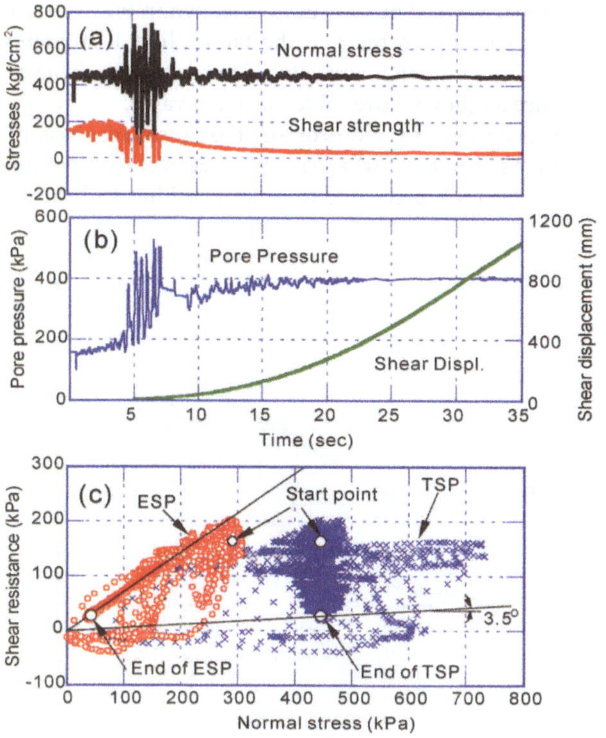

Fig. 6.15. Undrained cyclic tests on saturated sand layer ($B_D = 0.99$, initial dry density = 15.2 kN m^{-3}). **a** Time series data for normal stress and shear stress; **b** time series data for pore-water pressure and shear displacement; **c** stress path

disturbed by tectonic, seismic, and possibly gravity stresses in the past. Therefore, some parts of the slope are disturbed, namely well-softened, and some parts are over-consolidated and stiff. While conducting slope stability analysis in the course of planning for works to prevent a possible sliding surface liquefaction phenomena in this site, it would be dangerous to assume that the sliding surface will pass into the stiff part of soil layer. A more prudent assumption is that the potential sliding surface will be formed in the normally consolidated layer. This assumption was adopted in the present study.

In the APERITIF project, a series of naturally drained tests was also conducted, where the upper valve of the shear box was kept open during the test. Because the permeability of the sands and gravels was not small, generated pore-pressure could be dissipated partly during the short time period of the main shock. This type of tests could be more practical. A mitigation measure for the prevention of landslides by lowering the groundwater to cancel out the pore-pressure generation during earthquake loading has been proposed (Sassa et al. 2003). Quantitative estimation of pore-pressure generation during earthquakes is necessary for reliable prediction of earthquake-induced landslides. In this regard, this series of ring shear tests provided an effective approach for the prediction of earthquake-induced landslides and design of the remedial measures.

Fig. 6.16 Aerial oblique view of the 2003 Minamata debris flow

6.7 Slide-triggered Debris Flow

On 20 July 2003, a landslide occurred in an andesitic weathered lava layer on a mountain slope of 31–32 degrees in Minamata City, Kumamoto Prefecture, Kyushu Island, Japan. It was triggered by a heavy rainstorm with 314 mm total rainfall and a maximum rate of rainfall of 91 mm hr^{-1}. The slide mass entered a torrent, where it was transformed into a debris flow that struck a village along the torrent, destroying 15 houses, killing 15 people, and injuring an additional six people. Figure 6.16 presents a view of the debris flow. The initial slide can be seen at the head of the debris flow. Apparently, this debris flow was triggered by the slide, and the landslide mass flowed downstream along the torrent, increasing its volume by entraining material from the channel and weathered surface soils of the mountain slopes on both sides of the channel. Figure 6.17 shows the central section through the initial slide surveyed by a non-mirror total station, and Fig. 6.18 is a photo presenting the sampling point of the weathered andesitic lava in the source area. Based on a topographic survey made after the landslide occurred, the initial slide was estimated to have occurred along a failure surface with an inclination of 26.5 degrees and depth of approximately 10–12 m.

Two samples were collected, one from andesitic lava, one from tuff breccia. The andesitic lava sample was taken under the sliding surface near the head scarp at the mark of Sample A in Fig. 6.17. The tuff breccia sample was collected from the mountain slope near the channel in the flank of the initial landslide at the mark of Sample B (projected to the central cross-section) in Fig. 6.17. This initial landslide occurrence was geotechnically simulated using the DPRI-5 ring shear apparatus. The initial stresses on the sliding surface were reproduced in the apparatus; then, the pore pressure was gradually increased simulating the rise of ground-water level during rainfall. In natural slopes, the ground-water rise would not be rapid; thus, the undrained loading condition was not used. To simulate drained ground-water conditions, water pressure supplied to the shear box through the upper drainage valve was gradually increased. Thus, water pressure was controlled, but the water was free to move through the upper valve. Therefore, the sample was subjected to a natural drained condition. If pore pressure was generated in the shear zone during loading, it drained naturally through the upper valve. The test result is shown in Fig. 6.19. After the stress path reached the failure line, it suddenly dropped to a much lower value. Thereafter, the shear resistance slightly recovered to a certain value.

This action of rapid drop of shear resistance was interpreted as follows: when shear failure occurred, volume reduction took place due to grain crushing and the re-

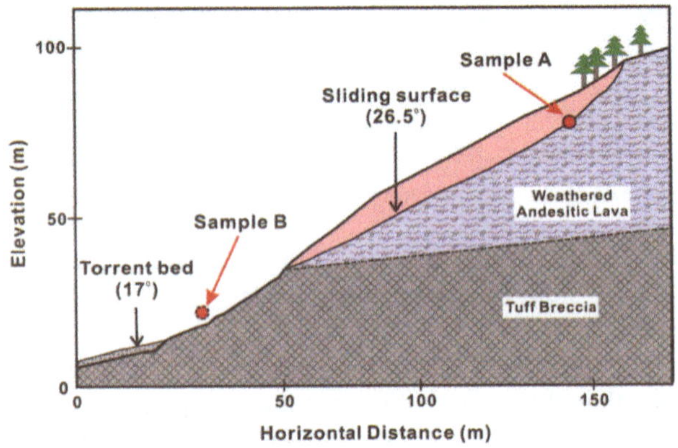

Fig. 6.17. Central section through the Minamata landslide in its original state

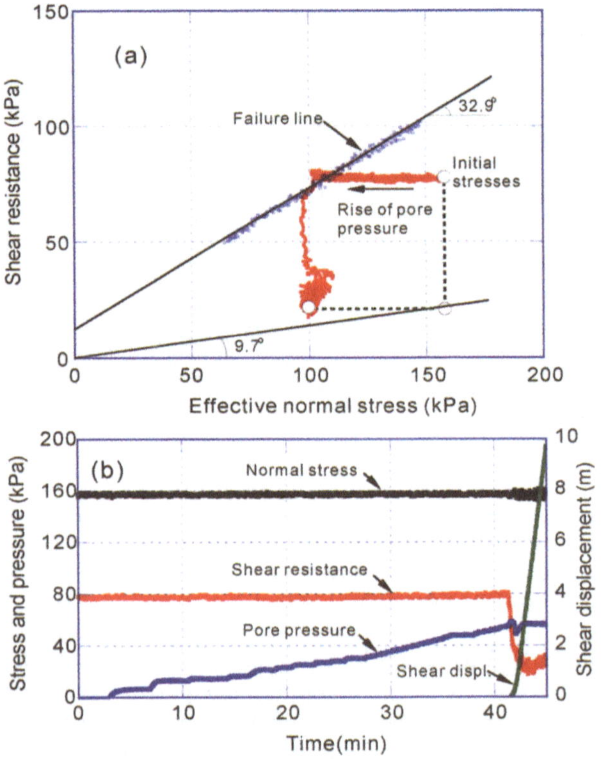

Fig. 6.19. Results of tests to simulate the initiation of the Minamata landslide ($B_D = 0.86$)

Fig. 6.18. Sampling of weathered andesitic lava in the Minamata landslide

sulting failure of the soil structure; this caused rapid excess pore-pressure generation. However, excess pore pressure was not monitored by the pore-pressure transducer because the upper valve was open and the transducer was not located at the shear zone, as illustrated in Fig. 6.5. Therefore, the stress path deviated from the failure line and dropped vertically.

The high pore pressure in the shear zone dissipated through the upper valve, but some excess pore pressure remained as the combined effect of pore-pressure generation within the shear zone and pore-pressure dissipation from the shear zone. The pore pressure generation speed was likely to have decreased from the immediate post-failure condition, while at the same time, the pore pressure dissipation speed was maintained. Therefore, the shear resistance recovered somewhat and settled to a certain value. The result gave an apparent friction angle of 9.7 degrees.

This test is called a naturally drained test because generated excess pore pressure is naturally drained from the shear zone to the upper valve. After the naturally drained test, the sample was again consolidated by dissipating the generated pore pressure. Then, slow shearing was applied. After reaching the failure line, the normal stress was decreased very slowly to maintain the drained condition, continuing slow movement without generating any pore-water pressure. This test determined the status of the failure line during motion, as shown by the blue line in Fig. 6.19. This test gave 32.9 degrees as the friction angle during motion.

These test results show that the andesitic lava deposit was subjected to high excess pore-pressure generation due to post-failure shearing and a low apparent friction angle of 9.7 degrees was mobilized. This means that the Minamata slide rapidly reached the torrent bed. The gradient of the torrent bed at the toe of the slope was 17 degrees; so, the landslide necessarily continued to move downstream and changed into a debris flow.

Loading by the failed slide mass onto the pre-existing torrent deposits was first modeled by Sassa et al. (1997) as shown in Fig. 6.20. The slide mass moved down the slope (I), and applied load onto the torrent deposits at the foot of the slope (II). Because a surface water stream or subsurface flow existed and some of the deposits were saturated, the torrent deposit was sheared by undrained loading and transported downstream together with the sliding mass (III). Let us consider a column of unit width, which is a part of the torrent deposit. In the position (I) of the sliding mass, the weight of the column (W_0) was in effect. When the sliding mass rode on to the torrent deposit (II) with a certain velocity, it provided dynamic loading of the column. Here, we assume that the applied stress on the torrent deposits was as the sum of the static stress, ΔW, (load due to the weight of the sliding mass) and the dynamic (impact) stress, Fd, working in the direction of motion of the sliding mass.

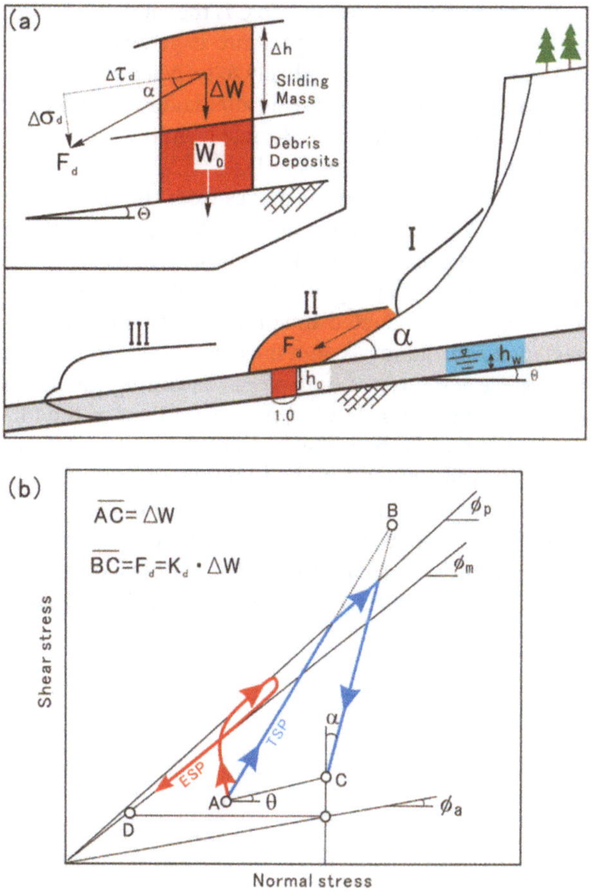

Fig. 6.20. Model of the landslide triggered debris flow (Sassa et al. 1997). **a** Illustration of the model; **b** stress path of the torrent deposit during loading. α: angle of thrust between the slope and the torrent bed; F_d: dynamic stress; k_d: dynamic coefficient ($F_D/\Delta W$)

The stress working on the bottom of the soil column is presented in Fig. 6.20b. The initial stress is expressed by the point "A", which corresponds to the position (I) of the sliding mass in Fig. 6.20a. If no excess pore pressure is generated during loading, the stress point moves to point (C) by adding the static stress (ΔW) to the initial stress. In addition, by adding the dynamic stress (F_d) to the static stress, the total stress moves to point (B). Therefore, the stress path in the actual field case tends to move from point A to point B. However, when the stress path reaches the failure line, it moves along the failure line as seen in Fig. 6.20b, because the stress path cannot exceed the failure line. At the point at which the dynamic stress reduces to zero, the total stress moves back to the stress point (C), namely the sum of W_0 and ΔW. Denoting the angle of thrust at collision with the torrent deposit as α and the dynamic stress as F_d, using a dynamic coefficient $k_d = (F_d/\Delta W)$, the dynamic shear stress and normal stress are expressed as:

$$F_d \cos\alpha = \tau_d, \quad F_d \sin\alpha = \sigma_d \tag{6.2}$$

The stress path from A to B to C is the total-stress path in the case where no pore-pressure is generated. However, excess pore pressure is likely to be generated during loading and also during shearing after failure. In this case, the effective-stress path will deviate from the total-stress path as a curved line from A to D.

When the landslide mass moves from the steep slope to a gentle slope, the angle α in Fig. 6.20 is great, but when the landslide mass (i.e., the debris flow) travels along the torrent (as shown in Fig. 6.21), the angle α is zero. Another test was conducted to simulate the landslide (debris) masses as it moved onto the torrent deposits and/or the surface soil layer of both slopes assuming the depth of moving mass was about 10 m, the gradient of the torrent bed was 15 degrees, the depth of the torrent deposit or surface soils was 2–4 m, and the dynamic coefficient was 0.9. Because of such rapid loading by the fast-moving slide mass (more than 10 m s^{-1}), the test was carried out under undrained conditions. The sliding surface was formed inside the torrent deposits which were composed of andesitic lava or tuff breccia. The surface soil layers of mountain slopes near the upper and middle part of torrent was mainly tuff breccia, though the lower part of the torrent was composed of non-volcanic sedimentary rocks. Therefore, two undrained loading tests were conducted for deposits of tuff breccia and those of andesitic lava.

The test result of Fig. 6.22 shows the case for tuff breccia, in which the torrent deposit and/or the surface soils were sheared and moved together with the original slide mass. Only 18.5 kPa was necessary as additional shear stress to cause shear failure and the mobilized apparent friction angle was only 1.9 degrees in this rapid and undrained loading condition as estimated from Fig. 6.22b. The test result of Fig. 6.23 shows the case for andesitic lava. This is stronger, the additional necessary shear stress was 45.5 kPa and the mobilized apparent friction angle at the steady state was 3.4 degrees. Both cases suggested that such saturated deposits were scraped and included in the moving mass. However, if the material was not fully saturated, it probably would not be scraped as the case

Fig. 6.21. Illustration of moving landslide mass along a torrent

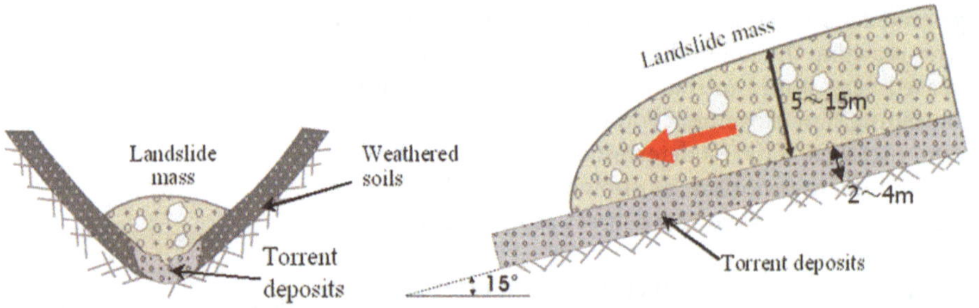

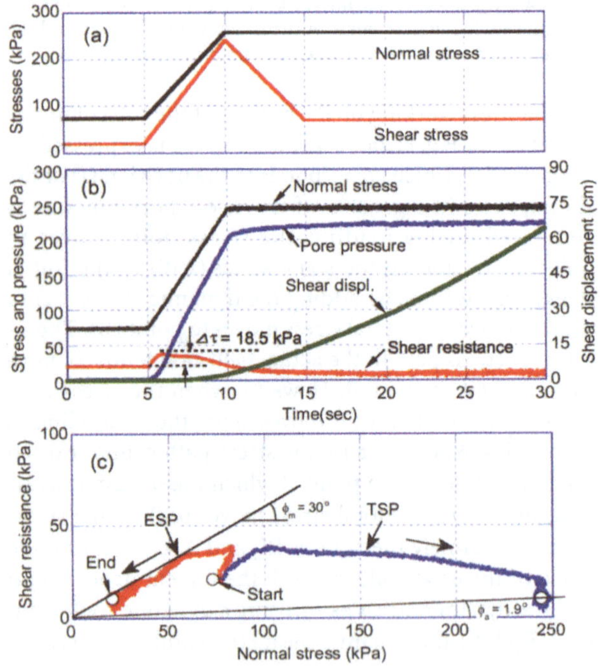

Fig. 6.22. Results of a test to simulate undrained loading of the tuff breccia deposits in the Minamata landslide ($B_D = 0.89$)

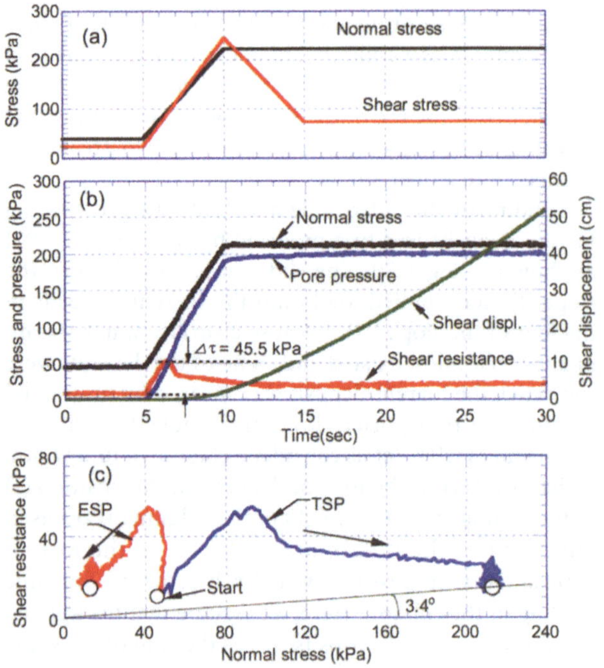

Fig. 6.23. Results of a test to simulate undrained loading of the weathered andesitic lava in the Minamata landslide ($B_D = 0.97$)

study in the Kameyama landslide, Hiroshima, Japan (Wang, Sassa and Fukuoka 2003).

To examine and interpret the actual phenomena, additional investigation and testing are necessary. However, this test has shown that this apparatus and its application provide an effective tool to study the mechanism of slide-triggered debris flows and the increase of landside mass by material entrainment during the flow.

6.8 Summary and Conclusions

This paper has presented the principles of design and construction of a series of undrained ring shear apparatus and its testing method developed and improved by Professor Sassa and his colleagues at DPRI, Kyoto University. These test machines enable three kinds of shear control, namely, the torque-, speed- or shear-displacement-controlled methods. But natural phenomena are under the stress control condition. Therefore, stress control tests to reproduce natural stress state during earthquakes and rainfalls were introduced. The controlling signals can be either edited signals of monotonic/cyclic loading or seismic records of earthquakes.

Introduction of a rubber edge and the gap controlling system enabled the system to be leakproof even when the sample was sheared at high speed (i.e., at the maximum rotating speed of 224 cm s^{-1} for DPRI-6 and 300 cm s^{-1} for DPRI-7). This leakproof system enabled examination of the undrained shear behavior of soils with high mobility at essentially limitless shear-displacement levels.

The assessment of landslide risk of the Nikawa slope area was performed on the basis of field investigation and laboratory ring shear simulation tests. Field investigation drilling at Nikawa revealed that shear failure could possibly occur along the boundary between a granitic gravel stratum and a layer of sand. The results of an undrained ring shear test on a sample taken from the granitic sand layer were presented in this paper. The results showed that sliding surface liquefaction could be triggered by an earthquake similar to the 1995 Kobe earthquake.

Application of the ring shear test to soils from the Minamata debris flow simulated a landslide triggered by the gradual increase of ground-water level during heavy rainfall. Sliding surface liquefaction resulted with the progress of shearing even in the naturally drained condition corresponding to the gradual increase of pore water pressure in the debris-flow materials. An undrained ring shear test was carried out simulating the undrained loading process that takes place in the pre-existing torrent deposits and the weathered surface soils on the valley slopes. This revealed that only a small increment in the shear stress due to the impacting by the displaced soil mass could have caused shear failure of these deposits/soils with the generation of high pore water pressure. In addition, the volume of the landslide mass was increased by entrainment of these deposits/soils. A new ring shear apparatus (DPRI-7) with a transparent shear box was developed in 2003. It enabled to observe the shear zone during the initiation and post-failure motion of landslides. DPRI-7 and the test results are introduced in other papers (Sassa et al. 2004; Fukuoka et al. 2006; and in this volume).

Acknowledgments

The first attempt at design and fabrication of an undrained stress-controlled dynamic loading ring shear apparatus was conducted with the support of Scientific-Grant-in-Aid (No. 03556021) of the Ministry of Education, Science, Culture and Sport of Japan in 1992 (Sassa 1994). This apparatus was designated as DPRI-3. The system and the testing procedures were improved during Ph.D. studies by Dr. Zieaoddin Shoaei, and by others at DPRI, Kyoto University. DPRI-5 and DPRI-6 were developed by means of financial support from the Ministry of Education, Science, Culture and Sports of Japan, for earthquake-disaster mitigation research after the Hyogo-Ken Nanbu earthquake of Japan in 1995, Japan. Apparatus DPRI-7 was developed with support from the project, "Areal Prediction of Earthquake and Rain Induced Rapid and Long-traveling Flow Phenomena" (APERIF), of the Special Coordinating Fund for Promoting Science and Technology of the Ministry of Education, Culture, Sports, Science and Technology of Japan (MEXT).

APERIF project was approved as a project of the International Programme on Landslides (IPL) by the International Consortium on Landslides (ICL) as IPL M-101 APERITIF project. These supports from MEXT and ICL are properly appreciated. Thanks also go to all colleagues of the Research Centre on Landslides of the DPRI of Kyoto University for their cooperation on this study.

References

Alarcon-Guzman A, Leonards GA, Chameau JL (1988) Undrained monotonic and cyclic strength of sands. J Geotech Eng-Asce 114(10):1089–1109

Ambraseys NN, Bommer JJ (1991) The attenuation of ground accelerations in Europe. Earthquake Eng Struc 20:179–1202

Bishop AW, Green GE, Garga VK, Andersen A, Brown JD (1971) A new ring shear apparatus and its application to the measurement of residual strength. Géotechnique 21(1):273–328

Bromhead EN (1979) A simple ring shear apparatus. Ground Eng 12(5):40–44

Bromhead EN (1986) The stability of slopes. Surrey University Press, London, 373 p

Castro G (1969) Liquefaction of sands. Ph.D. Thesis, Harvard University, Cambridge, Massachusetts

Castro G, Enos JL, France JW, Poulos SJ (1982) Liquefaction induced by cyclic loading. Report to National Science Foundation, Washington, DC, No. NSF/CEE-82018

Fukuoka H, Sassa K, Wang G, Sasaki R (2006) Observation of shear zone development in ring-shear apparatus with a transparent shear box. Landslides 3(3):239–251

Fukuoka H, Sassa K, Wang G (2007) Influence of shear speed and normal stress on the shear behavior and shear zone structure of granular materials in naturally drained ring shear tests. Landslides (in print) published online, DOI:10.1007/s10346-006-0053-0

Fukushima Y, Tanaka T (1990) A new attenuation relation for peak horizontal acceleration of strong earthquake ground motion in Japan. Bull Seismol Soc Am 80(4):757–83

Garga VK, Sendano JI (2002) Steady state strength of sands in a constant volume ring shear apparatus. Geotech Test J 25(4):414–421

Gibo S (1994) Ring shear apparatus for measuring residual strengths and its measurement accuracy. Journal of Japan Landslide Society 31(3):24–30

Hungr O, Morgenstern NR (1984) High velocity ring shear tests on sand. Géotechnique 34(3):415–421

Hvorslev MJ (1939) Torsion shear tests and their place in the determination of the shearing resistance of soils. Proc Am Soc Test Mater 39:999–1022

Ishihara K (1993) Liquefaction and flow failure during earthquakes. Géotechnique 43(3):349–451

Kramer KL, Seed HB (1988) Initiation of soil liquefaction under static loading conditions. J Geotech Eng-ASCE 114:412–430

Okada Y, Sassa K, Fukuoka H (2000) Liquefaction and the steady state of weathered granitic sands obtained by undrained ring shear tests: a fundamental study on the mechanism of liquidized landslides. J Natural Disaster Sci 22(2):75–85

Poulos SJ (1981) The steady state of deformation. J Geotech Eng-ASCE 107 (GT5):553–562

Poulos SJ, Castro G, France JW (1985) Liquefaction evaluation procedure. J Geotech Eng-ASCE 111(6):772–792

Rouse WC (1984) Flowslides. In: Brunsden D, Prior DB (eds) Slope instability. John Wiley & Son, New York, pp 491–522

Sassa K (1984) The mechanism starting liquefied landslides and debris flows. In: Proceedings of 4th International Symposium on Landslides, Toronto, June, 2, pp 349–354

Sassa K (1988) Geotechnical model for the motion of landslides. In: Special Lecture of 5th International Symposium on Landslides, "Landslides", 10–15 July, 1, pp 37–55

Sassa K (1994) Development of a new cyclic loading ring shear apparatus to study earthquake-induced-landslides. Report for Grain-in-Aid for Developmental Scientific Research by the Ministry of Education, Science and Culture, Japan (Project No. 03556021), 106 p

Sassa K (1995) Access to the dynamics of landslides during earthquakes by a new cyclic loading high-speed ring shear apparatus (keynote paper). In: 6th International Symposium on Landslides, "Landslides", A.A. Balkema. Christchurch, 10–14 February, 3, pp 1919–1937

Sassa K (1996) Prediction of earthquake induced landslides. In: Proceedings of 7th International Symposium on Landslides, A.A. Balkema, Trondheim, 17-21 June, 1, 115–132

Sassa K (1997) A new intelligent-type dynamic loading ring shear apparatus. Landslide News 10:33

Sassa K (1998) Recent urban landslide disasters in Japan and their mechanisms. In: Proceedings 2nd International Conference on Environmental Management, "Environmental Management", 1, pp 47–58

Sassa K (2000) Mechanism of flows in granular soils. In: Proceedings of the International Conference of Geotechnical and Geological Engineering, GEOENG2000, Melbourne, 1, pp 1671–1702

Sassa K (2002) Study on the mechanism of earthquake and rainfall induced rapid flow phenomena- Disasters and their mitigation in large-scale cities. In: Proceedings of the Symposium on Aerial Prediction of Earthquake and Rainfall Induced Flow Phenomena (APERIF), "New century of urban area landslide disaster mitigation," Tokyo. 31 August – 1 September 2002, pp 7–33 (in Japanese)

Sassa K (2004) Undrained dynamic-loading ring-shear apparatus and its application to landslide dynamics. Landslides 1(1): 7–19

Sassa K, Fukuoka H, Scarascia-Mugnozza G, Evans S (1996) Earthquake-induced-landslides: distribution, motion and mechanisms. Special issue for the great Hanshin Earthquake Disasters, Soils and Foundations 53–64

Sassa K, Fukuoka H, Wang FW (1997) Mechanism and risk assessment of landslide-triggered-debris flows: lesson from the 1996.12.6 Otari debris flow disaster, Nagano, Japan. In: Cruden DM, Fell R (eds) Landslide risk assessment. Proceedings of the International workshop on landslide risk assessment. Honolulu, 19–21 February, pp 347–356

Sassa K, Fukuoka H, Wang FW (1998) Mechanism of rapid long run-out motion in the Sumikawa reactivated landslide in Akita Prefecture and the Harihara landslide-debris flow in Kagoshima Prefecture, 1997, Japan. J Jpn Landslide Soc 35(2):29–37 (in Japanese)

Sassa K, Wang FW, Wang GH (1999) Mechanism of rapid landslides by the ring shear tests – on the long run-out rapid landslide in Hiegaeshi, Saigo Village, Fukushima Prefecture. In: Proceedings Symposium on1998 Slope Disasters and Sediment Disasters – Characteristics and Actual Conditions, Japan Landslide Society, pp 38–49 (in Japanese)

Sassa K, Wang G, Fukuoka H (2003a) Assessment of earthquake-induced catastrophic landslides in urban areas and their prevention planning. In: Proceedings of International Conference on Slope Engineering, December 2003, Hong Kong, 1, pp 26–49

Sassa K, Wang G, Fukuoka H (2003b) Performing undrained shear tests on saturated sands in a new intelligent type of ring shear apparatus. Geotech Test J 26(3):257–265

Savage SB, Sayed M (1984) Stresses developed in dry cohesionless granular materials sheared in an annular shear cell. J Fluid Mech 142:391–430

Shimamoto T, Tsutsumi A (1994) A new rotary-shear high speed frictional testing machine: its basic design and scope of research. Structural Geology 39:65–78 (in Japanese)

Shoaei Z, Sassa K (1994) Basic study on the shear behavior of landslides during earthquakes – excess pore pressure in the undrained cyclic loading ring shear tests. Bull Disaster Prev Res Inst Kyoto Univ 44(1):1–43

Sladen JA, D'Hollander, RD, Krahn J (1985) The liquefaction of sands, a collapse surface approach. Can Geotech J 22:564–578

Tika TM (1989) The effect of rate of shear on the residual strength of soil. PhD thesis, University of London (Imperial College of Science and Technology), 494 p

Tika TE, Hutchinson JN (1999) Ring shear tests on soil from the Vaiont landslide slip surface. Géotechnique 49(1):59–74

Vaid YP, Chern JC (1985) Cyclic and monotonic undrained response of saturated sands. Khosla V (ed) Advances in the art of testing soils under cyclic conditions. ASCE, pp 120–147

Vankov D, Sassa K (1999) Mechanism of earthquake-induced landslides on almost flat slopes studied with a ring shear apparatus. J Natural Disaster Sci 21(1):23–35

Voight, B (ed) (1978) Rockslides and avalanches 1: natural phenomena. Elsevier, Amsterdam

Voight, B (ed) (1979) Rockslides and avalanches 2: engineering sites. Elsevier, Amsterdam

Wang G, Sassa K, Fukuoka H (2003) Downslope volume enlargement of a debris slide-debris flow in the 1999 Hiroshima, Japan, rainstorm. Eng Geol 69:309–330

Zhang S, Tullis TE, Scruggs VJ (1999) Permeability anisotropy and pressure dependency of permeability in experimentally sheared gouge materials. J Struct Geol 21(7):795–806

Shear Behavior and Shear Zone Structure of Granular Materials in Naturally Drained Ring Shear Tests

Hiroshi Fukuoka* · Kyoji Sassa · Gonghui Wang

Abstract. "Sliding Surface Liquefaction" is a process causing strength loss and consequent rapid motion and long runout of certain landslides. Using a new ring shear apparatus with a transparent shear-box and digital video camera system, shear-speed-controlled tests were conducted on mixed grains (mixture of three different sizes of sand and gravel) and mixed beads to study shear behavior and shear zone development process under the naturally drained condition in which pore pressure is allowed to dissipate through the opened upper drainage valve during shearing. Higher excess pore water pressure and lower minimum apparent friction were observed in the tests where grain crushing was more extensive under higher normal stress and higher shear speed. Along with the diffusion of silty water generated by grain crushing, smaller particles were transported upward and downward from the shear zone. Concentration of larger grains to the central and upper part of the shear zone was confirmed by means of visual observation together with grain size analysis of sliced samples from several layers after the test. On the other hand, smaller particles were accumulated mostly below the layer where larger grains were accumulated. The reason why larger grains were accumulated into the shear zone may be interpreted as follows: grains under shearing are also subjected to vertical movement, the penetration resistance of larger grains into a layer of moving particles is smaller than that into the static layer. Therefore, larger grains tend to move into the layer of moving grains. At the same time, smaller particles can drop into the pores of underlying larger grains downward due to gravity.

Keywords. Naturally drained and speed-controlled ring shear tests, transparent shear box, shear zone development, concentration of grains and particles, velocity distribution profile, grain size distribution

7.1 Introduction

Every year, many rapid and long-travel landslides take place due to earthquakes and heavy rainfall. These landslides brought about casualties and property damage, worsened by the widespread development of towns and cities into upland areas. To reduce such landslide disasters, the development of reliable risk assessment has the first priority. Practical assessment of future landslide runout distance towards development areas requires precise prediction of the friction characteristics of the materials. Based on geotechnical and experimental studies using an undrained ring shear apparatus on the January 1995 Nikawa landslide in Japan, Sassa and colleagues found that granular materials show very high mobility under saturated undrained conditions. They named the mechanism as "Sliding Surface Liquefaction", (Sassa 1996, 2000; Sassa et al. 1996) hereafter often abbreviated as SSL.

Studies of the friction of granular materials for the purpose of landslide runout mechanism were conducted by many researchers using geological, geotechnical, and geophysical approaches. The effect of shear speed and normal stress on shear characteristics of granular materials had been widely examined (Novosad 1964; Scarlett and Todd 1969; Bridgwater 1972; Hungr and Morgenstern 1984; Sassa 1984; Vibert et al. 1989; Fukuoka et al. 1990; Fukuoka 1991; Tika 1989; Tika and Hutchinson 1999; Lemos 2003). These experimental studies were conducted by ring shear apparatus or torsion shear apparatus. Existing apparatuses have not allowed the experimenters to observe the grains of the sample during the test, nor to detect the development of a shear zone during shearing. Lang et al. 1991 tried to use a transparent shear box for the low-stress ring shear apparatus which was developed by Sassa 1984 and used a video image processing system to track the movement of the samples' grains. They conducted a series of tests on glass beads of 2–6 mm diameters under low normal stresses of 2.9 kPa and 29 kPa and shear speeds of 1 cm s^{-1} and 10 cm s^{-1}. They found that shear zone thickness decreased under higher normal stress in the test on beads of smaller diameter, and that the shear zone thickness was slightly greater during faster shear. However, these tests used only glass beads and did not use natural materials. Moreover, their apparatus was not capable of maintaining undrained conditions, or to measure pore water pressure.

Sassa and his colleagues have developed seven designs of ring shear apparatus since 1984 (Sassa et al. 2004a). The first apparatus (DPRI Ver.1) was used in the study of Vibert et al. (1989), Lang et al. (1991), and DPRI Ver.2 was used by Fukuoka (1991). Wang and Sassa 2002 examined the shear zone structure by exposing sections of silica sand samples in the ring shear apparatus DPRI Ver.6 after undrained tests of large shear displacement. However, their sample box of DPRI Ver.6 was metallic and they could not observe the sample during shear. Wafid et al. (2004) investigated the development of shear zones in undrained ring shear tests by observing the sections of sand samples stopped at different stages of shear displacement from the initiation of failure to the steady state. They found coarse grains accumulated in the shear zone and proposed a segregation model for shear zone development under undrained conditions. However, continuous monitoring from outside of the shear box and detection of horizontal and vertical movement of the grains were still impossible.

The authors deployed the DPRI Ver.7 with a transparent shear box and conducted shear-speed-controlled tests on coarse grained silica sands to study the shear zone formation process in granular materials (Fukuoka et al. 2005a,b). Velocity distribution profiles of grains under shear at various stages in the ring shear tests were observed through processing the video image. They found that after shear resistance reached peak strength, the thickness of the shear zone tends to decrease. However, these two studies focused on the variation of the velocity distribution profiles, and the variation of mechanical properties was not examined in thorough detail.

In this study, the main purpose is to study the influence of shear speed and normal stress on the shear behavior from an approach of shear zone development process observation through examining mechanical properties and variation of grain size distribution in the sample.

Fig. 7.1. Overview of the ring shear apparatus with a transparent shear box (DPRI Ver.7)

7.2 Ring Shear Apparatus and Observation System

The ring shear apparatus DPRI Ver.7, the latest model of undrained ring shear apparatus, has a transparent shear box which enables the observation of sand grains during slow to high speed shearing (Sassa et al. 2004a). The overview of the ring shear apparatus DPRI Ver.7 is shown in Fig. 7.1 and its transparent shear box is shown in Fig. 7.2. The basic structure of this apparatus is the same as DPRI-5 and DPRI-6 which are introduced by Sassa et al. (2003, 2004).

The basic structure of the DPRI Ver.7 is similar to its predecessors DPRI Ver.5 and DPRI Ver.6, introduced by Sassa et al. (2003, 2004a). The specifications of the DPRI Ver.7 apparatus are given by Sassa et al. (2004a) and Fukuoka et al. (2005a,b). The most important detail of the DPRI Ver.7 is the transparent sample box which allows direct observation of grains inside the shear box during shearing from outside.

The schematic illustration of the samples in the apparatus and set-up of two digital video cameras are shown in Fig. 7.3a. Movies of the sample during shear are taken by two digital video cameras (hereafter, abbreviated as DV cameras) 1 and 2. Initially, only a floor-mounted DV camera 1 was used. The image obtained by DV camera 1 is shown in the middle photo of Fig. 7.3b. As shown in the left photo of Fig. 7.3b, the image of the upper and lower shear box sample is clear enough to distinguish each sand grain before shear. However, during shear, the image of the lower shear box (inside the blue box in the middle photo of Fig. 7.3b) was severely blurred and it became impossible to distinguish each grain. Another digital video camera (fixed on the rotating table of the lower shear box) shown as "DV camera 2" in Fig. 7.3a was then installed. In the right photo of the Fig. 7.3b is a sample image taken by DV camera 2 during shear. Although grains in the upper shear box (in the blue box) cannot be distinguished in this photo because the upper shear box was

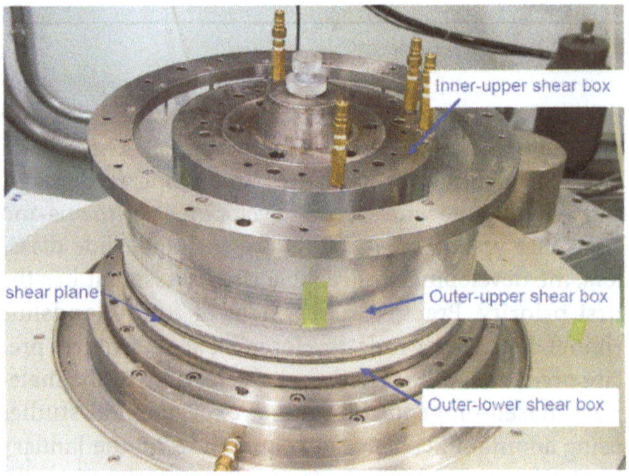

Fig. 7.2. Transparent shear box of DPRI Ver.7

rotating during shear, the image of the grains below the shear plane is clear enough for them to be distinguished.

A 2.0-mm-thick stainless steel plate is attached to the bottom of the upper half of the outer shear box contacting the rubber edge. This plate is necessary to protect the acrylic shear box from damage due to heating during high speed shearing. This system did not allow the observation of the exact shear plane and its vicinity, because the shear zone is shaded by invisible contacting parts consisting of the rubber edges and the metal plate shown as the "invisible part" in Fig. 7.3b. Pore pressure transducers were installed in the upper and inner sample box (Sassa et al. 2004a; Fukuoka et al. 2005b).

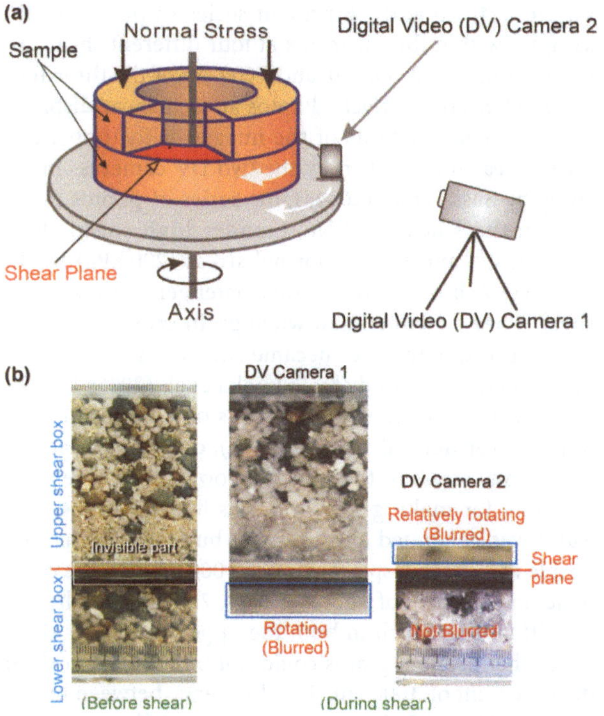

Fig. 7.3. a Schematic illustration of the samples in the ring shear apparatus DPRI Ver.7 and the digital video cameras 1 and 2 for video shooting; **b** shooting area of the digital video (DV) cameras 1 and 2. The image of the lower shear box taken by DV camera 1 is blurred. DV camera 2 installed on the rotating table of the lower shear box can take clear video images of the samples

7.3 Samples and Their Physical Properties

In this study, two types of samples were used: the first sample is mixed sands consisting of gravels and silica sands No. 1 (S1) and No. 4 (S4). The second sample is mixed beads. The properties of the samples are summarized in Table 7.1. Gravel used in this study consists of not completely round grains, but with smooth surfaces. Basic properties of the gravels are as follows: mean diameter $D_{50} = 6.5$ mm, uniformity coefficient $U_c = 1.47$, and specific gravity $G_s = 2.65$. Silica sand is produced artificially by crushing of natural rocks and sieving. It is commonly used as construction material. This silica sand consists of roughly round grains, of which 92–98% are quartz and a little amount of feldspar. Basic properties of silica sands S1 and S4 are as follows: silica sand S1 has a mean diameter of $D_{50} = 3.01$ mm, uniformity coefficient of $U_c = 1.64$, and specific gravity of $G_s = 2.64$. Silica sand S4 has $D_{50} = 1.05$ mm, $U_c = 2.35$, and $G_s = 2.64$.

Mixed beads consist of glass beads, plastic beads, and polystyrene beads. The basic properties of glass beads are $D_{50} = 2.0$ mm, $U_c = 1.0$ (single-size grains), and $G_s = 2.52$; plastic beads are $D_{50} = 6.0$ mm, $U_c = 1.0$, and $G_s = 1.77$; polystyrene beads are $D_{50} = 1.0$ mm, $U_c = 1.0$, and $G_s = 1.00$. Each of the three types of beads is mixed by equal percentage in bulk volume.

In this study, the tests were performed on the mixed sands and mixed beads, under different normal stresses and shear speeds, then the dynamic behaviours were observed.

7.4 Testing Conditions and Procedure

The test series consists of three ring shear tests on mixed sands and one test on mixed beads. The conditions of the four tests are summarized in Table 7.2. In the all tests, the speed-controlled mode is employed to examine shear speed dependency of shear resistance. In the initial stage of each test, shear speed was increased at a constant acceleration rate until shear speed reached the target speed. Stress and shear speed conditions of tests S1, S2, and S3 on mixed silica sand were selected from combinations of high normal stress (200 kPa) and low normal stress (22 kPa) conditions and high shear speed (150 cm s^{-1}) and low

Table 7.1.
Properties of tested samples

		Grain size D_{50}	Coefficient of uniformity U_c	Specific gravity G_s
Gravels		6.50	1.47	2.65
Silica sands No. 1		3.01	1.64	2.64
Silica sands No. 4		1.05	2.35	2.64
Mixed sands		3.0	4.38	2.64
Mixed Beads	Glass beads	2.0	1.0	2.52
	Plastic beads	6.0	1.0	1.77
	Polystyrene beads	1.0	1.0	1.00

Table 7.2. Test program

Test No.	Sample	Normal stress (kPa)	Shear speed (cm s^{-1})	Acceleratio n (cm s^{-2})	Drainage	Initial void ratio
S1	Mixed sands	22	150	50	Naturally-drained	0.67
S2	Mixed sands	200	150	50	Naturally-drained	0.62
S3	Mixed sands	200	9	2	Naturally-drained	0.61
B1	Mixed beads	22	150	50	Naturally-drained	–

shear speed (9 cm s^{-1}) conditions. Test S1 was conducted under low normal stress and high shear speed condition, test S2 was under high stress and high shear speed conditions, and test S3 was under high stress and low shear speed conditions. Test B1 was conducted on mixed beads under low normal stress and high shear speed.

In this study, all of the tests were conducted under naturally drained condition keeping the valve of the drainage in the upper sample box open during shear. In this condition, pore pressure during shear is not always zero. In the actual landslides, shearing along the sliding surface is neither completely undrained condition (no pore pressure dissipation), nor completely drained condition (no excess pore pressure) as discussed by Sassa et al. (2004b) where the effects were examined for the Nikawa landslide triggered by the Hyogoken-Nambu earthquake. Pore pressure build-up is a combined effect of the pore pressure generation rate in the shear zone and pore pressure dissipation rate which are functions of shear speed, permeability of soil layers, drainage path, etc. The purpose of the ring shear tests is to mechanically simulate shearing in real landslides. In real landslides, upward drainage is allowed. Therefore, the shearing was conducted keeping the upper drainage valve open allowing the pore pressure dissipation only upward. This test procedure may not reproduce the natural condition in a quantitative way, but it may illuminate the pore pressure build-up behavior in a qualitative manner.

The testing procedure was as follows: (*1*) samples were placed into the sample box by dry deposition and CO_2 gas was supplied to expel the air. Then, de-aired water was infiltrated from the bottom slowly under a pressure of about 1 m head to saturate the sample; (*2*) a targeted normal stress was loaded slowly under the naturally drained conditions, keeping the upper drainage valve open; (*3*) the sample was then sheared under the shear-speed-controlled condition and the naturally drained condition as shown in Table 7.2 until the shear speed reached the target value; (*4*) shearing was continued at constant shear speed until the shear displacement reached 3000 cm (30 m). During shearing, DV cameras 1 and 2 were monitoring the upper and lower shear box, respectively; (*5*) after the termination of shearing, the upper sample box was dismounted to expose the sample; (*6*) samples were taken from aeach layer of about 1 cm thickness from top to bottom of the samples within the sample box; finally, (*7*) the grain size distribution of each layer was measured by sieving.

7.5 Test Results on Mixed Sands

7.5.1 Video Observation of Shear Zone Development with Progress of Shear Displacement

Figures 7.4, 7.5 and 7.6 present series of photos of the samples within the shear box at four different shear displacements, i.e., 0, 30, 300, and 3000 cm of the three tests of S1, S2, S3, respectively. Photos at each shear displacement are combinations of the images of the upper and lower shear boxes taken by the two DV cameras. In the study by Sassa et al. (2004a), the series of photos of the test on the silica sand No. 1 under high speed shear (200 cm s^{-1}) and higher normal stress (200 kPa) by the DPRI Ver.7 showed obvious silty water generation during shear. They explained that when grain crushing created small grains, pore water became silty and gradually diffused later. In the study by Fukuoka et al. (2005a,b), silty water was not obvious in their tests on silica sand No. 1 under lower normal stress (22 kPa), due to the fact that the normal stress of the tests was too low to cause grain crushing for making the samples silty. In Fig. 7.4, silty water was generated and diffused, but some grains still visible at a shear displacement of 3000 cm. On the other hand, in the series of photos in Figs. 7.5 and 7.6, silty water diffused so much in both the upper and lower shear boxes that original grains could not be visible at a shear displacement of 3000 cm. The difference between test S1 and tests S2 and S3 is mainly due to the different normal stresses, i.e., higher normal stress in tests S2 and S3 promotes more intensive grain crushing and creates more amounts from smaller grains to make the pore water silty.

7.5.2 Variation of Stress and Pore Pressure with Progress of Shear Displacement

Variation of the shear resistance, pore pressure, and sample height increment against the shear displacement of tests S1, S2, and S3 are shown in Figs. 7.7, 7.8 and 7.9. In these tests, the sample height continued reducing from the beginning to the end. In test S1 (shear speed: 150 cm s^{-1}, normal stress: 22 kPa) (Fig. 7.7), the sample height reduction at shear displacement of 3000 cm was about 1.1 mm. On the other hand, tests S2 and S3 under high normal

Fig. 7.4.
A series of photos of the upper and lower shear box of the ring shear test S1 on mixed silica sands under the naturally drained condition. Normal stress = 22 kPa, shear speed= 150 cm s^{-1}, acceleration at the initial stage was 50 cm s^{-2}. Video images of the upper shear box were taken by DV camera 1 which was located on the laboratory floor, and the images of lower shear box were taken by DV camera 2 which was fixed on the rotating table of the lower shear box

Fig. 7.5.
A series of photos of the upper and lower shear box of the ring shear test S2 on mixed silica sands under the naturally drained condition. Normal stress = 200 kPa, shear speed= 150 cm s^{-1}, acceleration = 50 cm s^{-2}

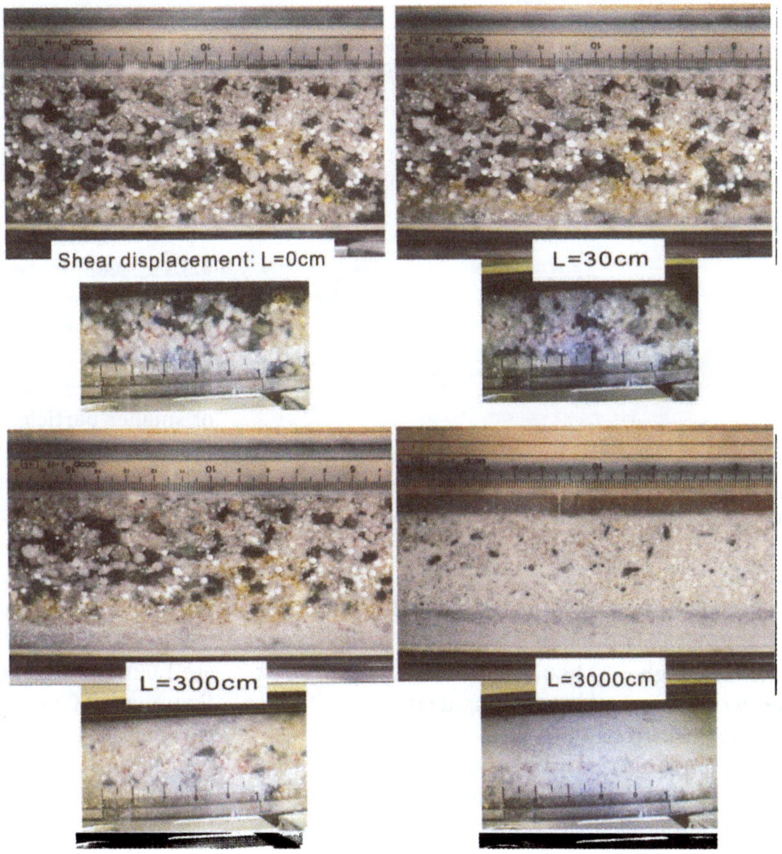

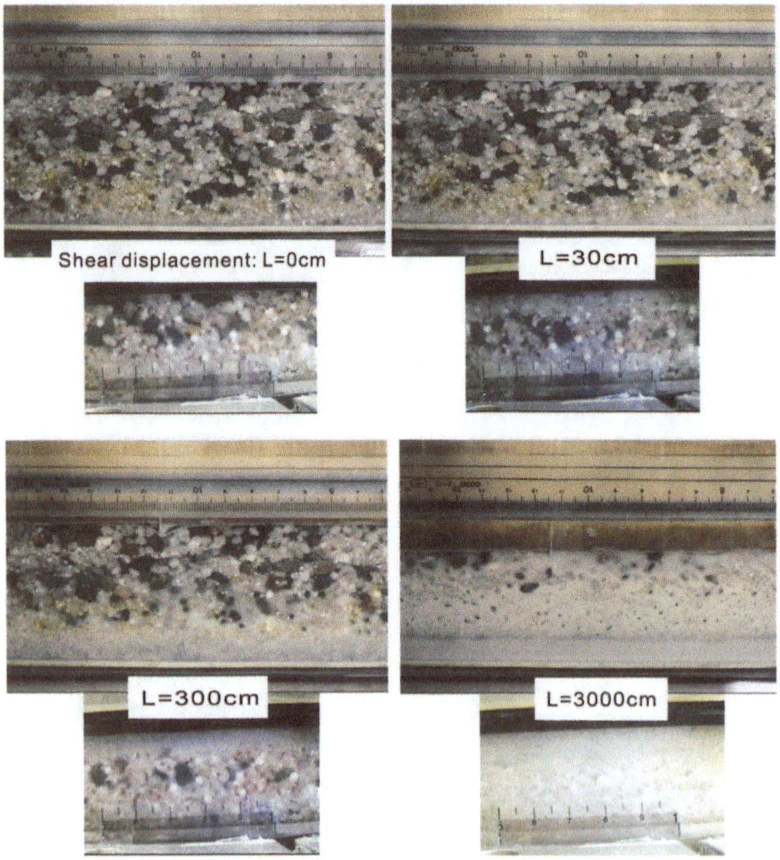

Fig. 7.6. A series of photos of the upper and lower shear box of the ring shear test S3 on mixed silica sands under the naturally drained condition. Normal stress = 200 kPa, shear speed = 9 cm s^{-1}, acceleration = 2 cm s^{-2}

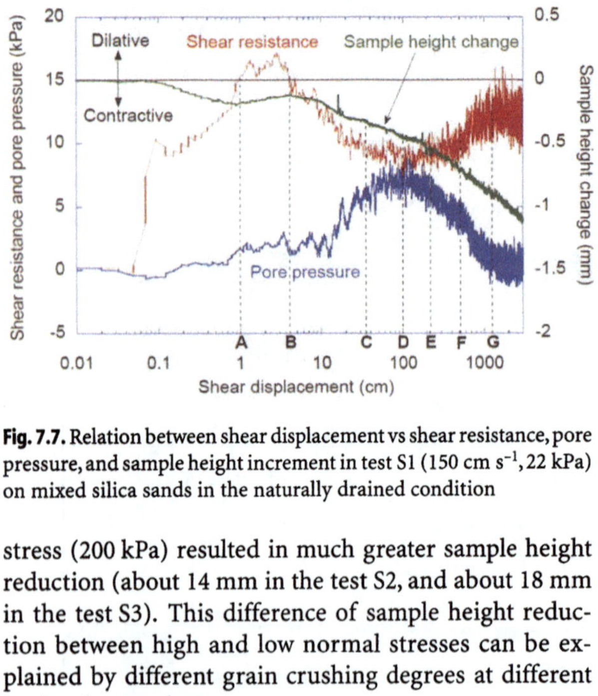

Fig. 7.7. Relation between shear displacement vs shear resistance, pore pressure, and sample height increment in test S1 (150 cm s^{-1}, 22 kPa) on mixed silica sands in the naturally drained condition

stress (200 kPa) resulted in much greater sample height reduction (about 14 mm in the test S2, and about 18 mm in the test S3). This difference of sample height reduction between high and low normal stresses can be explained by different grain crushing degrees at different levels of normal stress.

Because all of the tests were conducted under the naturally drained condition, pore pressure could change during the test. It is noted that excess pore pressure was generated after shear resistance reached peak strength, i.e., after the shear failure, in all of the tests of S1, S2, and S3. The shear resistance dropped due to the generated excess pore pressure and then increased again along with the dissipation of the excess pore pressure.

7.5.3 Concentration of Larger Grains in the Shear Zone during Shearing and Resulting Variation of Grain Size Distribution

In Figs. 7.4, 7.5 and 7.6, it can be seen that the proportion of smaller particles was increased with shearing, due to the grain crushing in the shear zone. To compare the results of the crushable sands with the uncrushable samples, another test B1 was performed on the mixed beads, where the test conditions were the same as those in test S1.

The series of photos at four stages from before shearing until the end of shearing taken in test B1 are presented in Fig. 7.8. White beads are 6.0 mm in diameter, blue beads are 1.0 mm, and transparent beads (not so visible in this figure) are 2.0 mm. In this figure, before shearing, white and blue beads are distributed almost uniformly. However, a blue zone appeared at shear displacement of $L = 30$ cm just above the shear zone in the upper shear box. Thereafter, the blue band of small beads appeared

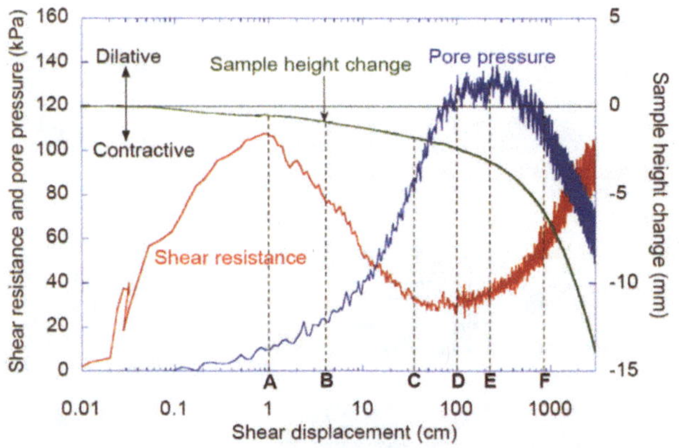

Fig. 7.8. Relation between shear displacement vs shear resistance, pore pressure, and sample height increment in test S2 (150 cm s^{-1}, 200 kPa) in the naturally drained condition

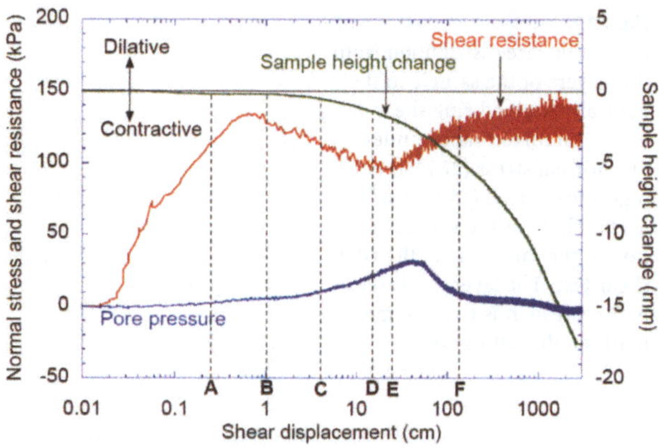

Fig. 7.9. Relation between shear displacement vs shear resistance, pore pressure, and sample height increment in test S3 (9 cm s^{-1}, 200 kPa)

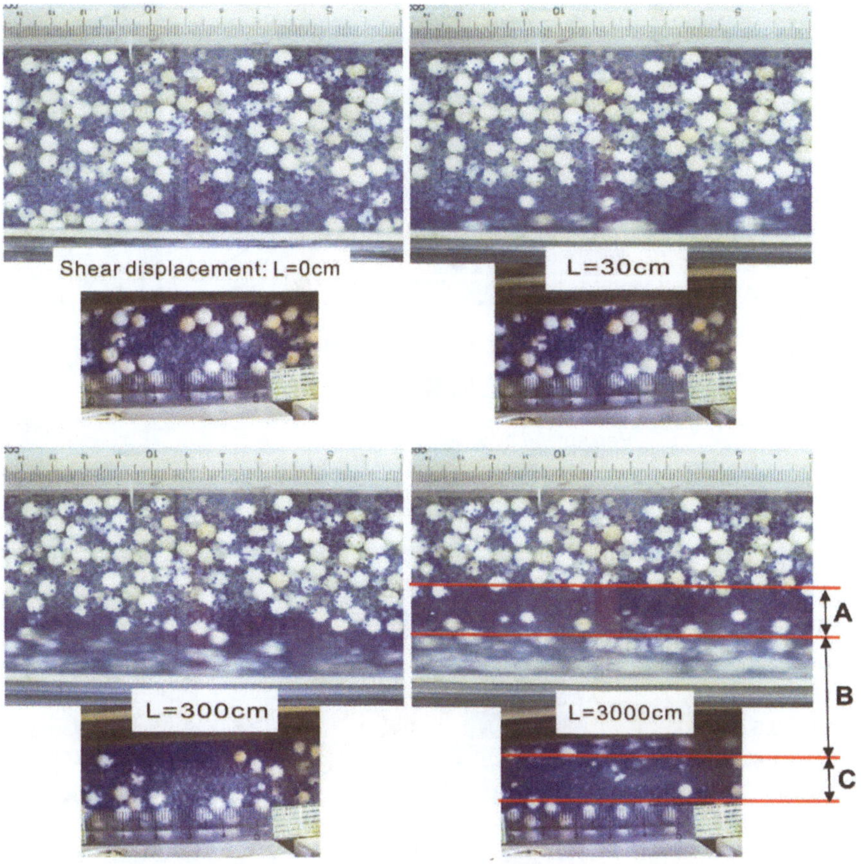

Fig. 7.10. Series of photos in the ring shear test B1 on mixed sands under low normal stress (22 kPa) and high shear speed (150 cm s^{-1}) in the naturally drained condition. *L* is the shear displacement at each photo. In zones *A* and *C*, fine blue beads are concentrated, while white coarser beads are concentrated in the shear zone *B*. The concentration of coarse grains in the shear zone was also confirmed in the vertical section after the test

also in the lower shear box as shown in Fig. 7.10 at the shear displacement of $L = 300$ cm and 3 000 cm, which are marked as A and C. On the other hand, comparing the images of the shear zone at $L = 300$ and 3 000 cm, the number of greater white beads seemed to increase in the 3 000 shear displacement, forming a layer, which is marked as B in the photo, while small blue beads concentrated in the two layers of A and C, located just above and below the layer B.

Figures 7.11, 7.12 and 7.13 show the results of grain size analysis performed for sand samples from different layers for tests S1, S2, and S3, respectively. On the left of each figure, the grain size distribution of each sliced part, which is marked as A–F or A–G, is presented in color bars. The corresponding photo of a vertical section view of the sample taken after ring shear test is shown on the right of each figure. The red dotted lines in these photos show the location of the shear plane, i.e., the boundary between

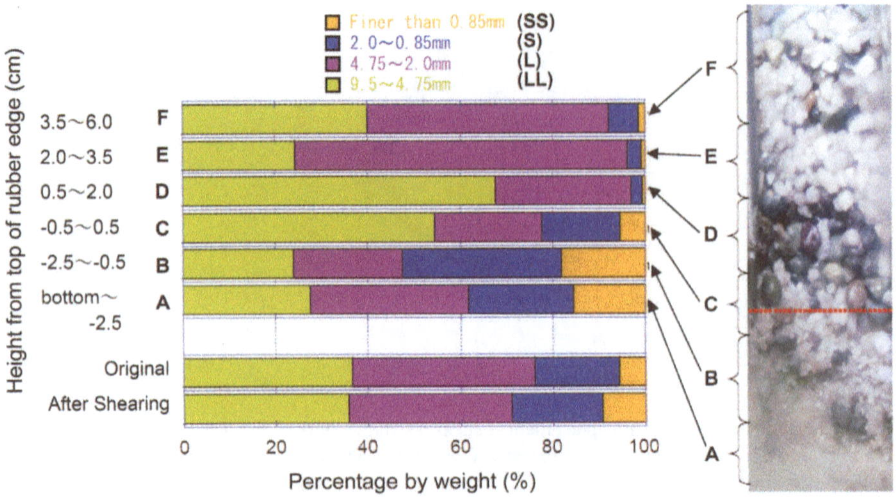

Fig. 7.11.
Left: Grain size distribution from thin layers of the sample in the naturally drained ring shear test S1 on mixed sands under low normal stress (22 kPa) and high shear speed (150 cm s^{-1}). *Right*: Photo of the vertical section of the sample after the ring shear test. The layer marked as A–F corresponds to the same mark on the left figure

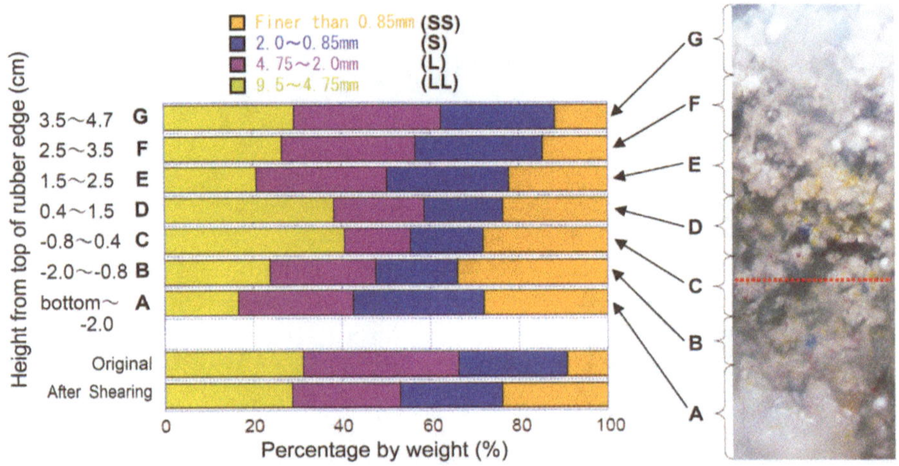

Fig. 7.12.
Left: Grain size distribution from thin layers of the sample in the naturally drained ring shear test S2 (200 kPa, 150 cm s^{-1}) on mixed sands. *Right*: Photo of the vertical section of the sample after the ring shear test. The layer marked as A–F corresponds to the same mark on the left figure

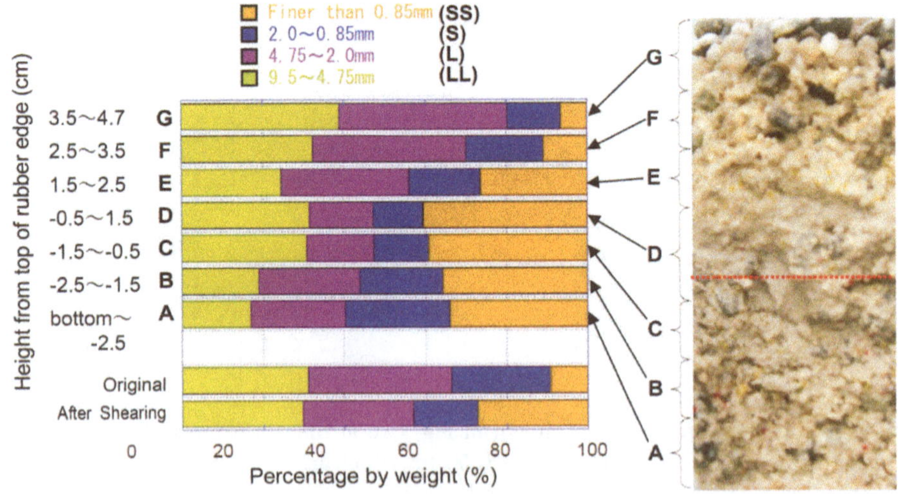

Fig. 7.13.
Left: Grain size distribution from thin layers of the sample in the naturally drained ring shear test S3 (200 kPa, 9 cm s^{-1}) on mixed sands. *Right*: Photo of the vertical section of the sample after the ring shear test. The layer marked as A–F corresponds to the same mark on the left figure.

the upper shear box and the lower shear box of the apparatus. Hereinafter, grains and particles of which the size is between 9.5 mm and 4.75 mm are denoted as "LL-grains" (percentage in weight is presented as green bars), 4.75 mm and 2.0 mm as "L-grains" (purple bars), 2.0 mm and 0.85 mm as "S-particles" (blue bars), and smaller than 0.85 mm as "SS-particles" (yellow bars). The color bar marked as "Original" presents the grain size distribution of the original samples before the ring shear test and the one at the bottom of the figure marked as "After shearing" presents the average grain size distribution of the samples after the ring shear test, which were collected from top to bottom within the samples.

Figure 7.11 shows the grain size distribution of the sample in the test S1. In the right photo, LL-grains of darker color concentrated above the shear plane. The percentage of LL-grains in the color bars of C and D is much larger than the original. Especially LL-grains of D is almost double the original. L-grains concentrated in the bars of E and F above the LL-grain concentrating zone. S-particles concentrated in the two bars A and B located below the LL-grain dominant zone. It was observed that S-particles and SS-particles settled downward under this low normal stress test (S1) immediately after shearing was stopped at shear displacement of 3 000 cm. Therefore, the amount of S-particles and SS-particles in the bars of D, E, and F are smaller than the original. During shear, the contents of S-particles and SS-particles must be a little more than that presented in the upper shear box and a little less in the lower shear box. The total amount of new SS-particles generated during shear by grain crushing is presented by the difference between the colored bars of "Original" and "After shearing" and it is only about 4%. Consequently, the S-particles and SS-particles concentrated into the lower box, and the LL-grains and L-grains concentrated around and above the shear plane.

Figure 7.12 shows the results for test S2. Because the normal stress of test S2 is much higher than test S1 in Fig. 7.11, more SS-particles were generated by grain crushing as shown by the difference of SS-particles content between the colored bars of "Original" and "After shearing." In the photos of Fig. 7.12 for the higher normal stress tests (200 kPa), the concentration of LL-grains in the vicinity of the shear plane was not so visible. SS-particles, most of which may be generated by grain crushing, are distributed throughout the vertical section. The grain size distribution in Fig. 7.12 shows that SS-particles concentrated in the three colored bars of A, B, and C, implying occurrence of deposition toward the bottom of the shear box. This figure also shows that LL-grains concentrated around the shear surface of the two bars of C and D. However, their percentage was not so high as the bars of C and D in Fig. 7.11 for lower normal stress test (22 kPa). The reason for the fact that accumulation of LL-grains occurred extensively under lower normal stress must be the only difference in their test condition of the normal stress between test S1 (22 kPa) and test S2 (200 kPa). The contents of S-particles and SS-particles within the top two colored bars of F and G in Fig. 7.12 were even greater than the original sample. Considering that in Fig. 7.5 sand grains near the top of the sample beneath the loading plate, corresponding to the two colored bars of F and G in Fig. 7.12 showed that most of the grain did not move up to 300 cm of shear displacement, and there is a possibility that the new S-particles and SS-particles were not created at the same height area. Therefore, it was estimated that fine particles generated within the shear zone moved upward along with the diffusion of silty water due to high excess pore water pressure in the shear zone.

Figure 7.13 for high normal stress (200 kPa) and low shear speed (9 cm s^{-1}) test shows the grain size distribution and a photo of vertical section of the sample after the test S3. In the photo, the concentration of LL-grains was not so visible as Fig. 7.11. When the increase of SS-grains amount is compared to the differences of "Original" and "After shearing" with those in Figs. 7.11, 7.12 and 7.13, obviously the greatest amount of new SS-particles was generated by grain crushing in Fig. 7.13 for test S3. The difference of conditions between tests S2 (Fig. 7.12) and S3 (Fig. 7.13) was only shear speed. S3 (9 cm s^{-1}) took 15 times longer time to reach 30 m shearing displacement than S2 (150 cm s^{-1}). Therefore, S-particles generation will be not only the function of shear displacement and normal stress, but also affected by shearing time.

The SS-particles concentrated around the shear plane and below the shear plane in the lower shear box as shown in the four colored bars of A, B and C in Figs. 7.11 and 7.12 and A, B, C and D in Fig. 7.13. On the other hand, the LL-grains concentrated in the two colored bars of C and D in Figs. 7.11 and 7.12. However the observed amount of the LL-grains of the two colored bars of C and D was almost same as the "Original" sample in Fig. 7.13. The amount of LL-grains of the three colored bars of A, B and E, located above and below the slices of C and D in the vertical section, was smaller than C and D in Figs. 7.11, 7.12 and 7.13. This result suggests a possibility that there is a mechanism which lets the LL-grains concentrate around the shear plane.

7.6 Discussion

7.6.1 Influence of Shear Speed and Normal Stress on the Mobilized Shear Resistance

In Fig. 7.14a, variations of apparent friction and pore pressure ratio of the mixed sands tests S1, S2, and S3 are shown against the shear displacement from 0.1 cm to 3 000 cm. In Fig. 7.14b, the same parameters were plotted vs elapsed time. Shear displacement and elapsed time are presented in log scale. Apparent friction, which is represented by red lines, is calculated by mobilized shear resistance τ divided by total normal stress σ. Apparent friction defined as $\tan^{-1}(\tau/\sigma)$ by Sassa (1988) is used as the most important index for landslide mobility. When apparent friction is reduced in the shear zone, higher mobility is expected.

Pore pressure ratio r_u, which is represented by blue lines, is calculated by generated excess pore pressure u divided by total normal stress σ. Values of r_u are not

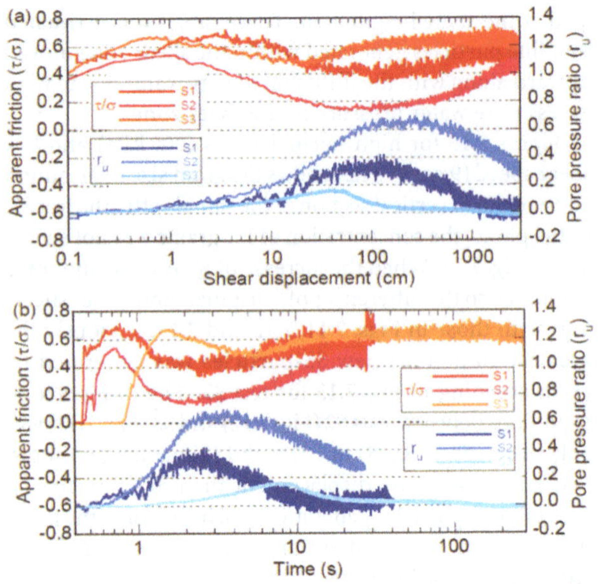

Fig. 7.14. Variation of apparent friction, pore pressure ratio vs (**a**) shear displacement, (**b**) elapsed time in log scale in the naturally drained test of mixed sands

greater than 1.0. When the sample nears a state of liquefaction or sliding liquefaction, r_u approaches 1.0. As explained in previous sections, the ring shear tests of this study were conducted under the naturally drained condition. Excess pore pressure was generated during shearing of each test, but was naturally dissipated through the upper drainage line of the shear box.

In Fig. 7.14a, apparent friction once increased as the shear resistance reached the peak strength in red plots of S1 and S2. The plots of S3 reached peak before shear displacement of about 0.7 cm. But after reaching the peak strength, shear resistance of each test turned to be reduced and showed a minimum value as the excess pore pressure was generated. The characteristics of this figure were: (*1*) order of the value of the peak r_u is S2 > S1 > S3; (*2*) order of the local minimum value of apparent friction in each test is S2 < S1 < S3. In general, when the excess pore pressure ratio is higher, the apparent friction is lower. The shear displacement at which the peak value of r_u appears in each test is larger than the one at which the smallest value of apparent friction appeared.

The drainage path of the apparatus is described in Sassa et al. (2004a). Pore pressure is measured at the annular gutter just 2 mm above the shear plane and filled with metal filters and felting cloth. Pore pressure transducers are installed at the exit of the pipes extending from the annular gutter. Although the samples are sandy material and permeability is higher than clayey materials, slight delay of measuring excess pore pressure in the shear plane may take place due to the 2 mm distance from shear plane in these naturally drained tests.

The effect of shear speed is obvious by comparing the results of tests S2 and S3, where the applied normal stress were 200 kPa, but the shear speed was 150 cm s^{-1} for S2, and 9 cm s^{-1} for S3, respectively. The peak excess pore pressure ratio of S2 is about 0.65, and that of S3 is about 0.16. The peak excess pore pressure of test S2 was almost four times that in test S3. The large difference of the peak excess pore pressure may be caused by the large difference in the shear speed. The difference in the build-up rate of excess pore pressure is about 15 times between S2 and S3. However, the dissipation rate is almost the same among tests S1, S2, and S3, because dissipation depends on pore pressure difference, dissipation path length and permeability. The values of these factors are almost the same (except the first factor). Thus, the peak excess pore pressure which was built up under the naturally drained condition is higher in the tests under higher shear speed. Thus, it is reasonable that the observed peak excess pore pressure in test S2 was much greater than S1. Consequently, in the naturally drained ring shear tests under higher shear speed and higher normal stress condition, higher excess pore pressure and lower minimum apparent friction can be expected.

On the other hand, when the normal stress condition is compared, S1 was under 22 kPa and S2 was under 200 kPa, and both tests were conducted under the same shear speed of 150 cm and the same acceleration of 50 cm s^{-2}. The peak excess pore pressure in test S2 is about 0.65 and that in test S1 is about 0.31. The peak excess pore pressure of test S2 showed about two times of that in test S1. The final sample height reduction in test S1 was 1.1 mm (Fig. 7.7), but that in test S2 was 14 mm (Fig. 7.8), more than ten times the value in S1. Therefore, the difference of pore pressure build-up was caused by the difference of grain crushing and the resulting pore pressure generation due to a difference of normal stress.

7.6.2 Influence of Normal Stress and Shear Speed on Grain Crushing and Concentration of the Coarse Grain in the Shear Zone

In Fig. 7.15, the variation in the amount of LL-grains and SS-particles in each layer of the sample is plotted. The horizontal axis is the height from the shear plane. The location of the shear plane is represented by the vertical straight line at height = 0 cm. The positive side of the height from shear plane corresponds to the upper shear box, while the negative side corresponds to the lower shear box. Blue plots and lines represent the variation of SS-particles amount, and red plots and lines represent variation of LL-grains amount for each test of S1, S2, and S3. Variation of the grains' proportions in each layer of

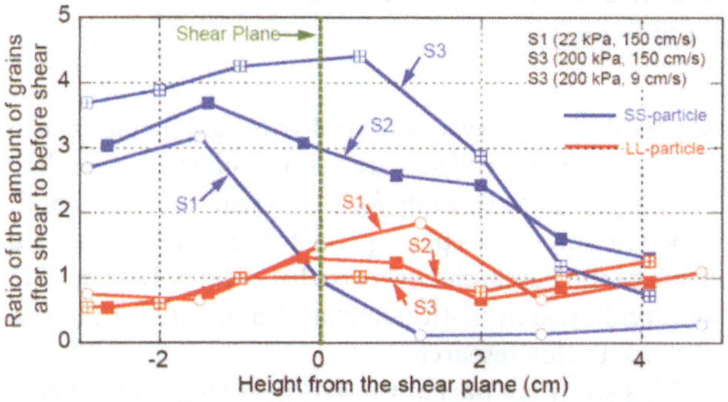

Fig. 7.15. Variation of the rate of sample height reduction (dh/dt) and its acceleration (d^2h/dt^2) for the shear displacement from 0.1 cm to 3 000 cm in log scale. Variations in the amount of LL-grains and SS-particles in each layer of the sample, represented by the ratio between before and after the tests of S1 (22 kPa, 150 cm s^{-1}), S2 (200 kPa, 150 cm s^{-1}), and S3 (200 kPa, 9 cm s^{-1}). *Blue plots and lines* represent variation of SS-particle amounts, and *red plots and lines* represent variation of LL-grain amounts

the sample is represented by the ratio between those before and after the tests. When the ratio is greater than 1.0, it means that the amount of grains increased and when it is smaller than 1.0, it means that the amount decreased.

As for the effect of normal stress on grain crushing, the order of the peak value of SS-particles distribution was S3 > S2 > S1. This means that higher normal stress in tests S2 and S3 causes generation of greater amount of SS-particles than the low normal stress in test S1. On the other hand, higher shear speed in test S2 causes less amount of SS-particles than the low shear speed in test S3. This is probably due to the fact that test S2 had greater excess pore pressure during shearing, which led to lower effective normal stress, thus less progress in grain crushing.

A general tendency is that greater amount of SS-particles is observed in the lower shear box in the three tests. Although deposition of SS-particles was observed when shearing was stopped, the ratio was still greater than 1.0 at heights of about 3 cm and 4 cm above the shear zone in test S2. This suggests that the finest particles generated by grain crushing were not only deposited downward, but also moved upward. Sassa et al. (2004a) observed that silty water generated by grain crushing during ring shear test under high speed high normal stress was diffused both upward and downward from the shear zone. The smallest particles floating in the pore water have also diffused. This process may contribute to the diffusion of the finest particles.

As for the variation of LL-grains distribution represented in red lines, the ratio shows greater than 1.0 for tests S1 and S2 in the vicinity of the shear plane (more precisely described as at the level of the shear plane and around 1 cm above the shear plane in Fig. 7.15). It means that LL-grains concentrated in the vicinity of the shear zone under high speed tests. This tendency is consistent with the results of test B1 shown in Fig. 7.10, but in test S3, the maximum ratio of the LL-grains was almost 1 probably affected by greater grain crushing and the resulting less LL-grains.

One possible interpretation why large grains were accumulated in the vicinity of the shear zone is as follows: grains under shearing are subjected to vertical force due to dilatancy. Small particles can enter into the pores between the large grains, however, large grains can move vertically only by the penetration into a layer by moving other particles and making spaces. If a layer of moving particles such as the shear zone will exist on one side and a layer of static layer exist on another side, the penetration of large grains into the layer of moving particles will be much easier, because there is no need to spend energy to force moving other particles. This mechanism can be understood by a simple experiment that a finger can be inserted into sands under shaking condition much easier than into sands under static condition. Melosh (1986) tested a series of such experiments in which sands were excited by a 37 kHz acoustic field and an aluminum rod was pushed into the sands under various strain rate to measure shear penetration resistance variation compared with the one under static condition. The test results showed obvious reduction in the 37 kHz acoustic condition.

Therefore, large grains tend to move into the layer of moving particles, namely, large grains above the shear zone tend to move downward to the shear zone, and large grains below the shear zone tend to move upward to the shear zone. In this way, coarse grains tend to concentrate in the vicinity of the shear zone. Consequently, small particles tend to remain above and below the shear zone, and at the same time small particles tend to drop into the pores between the larger grains due to gravity. In addition to this process, some of the small particles dropped through voids at the moment shearing was terminated. Distribution of small particles after the test is the result of these effects.

Wafid et al. 2004 studied the segregation process during ring shear tests. In their study, undrained ring shear tests were conducted on Silica sand No. 8 (D_{50} = 0.048 mm) under normal stress of 180 kPa by using the shear-stress-controlled mode. The maximum shear speed was limited to 5.4 cm s^{-1} and the samples were sheared up to a displacement of 1 000 cm. They found that large grains concentrated in the upper part of the shear zone and small particles accumulated at the bottom of the shear zone when the sample reached the steady state. They called this phenomenon segregation. When precisely observing the distribution of large grains in Fig. 7.15, the large grains seem to be accumulated in the upper part of the shear zone. The result appears consistent with that presented by Wafid et al. 2004.

7.7 Conclusions

By using a newly developed ring shear apparatus with a transparent shear box, a series of naturally drained monotonic shear-speed-controlled tests were conducted on mixed sand and mixed beads to study the process of shear zone development. Movement of grains and silty water due to grain crushing in the samples was monitored by video cameras. The variation of pore water pressure and shear resistance were monitored. The variation of grain size distribution in the sample was examined for sliced samples after tests for quantitative analysis of vertical displacement of grains and particles as well as creation of fine particles due to grain crushing. The following results were obtained:

1. In the naturally drained ring shear tests under greater shear speed and greater normal stress conditions, greater excess pore pressure and lower apparent friction were observed. Comparing the test results with different conditions of normal stress (200 kPa and 22 kPa) and different shear speeds (150 cm s^{-1} and 9 cm s^{-1}), the greatest pore pressure build-up and the minimum apparent friction were obtained in the combination of high normal stress and high speed shearing test. It is probably because high normal stress contributes to great grain crushing and the resulting volume reduction, and high speed contributes to a high pore pressure generation rate and the resulting greater pore pressure build-up.
2. Grain size analysis of the samples' sliced layers after the tests revealed that greater amounts of small particles were generated under greater normal stress. Along with the diffusion of silty water generated by grain crushing, small particles were transported upward and downward from the shear zone.
3. The concentration of large grains from the center to the upper part of the shear zone was confirmed by means of visual observation through the transparent shear box together with grain size analysis of the sliced samples from several layers after the test. On the other hand, small particles were accumulated mostly below the layer where large grains were accumulated. The reason why large grains were accumulated near and in the shear zone could be interpreted as follows: grains under shearing are subjected to vertical movement, the penetration resistance of larger grains into a layer of moving particles is smaller than that into the static layer. Therefore, both lager grains below the shear zone and larger grains above the shear zone tend to move into the layer of moving grains in the shear zone. At the same time, smaller particles can drop into the pores between the larger grains downward due to gravity.

Acknowledgments

Mr. Naohide Ishikawa, and Mr. Ryo Sasaki former master course graduate students are acknowledged for conducting the preliminary study for this research in our laboratory. The authors deeply appreciate the members and students of the Research Centre on Landslides, Disaster Prevention Research Institute, Kyoto University for their assistance to this research.

This study is a part of the M101 Project, "Areal Prediction of Earthquake and Rain Induced Rapid and Long-traveling Flow Phenomena" (APERITIF, proposer: K. Sassa), of the International Programme on Landslides (IPL) operated by the International Consortium on Landslides (ICL) since 2002, and is also a part of the Japanese project "Aerial Prediction of Earthquake and Rain-Induced Flow Phenomena" (APERIF project, represented by K. Sassa), which was supported by the Special Coordinating Fund for Promoting Science and Technology of the Ministry of Education, Culture, Sports, Science and Technology (MEXT) of the Japanese Government for 3 years from 2001–2004.

References

Bridgwater J (1972) Stress–velocity relationships for particulate solids. ASME Paper No. 72-MH-21, 7 p

Fukuoka H (1991) Variation of the friction angle of granular materials in the high-speed high-stress ring-shear apparatus, influence of re-orientation, alignment and crushing of grains during shear. Annu Disaster Prev Res Inst Kyoto Univ 41(4):243–279

Fukuoka H, Sassa K, Shima M (1990) Shear characteristics of sandy soils and clayey soils subjected to the high-speed and high-stress ring shear tests. Annu Disaster Prev Res Inst Kyoto Univ 33(1): 179–190 (in Japanese)

Fukuoka H, Sassa K, Wang GH, Sassaki R (2005a) Measurement of velocity distribution profile in ring-shear apparatus with a transparent shear box. In: Sassa K, Fukuoka H, Wang GH, Wang F (eds) Landslides, risk analysis and sustainable disaster management. Springer-Verlag, Berlin Heidelberg New York, pp 149–156

Fukuoka H, Sassa K, Wang GH, Sassaki R (2005b) Observation of shear zone development in ring-shear apparatus with a transparent shear box. Landslides (in contribution)

Hungr O, Morgenstern NR (1984) High velocity ring shear tests on sand. Géotechnique 34(3):415–421

Ishikawa N (2004) An experimental study on the shear behavior of granular materials by means of a new ring shear apparatus with a transparent shear box and image processing. Master's thesis, Kyoto University

Lang Y, Ote K, Fukuoka H, Sassa K (1991) Image-processing the velocity distribution of grains in ring shear tests. In: Proceedings of the Conference of Japan Society of Erosion Control Engineering 1991, pp 302–305 (in Japanese)

Lemos LJL (2003) Shear behaviour of pre-existing shear zone under fast loading – insights on the landslide motion. In: Proceedings International Workshop on occurrence and mechanisms of flow-like landslides in natural slopes and earthfills, pp 229–236

Melosh HJ (1986) The physics of very large landslides. Acta Mech 64:89–99

Novosad J (1964) Studies on granular materials. II. Apparatus for measuring the dynamic angle of internal and external friction of granular materials, Collection Czech Chem Commun 29: 2697–2701

Sasaki R (2005) An experimental study on the process of shear-zone development using the ring shear apparatus with a transparent shear box. Master Thesis of Graduate School of Science, Kyoto University, 74 p

Sassa K (1984) The mechanism starting liquefied landslides and debris flows. In: Proceedings of 4th International Symposium on Landslides, Toronto, June, 2, pp 349–354

Sassa K (1988) Geotechnical model for the motion of landslides. In: Special Lecture of 5th International Symposium on Landslides, "Landslides", 10–15 July, 1, pp 37–55

Sassa K (1996) Prediction of earthquake induced landslides. In: Proceedings of 7th International Symposium on Landslides, A.A. Balkema, Trondheim, 17–21 June, 1, 115–132

Sassa K (2000) Mechanism of flows in granular soils. In: Proceedings of the International Conference of Geotechnical and Geological Engineering, GEOENG2000, Melbourne, 1, pp 1671–1702

Sassa K, Fukuoka H, Lee JH, Zhang DX (1992) Measurement of the apparent friction angle during rapid loading by the high-speed high-stress ring shear apparatus; interpretation of the relationship between landslide volume and the apparent friction during motion. In: Proceedings 6th International Symposium on Landslides, pp 545–552

Sassa K, Fukuoka H, Scarascia-Mugnozza G, Evans SG (1996) Earthquake-induced-landslides: distribution, motion and mechanisms. Special issue for the great Hanshin Earthquake Disasters, Soils and Foundations 53–64

Sassa K, Wang G, Fukoka H (2003) Performing undrained shear tests on saturated sands in a new intelligent type of ring shear apparatus. Geotech Test J ASTM 26(3):257–265

Sassa K, Fukuoka H, Wang GH, Ishikawa N (2004a) Undrained dynamic-loading ring-shear apparatus and its application to landslide dynamics. Landslides 1:7–19

Sassa K, Wang G, Fukuoka H, Wang F, Ochiai T, Sugiyama M, Sekiguchi T (2004b) Landslide risk evaluation and hazard zoning for rapid and long-travel landslides in urban development areas. Landslides 1(3):221–235

Scarlett B, Todd AC (1969) The critical porosity of free-flowing solids. J Eng Ind Ser A 91(1):478–488

Tika TM (1989) The effect of rate of shear on the residual strength of soil. PhD thesis, University of London (Imperial College of Science and Technology), 494 p

Tika TE, Hutchinson JN (1999) Ring shear tests on soil from the Vaiont landslide slip surface. Géotechnique 49(1):59–74

Vibert C, Sassa K, Fukuoka H (1989) Friction characteristics of granular soils subjected to high speed shearing, In: Proceedings the Japan-China Symposium on Landslides and Debris Flows, 1, pp 295–299

Wafid MA, Sassa K, Fukuoka H, Wang GH (2004) Evolution of shear-zone structure in undrained ring-shear tests. Landslides 1(2):101–112

Wang GH, Sassa K (2002) Post mobility of saturated sands in undrained load-controlled ring shear tests. Can Geotech J 39(4):821–837

Chapter 8

Rockslides and Their Motion

Mauri McSaveney* · Tim Davies

Abstract. The motion of landslides sourced from mostly bedrock (called rockslides) is controlled by the phenomenon of grain flow, and the frictional resistance of the constituent rock grains and their interstitial fluids. Modern understanding of grain-flow dynamics recognises that the important interactions between grains are irregularly distributed within the grain mass, with fortuitous alignments of grains carrying most of the stress in "force chains", while other grains are only weakly stressed. In rapidly shearing grain flows, under substantial confining stress, force-chain stresses rise high enough to crush grains. Such comminuting grain flows develop a distinctive grain-size distribution that is fractal over many orders of magnitude of grain size down to sub-micron sizes. In the moment of crushing, grains are not solids, and behave as high-pressure fluids. As the grain fragments are injected into lower pressure surroundings, they behave as would any other fluid, lowering the effective stress on other grains, and thereby lowering frictional resistance to flow. We show how this affected the blockslide component of New Zealand's prehistoric giant Waikaremoana rockslide; New Zealand's Falling Mountain rock avalanche triggered by an earthquake in March 1929; and a small prehistoric New Zealand rockslide that was too small to be a comminuting grain flow, but which fell on and mobilized a fine, saturated substrate. We use grain-flow dynamics to explain the motion of these rockslides determined through field studies and physical and numerical modeling.

Keywords. Rockslide, rock avalanche, grain flow, comminution, fragmentation, fracture energy, force chain, grain bridge, frictional resistance

8.1 Outline

The larger landslides on the rocky planets invariably involve bed rock, and are generally called rockslides, which are the largest natural manifestations of the phenomenon of grain flow. Here, we first describe what is meant by the term rockslide, and discuss modern understanding of grain-flow dynamics as a process governing the motion of landslides. Of particular interest in large rockslides is the distinctive grain-size distribution that results from comminuting grain flow. We discuss this grain-size distribution, how it arises, and its significance in the energetics of rockslide motion. Last we look at how comminuting-grain-flow dynamics has affected some real rockslides: the blockslide component of New Zealand's prehistoric giant Waikaremoana rockslide; New Zealand's Falling Mountain rock avalanche triggered by an earthquake in March 1929; and a small prehistoric New Zealand rockslide that was too small to be a comminuting grain flow, but which fell on and mobilized a fine, saturated substrate. Field and physical and numerical modeling studies have been used to gain understanding of the motion of these rockslides during emplacement.

8.2 What is a Rockslide?

David Varnes (Varnes 1978, updated in Cruden and Varnes 1996) presented the most widely adopted landslide classification. It is useful because it is simple and adaptable, and not because it is completely accurate; no classification can be perfect, but a common classification is essential if people are to know that they are discussing similar things. The terms rock slide and rockslide are synonyms. They imply a mass, or contiguous group of masses of mostly bedrock that moved or is moving downslope, apparently dominantly on surfaces of rupture, or on relatively thin zones of intense shear strain, so that the displaced mass of bedrock appears to have slid. The displaced mass may be somewhat disrupted, and there may also be evidence of some distributed deformation (flow) within it (Fig. 8.1). Rockslides become debris slides when the displaced mass appears completely disrupted, and this transition can occur in time or in space on the same landslide, so that what might technically be debris slide in the strict Varnes classification may still be called a rockslide in practice. There also can be transitions from sliding to flow or spreading (translation of blocks on a mass with distributed flow). Since the transitions can be gradational, and not necessarily readily recognisable, what may be a rockslide to some, may be a debris flow or a rock spread to others depending on the details of their perception of what is important. This seldom presents any particular technical difficulty in understanding as it has no bearing on the mechanical processes of the motion.

At the larger end of the rockslide scale, are what are often termed rock avalanches or rockslide avalanches in the literature (Fig. 8.2). Avalanche is not in the Varnes classification, but is acceptable in English usage. The term avalanche has never necessarily implied snow or ice, but in the context of snow avalanches, the qualifying adjective "snow" is normally readily apparent in context and is

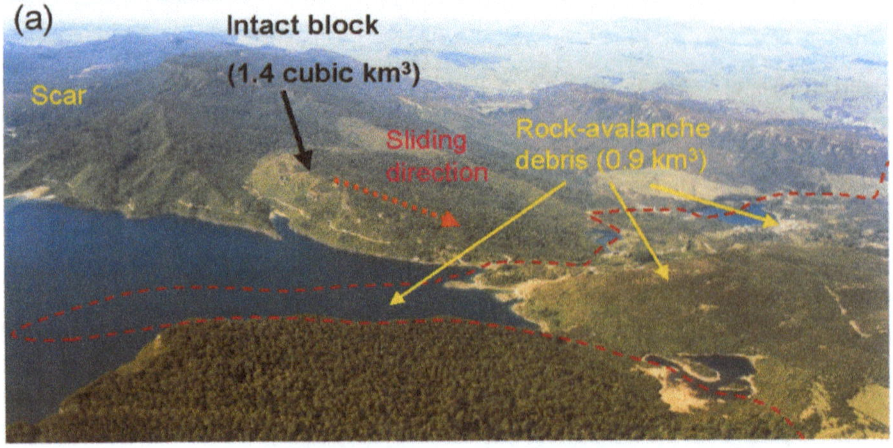

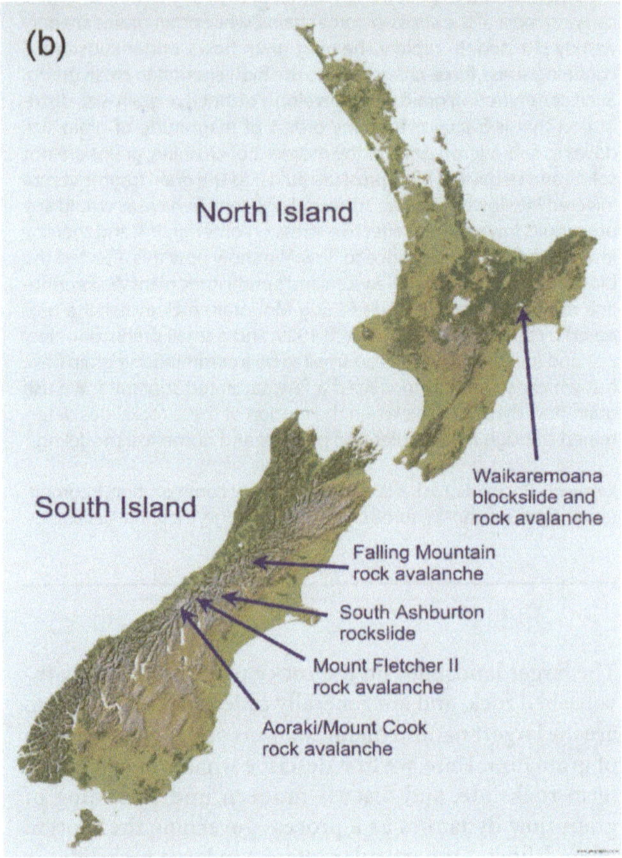

Fig. 8.1.
Aerial view of the Waikaremoana landslide and a relief map. **a** Aerial view of the Waikaremoana landslide showing where the blockslide component plowed into the moving rock avalanche and pushed up a 150 m high mound of debris as it came to rest from a speed of 40 m s^{-1} (photograph by Lloyd Homer); **b** relief map of New Zealand showing locations of the main rockslides described in the text

dropped. If the greater part of the displaced mass translates as an apparently intact block, it may be called a block slide or blockslide (Fig. 8.1), or even a block glide. These are just combinations of simple English words, each with precise meaning given by context. Rockslide avalanches, and blockslides are all encompassed by "rockslide".

The apparent sliding movement can be rotational (moving on a concave curved surface) or translational (moving on a planar surface). If rotation is along a circular arc, it can be accomplished with minimal break-up of the rock mass, but more usually the sliding surface is neither circular nor planar, and some amount of rock-mass break-up is inevitable due to flexing of the mass.

Although frictional sliding often appears to be the mechanism of displacement of rockslides, this is never true. The displacement mechanism is always grain flow, and the deformation takes place across a zone of granular material of finite thickness. For this reason, as with all landslides, a key to understanding rockslide motion is understanding grain-flow mechanics.

Fig. 8.2.
The Falling Mountain rock avalanche triggered by an earthquake in 1929 (photograph by Lloyd Homer). The landslide is believed to have first begun to move as a blockslide, which rapidly accelerated as it disintegrated to joint-bounded blocks on flexing during evacuation of the scar, before comminuting to mostly sand and silt in its 4 km runout

8.2.1 Force Chains and Their Role in Grain-flow Mechanics

The deforming part of all landslides is always made up of granular material, with an interstitial fluid of some kind, most often air or water. It is through deformation of this granular material that all landslides move, and it is through the interaction of the grains and their interstitial fluids that granular materials gain their frictional properties exhibited during flow. Granular friction resists motion, so frictional properties determine whether a given landslide subject to a given force speeds up or slows down. Grain flows are of great industrial importance; they are much researched, but not generally well understood (Campbell 2006). The literature is diverse and spread across many disciplines, in many languages, and with diverse technical jargon to inhibit cross-discipline communication.

Long-recognised complexities of industrial grain flows, are that flow can lock up unpredictably, and be difficult to restart, or the stopping of flow can transfer the flow's momentum to the container and rupture it. These occur because, although the grain mass as a whole behaves as a fluid, it is made up of rigid grains, and when deformation randomly assembles a series of grains into perfect alignment in direct contact, the chain of grains is a solid rod that resists deformation until it fails for some reason. These chains are called force chains, or grain bridges (Campbell 2006; Liu et al. 1995; Sammis et al. 1987). A well known example of a usually static force chain is the stone arch. Force chains exist in all grain flows where grains are moving towards one another; they cannot exist where grains are diverging, because this brings grains out of contact and breaks the force chain. Force chains can not normally be seen. Experimental "two-dimensional" grain flows using uniform discs that polarize transmitted light under stress have been used to image force chains (Fig. 8.3) (Howell, et al. 1999a,b), and they have been modeled numerically in four dimensions for materials of uniform and widely graded grain sizes (Hazzard and Mair 2003; Mair et al. 2001).

Present in any grain mass, whether moving or static, force chains form a 3-dimensional mesh of connected more highly stressed grains forming relatively short beams, columns and braces (to use a building analogy). As the range of grain sizes present in a grain flow increases, force chains become more diffuse, tending to proliferate around larger grains, but they still exist. All experiments show that much of the resisting stress in a shearing grain flow is carried by force chains. The principle compressive stress within each grain is highly nonuniformly distributed among grains, and strongly controlled by the shear rate. Some grains carry little of the average principle compressive stress, while others carry much of it, so some grains are highly stressed. For shear strain to persist, the force chains must fail, and force chains are constantly forming, stressing, failing and reforming among other grains – in all grain flows. Where the flowing grains are more heterogeneous in size, shape, or composition, the stress field of force chains is more varied.

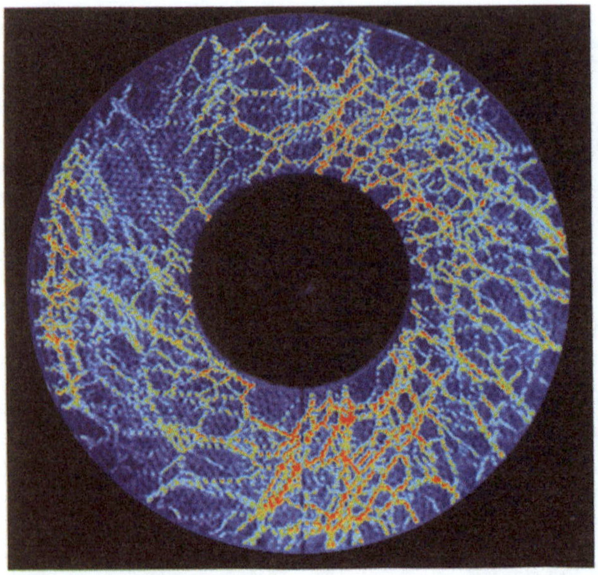

Fig. 8.3. Transparent plastic discs shearing under stress viewed with transmitted circularly polarised light to show force-chain stress concentrations (*http://www.phy.duke.edu/~bob/*). This false-color image is from Dan Howell (Howell et al. 1999a,b). This is a 2D experiment in which a collection of disks undergoes steady shear. When viewed through circular polarizers, high forces show up as bright regions. When the force is very large, the polarization is rotated through multiple phases of π, showing fringes. The experiment shows that forces in granular systems are inhomogeneous and intermittent if the system is deformed. The forces are detected by means of photoelasticity: when the grains deform, they rotate the polarization of light passing through them. In granular materials, force is rarely transmitted uniformly, but rather preferentially along a network forming force chains – often right next to regions where there is little or no force. The illustrated disks are under a large load the high-force areas are highlighted in *red* (false color). Note that some of the disks are *not coloured* at all (no force), while disks right next to them are highlighted in *red* (high force)

Individual force chains are usually short. There is no theoretical reason for any particular length, although lengths of one and two grains are meaningless and prohibited by semantics. Although the meshwork must straddle the grain mass, the alignments are fortuitous, and so long perfect alignments are rare, and very short chains are indistinguishable from nodes in the mesh. In modeled grain flows, chain lengths are variable, but generally ~5–15 grains.

Force chains can fail by several mechanisms; most fail by inter-granular slip, grain rotation, or buckling of the whole chain, some can fail by grain divergence, but the weaker and more highly stressed grains can fail by crushing (grain fracture) when force-chain stress exceeds grain strength. At failure, the elastic strain stored in all the grains participating in a force chain is released, at least in part, and this drop in elastic potential energy appears as kinetic energy in the direction of failure as rotating and (or) translating grains, or if a grain has fractured, as the kinetic energy of its fragments moving away from its former center of mass.

Several elements of force-chain failure are of particular interest in landslide behavior. If the granular mass has significant cohesion through the presence of clays and water, force chains can exist in tension, and so the divergence of grains (as well as their convergence) is resisted. Hence such cohesive materials can provide more resistance to flow than were they dry and non-cohesive. In addition, the platy character of clay minerals makes for particularly weak force chains as grains become aligned by shear displacement, and slip is facilitated by water films and grain alignments.

Force-chain induced fracture of brittle grains is important in rockslides. The strength of materials increases as the rate of application of stress increases; the faster a landslide mass deforms, the more stress its grains can bear before failure, and the transfer of stress at grain failure becomes higher. In part, this is because as the strain rate increases, the amount of super stress above the grain yield strength increases because of the finite time for fracture crack initiation and growth. This increases the probability of rapid, catastrophic failure, and decreases the mean fragment size when fracture occurs. When individual grains crush, they momentarily cease being solids; they can bear no directed stress, and briefly behave as a fluid under pressure, transmitting an isotropic stress (pressure) to their surroundings. The value of this pressure is the applied directed stress causing grain fracture – at the instant of grain failure, it is always a little higher than the compressive strength of the grain material under the ambient conditions of confining stress, temperature, and strain rate, but it diminishes as the crushing grain expands into its lower-pressure, surrounding grain mass beyond the immediate failing force chain. Fragments from the crushing grain are pushed out of the shortening force chain, and injected into surrounding lower-pressure regions, between other, less stressed grains, thereby forcing them apart, and transferring stress from the force chain to surrounding grains.

When ephemeral force chains fail, their temporarily stored elastic energy returns to kinetic energy, but as in all such things, some energy is lost to heat. With slip, rotation and buckling, kinetic energy is returned only in the direction of slip, rotation or buckle. In grain breakage, it is returned in all lower-pressure directions radial to the fragmenting grain. That is, momentarily, the crushing grain transfers a stress equal to its compressive strength under the ambient loading conditions, from the direction of the principle compressive strain to all other stress components, resulting in overall increases in the mean regional stresses in these directions in proportion to the volume of grains fragmenting, and the compressive strength of the rock (which varies with confining stress and rate of application of load). By adding to the mean stress that supports the overburden weight, the effective overburden weight carried by other grains is reduced, and they can move with reduced frictional resistance. Lateral and longitudinal mean stresses are also increased, and these also can significantly affect motion in thick crushing grain flows.

The effect of grain crushing is simplest understood through the well known Mohr-Coulomb relationship (Eq. 8.1) which applies also to non-fragmenting granular flows. It relates shear stress, τ, to grain contact stress through an internal friction angle φ (Eq. 8.2) or coefficient of internal friction μ such that:

$$\tau = \mu(\sigma - p) - S_0 \tag{8.1}$$

and

$$\mu = \tan\varphi \tag{8.2}$$

for intergranular contact (effective) stress $(\sigma - p)$ (σ is overburden load and p is pore-fluid pressure) and cohesive strength S_0. The crushing grains add an ephemeral high "fluid" pressure p to the grain mass and reduce the effective confining stress on grains in proportion to the crushing strength of the material and the proportion of the grain mass that is crushing simultaneously. Since the crushing strength is usually very much higher than the overburden stress (for typical brittle crustal rocks, the pressure of a fracturing grain can be of the order of 0.5 to 5 GPa, and much more than the overburden stress in even the largest landslides) only a small proportion of simultaneous crushing grain mass is needed to significantly lower the frictional resistance to flow.

Crack growth and hence grain fracture are catalysed by water, and therefore occur at slightly lower stresses than in dry rock, a difference noticeable mostly on the waterless planetary bodies of Mercury and the Moon. Of course, under water-saturated conditions, the pore-water pressure is additional to that from grain fragmentation, and a lower proportion of simultaneously crushing grains will produce the same reduction in frictional resistance.

8.2.2 Grain-size Distribution in Rockslides

In the conventional measure of grain-size distribution (distribution by mass) the grain-size distribution in many landslides appears to follow a two-parameter Weibull distribution (Eq. 8.3), the so called "Rosin's Law of crushing", or Rosin-Rammler distribution) well enough to pass some statistical fitting tests, but its usually slight misfit appears to be systematic across the range of grain sizes (McSaveney 2002; Dunning 2004). In the Weibull distribution, the proportion by weight of rock debris (P_f) finer than a given grain size x, is given by

$$P_f = 1 - e^{-ax^b} \tag{8.3}$$

were a and b are empirical constants.

Distribution by mass, however, is deceptive and is an inappropriate measure when grain mass is not of interest. Of interest in grain-to-grain interactions are the size of the average grain and the probabilities of grains encountering grains of similar or different sizes. Probability of grain encounters is linearly proportional to grain surface area, not grain volume: the more total grain surface area is contained on grains of a given size, the more likely that a given grain of any size will make contact with a grain of that size. The total surface area on grains of a given size is determined by the average surface area of such grains, and their total number. If we assume that all grains have a similar range of shapes, regardless of size, then grain surface area (a_d) of a given grain size (diameter d) will be given by:

$$a_d = s\pi d^2 \tag{8.4}$$

where s is a shape factor >1.0 relating the surface area of the average grain shape to the area of a sphere of the same mass. As long as our interest is only in relative areas, and all grain sizes have similar ranges of shapes, we do not need to know this shape factor. The total area (A_d) contained on all grains of a given size per unit volume of deposit is given by:

$$A_d = n_d a_d \tag{8.5}$$

where n_d is the number of grains in that volume of a given size d. The total grain surface area A then is:

$$A = \sum_{d=0}^{d=d_{max}} n_d \, a_d \tag{8.6}$$

Mean grain size (d_m) per unit volume is given simply by:

$$d_m = \rho / \sum_{d=0}^{d=d_{max}} n_d \tag{8.7}$$

where ρ is the ratio of densities of deposit and grain material (ignoring for simplicity, density differences between component mineral grains which become more prominent at the smallest sizes).

Critical to accurate estimation of mean grain size, and relative proportions of surface area is accurate knowledge of the complete grain-size distribution, particularly in the smallest grain sizes where most of the grains are. A "grinding limit" of about 1 μm for quartz has been suggested (e.g., An and Sammis, 1994). Electron-microscopy on a number of rock-avalanche samples (A.L. Strom, RAS Moscow, personal communication 2005), includes several recent New Zealand examples supplied by us (1991 Mount Cook, and 1992 Mount Fletcher II rock avalanches (McSaveney 2002)), and provides convincing evidence of grains in sub-micron sizes down to the resolution of the electron microscope used for counting (~0.077 µm).

Grain counts across two size ranges (~0.077–2.75 µm and ~1.0–10 µm) have unknown quantitative relationships to each other and to grain counts calculated from conventional weight-percentage particle-size analysis over the range 0.002–100 mm. By assuming that the sub-samples are parts of the same grain-size distribution, however, the three distributions were aligned by simple multipliers to approximate the complete distribution (Fig. 8.4). Both the Mount Cook and Mount Fletcher II samples are sufficiently similar over overlapping portions of the range of sizes, and so no error is likely to have arisen by assuming that the distributions should be continuous from one sub-sample to the next.

Above about 0.6 microns, both distributions closely follow a simple Power-law distribution with an exponent of ~-2.58, to the largest grains present in the deposits with no systematic departures. That is, the distributions are fractal across most of the range of sizes present. The fractal nature of comminution by fragmentation is well established (Turcotte 1986). The fractal character has significant implications as to the origins of deposits and the dynamics of the processes that created them.

The sub-micron portions of both distributions are non-fractal (Fig. 8.4). There are systematic decreases in proportions of grains below 0.6 µm in both deposits. Despite

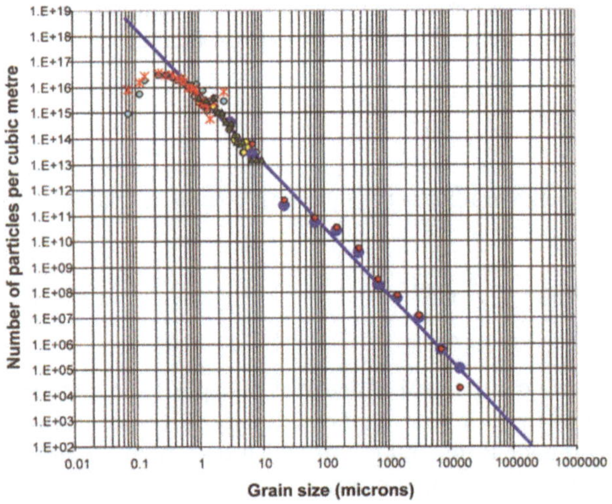

Fig. 8.4. Distribution of grain numbers by grain size in two recent New Zealand rock avalanches (Aoraki/Mount Cook of December 1991 and Mount Fletcher II of September 1992, McSaveney 2002). Grain counts courtesy of Dr. A.L. Strom, RAS, Moscow. The line is drawn with a power-law exponent of -2.58; it has not been fit to the data. We explain the departure of the data from the power-law at the finest sizes as resulting from the finest grains selectively adhering to other grains through Van der Waals forces and being selectively depleted during grain-size analysis

major differences in sample histories, both samples are identically self-similar across the range of grain sizes above 0.6 µm. (and almost identically non-fractal below 0.6 µm). Electron-microscope images of debris-avalanche grains (such as Fig. 8.9c,d of Belousov et al. 1999) reveal a possible explanation for the change in distribution trend below 0.6 µm. Such images always show sub-micron fragments clinging to sand grains. Increasingly smaller grains are increasingly more influenced by Van der Waals forces, and adhere strongly to any other grain; hence they are removed on larger grains if these are selectively removed in analysis. Below an undetermined size smaller than 0.077 µm, smaller sizes are incapable of maintaining independent existence; they adhere too strongly to other grains, and so are not measurable after the event. If these are the causes of the reversal in trend at the smallest sizes in these grain-size distributions, it implies that a "grinding limit" recognised in industrial grinding is not the smallest size of fragments capable of being created, but the smallest size capable of maintaining independent existence. Much smaller sizes are possible, and may be produced, but they can not be verified by this technique. Hereafter, we assume that grain comminution is identically self-similar across the entire range of particle sizes present in grain flows, down to and including the smallest sizes capable of maintaining independent existence. This suggests that at 0.077-µm nominal grain diameter 99.82–99.98% of the grains of this size might have been missed in counting, for one reason or another, but if so, the missed grains account for less than 0.1% of the mass (<1 g kg^{-1}). The assumption is one of convenience; it is not essential to understanding the role of comminution.

Grain flows in which grain crushing has been important are thus very distinctive. Not only are they much finer grained than their parent materials, they also exhibit a distinctive grain-size distribution that makes them look as if their grains have never been sorted by mass or volume. Since all sizes are present, they can never have been selected for size in their making, other than to have become as small as possible in the time available, but the apparent lack of sorting is deceptive, because it demonstrates that a sorting mechanism has operated (see below).

Although crushing-grain-flow deposits often appear to be what is termed "matrix-supported" (the classical Blocks-in-Matrix or BIM rocks), apparently matrix-supported grain masses can grade laterally and vertically to apparently clast-supported debris, usually with no apparent reason, and sometimes with identifiable inherited source structures traceable across the gradational boundaries. On detailed sampling and size analysis, relative-size relationships between larger and smaller grains are self-similar at most scales, and geometrically the deposits either are all matrix, bar the largest few grains, or have no matrix. The difference between matrix- and clast-supported structures in comminuted grain masses is merely one of perception and degree of comminution, which often is spatially very highly variable in these deposits. Locally lesser comminution can arise because: the locally viewed grain mass is from a larger grain crushed late in the flow; the rate of last grain crushing was slow; the local shear rate was always slow; or any combination of these only partly independent variables. The number of fragments from a single crushing is a positive power-law function of strain rate (Zhang et al. 2004).

In theoretical and experimental studies, Sammis et al. (1987) found fault-gouge grain-size distributions to be characterized by a fractal dimension of 2.58 (based on a theoretical three-dimensional geometry with no two grains of the same size adjacent to one another, although this arrangement is impossible to demonstrate by direct observation). Other theoretical models with different fractal dimensions exist (1.97, Turcotte, 1986; 2.84, Allegre et al. 1982). In addition to the two samples discussed above, we (Davies and McSaveney, in press) find that fractal dimensions of a number of grain-size distributions from pervasively crushed samples from many rockslide avalanches and two fault gouges (Fig. 8.5) are well approximated by 2.58. In more detailed sampling of a few rock-avalanche deposits, some in common with the data of Fig. 8.5, Dunning (2004) finds a variety of lower fractal dimensions that appear to approach and reach 2.58 with sample distance from source. Crosta et al. (in press) find a variety of fractal dimensions ranging from 1.2 to 2.9 in the Italian Val Pola rock-avalanche deposit. Possibly of most significance in these studies is the conclusion that grain-size distributions by number are almost always fractal across most of the range of grain sizes present in such deposits – they do not follow Rosin's Law, but they are formed by crushing.

The Val Pola evidence may imply that several mechanism are causing fragmentation, and acting to greater or lesser extents depending on local conditions, such as grain confinement, and grain material. Possible fracturing mechanisms are crushing between grains, grain-to-grain collisions (impacts), attrition of grain edges, and grain bending moments. All involve locally exceeding the strength of the material. At low confining stress, large voids additionally allow gravitational mechanical sieving to change grain-size distributions, but in general, fractal size distributions and high confining stresses strongly inhibit gravitational sieving. Immense dust clouds are usually seen with rockslides, and it is likely that this is expelled from the weakly confined, usually coarser outer part of the rockslide, that we have come to call the "carapace" (Dunning 2004). The extremely dangerous environment during rock avalanching precludes direct field observation of any mechanism, prohibiting evaluation of relative roles.

The simplest theoretical model which has no two grains of the same size adjacent to one another is one in which no two grains anywhere have any reason to be exactly the same

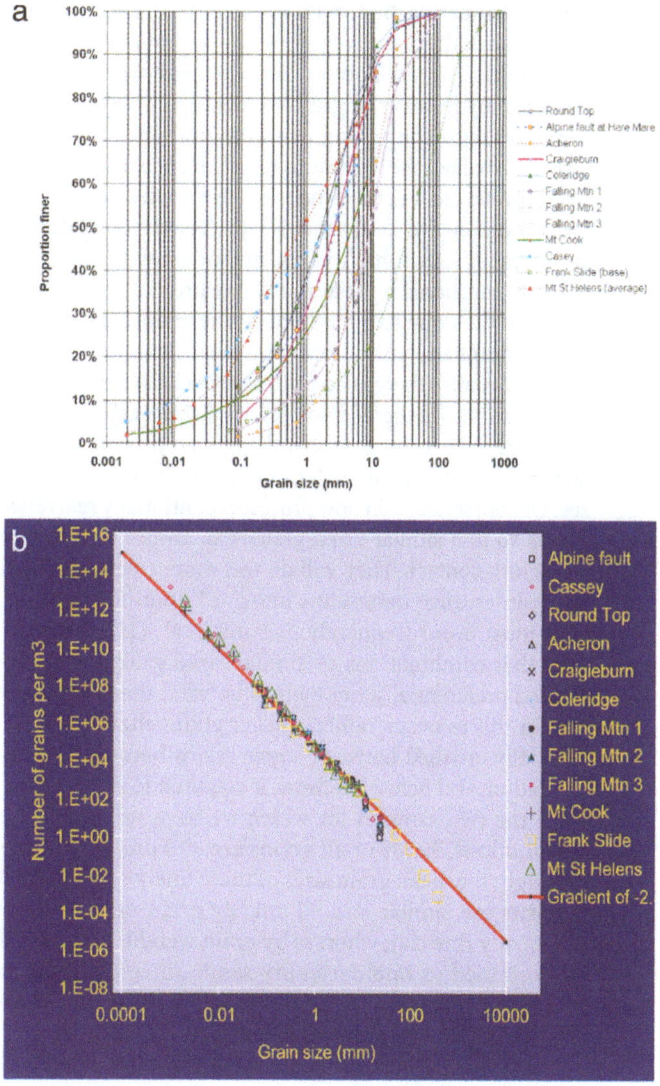

Fig. 8.5. Grain size distributions. **a** Grain-size distributions of a number of rockslide deposits expressed conventionally as proportion by weight finer than a given grain size. Alpine Fault and Casey are samples of fault gouge. **b** Grain-size distributions of the same rockslide deposits expressed numbers of grains per unit volume of a given grain size. The power-law line is drawn with an exponent of –2.58

size, and all are therefore likely to be of different sizes. Grain-crushing appears to be such a mechanism, but it can not directly produce more finer grains than coarser grains in a single fracturing event, this can only evolve through many events and with other processes operating.

Comminution is progressive, and there is a fixed lower limit below which grains do not maintain independent existence, so the total number of grains in any given volume of debris must increase with continuing comminution by increasing the numbers of smaller grains at the expense of the largest grains (this is axiomatic from the definition of grain comminution). As the number of grains increases, the proportion represented by the largest grain must decline, even as the largest grain size declines (since there is always only one largest grain). Hence one might expect the power-law exponent of the size distribution by number of grains to decrease (become more negative) with continuing comminution once the modal size approaches the "grinding limit". Hence, there should be no preferred fractal dimension that such size distributions could evolve to. If there is a preferred fractal dimension around 2.58, it has to arise in tandem with the grain-creation mechanism. That is, if there are no two adjacent grains of the same size, it is because the grain-creation process is simultaneously destroying such fortuitous combinations because there is no process especially creating such combinations (ie there is no preferred size, and so all sizes are made). We know also that there is no process creating larger grains, except at the "grinding limit" which we do not see. Of course, all grain flow processes continually rearrange grains so that similar-sized grains can always fortuitously come into contact. They will do this more frequently than they can arrange themselves into 5–15 grain force chains, so it must occur frequently. Sammis et al. (1987) considered that combinations of similar-sized grains were destroyed preferentially, but there is no valid theoretical reason for this to occur, rather, smaller grains should be preferentially crushed between larger grains because less deformation, and hence less work is required to crush them.

In the two samples for which we have full grain size distributions, 99.5% of all grains are <10 µm, so not surprisingly, the mean grain sizes of these finely comminuted deposits are similar and ~1 µm. By grain number, they are mostly fine clay, whereas by grain weight, they would be described as bouldery, silty sand. Since boulders to more than 10 m across are present in the deposits, clearly smaller grains have been crushed preferentially.

Although we lack electron microscopy for our other samples, they are all very similar across the range of sizes determined (down to 2 µm for several samples), strongly suggesting that they too might extend similarly to below 1 µm (but their distributions below 0.6 µm are as yet unknown). Few if any deposits have no visible dust, and every dry rockfall is accompanied by a prominent dust cloud that limits detailed observation. Calculation of accurate mean grain size and accurate proportions of surface area, are critically dependent on knowing the distribution in the range below 1 µm. In any event, the deposit mean grain sizes are likely to be very small, and most (>99%) of the grains in these comminuted-grain-flow deposits are silt and clay-sized particles. This is well hidden from view by the <1% of larger grains that occupy almost the entire volume, and lead such deposits to be described as bouldery, at least until their interiors are exposed by erosion.

Any model of how this peculiarly regular grain-size distribution comes to exist must entail interactions between grains, either by considering real grain interactions or considering them by proxy. The probability of grain interactions is proportional to grain-surface area (Fig. 8.6), and most grain interactions most of the time are between the grain sizes that represent most of the grain-surface area, and not with the grains that represent most of the volume. Since grain-flow deformation takes place by relative motion between grains, it follows that during much of the flow duration, the deformation in these comminuting grain flows takes place between smaller grains – at all scales.

Because the comminution process appears to be identical at all scales, we do not necessarily need to study the process at micron and sub-micron scales; it suffices to comprehend it at any macro-scale, and to verify that scale is not a relevant parameter at macro-scales. It also follows from the fractal distribution that every grain, even in what appears to be merely the "dust" among the larger grains, is as important to the flow dynamics as is any other grain in the flow, and the dust should not be ignored simply because it is unseen.

The least relevant grains are the large boulders that often litter the deposit surface. They are less relevant because they lack the same opportunities to interact with as many surrounding grains as do grains in the interior. They participate mostly by providing overburden pressure. A carapace of larger grains is commonly found on the upper surface of many rockslide deposits. Carapaces can vary in thickness, but generally are about 10 m thick. Individual carapace boulders can range up to megaclasts hundreds of meters across. Large "Toreva" blocks of original volcanic edifice sometimes kilometers across are often recognised on top of volcanic debris avalanches, and some large blockslides cover tens of km^2 with mostly intact blocks hundreds of meters thick, and only a relatively thin layer (a few meters thickness) of comminuted grains at their base.

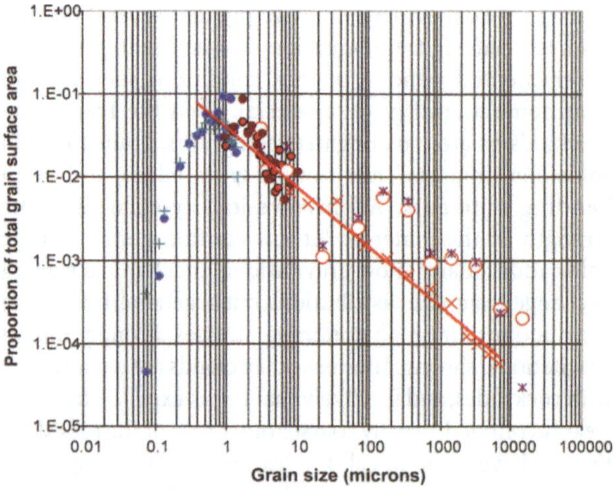

Fig. 8.6. Distribution of total grain surface area by grain size for the rockslide deposits of Fig. 8.4. The power-law line is fit to the Mount Cook data of Fig. 8.5 which was determined from a much larger sample than those available for Fig. 8.4, and so is considered to be more representative of the larger fraction of the Aoraki/Mount Cook rockslide avalanche

8.3 Size Distribution from Grain Crushing

It is self-evident that mass is conserved in grain crushing, so the sum of fragment volumes must remain the original volume of the crushed grain. It is observed empirically that the number of fragments is a positive power-law function of strain rate (Zhang et al. 2005). Fragments in crushing are defined by the intersections of cracks as they propagate through the crushing grain. There appear to be no other contraints. The largest fragment must be smaller than the parent; the smallest fragments may be incapable of maintaining independent existence, and incapable of verification, but there appear to be no other size requirements. Geometric logic dictates that the total surface area on all fragments is the original grain surface area plus the twice the total area of cracks generated. Thus grain numbers scale linearly with total crack area, and hence with the square of mean diameter (nominal grain size). If grain fragments form without regard for size, it follows that they will form randomly with the square of grain size, and there will be more larger grains than smaller grains among the fragments, with a Power-law gradient of +1.0, not –2.58.

If fragments are being formed by grain crushing, then grains are being crushed between other grains. If they all are of similar strength, then those requiring less deformation (ie smaller grains between larger grains) are preferentially crushed. Since the comminution process cannot make grains any larger, continued comminution leads to more and more smaller grains relative to larger grains, across all but the finest scales of grains that have been created. The tail of diminishing grain numbers at the finest end of the distribution will be continually pushed to finer and finer sizes until grains reach the limit beyond which they cannot maintain independent existence.

How then, does a power-law gradient of –2.58 come to be so common? If generally no two adjacent grains of the same size are needed to fulfill the theoretical requirements for a fractal dimension of 2.58, it cannot be because it gives all grains equal and minimum probability of being further fractured as Sammis et al. (1987) believed.

If we consider the fate of the fragments formed by grain crushing in force chains, we can see how the system can self arrange to eliminate combinations of similar-sized grains. When crushing-grain fragments are pushed from force chains, they are injected into lower pressure surroundings. The lowest pressure surroundings are the pore spaces between grains, and so smaller grains are being injected between larger grains at all scales, destroying combinations of similar-sized grains which require larger void space

As shown above, no particular fractal dimension should be maintained once the modal grain size approaches the "grinding limit"; and so the significant information is that the distribution is self-similar at all scales, and presumably formed by the same processes at all scales. Also, injection of finer grains between larger grains confirms that the fragmentation process can be viewed as introducing a high-pressure fluid into the pore space; this fluid competes for pore space with every other fluid that might be present.

8.3.1 Energy Loss in Rock Comminution

Such intense rock comminution as described above is found in all large rapid rockslides and beneath blockslides. There are a variety of concepts of the role comminution may play in their motion. Many studies ignore the comminution, assuming that it occurs before the landslide, or in the initial moments, and giving it no part in the motion. In studies that consider comminution, opinions are divided as to whether it aids or hinders motion. Many see it as an energy sink, whereas we see it as a temporary energy diversion. For certain, it is positive evidence for resistance to motion, which we call friction, and proof that energy has been lost to friction, but is the loss to friction more, less, or the same as the loss that would occur if comminution had not occurred during the grain flow? Assuming that it occurred before leaves unanswered the question of the means by which it occurred. Assuming that it occurred in the initial moments requires an explanation of where the energy came from to achieve it, and why the huge amount of elastic strain energy released in that instant is not all too abundantly evident? We know how rock masses respond to sudden huge releases of elastic strain energy. Where is the required ballistic ejecta sheet that has never been reported?

Fragmentation of rock is of great industrial importance, and has been much researched for some centuries, even before modern concepts of energy had been developed. There are applications desiring only large fragments, and others calling for intense fragmentation. Whatever the application, fragmentation is accomplished at a cost. It takes input of energy to fragment rock, and controlling the rate of input of energy can cause it to fragment in a specific, desired way. In industry, it costs money to apply the amount of energy that will break rock, and so the concept of "fragmentation energy" arose; although it is much researched; it is not well understood, and generally is misunderstood. All methods for fragmenting rock use the same process – a mass is deformed until it breaks; if it is not deformed in any way, it will never break.

Even in a granular mass, work is done to accomplish deformation; for the same amount of deformation, if the same force is applied, the same work is done, whether or not grains fragment. This work done is energy expended, not withstanding that the elastic component is recovered. After comminution, the energy expended can be divided by the new surface area created by the rock breakage, and "fracture surface energy" is calculated this way. In industrial applications, this is an energy loss to be paid for. But the work done is the deformation of the granular mass (aside from energy that is lost to heat, which is lost whether or not there is grain breakage). Deformation of the grain mass is an integral part of landslide motion, and so is *not* an energy loss in landsliding. There is no actual energy associated with the surface itself – the energy is *expended* in the process of creating new surfaces. Van der Waals forces are associated with proximity to molecules, not proximity to surfaces. They are present internally, but cannot be accessed. As granular solids are crushed finer and finer, there is no net increase in the total energy of the mass, other than any thermal energy from the friction – there is no energy trapped on the surface. Fracture surface energy is not an energy loss in landslide motion, and is associated with the fracturing deformation and not with the resulting surfaces. Its association with surfaces is through historical accident involving the measuring process, and many theoretical studies of its significance are spurious. It does, however, have much useful application to industry. To associate it with the surfaces is a misinterpretation of its significance; just as vehicle fuel consumption is associated with the engine and the journey, and not associated with the fuel, surface energy is associated with the material and its deformation (journey); it is not associated with the surfaces.

Fracture-surface-creation energy returns to grain flows as grain kinetic energy and heat, and there is a major logical error in accounting for it as a separate entity. This logical error is perpetuated in recent literature (eg Locat et al. 2006), but had its origins in the 18th and 19th centuries (Belidor 1725; Rittinger 1867, cited in Bond 1952). A short quote from a historical summary of how rocks break (Bond 1955) reveals how easily the error arose in the transfer from industrial to scientific application: "*Rock breakage is produced by deforming the rock, commonly under pressure, until the resulting stress locally exceeds the breaking strength and crack tips form, usually on the surface. The surrounding strain energy then flows to the new crack, which is thereby extended to split the rock. When the rock breaks, or the strain is otherwise released, the mechanical energy input is transformed into heat.*" We now recognize that cracking releases strain energy which "flows" (concentrates) away from the crack which is thereby extended. When the strain is released, it is ultimately transformed into heat, but the post-fracture transformation process (deformation) is more relevant to deformation studies than either the fracture process, or the heating. In blasting and crushing, however. post-fracture deformation is irrelevant, and the ultimate fate of released mechanical energy is frictional heat. In landslides, the post-failure deformation is the landslide, but the ultimate fate of all the released mechanical energy is still frictional heating.

8.3.2 Dynamic Fragmentation and Collapse

Comminution is a major physical-weathering process that creates smaller particles of rock from larger particles. Although the landslide source rock may sometimes be described as "intact", it contains many unhealed joint systems, formed by a variety of stresses endured between emplacement of the geological formation, and its exposure in outcrop. There are never landslides in completely intact rock. Some of the comminution is collapse of the original rock mass along such defects, with breakage caused by flexing or contact stresses during motion. Inspection of finely fractured debris, however, shows that in addition to blocks defined by pre-existing joints, there also many fragments formed by breakage well below any joint spacing (Fig. 8.7).

Separation along pre-existing joints may occur at relatively low stresses, depending on the degree of development of the joints or secondary cementation, but breakage of intact rock between joints requires high stresses, and hence occurs most commonly (but not exclusively) where high dynamic stresses develop – in, for example, large-volume, rapid, rockslides and those involving large fall height or high impact velocity. Breakage of unjointed rock by dynamic mechanical stress is termed "dynamic fragmentation" (Grady and Kipp 1987; McSaveney and Davies 2006).

To distinguish the high- and low-stress modes of comminution of a parent body, Davies et al. (1999) used the term *collapse* for separation along joints, and *fragmentation* for the further comminution below that dictated by joint spacing. In this context, collapse and fragmentation have different effects on the behavior of the developing granular mass. The distinction is useful in concept, but collapse and fragmentation are merely extreme end mem-

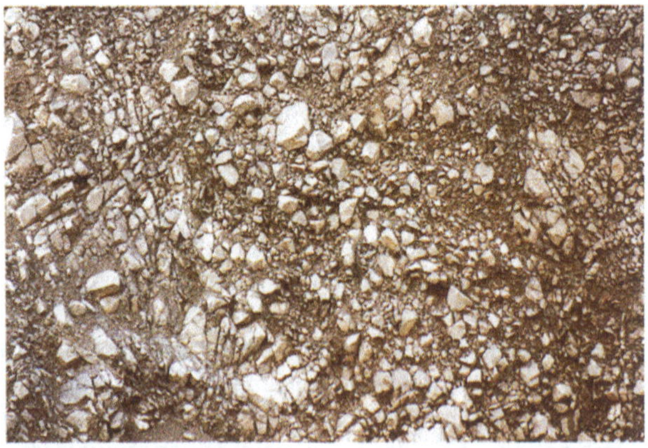

Fig. 8.7. Finely fractured debris in the Falling Mountain rock avalanche showing breakage of grains to well below any pre-existing joint spacing. Field of view is about 2 m across, but the fractal nature of the grain-size distribution indicates that it would appear essentially the same if viewed at any scale

bers of a continuum of behavior of rock-mass defects in the break-up of large rock masses. Our interest mostly is with the behavior of brittle rock masses at rapidly applied high stresses where the grains in the deforming mass are broken and re-broken many times, so that no especially weak defects remain, other than between grains.

Since joint spacings are often fractal, the size distribution of joint-bounded blocks is often fractal. Dunning (2004) has shown that granular masses formed mostly from collapse can not be distinguished from ones that are much more intensely fragmented solely on the basis of power-law gradient of the grain-number distribution by grain size.

The elastic-strain energy (W) released by failure of a unit volume of stressed rock is given by:

$$W = Q^2/2E \tag{8.8}$$

(Herget 1988), where Q is the rock strength and E is its elasticity (Q and E assumed to be isotropic). It is available to do work only while it is being released. This is why it is absurd to assume that all, or most of the comminution occurs in the initial moments of the landslide; sudden release of so much energy could not pass unnoticed. Nor is there the means to accumulate this much energy in a static grain mass.

Note that the work achievable from the energy release is that due to a single failure episode; the same grain mass can repeatedly absorb and release the same amount of energy each time that its increasing number of grains are stressed to failure (as distinct from relict tectonic or overburden stresses that can only be released once).

In dynamic fragmentation, a grain begins to fail when its material Hugoniot elastic limit (HEL) is exceeded. The rate of progress to total failure depends on intrinsic material properties, the rate of applied strain and the confining stress. The static unconfined compressive strength (UCS) of the material is the lowest possible HEL. For many types of brittle rock in outcrop, the HEL is of the order of 1 GPa (Melosh 1997). It has been stated (Knoeberl 1997) that "The only known process that produces shock pressures exceeding the HELs of most crustal rocks and minerals in nature is impact cratering. Volcanic processes are not known to exceed 0.5 to 1 GPa." But such thinking arises through misunderstanding the HEL, which includes, but is not restricted to, shock compression. Although the shock fronts associated with the formation and failure of force chains in rockslides are in the lower range of what might commonly be perceived as "shock" compression, it is nevertheless HEL exceedence. In the moment of grain failure, the crushing grain should not be described as a solid, above their HELs all materials are high-pressure vapours, and the crushing grain effectively is a high-pressure pore fluid in the moment of crushing – a very ephemeral fluid.

8.3.3 Effect of Dispersive Stresses Caused by Dynamic Grain Fragmentation

Dynamic grain fragmentation always generates a high local pressure, but the proportion of the total grain mass being fragmented at any instant is often very small. For example, the formation of a scree slope by intermittent rockfall results from and generates energetic breakage of grains in occasional, widely spaced events. Likewise, a small rockslide may have very few grains simultaneously fragmenting because most grains are free to slip or rotate in response to applied stresses, rather than being confined and forced to deform and fragment. Spatially and temporally widely separated dispersive stresses, even though individually of high magnitude, need not pervasively affect the overall motion of the associated grain mass, and thus may not result in noticeable morphological characteristics. A part of the reason for this is that grain masses are very effective at absorbing and redistributing locally high stresses; this is why soldiers can be protected from individual high-velocity bullets by sandbags.

Where fragmentation events are more frequent and more closely-spaced, however, deposit morphologies begin to exhibit characteristics indicative of enhanced mobility during emplacement. This enhanced mobility comes from the added "fluid" pressure of fragmentation as illustrated in the Mohr-Coulomb relationship (Eq. 8.1). It takes its most dramatic form in impact cratering when 100% of the rock mass is fragmenting in an instant, but ~1 GPa pressures can dramatically alter mobility characteristics by reducing frictional resistance when less than 1% of grain mass is instantaneously fragmenting (1% of 1 GPa is equivalent to a piezometric head of 300 m of water in the pore space).

8.3.4 Fragmentation in Confined Layers

We first consider rockslides in which rock is fragmented in a relatively thin layer beneath a large intact rock mass. This situation is very similar to that frequently investigated in laboratory tests. Common landslides of this type are long-runout, low-angle block slides in which low sliding friction between rock surfaces is necessary to explain observations.

Shear banding is commonly reported in blocksliding (e.g., Anders et al. 2000) and some volcanic debris avalanches (Clavero et al. 2002), and less commonly in non-volcanic debris (rock) avalanches (Yarnold and Lombard 1989). In our experience, shear bands are not absent in non-volcanic debris avalanches, but they are not visually dominant.

Consider a thin, granular, shearing, fragmenting stratum between a moving non-deforming layer and a sta-

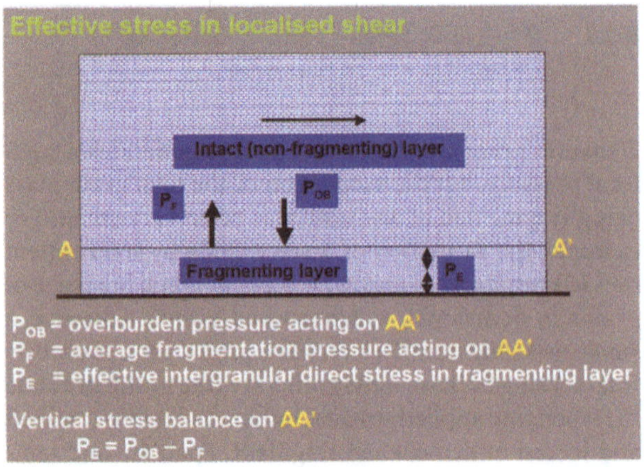

Fig. 8.8. Definition sketch of stresses operating in a thin, granular, shearing, fragmenting stratum between a moving non-deforming layer and a stationary base

tionary base (Fig. 8.8): in which P_{OB} is the overburden pressure at top of the fragmenting layer, P_F is the average fragmentation pressure in the fragmenting stratum and P_E is the effective intergranular direct stress in it. Stresses and strain rate are assumed time-invariant.

Applying a vertical stress balance at the top of the fragmenting stratum:

$$P_E = P_{OB} - P_F \tag{8.9}$$

Each fragmenting grain generates a pressure = Q, so

$$P_F = k_1 F_f \tag{8.10}$$

where F_f = proportion of simultaneously fragmenting grain mass; $k_1 = Q \sim 10^9$ Pa.

The simultaneously fragmenting proportion (at a given shear rate) will depend on the amount by which the effective intergranular direct stress P_E exceeds the minimum needed to cause fragmentation P_C:

$$F_f = k_2(P_E - P_C) \tag{8.11}$$

assuming linear dependency, which seems reasonable.

In a simulation of the Falling Mountain non-volcanic debris avalanche (Davies and McSaveney 2002), the required mean fragmentation stress explaining the runout distance was ~2.5 MPa, implying $F_f \sim 0.01$. This simulation simply used the longitudinal component of dispersive pressure to provide the force needed to generate the observed runout, without affecting effective stress, so that $P_{OB} = P_E$. The mean overburden pressure resulting in this stress was ~0.75 MPa. The minimum stress causing fragmentation (P_C) was that at the base of the relatively unfragmented 10-m deep carapace, ~0.3 MPa. This observation suggests that $P_C \sim 10^{-3} \times$ Unconfined Compressive Strength.

Hence in Eq. 8.11, $0.01 = k_2(0.75 - 0.3) \times 10^6$, so $k_2 = 2 \times 10^{-8}$.

From Eqs. 8.9, 8.10 and 8.11,

$$P_F = k_1 k_2 (P_{OB} - P_C)/(1 + k_1 k_2) \tag{8.12}$$

$$k_1 k_2 = 10^9 \times 2 \times 10^{-8} = 20 \quad , \text{ so}$$

$$P_F = 20(P_{OB} - P_C)/21 = 0.95(P_{OB} - P_C) \tag{8.13}$$

Hence

$$P_E = P_{OB} - P_F = 0.05 P_{OB} + 0.95 P_C \tag{8.14}$$

In terrestrial landslides P_{OB} may be up to ~7 MPa (200–300 m deep), while P_C is ~0.2 MPa, in which (extreme) case $P_E = 0.35 + 0.19 = 0.37$ MPa ~ $2P_C$, so:

$$P_E < 2P_C \approx O(P_C) \tag{8.15}$$

and the steady-state or time-averaged effective stress in the fragmenting layer of a landslide is of the order of the minimum stress needed to maintain fragmentation.

8.4 Block Slides

We apply this theory first to blockslides, which obviously conform closely to our model of confined granular flow.

8.4.1 The Problem

Blockslides are relatively intact blocks that lie some distance from their original locations, having moved on gentle slopes. Well-known examples are the Heart Mountain slide, Montana and Wyoming, USA (Prostka 1978); the Horse Creek and South Creek slides, Idaho, USA (Beutner 1972), the Bearpaw Mountains slide, Montana (Gucwa and Kehle 1978) and the Dakota Group rockslides, Colorado, USA (Braddock 1978). These have volumes of tens to hundreds of cubic kilometers, thicknesses of about a kilometer, and moved typically tens of kilometers on slopes as gentle as 2°. How do they move so far on such low-angle slopes? Is their motion fast or slow?

8.4.2 Our Approach to a Solution

New Zealand's prehistoric 2.1 km³ Waikaremoana landslide, (Read et al. 1991) is a very useful blockslide example. About two thousand years ago it dammed Waikaretaheke River to form Lake Waikaremoana, a large lake that today forms the centrepiece of Urewera National Park. It has both blockslide and rock-avalanche components. Following 2 km of displacement on a 5.5–8° slope, its 1.4 km^3 blockslide portion ploughed into rock-avalanche debris, raising a 150 m high debris mound (Fig. 8.1). Using an energy-balance approach and assuming 80% efficiency, Beetham (1984) estimated an impact speed of 26 m s^{-1}. Davies et al. (2006) sought confirmation of this assumption through a small physical model. When the model indicated an even higher impact speed, and drilling showed highly comminuted rock on the slide plane, we interpreted the remarkably low coefficient of friction in terms of the ephemeral fluid pressure due to dynamic fragmentation.

8.5 The Waikaremoana Blockslide Model

8.5.1 Preparing the Physical Model

To assess how fast the Waikaremoana block slid to push up the mound, we built a small (1 : 10 000 linear scale) laboratory model. The crucial design criterion was that the ratio of inertial to gravitational forces in the model was the same as in the landslide, so that the ability of the kinetic energy or momentum in the model block to displace granular material would be similar to that in the field.

We satisfied the criterion by scaling the block and grain masses in proportion to the cube of the linear scale, by using sand of the same bulk density (about 2 t m^{-3}) and internal friction coefficient (about 0.6) as found in the prototype, and by adjusting the model block weight to correspond to that of the Waikaremoana block. Since the acceleration due to gravity is identical in model and prototype, the two situations are dynamically similar. With this similarity of forces, the velocity scale is equal to the square root of the linear scale (Yalin 1971) – in this case $(1:10\,000)^{0.5} = 1:100$.

The model apparatus (Fig. 8.9) conceptually represented only a small part of the total Waikaremoana landslide, specifically where the block stopped against rock-avalanche debris at the edge of a proto-valley, and the mound was formed. It was not required to model any other landslide part. The model block was swung from four suspension wires to hit the grain mass (sand) in the correct attitude at a range of speeds, and we observed the repositioning of the sand using a high-speed video camera (200 fps) which allowed us to measure block and sand speeds.

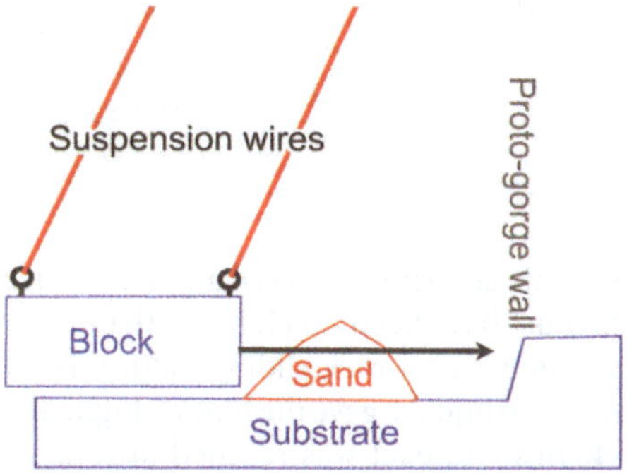

Fig. 8.9. Diagrammatic sketch of the model used to determine the speed of the Waikaremoana blockslide at the moment of impact with the opposing valley wall

8.5.2 Model Results

To achieve a satisfactory model mound, two conditions were needed. First, the sand needed to move with the block as they both impacted the proto-valley side. We achieved this by putting the sand some distance from the barrier, and allowed the block to swing into it faster than required; it then set the sand in motion, so that both were

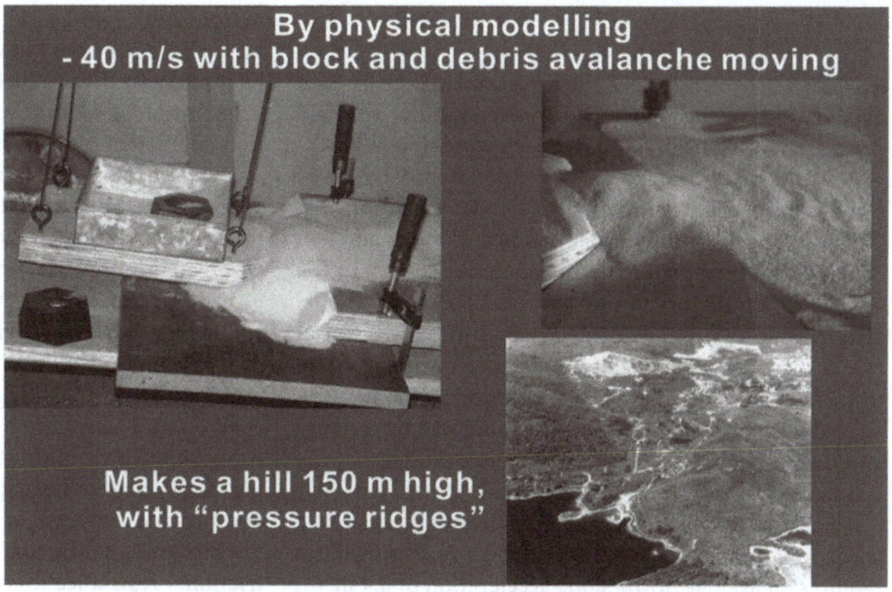

Fig. 8.10. Prototype and model debris mounds

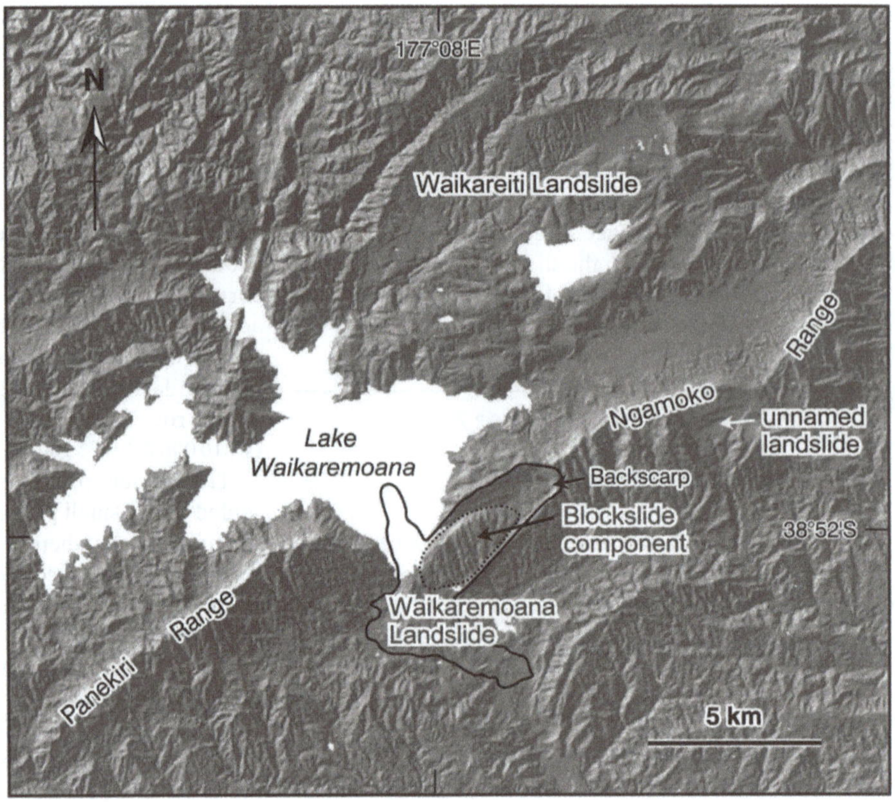

Fig. 8.11.
Shaded relief map of the Waikaremoana landslide at Lake Waikaremoana, showing the relationship between the blockslide and rock avalanche components

moving at a lower velocity when they hit the barrier. Second, the block and the sand needed to hit the barrier at about 0.4 m s^{-1}.

These conditions produced a mound about 15 mm high, correctly representing the 150 m height in the field. As well, the morphology of the mound, with "pressure" ridges, was strikingly similar to that in the field (Figs. 8.10 and 8.11).

8.5.3 Applying the Model Results to the Field

The model correctly scaled the forces during the impact, and so valid conclusions can be drawn regarding the field processes. We concluded that the Waikaremoana blockslide was rapid, with an impact speed of ~40 m s^{-1}. It could not emplace the mound with an impact speed of only 26 m s^{-1}. The 33% higher speed indicates a less efficient conversion of kinetic energy to potential energy in the collision than the previously assumed 80%. The model also showed that the deformed debris mass was not at rest at the time of impact; it had to have been in motion to form a mound with "pressure ridges". This demonstrated that the rock avalanche was generated by collapse and disintegration of the front of the block and that the two landslide parts were synchronous.

The block traveled 2 km before stopping against a gorge wall and debris. To reach 40 m s^{-1} in this distance requires an average down-slope acceleration of 0.4 m s^{-2}. To achieve this acceleration, basal and lateral friction coefficients must have been much lower than normally associated with rock-on-rock sliding (0.6). It is inconceivable that an earthquake gave the block a high initial speed and its 40 m s^{-1} impact was reached after decelerating under normal friction. This requires an impossible initial speed of ~600 m s^{-1}. We inferred that the intact block accelerated to a maximum 40 m s^{-1} during its 2-km travel while sliding on the 5.5–8° basal surface and its left-lateral margin (the right-lateral margin was unconfined). There has to have been very low frictional resistance to motion of the block. If the block started from rest, an average acceleration of 0.4 m s^{-2} is needed, requiring a mean friction coefficient of 0.06 – about 1/10th of normal friction.

A number of possibilities, including high pore-water pressures and undrained loading of a saturated substratum, were investigated for explaining this low friction and eliminated for a variety of reasons (Beetham 1983; Read et al. 1991; Davies et al. 2006). Rock fragmentation at the sliding interface was inevitable, and a thin (300 mm) layer of finely-ground rock at the base of the displaced block was indicated by drilling through this horizon (Davies et al. 2006). This was interpreted as fragmented sandstone. A high ephemeral fluid pressure, generated by fragmenting clasts, can readily substitute for pore-water pressure and provide sufficient vertical force to support much of the block weight during motion, and hence reduce the frictional resistance to motion.

The rock above and below the fragmenting material also are available to fragment, and so the system cannot run out of pore fluid until it runs out of fragmentation-inducing motion.

The static unconfined compressive strength of the sandstone at Waikaremoana is about 50 MPa (Read 1979), so its dynamic Q may be about 100–200 MPa (or even greater at the ~100 s^{-1} shear rate required for a velocity differential of ~30 m s^{-1} to take place across 0.3 m or less). Applying Eq. 8.14, the frictional retarding stress in shear bands should be $\sim 0.95 \times 0.2 + 0.05 \times 6.6 = 0.52$ MPa. The geostatic stress due to the weight of the 275-m deep block is ~6.6 MPa, so to support most of this requires about one part in ~30 of the participating grain mass to be fragmenting at any time to support the weight of the block. If each force chain across the width of a shear band contains about 10 grains, then about 1 part in 3×10^{-3} of a 100 mm thick shearing layer needs to be fragmenting simultaneously. The net downslope stress (gravity component minus friction) is then $6.6 \sin(7°) - 0.52 = 0.28$ MPa. The block mass per unit area is 0.69 kt m^{-2}, so the downslope stress would cause block acceleration of 0.4 m s^{-2}. Two kilometers of travel at this acceleration would develop a velocity of about 40 m s^{-1}.

8.5.4 A Fragmentation Model for Initiation of the Waikaremoana Landslide

Davies et al. (2006) showed that fragmentation in a confined layer could explain the high-speed emplacement of the blockslide with low frictional resistance, but we did not explain how the blockslide could develop the initial shear deformation rate to induce widespread fragmentation at its base. There is reason to believe that it started during strong ground shaking in an earthquake (Beetham 1983). It was once thought to have slid on the interface between mudstone and sandstone strata (Beetham 1983), but drilling showed it to have failed entirely in sandstone (Davies et al. 2006). Although part of the failure surface is thought to have been a bedding plane, it cannot all have been a bedding plane, because the bedding-plane geometry is impossible for failure to occur that way (Beetham 1983; Riley and Read 1992). The failure surface had to climb up through the stratigraphic pile before slip could occur. One block of the landslide is essentially intact, and is some 3.5 km long. If it was differentially flexed in its 2 km journey along the failure surface, flexure can only have been very minor, and flexed over a rise (opening joints near the surface) and not through a hollow (opening joints near the base and compressing the surface). Hence, we infer that there is little curvature on the failure surface in the direction of sliding. We can reasonably infer that it was like this from the start.

In material-failure theory, all failure utilizes pre-existing defects. As stress grows in a body, suitably oriented pre-existing defects grow and coalesce, until a through-going defect is formed to permit failure. When imperfectly elastic rock masses are deformed by folding, some slip between some beds is a kinematic requirement, and so bedding-plane shears develop in folded rocks. These are often exploited in landslides. At Waikaremoana, the axis of shortening due to folding is not easy to define with any precision because the folding is complicated, and it may result from an aggregation of episodic deformation events from slip on a number of different faults. Each separate event may have its own orientation of bedding-plane shear. The gross pattern of folding would not have produced bedding-plane shear parallel to the direction of eventual landslide motion.

A clue to how a failure surface may have developed at 275 m depth beneath the Waikaremoana landslide perhaps can be found elsewhere in the local structural geology. The SE flanks of the adjacent hills are dip slopes. In the vicinity of the headscarp of the landslide, the dip slope averages 19.1° while beyond the blockslide toe it averages 14.7°. These correspond closely to nearby measured dips on exposed bedrock and are probably more precise estimates of the bedding attitude than direct measurements at small outcrops. They indicate a twist in the bedding in the vicinity of the landslide. The left lateral margin of the Waikaremoana blockslide is inferred by Beetham (1983) to follow a fault. It is normal to the expected direction of principal tension from the twist motion, however, and we could infer that the left lateral release surface for the blockslide might be a tension joint formed by the twist folding.

Twist-derived tensional jointing offers a simple explanation for that major release surface, but the same twist deformation offers several possible explanations for the other major release surface, the basal slide plane. First, the direction of bedding-plane slip in the twist is parallel to the direction of eventual slip of the blockslide.

The local bedrock involved in the landslide is a strongly cemented calcareous sandstone. Its behavior under impulse loading could be highly elastic, especially in the near-surface environment. When a thin elastic sheet is twisted, it has two stable modes: in one, the center of the sheet is down; and in the other, it is up. If it is done experimentally, a small sheet can be flipped from one to the other mode with little effort being required. At field scale, the "up" mode is unlikely to be adopted at depth because it requires formation of void space beneath the elastic sheet, but for surface strata, the "up" mode is a transient option under impulse loading. Perhaps, the hard sandstone that forms the crest and dip slope of the range was coseismically twisted some 2200 years ago, and the block at the SW end of the range popped up, with a ground acceleration normal to the bedding plane of >1g, while below it, the rock mass popped down. With rock masses accelerating rapidly in opposite directions, a zone of rock mass would be put in tension, a myriad of favourably oriented defects would coalesce, and a surface of separation would form in tensional failure. This offers one potential explanation of the separation.

Another explanation also uses the twist, but in combination with the local geomorphology. The range is a cuesta (Fig. 8.11), with one flank covered by a thick strongly cemented sandstone. During coseismic twisting of the underlying rock mass, a portion of the relatively thin skin, in close proximity to two free edges of the beds, over the twist may have failed to fold (due to inertia) as the rock underlying it folded. A zone at the base of the strong mass was put in tension, a myriad of favourably oriented defects (pre-existing bedding-plane shears) coalesced, and a surface of separation formed in tensional failure, delaminating a slab of strong rock from the underlying beds. This could leave a rock slab a little larger than 2.1 km^3 unsupported above a double-wedge-shaped gap. The slab then could fracture along pre-existing planar defects (an orthogonal set of tensional joints from earlier folding episodes. The portion of this rock mass towards the lower-dip end of the twist (the SE end of the range, towards the proto-gorge) would be raised highest above its substrate and fall furthest on failure. The delaminating gap would decrease progressively NE along the range. In the

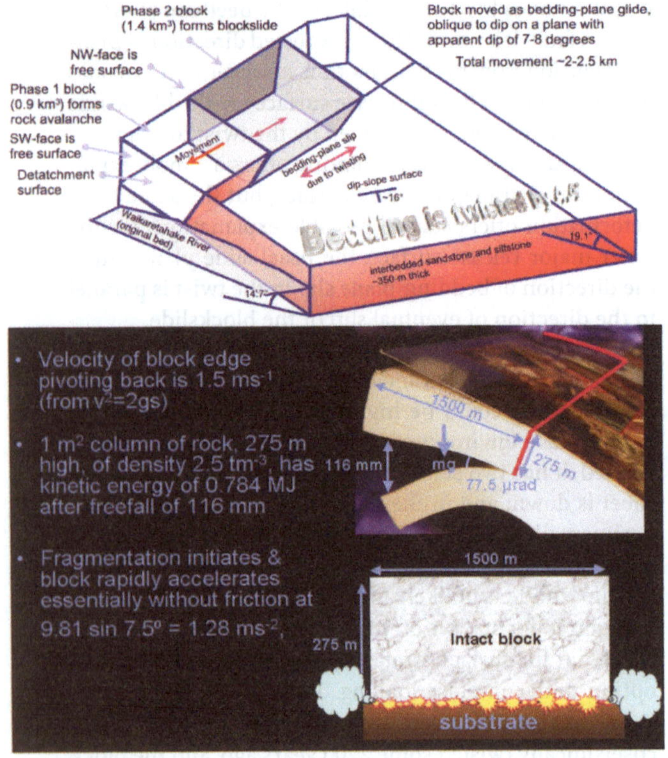

Fig. 8.12. Our concept of how the Waikaremoana landslide may have been initiated as an elastic response to coseismic folding of near-surface strata

vicinity of the headscarp, the fallback may have been trivial and unable to destabilize the rock mass beyond there.

The total twist is ~4.4° (0.0775 radians). If this has occurred over ~5 million years, the annual rate is ~1.55×10^{-8} radians per year, or ~77.5 µrad per event, assuming a recurrence interval of perhaps 5000 years for the coseismic twisting (we have no data on this). Over the 1500-m width of the Waikaremoana blockslide, the widest gap(s) would be ~116 mm. The geometry is shown diagrammatically in Fig. 8.12. The velocity, v, of the outer edge of the block as it pivots back in free fall under gravity (g) is given by $v^2 = 2gs = 1.5$ m s^{-1}. A 1-m^2 column of rock, 275 m high, of density 2.5 t m^{-3}, would have kinetic energy of 0.784 MJ after a freefall of 116 mm under gravity.

Either of these alternatives would have left 2.1 km^3 of rock mass literally up in the air (though possibly above a vacuum). We might suppose that the tensional failure was not a single clean break, but a multitude of propagating cracks, so that many small pieces of rock were torn between the two parted surfaces. In addition, the mass would not fall directly back on its pre-failure position. Thus, the rock mass would fall back with much stress concentration at asperities, which could be instantly crushed. In addition to removing asperities from the failure surface, however, the instantaneous crushing would also remove most frictional resistance to downslope motion. The fragmenting rock would behave as a heavy vapor at a pressure of the Hugoniot elastic limit (HEL) of the rock under the local environmental conditions (of confining pressure, temperature, and rock strain rate). The HEL of the cemented calcareous sandstone is greater than the static unconfined compressive strength (~50 MPa) and probably in the region of 100–200 MPa, whereas the overburden pressure beneath the blockslide at Waikaremoana was ~6.6 MPa. Hence we can suppose that the block proceeded to rapidly accelerate downslope at ~9.81 sin(7.5°) = ~1.28 m s^{-2}, initially essentially with no frictional resistance. The fragmentation process, however, is self-regulating: if there is no effective normal load in the fragmenting layer, the grains will not be stressed, and so will not continue to fragment. A balance should be quickly achieved in which the load carried by the ephemeral pressure of fragmenting grains leaves sufficient load to maintain some fragmenting grains. We surmise that fragmentation of rock at the base of the block was initially close to 100% of all rock particles, but quickly dropped to ~0.03% of the shearing layer fragmenting at any instant, and continued at that level until block motion abruptly stopped when it hit the opposite valley wall. In order to hit the wall at ~40 m s^{-1}, an average acceleration of 0.3 m s^{-2} was required, so it is not likely that an initial downslope acceleration of 1.28 m s^{-2} was maintained more than momentarily. What is more important than the duration of high acceleration is that the mass may have had its highest downslope acceleration when starting to "slide", declining to some steady-state value as the mass gained velocity. In this model, the initial lowest friction was caused by massive fragmentation of rock at the base of the block as it fell back in a 1-g freefall from being raised (or left) into the air, being a near-surface elastic response of a rock mass to coseismic folding. Although it appears to be a feasible model, we do not know of a means to test whether it occurred.

8.5.5 Other Rockslide Phenomena

We now consider rockslides involving fragmentation throughout most the moving debris mass, other than within the upper few meters where low confining stresses readily permit inter-granular slip and rotation. Large rock avalanches have a fascinating morphological characteristic; their runout lengths are proportionally greater for their volume than those of their smaller counterparts, to a degree that increases markedly with the volume of the deposit. This is the so-called "size effect" (e.g., Hsü 1975, 1978; Scheidegger 1973). We have investigated this anomaly (McSaveney 1978; Davies 1982; Davies and McSaveney 1999, 2002, 2003; Davies et al. 1999; McSaveney and Davies 2002, 2003; McSaveney et al. 2002), and conclude that, rather than requiring some novel physical mechanism (e.g., Shreve 1968; Goguel 1978; Melosh 1979; Campbell 1989; Campbell et al. 1995; Collins and Melosh 2003), or extensive occurrence of some particular substrate condition (e.g., Sassa et al. 2004; Abele 1997; Hungr and Evans 2004) to generate low basal friction beneath the mass of translating debris, the continuing process of dynamic fragmentation during the runout generates sufficient internal "fluid" pressure that, under conditions of normal intergranular and basal friction, the debris can spread to a greater extent than would occur without fragmentation. This extra spreading results in the extended distal runout observed in rock-avalanche deposits, and is accompanied by reduced runout of proximal material.

8.5.6 Transition from Blockslide to Rock Avalanche

We have shown above that dynamic-fragmentation is capable of explaining the extraordinarily rapid acceleration needed in some blockslides to cause their subsequent motion and effects. The Avalanche Lake rock avalanche in Canada, began as a blockslide, and apparently needs a basal friction coefficient of 0.04 to explain its motion (Hungr 1995). The acceleration of the initial blocks of the 1980 Mt. St. Helens debris avalanche, computed from the Rosenquist series of photographs, requires a basal friction coefficient of 0.1 (Voight 1981). The very rapid motion of the failed blocks of Mt. Toc in the Vaiont disaster of 1963 has been attributed to low-strength clay (Kilburn and Petley 2003), but it seems possible that rock fragmentation may have played some role there too.

Some of the anomalously low travel angles exhibited by rock-avalanche deposits (e.g., Cruden 1980) can be explained by the notion that some rock avalanches move initially as blockslides. An initial blockslide phase at low friction allows faster acceleration than under normal frictional resistance. But not all rock avalanches need have an initial blockslide phase of any significant duration.

8.5.7 Dry Rock Avalanches

Rock avalanches generally move as dry granular flows, because the large volume (~20%) of void space created by the observed pervasive fragmentation of initially massive rock is several orders of magnitude more than sufficient to prevent saturation of the debris mass by any water present in the parent rock or along the runout path

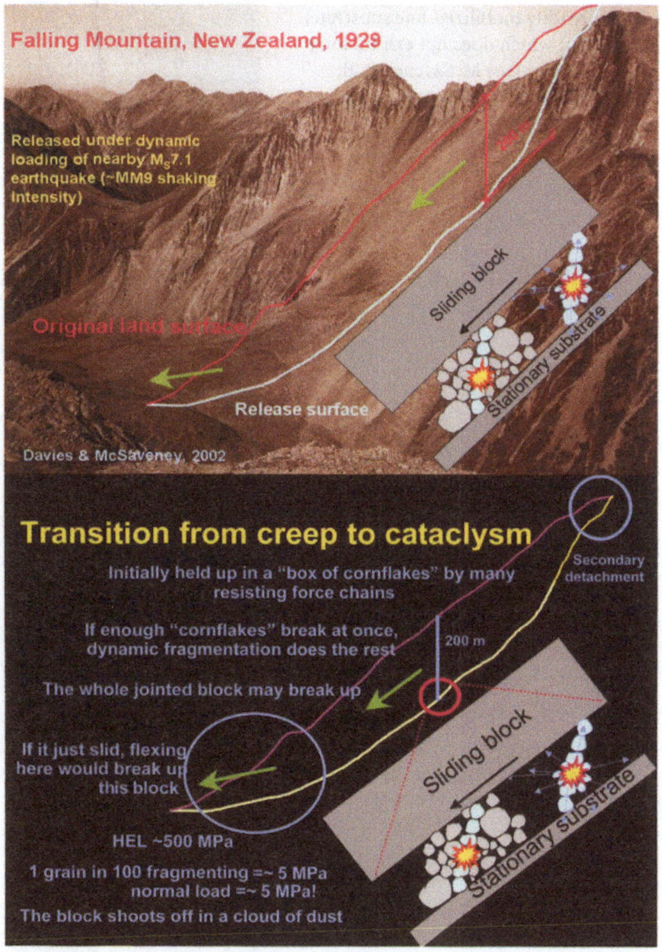

Fig. 8.13. Our concept of how the Falling Mountain rock avalanche may have initiated as a rockslide. The "cornflake" analogy is because force chains are what often prevent breakfast cereals from pouring from the box, requiring shaking to disrupt the force chains

(McSaveney and Davies in press). They are, however, highly erosive, tending to incorporate any readily deformable substrates encountered along their paths and to deposit them in the distal bases of their deposits – this can include rivers, lakes and groundwater systems.

Our concept of grain conditions in rock-avalanche motion is illustrated in Fig. 8.13; the stars represent grains of various sizes fragmenting simultaneously.

8.5.8 Falling Mountain

In the case of the Falling Mountain deposit (Fig. 8.2) investigated by Davies and McSaveney (2002), the rock is a highly indurated greywacke sandstone with a static, unconfined, intact compressive strength of ~280 MPa. We do not know the appropriate dynamic strength at the local strain rates (>10 s^{-1}) and overburden pressures (>0.3 MPa) causing fragmentation during the rock avalanche but estimate that it is likely to be at least ~500 MPa (following Kobayashi 1970). Our previous simulation (Davies and McSaveney 2002) showed that the avalanche runout could be represented by adding an isotropic dispersive stress equivalent to 0 5% of the grain mass fragmenting simultaneously in a 1-dimensional simulation (DAN (Hungr 1995)). We now modify this by first calculating the resisting stress in the shearing layer, which is equal to the basal resisting stress. Since shear takes place in localized bands, this is given by Eq. 8.15, and is 0.34 MPa. This results in an apparent friction angle of 20°. The depth-averaged overburden pressure resisted by fragmentations to leave this residual shear resistance is

Fig. 8.14.
Aerial view of the small prehistoric South Ashburton rockslide which fell onto and eroded a wet, readily mobilized fine substrate, but which does not exhibit long runout (see McSaveney et al. 2000; McSaveney and Davies 2005, 2006)

0.5 MPa, and the spatial density of simultaneous fragmentations needed to achieve this is about 0.01%. Rerunning the DAN simulation model (of Davies and McSaveney 2002) with a friction angle of 20° and an isotropic fragmentation-induced dispersive pressure of 0.5 MPa yields exactly the correct runout of the avalanche.

Our previous simulation used trial-and-error to obtain the correct runout by varying the spatial density of simultaneous fragmentations; the present simulation was based on calculated internal friction resulting from fractally distributed, isotropically oriented shear bands below a passive carapace, and their effect on internal friction calculated by Eq. 8.13. The success of the present simulation independent of any calibration is remarkable.

If all the fragmentation observed in the rock-avalanche deposit were to occur at the start of the motion, as often has been suggested, the concentration of GPa-range forces acting simultaneously would cause an instantaneous explosive fragmentation of the whole mass, and its dispersion as a (probably mushroom-shaped) dust cloud, with no evident translational mass movement – this does not occur. Equation 8.8 suggests that complete fragmentation of the 55×10^6 m^3 volume of the Falling Mountain rock avalanche in 1 second at the start of its motion would have released elastic-strain energy equivalent to a nuclear explosion in the several-megaton range, if complete fragmentation took place in 1000 stages. But the energy for fragmentation is fed from the kinetic energy of shear-strain deformation, and there obviously is insufficient kinetic energy available initially to cause such an explosion at the start of the landslide. The concept of near-instantaneous, initial fragmentation is thus an absurdity.

8.5.9 South Ashburton Landslide

New Zealand's South Ashburton landslide (Fig. 8.14; McSaveney et al. 2000) is an unusual and instructive rockslide; it fell onto, and ran out over a fine-grained, saturated, and deformable substrate, but it does not exhibit classical long runout (McSaveney and Davies 2005, 2006). Because the landslide was too small to utilize dynamic fragmentation to any significant degree, its runout could be modeled with dynamic similarity using sand (McSaveney et al. 2000). Our physical modeling of it (Fig. 8.15) suggested that it fell first as a block, which then disintegrated and spread at the foot of the slope.

8.6 Conclusion

The motion of large rockslides is most readily understood as being controlled by comminuting-grain-flow dynamics, where grains crushing under large dynamic stresses in force chains form an ephemeral high-pressure fluid in the pore space between larger grains at all scales and reduce the overall resistance to motion in proportion to the fraction of the grain mass that is being crushed at any instant. The motion of small rockslides where grain comminution is unimportant is controlled by grain-to-grain friction and intergranular stresses determined by the non-fragmenting grains and any other pore fluid which may be present.

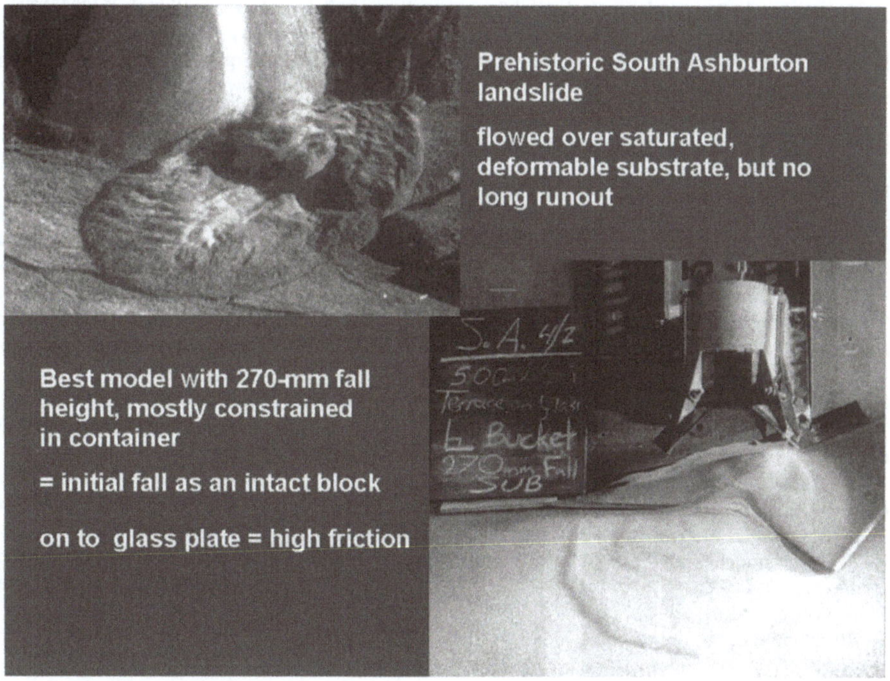

Fig. 8.15. Prototype and physical model of the South Ashburton rockslide. The most satisfactory model was with the debris mass gaining momentum in a container, before flowing onto a high-friction glass plate coated with a thin layer of loose sand forming an erodible substrate. This eroded substrate was carried at the base of the rockslide front and formed the raised distal rim. Without the erodible substrate, the raised rim could not be formed. Similarity of model and prototype indicate that the prototype flowed with normal intergranular friction similar to that of the sand

Acknowledgment

This research has been supported by the New Zealand Foundation for Research, Science and Technology. Alexander Strom, Stuart Dunning, and Giovanni Crosta helped us develop our ideas on particle-size distributions in comminuting grain flows. We gratefully acknowledge many other colleagues, both believers and doubters, whose discussions have helped mature our concepts.

References

Abele G (1997) Rockslide movement supported by the mobilization of groundwater-saturated valley floor sediments. Z Geomorphol 41(1):1–20

Allegre CJ, Mouel JLL, Provost A (1982) Scaling rules in rock fracture and possible implications for earthquake prediction. Nature 297:47–49

An LJ, Sammis CG (1994) Particle size distribution of cataclastic fault materials from Southern California: a 3-D study. Pure Appl Geophys 143:203–227

Anders MH, Aharonov E, Walsh JJ (2000) Stratified granular media beneath large slide blocks; implications for mode of emplacement. Geology 28:971–974

Anthony JL, Marone C (2005) Influence of particle characteristics on granular friction. J Geophys Res 110: B08409 Doi: 10.1029/2004 JB003399

Beetham RD (1983) Seismicity and landsliding with special attention to New Zealand. Unpubl. MSc thesis, University of London

Belidor BF (1725) Nouveau cours de mathématique à l'usage d'artillerie et du génie. Paris, 505 p

Belousov A, Belousova M, Voight B (1999) Multiple edifice failures, debris avalanches and associated eruptions in the Holocene history of Shiveluch Volcano, Kamchatka, Russia. B Volcanol 6:324–342

Beutner EC (1972) Reverse gravitative movement on earlier overthrusts, Lemhi Range, Idaho. Geological Society of America Bulletin 83:839–846

Beutner EC, Gerbi GB (2006) Catastrophic emplacement of the Heart Mountain blockslide, Wyoming and Montana, USA. GSA Bull 117:724–735; doi: 10.1130/B25451.1

Bond FC (1952) The third theory of comminution. T Am I Min Met Eng 193:494–496

Bond FC (1955) How does rock break? The Scientific Monthly 81: 196–198

Braddock WA (1978) Dakota Group rockslides, northern Front Range, Colorado, U.S.A. In: Voight B (ed) Rockslides and avalanches, I. Developments in Geotechnical Engineering, 14A, Elsevier, Amsterdam, pp 439–479

Campbell CS (1989) Self-lubrication for long-runout landslides. J Geol 97:653–665

Campbell CS (2006) Granular material flows – an overview. Powder Technol 162:208–229

Campbell CS, Brennen CE (1985) Computer simulation of granular shear flows. J Fluid Mech 151:167–188

Clavero JE, Sparks RJS, Huppert HE, Dade WB (2002) Geological constraints on the emplacement mechanism of the Parincota debris avalanche, northern Chile. B Volcanol 64: 40–54. doi: 10.1007/s00445-001-0183-0

Collins GS, Melosh HJ (2003) Acoustic fluidization and the extraordinary mobility of Sturzstroms. J Geophys Res 108(B10): 2473, doi:10.1029/2003JB002465

Crosta GB, Frattini P, Fusi N (2007) Fragmentation in the Val Pola rock avalanche, Italian Alps. J Geophys Res (in press)

Cruden DM (1980) The anatomy of landslides. Can Geotech J 17: 295–299

Cruden DM, Varnes DJ (1996) Landslide types and processes. In: Turner AK, Schuster RL (eds) Landslides: investigation and mitigation. Special Report 247, Transportation research board, US National research council, Washington, D.C., pp 36–75

Davies TRH (1982) Spreading of rock avalanche debris by mechanical fluidisation. Rock Mech 15:9–24

Davies TRH, McSaveney MJ (1999) Runout of dry granular avalanches. Can Geotech J 36(2):313–320

Davies TRH, McSaveney MJ (2002) Dynamic simulation of the motion of fragmenting rock avalanches. Can Geotech J 39:789–798

Davies TRH, McSaveney MJ (2006) Runout of rock avalanches and volcanic debris avalanche. In: Picarelli L (ed) Proceedings of the International Conference on fast slope movements: prediction, and prevention for risk mitigation. Naples, May 11–13, 2003, vol. 2

Davies TRH, McSaveney MJ, Hodgson KA (1999) A fragmentation-spreading model for long-runout rock avalanches. Can Geotech J 36:1096–1110

Davies TRH, McSaveney MJ, Beetham RD (2006) Rapid block glides – slide-surface fragmentation in New Zealand's Waikaremoana landslide. Q J Eng Geol Hydroge 39:115–129

Dunning SA (2004) Rock avalanches in high mountains – A sedimentological approach. University of Luton, unpublished PhD Thesis

Evans SG, Hungr O, Enegren EG (1994) The Avalanche Lake rock avalanche, Mackenzie Mountains, Northwest Territories, Canada: description, dating and dynamics. Can Geotech J 31:749–768

Goguel J (1978) Scale-dependent rockslide mechanisms, with emphasis on the role of pore-fluid vaporization. In: Voight B (ed) Rockslides and avalanches, I, : developments in geotechnical engineering. 14A, Elsevier, Amsterdam, pp 693–705

Grady DE, Kipp ME (1987) Dynamic rock fragmentation. In: Atkinson BK (ed) Fracture mechanics of rock. Academic Press, London, pp 429–475

Gucwa PR, Kehle RO (1978) Bearpaw Mountains Rockslide, Montana, U.S.A. In: Voight B (ed) Rockslides and avalanches, I: developments in geotechnical engineering. 14A, Elsevier, Amsterdam, pp 393–421

Hazzard JF, Mair K (2003) The importance of the third dimension in granular shear. Geophsy Res Lett 30, doi: 10.1029/2003GL017534

Herget G (1988) Stresses in rock. Balkema, Rotterdam, 179 p

Howell DW, Behringer RP, Veje CT (1999a) Fluctuations in granular media. Chaos 9:559–572

Howell DW, Behringer RP, Veje CT (1999b) Stress fluctuations in a 2D granular Couette experiment: a continuous transition. Phys Rev Lett 26:5241–5244

Hsü KJ (1975) Catastrophic debris streams (Sturzstroms) generated by rockfalls. Geol Soc Am Bull 86:123–140

Hungr O (1995) A model for the runout analysis of rapid flow slides, debris flows and avalanches. Can Geotech J 32(4):610–623

Hungr O, Evans SG (2004) Entrainment of debris in rock avalanches: an analysis of a long run-out mechanism. Geol Soc Am Bull 116:1240–1252

Kilburn CRJ, Petley DN (2003) Forecasting giant, catastrophic slope collapse: lessons from Vajont, Northern Italy. Geomorphology 54(1–2):21-32

Knoeberl C (1997) Impact cratering: an overview of mineralogical and geochemical aspects http://www/univie.ac.at/geochemistry/impp.html

Kobayashi R (1970) On mechanical behaviours of rocks under various loading rates. Rock Mech Jpn, 1:56–58

Liu CH, Nagel SR, Schecter DA, Coppersmith SN, Majumdar S, Narayan O, Witten TA (1995) Force fluctuations in bead packs. Science 269:513–515

Locat P, Couture R, Leroueil S, Locat J, Jaboyedoff M (2006) Fragmentation energy in rock avalanches. Can Geotech J 43:830–851

Mair K, Frye KM, Marone C (2001) Influence of grain characteristics on the friction of granular shear zones. J Geophys Res 107: 2219 Doi:10.1029/2001BJ000516

McSaveney MJ (1978) Sherman Glacier rock avalanche, Alaska, USA. In: Voight B (ed) Rockslides and Avalanches 1, Natural Phenomena. Developments in Geotechnical Engineering 14A(6):197–258

McSaveney MJ (2002) Recent rockfalls and rock avalanches in Mount Cook National Park, New Zealand. In: Evans SG, DeGraff JV (eds) Catastrophic landslides: occurrence, mechanisms and mobility. Geological Society of America Reviews in Engineering Geology 15:35–70

McSaveney MJ, Davies TRH, Hodgson KA (2000) A contrast in deposit style and process between large and small rock avalanches. In: Bromhead E, Dixon D, Ibsen M-L (eds) Landslides in research, theory and practice. Thomas Telford Publishing, London, pp 1053–1058

McSaveney MJ, Davies TRH, McSaveney MJ (2006) Rapid rock-mass flow with dynamic fragmentation. In: Evans SG, Scarascia-Mugnozza G, Strom A and Hermanns RL (eds) Advanced research workshop: landslides from massive rock slope failure. NATO Science Series, IV Earth and Environmental Sciences, June 16–21, 2002, Celano, Italy, 49, pp 285–304

Melosh HJ (1997) Impact cratering. In: Shirley JH, Fairbridge RW (eds) Encyclopedia of planetary sciences. Chapman and Hall, London, pp 326-335

Prostka HJ (1978) Heart Mountain fault and Absaroka volcanism, Wyoming and Montana, U.S.A. In: Rockslides and avalanches, I.Voight B (ed): Developments in geotechnical engineering. 14A, Elsevier, Amsterdam, 423–437

Read SAL (1979) Lake Waikaremoana outlet – engineering geological studies of factors related to leakage through the natural dam. NZ Geological Survey report EG336

Read SAL, Beetham RD, Riley PB (1991) Lake Waikaremoana barrier – a large landslide dam in New Zealand. In: , Bell DH(ed) Landslides. Balkema, Rotterdam, pp 1481–1487

Riley PB, Read SAL (1992) Lake Waikaremoana: present day stability of landslide barrier. In: Bell DH (ed) Landslides : Proceedings of the sixth International symposium 10–14 February 1992, Christchurch, Balkema, Rotterdam, pp 1249–1256

Sammis C, King G, Biegel R (1987) The kinematics of gouge formation. Pure Appl Geophys 125:777–812

Sassa K, Fukuoka H, Wang G, Ishikawa N (2004) Undrained dynamic-loading ring-shear apparatus and its application to landslide dynamics. Landslides 1:7–19

Scheideigger AE (1973) On the prediction of the reach and velocity of catastrophic landslides. Rock Mech 5:231–236

Sheve RL (1968) The Blackhawk landslide. Geological society of America, Special paper, 108, 47 pp

Turcotte DL (1986) Fractals and fragmentation. J Geophys Res 91: 1921–1926

Varnes DJ (1978) Slope movement types and processes. In: Schuster RL, Krizek RJ (eds) Special report 176: Landslides: analysis and control. Transportation research board, National research council, Washington, D.C., pp 11–33

Voight B (1981) Time scale for the first moments of the 1980 eruption. In: Lipman PW, Mullineaux DR (eds) The 1980 eruptions of Mt St Helens, Washington. , US Geological Survey Professional Paper, 1250, pp 69–134

Voight B, Glicken HX, Janda RJ, Douglas PM (1981) Catastrophic rockslide-avalanche of May 18 1980. In: Lipman PW, Mullineaux DR (eds) The 1980 eruption of Mt. St. Helens, Washington. US Geological Survey Professional Paper, 1250, pp 347–377

Yalin MS (1971) Theory of hydraulic models. Macmillan, London

Yarnold JC, Lombard JP (1989) A facies model for large rock avalanche deposits formed in dry climates. In: Colburn IP, Abbott PL, Minch J (eds) Conglomerates in basin analysis: a Symposium dedicated to A. O. Woodward. Pacific Section of Society of Economic Paleontologists and Mineralogists, 62:9–31

Zhang YQ, Lu Y, Hao H (2004) Analysis of fragment size and ejection velocity at high strain rate. Int J Mech Sci. 46:27–34

Chapter 9

Residual Shear Strength of Tertiary Mudstone and Influencing Factors

Binod Tiwari

Abstract. Extensive research done on the residual shear strength of the Tertiary mudstone showed that mudstones are rich in expansive clay minerals. Those clay minerals are responsible for making the mudstone highly weathering susceptible, which is the main cause for excessive landslides in the mudstone formations. Because of high brittleness, residual shear strength is more important in analyzing the landslides occurred in such formations. Research results showed that average liquid limit, plasticity index, activity, and clay sized fractions in the mudstone from the Niigata Prefecture of Japan are 72%, 36%, 1.4, and 26%, respectively. Dominating clay mineral in mudstone is smectite with an average proportion of 15%. Dominating clay mineral oxide of mudstone was aluminum oxide with an average value of 18%. Average residual friction angle was observed to be 13°. At the presence of saline pore water the strength can increase up to 40%, depending on the nature of soil and site condition. If it is hard to get the soil specimens in sufficient quantity from the mudstone area, residual friction angles can be approximately estimated with index properties. However, if the mineralogical composition is known, residual friction angle can be estimated with up to 90% accuracy, using the diagram proposed by Tiwari and Marui (2005). This paper deals with the chemical, mineralogical, and mechanical properties of the Tertiary mudstone from the Niigata Prefecture of Japan, as well as the methods to measure residual shear strength in conventional soil testing devices.

Keywords. Landslides, mudstone, mineralogical composition, chemical composition, weathering index, residual shear strength, index properties

9.1 Background

Tertiary mudstones are over-consolidated sedimentary rocks made of clays and silts of sedimentary origin. Because of the tectonic activities and unequal settlements, these rocks are highly fissured. Due to their low strength and easily weathering nature, they are also referred to as soft rocks. As they have low strength, have low permeability, are expansive in nature, and are very brittle, mudstones ate susceptible to shear failure. We have evidenced a large number of progressive failure cases on the slope formed on mudstones. In Japan, several places are formed with this type of Tertiary mudstone. Natural hazards such as landslides, and slope failures are frequently occurred on those locations. Niigata Prefecture of Japan is one among the prefectures that have numerous dormant landslides triggered on the mudstone formations. Shown in Fig. 9.1 is the county-wise distribution of landslides in Niigata Prefecture. Higashi Kubiki County of the Niigata Prefecture has the highest number of landslides (Tiwari et al. 2004). Approximately 91% of the county is covered with the mudstone formation. More than 50% of the entire county was observed under dormant landslides. Those landslides were observed in two types of mudstone formations – Sugawa and Taruda formations. This paper deals with the properties of those mudstones pertinent to the occurrence of landslides.

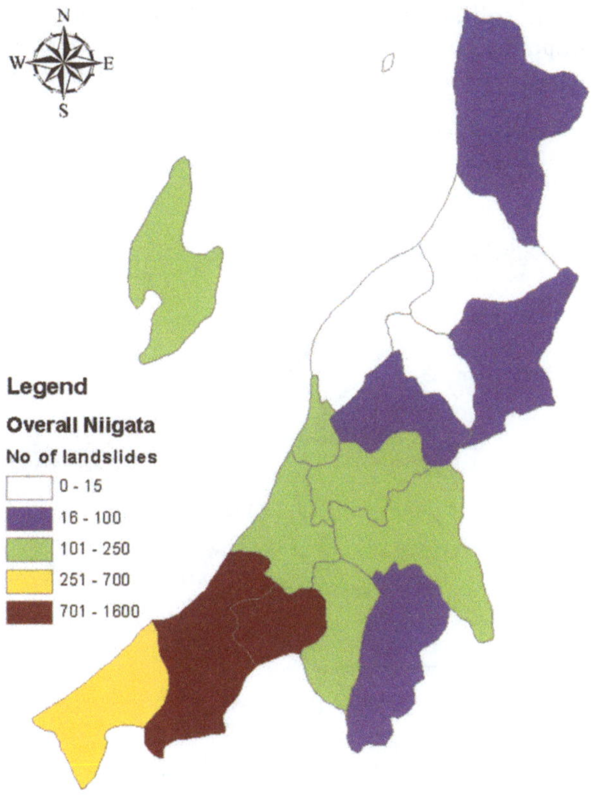

Fig. 9.1. County-wise Distribution of Landslides in the Niigata Prefecture of Japan

9.2 Properties of Tertiary Mudstone

Mudstones are made of expansive clay minerals that include smectite. Although they are stiff in nature, they soften easily under the presence of water. Therefore, proportions of expansive minerals play an important role in the properties of the mudstone. To measure the properties of the Tertiary mudstone, numerous soil samples were collected from a number of landslide sites in Higashi Kubiki County (Fig. 9.2). Shown in Fig. 9.3 is the geological map of the soil sampling area. Index properties of those mudstone specimens are presented in Table 9.1. The liquid limit, plasticity index, and clay fractions ranged from 51 to 120%, 15 to 68%, and 3 to 45%, respectively, and the average values were obtained to be 72.3%, 36.2%, and 25.5% respectively.

Main dominating clay minerals in the Tertiary mudstone are smectite and kaolinite. Nature of the soil, liquid limit, activity, and shear strength of the soil depend on the proportion of smectite in it. Shown in Table 9.2 are the proportions of various minerals present in the studied Tertiary mudstones. Those proportions were measured by X-ray diffraction method. For details of the X-ray diffraction, please refer Tiwari and Marui (2005). Table 9.2 shows the range of minerals in the soil specimens. According to the Table 9.2, proportion of smectite, kaolinite, illite, and massive minerals (quartz and feldspar) ranged from 7 to 24%,

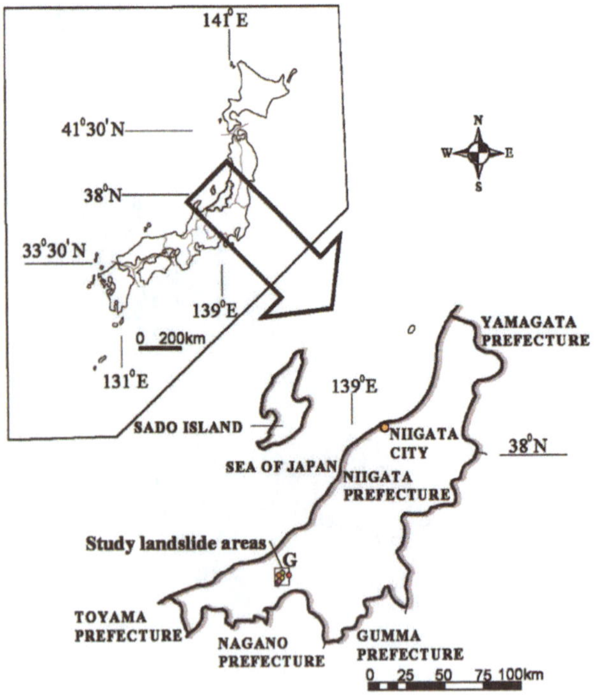

Fig. 9.2. Location of the mudstone sampling area

Table 9.1. Index properties of the mudstone specimens

Landslide sites	LL (%)	PI (%)	CF (%)
Okimi 1	75.0	38.0	30.2
Okimi 2	82.0	39.1	32.2
Okimi 3	78.0	38.0	30.1
Okimi 4	66.7	21.9	17.5
Okimi 5	71.0	35.0	30.2
Okimi 6	72.3	37.1	31.1
Okimi 7	76.0	36.0	30.8
Okimi 8	68.1	34.6	28.8
Okimi 9	84.0	52.9	45.2
Okimi 10	69.2	31.0	26.3
Okimi 11	81.4	43.9	37.5
Okimi 12	78.2	25.8	20.5
Okimi 13	75.5	25.6	21.5
Okimi 14	77.2	49.5	38.2
Okimi 15	66.0	31.0	27.7
Okimi 16	55.0	18.0	19.5
Yosio 1	86.3	47.1	28.2
Yosio 2	89.0	41.9	37.5
Yosio 3	73.0	35.3	22.1
Yosio 4	64.0	34.1	19.2
Yosio 5	68.0	38.2	28.2
Yosio 6	69.0	40.0	27.2
Yosio 7	51.0	19.0	21.8
Mukohidehara 1	71.5	36.4	21.2
Mukohidehara 2	55.8	32.6	20.5
Mukohidehara 3	70.0	38.0	24.0
Mukohidehara 4	65.0	34.0	22.0
Mukohidehara 5	61.0	15.0	20.0
Mukohidehara 6	53.0	26.0	14.0
Engyoji 1	91.3	50.5	32.8
Engyoji 2	94.6	62.4	33.2
Engyoji 3	69.0	38.0	27.5
Engyoji 4	62.0	29.0	19.9
Engyoji 5	59.0	19.0	20.0
Iwagami 1	96.2	48.0	32.2
Iwagami 2	94.7	59.2	33.5
Iwagami 3	63.0	32.0	24.8
Iwagami 4	71.0	38.0	26.2
Iwagami 5	64.0	21.0	23.0
Tsuboyama 1	79.8	36.3	40.2
Tsuboyama 2	100.0	68.3	42.8
Tsuboyama 3	120.0	65.5	42.0
Tsuboyama 4	108.0	66.2	41.2
Tsuboyama 5	58.0	29.2	21.3
Hirayama 1	57.0	17.0	10.0
Hirayama 2	57.0	23.0	2.8
Takao 1	61.0	24.0	18.0
Takao 2	55.0	27.0	6.5
Hotenji 1	82.0	51.0	31.0
Hotenji 2	84.0	31.0	22.0
Heito 1	54.0	19.0	10.0
Heito 2	56.0	28.0	5.4
Utsunomata 1	83.0	54.0	51.0
Utsunomata 2	67.7	27.5	17.0
Utsunomata 3	72.3	29.9	17.0
Utsunomata 4	55.6	38.6	12.0
Utsunomata 5	56.8	34.3	16.0

Table 9.2. Mineralogical compositions of the mudstone specimens

Landslide sites	Massive minerals (%)	Kaolinite (%)	Illite (%)	Smectite (%)
Okimi 1	81.0	3.9	0.1	15.0
Okimi 2	82.0	2.8	0.2	15.0
Okimi 3	81.0	3.9	0.1	15.0
Okimi 4	84.0	1.9	0.1	14.0
Okimi 5	79.0	4.9	0.1	16.0
Okimi 6	79.0	4.8	0.2	16.0
Okimi 7	82.0	2.9	0.1	15.0
Okimi 8	78.0	5.9	0.1	16.0
Okimi 9	75.0	2.9	0.1	22.0
Okimi 10	78.0	6.9	0.1	15.0
Okimi 11	77.0	2.9	0.1	20.0
Okimi 12	76.0	9.0	0.1	15.0
Okimi 13	77.0	7.9	0.1	15.0
Okimi 14	77.0	4.9	0.1	18.0
Okimi 15	79.4	4.5	0.6	15.5
Okimi 16	85.3	3.5	1.0	10.2
Yosio 1	77.0	5.0	0.0	18.0
Yosio 2	73.0	8.0	0.0	19.0
Yosio 3	74.9	7.7	1.6	15.8
Yosio 4	74.2	6.9	1.2	17.7
Yosio 5	75.6	5.4	0.4	18.6
Yosio 6	76.4	7.0	0.7	15.9
Yosio 7	80.0	6.1	0.2	13.7
Mukohidehara 1	84.0	7.0	0.0	9.0
Mukohidehara 2	85.0	6.0	0.0	9.0
Mukohidehara 3	72.6	11.2	0.9	15.3
Mukohidehara 4	72.3	7.3	0.8	19.6
Mukohidehara 5	84.6	2.8	0.5	12.1
Mukohidehara 6	85.7	2.0	1.9	10.4
Engyoji 1	73.0	10.0	0.0	17.0
Engyoji 2	74.0	8.0	0.0	18.0
Engyoji 3	75.0	4.5	2.5	18.0
Engyoji 4	78.8	5.2	2.6	13.4
Engyoji 5	85.1	3.5	1.4	10.0
Iwagami 1	78.0	6.0	1.0	15.0
Iwagami 2	78.0	6.0	0.0	16.0
Iwagami 3	75.9	14.5	1.9	7.7
Iwagami 4	77.8	10.8	0.3	11.1
Iwagami 5	70.5	3.9	1.7	23.8
Tsuboyama 1	75.0	8.0	0.0	17.0
Tsuboyama 2	76.0	6.0	0.0	18.0
Tsuboyama 3	73.7	3.9	1.7	20.7
Tsuboyama 4	73.6	8.7	2.2	14.4
Tsuboyama 5	86.5	3.8	0.4	9.3
Hirayama 1	76.9	3.1	1.8	18.1
Hirayama 2	75.0	4.4	1.8	17.7
Takao 1	85.4	3.2	1.4	10.0
Takao 2	84.0	4.1	0.9	11.0
Hotenji 1	82.0	6.0	2.2	9.8
Hotenji 2	84.8	3.2	1.7	10.3
Heito 1	74.6	6.9	1.8	16.5
Heito 2	80.7	11.7	1.0	6.6
Utsunomata 1	76.4	7.0	1.4	15.2
Utsunomata 2	79.3	3.5	0.2	17.0
Utsunomata 3	81.5	4.1	0.3	14.1
Utsunomata 4	72.1	6.2	0.5	21.2
Utsunomata 5	70.2	7.0	0.4	22.4

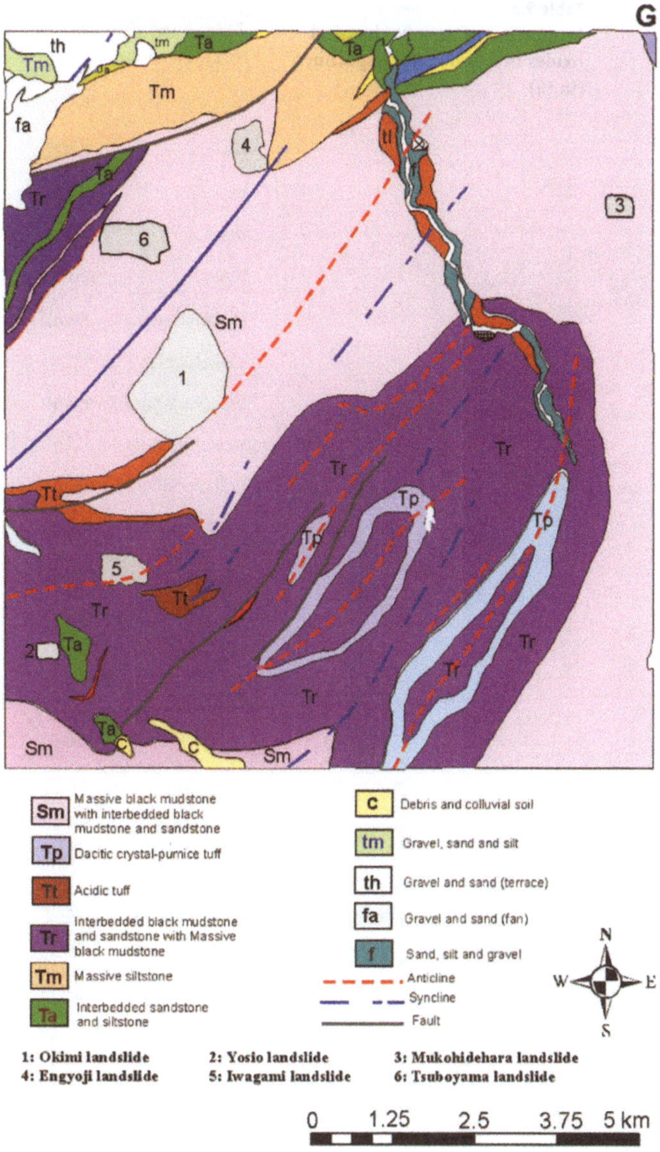

Fig. 9.3. Geological map of the study area

2 to 15%, 0 to 3%, and 70 to 87%, respectively with the average value of 15.2%, 5.8%, 0.7%, and 75.6%, respectively.

Mudstones are highly susceptible to physical weathering. It is important to know the chemical compositions of the soil specimens to predict the physical and chemical weathering behavior. Proportions of the major chemical oxides of several mudstone specimens are presented in Table 9.3. According to the Table 9.3, Al_2O_3 is the major chemical oxide pertinent to clay minerals, obtained in the Tertiary mudstone. CIA (Chemical Index of Alteration), CIW (Chemical Index of Weathering), PFA (Plagioclase Index of Alteration), V (Vogt's Residual Index), R (Ruxton Ratio), WIP (Weathering Index of Parker), and

Table 9.3.
Proportion of various chemical oxides in the mudstone specimen (in %)

Sampling site	SiO_2	TiO_2	Al_2O_3	Fe_2O_3	MnO	MgO	CaO	Na_2O	K_2O	P_2O_5
Okimi 1	66.79	0.78	17.92	5.62	0.06	1.94	0.58	1.46	2.90	0.18
Okimi 2	66.37	0.78	18.42	7.07	0.08	1.91	0.68	1.96	3.08	0.18
Yosio 1	67.26	0.74	18.44	6.41	0.04	2.15	0.75	1.32	2.73	0.11
Yosio 2	70.02	0.75	17.02	5.79	0.02	1.34	0.86	1.39	2.69	0.10
Yosio 3	70.17	0.76	16.45	5.76	0.03	1.32	0.90	1.35	2.61	0.10
Yosio 4	68.43	0.72	17.58	5.97	0.04	2.05	0.61	1.13	2.51	0.12
Mukohidehara 1	66.87	0.73	17.58	8.39	0.05	2.06	0.95	1.37	2.26	0.17
Mukohidehara 2	68.43	0.73	17.04	6.25	0.05	2.30	1.50	1.32	2.17	0.10
Mukohidehara 3	68.96	0.72	17.02	5.63	0.04	2.07	1.01	1.84	2.31	0.12
Mukohidehara 4	67.65	0.72	17.42	5.77	0.05	2.32	1.10	2.22	2.35	0.13
Mukohidehara 5	67.45	0.74	17.68	6.65	0.05	2.42	0.70	1.57	2.51	0.10
Mukohidehara 6	69.12	0.68	17.36	5.20	0.04	2.02	1.30	1.96	2.36	0.12
Engyoji 1	69.70	0.67	17.86	5.97	0.03	1.66	0.55	1.24	2.29	0.12
Engyoji 2	68.15	0.73	17.37	6.00	0.06	2.40	1.06	1.77	2.51	0.13
Engyoji 3	68.14	0.57	17.70	5.85	0.04	2.35	0.93	1.95	2.34	0.12
Engyoji 4	66.03	0.82	18.83	7.36	0.02	2.25	0.68	1.01	2.92	0.08
Iwagami 1	68.51	0.77	17.03	5.32	0.04	1.91	0.53	1.89	3.06	0.10
Iwagami 2	68.85	0.72	17.95	5.14	0.04	1.79	0.68	1.90	2.01	0.18
Iwagami 3	69.95	0.76	16.56	4.99	0.03	1.72	0.76	2.69	2.06	0.09
Tsuboyama 1	68.04	0.79	17.99	7.00	0.02	1.70	0.17	0.92	2.61	0.10
Tsuboyama 2	68.68	0.68	17.29	5.63	0.03	1.79	0.52	1.47	2.76	0.12
Tsuboyama 3	68.31	0.79	17.54	7.05	0.02	1.73	0.54	1.24	2.49	0.10
Tsuboyama 4	67.93	0.68	18.29	7.56	0.03	1.70	0.24	0.87	2.67	0.12
Hirayama 1	66.57	0.70	16.55	10.04	0.07	2.07	0.46	1.20	2.64	0.19
Hirayama 2	65.69	0.72	16.99	8.52	0.05	2.35	0.68	1.42	2.69	0.16
Takao 1	65.80	0.72	18.88	8.38	0.15	1.88	0.65	0.98	2.34	0.22
Takao 2	69.13	0.71	17.75	5.46	0.03	2.04	0.71	1.46	2.58	0.11
Heito 1	65.33	0.73	18.94	9.03	0.04	1.79	0.52	1.16	2.29	0.18
Heito 2	65.04	0.70	18.02	8.69	0.04	2.69	0.81	1.42	2.48	0.10

STI (Silicate Titanium Index) of those specimens were also calculated and are presented in Table 9.4. These parameters are referred to as the weathering indices. CIA is calculated as

$$\frac{Al_2O_3}{Al_2O_3 + CaO + Na_2O + K_2O} \times 100\%$$

CIA value of less than 50% shows that the rock is still fresh, whereas CIA value of 100% shows the rock is in the verge of complete weathering (Nesbitt and Young 1982). CIW is calculated as

$$\frac{Al_2O_3}{Al_2O_3 + CaO + Na_2O} \times 100\%$$

CIW value of less than 50% shows that the rock is still fresh, whereas CIW value of 100% shows the rock is in the verge of complete weathering. Likewise, PIA is calculated as

$$\frac{Al_2O_3 - K_2O}{Al_2O_3 + CaO + Na_2O - K_2O} \times 100\%$$

PIA value of less than 50% shows that the rock is still fresh, whereas PIA value of 100% shows the rock is in the verge of complete weathering. V is calculated as

Table 9.4.
Values of various weathering indices for the mudstone specimens

Sampling site	CIA	CIW	PIA	V	R	WIP	STI
Okimi 1	73.11	83.85	81.07	2.52	6.33	44.91	82.40
Okimi 2	70.27	80.52	77.19	2.34	6.12	51.21	82.12
Yosio 1	73.96	82.90	81.40	2.38	6.19	43.19	82.44
Yosio 2	71.56	81.54	78.55	2.75	6.98	41.55	83.37
Yosio 3	71.10	80.99	77.92	2.68	7.24	40.54	83.56
Yosio 4	74.54	85.53	83.32	2.49	6.60	38.95	83.10
Mukohidehara 1	73.28	81.59	79.24	2.18	6.45	39.85	82.78
Mukohidehara 2	70.13	77.64	74.95	1.81	6.82	40.81	83.26
Mukohidehara 3	69.80	77.77	74.90	1.93	6.88	44.83	83.36
Mukohidehara 4	67.97	74.47	72.43	1.73	6.59	49.67	83.00
Mukohidehara 5	72.91	6.47	44.23	2.05	82.12	79.53	82.81
Mukohidehara 6	68.08	75.65	72.61	1.86	6.76	46.94	83.46
Engyoji 1	76.39	85.45	83.49	2.81	6.62	36.86	83.41
Engyoji 2	69.68	78.20	75.16	1.84	6.66	46.96	83.07
Engyoji 3	70.42	78.31	75.57	1.87	6.53	46.70	83.74
Engyoji 4	75.65	86.65	84.38	2.56	5.95	42.05	81.72
Iwagami 1	69.79	80.75	77.16	2.29	6.83	49.95	83.04
Iwagami 2	73.29	80.43	78.32	2.26	6.51	41.27	82.99
Iwagami 3	67.34	74.05	71.18	1.85	7.17	48.97	83.45
Tsuboyama 1	79.47	90.80	89.27	3.40	6.42	35.77	82.52
Tsuboyama 2	73.13	83.71	80.95	2.57	6.74	43.25	83.43
Tsuboyama 3	75.41	85.30	83.08	2.73	6.61	38.73	82.73
Tsuboyama 4	79.35	90.73	89.18	3.43	6.30	35.99	82.91
Hirayama 1	74.55	85.54	83.04	2.42	6.82	40.27	83.33
Hirayama 2	72.40	82.65	79.79	2.09	6.56	44.12	82.91
Takao 1	77.99	87.12	85.41	2.83	5.91	35.74	82.14
Takao 2	73.20	82.74	80.15	2.32	6.61	42.87	83.16
Heito 1	78.06	86.95	85.27	2.90	5.85	36.38	81.97
Heito 2	73.48	82.53	80.08	1.95	6.12	43.71	82.47

$$\frac{Al_2O_3 + K_2O}{MgO + CaO + Na_2O} \times 100\%$$

For fresh rock, V is less than 1, where as for the rock close to complete weathering, V is infinite. Similarly, R is calculated as

$$\frac{SiO_2}{Al_2O_3}$$

R values of fresh rock and completely weathered rock are more than 10, and 0, respectively. WIP is calculated as

$$\left(\frac{Na_2O}{0.35} + \frac{MgO}{0.9} + \frac{2K_2O}{0.25} + \frac{CaO}{0.7}\right) \times 100\%$$

WIP value is more than 100% for the fresh rock, whereas WIP value is 0% for the completely weathered rock. Likewise, ATI is calculated as

$$\left(\frac{\dfrac{SiO_2}{TiO_2}}{\dfrac{SiO_2}{TiO_2} + \dfrac{SiO_2}{Al_2O_3} + \dfrac{Al_2O_3}{TiO_2}}\right) \times 100\%$$

STI value of more than 90% shows that the rock is still fresh, whereas STI value of 0% shows the rock is in the verge of complete weathering.

CIA, CIE, PFA, V, R, WIP, and STI in those specimen varied from 67-79%, 74-90%, 71-80%, 1.8-3.4, 5.8-7.2, 36-51%, and 81.7-83.6%, respectively. The average value of SiO_2, TiO_2, Al_2O_3, Fe_2O_3, MnO, MgO, CaO, Na_2O, K_2O, and P_2O_5 are 68%, 0.7%, 18%, 6.6%, 0.04%, 2%, 0.7%, 1.5%, 2.5%, and 0.13%, respectively. Likewise, average values of CIA, CIW, PIA, V, R, WIP, and STI were 73%, 79.7%, 78.6%, 2.4, 9.1, 43.8%, and 82.9%, respectively.

9.3 Shear Strength Properties

As a result of brittleness in mudstone due to the overconsolidation effect, there is a sharp drop in the strength from peak to residual within a short displacement amount (Skempton 1985; Tiwari and Marui 2005; Duncan and Wright 2005; Mitchell and Soga 2005). Shown in Fig. 9.4 is a typical stress-strain diagram for the undisturbed and remolded soil specimens from Okimi landslide. Due to a number of fissures existing in the rock mass, fully softened or residual shear strength is mobilized during sliding (Morgenstern 1978) in many cases. As a result of the expansive nature of the mudstone and due to the existence of many dormant landslides in the mudstone areas, it is important to evaluate the residual shear strength of the soil than the peak strength. Table 9.5 shows the ranges of peak, fully softened, and residual shear strength of a number of mudstone specimens. Peak shear strength was obtained by shearing the undisturbed specimen and the fully softened shear strength was obtained by shearing the remolded specimen both in the CD-triaxial tests. Residual shear strength was measured in the ring shear device.

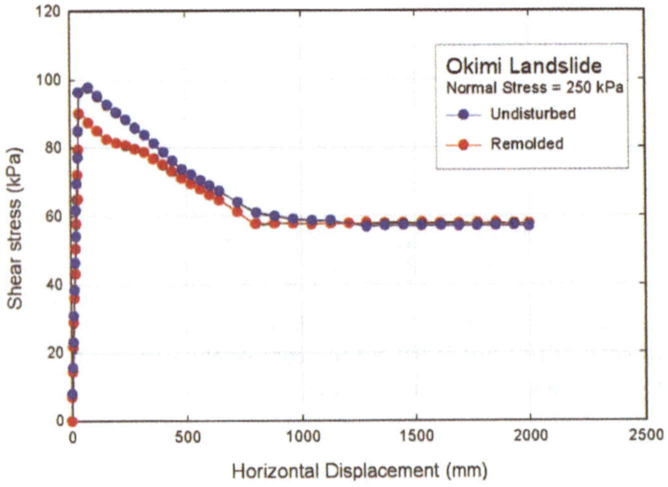

Fig. 9.4. Typical Stress-strain diagram for the tested mudstone specimens

Table 9.5. Friction angles of the mudstone specimens (Degree)

Landslide sites	Peak	Fully softened	Residual
Okimi 1	N/A	N/A	15.0
Okimi 2	N/A	N/A	16.0
Okimi 3	N/A	N/A	16.0
Okimi 4	N/A	N/A	22.0
Okimi 5	N/A	N/A	14.0
Okimi 6	N/A	N/A	14.0
Okimi 7	N/A	N/A	16.0
Okimi 8	N/A	N/A	12.0
Okimi 9	N/A	N/A	11.0
Okimi 10	N/A	N/A	12.0
Okimi 11	N/A	N/A	11.0
Okimi 12	N/A	N/A	12.0
Okimi 13	N/A	N/A	12.0
Okimi 14	N/A	N/A	10.0
Okimi 15	N/A	N/A	12.5
Okimi 16	N/A	N/A	18.8
Yosio 1	25.0	15.0	10.0
Yosio 2	29.0	16.0	10.0
Yosio 3	37.0	14.0	10.1
Yosio 4	24.0	19.0	12.7
Yosio 5	N/A	N/A	9.8
Yosio 6	N/A	N/A	9.8
Yosio 7	N/A	N/A	10.7
Mukohidehara 1	37.0	21.0	19.0
Mukohidehara 2	34.0	20.5	18.0
Mukohidehara 3	25.0	15.0	8.9
Mukohidehara 4	20.0	14.0	10.7
Mukohidehara 5	30.0	18.0	16.2
Mukohidehara 6	35.0	21.0	19.3
Engyoji 1	15.0	12.0	10.0
Engyoji 2	N/A	N/A	10.0
Engyoji 3	15.0	12.0	9.8
Engyoji 4	22.0	15.0	12.8
Engyoji 5	32.0	20.0	17.6
Iwagami 1	N/A	N/A	12.0
Iwagami 2	N/A	N/A	12.0
Iwagami 3	N/A	N/A	17.1
Iwagami 4	N/A	N/A	11.2
Iwagami 5	N/A	N/A	11.1
Tsuboyama 1	22.0	13.0	10.0
Tsuboyama 2	25.0	16.0	10.0
Tsuboyama 3	30.0	18.0	10.9
Tsuboyama 4	20.0	16.0	14.7
Tsuboyama 5	N/A	N/A	19.2
Hirayama 1	N/A	N/A	12.7
Hirayama 2	N/A	N/A	10.4
Takao 1	N/A	N/A	18.1
Takao 2	N/A	N/A	17.6
Hotenji 1	N/A	N/A	15.2
Hotenji 2	N/A	N/A	14.0
Heito 1	N/A	N/A	12.9
Heito 2	N/A	N/A	25.7
Utsunomata 3	N/A	N/A	11.0
Utsunomata 6	N/A	N/A	14.4
Utsunomata 7	N/A	N/A	17.2
Utsunomata 8	N/A	N/A	8.3
Utsunomata 9	N/A	N/A	7.8

As residual shear strength is more important in the natural mudstones, considerable amount of time and effort was spent to conduct research on the residual shear strength of the soil from mudstone area. In the case of residual shear strength, there is no difference whether the specimen is remolded or undisturbed (Tiwari et al. 2005). The detailed outcome from those research works is presented below.

Soil samples were collected from 11 landslide areas, all of them from the mudstone formation. Soil samples were collected from the shallow sliding zones and deep sliding zones in the same sliding block of several landslides and were sheared in a ring shear device to obtain the residual shear strength. Shown in Fig. 9.5 is the Imperial College and NGI type ring shear device (Bishop et al. 1971) used for the study. Shown in Fig. 9.6 is a special trimmer fabricated to trim the sample for undisturbed soil testing in the ring shear device. Detailed description of the device and testing methodology can be obtained from Tiwari and Marui (2003, 2005). Tiwari et al. (2005) showed that there is no difference between the residual shear strength of the soil from deep and shallow sliding surfaces as well as intact and remolded specimens if the landslides are triggered in a homogenous mudstone formation, especially with respect to the mineralogical composition. Residual cohesion is negligible (less than 5 kPa) if not zero. The shear envelope of those specimen curves sharply near the origin i.e. at lower normal stress ranges. Shown in Fig. 9.7 is the typical shear envelope obtained from the mudstone of Okimi landslide. Values of residual friction angles for the studied soil specimens are presented in Table 9.5. The range of friction angle was from 8° to 26° with an average of 13.2°.

Fig. 9.5. Imperial College and NGI type Ring shear device used for the study

Fig. 9.6. Specially designed cutter to cut the undisturbed sample for ring shear test

As ring shear tests were conducted to obtain residual friction angle of the tested specimen and it is very hard to observe undrained shear strength of clays in the ring shear device, Tiwari et al. (2006) conducted extensive research works to obtain undrained shear strength of the soil from triaxial compression. A 37 mm diameter and 87 mm tall clay specimen was trimmed and was pre-cut at an angle close to $(45° + \phi'/2)$ in a special mold. Both halves of the specimen were polished against the frosted glass surface for the displacement of about 1 m. Shown in Fig. 9.8 is the mold to prepare the pre-cut and polished specimens. Those halves were reassembled and sheared in a triaxial device having "both side free" platen. Shown in Fig. 9.9 is the special device used for the residual shear strength measurement in triaxial compression. Shown in Fig. 9.10 is the specimen obtained after the shearing. It clearly shows that the specimen shears along the shearing plane. The research result showed that with the modified platen, residual shear strength can be obtained with reasonable accuracy in a triaxial compression. Residual friction angle by triaxial compression using the above mentioned method was approximately 10% higher than that by the ring shear device. This was very reasonable compared to the results presented by Bishop et al. (1971). Similar platen assembly was used to obtain the undrained residual shear strength of the specimen. Undrained shear strength of the same specimen was about 5% lower than the drained strength. For detail please refer to Tiwari et al. (2006). Chandler (1966) also conducted the similar tests to measure the residual shear strength of Keuper Marl.

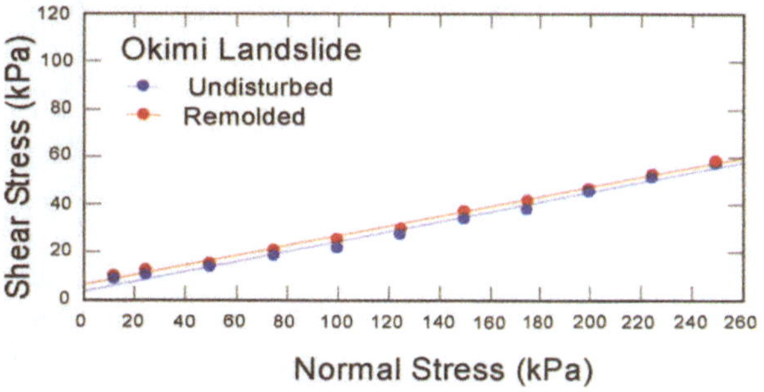

Fig. 9.7. Typical shear envelop for the residual shear strength of undisturbed and remolded soil specimens

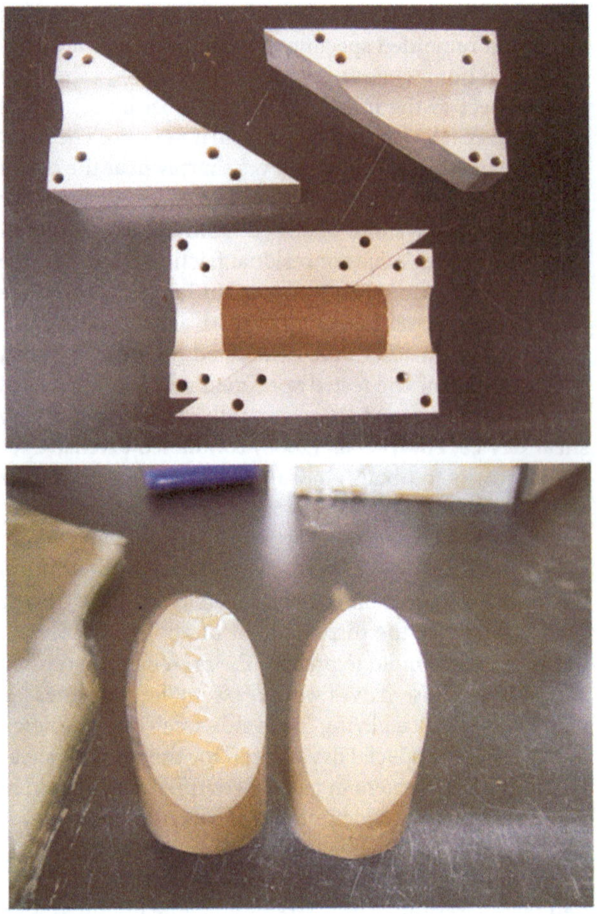

Fig. 9.8. Pre-cut mold and polished specimen used for the residual shear strength measurement at triaxial device

Fig. 9.10. Pre-cut soil specimen after shearing in triaxial compression

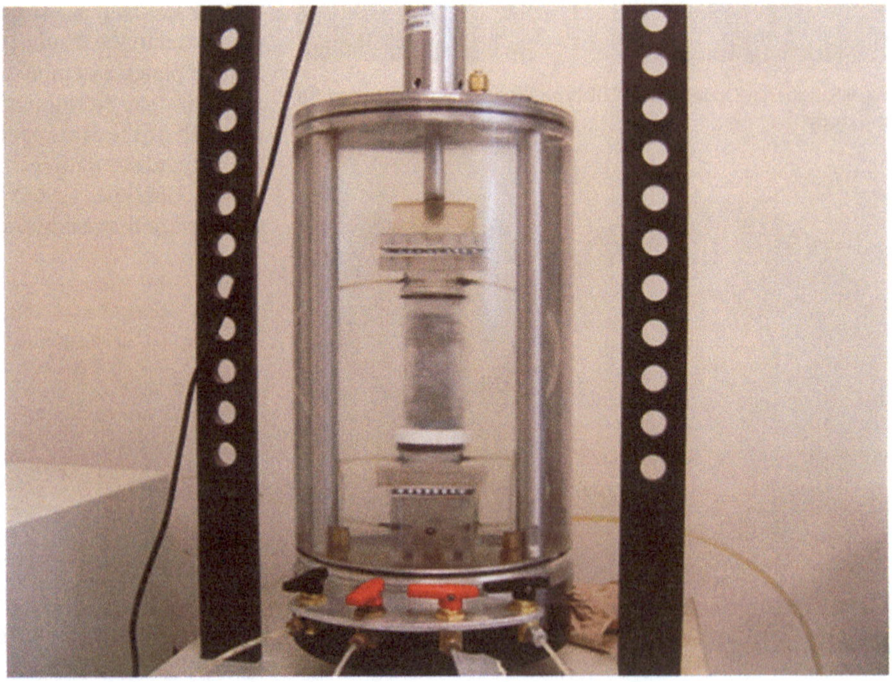

Fig. 9.9.
Specially fabricated free platen for the triaxial testing to measure residual shear strength

9.4 Factors Influencing the Residual Shear Strength of Mudstone

As majority of the minerals in the soil from the mudstone are smectite, kaolinite, illite, feldspar, and quartz, Tiwari and Marui (2005) conducted a separate research to observe the quantitative influence of mineralogical composition on the residual shear strength. Tiwari and Marui (2005) compared the residual shear strength of the soil specimens with index properties that include clay fraction, liquid limit, and plasticity index, as well as the proportion of smectite and specific surface area. Shown in Figs. 9.11, 9.12, and 9.13 are those relations. The result shows that each type of dominant clay mineral gives unique correlation curve, and the vertical position of the curve depends qualitatively on the uniformity coefficient. Therefore, although residual shear strength could be estimated approximately with the index properties, precise estimation requires the exact proportion of constituent minerals. Based on the soil test results from more than 120 natural and laboratory prepared soil specimen, Tiwari and Marui (2005) proposed a triangular estimation chart, as shown in Fig. 9.14, which gives better estimation of residual friction angle of the soil specimen based on the mineralogical composition. Position of each soil specimen is plotted in the triangular estimation chart based on the proportions of massive, low plasticity clay, and high plasticity clay minerals and residual friction angle is estimated by interpolating the corresponding contour lines. Shown in Fig. 9.15 are seven different zones, where the residual shear strength behavior changes with the domination of individual clay mineral. For the proportion of smectite and kaolinite up to 8% each, residual friction angle is controlled by the quartz and feldspar, which is approximately 29°. However, smectite proportion in the soil, if is more than 8%, controls the behavior of the soil significantly. If the proportion of smectite in the soil specimen is more than 42%, residual shear strength of the soil is approximately 4°, irrespective of the proportion of the other minerals. For other mineralogical composition, the residual friction angle of a soil specimen depends on the proportion of massive minerals (quartz, feldspar, calcite etc), low plasticity clay minerals (kaolinite), medium plasticity clay minerals (illite and chloride) and high plasticity clay minerals (smectite). Approximate values of the residual friction angle for different combinations can be obtained by interpolating the contours in the chart presented in Fig. 9.14. Tiwari and Marui (2005) could estimate the residual friction of mudstone specimens with 90% accuracy using the Fig. 9.14. Shown in Fig. 9.16 is the domination of kaolinite and smectite in each zone. Both smectite and kaolinite are dominant in zone A and zone B. Kaolinite has less effect in zone C, D, and E.

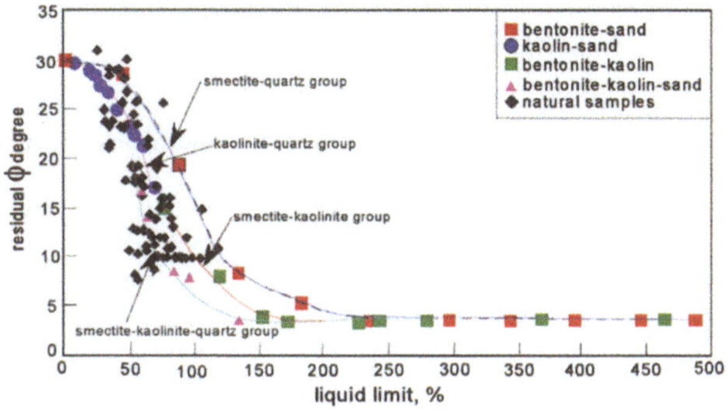

Fig. 9.11. Relation between residual friction angle and liquid limit

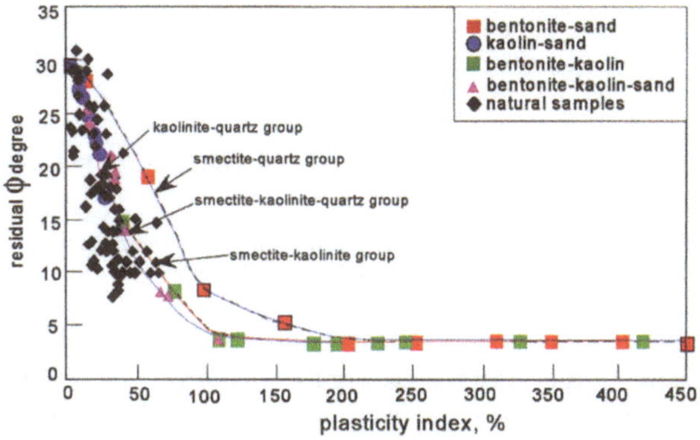

Fig. 9.12. Relationship between residual friction angle and plasticity index

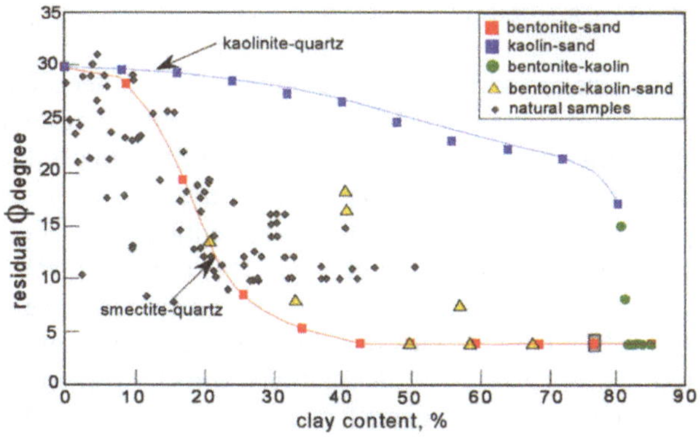

Fig. 9.13. Relationship between residual friction angle and clay fraction

Other than the mineralogical composition, pore water chemistry plays an important role in the residual shear strength of mudstone. When residual shear strength of different minerals were measured with sea water as a pore fluid, smectite showed a significant increase in residual friction angle, i.e. 10° from the value of 4° (with distilled water) (Tiwari et al. 2005). However, kaolinite did not show much influence of sea water on residual shear strength. This shows that pore water chemistry plays an important role on the residual shear strength of mudstone. Figure 9.17 shows the increase in residual friction angle due to the presence of sea water with the proportion of smectite. The increase in the residual friction angle with

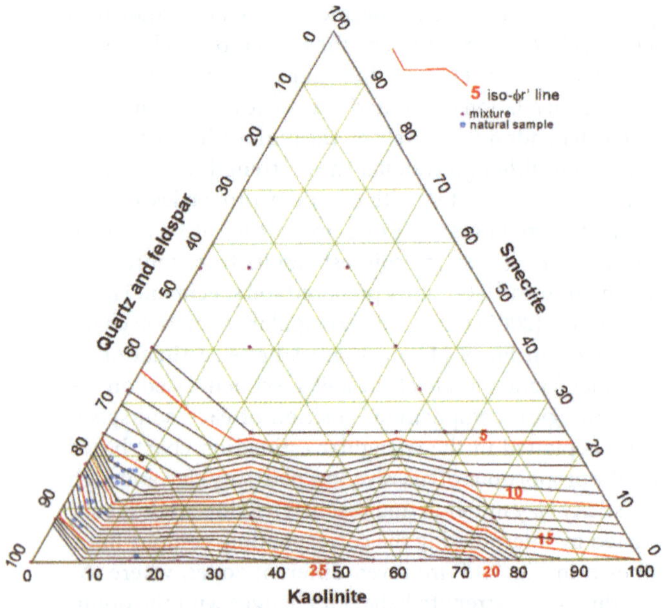

Fig. 9.14. A triangular diagram for the estimation of residual friction angle with mineralogical composition

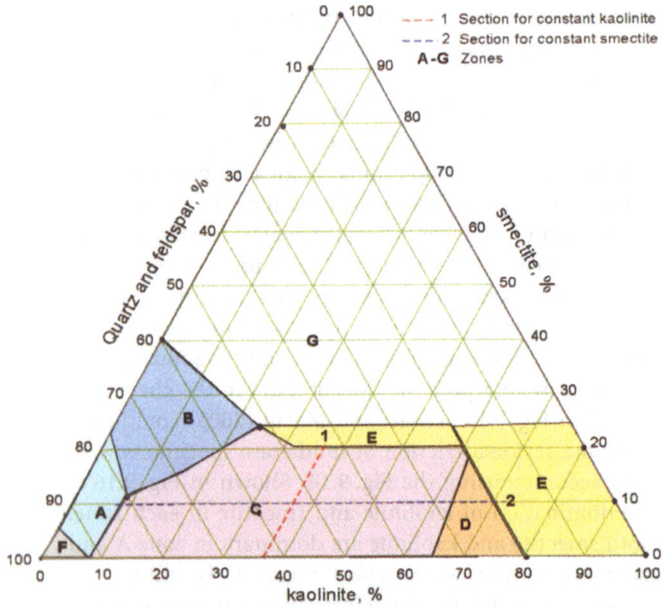

Fig. 9.15. Different zones that have influence in residual friction angle of soil

saline water depends on the thickness of diffused double layer (Tiwari et al. 2005; Mitchell and Soga 2005). Due to its formation process, mudstones near the coastal area are rich in NaCl concentration. A significant proportion of Na^{+} ion leaches out through the interaction with fresh water after being crushed out during sliding and residual shear strength decreases continuously. This degrades the stability of the slope. Therefore, we should be very care-

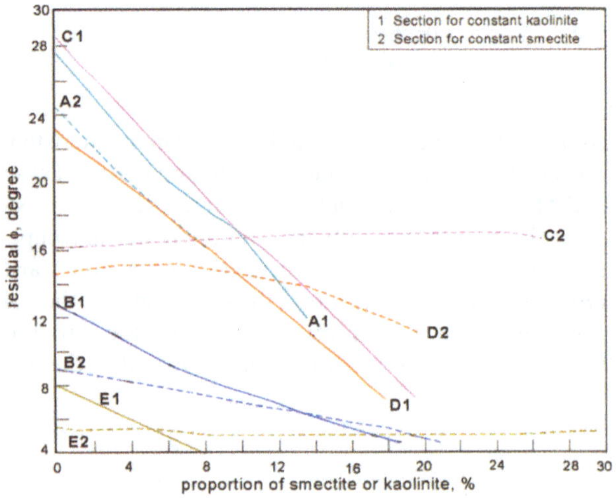

Fig. 9.16. Influence of smectite and kaolinite in residual shear strength for various zone shown in Fig. 9.15

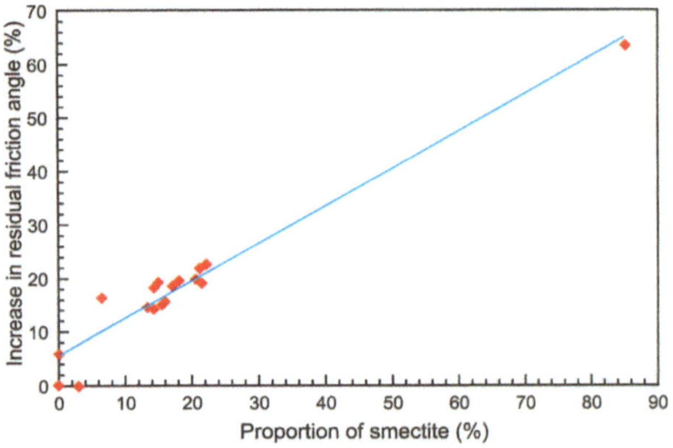

Fig. 9.17. Increase in residual friction angle with proportion of smectite while testing the specimen at saline water

ful while assessing the long-term stability of a mudstone slope in the coastal area. Figure 9.18 shows the residual friction angle of a number of mudstone specimens under sea water, distilled water and after leaching.

9.5 Conclusion

Extensive research was conducted for about a decade to measure the residual shear strength of mudstones and the influencing factors on the residual shear strength. The research showed that Tertiary mudstones are rich in expansive minerals and are highly susceptible to strength softening in the presence of water. Due to its brittleness characteristics, residual shear strength plays vital role in the stability of the mudstone slope. Average values of the liquid limit, plasticity index, activity, and clay fraction of the mudstone from Niigata Prefecture of Japan were 72%, 36%, 1.4, and 26%, respectively. Dominating clay mineral

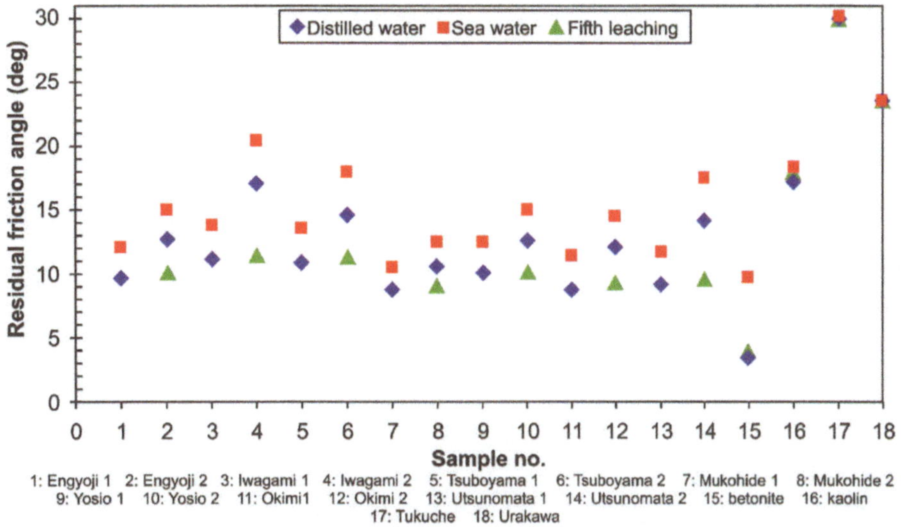

Fig. 9.18. Residual friction angle of the soil specimen with distilled water, saline water, and after leaching the Na^+

of the Tertiary mudstone was smectite with an average proportion of 15%. Average residual friction angle of those mudstones was 13°. The shear strength of the mudstone in the field, where saline pore fluid was observed was about 40% higher than the mudstone after leaching Na^+ ion. The research also shows that the residual shear strength can be estimated approximately with liquid limit, plasticity index, and specific surface area. However, better estimation can be performed using the diagram proposed by Tiwari and Marui (2005) if the mineralogical composition is known. Quantity of soil specimen obtained from boring core is sufficient to estimate the mineralogical composition.

References

Bishop AW, Green GE, Garga VK, Andresen A, Brown JD (1971) New ring shear apparatus and its application to the measurement of residual strength. Géotechnique 21(4):273–328

Chandler RJ (1966) The measurement of residual shear strength in triaxial compression. Géotechnique 16(3):81–186

Duncan JM, Wright S (2005) Soil strength and slope stability. John Wiley & Sons Inc., New York

Mitchell JK, Soga K (2005) Fundamental of soil behavour. 3rd ed. John Wiley & Sons, Inc., USA

Morgenstern N (1977) Slopes and excavations in heavily over consolidated clays, In: Proceeding of 9th International Conference of Soil Mechanics and Foundation Engineering, 2, pp 567–581

Nesbitt HW, Young GM (1982) Early proterozoic climates and plate motions inferred from major element chemistry of lutites. Nature 279:715–717

Skempton AW (1985) Residual strength of clays in landslides, folded strata and the laboratory. Géotechnique 35(1):3–18

Tiwari B, Marui H (2003) Estimation of residual shear strength of Toyoura sand-Bentonite-Kaolin Mixture. J Jpn Landslide Soc 40(2):124–133

Tiwari B, Marui H (2005) A new method for the correlation of residual shear strength of the soil with mineralogical composition. J Geotech Geoenviron 131(9):1139–1150

Tiwari B, Marui H, Aoyama K, Bhattarai P, Tuladhar G (2004) Preparation of geotechnical database of mountainous cities with ArcGIS. In: Proceedings of ESRI's 24th Annual International User's Conference, San Diego, 1569, pp 1–30

Tiwari B, Brandon T, Marui H, Tuladhar G (2005a) Comparison of residual shear strength from back analysis and ring shear tests on undisturbed and remolded specimens. *J Geotech Geoenviron* 131(9):1071–1079

Tiwari B, Tuladhar G, Marui H (2005b) Variation in residual shear strength of the soil with pore water chemistry. J Geotech Geoenviron 131(12):1445–1456

Tiwari B, Duncan JM, Brandon TL (2006) Evaluation of undrained shear strength of clays in triaxial compression, submitted to National Science Foundation, Virginia Tech

Chapter 10

On Failure of Municipal Waste Landfill

Ikuo Towhata

Abstract. One of the serious problems in mega cities in developing countries is the management of municipal waste. Due to priority of economic development or insufficient attention to waste problems, many mega cities simply dump waste in the field without provision for environmental and mechanical risks. An example of problems of this type occurred in Bandung City of Indonesia in February, 2005, in which a large waste landfill collapsed after rain fall and killed more than one hundred people. Similar accident occurred earlier in the Philippines as well. This text therefore makes a brief report on this event and shows the need for more care for landfill operation and management.

Keywords. Landfill, slope failure, municipal waste, case study

10.1 Introduction

In the recent decade, many countries which used to be called developing countries have made remarkable success in economic development. Accordingly, their population increased and the increased people started to live in big cities. This situation resulted in emerging mega cities which have millions of population. The problem lying in those new mega cities is that economic development is given with the first priority and that there is not sufficient provision for good environment. Hence, air and water are polluted and poor traffic system worsens the congestion and air pollution. Environmental pollution made by industrial waste is another problem.

A problem which is similarly serious is the management of municipal waste which is produced by people's daily life as well as offices and small industries. Since budget is not sufficient for waste treatment, the municipal waste is often dumped without taking care of accompanying risks. While the risk of environmental pollution is frequently mentioned, the present text concerns slope instability problem. The failure of Payatas landfill in Metro Manila is famous as a rainfall-induced failure of waste deposit (Merry et al. 2005). A similar accident, however, has occurred elsewhere.

Fig. 10.1. Location of Bandung City

10.2 Failure of Waste Leuwigajah Landfill in Bandung, Indonesia

Bandung, Indonesia, is a city which is located in West Java Island and has six-million population (Fig. 10.1). In this city, the Leuwigajah Landfill was the biggest and most important site for dumping waste. It was situated in a narrow valley whose original gradient was 5–10% and its subsurface conditions consisted of bedrock covered by 1-m thick clay. The dumped landfill formed a waste deposit of 60 to 70 m in thickness. It appears that all kinds of municipal waste of the city were simply dumped without incineration or separation (Fig. 10.2). Hence, organic waste and plastics are mixed in the fill. The landfill was constructed since 1992 by simply dropping garbage from the top over an edge and poor compaction was made by crawlers.

The Leuwigajah landfill collapsed suddenly at 2 A.M. on February 21st, 2004, over an area of 0.13 km^2 (Koelsch et al. 2005). The author visited the site in February, 2006, which was one year after the accident. The present text is written based on knowledge obtained during this visit.

Fig. 10.2. Appearance of remaining part of Leuwigajah landfill

Fig. 10.3. Head scarp of landfill failure

Fig. 10.4. Widespread deposit of failed waste

The failure created a cliff at its top part (Fig. 10.3). Being nearly vertical and having been stable for one year after the failure, this cliff suggests that the waste material has some shear strength, although the downslope part failed. Figure 10.4 on the contrary demonstrates the view towards the downstream direction. The failed waste flowed over a long distance and covered an area of $300 \times 900\ m^2$. The apparent angle of friction as determined from the vertical and horizontal travel distances was 6 degrees. The death toll due to this accident was 146. Local people told the author that there was a heavy rainfall for three days prior to the failure. Since the landfill was located in valley topography, probably much ground water came into the waste deposit and affected the stability.

10.3 Significance of Landfill Failure

Similar to other landfills in South East Asia, many scavengers were living in the downslope area of the Leuwigajah landfill. Their houses were too close to the landfill (Fig. 10.5) and many of them were buried under the failed waste mass. Scavengers pick up many kinds of waste (Fig. 10.6) and sell for money. For example, 1 kg of plastics are sold at 750 Rupia in the local currency; approximately equal to 9 Japanese Yen or 6 US Cents. Although there was no eyewitness of this midnight slope failure, the local people mentioned that the landfill had been moving for two days prior to the ultimate failure. This implies the effects of rainfall which occurred in the meantime.

Figure 10.4 showed that the failed waste reached the area of rice field. Although the quality of irrigation water is affected by the waste, there is no effort to prevent the pollution problem.

Fig. 10.5.
Scavengers' houses

Fig. 10.6. Scavenger who pick up waste

Fig. 10.7. Houses which survived the failure of landfill

10.4 Proposals for Better Conditions

The ultimate solution for environmental and stability problems is obviously a construction of dam on the downstream side of the dumping site. Such a measure is, however, ruled out for economic reasons in many developing countries. Hence, alternative ideas are needed.

Firstly, a control of habitation is necessary. In the Leuwigajah case, houses located in the valley slope survived the accident (Fig. 10.7). It seems possible to make a rough estimate of the run-out area of a possible landfill failure by using the aforementioned apparent friction angle of 6 degrees. Houses should be located in areas which are not affected by the waste flow. Secondly, the dumping of waste should make a gentle slope gradient; formation of a steep slope should be avoided. Warning system for slope instability is a useful idea, for which needed cost is not high. Finally, good drainage system should be installed and the leachate water should not affect the habitation and agricultural areas. A water treatment facility seems to be a good measure to achieve this goal. Although such a facility may be called expensive, it is a reasonable target of international development aids. Mitigation of many mega city problems in developing countries should be considered as a useful and meaningful point of international aids.

10.5 Conclusions

A brief report was made of a recent collapse of municipal waste landfill in Indonesia. It was seen that safe but expensive measures are not accepted due to economic reasons. Alternative safety measures are the location of houses in safer area, making gentle slope angle during waste dumping, and installation of warning system and drainage as well as waste treatment facility.

Acknowledgments

The author's visit to the failed landfill was made easy by assistance of Dr. F. Koelsch of Technical University of Braunschweig, Germany, and Dr. Ilyas Suratman of Bandung Institute of Technology. The author expresses his sincere thanks to their collaboration.

References

Koelsch F, Fricke K, Mahler C, Damanhuri E (2005) Stability of landfills, the Bandung dumpsite disaster. In: Proceedings of the 10th International Landfill Symposium

Merry SM, Kavazanjian E, Fritz WU (2005) Reconnaissance of the July 10, 2000, Payatas landfill failure. Journal of Performance of Constructed Facilities, ASCE 19(2):100–107

Experimental Study with Ring Shear Apparatus on the May 2004 Landslide–Debris Flow at Bettou-dani Valley, Haku-san Mountain, Japan

Fawu Wang* · Kyoji Sassa

Abstract. In May 2004, a landslide occurred at the right flank of the Jinnosuke-dani landslide, and transformed into a debris flow after fluidization. By analysis of the monitored video images of the debris flow, field investigation on the source area of the landslide, and a series of simulation tests with a ring-shear apparatus on the initiation of the rainfall-induced landslide and its traveling process, the initiation and traveling mechanisms of the debris flow traveling in the valley were investigated. It is shown that concentrated groundwater flow was the main reason for the landslide initiation, and a rapid decrease of the mobilized shear resistance even under naturally drained condition caused the rapid landslide motion. During the debris motion in the valley, high potential for grain-crushing of deposits in upstream and lower potential for the downstream deposits controlled the traveling and depositing process of the debris flow. Different grain-crushing potential of the valley deposits played an important role in the debris flow traveling and depositing processes.

Keywords. Landslide, case study, fluidization, groundwater, grain-crushing, ring-shear test

11.1 Introduction

Haku-san Mountain is located at the boundary between Ishikawa Prefecture and Gifu Prefecture in Hokuriku district, Japan (Fig. 11.1). It is an active volcano with a summit 2703 m in elevation, and the whole mountain is a national park. This park is famous for its beautiful scenery. About 50000 mountain climbers visit this mountain in the period from 15 May to 15 October every year. Tedori River, the largest river in Ishikawa Prefecture, originates in this area. The Jinnosuke-dani landslide (Fig. 11.2) is a giant landslide located on the southwestern slope of Haku-san Mountain ("dani" means valley or torrent in Japanese.). It was also the first landslide designated as a "Landslide prevention area" by the "Japanese Landslide Prevention Law" in 1958. Landslides frequently occur in this area, and commonly trigger debris flows that travel long distances and damage properties in the downstream valley of the Tedori River. In the photograph of Fig. 11.2, the areas not covered by vegetation are local slope failures. For example, at the left side of the photograph, the main scarp and sliding surface of the Bettou-dani failure, which occurred in 1934, is visible. In that event, a debris flow initiated by a landslide reached the Japan Sea after traveling for 72 km. The upper part of the central ridge sandwiched by the Bettou-dani at the left side and the Jinnosuke-dani at the right side is an active landslide, according to data obtained by monitoring, and is called the "Central Ridge Block" of the Jinnosuke-dani landslide (Fig. 11.3). The width and length of this block are 500 m and 2000 m respectively. Besides the "Central Ridge Block", there are many other active landslide blocks in this area. In Fig. 11.3, the blocks with arrow inside are active landslide blocks. In recent decades, accompanying the deformation of the Central Ridge Block, local landslides with different volumes occurred at both boundary valleys of the Central Ridge Block and caused damage, although many countermeasure works have been conducted in this area for more than 50 years.

The Electronic Distance Measuring (EDM) method and Global Positioning System (GPS) monitoring method have been used at this site to monitor slope displacements. Figure 11.4 shows the initial locations of the monitoring points in the regulated "Landslide Prevention Area" and the displacement vectors of the slope surface from 1994 to 2001 (Wang and Sassa 2007). There are 9 survey points (A1 to A9) in the lower block of the landslide and 12 survey points (B1 to B12) in the upper block. Six points (C1 to C6) were located outside of the Jinnosuke-dani land-

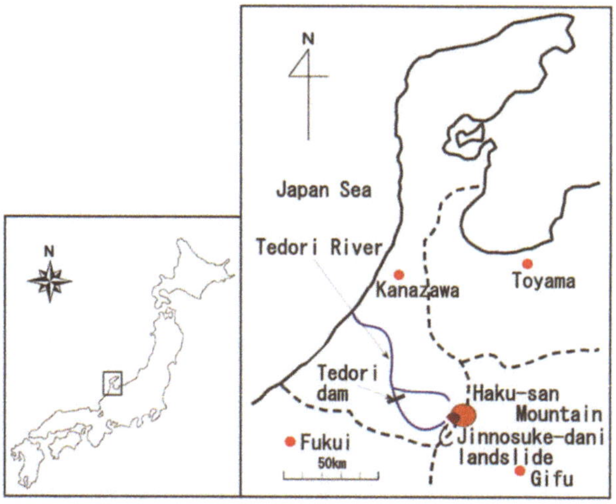

Fig. 11.1. Location map of the Jinnosuke-dani landslide

Fig. 11.2.
Aerial photograph of the "Landslide Prevention Area" of the Jinnosuke-dani landslide (photo courtesy of Kanazawa Office of Rivers and National Highways, MLIT)

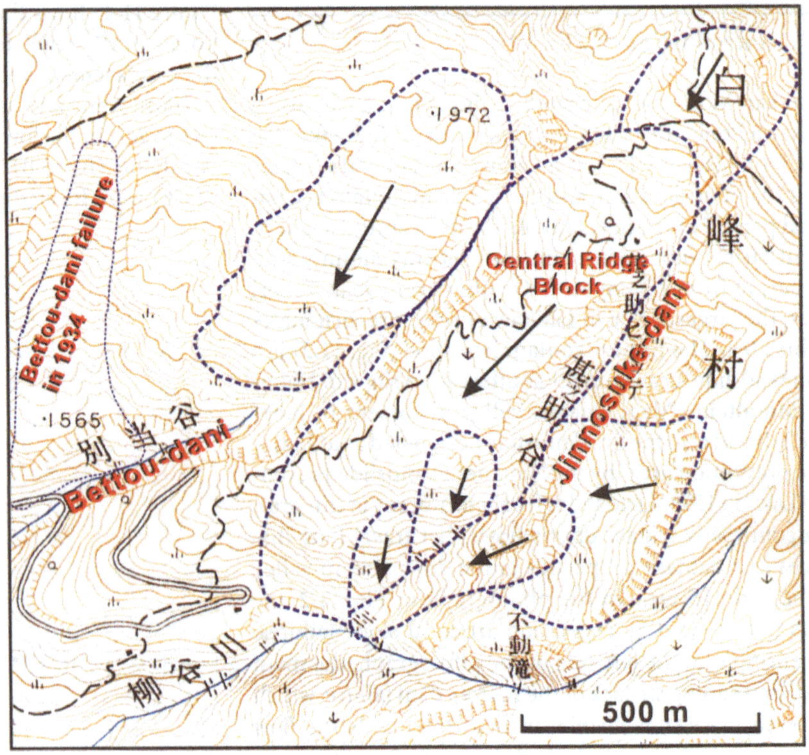

Fig. 11.3.
Active landslide blocks in the Haku-san mountain area around the "Central Ridge Block" of the Jinnosuke-dani landslide (based on Kanazawa Office of Rivers and National Highways, MLIT 2004a)

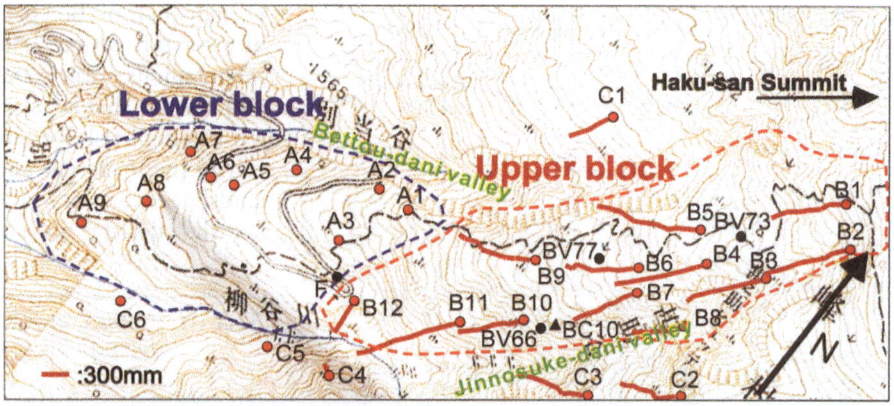

Fig. 11.4. Cumulative displacements of the landslide from 1994 to 2001. All the points of *A*, *B*, and *C* are slope surface displacement monitoring points

Fig. 11.5. Photo of debris retention dams constructed in the Jinnosuke-dani valley

slide (It is used to monitor the activity of the other landslide blocks in the regulated landslide prevention area). The monitored results show that the upper block displaced quite actively; the cumulative displacements of survey points B5 and B11 exceeded 1 100 mm in the 7 years. The main features of the monitoring results in the upper block are: (*1*) the points at the central part almost moved along the downslope direction; (*2*) the points near valleys had a component to the valley side, besides along the downslope direction. However, the lower block has been relatively stable because the monitored displacement at A1 to A9 at a rate of 3–15 mm yr^{-1}. While, C1, C2, C3 and C4 which located on different landslide blocks also indicated the motion of the corresponding blocks. The boundary between the upper block and the lower block is not clear at the slope surface. So, the landslide is divided to blocks just according to the surface displacement. The average movement direction of the upper block of the Jinnosuke-dani landslide is S 36° W. This direction of movement corresponds well with the dip direction of the Tedori Formation.

Figure 11.5 shows the debris retention dams constructed in the Jinnosuke-dani valley. In May 2004, a landslide occurred at the upper part of the Bettou-dani from the Central Ridge Block. This landslide was transformed into a debris flow that traveled more than 2 km after it slid into the Bettou-dani. A suspension bridge was completely destroyed, and a local road with a simple bridge utilized for debris-retention dam construction at the middle of Bettou-dani was heavily damaged. Fortunately, nobody was injured because there were not many mountain climbers passing through the valley when the landslide–debris flow occurred. However, because landslides with similar behavior frequently occur in this area, the risk for further landslide and debris flow activity still exists. As a national park, it should be absolutely safe for the tourists. Even if some large landslides cannot be completely stabilized, understanding their potential risk, es-

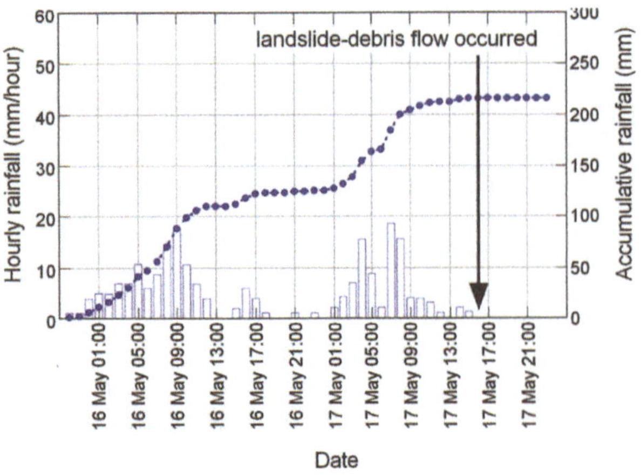

Fig. 11.6. Hourly rainfall before the landslide–debris flow occurred (The rainfall gauge was located at the central ridge block of Jinnosuke-dani landslide, and was measured by Kanazawa Office of Rivers and National Highways, MLIT)

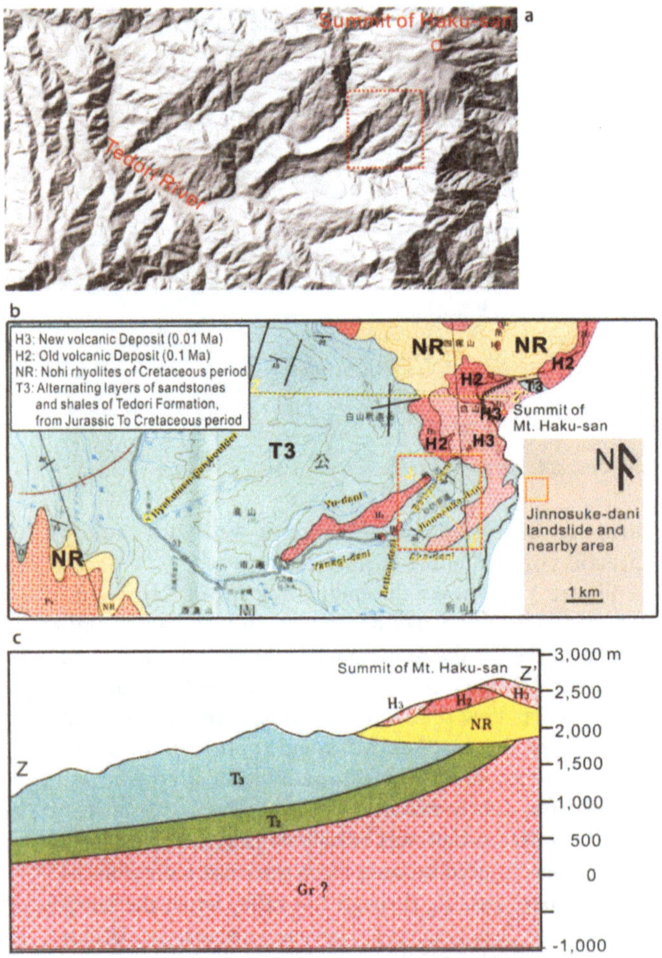

Fig. 11.7. Topography and geological condition of the Haku-san and nearby area. **a** DEM model; **b** geological map (after Kaseno 1993); **c** longitudinal section of Z–Z' in (**b**) (after Kaseno 1993). *T2*: Sandstone and shale in Kuwashima group (also belong to the Tedori Formation); *Gr?*: possible granite

pecially their motion behavior is also very important for disaster mitigation. Looking at the large parks in the mountainous areas in the world, it is easy to find a common feature, that is, steep topography with nice views often has high risk for landslide. This paper attempts to clarify the initiation and traveling mechanisms of the landslide–debris flow, aiming to supply insight for future landslide disaster prevention in similar area.

11.2 General Conditions of the Jinnosuke-dani Landslide on Haku-san Mountain

The Haku-san Mountain area is characterized by heavy precipitation and the Tedori River is characterized by its steep gradient (Wang et al. 2004; Okuno et al. 2004). In winter, due to the strong influence of monsoons from Siberia, the accumulative snowfall may exceed 12 m in the Haku-san Mountain area. In other seasons, half of the days are rainy. For this reason, local annual average precipitation is 3295 mm, about two times the national average of 1700 mm for Japan. In this area, snowmelt generally begins in the middle of March and finishes at the end of May. In May 2004 when the landslide occurred, snow melting was on-going, and the accumulative rainfall for three days before the landslide occurred was 216 mm (Fig. 11.6). It is also reasonable to consider that the effective water that infiltrated into the slope should be more than this value. Because the movement was recorded by a video camera of the Kanazawa Office of Rivers and National Highways, Ministry of Land Infrastructure and Transport, Japan (KORNH- MLIT) (2004b), the actual failure time was also exactly recorded, and the video image is very important for the study on the motion mechanism of the landslide and debris flow.

The basal bedrock in the Haku-san Mountain area is the lower Paleozoic Hida gneiss. As a part of the 1 : 50000 geological map of Haku-san Mountain area, a geological map of the Jinnosuke-dani landslide and nearby area was completed by Kaseno (2001). From the Jurassic to Early Cretaceous periods, the Haku-san Mountain area was a lake near the sea. The series of lacustrine sediments deposited in that period is called the Tedori Formation deposits. The deposits are sedimentary strata consisting of shale, sandstone, and conglomerate layers that have undergone hydrothermal alteration during the mountain-building process of Haku-san Mountain. General descriptions of the geology can be found in Kaseno (1993). Figure 11.7a is a DEM model for a large area around the Haku-san mountain area. The model is built on the elevation data with contour difference of 50 m. The image of lava deposit distribution around the summit of Haku-san (vent of the Haku-san volcano) is visible. Figure 11.7b is the geological map of the corresponding area to Fig. 11.7a. The Nohi Rhyolite (NR) of the Cretaceous period is distributed at the upper right corner in the figure and the alternating layers of sandstone and shale of the Tedori Formation (T3) are distributed below and to the left. Both units form the bedrock of the Haku-san Mountain area. Volcanic lava deposits consisting of calc-alkaline orthopyroxene and hornblende andesite, which erupted 100 000 and 10 000 years ago (H2 and H3, respectively), overlie the strata of the Tedori Formation and the Nohi Rhyolite. Figure 11.7c shows an estimated longitudinal section of the Z–Z' section (shown in Fig. 11.7b). The deposition subsequence of the strata and volcanic deposit is clear. The Jinnosuke-dani landslide area is indicated in the dotted line box. At the northwest corner of the dotted line area, the disconnection of H2 old volcanic deposit can be observed. The reason for this phenomenon maybe caused by erosion and landslide.

Figure 11.8 shows the distribution of strata in the area adjacent to the Jinnosuke-dani landslide. Alternating layers of sandstone and shale of the Tedori Formation are distributed at the left side of the figure, and the Cretaceous Nohi Rhyolite are distributed at the right side. Both form the bedrock of this area. Volcanic deposits, which erupted 100 000 years ago and 10 000 years ago, overlie the strata of the Tedori Formation and the Nohi Rhyolites.

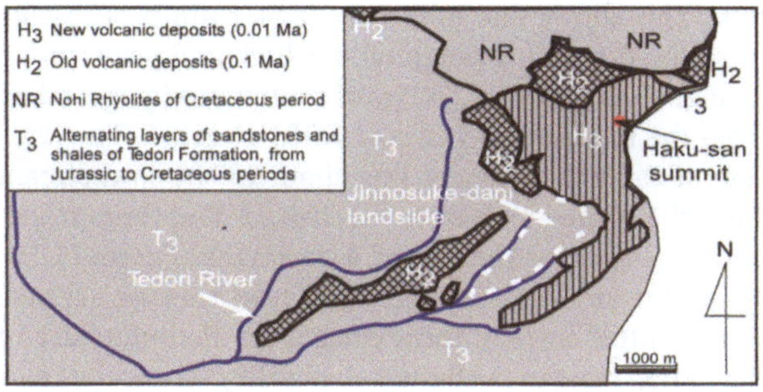

Fig. 11.8. Geological map of the area adjacent to the Jinnosuke-dani landslide (modified from Kaseno 2003)

11.3 The May 2004 Landslide–Debris Flow

As mentioned earlier, the landslide occurred on 17 May 2004 after continuous intense rainfall for two days. The elevation of the source area of the landslide was about 1 900 m, and the elevation of the toe part of the deposit of the debris flow caused by the landslide was about 1 200 m. Figure 11.9 shows two aerial photographs taken before the event (Fig. 11.9a: in the fall of 2003), and after the event (Fig. 11.9b: on 24 May 2004, 7 days after the landslide-debris flow), and Fig. 11.9c, the trace of the debris flow with elevations at some key points. As shown in Fig. 11.9a,

Fig. 11.9. The May 2004 landslide–debris flow, which occurred in the Bettou-dani from the Central Ridge Block of the Jinnosuke-dani landslide. **a** Aerial photograph taken before the slope failure (in the fall of 2003); **b** aerial photograph after the landslide (taken on 24 May 2004); **c** trace of the debris flow (photos courtesy of the Kanazawa Office of Rivers and National Highways, MLIT)

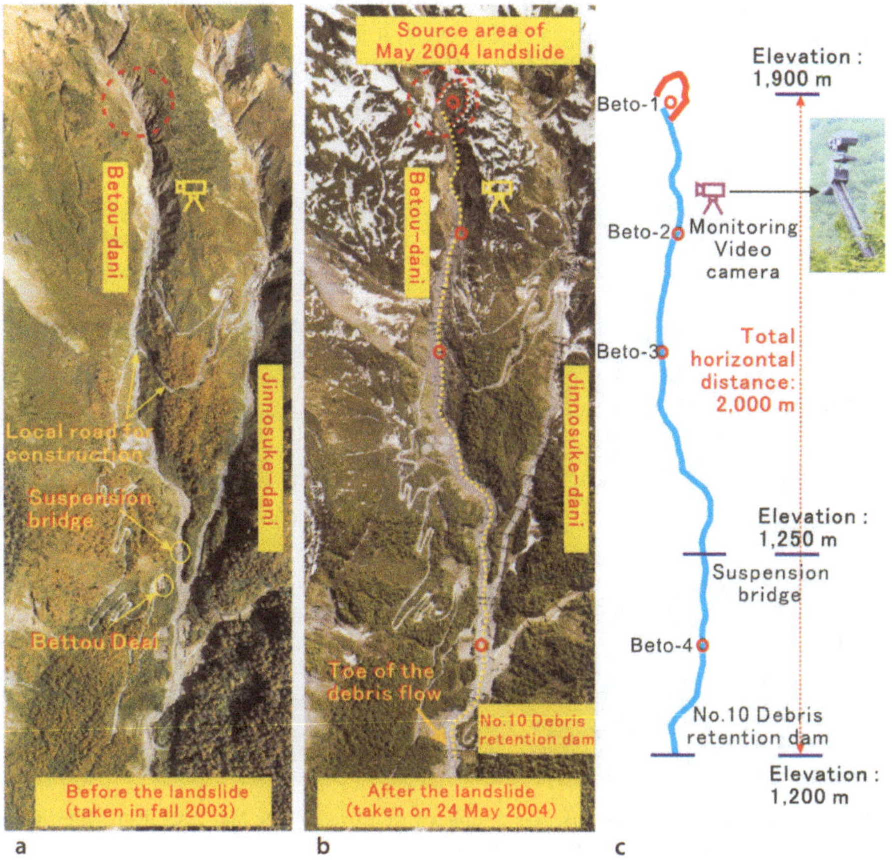

the source area is a steep cliff and there was not any vegetation on the lower segment of the landslide; however, at the upper part, the slope is relatively gentle and is covered by vegetation. For mountain climbers, after leaving Bettou Deai, which has facilities such as parking areas, rest rooms, simple restaurants, and a bus stop, most of the climbers have to cross the suspension bridge and access the Central Ridge Block of the Jinnosuke-dani landslide to get to the summit of Haku-san Mountain. At the middle of the Bettou-dani, an access road for construction of debris-retention dams crosses the valley and enters the Central Ridge Block. As shown in Fig. 11.9b, both the roads and the bridge were badly damaged when the debris flow hit them. The entire flowing process of the debris flow was recorded by a video camera (see Fig. 11.9c for the photo of the video camera set at the site), which was set at an elevation of about 1 860 m for the purpose to monitor the debris flow in the Bettou-dani valley by the KORNH-MLIT (2004c). It Japan, under the leadership of the Ministry of Landslide, Infrastructure and Transport, monitoring video cameras are set at different parts of important rivers, especially at the parts which have high risk for landslide and debris flow. The video camera monitoring was conducted with a remote control system, and the video image can be reviewed in real time on the internet. Because Bettou-dani is a valley with high potential for landslide and debris flow, it has been under control and monitored for 24 hours a day. Through analyzing the recorded video images of this event, it is estimated that the velocity of the debris flow may have reached a maximum of 20 m s^{-1}. As shown in Fig. 11.9c, the relative height difference between the source area and the toe of the deposits of the debris flow (near No. 10 debris retention dam) was 700 m, and the horizontal traveling distance was about 2 000 m. Based on these data, the apparent friction angle ϕ_a (defined as $\tan \phi_a = H/L$, where H is the difference of elevation between head and toe of a landslide, and L is the horizontal distance from head to toe) of the debris flow is estimated to be 19.3 degrees.

Figure 11.10 shows the situation when the suspension bridge was completely destroyed. Large boulders 3–4 m in diameter were transported and deposited near the bridge site. Some thin debris were deposited on the top of the left pillar of the bridge, about 10 meters above the valley bottom. This shows that even near the terminus of the debris flow, the sliding potential was high and powerful.

Figure 11.11 shows a series of continuous images taken from the monitoring video of KORNH-MLIT. The location of the video camera was about 250 m downstream from the source area of the landslide. The time in seconds is shown at the top of each image. Figure 11.11a shows the situation just before the debris flow arrived. White dotted lines in these images are the boundary of the sliding mass in the Bettou-dani valley. The white color in the images is snow. The debris flow passed through the video from 16:32:37 to 16:33:16 o'clock; thus, the entire process continued for only 40 seconds in front of the video camera. By analysis of the video images shown in Fig. 11.11, the debris flow can be divided into four separate waves. The first wave was from (b) to (g), which con-

Fig. 11.10. The suspension bridge that was completely destroyed by the May 2004 debris flow in the Bettou-dani (photo courtesy of the Kanazawa Office of Rivers and National Highways, MLIT)

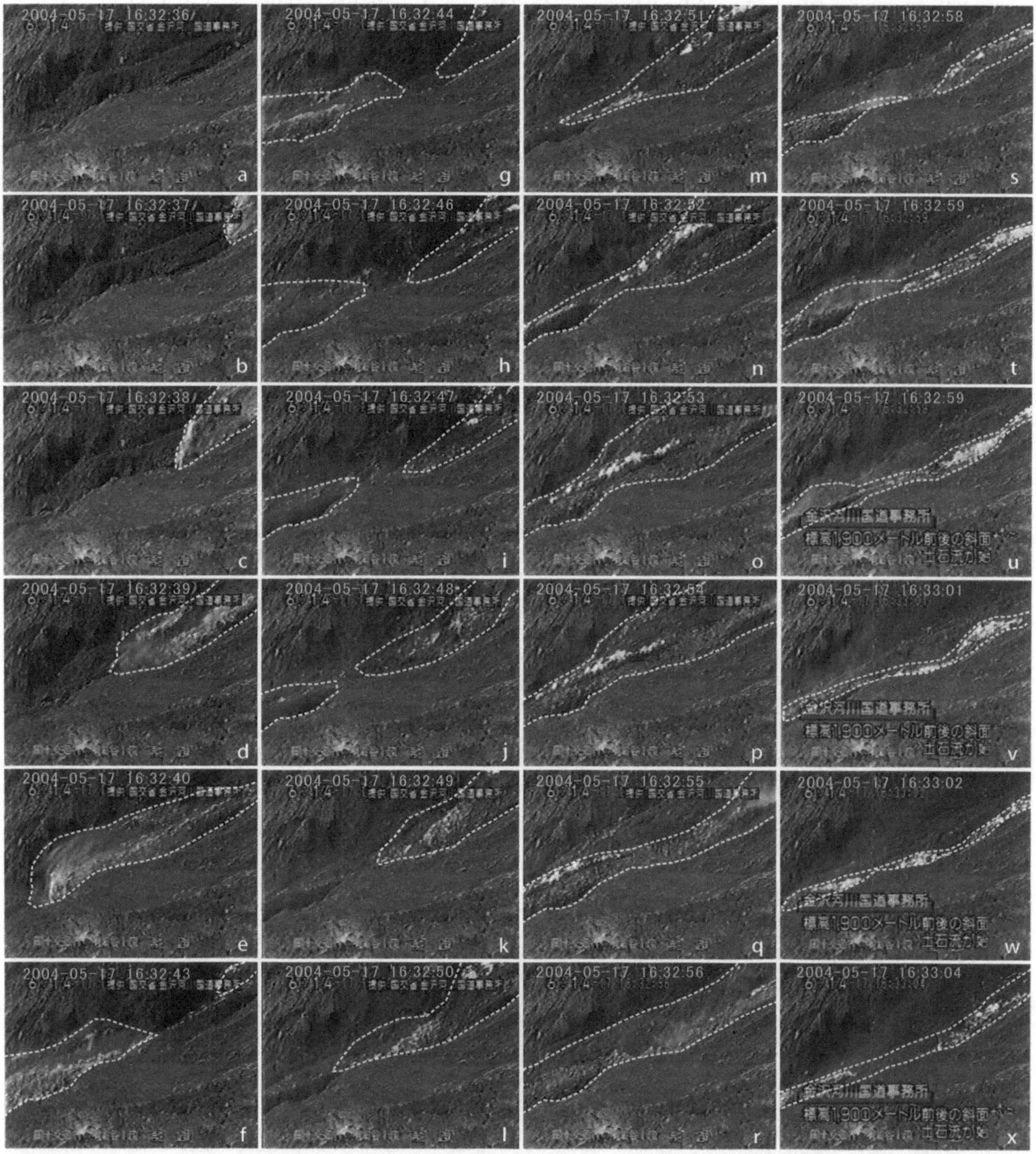

Fig. 11.11. Continuous images of the May 2004 debris flow in the Bettou-dani (video courtesy of the Kanazawa Office of Rivers and National Highways, MLIT). Sliding mass is shown in *white dotted lines*. *White color* in the white dotted lines is snow

tinued for 7 seconds. As a frontier of the sliding mass, there was fog going along with the sliding mass, indicating a high traveling speed of the sliding mass. The second wave was from (g) to (i), which continued for 3 seconds. In this short period, a relative small sliding mass passed the video very quickly. The third wave was from (i) to (s), which continued for 11 seconds. During this long period, the sliding mass passing before the video was very high and wide, indicating a large volume. As shown in Fig. 11.9s, an interrupt can be seen in the middle of the sliding mass. Actually, the flow of the sliding mass should keep continuous at the lower position of the valley (From the camera location, the bottom of the valley cannot be observed). From 16:32:58, the dimension of the flow became smaller, but showing a continuous flow. For this reason, the fourth wave was defined from (r) to the end of the motion (at 16:33:16), which continued for 18 seconds. The images after (x) were not presented here because the direction of the video camera was changed to observe the situation of the upstream part. For all of the images including the sliding mass, it is obvious that all of the debris included snow, and muddy fog can be recognized in Fig. 11.11d,e,r,s and 11.11t, indicating rapid motion during downstream traveling.

Figure 11.12 shows the situation at the source area of the landslide on 11 September 2005, more than one year after the landslide event. A man in the enlarged box can be seen as a scale. According to the report of investigation conducted soon after the event by KORNH-MLIT (2004b), the average slope angle was about 28 degrees, and the average thickness of the sliding mass was estimated as 30 m. In this figure, L1, L2, and L3 show the rear boundaries of three different sliding blocks, which moved for a limited distance from the main scarp; however, most of these blocks did not move so far, but just rested on the slope. At the middle block, between L1 and L3, most of the debris material slid out of the source area, entered into the Bettou-dani, and joined the debris flow. Also, at the lower part of L2, most of the debris slid out into the valley. A common phenomenon at the source areas of these sliding blocks is that concentrated groundwater flow exited at W1, W2, and W3 at relatively high positions. In the other hand, at the lower part of L3, there were three groundwater exits W4, W5 and W6. These groundwater exits were located in relative low position comparing with W1, W2 and W3. It looks like that the groundwater level was different at the left and right parts, and the exits indicated the different groundwater veins. It is estimated that the phenomenon of the sliding mass in L3 not moving so far is due to that the sliding mass at L3 was not fully saturated by the groundwater flow when the landslide occurred. This fact ensured that the debris, especially in the potential sliding zone, was fully saturated and that high water pressure was supplied to the back of the debris to make the slope unstable. The groundwater exiting at a high position at the head of the debris was a major triggering factor for the landslide which fluidized into a debris flow. It needs to be mentioned that, when the field investigation was conducted in September 2005, more than one year had passed since the landslide occurrence. Some situation may be changed by rainfall erosion and other factors. However, the basic condition should be the same because there was no large change in natural condition and no any engineering activity on this site. The difference is that, landslide occurred in the snow melt season combining with heavy rainfall, and September is not rainy season. Considering this difference, it can only be estimated that the flow at the groundwater exits should be stronger than that at the investigation.

To investigate the initiation and traveling mechanism of the landslide-debris flow, soil samples were obtained at the source area and along the travel path in the valley

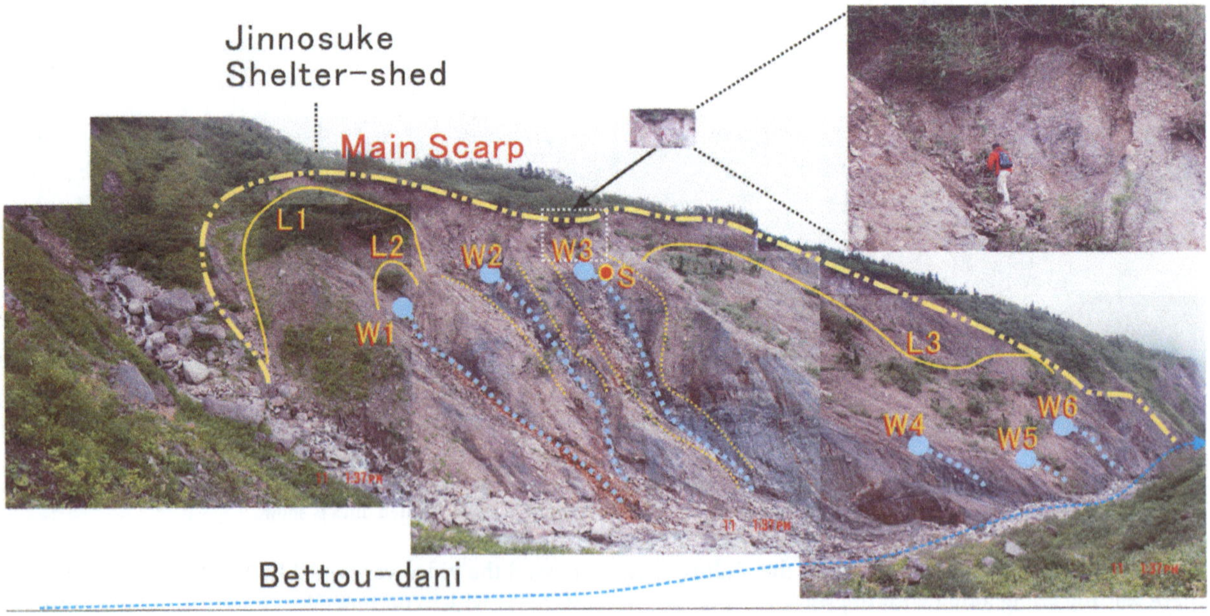

Fig. 11.12. Source area of the May 2004 landslide-debris flow in the Bettou-dani. *Blue points* represent the groundwater exits. *Solid lines* are the rear boundaries of the sliding blocks (photo taken by F.W. Wang on 11 September 2005). Points "*W1*"–"*W6*" shows groundwater exits, and "*S*" is the sampling point. A man in the enlarged box can be seen as a scale

of the Bettou-dani. Soil sample Beto-1 was taken from point "S" (Fig. 11.12) at the source area. This sample was subjected to ring-shear test, simulating a rainfall-induced landslide, to investigate the initiation mechanism of the landslide. Soil samples Beto-2, Beto-3, and Beto-4 were taken from the surface of the valley deposits along the traveling path of the debris flow, and were subjected to ring-shear tests that simulated dynamic loading of the landslide mass on the valley deposits and the dynamic loading of the debris flow on the valley deposits to investigate the traveling mechanism of the debris flow traveling down the valley.

In the soil sampling, coarse grains larger than 20 mm were excluded because it is impossible to use them in the laboratory tests. Figure 11.15 shows the grain-size distribution of the four soil samples. Sample Beto-1 is the finest sample among them, and all of the samples show a similar gradation. For sample Beto-2, 3, 4, Beto-3 has the least amount of fines. The grain-size distribution of the samples may indicate the potential of the water transport in the valley, i.e., upstream of the Beto-3 sample site, the fine part dominates because of the supply of the weathered material; while the downstream part is rich in fines, because of their transport by water. The average grain-sizes of the soil samples Beto-1, Beto-2, Beto-3 and Beto-4 were 2.7 mm, 6.0 mm, 4.8 mm and 6.0 mm, while the uniformity coefficient was 20.9, 25.7, 7.6 and 34.7, respectively. The specific gravity G_s of the samples was 2.71.

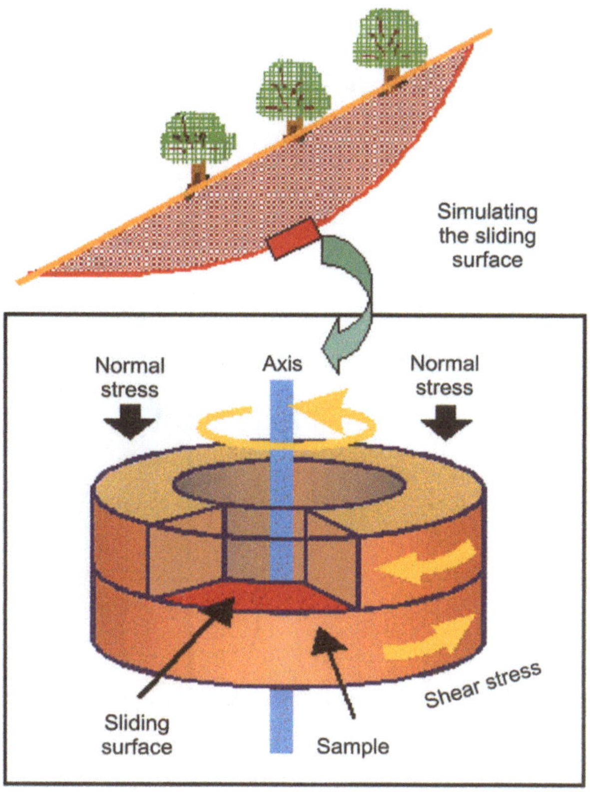

Fig. 11.13. Design concept of ring shear apparatus (from Sassa et al. 2004a)

11.4 Ring Shear Tests on the Initiation and Traveling Mechanisms of the Landslide–Debris Flow

Figure 11.13 shows the design concept of the ring shear apparatus, by which the initial normal stress, initial shear stress and the initial pore pressure acting on a soil element at sliding zone can be simulated. The major purpose and the design concept is to geotechnically simulate the formation of the shear zone and the post-failure mobility of high-speed landslides and observes the consequence of mobilized shear resistance, as well as the post-failure shear displacement and generated pore-water pressure (Sassa et al. 2004a). In this study, ring-shear tests were conducted using ring-shear apparatus DPRI-5, which was developed by Sassa in 1996 (Sassa et al. 2003).

One of the most important features in the development of the ring shear apparatus is the development of an effective and durable pore-pressure monitoring system. To have a large inlet section and provide an average pore-pressure value throughout the soil sample, pore-pressure transducers are connected to a gutter (4×4 mm) extending along the entire circumference of the inner wall of the outer ring in the upper box, as shown in Fig. 11.14. The gutter is located 2 mm above the shear surface and is covered by two metal filters, with a filter cloth between them. This system is quite durable in regard to shearing and is sensitive to pore-pressure monitoring, although the monitoring point is not at the center of the shear zone (Sassa et al. 2004a).

The diameters of the outer ring and inner ring are 180 mm and 120 mm, respectively. The sample, after placement in the shear box, had a donut shape with a width of 30 mm. To avoid possible grain-size effects on the shearing behavior, only grains with diameter smaller than 4.75 mm were included in the tested samples.

11.4.1 Grain-crushing Susceptibility of the Valley Deposits in Different Parts of the Bettou-dani

It is believed that the difference in grain-crushing susceptibility should cause the difference in the traveling process of the landslide–debris flow. When the soil is easy to be crushed, excess pore pressure will be easily generated during the shearing process under rapid motion (means undrained condition), and in turn, the shear resistance will decrease and result in high mobility of landslides. To check the grain-crushing susceptibility of the valley deposits, dry ring-shear tests were conducted on the samples. The test conditions were: consolidate

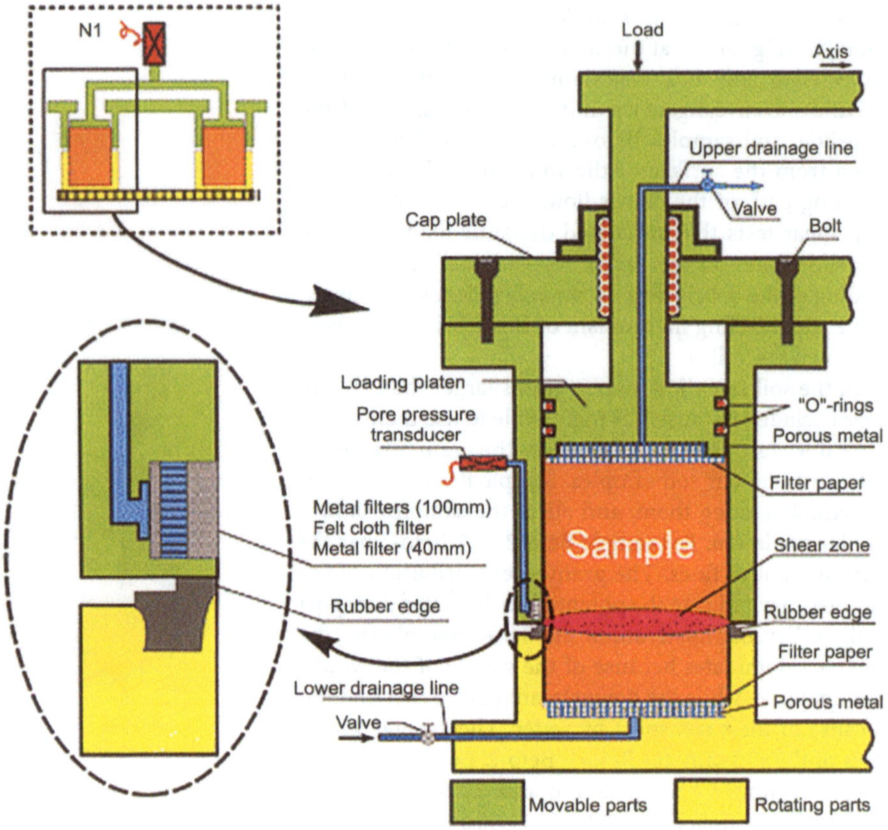

Fig. 11.14. A half section of the shear box and the close-up diagram of the edges (from Sassa et al. 2004a)

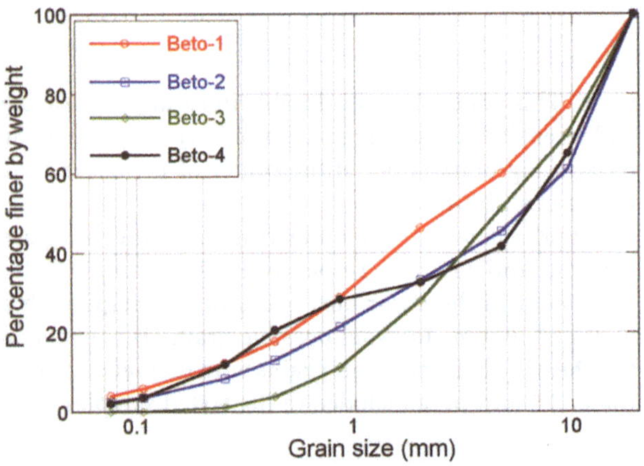

Fig. 11.15. Grain-size distribution of soil samples taken from the source area of the 2004 landslide and from the traveling path of the debris flow

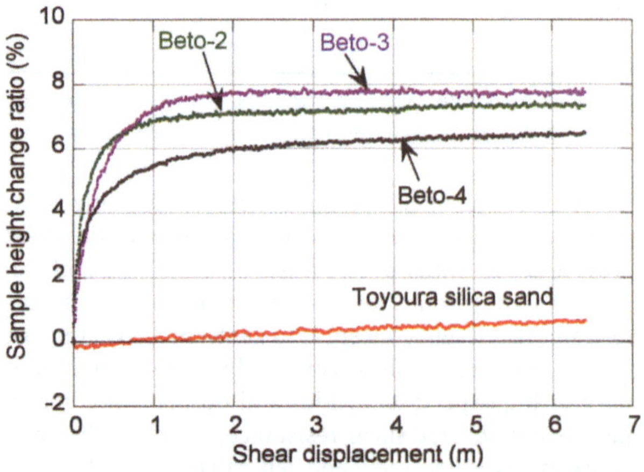

Fig. 11.16. Sample-height change with shear displacement during constant-shear-speed dry ring-shear tests on samples Beto-2, Beto-3, Beto-4, and Toyoura silica sand. Normal stress = 300 kPa, Shear velocity = 10.0 mm s^{-1}. The void ratios for the four samples after consolidation (before shearing) are 0.687, 0.760, 0.753 and 0.901, respectively

the sample at 300 kPa normal stress which corresponds to the actual stress level of the landslide, and shear it under constant speed of 10.0 mm s^{-1} until the shear displacement reaches 6.4 m. For comparison, Toyoura silica sand which is known as a standard sandy soil that is difficult to crush, was also sheared under the same test conditions.

Figure 11.16 shows sample-height change during the dry ring-shear tests. Soil with high grain-crushing susceptibility generally has large sample-height change (contraction) during dry shear. The sample-height changes that occurred in the samples taken from Bettou-dani were quite a bit larger than that of Toyoura silica sand. Among the samples taken from Bettou-dani, Beto-4 has the smallest sample-height change during shearing, showing a relatively lower grain-crushing susceptibility.

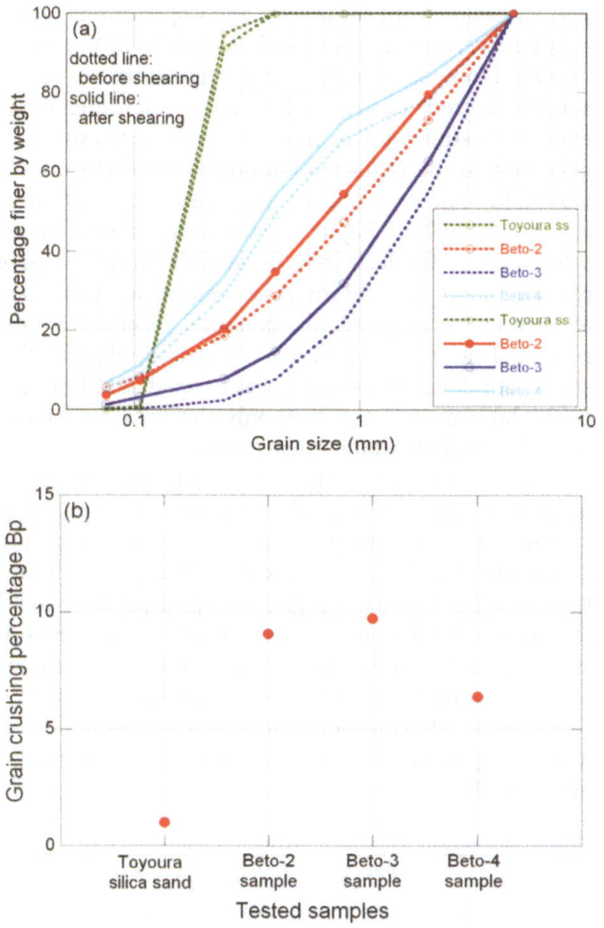

Fig. 11.17. Grain crushing occurred in the dry constant-shear-speed ring-shear tests on samples Beto-2, Beto-3, Beto-4, and Toyoura silica sand. **a** Grain-size distribution of the tested samples before and after shearing; **b** Marsal's grain crushing susceptibility B_p (defined by Marsal 1967)

Figure 11.17a,b shows the grain-size distribution of the tested samples before and after shearing, and the grain-crushing percentage B_p of all samples (Marsal 1967), respectively. B_p is the summation of the difference of grain-size distribution at each sieve size of the sample before and after dry shear (taken from shear zone), and it indicates the grain-crushing susceptibility of the soil. It is obvious that grain-crushing susceptibility becomes lower from the upstream part to the downstream part of the Bettou-dani.

11.4.2 Ring-shear Tests on Soil Samples Taken from the Source Area

The purpose of this test is to simulate the initiation of the landslide by water pressure. The initial slope condition was simplified as being 30 m in thickness (h) and 28 degrees in slope angle (θ). The landslide was triggered in the ideal slope by rainfall and snowmelt water. Through $\sigma_0 = \gamma_t h \cos^2\theta$ and $\tau_0 = \gamma_t h \sin\theta \cos\theta$, where γ_t is the united weight of the sliding mass, a average value of 18 kN m^{-3} was assumed in the test. Assuming the groundwater table near the sliding surface, the initial effective normal stress (σ_0) and shear stress (τ_0) acting on the sliding surface were 420 kPa and 224 kPa, respectively. Then, pore-water pressure acting on the element increased as the result of rainfall and snowmelt. The test was conducted under the following procedure:

1. saturate the soil sample to a high degree of saturation with carbon dioxide and de-aired water; it was confirmed that the B_D value (B_D is a pore pressure parameter, related to the degree of saturation in the direct-shear state. It was proposed by Sassa (1988)) reached 0.96, showing a high degree of saturation;
2. consolidate the sample under normal stress of 420 kPa;
3. apply the initial shear stress of 224 kPa gradually at 41.7 Pa s^{-1} to avoid pore-water pressure generation;
4. increase the pore-water pressure gradually at the rate of 0.5 Pa s^{-1} through the upper drainage line directly connecting to the upper surface of the sample until failure occurs;
5. measure the residual friction angle of the soil with constant shear speed, while increasing the normal stress gradually from a low stress level (about 90 kPa) to a high stress level (about 400 kPa).

Figure 11.18a–c presents the test results. Fig. 11.18a shows the time-series data for the whole test series. From the beginning to nearly 200 s, the normal stress and shear resistance were kept constant, while the pore pressure was increased gradually. From 200 s to 330 s, small displacement occurred, and the shear resistance mobilized a little bit higher, although the shear stress was kept constant. The void ratio at 250 s was 0.350. After 330 s, rapid failure occurred, which can be confirmed by the acceleration of the shear displacement. Corresponding to the rapidly increasing shear displacement, the shear resistance decreased rapidly to about 110 kPa. At that point, the apparent friction angle became 15.1 degrees, which is shown by the total stress path and effective stress path in Fig. 11.18b. The residual friction angle of the soil was measured under complete drained condition as 33.7 degrees, which is shown in Fig. 11.18c, when the above test was finished. The procedure was to increase the normal stress gradually (5.6 Pa s^{-1}) from about 95 kPa to 405 kPa when keeping a constant shear velocity. The shear velocity was set as slow as 0.02 mm s^{-1} to avoid excess pore pressure generation in the shear zone. From the above test results, it is easy to calculate that the slope can remain stable at its initial slope angle of 28 degrees, if there is no increase in pore pressure at the sliding surface. In addition, from Fig. 11.18b, it can be seen that the peak friction angle of the soil at initial failure is much higher than 33.7 degrees.

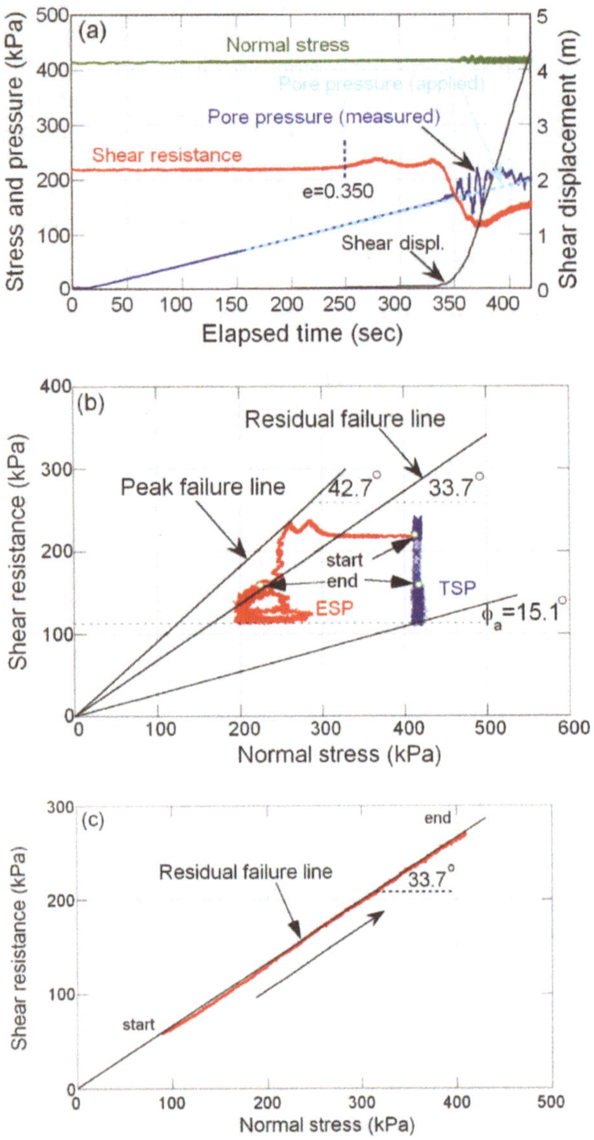

Fig. 11.18. Simulation test results of landslide initiation triggered by rainfall under naturally drained condition. $B_D = 0.96$, pore water pressure increasing rate = 0.5 kPa s^{-1}. **a** Time-series data; **b** total stress path (TSP) and pseudo effective stress path (ESP); **c** residual friction angle of the tested soil sample Beto-1 from the source area

In Fig. 11.18, there are some concepts that need to be clarified. At first, naturally drained condition means that in the test procedure drainage is not prevented and excess pore-water pressure can generate depending on material behavior and loading rate, and it was referred to as naturally drained conditions by Sassa et al. (2004b). Under the naturally drained condition, it is easy to find out in Fig. 11.18a, that there is a difference between the pore pressure (applied) and pore pressure (measured). As shown in Fig. 11.14, the pore pressure transducer is connected with the shear zone, also with the upper drainage line, which pore pressure was supplied. When the pore pressure value from shear zone and upper drainage line is difference, the high value will be recoded by the transducer. It is obvious that the high portion of the pore pressure (measured) was from shear zone. From this difference, it is estimated that high excess pore pressure was generated in the rapid motion process after the sample failure. Related to the above concept, the pseudo effective stress path (EPT) in Fig. 11.18b is not the actual effective stress state in the shear zone, because the measured pore pressure is not directly from the shear zone. It was also affected by the supplied pore pressure. When pore pressure at shear zone is correctly measured, the effective stress path should move along peak or residual failure line. In the later part of this test, only the shear resistance should be replied on for data explanation.

Through the above test aiming to simulate the failure process of a natural slope when the pore pressure acting on the potential sliding surface was increased by the groundwater table rising caused by heavy rainfall and snowmelt, it is found that high excess pore pressure could have resulted, and in turn, the shear resistance of the soil at sliding surface could drop down rapidly. The resultant rapid drop-down of the shear resistance should cause acceleration of the landslide motion when it moved down to the valley, and resulted in a high velocity when it rushed into the valley.

11.5 Ring-shear Tests on Soil Samples Taken from the Landslide Travel Path in the Bettou-dani

The above test simulated the initiation of rapid landslide from natural slope. What will happen when the failed sliding mass rushed into the Bettou-dani valley, where thick torrent deposits were distributed? To simulate the landslide motion in the Bettou-dani, three other samples (Beto-2, Beto-3, Beto-4) from different portion of the Bettou-dani were sampled and used in ring-shear tests to show the fluidization process of the landslide (see Fig. 11.9b). Beto-2 was taken near the uppermost debris-retention dam in the Bettou-dani; Beto-3 was taken near the destroyed bridge; and Beto-4 was taken below the suspension bridge and near the toe of deposit of the debris flow.

Figure 11.17 is a model proposed by Sassa et al. (1997) to simulate the undrained-loading behavior of valley deposits by a rapidly sliding mass. The slide mass moved down the slope (I), and applied load onto the torrent deposits at the foot of the slope (II). Because a surface water stream or subsurface flow existed and some of the deposits were saturated, the torrent deposit was sheared by undrained loading and transported downstream together with the sliding mass (III). Here, it is assumed that a column with unit length of its sliding surface, which is a part of the torrent deposit. In the position (I) of the sliding mass, the weight of the column (W0) was in effect.

When the sliding mass rode on to the torrent deposit (II) with a certain velocity, it provided dynamic loading of the column. It is assumed that the applied stress on the torrent deposits was as the sum of the static stress, W, (load due to the weight of the sliding mass) and the dynamic (impact) stress, F_d, working in the direction of motion of the sliding mass.

At the Bettou-dani, the sliding mass moved down the slope (I), and applied a load to the valley deposits at the foot of the slope (II). Because a surface-water stream or subsurface flow existed and some of the deposits were saturated, the valley deposit was sheared by undrained loading and transported downstream together with the sliding mass (III) (Sassa et al. 2004a). The above test shown in Fig. 11.18 corresponds to the landslide that occurred at slope (I).

To simulate the succeeding process, sample Beto-2 was used to simulate the situation at slope (II), while samples Beto-3 and Beto-4 were used to simulate the behavior at slope (III), and the local slope angles of the valley at the sampling points (Beto-3 and Beto-4) were considered.

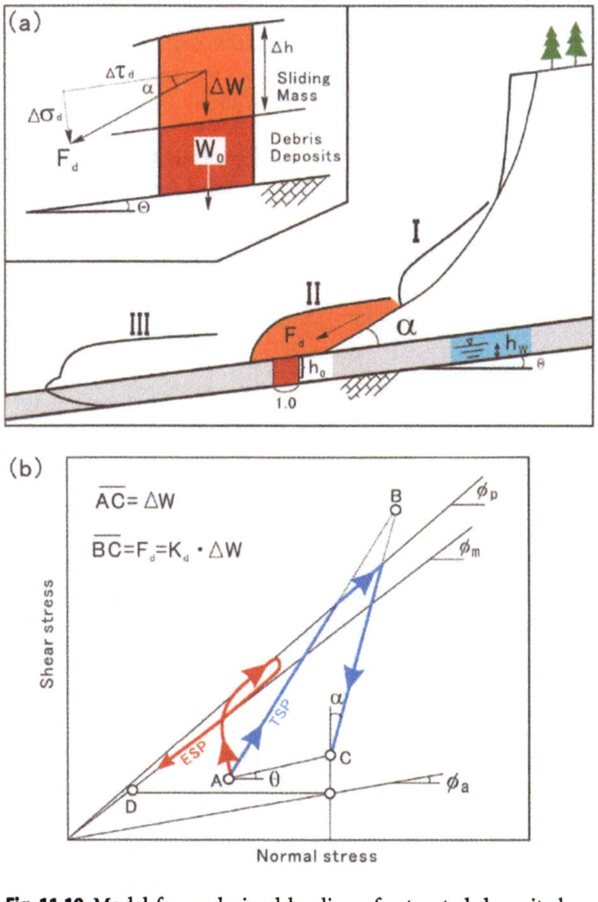

Fig. 11.19. Model for undrained loading of saturated deposits by a displaced sliding mass (Sassa et al. 1997). **a** Illustration of the model; **b** stress path of the torrent deposit during loading. α: Angle of thrust between the slope and the torrent bed; F_d: dynamic stress; k_d: dynamic coefficient ($F_d/\Delta W$)

In Fig. 11.19, the valley deposit has a thickness of h_0, an initial slope angle of the valley θ, and groundwater thickness on the sliding surface h_w. The undrained loading from a rapidly moving displaced landslide has a thickness of Δh, an intrusion angle of α, and a dynamic (impact) coefficient of K_d. Based on Sassa et al. (1997) and Sassa et al. (2004a), it is reasonable for K_d to take a value of unity. Then, the increment of normal stress and shear stress from the rapidly sliding mass to the deposits can be determined. The initial conditions for the three sampling points, which were employed in the ring-shear tests, are summarized in Table 11.1. From the upstream to downstream, the slope angle was changed from 18 degrees to 5 degrees near the No. 10 debris retention dam. The initial thickness of the torrent deposits at the valley was assumed to be the same value of 5 m, and the thickness of groundwater was assumed to be 3 m (groundwater table was 2 m under below the torrent surface). The intrusion angle was assumed as 10 degrees for Beto-2 sample near the source of the landslide from natural slope. It is the difference between the slope angle of the initiated landslide and the slope angle of the valley bed. When sliding mass moved along the Bettou-dani valley, the intrusion angle became to zero for Beto-3 and Beto-4. Because of the effect of the debris retention dam, the thickness of the sliding mass became thinner and thinner. Considering this phenomenon, the thickness of the sliding mass was assumed to be 30 m, 20 m and 5 m from upstream to downstream.

Figure 11.20 shows the results of the simulation test on slope (II) using sample Beto-2. Fig. 11.20a shows the input stress signals of normal stress and shear stress before (0–5 s), during (5–15 s), and after the dynamic impact process (15–30 s). The signal was loaded on the sample under the undrained condition to simulate the rapid loading of sliding mass on the torrent deposits, which had initial normal stress and initial shear stress acting on it.

As can be seen in the time-series data (Fig. 11.20b), pore-water pressure was generated at the same rate as the applied normal stress, and failure occurred as soon as the loading was applied. Shear resistance reached its peak strength at 7 s, and arrived at its steady-state strength at 15 s. As shown by the stress paths (Fig. 11.20c), the apparent friction angle is only 2.6 degrees, showing a high

Table 11.1. Initial condition for undrained loading on the valley deposits from a rapid sliding mass, Bettou-dani landslide

	θ (deg)	h_0 (m)	h_w (m)	α (deg)	Δh (m)
Beto-2	18	5	3	10	30
Beto-3	18	5	3	0	20
Beto-4	5	5	3	0	5

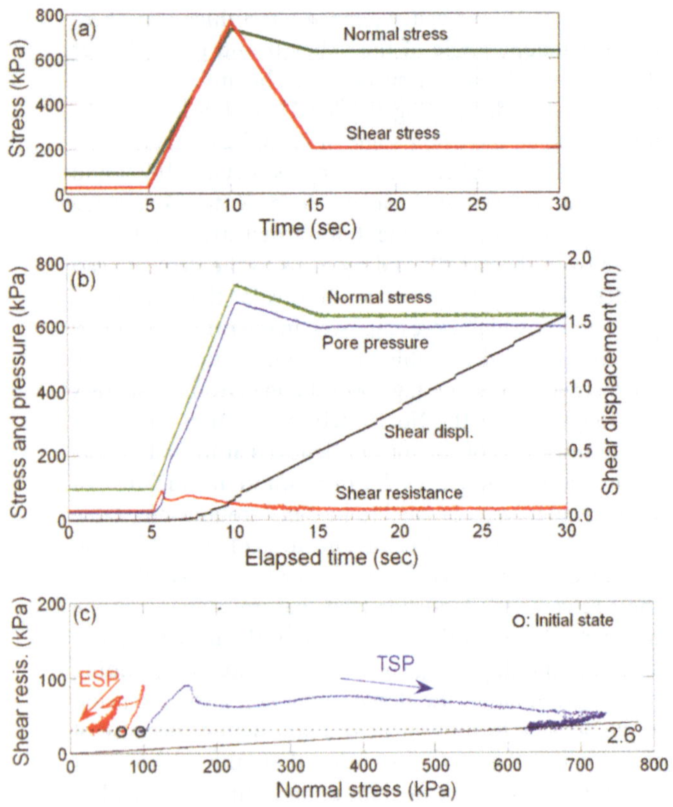

Fig. 11.20. Simulation test results on sample Beto-2 when sliding mass rushed into the valley and loaded on the torrent deposits. **a** Applied-stress signals (normal-stress and shear-stress increments); **b** time-series data; **c** effective stress path (ESP) and total stress path (TSP). $B_D = 0.96$

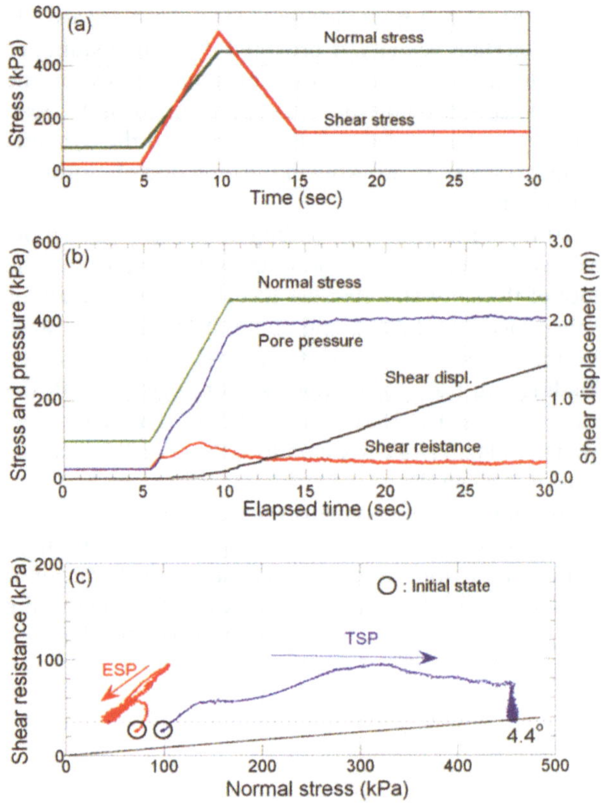

Fig. 11.21. Simulation test results on sample Beto-3 when debris flow traveled along the valley in the middle part of the Bettou-dani valley. **a** Applied-stress signals (normal stress and shear stress increments); **b** time-series data; **c** effective stress path (ESP) and total stress path (TSP). $B_D = 0.99$

Table 11.2. Summary of ring-shear test results on sample Beto-2, 3, and 4

Sample	Shear resistance at steady state under the undrained condition (kPa)	Total normal stress (kPa)	The minimum apparent friction angle (deg)
Beto-2	About 35	620	2.6
Beto-3	About 35	460	4.4
Beto-4	About 35	205	5.0

mobility of the valley deposit after the dynamic loading. All the series test results show that the valley deposits fluidized after the dynamic loading of the rapidly moving displaced sliding mass in an undrained condition.

Figure 11.21 shows the results of the simulation test on sample Beto-3 at slope (III). The intrusion angle was assumed to be zero because the displaced sliding mass came from the upper part of the same valley with the same slope angle. Fluidization also occurred, and the apparent friction angle mobilized at the steady state came to 4.4 degrees, slightly higher than that of the Beto-2 sample.

Figure 11.22 shows the simulation-test results on sample Beto-4, taken near the toe of the deposits. The apparent friction angle of Beto-4 was 5.0 degrees, the same value as the slope angle of the Bettou-dani at this part. For this reason, the shear displacement generated in this test was only 28 mm when the loading was completed.

As a summary of the above dynamic tests, the shear resistance at steady state under the undrained condition, the minimum apparent friction angle to show the mobility of the torrent deposit loaded by rapid moved sliding mass, and the residual friction angle of the three soil samples taken from the Bettou-dani are presented in Table 11.2. Because all of the possible grain-crushing should have been completed when the shearing reached the steady state, the shear resistances at the steady state were almost the same as about 35 kPa. When the normal stress became smaller from 620 kPa to 205 kPa, the ap-

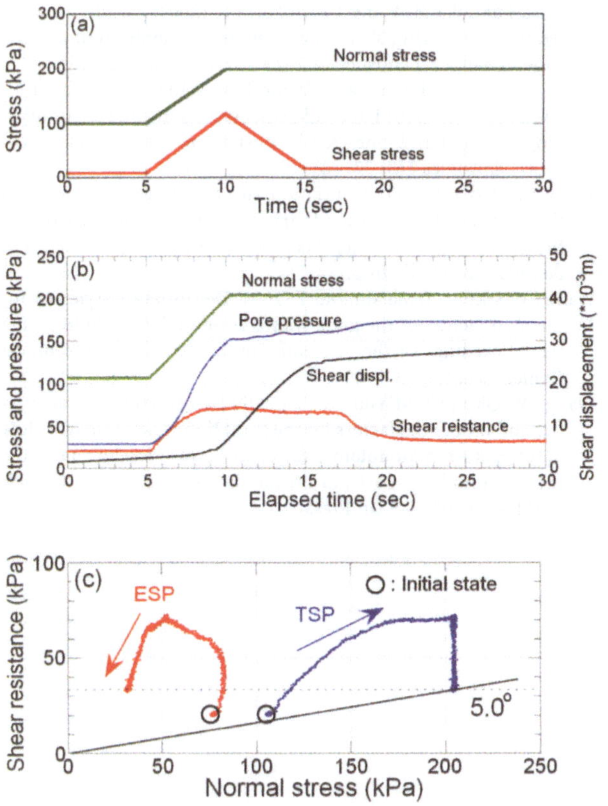

Fig. 11.22. Simulation test results on sample Beto-4 when debris flow traveled to and deposited near No. 10 debris retention dam. **a** Applied-stress signals (normal stress and shear stress increments); **b** time-series data; **c** effective stress path (ESP) and total stress path (TSP). $B_D = 0.95$

parent friction angle became larger from 2.6 degrees to 5.0 degrees; and when it became larger than the slope angle (it was about 5.0 degrees at the position near No. 10 debris retention dam), the debris flow should have decelerated and finally came to a stop.

From the above simulation tests that reproduced rapid loading on valley deposits, the impact process, the traveling process, and depositing process of the debris flow that occurred in May 2004 were well reproduced in the laboratory.

11.6 Conclusions

The May 2004 landslide–debris flow that occurred in the Bettou-dani of the Jinnosuke-dani landslide, Haku-san Mountain, showed a fluidization process from landslide to debris flow. By analysis of the monitored video images of the debris flow, field investigation of the source area of the landslide, and laboratory ring-shear tests that simulated the rainfall triggering mechanism and the fluidization mechanism during the process of downstream travel, the following were concluded:

1. Concentrated groundwater flows were a main triggering factor for the landslide initiation by increasing water pressure in the slope;
2. In the ring-shear simulation test of the landslide initiation, it was shown that even under naturally drained conditions, the mobilized shear resistance of the weathered soil in the source area showed a rapid decrease after landslide initiation, and this should be the instinctive factor for rapid landslide motion after its initiation;
3. In the ring-shear simulation test of dynamic loading on the valley deposits, it was shown that high potential for grain-crushing of upstream deposits and lower potential of the downstream deposits controlled the traveling and depositing process of the debris flow;
4. The shear resistance at steady state under undrained conditions is the same for the soil samples taken from different parts of the valley (Sample Beto-2, 3, 4). A possible reason is that although the initial grain gradations of these samples differ at the initial state, the soil at the shear zone would become the same when the shearing process reached the steady state, when all of the possible grain-crushing is completed.

Acknowledgments

Deep thanks are given to the Kanazawa Office of Rivers and National Highways, MLIT, for cooperation in the field work and as a source of information on the May 2004 landslide–debris flow. Financial supports by research grants (No. 15310127, Representative: F.W. Wang) from the Ministry of Education, Culture, Sports, Science, and Technology of Japan (MEXT), and 16G-03 Joint Research Programme of Disaster Prevention Research Institute (DPRI), Kyoto University are highly appreciated. Dr. Huabin Wang, Mr. Ryuta Saito, Mr. Jozef Jurko, and Mr. Taichi Minamitani of the Research Centre on Landslides (RCL), Disaster Prevention Research Institute, Kyoto University, joined the sampling and field investigation.

References

Kanazawa Office of Rivers and National Highways, Ministry of Land, Infrastructure and Transport of Japan (2004a) Investigation report on Jinnosuke-dani landslide (in Japanese)

Kanazawa Office of Rivers and National Highways, Ministry of Land, Infrastructure and Transport of Japan (2004b) Debris flow occurred on 17 May 2004 in Bettou-dani. Newsletter Sabo Hakusan 6:1–4 (in Japanese)

Kanazawa Office of Rivers and National Highways, Ministry of Land, Infrastructure and Transport of Japan (2004c) http://210.131.8.12/~kanazawa/mb5_kouhou/press/news.html

Kaseno Y (1993) Geological mapping of Ishikawa. Geology Bulletin (in Japanese)

Kaseno Y (2001) Supplement of geological mapping of Ishikawa. Geology Bulletin (in Japanese)

Marsal RJ (1967) Large scale testing of rockfill materials. J Soil Mech Found Div-ASCE 93(SM2):27–43

Okuno T, Wang FW, Matsumoto T (2004) The deforming characters of the Jinnosuke-dani landslide in Haku-san mountainous area, Japan. In: Lacerda W, Ehrlich M, Fontoura S, Sayao A (eds) Landslides: Evaluation & Stabilization. Proceedings IX International Symposium on Landslides, Rio de Janeiro, 2, 1279–1285

Sassa K (1988) Geotechnical model for the motion of landslides. In: Special Lecture of 5th International Symposium on Landslides, "Landslides", 10–15 July, 1, pp 37–55

Sassa K, Fukuoka H, Wang FW (1997) Mechanism and risk assessment of landslide-triggered-debris flows: lesson from the 1996.12.6 Otari debris flow disaster, Nagano, Japan. In: Cruden DM, Fell R (eds) Landslide risk assessment. Proceedings of the International workshop on landslide risk assessment. Honolulu, 19–21 February, pp 347–356

Sassa K, Wang G, Fukuoka H (2003) Performing undrained shear tests on saturated sands in a new intelligent type of ring shear apparatus. Geotech Test J 26(3):257–265

Sassa K, Fukuoka H, Wang G, Ishikawa N (2004a) Undrained dynamic-loading ring-shear apparatus and its application to landslide dynamics. Landslides 1:7–19

Sassa K, Wang G, Fukuoka H, Wang FW, Ochiai T, Sugiyama M, Sekiguchi T (2004b) Landslide risk evaluation and hazard mapping for rapid and long-travel landslides in urban development area. Landslides 1(3):221–235

Wang FW, Sassa K (2007) Initiation and traveling mechanisms of the May 2004 landslide–debris flow at Bettou-dani of the Jinnosuke-dani landslide, Haku-san Mountain, Japan. Soils and Foundations 47(1) (in press)

Wang FW, Okuno T, Matsumoto T (2004) Deformation style and influential factors of the giant Jinnosuke-dani landslide in Japan. In: Proceedings of the Fifteenth Southeast Asian Geotechnical Conference, 1, pp 399–404

Wang FW, Okuno T, Matsumoto T (2007) Deformation characteristics and influential factors for the giant Jinnosuke-dani landslide in the Haku-san Mountain area, Japan. Landslides: Journal of the International Consortium on Landslides 4(1), DOI 10.1007/s10346-006-0049-9 (in Press)

Chapter 12

On the Pore-pressure Generation and Movement of Rainfall-induced Landslides in Laboratory Flume Tests

Gonghui Wang* · Kyoji Sassa

Abstract. Using a small flume, a series of tests was conducted to trigger rainfall-induced landslides. Based on monitoring of sliding distance and pore pressures, the process of pore-pressure generation in relation to sliding distance was examined. By performing tests on sands of different grain sizes (silica sand no. 7 (D_{50} = 0.14 mm) and no. 8 (D_{50} = 0.057 mm)) at different initial dry densities or different thicknesses, the effects of these factors on pore-pressure generation and failure behavior of a landslide mass were analyzed. Results from tests of different initial densities showed that for each sample there was an optimal density index, at which the pore pressure build-up after failure reached its maximum value. This optimal density index varied with the thickness of sample and the grain size of samples. Moreover, observed failure phenomena showed that the failure mode also depended greatly on the grain size and sample thickness. In fact, flowslides were initiated in the tests on finer silica sand (no. 8), whereas retrogressive sliding occurred in the tests on silica sand no. 7. Results of tests on mixtures of silica sand no. 8 with different contents of loess by weight showed that the existence of fine-particle soil (loess) could significantly change the flow behavior of a landslide mass during motion: the flow behavior of soils with 20 and 30 percent loess was different from these two silica sands and the mixture with 10 percent loess, showing greater velocity without deceleration. This suggests the existence of a mechanism that maintains high pore pressures during motion for these soils. In addition, by rotating saturated samples in a double-cylinder apparatus, a mechanism was examined in which pore pressure in saturated soils during motion was maintained. The results showed that the pore pressure of the saturated mixture increased with velocity because of the "floating" of sand grains that accompanied the movement for each test. In addition, the sample with finer grain sizes or greater fine-particle (loess) contents floated more easily, and high pore pressure could be maintained during motion. The floating ratios of grains reached a high value (>0.85) at a very slow velocity for samples with 20 and 30 percent loess. Based on these test results, it is concluded that grain size and fine-particle contents can have a significant impact on the mobility of rainfall-induced landslides.

Keywords. Landslide; liquefaction; laboratory tests; pore pressure; grain size; fine-particle content

12.1 Introduction

Rainfall-induced landslides pose significant hazards in many parts of the world especially in mountainous areas in rainy environments, because these circumstances occur frequently. Among hazardous rainfall-induced landslides, fluidized landslides are often the most dangerous and damaging because they usually occur unexpectedly and are characterized by rapid movement and long runout distance. It is generally recognized that rainfall-induced landslides are caused by increased pore pressures and seepage forces during periods of intense rainfall (Terzaghi 1950; Sidle and Swanston 1982; Sitar et al. 1992; Anderson and Sitar 1995). It is the increased pore pressure that decreases the effective stress in the soil and thus reduces the soil shear strength, eventually resulting in slope failure (Brand 1981; Brenner et al. 1985). Further studies have illustrated that in some cases of rainfall-induced landslides, movement along the sliding surface leads to crushing of the soil grains, which results in the liquefaction along this surface, finally resulting in rapid movement and long runout distance (Sassa 1996; Sassa 1998a, b); thus, high pore pressure is a result of shearing along the sliding surface.

Liquefaction, a process by which the soil suddenly loses a large proportion of its shear resistance due to the generation of high pore pressure, is always the reason for fluidized landslides. Liquefaction triggered by dynamic effects, such as earthquakes, or by static effects, such as rainfall, snowmelt, etc., has been studied extensively (e.g., Terzaghi 1956; Seed 1966, 1979; Bishop 1967, 1973; Castro 1969; Casagrande 1971; Castro and Poulos 1977; Sassa 1984, 1996, 1998a, 1998b; Eckersley 1985, 1986; Hird and Hassona 1990; Ishihara et al. 1990; Ishihara 1993), and much knowledge has been accumulated about the mechanism of liquefaction. However, most of the understanding is based on simple element testing of small specimens under idealized conditions (e.g., saturated, undrained, and stress isotropy). More complicated fabric effects, stress anisotropy, entrapped air, deformation of thin shear zones, and some currently unknown processes, which may also have significant influences on the development of liquefaction, have not been thoroughly studied (Eckersley 1990). Thus, it does not seem suitable to use the results inferred from the simple element tests alone to explain the observed field phenomena of fluidized landslides.

In addition, in recent decades, many small-scale landslides have been triggered under laboratory condition. For example, Sassa (1972, 1974) performed a series of flume tests on Toyoura sand (Japanese standard sand) and concluded that the changes in rigidity of sand and upper yield strain within a slope are essential to the analysis of slope stability. Further studies on liquefied landslides revealed that the increasing pore pressure due to an undrained soil layer caused slope failure during rainfall and that the generation of pore pressure was a result of sudden initiation of subsidence (Sassa and Takei 1977a, 1977b; Sassa 1984). Eckersley (1990) triggered flowslides in coking-coal stockpiles by raising the water table, and inferred that excess pore pressures were generated during, rather than before, the movement. However, in this work, the sliding displacement was recorded only by video cameras; observations of deformation within the stockpiles were not carried out. Another laboratory flowslide study was conducted on loose saturated fine quartz sands, in which the motion of liquefied sands and pore pressures during motion were analyzed (Spence and Guymer 1997). Iverson and his colleagues triggered landslides and debris flows in a large flume with emphasis on examination of landslide movement and intergranular pore pressures (Iverson and LaHusen 1989; Iverson 1997; Iverson et al. 1997, 2000; Major and Iverson 1999). Those studies showed that landslide rates are significantly dependent on the initial soil porosity, and rapid fluidized landsliding involves partial or complete liquefaction of the mass by high pore-fluid pressure. It was also concluded that the magnitude of pore pressure change induced by porosity change during landsliding depends not only on initial porosity, but also on the relative time scales for soil deformation and pore pressure diffusion; pore pressures fluctuated dynamically during rapid steady shear deformation of water-saturated granular materials, and dissipated significantly only during post depositional sediment consolidation.

Although great effort has been devoted to the study of initiation, motion, and deposition of fluidized landslides with emphasis on pore-pressure generation and dissipation, the understanding of the fluidized landslide process is as yet limited. How the excess pore pressure generates and is maintained, and then how the generated pore-pressure affects the movement of displaced landslide material requires further scrutiny. In addition, by now almost all of the experimental studies have been performed on clean sands, i.e., little work has been performed on silt and silt-clay mixture, although there have been a few exceptions (e.g., Bishop 1973; Kramer 1988; Eckersley 1990; Ishihara, et al. 1990; Thevanayagam 1998; Youd and Gilstrap 1999; Vallejo and Mawby 2000). It is highly desirable to examine how the grain size and fine-particle content affect pore-pressure generation and the resulting movement of the landslide mass.

Therefore, in this research, to study those rainfall-induced landsides where silty soils control the triggering and movement of the landslide mass, a series of flume tests was initially performed on sands of different grain sizes and on mixtures of sandy silt with loess by different weight, with emphasis on pore-pressure generation and on movement of the displaced landslide mass. Due to ineffectiveness of the flume apparatus in measuring the pore-water pressure of landslide mass in motion, a series of double-cylinder rotating tests was then performed to examine the maintenance of pore pressure within the soil mass during movement. Based on the test results, the effects of grain size and fine-particle content on the pore-pressure generation/maintenance and the movement of the landslide mass are discussed.

12.2 Properties of the Samples

To examine the effects of grain size in this study, fine silica sand no. 7 (S7) and no. 8 (S8) were selected as the samples. Each was comprised of 92–98 percent of subangular to angular quartz, and a small percentage of feldspar. To study the influence of finer-grain content upon slope failure behavior, loess was also used in this research. This loess, which was composed mainly of fine sand and silt, was collected from a landslide at Lishan, Xi'an, China. In the present work, a series of tests was conducted on the mixture of S8 and loess with loess content of 10, 20, and 30 percent by weight, which were termed M10, M20 and M30, respectively. The grain-size distributions of S7, S8, and loess are shown in Fig. 12.1. Note that those grains smaller than 0.005 mm existing in loess could be regarded as clay, while those in S7 and S8 could be classified as fine silt or rock flour, because the

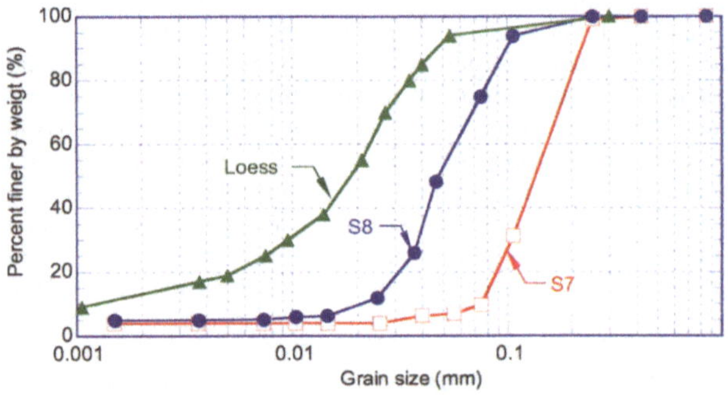

Fig. 12.1. Grain-size distribution curves for silica sand no. 7, no. 8, and loess (*S7*: Silica sand no. 7; *S8*: Silica sand no. 8)

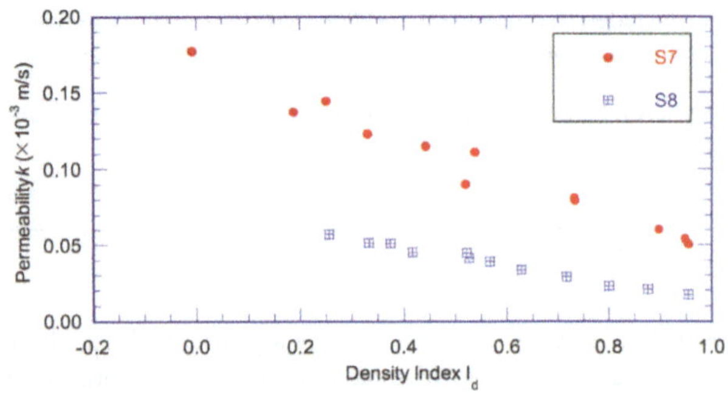

Fig. 12.2. Variation of permeability of silica sand no. 7 and no. 8 versus density index ($k = 0.013024 + 0.19645\,e^3/(1+e)$, correlation coefficient $R = 0.985$ for S7; $k = 0.0010681 + 0.04621\,e^3/(1+e)$, correlation coefficient $R = 0.989$ for S8)

grains are formed by mechanical grinding. Figure 12.2 plots the permeability of S7 and S8 against density index (density index, I_d, is used to express the density in this paper; $I_d = (e_{max} - e)/(e_{max} - e_{min})$, where e_{max} and e_{min} are the maximum and the minimum void ratio of dry sand, respectively, and e is the initial void ratio). The variation of permeability is given as a regression function of void ratio from the Kozeny-Carmen equation, based on the test results. Some other properties of these samples are listed in Table 12.1.

12.3 Flume Test Apparatus and Test Procedures

12.3.1 Flume Test Apparatus

Figure 12.3 shows the employed experimental apparatus. The flume, with transparent sides, was 180 cm long, 24 cm wide, and 15 cm high. The flume angle is made changeable for different test requirements. To assure the same friction between the sand particles and the base of the flume as of that of sands inside the flume, silica-sand grains were glued to the surface of the flume base. At longitudinal distance $x_1 = 45$ cm from the upstream end of the flume, a hole 1.5 cm in diameter was drilled along the central line, as seen in Fig. 12.3 (the left part of this figure shows the placing of transducers on the flume bottom). A vinyl tube was inserted into the hole with one end being flush with the flume bed (end A in Fig. 12.3, which was covered by filter paper to prevent sand ingress), and the other end (end B) was connected to the pressure transducer to measure the pore-water pressure. On the side of the hole at position x_1, a normal-stress transducer and a shear-stress transducer were installed in an attempt to measure the stresses in the soils. Adjacent to and downstream from x_1, a styrene foam ball 2 cm in diameter was laid and connected with a linear displacement transducer by means of a wire. During the test, the ball buried in the sample moves together with the sample; in this manner, the time series of sliding distance can be monitored. Because of the resolution of this linear displacement transducer with rated capacity (100 cm), displacements smaller than 1.0 mm cannot be measured correctly. Therefore, to get precise measurement, a laser displacement sensor with resolution of 0.015 mm and rated capacity of 15 mm was

Table 12.1.
Properties of employed samples, S7, S8, M10, M20, M30, and loess

Sample	D_{50} (mm)	D_{10} (mm)	U_c	e_{max}	e_{min}	G_s
S7	0.13	0.050	0.0185	0.047	0.043	0.040
S8	0.074	0.018	0.0012	0.0118	0.0084	0.0057
Loess	2.1	3.7	19.0	4.6	6.0	8.3
M10	1.23	1.66	–	1.51	1.52	1.56
M20	0.70	0.85	–	0.73	0.73	0.73
M30	2.63	2.63	2.72	2.64	2.65	2.66

Note: D_{50} (mm) and D_{10} (mm): Mean and effective grain size, respectively; U_c: Uniformity coefficient; e_{max} and e_{min}: maximum and minimum void ratio, respectively; –: was not measured. Properties of M10, M20, and M30 were obtained from S8 and loess, using the weighted average method.

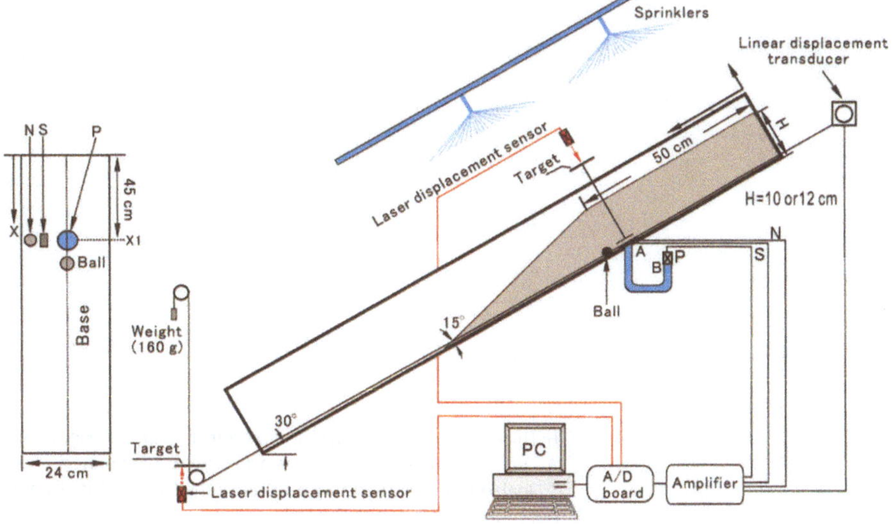

Fig. 12.3.
Arrangement of the experimental apparatus for rainfall-induced slope failure

also used by fixing a target on the wire and shining a laser beam upon the target. A 160-g weight was attached to the other end of the wire to balance the pulling resistance of the linear-displacement transducer. Above the flume two spray nozzles were placed. By keeping the supplied water pressure constant by means of a pump, and adjusting the distance to the nozzles, uniform artificial rainfall was assured. A video camera was used to monitor the entire test process from one side of the flume.

12.3.2 Test Procedures

All the transducers were installed at first. During installation, normal-stress and shear-stress transducers were positioned flush with the surface of the flume base. Pore-pressure transducer (P) was installed with end B (Fig. 12.3) 1 cm lower than end A. Tube AB was filled with de-aired water and air was completely evacuated from end B to ensure reliable measurement of pore-water pressure. After the installation of the transducers, the soil sample was prepared and placed in the flume. To obtain loose samples with different void ratios, water was first added to the oven-dried samples to raise the initial water content to 5 percent, and then the sand was stirred to a homogeneous consistency. After that, the sample was packed into the flume. To make the sample uniform, while packing, the sample was placed in a series of 2-cm layers parallel to the flume base, and each layer was tamped. Finally, the superfluous parts of the placed sample were removed and the sample was shaped as in Fig. 12.3.

Table 12.2.
Summary of flume test data

Test	Test series	Sample	H (cm)	e	I_d	Δu (kPa)	ΔS (cm)	V_p (cm s^{-1})
$S7_{I,1}$	I	S7	10	0.97	0.49	0.14	1.1	0.7
$S7_{I,2}$				1.01	0.41	0.03	*	*
$S7_{I,3}$				1.14	0.17	0.21	3.2	4.5
$S7_{I,4}$				1.22	0.01	0.72	8.3	14.5
$S7_{I,5}$				1.24	−0.06	0.54	5.3	10.6
$S7_{I,6}$				1.35	−0.23	0.23	*	*
$S7_{I,7}$				1.41	−0.34	0.32	5.5	7.6
$S7_{I,8}$				1.49	−0.49	0.17	8.2	9.5
$S7_{I,9}$				1.50	−0.51	0.18	2.3	7.3
$S8_{II,1}$	II	S8	10	1.29	0.46	0.10	13.4	4.1
$S8_{II,2}$				1.41	0.31	0.48	33.9	8.6
$S8_{II,3}$				1.46	0.25	0.53	*	*
$S8_{II,4}$				1.46	0.25	0.62	*	*
$S8_{II,5}$				1.48	0.22	1.01	81.2	36.8
$S8_{II,6}$				1.50	0.20	1.02	79.0	35.0
$S8_{II,7}$				1.51	0.19	0.99	29.6	8.5
$S8_{II,8}$				1.58	0.08	0.64	45.6	29.8
$S8_{II,9}$				1.68	−0.03	0.51	58.6	39.6
$S8_{II,10}$				1.70	−0.05	0.13	11.8	16.1
$S8_{II,11}$				1.71	−0.06	0.33	24.0	31.1
$S8_{II,12}$				1.77	−0.14	0.20	39.3	18.1
$S8_{III,1}$	III	S8	12	1.86	−0.25	0.51	38.0	13.4
$S8_{III,2}$				1.82	−0.20	0.50	39.0	16.4
$S8_{III,3}$				1.81	−0.18	0.39	23.4	22.2
$S8_{III,4}$				1.73	−0.08	0.56	*	*
$S8_{III,5}$				1.70	−0.05	0.97	83.4	43.4
$S8_{III,6}$				1.66	0.00	1.09	84.6	47.4
$S8_{III,7}$				1.63	0.04	1.44	27.8	9.5
$S8_{III,8}$				1.58	0.10	0.86	*	*
$S8_{III,9}$				1.53	0.16	0.87	48	23.0
$S8_{III,10}$				1.49	0.21	0.38	57	18.9
$S8_{III,11}$				1.45	0.26	0.40	9.5	2.3
$M10_{IV,1}$	IV	M10	10	1.61	−0.13	0.55	90.7	59.3
$M20_{IV,1}$		M20	10	1.50	0.01	0.54	>100	68.2
$M30_{IV,1}$		M30	10	1.58	−0.02	0.58	>100	77.6

H: Thickness of the soil layer in flume test shown in Fig. 12.3; V_p: Peak value of velocity for each test; ΔS: Runout distance during main failure; Δu: Generated excess pore pressure during main failure; *: Unable to measure distance, because the "ball" of the distance measurement system in Fig. 12.3 did not properly move with the soil.

Initial dry density was determined from the oven-dried weight of the used mass and the volume of the sample. When water sprinkling began, the data-logging system and video camera were activated. To successfully obtain the rapid change of pore pressure during quick failure, both pore pressure and displacements (parallel and normal to the slope) were logged at 0.05-second intervals.

12.3.3 Test Condition

Table 12.2 presents the test conditions and some test data in flume tests. In all the tests, the flume angle was kept constant at 30 degrees. The limits of the relative density index, which were defined by Spence and Guymer (1997), are used in this paper, i.e., extremely loose: $I_d < 0$, very loose: $0 = I_d < 0.15$, loose: $0.15 = I_d < 0.35$, medium: $0.35 = I_d < 0.65$, dense: $0.65 = I_d < 0.85$, very dense: $0.85 = I_d = 1$. Rainfall intensity was kept constant as 1.7 mm min^{-1} in all the tests. The flume tests were carried out in four series: (*a*) Test series I: tests on S7 with thickness being 10 cm; (*b*) test series II: tests on S8 with thickness being 10 cm; (*c*) test series III: tests on S8 with thickness being 12 cm; (*d*) test series IV: tests on M10, M20, and M30 with thickness being 10 cm.

12.4 Observed Phenomena in Flume Tests and Discussion

12.4.1 Movement of Failed Landslide Mass

The movement of landslide soil depends greatly on the sample itself. In the series tests on S7 (Series I), although the test phenomena differed from each other when the initial densities were different, retrogressive sliding occurred in each test. However, in the series tests on S8 (Series II and III) and the mixtures of M10, M20, and M30 (Series IV), flowslides were initiated. According to the video recordings, it was observed that the movement of the displaced landslide was affected greatly by the initial density and samples. Although in this research, two series of tests were performed on S8 with initial thickness being 10 cm and 12 cm, respectively, there was no obviously differing in their failure mechanisms through the video recordings, probably due to the little difference (2 cm) between their thicknesses. To have a brief understanding of the entire failure process, the failure modes are summarized in Fig. 12.4 as Types A and B for the tests on S7, and Types C and D for the tests on S8, M10, M20, and M30.

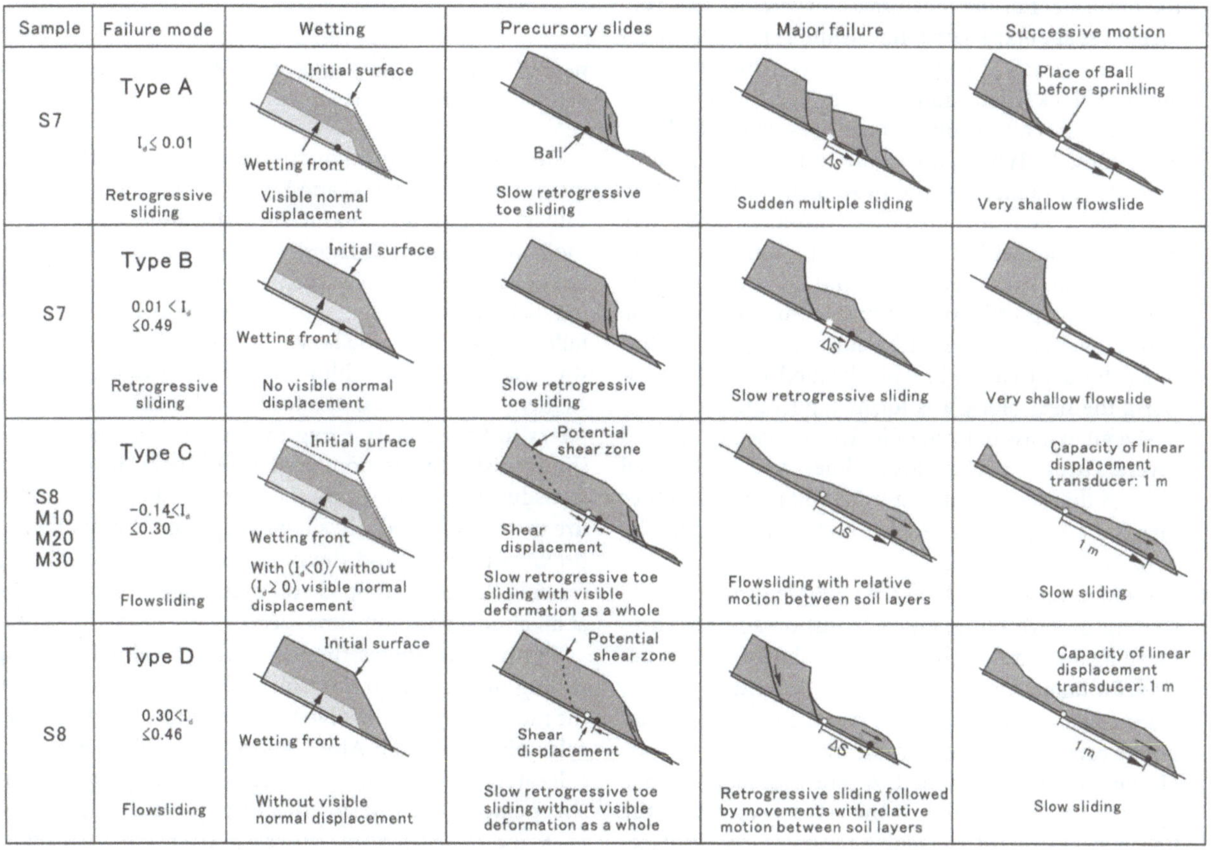

Fig. 12.4. Summarized failure modes

In each type of failure mode, the failure process was divided into four periods:

1. *Wetting*: After sprinkling, the water gradually flowed toward the base, and the wetting front was approximately parallel to the base. During this period, there appeared obvious normal displacement in the tests on S7 of failure-mode Type A and in those tests on extremely loose sample (S8) of Type C, while there was no visible normal displacement in other tests.
2. *Precursory slides*: After the wetting front reached the base, with the rising of saturation degree, retrogressive compound shallow sliding appeared at the toe part of slope. After sliding, the masses became very thin and then flowed downward. In this period, obvious shear deformation occurred within the whole layer for Type C, while there was no obvious deformation for other types.
3. *Major failure*: Following the retrogressive compound shallow sliding, major failure occurred. The displaced soil slid a certain of distance (ΔS, as shown in Fig. 12.4), with the styrene foam ball within it, and then stopped. During this period, the sliding behaviors were quite different in these four types. Failure-mode Types A and B showed retrogressive sliding only, whereas Types C and D behaved as flowslide motion. In Type A, during the movement of the failed mass, several blocks were formed at about the same time, while Type B was characterized by slow retrogressive sliding. After one (usually big) block failed and slid downward, the next block was formed. This retrogressive process continued several times. In Type C, according to the video recordings, it was found that the movement at the surface was faster than that at the flume bottom, i.e., there was relative motion within the failed soil at different depth, showing the characteristics of flows, although the vertical velocity profile was not measured. Failure-mode Type D was characterized by slow retrogressive sliding. After one (usually big) block failed and slid downward, the next block was formed, pushing the previously failed mass in front of it. This process continued several (usually two or three) times, and, in the final stage of this period, the displaced soil moved together forming a large single block, exactly as illustrated by Type D.
4. *Successive movement*: After the major failure, with continuing of sprinkling, failed landslide soil began to move again and flowed downward slowly. In this period, for failure-mode Types A and B, because the displaced soil became so shallow that the foam ball could not be carried to move with the mass, the moving displacement was not obtained. Whereas for failure-mode Types C and D, because of the rated capacity of the linear-displacement transducer, the measured displacement for each test was 1 m at the maximum capacity.

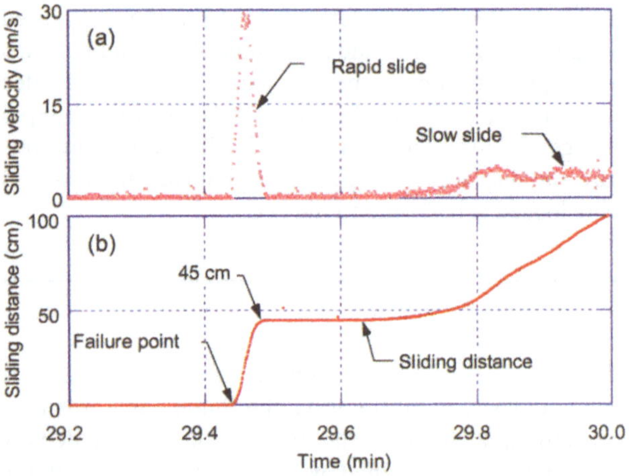

Fig. 12.5. Variation of sliding distance and sliding velocity for test $S8_{II.8}$

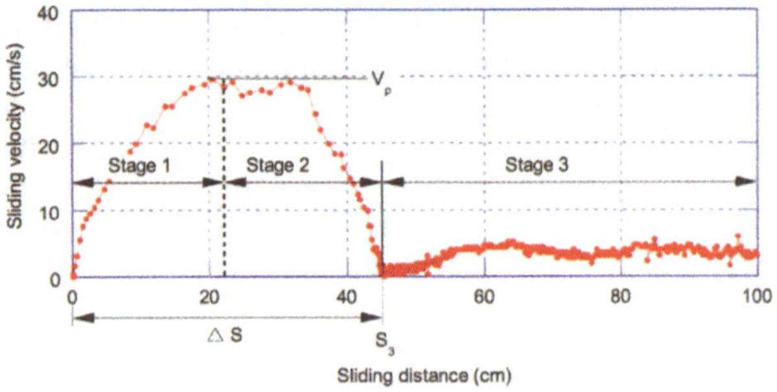

Fig. 12.6. Sliding velocity against sliding distance for test $S8_{II.8}$ (*S3*: Beginning of Stage 3; ΔS: rapid runout distance)

To examine the movement characteristics of flowslides quantitatively (Types C and D), here the results of a typical test $S8_{II.8}$ are used to interpret the motion process (in the major failure). This test was carried out with an initial density index of 0.08, and the time series data of sliding distance and sliding velocity just before and after the failure of the test are shown in Fig. 12.5. Because there was relative motion within the soil layer during flowing as mentioned above in the tests, the "velocity" used here refers to the velocity along the flume base because the sliding distance was monitored by the linear-displacement transducer installed along the base. As shown, after failure the mass flowed and showed rapid movement for approximate 45 cm, and then it stopped. With continuing rainfall, the once-stopped soil mass then began to move at an almost constant slow speed.

Figure 12.6 shows the variation of velocity during sliding in relation to the sliding distance. As indicated, the moving process involved three stages: (*1*) Stage 1 (Accelerating process): After failure, the sliding mass accelerated until reaching a certain velocity, which was the peak value during the entire sliding process; (*2*) Stage 2 (Decelerating process): After velocity reached its peak value, the sliding mass decelerated; after a certain distance (ΔS) of movement, the velocity reduced to zero; (*3*) Stage 3 (Steady slow sliding): The mass began to accelerate again and finally shifted to a slow motion at an almost constant value. For simplification, the sliding distance of movement during the first two stages is termed the "rapid runout distance" (ΔS) in the following sections.

12.4.2 Pore-pressure Generation

To study the effect of pore water pressure on the motion of failed mass, pore water pressure was monitored in the source area. And here the results of one test are used to illustrate the pore water pressure generation process.

Figure 12.7 shows the results of test $S8_{II.8}$. Fig. 12.7a illustrates the time series of pore pressure and sliding distance immediately before and after the failure. Fig. 12.7b is the enlargement of Fig. 12.7a for a very short period of failure. The precise monitoring of transient pore pressure in relation to slope movement demonstrated that high pore water pressure in the shear zone seemed not to exist prior to the initiation of failure (Fig. 12.7b). Checking on the possible time lag in the pore pressure monitoring system through a tube and a diaphragm of the transducer revealed that the time lag in the pore pressure measurement system was at least less than 0.005 s interval (Wang and Sassa 2001). Therefore, the time resolution of 0.05 s data logging in Fig. 12.7 is reliable enough.

Figure 12.7 shows that there was gradual increase of pore pressure, u_i, before the failure. This u_i could be divided into two components: one is the static water pressure due to the rise of water surface in the sample caused by rainfall; and the other was the excess pore water pressure due to the slow shearing. It could be seen that small shear displacement was initiated before the failure; meanwhile, as mentioned above, there was obvious shear deformation within the whole layer shortly prior to the failure (according to the video recordings observation). After failure, the pore water pressure rose rapidly; this probably have been due to the influence of excess pore-pressure built-up accompanying the quick shearing, because rainfall intensity itself was kept constant. Soon after the rapid pore pressure rise, the pore water pressure dropped down (Fig. 12.7a). It would be due to the following possible reasons: (*1*) after the major failure (see Fig. 12.4) happened, the failed mass would move downward, and then the mass above the place of pore pressure transducer (*P* in Fig. 12.3) would decrease in mass height or moved down completely; (*2*) with increase in elapsed time, dissipation of pore pressure could result in the reduction in pore pressure also. It should be noted that the maximum value of pore water pressure (u_m) measured in the major failure was much greater than u_i.

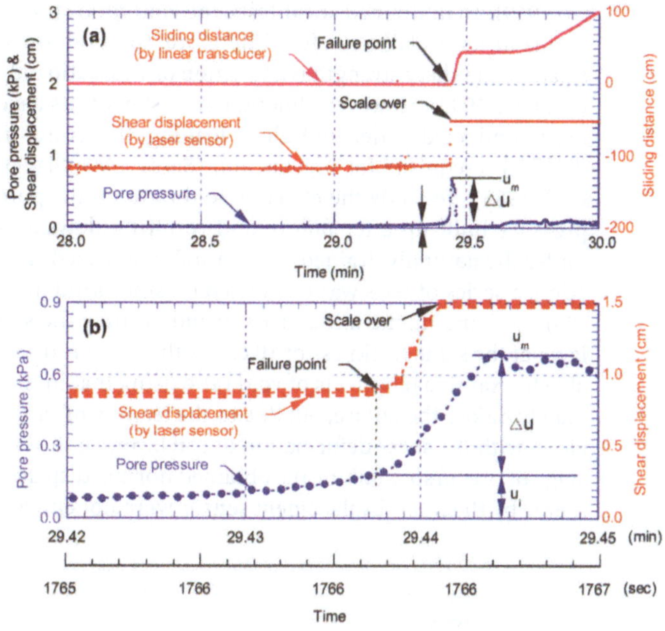

Fig. 12.7. Time series of pore pressure and sliding distance for test $S8_{II.8}$ (capacities of linear-displacement transducer and laser-displacement sensor are 100 cm and 15 mm, respectively; displacements greater than the capacities were not determined). **a** Time series of pore pressure and sliding distance immediately before and after failure; **b** enlargement of Fig. 12.6a during the very short period of failure

12.4.3 Pore Water Pressure and Flowslide Motion

Flowslide motion presented in Fig. 12.6 could be interpreted based on the monitored pore pressure behavior as follows. At stage 1, excess pore pressure was generated which reduced the shear resistance, and then an unbalanced driving force was resulted in, which caused the acceleration of sliding. At stage 2, accompanying the dissipation of pore water pressure along with elapsed time, and meanwhile, due to the possible reduction in driving force when the displaced mass moved and the geometry changed, the shear resistance became greater than the driving force, and then deceleration was resulted in. Finally, in stage 3, the driving force and shear resistance reached a balance, and the sliding velocity tended to the same value. To examine the relationship of pore water pressure built-up after failure (Δu in Fig. 12.7b) and initiated "rapid runout distance" (ΔS), these two values for different tests were examined and summarized in Table 12.2.

12.4.4 Effects of Initial Void Ratio

The initial void ratio plays an important role in liquefaction. Saturated sands with considerably greater void ratio than those on the steady-state line were found to liquefy when subjected to undrained monotonic shear (Castro 1969). In undrained shear tests, the void ratio still remains at its prefailure value. Therefore, the undrained steady-state shear strength is a function of void ratio only (Ishihara 1993). However, although most shear tests have been conducted under undrained loading, this is not in itself a prerequisite for liquefaction (Sassa 1985; Sladen et al. 1985). To study the effect of void ratio on the pore pressure built-up after failure and resulting flowslides under the naturally drained and naturally saturated situation, a series of tests was conducted on silica sand no. 8 (S8) by changing the initial density and keeping the soil layer at the same thickness of 10 cm. Although it is desirable to examine the effects of void ratio using that immediately before the failure, but due to the leaning of target along with the shear deformation of soil layers in the last stage of precursory failure, the obtained normal displacements by the laser displacement sensor were not enough

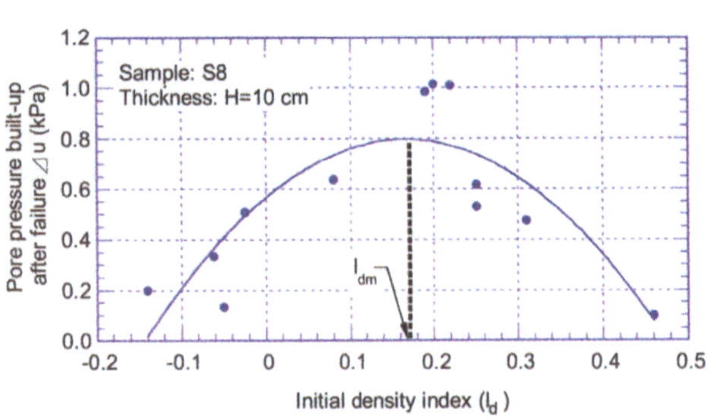

Fig. 12.8. Pore pressure build-up after failure versus initial density index for tests on S8 with thickness being 10 cm

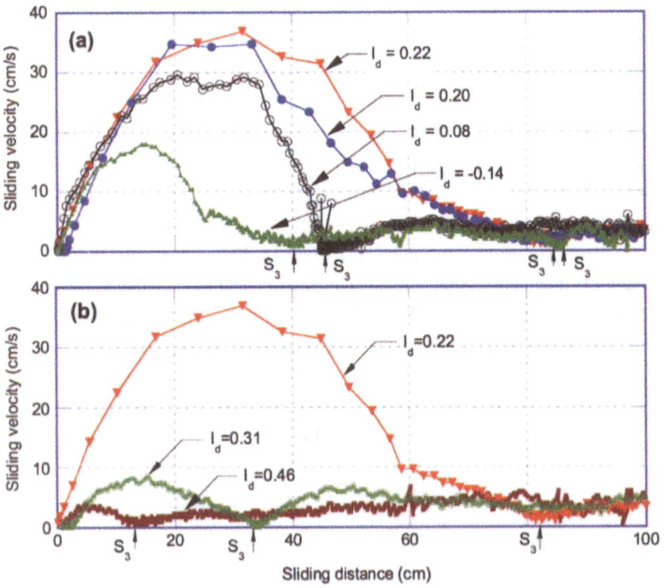

Fig. 12.9. Variation of sliding velocity for tests on S8 with different initial density indexes. **a** For tests with initial density index $I_d \leq 0.22$ ($S8_{III.5}$, $S8_{III.6}$, $S8_{III.8}$, $S8_{III.12}$); **b** for tests with initial density index $I_d \geq 0.22$ ($S8_{III.1}$, $S8_{III.2}$, $S8_{III.5}$)

credible in this series tests. According to another series of tests presented in Wang et al. (1998), it was found that the sand with a greater initial void ratio was subjected to a greater volume reduction, but it still kept a greater void ratio immediately before the failure. Therefore, the initial void ratio was used in the following analyses.

12.4.4.1 Effect on Pore Water Pressure Built-up after Failure

A quick increment of pore water pressure was generated after failure of the slope, as shown in Fig. 12.7b. To analyze the relationship between pore water pressure and failure motion, the pore pressure built-up after failure (Δu) of each test was estimated by regarding it as the difference between u_m and u_i (Fig. 12.7), and then, the relationship of Δu with initial density (expressed as initial density index I_d, $I_d = (e_{max} - e)/(e_{max} - e_{min})$) was examined.

The test results show that the pore pressure built-up after failure for each test greatly depended on the initial density index. Figure 12.8 plotted the pore pressure built-up after failure (Δu) against density index for these tests on S8 with the initial thickness being 10 cm. As presented, with the increase of density index, Δu increased until a certain value of I_d (I_d value at which Δu reached its peak value is denoted as I_{dm}), and thereafter, it decreased.

Usually, for saturated samples in undrained condition, it could be expected that greater pore pressure would be built-up in looser materials. In this natural (drained) slope condition, although many factors are involved in affecting the pore water pressure built-up during flowsliding, the main reasons for the existing of I_{dm} as the synthetic results of various factors could be interpreted as follows. For the tests in which I_d is smaller than I_{dm}, the permeability is greater (inferred from the variation tendency of permeability against density index shown in Fig. 12.2); thus, the dissipation must be quicker. Accordingly, the effect of greater pore pressure dissipation rate is likely to exceed the effect of greater pore pressure generation in the range of I_d smaller than I_{dm}. For the tests in which I_d is greater than I_{dm}, the volume reduction during shearing is small because of the smaller void ratio; thus, less pore-water pressure generation is likely to accompany the failure. Therefore, it is concluded that there is an optimal density index for pore water pressure built-up for the flowslides in this research, while given the initial thickness of the sample being the same.

12.4.4.2 Effect on Flowslide Motion

Variation of sliding velocity in relative to sliding distance in the tests with different initial densities is shown in Fig. 12.9. All cases show the accelerating, decelerating, and steady slow-movement stages. During the rapid-movement stage (accelerating period and decelerating period), the peak values of velocity and rapid runout distance (ΔS) differ for the samples with different initial density indexes. However, during the final steady slow-movement stage, there appeared no obvious difference in their sliding velocities.

Although there were an exception ($S8_{II.10}$ shown in Table 12.2, where both the peak value and rapid runout distance were very small, due to the difficulties with test-sample preparation), the peak value of velocity and rapid runout distance (ΔS) of flowslides became greater with increase of the initial density index (Fig. 12.9a). Also there was a density index at which both of peak velocity and rapid runout distance (ΔS) reached their maximum values. After that, with increase of initial density index, they tended to decrease (Fig. 12.9b). This variation was in accord with the tendency of pore water pressure built-up after failure in relation to initial density index (Fig. 12.8). Comparing both results of Figs. 12.8 and 12.9, the test with the greater pore water pressure built-up after failure had greater peak velocity and longer rapid runout distance. Therefore, high pore water pressure built-up after failure could be considered to be responsible for the rapid motion.

12.4.5 Effects of Initial Thickness

Triaxial laboratory studies of static liquefaction have shown that the level of initial shear stress existing under the drained condition strongly affects static liquefaction resistance (Hird and Hassona 1990; Ishihara 1993; Kramer and Seed 1988). For samples anisotropically consolidated to high levels of initial shear stress, the increase in shear stress under the undrained condition required to initiate liquefaction may only be a small percentage of the existing initial shear stress. By means of the experimental study on flowslides initiated in coking-coal stockpiles, Eckersley (1990) found that the built-up of pore pressures might be affected by shear deformation, dissipation, thickness of mass overlying the sliding surface, and other factors that are not completely understood. But in Eckersley's (1990) research, the detailed study of the relative importance of these factors had not been carried out. Thus, to make an insight into the effects of soil-layer thickness on pore-pressure built-up and flowslide motion, tests with different initial thicknesses of soil layer were conducted on S8 silica sand. Due to the limitation of flume height, the sample thickness was changed only from 10 cm to 12 cm. The samples with a thickness of 12 cm were made following the same steps for preparing the samples of 10-cm thickness, but the tests were conducted using differing initial void ratios.

12.4.5.1 Effect on Pore Water Pressure Built-up

Pore water pressure built-up after failure versus initial density index for the tests in which $H = 10$ cm and $H = 12$ cm have been plotted in Fig. 12.10. As presented,

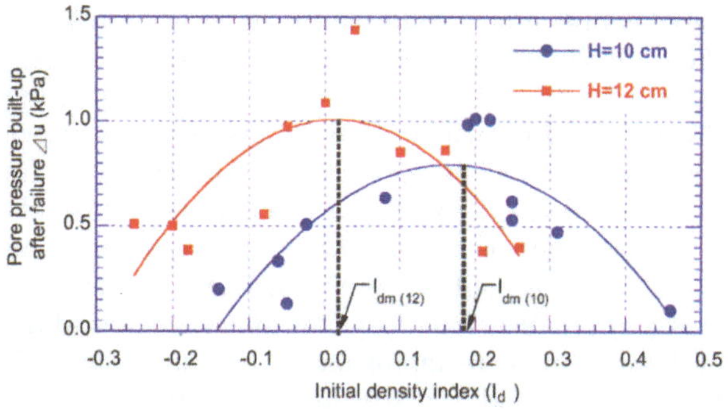

Fig. 12.10. Relationship between pore pressure built-up after failure and initial density index for samples with different thicknesses

the test results for $H = 12$ cm showed the same tendency as those obtained in the tests in which $H = 10$ cm: with increase of initial density index, the pore water pressure built-up after failure increased until a peak value was reached; thereafter, it decreased. However, it can easily be seen that the maximum value of pore pressure for the tests in which $H = 12$ cm is greater than that for $H = 10$ cm. The optimization density index for the tests on samples that are 12 cm thick is approximately 0.0, while that for the 10-cm samples is approximately 0.2. The greater thickness means a smaller pore pressure dissipation rate at a certain permeability and greater possibility of pore water pressure generation due to the greater initial (total/effective) stress state. When assuming the optimum density index as the combined results of pore pressure generation rate and pore pressure dissipation rate, it is reasonable that the greater thickness of soil layer gives the higher pore pressure built-up and the smaller I_{dm} value. Thus, the test result of Fig. 12.10 is conformable to the interpretation of the existence of optimal density index (I_{dm}) in Fig. 12.8.

12.4.5.2 Effect on Sliding Velocity and Rapid Run-out Distance

The variations of sliding velocity in relation to sliding distance for tests on samples that were 12 cm thick are shown in Fig. 12.11. All initiated flowslides had accelerating, decelerating, and steady slow-sliding periods, similar to those shown in Fig. 12.9. Samples with different initial density indexes had different peak velocity values. During the final steady slow-moving stage, there appeared to be no obvious differences between their sliding velocities. The peak velocity value and rapid run-out distance (ΔS) of flowslides became generally greater with increase of the initial density index until $I_d = 0.0$ was reached (Fig. 12.11a). After that, it decreased with initial density index (Fig. 12.11b). The effects of sample initial thickness on the flowslide motion could be compared in Fig. 12.11c and d. The flowslide motion with the greatest peak velocity from the series tests on 12-cm thick samples is compared with that obtained from the series tests on 10-cm thick samples in Fig. 12.11c. As shown evidently, the peak velocity for the sample of 12 cm thickness is greater than that for the sample of 10 cm thickness (the peak velocities are 37 cm s^{-1} for $H = 10$ cm, and 48 cm s^{-1} for $H = 12$ cm, respectively). Figure 12.11d superimposes the flowslide motions of samples with approximately the same initial density but different initial thickness. It can be seen in Fig. 12.11d that even though the initial density indexes of these samples are approximately the same ($I_d = -0.05$ for $H = 12$ cm, and $I_d = -0.03$ for $H = 10$ cm), their peak velocity values and rapid run-out distances (ΔS) differ. This is probably due to the higher pore water pressure built-up due to less pore pressure dissipation rate in greater initial thickness.

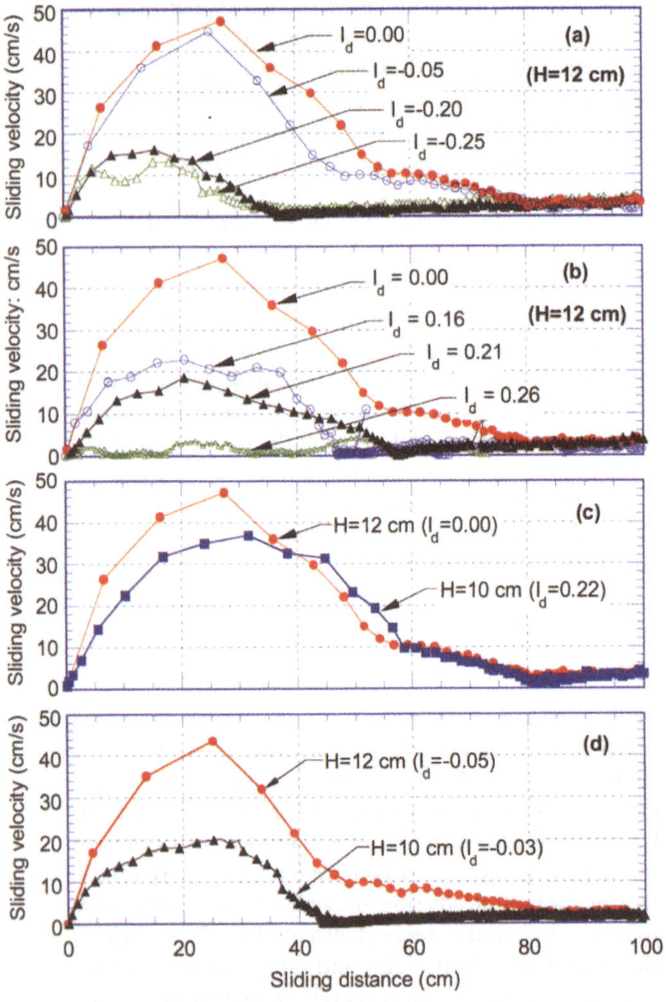

Fig. 12.11. Variation of sliding velocity for samples with different initial density indexes. **a** For tests with initial density index $I_d \leq 0.00$ (sample thickness: 12 cm); **b** for tests with initial density index $I_d \geq 0.00$ (sample thickness: 12 cm); **c** for the maximum velocities obtained in the series of tests on samples that were 12 cm thick and 10 cm thick; **d** for the velocities in which the initial density indexes for the samples were approximately the same but their thicknesses were different

12.4.6 Grain Size, Pore-pressure Build-up and Failure Modes

12.4.6.1 Grain Size and Pore-pressure Build-up

When we focus on the variation tendency of pore-pressure build-up after failure with initial density index for the tests on S7 and S8 (Fig. 12.12), it is noticed that S7 and S8 had the same changing tendency with initial density index, as mentioned previously. However, an observation on these two Δu–I_d curves can find that the pore pressure build-up after failure for S8 is generally greater than that for S7, indicating that higher pore pressure was easier to be built up in finer sand S8. This could be a result of the greater permeability of S7 (see Fig. 12.2). In the analysis of pore pressure generation during slope failure and sliding, Iverson and LaHusen (1989) and Iverson et al. (1997) pointed out that the pore pressure build-up mainly depends on the rate of landslide motion and soil deformation as well as the permeability of soils. If the moving velocity of failed landslide mass is the same, the soil with greater permeability will have quicker dissipation, and thus smaller pore-pressure build-up.

Furthermore, it was found that the optimal density index (I_{dm}) for S7 differs from that of S8. This shift of I_{dm} can be interpreted by the volume-reduction potential, which is directly related to the generation of pore-water pressure for a saturated sample under undrained conditions during shearing. In fact, if S7 has the same volume reduction potential as S8, it should have a smaller density index than S8 soil. This was certified by the results of two undrained ring-shear tests on loose S7 and S8 soil performed on a newly developed undrained ring-shear apparatus (Wang and Sassa 2003). Moreover, a smaller density index would result in quicker dissipation of generated pore-water pressure in the naturally drained condition of the flume test. Therefore, the combination of two aspects would necessarily make the optimal density index for S7 become smaller (looser), shifting leftward, as shown in Fig. 12.12.

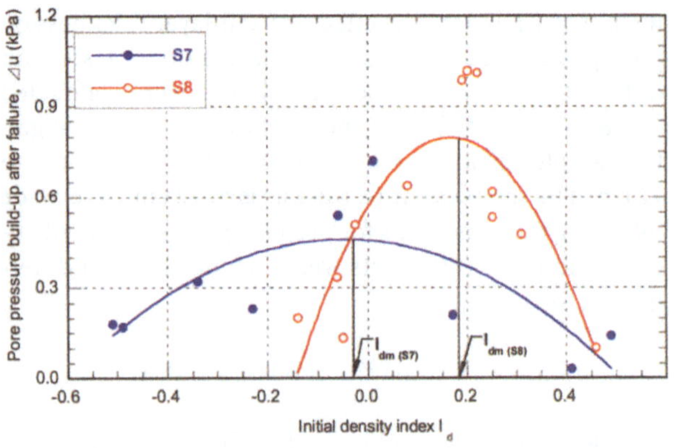

Fig. 12.12. Pore pressure build-up after failure versus initial density index for tests on S7 and S8, respectively (sample thickness: 10 cm)

12.4.6.2 Grain Size and Movement of Landslide Mass

The influence of grain size on the movement of the landslide mass was significant. As introduced in Fig. 12.4 for the tests on S7, retrogressive slides were initiated, whereas in the tests on S8, rapid flowslides occurred, showing completely a different failure mode. As the main reasons for the different failure modes for S7 and S8, the following can be considered: (*1*) the pore-pressure build-up after failure is normally smaller for tests on S7, as shown in Fig. 12.12; (*2*) the generated pore-water pressure is dissipated very quickly for S7 during movement because of its greater permeability, as shown in Fig. 12.2; (*3*) S8 is more likely to float during movement because of its smaller grains (see Fig. 12.16).

12.4.6.3 Fine-particle (Loess) Content and Flowslide Motion

A series of tests was performed on S8 with different loess contents (10 percent, 20 percent, and 30 percent) to examine the effects of fine-particle (loess) content on the movement of the landslide mass. To minimize the effects of variation of initial void ratio, the samples were formed by using the same tamping method. Due to difficulties in sample preparation, the initial densities were not the same, but were 1.01, 1.06, and 1.03 g cm^{-3}. Their initial density indexes were −0.13, 0.01, and −0.02, respectively (Table 12.2). From these limited test results, there was no obvious difference in pore-pressures build-up after failure (Δu), 0.56, 0.55, and 0.58 kPa for M10, M20, and M30, respectively. However, the flow behavior varied remarkably. Thus, attention was paid to the effects of fine-particle (loess) content on the movement of the initiated flowslides.

The variation of sliding velocity with sliding distance for samples with different fine-particle (loess) contents is plotted in Fig. 12.13. As shown in Fig. 12.13a, for the tests on S8 and M10, the sliding mass accelerated until reaching a certain velocity, and then decelerated until cessation of movement at point S_3. Thereafter, the mass resumed movement at an almost constant velocity. Unlike the phenomena observed for S8 and M10, however, in the tests on M20 and M30, the landslide mass continued to move without deceleration, at least within the range of available sliding distance measurement (100 cm), as shown in Fig. 12.13b. This suggests that a mechanism for maintaining the generated pore pressure was at work during movement for M20 and M30. This mechanism was proposed by Sassa (1985, 1988a, 2000), where the maintaining of pore pressure was explained by the flotation of finer grains and the resultant increase of specific gravity of pore liquid. To examine the variation of pore pressure along with movement, double-cylinder rotating tests were performed on water-saturated samples. The pore-pressure maintaining mechanism can then be discussed as follows.

12.5 Double-cylinder Apparatus and Test Procedures

12.5.1 Double-cylinder Apparatus

Figure 12.14 presents a schematic illustration of the employed double-cylinder rotating apparatus (designed by Sassa 1988a). This apparatus is composed of a rotating system, a double cylinder (14 cm in inside diameter, 30 cm in outside diameter, and 28 cm in depth), and a data-recording system. The pore-water pressure within the moving mass is measured by drawing a vinyl hose through a

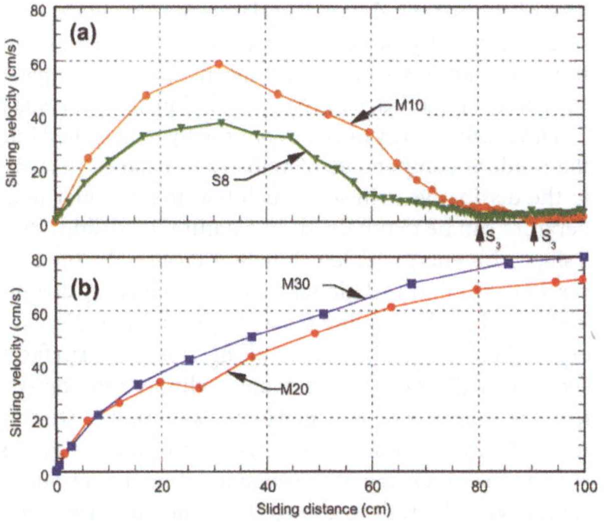

Fig. 12.13. Sliding velocity against sliding distance. **a** For S8 and M10; **b** for M20, and M30 (M10, M20, and M30: mixtures of S8 with 10, 20, and 30 percent loess by weight, respectively)

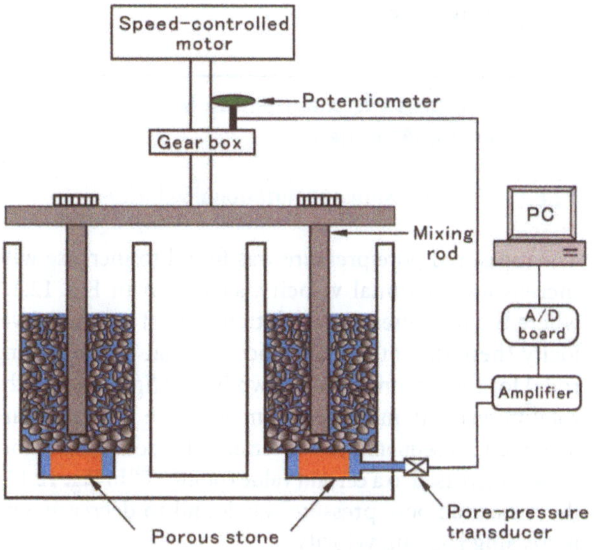

Fig. 12.14. Arrangement of double-cylinder rotating apparatus (after Sassa 1988b)

filter from the bottom of cylinder and installing a pore-pressure transducer into the hose. During tests, the mixture of the sample and water inside the cylinder is rotated by four mixing rods that are connected to a rotating axis, and the rotating velocity is measured with the help of a potentiometer. All the tests presented in this research were performed with the height of sample after saturation (h_s) being the same as that of the water (h_w).

12.5.2 Test Procedures and Test Conditions

An oven-dried sample was weighed and set into the cylinder until the sample height reached 15 cm. Then de-aired water was sprinkled from the upper opening mouth of the cylinder until the water level reached the same height as the sample. Note that the sample height decreased during the sprinkling of water. Finally, the water surface was kept at the same height as the sample for a long time (usually 24 hours) to obtain a highly saturated sample. The saturated sample was rotated to move with the mixing rods, and then the data of pore pressure and potentiometer were recorded by a computer. The velocity was increased gradually by steps; at each step, rotation was continued for 10 to 15 minutes, after which, the pore-water pressure and rotating velocity were logged.

According to Casagrande (1936), during shear deformation, the volume change of sand in loose state and in dense state tends to produce the same "critical density" or "critical void ratio". Therefore it can be inferred that the tested sand will tend to arrive at the same void ratio after reaching steady-state movement, no matter whether the sand was initially in loose or dense state. Hence, in this test series, all the samples were placed by means of dry deposition without tamping, and the effects of initial density were not examined.

12.6 Pore-pressure-maintaining Mechanism during Movement

12.6.1 Pore Pressure and Rotational Velocity

The measured pore pressure was found to increase with increasing rotational velocity, as shown in Fig. 12.15, where the pore pressure is plotted against rotational velocity (here the rotational velocity is that of the mixing rods) for the test on S8. As shown in this figure, when the rotating velocity increased from zero, the pore pressure increased consequently; however, after the rotational velocity was increased to a certain value (point "T" in Fig. 12.15), the measured pore pressure was found to decrease with increasing rotating velocity.

The increase of pore-water pressure with increasing rotational velocity can be interpreted using a phenomenon known as granular temperature, which measures the degree of agitation of solid grains and is determined by the ensemble average of grains' velocity fluctuation about their mean velocity (Ogawa 1978; Campbell 1990; Iverson et al. 1997). When the saturated sample is rotated by the rods, agitation by the rotating rods influence not only the mean moving velocity of grains, but also the ability of grains to move past one another. This leads to the increase of granular temperature. With increase in the velocity of rotating rods, the granular temperature becomes higher, i.e., the random moving velocity becomes greater. Interparticle collisions of grains with greater random moving velocity can favor those grains of greater diameter to fluctuate and float on the pore fluid, and then elevate the pore water pressure consequently (Sassa 1988a; Iverson and LaHusen 1989). If every grain floats in water, it means that the sample is 100 percent liquefied, namely, that full liquefaction results.

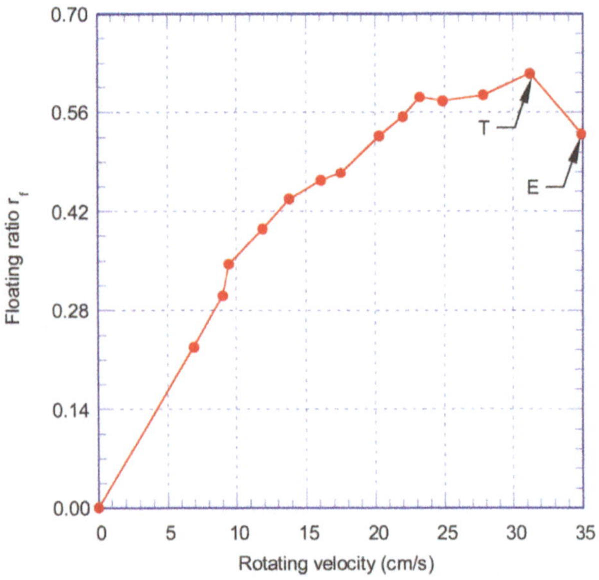

Fig. 12.15. Variation of pore pressure in relation to rotational velocity for test on saturated S8

Obviously, energy must be continually replenished into the random components of velocity to balance that lost to the dissipative collisions, such that the granular temperature can be maintained. In a natural landslide down a slope, this energy replenishment is ensured by the conversion of bulk translational energy (supplied by downslope travel of the moving mass) to grain fluctuation energy when the grains shear along irregular surfaces (Iverson et al. 1997). In this double-cylinder rotating system, the energy is replenished by the agitation of stirring rods. Meanwhile, due to the circular motion, the grains close to the inner wall of the outside cylinder continue to collide with the wall to change their moving directions; this also stimulates the fluctuation of grains, i.e., consequently elevates the granular temperature. Though the method of energy replenishment in this double-cylinder rotating test differs from that of natural downslope landslide movement, this double-cylinder rotating test is believed to be an effective approach for gaining an understanding of the pore-water pressure within saturated soils during steady-state motion.

Beyond point "T" in Fig. 12.15, because the mixing rods rotated too fast (relative to the moving mixtures) so that the mixtures could not move with the mixing rods together, the mixtures were subjected to shear; therefore, a reduction in pore pressure evidently occurred as a consequence, as shown in Fig. 12.15. Note that the test was terminated when point "E" was reached.

Though samples for this series of tests were formed to the same initial height (15 cm) in the dry state, the heights after saturation differed for different samples because of the various settlements that accompanied the sprinkling of water. To compare pore-water generation during motion for different samples, test results were normalized by the total normal stress (σ_t), and were presented by floating ratio (r_f). Floating ratio is a kind of pore-pressure ratio (identical to excess pore-pressure ratio in value) that has been defined by Sassa (1988a), and is formulated as:

$$r_f = \frac{u - u_s}{\sigma_t - u_s} \tag{12.1}$$

Here, u: measured pore pressure; σ_t: total normal stress ($\sigma_t = \gamma_s h_s$), h_s is the height of sample after saturation; u_s: static water pressure ($u_s = \gamma_w h_w$), is regarded as the initial water pressure that was measured before the rotating, h_w is the height of water, $h_s = h_w$; γ_w is unit weight of water and was regarded as 1.0 gf cm^{-3} in the calculation; γ_s is unit weight of saturated soil, which was obtained from the calculated dry unit weight (γ_d) of the soil in cylinder and the corresponding void ratio (e), namely,

$$\gamma_s = \gamma_w \frac{e}{1+e} + \gamma_d \tag{12.2}$$

If every grain floats in water, namely 100 percent liquefied, $r_f = 1.0$.

12.6.2 Grain Size, Fine-particle (Loess) Content, and Pore Pressure

To examine the effects of samples, the test results for all of these samples are shown in Fig. 12.16 in the form of floating ratio versus the velocity of the stirring rods. From Fig. 12.16, it can be seen that although the floating ratio increases with increasing rotating velocity for all of the samples, the changing tendency along with rotational velocity differs for different samples. At a given rotational velocity, when the sample becomes finer in grain size (from S7 to S8) or greater in fine-particle (loess) content (from M10 to M30), the floating ratio becomes greater. Note that for the tests on M20 and M30, the floating ratios increased very quickly, and the floating ratio reached a high value at a very small rotational velocity, thereafter kept almost constant (about 0.80 and 0.95 for M20, and M30, respectively).

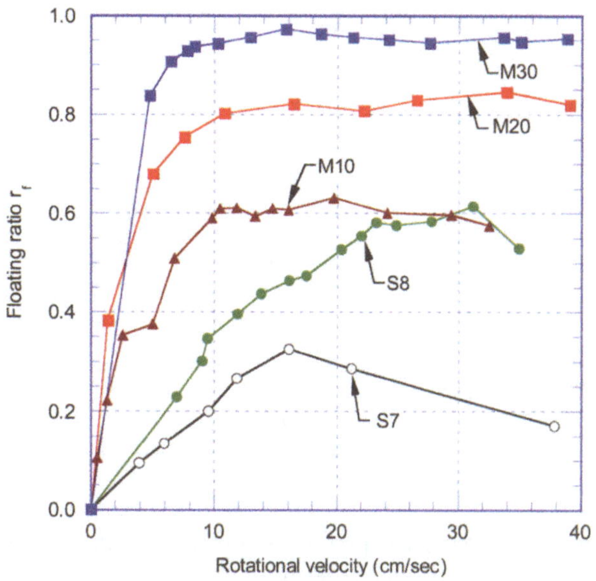

Fig. 12.16. Floating ratio versus rotational velocity for different samples

This phenomenon could be interpreted by the main factors controlling the pore pressure during motion. As mentioned above, the increase in pore pressure is due to the floating of grains in the fluid that would increase the mobilized specific gravity of fluid. Finer grains are easier to float. Therefore, corresponding to a certain rotational velocity, there will be more grains involved in floating for the finer sample; consequently, the resultant pore-pressure increment will be greater. This is evidenced clearly by the changing tendency of M20 and M30 shown in Fig. 12.16.

It is of interest to note that after the rotating was stopped, the dissipation of generated pore pressure was observed. The test results revealed that the dissipation of generated pore pressure for S7 in greater grain size showed a quick dissipation (within about 27 seconds), and when the grain size became finer or fine-particle content became greater, the dissipation became remarkably slower, especially for the test on M30. It took approximately 24 hours before the pore pressure became steady (as shown in Fig. 12.17), indicating that these fine particles played a most important role in the dissipation of generated pore pressure. This is of great importance for better understanding the deposition of debris flows, because, as stated by Major and Iverson (1999), many observations of debris flows in the field have revealed that the deposited liquefied materials kept moving or deformed for days to weeks (e.g., Fryxell and Horberg 1943; Curry 1966; Broscoe and Thomson 1969).

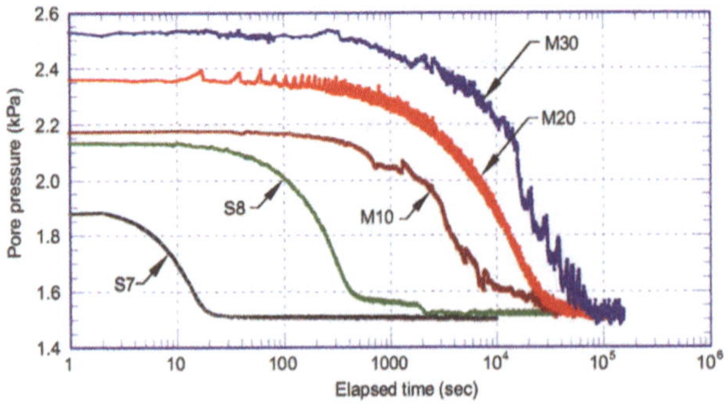

Fig. 12.17. Time series of floating ratio after the rotation ceased for different samples

12.7 Conclusions

Based on test results presented in this paper, the following conclusions can be drawn:

1. High pore water pressure was built up during movement of the failed mass. The pore pressure built-up after failure controls the velocity of movement and the runout distance of flowslides.
2. There is an optimal density index for pore water pressure built-up after failure as the combined results of generation and dissipation of pore water pressure for a slope in a natural condition. The peak value of velocity and the rapid runout distance depend largely on the initial density index. The relationship of the peak value of velocity and the rapid runout distance to the initial density index are the same as that of the pore water pressure built-up after failure to the initial density index.
3. Pore pressure built-up after failure depends on the initial sample thickness. With an increase of the initial thickness of the soil layer, the pore pressure built-up after failure became greater (at least within the tested range of initial density index, $I_d < 0.15$). The optimization density index for pore-pressure built-up also was found to change with initial thickness. The sample with greater thickness for the same density index had greater peak velocity and longer rapid run-out distance.
4. Grain size plays an important role in movement of a landslide mass and pore-pressure build-up after failure. The optimal density index is different for samples with different grain sizes. Usually the generated pore pressure was smaller for coarser sand (S7) and retrogressive slope failure occurred in tests on coarser sand (S7), while finer sand (S8) was prone to suffer rapid flowslide failure, probably due to the smaller permeability of S8 and the fact that it is easier for S8 of finer grains to float during motion (see Fig. 12.16).
5. If the sand included differing amounts of fine particles (loess), the flowslides were quite different in character. The tests showed that the peak velocity of flowslides became greater with increase of loess content from 10 percent (M10) to 30 percent (M30). In addition, there was no decelerating stage or steady slow-moving stage in tests on M20 and M30 soils.
6. Pore pressure within saturated sand increased with increasing moving velocity. In addition, corresponding to a certain moving velocity, the increment of pore pressure was greater for sand of finer grain size. Namely, finer grains started to float at slower moving velocity.
7. With increasing fine-particle (loess) content, the pore pressure increased more readily at lower moving velocity. In fact, for the tests on M20 and M30 soils, the floating ratios increased very quickly, and the floating ratio reached a great value (about 0.80 and 0.95 for M20, and M30, respectively) at a very small rotational velocity. Thereafter, the floating ratio remained approximately steady. In addition, the dissipation of generated pore pressure after cessation of rotation was very slow for tests on M20, and M30 soils.

Finally, it is noted that this report is mainly based on our two publications (Wang and Sassa 2001, 2003), from which more detailed information on the test results could be obtained.

References

Anderson SA, Sitar N (1995) Analysis of rainfall-induced debris flow. J Geotech Eng-Asce 121(7):544–552

Bishop AW (1967) Progressive failure-with special reference to the mechanism causing it. In: Proceedings of Geotechnical Conference, Oslo, Norway, 2, pp 142–150

Bishop AW (1973) The stability of tips and spoil heaps. Q J Eng Geol 6:335–376

Brand EW (1981) Some thoughts on rainfall induced slope failures. In: Proceedings of 10th International Conference on Soil Mechanics and Foundation Engineering, pp 373–376

Brenner RP, Tam HK, Brand EW (1985) Field stress path simulation of rain-induced slope failure. In: Proceedings of 11th International Conference on Soil Mechanics and Foundation Engineering, 2, pp 373–376

Broscoe AJ, Thomson S (1969) Observation on an alpine mudflow, Steele Creek, Yukon. Can J Earth Sci 6:219–229

Campbell CS (1990) Rapid granular flows. Annu Rev Fluid Mech 22:57–92

Casagrande A (1936) Characteristics of cohesionless soils affecting the stability of slopes and earth fills. Journal of the Boston Society of Civil Engineers, reprinted in Contributions to Soil Mechanics, 1925 to 1940, Boston Society of Civil Engineers, pp 257–276

Casagrande A (1971) On liquefaction phenomenon. Géotechnique 21(3):197–202

Castro G (1969) Liquefaction of sands. Ph.D. Thesis, Harvard University, Cambridge, Massachusetts

Castro G, Poulos SJ (1977) Factors affecting liquefaction and cyclic mobility. J Geotech Eng-ASCE 103:501–516

Curry RR (1966) Observation of alpine mudflows in the Tenmile Range, central Colorado. Geol Soc Am Bull 77:771–776

Eckersley JD (1985) Flowslides in stockpiled coal. Eng Geol 22: 13–22

Eckersley JD (1986) The institution and development of slope failures with particular reference to flowslides. PhD thesis, James Cook University of North Queensland

Eckersley JD (1990) Instrumented laboratory flowslides. Géotechnique 40(3):489–502

Fryxell FM, Horberg L (1943) Alpine mudflows in Grand Teton National Park, Wyoming. Geol Soc Am Bull 54:457–472

Hird HH, Hassona FAK (1990) Some factors affecting the liquefaction and flow of saturated sands in laboratory tests. Eng Geol 28:149–170

Ishihara K (1993) Liquefaction and flow failure during earthquakes. Géotechnique 43(3):349–451

Ishihara K, Okusa S, Oyagi N, Ischuk A (1990) Liquefaction-induced flowslide in the collapsible loess deposit in Soviet Tajik. Soils and Foundations 30(4):73–89

Iverson RM (1997) The physics of debris flows. Rev Geophys 35(3):245–296

Iverson RM, LaHusen RG (1989) Dynamic pore-pressure fluctuations in rapidly shearing granular materials. Science 246: 796–799

Iverson RM, Reid ME, LaHusen RG (1997) Debris-flow mobilization from landslides. Annu Rev Earth Planet Sci 25:85–138

Iverson RM, Reid ME, Iverson NR, LaHusen RG, Logan M, Mann JE, Brien DL (2000) Acute sensitivity of landslide rates to initial soil porosity. Science 290:513–516

Kramer KL (1988) Triggering of liquefaction flow slides in coastal soil deposits. Eng Geol 26:17–31

Kramer KL, Seed HB (1988) Initiation of soil liquefaction under static loading conditions. J Geotech Eng-ASCE 114:412–430

Major JJ, Iverson RM (1999) Debris-flow deposition: Effects of pore-fluid pressure and friction concentrated at flow margins. Geol Soc Am Bull 111(10):1424–1434

Ogawa S (1978) Multitemperature theory of granular materials. In: Proceedings of U.S.-Japan Seminar on Continuum-Mechanics and Statistical Approaches to the Mechanics of Granular Materials, Gukujutsu Bunken Fukyukai, Tokyo, pp 208–217

Sassa K (1972) Analysis on slope stability-I: mainly on the basis of the indoor experiments using the standard sand produced in Toyoura, Japan. J Jpn Soc Erosion Control Eng 25(2):5–17 (in Japanese with English abstract)

Sassa K (1974) Analysis on slope stability-II: Mainly on the basis of the indoor experiments using the standard sand produced in Toyoura, Japan. J Jpn Soc Erosion Control Eng 26(3):8–19 (in Japanese with English abstract)

Sassa K (1984) The mechanism starting liquefied landslides and debris flows. In: Proceedings of 4th International Symposium on Landslides, Toronto, June, 2, pp 349–354

Sassa K (1985) The mechanism of debris flows. In: Proceedings of XI International Conference on Soil Mechanics and Foundation Engineering, San Francisco, 3, pp 1173–1176

Sassa K (1988a) Motion of landslides and debris flows-prediction of hazard area. In: Report for Grant-in-Aid for Scientific Research by Japanese Ministry on Education, Science and Culture (Project No. 61480062), pp 1–15

Sassa K (1988b) Geotechnical model for the motion of landslides. In: Special Lecture of 5th International Symposium on Landslides, "Landslides", 10–15 July, 1, pp 37–55

Sassa K (1996) Prediction of earthquake induced landslides. In: Proceedings of 7th International Symposium on Landslides, A.A. Balkema, Trondheim, 17–21 June, 1, 115–132

Sassa K (1998a) Recent urban landslide disasters in Japan and their mechanisms. In: Proceedings 2nd International Conference on Environmental Management, "Environmental Management", 1, pp 47–58

Sassa K (1998b) Mechanisms of landslide triggered debris flow. In: Sassa K (ed) Environmental forest science: Proceedings of IUFRO Div. 8 Conference, Kyoto, Kluwer Academic Publisher, pp 471–490

Sassa K (2000) Mechanism of flows in granular soils. In: Proceedings of the International Conference of Geotechnical and Geological Engineering, GEOENG2000, Melbourne, 1, pp 1671–1702

Sassa K, Takei A (1977a) Consider vertical subsidence in slope unstabilization – I: Stress fall phenomenon and bearing power of the sides. J Jpn Landslide Soc 14(2):19–26 (in Japanese with English abstract)

Sassa K, Takei A (1977b) Consider vertical subsidence in slope unstabilization – II: Some examples in the field. J Jpn Landslide Soc 14(3):7–14 (in Japanese with English abstract)

Seed HB (1966) Landslides during earthquakes due to soil liquefaction. J Soil Mech Found Div-ASCE 94(5):1055–1122

Seed HB (1979) Soil liquefaction and cyclic mobility evaluation for level ground during earthquakes. J Geotech Eng-ASCE 105: 201–255

Sidle RC, Swanston DN (1982) Analysis of a small debris slide in coastal Alaska. Can Geotech J 19:167–174

Sitar N, Anderson SA, Johnson KA (1992) Conditions leading to the initiation of rainfall-induced debris flows. Geotech. Engrg. Div. Specialty Conference: Stability and Perf. of slopes and Embankments-II, ASCE, New York, pp 834–839

Sladen JA, D'Hollander RD, Krahn J (1985) The liquefaction of sands, a collapse surface approach. Can Geotech J 22:564–578

Spence KJ, Guymer I (1997) Small-scale laboratory flowslides. Géotechnique 47(5):915–932

Terzaghi K (1950) Mechanism of landslides. In: Paige S (ed) Application of geology to engineering practice (Berkey Volume) Geological Society of America, New York, pp 83–123

Terzaghi K (1956) Varieties of submarine slope failures. In: Proceedings of 8th Texas Conference on Soil Mechanics and Foundation Engineering, pp 1–41

Thevanayagam S (1998) Effects of fines and confining stress on undrained shear strength of silty sands. J Geotech Geoenviron 124(6):479–491

Vallejo LE, Mawby R (2000) Porosity influence on the shear strength of granular material- clay mixtures. Eng Geol 58:125–136

Wang G, Sassa K (2001) Factors affecting rainfall-induced flowslides in laboratory flume tests. Géotechnique 51(7):587–599

Wang G, Sassa K (2003) Pore pressure generation and movement of rainfall- induced landslides: effects of grain size and fine particle content. Engineering geology 69:109–125

Wang GH, Sassa K, Fukuoka H, Wang FW (1998) Study on the excess pore pressure generation in laboratory-induced-landslides. In: Moore DP, Hungr O (eds) Proceedings of 8th Congress of IAEG, Vancouver, Balkema, Rotterdam, 6, pp 4237–4244

Youd TL, Gilstrap SG (1999) Liquefaction and deformation of silty and fine-grained soils. In: Proceedings of 2nd International Conference on Earthquake Geotechnical Engineering, Lisbon, Portugal, 3, pp 1013–1020

Ring Shear Tests on Clays of Fracture Zone Landslides and Clay Mineralogical Aspects

Netra P. Bhandary* · Ryuichi Yatabe

Abstract. Various investigations on tectonically-induced landslides in Shikoku Region of West Japan have been carried out, most of which conclude at tectonic activities through the major tectonic faults and enhanced rock mineral decomposition as being mainly responsible for the landslide occurrence. This paper looks into strength parameters of the landslide clays, as measured in ring shear apparatus, from clay mineralogical point of view. As a result of strength tests and X-ray diffraction analysis, it is found that the drop from peak to residual friction angles for the tested samples reaches as high as 20°, and the residual strength of the landslide clays was found to decrease with higher amount of expansive clay minerals, which was estimated as being relative to chlorite mineral.

Keywords. Expansive clay minerals, expansive mineral ratio, fracture zone landslides, ring shear test, X-ray diffraction test

13.1 Introduction

Landslides are common natural phenomena that occur mainly under the influence of external inducing factors like earthquake forces and rainfall and internal inducing factors like slope material properties and slope conditions. Although many large-scale landslides do also occur in plains such as near the banks of deep cut rivers, the landslides to be discussed in this paper occur predominantly on mountainous terrains mostly with creeping displacements. Slopes in mountainous terrains often suffer failures in the form of minor collapses like rain-induced surface layer failures to large-scale failures like deep-seated gravitational creeps or massive slope movements. Japan with its more than 70% mountainous terrain has recognized a large number of active and potential landslide sites that threaten a huge loss of life and infrastructure. Shikoku Region in west Japan, for example, has a large number of tectonically induced landslides, often known as fracture zone landslides, as indicated by landslide distribution map in Fig. 13.1 (MLIT 1997). As of March 1997, the number of landslides under the direct jurisdiction of the Ministry of Land, Infrastructure, and Transport (abbreviated hereinafter as MLIT) in Shikoku alone is 670, which is nearly 21% of the number all over the country (ibid.).

The large-scale creeping landslides are often known for no immediate damages, but a rapid progress in their

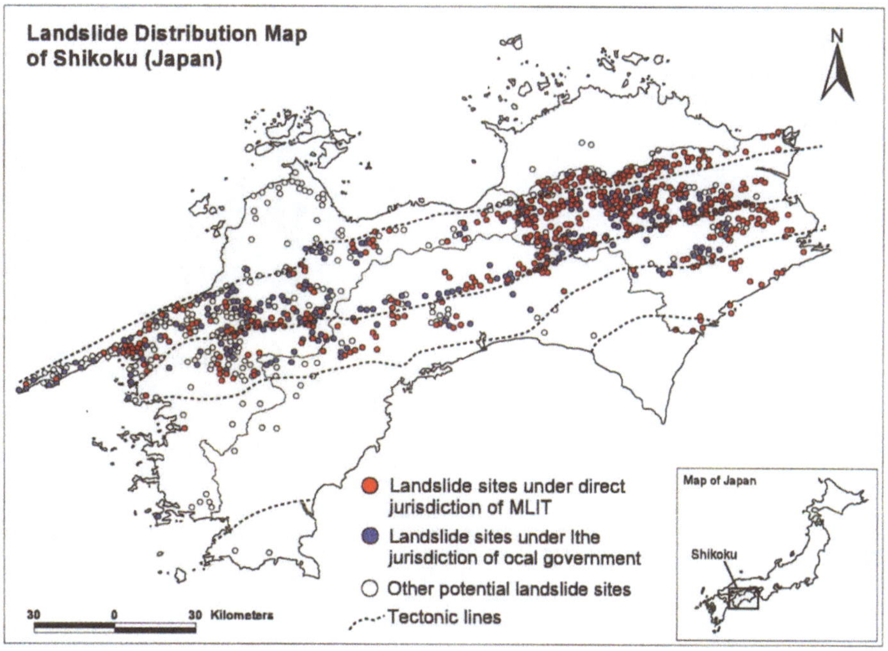

Fig. 13.1. Landslide distribution map of Shikoku (Fracture Zone Landslides, reproduced from MLIT 1997)

displacement may result in disastrous effects such as damming up of rivers, flow of human settlements, destruction of agricultural fields and forests, destruction of highways and roads, damage to dams, tunnels, highways and bridges, and transmission towers, and destruction of nature and environment. To overcome potential landslide threat, various landslide prevention and management projects are in execution throughout the Shikoku Region, some under the supervision of the local governments but most under the direct control of the MLIT. Such project works involve planning and design of proper and appropriate countermeasures, which require greater accuracy in the stability analysis. It is therefore important to have an in-depth understanding of the mechanism involved in creeping displacement of large-scale landslides.

Various field investigations carried out during the execution of landslide prevention and management projects have revealed that the landslides in Shikoku Region consist of clearly identifiable slip layers composed largely of fine clayey soils varying in thickness from a few centimeters to as thick as 20 centimeters (Yagi et al. 1990). As an attempt to elucidate the failure mechanism of slip layer soils and associated landslide displacement behavior, Yagi et al. (1990) and Yatabe et al. (1991a,b) have investigated shear characteristics of several clayey soil samples from the slip layers of landslides in various parts of Shikoku. Based on the laboratory shear tests, they found that most landslides in Shikoku Region are in pre-residual state of shear, which infers that the field angle of internal friction is greater than the residual angle of friction measured in the laboratory. Yagi et al. (1990) also compared the strength parameters of landslide clays in Shikoku with those of marine clays and found that the peak friction angles for the landslide clays were smaller by 3° to 8° than those for the marine clays in Japan. They mention that the fabric of landslide clay is different from that of marine clay and discuss that this difference can be attributed to the process and environment the clay soils are formed in. This paper is an extension of the work the above researchers carried out, and attempts further to look into shear strength characteristics of the clayey soils involved in the slip layers of fracture zone landslides based mainly on the ring shear tests. In addition, emphasis is put on investigating the influence of clay mineralogy on the angles of shearing resistance of the landslide clays, and its role in inducing the fracture zone landslides.

13.2 Features of Fracture Zone Landslides

13.2.1 Geological Background

Some of the major tectonic lines in Japan, often measuring hundreds of kilometers, pass through Shikoku Island. For example, Japan's longest tectonic fault system called *chuokōzōsen* (also known as median tectonic line and abbreviated as MTL) separates two major geological formations in the northern half of Shikoku Island. Likewise, two equally important tectonic lines known as Mikabu Tectonic Line and Butsuzo Tectonic Line (note: other minor tectonic lines have not been mentioned due to their insignificant role in causing landslides) pass across the island, and together with the MTL they divide Shikoku Region into four major strips of geological formations, namely Izumi Group, Sambagawa belt, Chichibu belt, and Shimanto belt including a few more minor geological formations (Fig. 13.2; GSJ 1992). The belts recognized as being particularly prone to landslides are Izumi Group,

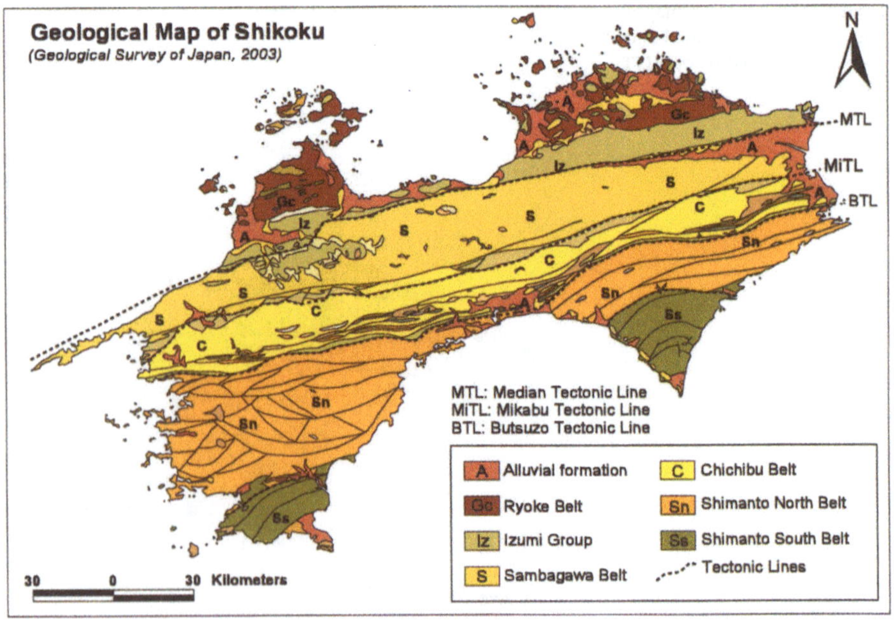

Fig. 13.2. Geological map of Shikoku showing major tectonic lines (GSJ 2003)

Sambagawa belt, Mikabu belt (a narrow strip along the Mikabu Tectonic Line separating Sambagawa belt and Chichibu belt), and Chichibu belt, which can be verified by superimposing Fig. 13.1 over Fig. 13.2.

The Izumi group is a sedimentary deposit composed primarily of sandstone with the intrusion of thin layers of shale, whereas the Sambagawa belt is a deposit of metamorphic rocks consisting mainly of green schist and black schist. Similarly, the narrow strip of Mikabu belt consists of greenstone (often known as Mikabu greenstone) as a metamorphic deposit. The rock type in the Chichibu belt is sedimentary composed of green schist, mudstone, conglomerate, etc.

13.2.2 Landslide Activation

The investigation and study on landslides in Japan began only half a century ago when the landslide prevention law was enacted in 1958 (JSSMFE 1985). While many landslides have remained active since long, a considerable number of relict landslides or landslide remnants in Shikoku were reactivated during the construction of express highways along the median tectonic line (MTL). Such reactivation is attributed mainly to huge slope cuts and tunneling. As also indicated in Fig. 13.1, the distribution of landslides in Shikoku can be found notably concentrated over two areas between the median and Mikabu tectonic lines.

In addition to other various factors, the internal factors causing landslides in Shikoku are flow behavior of groundwater, fluctuations in its level, and changes in soil strength properties. As reported by Scheidegger (1970) and supported by subsequent researchers such as Slivosky (1977), Carraro et al. (1979), and Varnes et al. (1989) the tectonic activity and earthquakes might also be related to the activation of landslides in the form of deep-seated gravitational creep. Particularly in Shikoku, the role of tectonic activities in inducing the landslides has been significant.

One highly favorable condition for land sliding in Shikoku is considered to be fractured state of the bedrocks, especially near the tectonic lines (MLIT 1997). It is believed that the fragile state of the bedrocks accelerates the rock mineral weathering by facilitating water ingress into the rock mass. Such weathering or decomposition of the bedrock minerals on a certain plane results in formation of clayey soils composed of weaker minerals such as chlorite, smectite, and vermiculite. Smectite and vermiculite minerals possess significantly small angles of shearing resistance at the residual state, which is because they absorb excess water and swell significantly upon wetting. So, a slight decrease in effective stress due especially to rise in groundwater level and simultaneous increase in driving shear stresses result in sliding of the slope material through the clay layer composed of weak minerals.

13.2.3 Topographical Features

Topographical features of creeping landslides, in general, consist of mild and gentle natural slopes, especially slopes that probably collapsed in the past and have attained a relatively gentle angle of inclination, i.e., in a range between 15° and 30°. The mountains of Shikoku compared to those in other regions are steep with slope inclinations ranging from 20° to 40°. It is also evidenced by a topographical feature that despite a narrow width of the island, some of the tallest mountains in west Japan lie in this region. Most mountain ranges form an alignment parallel with the fault lines extending in east-west direction, but minor cross alignments can be seen all throughout the fault lines due especially to cuts produced by streams and rivers. It is therefore obvious that the angles of slope for the landslide sites in Shikoku are high. Figure 13.3 (Ishii 1994) shows a comparative chart indicating that the landslide sites in Shikoku (represented by fracture zone landslides) have steeper slopes than those in Tertiary system deposits, which are dominant in the northern Japan and consist mainly of weathering-prone soft rocks such as Neogene sedimentary rocks, Palaeogene sedimentary rocks, and Tertiary volcanic rocks (MLIT 1997).

The average inclination of slope for the fracture zone landslides, as seen in Fig. 13.3, is much higher than that for the tertiary zone landslides, which more or less means that the size of a fracture zone landslide must be smaller than that of a tertiary zone landslide. This is evidenced by Fig. 13.4 (Ishii 1994), which compares the sizes of the fracture zone landslides and the tertiary zone landslides. This figure shows that most of the landslides in the fracture zone (represented by Sambagawa, Mikabu, and Chichibu in the figure) are smaller than those in the tertiary zone although the size ratio of the landslides in both the zones varies from $L = 0.5W$ to $L = 3.0W$, where L represents the slope length and W represents the width. The

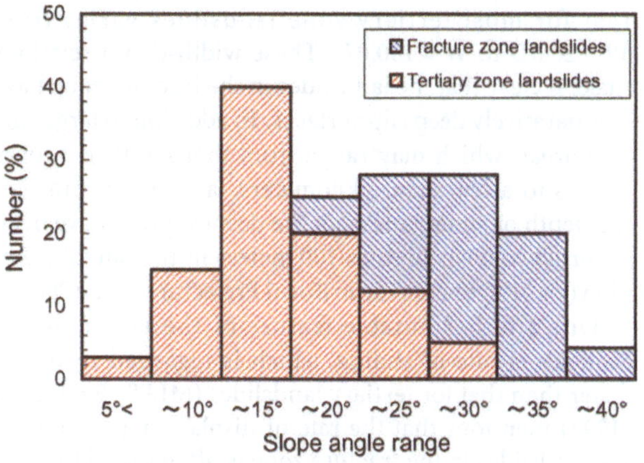

Fig. 13.3. Range of landslide slope angles for fracture zone landslides and tertiary landslides (from Ishii 1994)

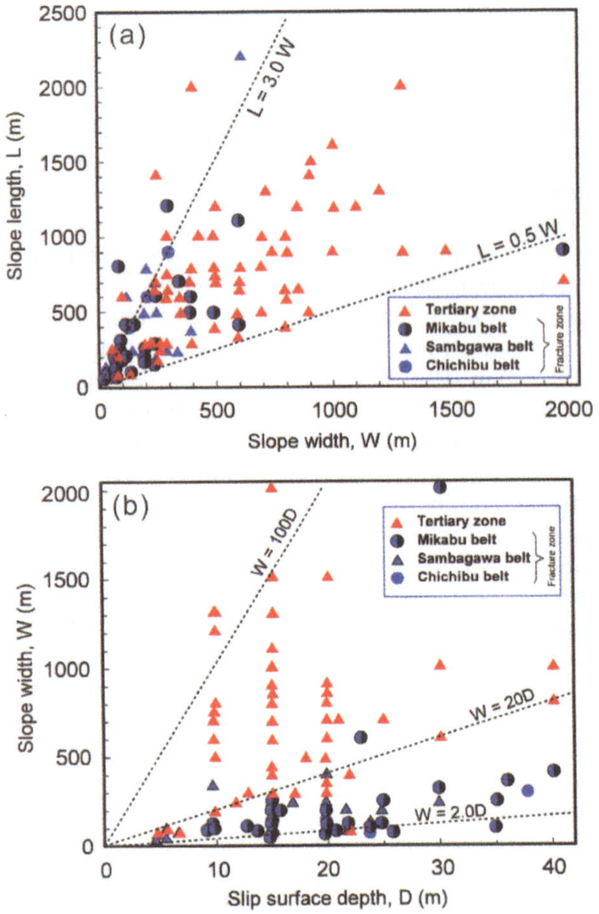

Fig. 13.4. Topographical features (slope width vs. slope length and slip surface depth vs. slope width) of fracture zone landslides (from Ishii 1994)

figure also shows that the average depth of sliding mass for the same width in the fracture zone is much higher than that in the tertiary zone. It is seen that the ratio of width to depth for most fracture zone landslides varies from $W = 2.0D$ to $20.0D$ (D represents the depth), whereas that for most tertiary zone landslides varies from $W = 20.0D$ to $W = 100.0D$. These width-depth relations make it clear that the landslides in the fracture zone have comparatively deep slip surfaces. In addition to large surface areas, which may range from thousands of square meters to a few square kilometers, landslide depths (or the depth of sliding mass) at the deepest point have been reported to be as high as 100 meters in the Shikoku Region (e.g., Kage landslide in Kochi Prefecture (Yube 2001)).

One another notable feature of the landslides in Shikoku is rate of sliding, which is regarded relatively faster than that for tertiary landslides (MLIT 1997). Ishii (1994) mentions that the rate of displacement of creeping landslides in the fracture zone is often found to go as high as several tens of centimeters a month, especially during rainy periods. With the lowering of groundwater level, however, the monthly rate of displacement comes down to only a few centimeters. Such slow displacements of the landslides take place under a condition when shear stress through the slip surface (i.e., slip layer) is slightly less than the shear strength of the slip layer soil, which causes a condition of slope instability with a factor of safety slightly greater than the unity. The failure of slip surface soil in such a condition is supposed to be due particularly to creeping behavior, in which a soil sample subjected to stresses equivalent to 90–95% of the shear strength in a longer time span deforms without resulting in complete failure.

13.3 Experimental Program

13.3.1 Material

The test samples were obtained from boring cores of over 20 landslide sites in Shikoku, most of which were from Sambagawa and Mikabu belts (Fig. 13.2) including a few samples from Chichibu belt. In addition to determination of physical properties of all collected samples, the laboratory investigation consisted of (*1*) shear strength measurement in ring shear apparatus and triaxial compression cell, and (*2*) identification of constituent minerals in the collected samples and quantitative evaluation of the constituent clay minerals with the help of X-ray diffraction analysis.

Figures 13.5 and 13.6 summarize a part of the physical properties of the tested clay samples (finer than 425 µm) in terms of grain size distribution, liquid limit, and plastic limit values. In Fig. 13.5, an overall picture of grain size distribution (in three gradations of finer than 2 µm, 2 to 20 µm, and 20 to 425 µm) of all collected samples has been plotted against the plasticity index values. As seen in the figure, the clay fraction (i.e., the percentage by weight of the soil particles finer than 2 µm) varies from a

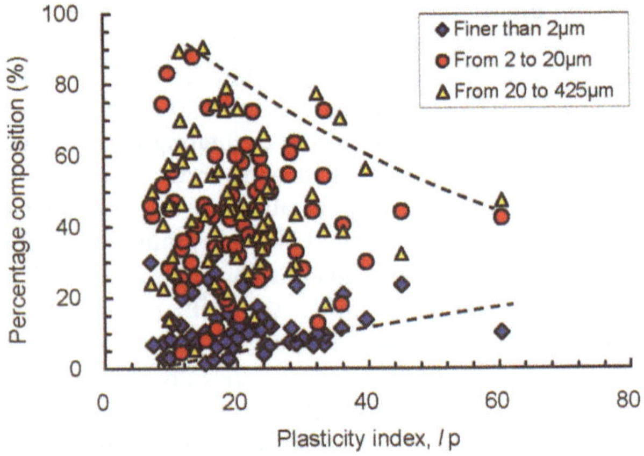

Fig. 13.5. Grain size distribution (finer than 425 µm) for tested landslide soil samples plotted against the plasticity index

few percent to 30 percent, and the amount of coarse particles (i.e., 20 to 425 μm) varies from a few to as high as 80 percent. Although in general the plasticity index value for a soil increases with higher percentage of clay fraction, this figure indicates that rather than due to increase in the percentage of clay fraction, the plasticity index seems to increase with the decrease in the percentage of coarse particles.

Figure 13.6, on the other hand, indicates the range for liquid limit and plastic limit for the collected samples. It is apparent that the plastic limit values remain within a range between 10% to 40%, while the liquid limit values vary from 25% to as high as 95%. Although there are slight indications that the liquid limit values increase with the increase in clay fraction value up to 15%, the scattered data indicate that the influence of clay fraction is less significant than other factors such as clay mineralogy.

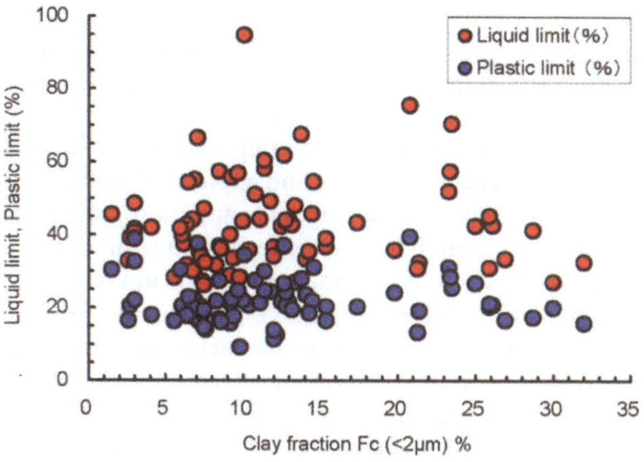

Fig. 13.6. Liquid limit and plastic limit values for the tested landslide soil samples plotted against clay fraction (<2 μm)

13.3.2 Method

13.3.2.1 Strength Tests

As stated above, ring shear tests and triaxial compression tests were planned for the strength measurement of the collected landslide clay samples. The ring shear tests were based on Bishop's ring shear apparatus (Bishop et al. 1971), and were performed on remolded reconsolidated specimens under drained conditions. The annular specimen in the ring shear apparatus measured 160 mm in outer and 100 mm in inner diameter with a thickness of 10 mm. Figure 13.7 shows a not-to-scale diagram of the ring shear apparatus used in the study, which was produced for the purpose of trial experimental studies. Similar to most other commonly employed ring shear apparatuses, it consisted of lower and upper parts. The shearing assembly was such that the lower part of the apparatus could be rotated with the help of speed regulated electric motor, while the upper part including the loading platen could only move but not rotate. The torque due to the rotation of the lower part could be transferred to the

Fig. 13.7. Ring shear apparatus employed in strength tests

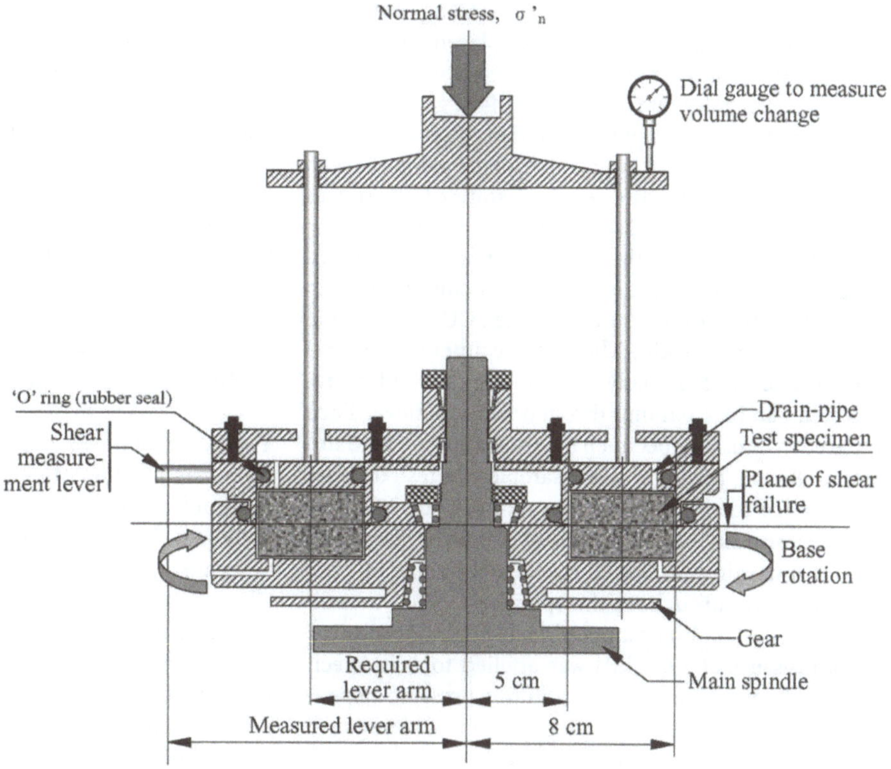

upper part through the specimen in the sample box and measured at the shear measurement lever in contact with an automatic load measuring device. Being a machine produced for the trial purpose, the expectations of mechanical friction, such as due to bearing contact, O-ring rubber seal contact, and contact between the loading platen rods and movable upper part of the apparatus, were high. However, it was confirmed that the friction remained below 30 N even during the application of a vertical pressure of 300 kPa. Moreover, sufficient drainage during shear (due to sample contraction) was ascertained by providing four drain holes through the loading platen as well as the lower base and using high permeability filter papers on top and bottom of the specimen.

The test specimens were prepared in the ring shear apparatus itself by one-dimensionally consolidating remolded pre-saturated samples passing through 425 μm sieve. Each specimen was sheared under three sets of effective pressure (i.e., vertical pressure, σ_v) of 98.1 kPa, 196.2 kPa, and 294.3 kPa. The rates of displacement for pre-peak and post-peak shears were set to be different because a faster shearing rate could produce excess pore-water pressure and a slower shearing rate could result in many days to achieve the displacement required for measuring the actual residual strength of the sample. Based on the method described by Garga (1971), the rate of shear until the peak strength value had been exhibited was set to be 0.044 deg min^{-1}. (i.e., 0.05 mm min^{-1}), which was then increased 10 times (i.e., 0.44 deg min^{-1} or 0.5 mm min^{-1}) for the post-peak shear until steady states of shear strength and volume change were confirmed. Depending on the state of consolidation and sample type, the displacement required to achieve the peak strength value ranged from 2 to 7 mm. Likewise, the angular rate of shear of 0.44 deg min^{-1} in the ring shear apparatus employed could produce a total displacement of 72 cm in 24 hours through the plane of shear, but for most samples the residual state of shear was achieved before a shear displacement of 50 cm.

The triaxial tests, on the other hand, were performed in isotropically consolidated-undrained conditions with pore-water pressure measurement (i.e., CU-bar tests) at controlled rate of strain. The test specimens were prepared by one-dimensionally consolidating remolded pre-saturated samples passing through 425 μm sieve. Each prepared cylindrical specimen measured 35 mm in diameter and 80 mm in height. Each sample was tested under three sets of total confining pressure (σ_3) of 98.1 kPa, 196.2 kPa, and 294.3 kPa. The rate of compression was set at 0.044 mm min^{-1}, at which the development of pore water pressure throughout the specimen would be uniform. To ascertain highest possible degree of saturation, a back pressure of 196.2 kPa was applied to each specimen, which raised the B-value (confining pressure-dependent coefficient of pore water pressure) up to 0.95 against a rise in confining pressure of 98.1 kPa.

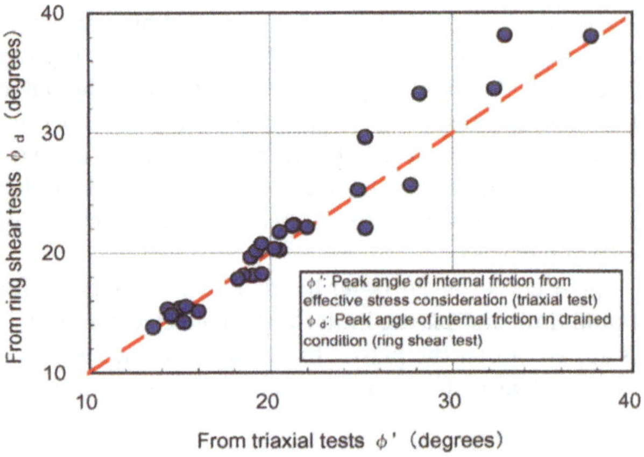

Fig. 13.8. Comparison of angle of effective internal friction (triaxial compression test) and angle of drained internal friction (ring shear test)

The main purpose of conducting triaxial compression tests was to ensure accuracy and applicability of the soil strength measured in the ring shear apparatus. Moreover, as the peak strength (drained strength) measured in ring shear apparatus could be erouneous due to unknown friction as well as unconfirmed excess pore water pressure development in the plane of failure during pre-peak shear, the results of triaxial compressions tests were employed to evaluate the peak shear resistance of the tested samples. Figure 13.8 shows a comparison of the angles of drained shear resistance for a few samples as measured in ring shear apparatus and the angle of effective shearing resistance as obtained from triaxial compression tests. The figure is evident that the drained shear strength (ring shear apparatus) and effective shear strength (triaxial compression test) are almost equal. Theoretically, the peak friction angle measured in ring shear apparatus at drained conditions must be close (if not equal) to the effective angle of friction measured in triaxial tests. It is because the total stress during fully drained conditions is equivalent to the effective stress. So, the choice was made of effective friction angle to represent the actual value of peak angle of friction for the tested samples.

13.3.3 X-ray Diffraction Analysis

Two basic methods employed to identify the constituent minerals of the collected soil samples were powder method and oriented aggregate method. The powder method of X-ray diffraction analysis, in which all size particles are crushed into fine powder, is meant for identifying all constituent minerals of the sample, while the oriented aggregate method, in which only finer than 2 μm particles are sampled from the soil-water suspension, helps to identify constituent minerals of clayey particles. Other relevant information in relation with the X-ray diffraction analysis method will be discussed in the following section.

13.4 Results and Discussion

13.4.1 Shear Characteristics

All strength test results have been expressed and presented in terms of angle of internal friction. In Fig. 13.9, the angles of internal friction as obtained from the triaxial compression tests (i.e. angle of effective internal friction, ϕ') and ring shear tests (i.e. residual angle of internal friction, ϕ_r) have been plotted against the clay fraction F_c. Various curves of F_c-ϕ_r relationship such as by Skempton (1964), Borowicka (1965), Binnie et al. (1967), and Blondeau and Josseaume (1976) have also been included for reference and comparison. No clear drop in internal friction angles with the increase in clay fraction value can be confirmed from this figure. This infers that unlike common natural clays, the landslide clays of fracture zone exhibit scattered correlations between strength and physical parameters.

In next attempt to see the influence of physical properties on strength behavior, the internal friction angles were plotted against the plasticity index (I_p), as shown in Fig. 13.10. For reference and comparison, various other data ranges (Vaughan et al. 1978; Bucher 1975; Kanji 1974; Seyček 1978; Fleischer 1972; Voight 1973) for I_p-ϕ_r relationship have also been included in this figure. The angles of residual internal friction for most tested samples lie within the range mentioned by Bucher (1975), but it is hard to draw a clear relationship between the plasticity index and angle of residual internal friction for the fracture zone landslide soils. The range of variation is very significant in compared with the results obtained for other natural clays such as by Vaughan et al. (1978), Kanji (1974), Seyček (1978) Fleischer (1972), and Voight (1973). As seen in the figure, the effective friction angles range from 20° to 40°, and the residual friction angles range from 10° to 25°. It is also worth noting that the drop from peak to residual values of friction angle is comparatively high. Figure 13.11 shows that such drops for the landslide soils in Shikoku range from a few degrees to as high as 20°. In general, the drop from peak to residual strength is considered to be higher for more plastic soils because higher drop means higher clay content and higher clay content means higher plasticity. However, it is interesting to note that the drops for the tested landslide soil samples do not definitely follow this pattern. Except for some, most samples exhibit scattered pattern.

In Fig. 13.12, the residual friction angles (ϕ_r) of only selected samples (the samples that were tested for constituent minerals, which are also plotted in Fig. 13.9) have been plotted against clay fraction, F_c. As also discussed above, the F_c-ϕ_r relation for these soil samples cannot be interpreted as what is generally regarded (such as Lupini et al. 1981; Skempton 1985). Rather than having a fine reduction in residual friction angle with the increase in

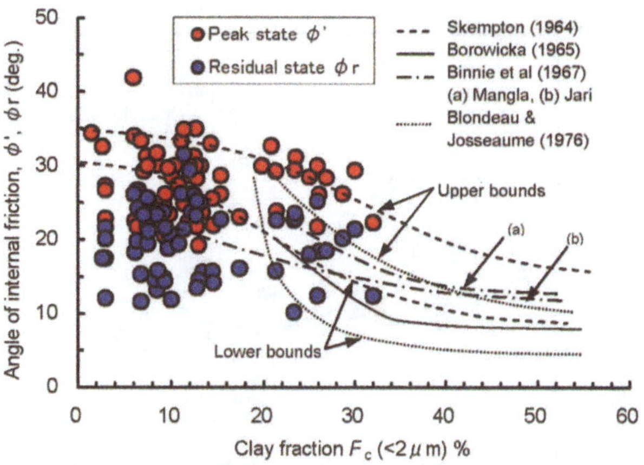

Fig. 13.9. Angles of internal friction for the landslide soil samples plotted against the clay fraction

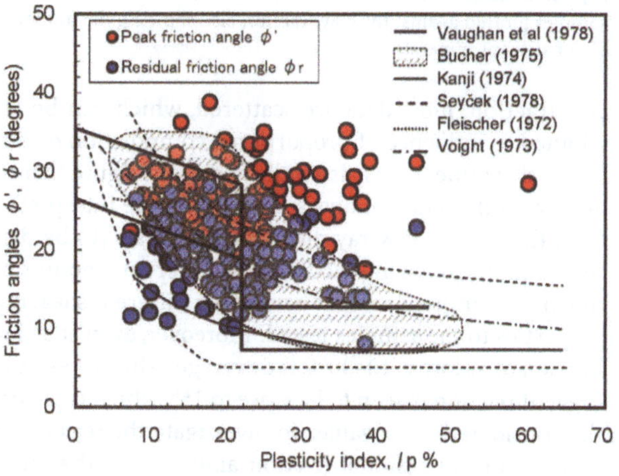

Fig. 13.10. Angles of internal friction for the landslide soil samples plotted against the plasticity index

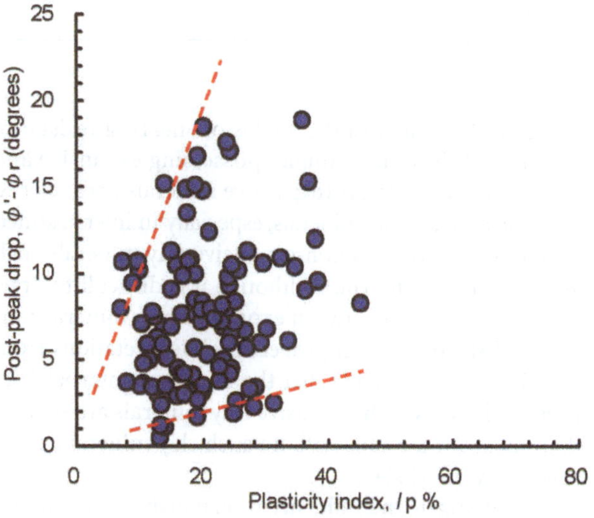

Fig. 13.11. Amount of drop from angle of peak internal friction to angle of residual internal friction

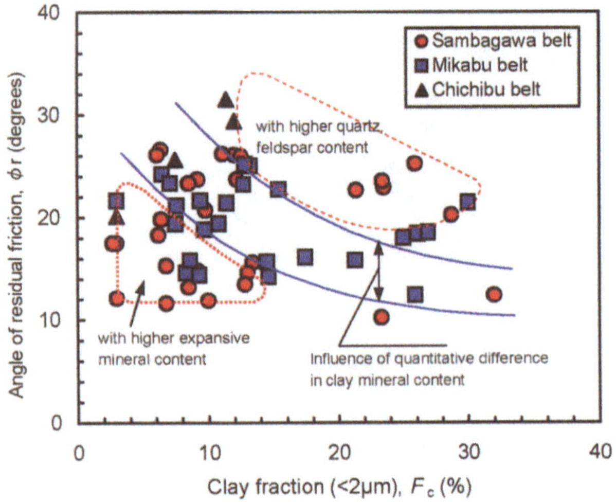

Fig. 13.12. Angles of internal friction for selected landslide soil samples plotted against the clay fraction also showing the influence of constituent minerals

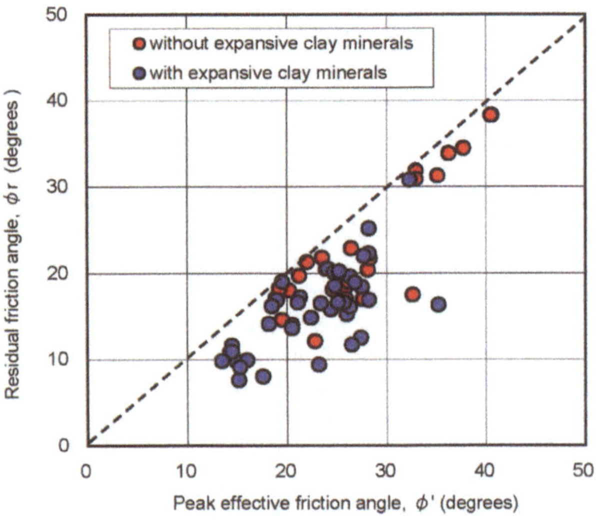

Fig. 13.13. Comparison of angles of internal friction for the samples with and without expansive clay minerals

clay fraction, most data are scattered, which can be attributed to influence of proportional composition of different clay minerals, as also indicated in the figure (based on X-ray diffraction analysis data). It can be interpreted from the results of X-ray diffraction analysis (to be discussed in the following section) that different amounts of constituent minerals might have given different shearing properties to the samples tested. Moreover, even at a clay fraction of as low as 10–30%, the average value of residual angle of friction is seen to be close to 15°, which is generally considered a low value. To investigate the reason for lower values of residual friction angle it was therefore essential to analyze the results of strength tests and X-ray diffraction analysis together, which is discussed in the following section.

13.4.2 Clay Mineralogy

Figure 13.13 compares the angles of shearing resistance for some of the tested samples possessing expansive and non-expansive minerals (expansive minerals here refer to smectite and like clay minerals, especially in interstratified state of expansive and non-expansive clay minerals such as smectite and chlorite). Although it is difficult to draw a clear relationship between expansive clay mineral content and shear strength properties, interpretations from this figure can be made that the angles of internal friction for the soils with expansive clay minerals are smaller than those for the soils with no such clay minerals, especially in residual state.

To investigate the influence of expansive clay mineral content on the strength, attempts were made to estimate the relative amount of the expansive clay minerals present. A method was adopted for the quantitative estimation of expansive minerals such as smectite and expansive chlorite. The term expansive chlorite here is used to represent a phase of chlorite mineral, in which certain portion of chlorite behaves like smectite in presence of water (i.e., interstratified chlorite-smectite mineral). Chlorites are formed generally by the alteration of smectite in presence of sufficient Magnesium (Mg^{2+}), which causes a crystalline brucite layer (octahedral structure with Mg) to replace the interlayer water (Mitchell 1976). It is however considered that the phase change also takes place from chlorite to smectite especially during mineral decomposition resulting in partial alteration of the mineral layers in chlorite to produce expansive chlorite, part of which behaves like smectite.

To estimate the relative amount of smectite in expansive chlorite, the use is made of X-ray diffraction patterns in powder method and glycol-treated oriented aggregate method. When an untreated sample containing expansive chlorite is placed on X-ray diffractometer, the peak for chlorite mineral, which has a basal spacing of 14 Å, is exhibited at a diffraction angle between 6.1°–6.4° and the peak for the expansive fraction of the mineral is exhibited near a diffraction angle of 6°. When treated with ethylene glycol, the position of the peak for the chlorite portion remains the same, whereas the position of the peak for the expansive fraction shifts to 5.2°–5.5°, which is a range of diffraction angle for glycolated montmorillonite (smectite). When a ratio of peak intensity for montmorillonite to that of chlorite near the diffraction angle of 6° is calculated, as illustrated in Fig. 13.14, it should represent the relative amount of smectite mixed with chlorite in a test sample (i.e., A/B in the figure). The data obtained from such calculations are plotted against the angle of shear-

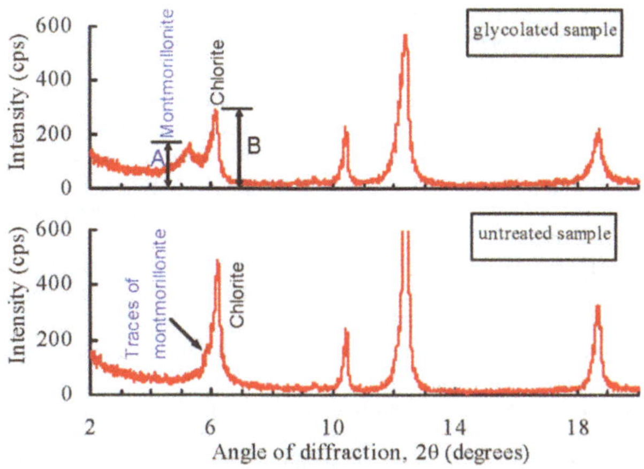

Fig. 13.14. Typical X-ray diffraction patterns for untreated sample and glycolated sample in oriented aggregate method showing the presence of smectite (montmorillonite) and the parameters (A, B) used in estimating relative amount of expansive clay mineral

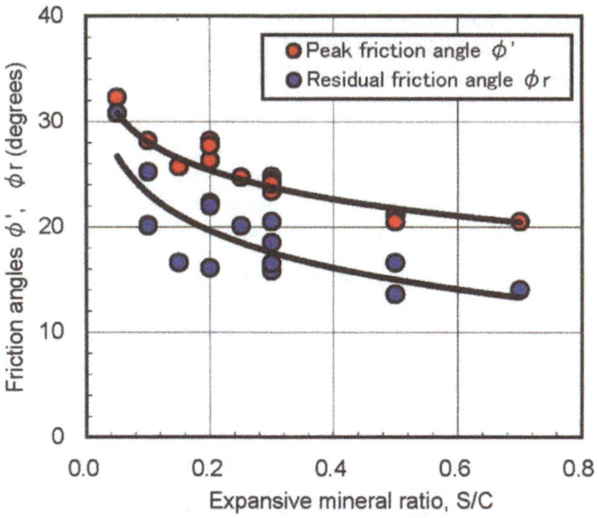

Fig. 13.15. Influence of expansive mineral ratio (S/C) on the angles of internal friction for selected landslide soil samples with smectite content

ing resistance for the tested samples in Fig. 13.15. The relative amount of expansive clay mineral is termed as expansive mineral ratio and is represented by a ratio S/C, in which S stands for smectite and C stands for chlorite.

The figure shows that the angles of shearing resistance for the tested samples decrease with the increasing values of the expansive mineral ratio. The data are slightly scattered, which can be attributed to experimental errors, especially with the ring shear tests. Moreover, the influence of some of the common constituent minerals of the tested samples such as mica, quartz, tremolite, and other non-clay minerals is not considered. So, if this effect is supposed to be minimal and the effect of expansive clay minerals is exaggerated, a drop of 10° in the angle of friction can be clearly seen against an increase of expansive mineral ratio by 0.5. Such a variation of angle of shearing resistance with the amount of smectite and similar minerals reveals that the degree of instability for a creeping landslide increases with increasing chances of other minerals turning into smectites. On the other hand, if the geochemical environment is favorable for smectite minerals to turn into other minerals like chlorite, the stability will increase. Therefore, it might be possible to stabilize creeping landslides to some extent chemically by introducing some substance in the slip surface soil to control the formation of weak clay minerals like smectites.

13.5 Concluding Remarks

This paper introduced the features of fracture zone landslides in Japan, where a large number of active and potential landslide sites have been identified and are under preventive efforts. The tectonic activities have led to formation of fracture zone underneath the Shikoku Region of Japan, which has given rise to abundant space for underground water that causes decomposition of the rock minerals leading to formation of weaker clay minerals in a layer of soil potentially acting as slip surface for the landslide. Smectites, chlorite, vermiculite, and illite are found as common clay minerals in the landslide clays in Shikoku. In this paper, certain attempts were made to analyze the strength properties of landslide clays based on the influence of expansive clay minerals such as smectite and expansive chlorite.

The peak effective angle of shearing resistance for the lab samples of landslide soils were measured to be 20° to 40°, and a significant drop of as high as 20° at the residual state was seen. Moreover, the drops from peak to residual values were not seen to have followed the increasing pattern with the increase in plasticity index. Mineralogy of the tested samples, on the other hand, was found to have significant influence in the residual strength of the tested samples. Semctite was detected in most samples but with great variation in the proportion. Samples composed of smectite were found to have lower angles of friction than those composed of non-expansive minerals. An analysis of angles of internal friction and relative amount of smectite in the tested samples showed a clear drop in the former with the increase in the latter. This implies that the angle of residual friction decreases with the increase in the amount of smectite minerals, and even an insignificant variation in the inclusion of expansive clay minerals significantly changes the strength properties. Finally, most fracture zone landslides, which have occurred in metamorphic deposits composed of black and green schist and sedimentary deposits composed of sandstone and shales, are probably active in creeping phase due to the influence of smectite like clay minerals.

Acknowledgment

The landslide clay samples were obtained from various landslide investigation and prevention authorities in Shikoku. Necessary information on landslide sites and preventive efforts were also made available by the local government authorities and consultants. The ring shear apparatus employed in the tests was developed with the help of Mr. Futagami, O., then Lab Technician of the Faculty of Engineering, Ehime University, who deserves special thanks from the authors.

References

Binnie MA, Clark JFF, Skempton AW (1967) The effect of discontinuities in clay bedrock on the design of dams in the Mangla Project. In: Proceedings of Trans 9th International Congress on large dams, Istanbul, 1, pp 165–183

Bishop AW, Green GE, Garga VK, Andresen A, Brown JD (1971) A new ring shear apparatus and its application to the measurement of residual strength. Géotechnique 21(4):273–328

Blondeau F, Josseaume H (1976) Measure de la resistance au cisaillement residulle en laboratoire. In: Bull Liaison Lab Ponts Chausśees Stabilite de talus 1, versants naturels, numero special II, pp 90–106

Borowicka H (1965) The influence of the colloidal content on the shear strength of clay. In: Proceedings of 6th International Conference on Soil Mechanics, Montreal, 1, pp 175–178

Bucher F (1975) Die Restscherffestigkeit natürlicher Böden. Ihre Einflussgrössen und Beziehungen als Ergebnis experimenteller Untersuchungen. Institut für Grundbau und Bodenmechanik Eidgenössische Technische Hochschule, Zürich, Report No. 103

Carraro F, Dramis F, Pieruccini U (1979) Large-scale landslides connected with neotectonic activity in the Alpine and Apennine ranges. In: Proceedings of 15th Plenary Meeting of the IGU-UNESCO Commission on Geomorphological Survey and Mapping, Modena (Italy), pp 213–230

Fleischer S (1972) Scherbruch und Schergleitfestigkeit von bindigen Erdstoffen. Neue Bergbautechnik 2(2):98–99

Garga VK (1970) Residual shear strength under large strains and the effect of sample size on the consolidation of fissured clay. PhD thesis, University of London.

GSJ (2003) Geological map of Japan 1:1,000,000, 3rd ed., CD-ROM version, digital geoscience map G-1. Geological Survey of Japan, AIST, 28 November, 2003, 2nd CD-ROM version

Ishii T (1994) Geotechnical study on stability and displacement features of the fracture zone landslides, PhD Thesis, Ehime University (unpublished, in Japanese)

JSSMFE (Japanese Society of Soil Mechanics and Foundation Engineering) (1985) Dictionary of words and phrases in Geotechnical Engineering, 6th ed. pp 126–127, 1991 (in Japanese)

Kanji MA (1974) The relationship between drained friction angles and Atterberg limits of natural soils. Géotechnique 24(4):671–674

Lupini JF, Skinner AE, Vaughan PR (1981) The drained residual strength of cohesive soils. Géotechnique 31(2):181–213

Mitchell JK (1976) Fundamentals of soil behavior. John Wiley & Sons, Inc., 442 p

MLIT (Ministry of Land, Infrastructure and Transport) (1997) Landslides in Japan. A booklet published by Sabo Publicity Center, Slope Conservation Division, Japan

Scheidegger AE (1970) Theoretical geomorphology. Springer-Verlag, Berlin Heidelberg

Seyček J (1978) Residual shear strength of soils. Bulletin of International Association of Eng Geol 17:73–75

Skempton AW (1964) Long term stability of clay slopes. Géotechnique 14(2):77–102

Skempton AW (1985) Residual strength of clays in landslides, folded strata and the laboratory. Géotechnique 35(1):3–18

Slivosky M (1977) Gravitational deformations of valley slopes in tectonically fractured rock masses. Bulletin of International Association of Eng Geol 16:114–118

Varnes DJ, Radbruch-Hall DH, Savage WZ (1989) Topographic and structural conditions in areas of gravitational spreading of ridges in the western United States, US Geological Survey professional papers, p 1496

Vaughan PR, Hight DW, Sodha VG, Walbancke HJ (1978) Factors controlling the stability of clay fills in Britain. Clay fills, Institution of Civil Engineers, London, pp 203–217

Voight B (1973) Correlation between atterberg plasticity limits and residual shear strength of natural soils. Géotechnique 23(2): 265–267

Yagi N, Enoki M, Yatabe R (1990) Consideration on mechanical characteristics of landslide clay landslides. In: Bonnard C (ed) Proceedings of the fifth International symposium on landslides, 10–15 July 1988, 1, pp 361–364

Yagi N, Yatabe R, Yokota K, Bhandary NP (1999) Strength of landslide clay from mineralogical point of view. In: Yagi N, Yamagami T, Jiang JC (eds) Proceedings of International Symposium on slope stability engineering, IS-Shikoku'99, Matsuyama, Japan, November 1999, 2, pp 701–704

Yatabe R, Yagi N, Enoki M, Nakamori K (1991a) Strength characteristics of landslide clay. J Jpn Landslide Soc 28(1):9–16 (in Japanese)

Yatabe R, Yagi N, Enoki M (1991b) Ring shear characteristics of clays in fractured-zone-landslide. J Jpn Soc Civil Eng, 436/III(16): 93–101 (in Japanese)

Yokota K, Yatabe R, Yagi N (1995) Strength characteristics of weathered serpentine. J Jpn Soc Civil Eng, 529/III(33):155–163 (in Japanese)

Yube M (2001) Mineralogical and geotechnical considerations on mechanism of landslides in mikabu belt, PhD Thesis, Ehime University, p 116 (unpublished, in Japanese)

Chapter 14

Landslides Induced by a Combined Effect of Earthquake and Rainfall

Kyoji Sassa* · Hiroshi Fukuoka · Fawu Wang · Gonghui Wang

Abstract. Many landslides are triggered by rainfall and many landslides are triggered by earthquakes. The probability that rainstorms (typhoons or hurricanes) and earthquakes attack same area is not high or rare. Combined effects of rainstorms and earthquakes have not been examined. The 2004 Mid-Niigata Prefecture earthquake (*M* 6.8) caused twelve landslides more than one million cubic meters, and many landslide dams were formed by large-scale displaced landslide masses. While, the 1995 Hyogo-ken Nambu earthquake (*M* 7.2) did neither cause any large-scale landslide, nor landslide dam although it has a greater magnitude and a similar depth of earthquake. One major difference is: a typhoon attacked Niigata Prefecture in three days before the earthquake, and the 1995 Hyogo-ken Nambu area was very dry before the earthquake. Combined effects of two triggering factors were examined for two cases which the authors investigated: the 2006 Southern Leyte landslide possibly triggered by a nearby small earthquake (*M* 2.6), and the Higashi-Takezawa landslide triggered by the 2004 Mid-Niigata Prefecture landslide.

Keywords. Earthquake-induced landslides, sliding-surface liquefaction, ring shear test, undrained test

14.1 Introduction

Major direct triggering factors of landslides are rainfall, earthquake, human activities which change pore-pressure or stress in the slope, although weathering and long-range creep may decrease the shear strength and cause landslides. Landslides are usually classified in types of movements and types of materials (Cruden and Varnes 1996). They can be classified by triggering factors as shown in Fig. 14.1, namely they are rainfall induced landslides, earthquake induced landslides, rainfall and earthquake induced landslides. Human activities such as mining and excavation or embankment cause landslides without rainfall or earthquake. Sometimes human activities, rainfall, earthquake are also combined. This table is not three dimensional, so human activities are not focused here because major interest is a combined effect of earthquake and rainfall. So far the combined effect triggering landslides have not been studied. However, the 2004 Mid-Niigata Prefecture earthquake (*M* 6.8) caused many of large-scale landslides and many landslide dams. Landslide dams posed the risk of dam failure resulting in debris flows. The Defense Agency and the Ministry of Land Infrastructure and Transport (MLIT) of Japan had to make great efforts to prevent failure of landslide dams for the protection of downstream towns. Similar magnitude of earthquakes relatively occurred in Japan, but no such risk was reported. Recent earthquakes in Japan occurred fortunately in winter, namely dry season. The 2004 Mid-Niigata Prefecture earthquake occurred late October, the rainy season and typhoon season passed by the month in ordinary years. But the year of 2004 was not usual, many typhoons attacked Japan until even late October. The typhoon No. 23 attacked this area in three days before the earthquake. Such large scale landslide disaster was likely due to the combined effect of earthquake and rainfall. Two recent landslides in Japan and Philippines are examined from the point of view of combined effect of earthquake and rainfall.

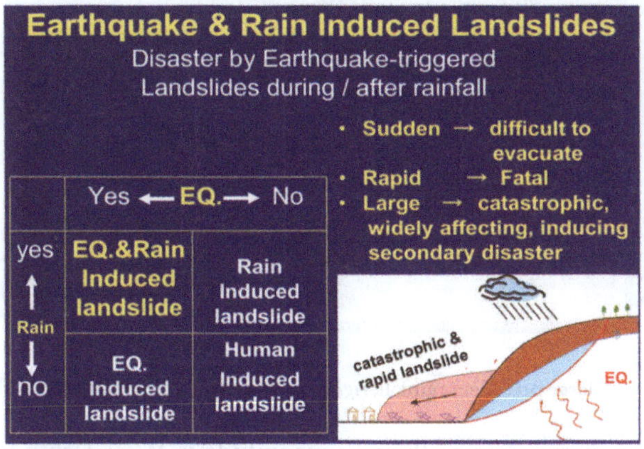

Fig. 14.1. Types of landslides classified by triggering factors and the characteristics of earthquake and rainfall induced landslides

14.2 Combined Effect of Rainfall and Earthquake

Figure 14.2 was made to present the significance of earthquake and rainfall induced landslides after the 2004 Mid-Niigata Prefecture earthquake disaster by focusing recent landslide disasters in Japan in 2003–2005. Figure 14.3 presents four examples of landslides in relation to rainfall, earthquake and types of induced landslides

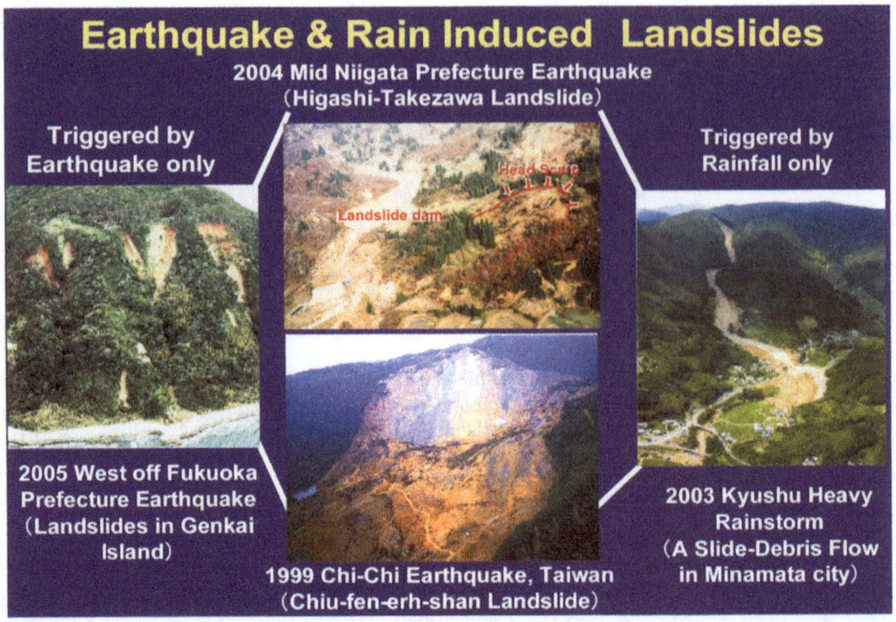

Fig. 14.2. Typical examples of earthquake-induced landslides (*left*), a rainfall-induced landslide (*right*), and two earthquake and rainfall induced landslides (*center*)

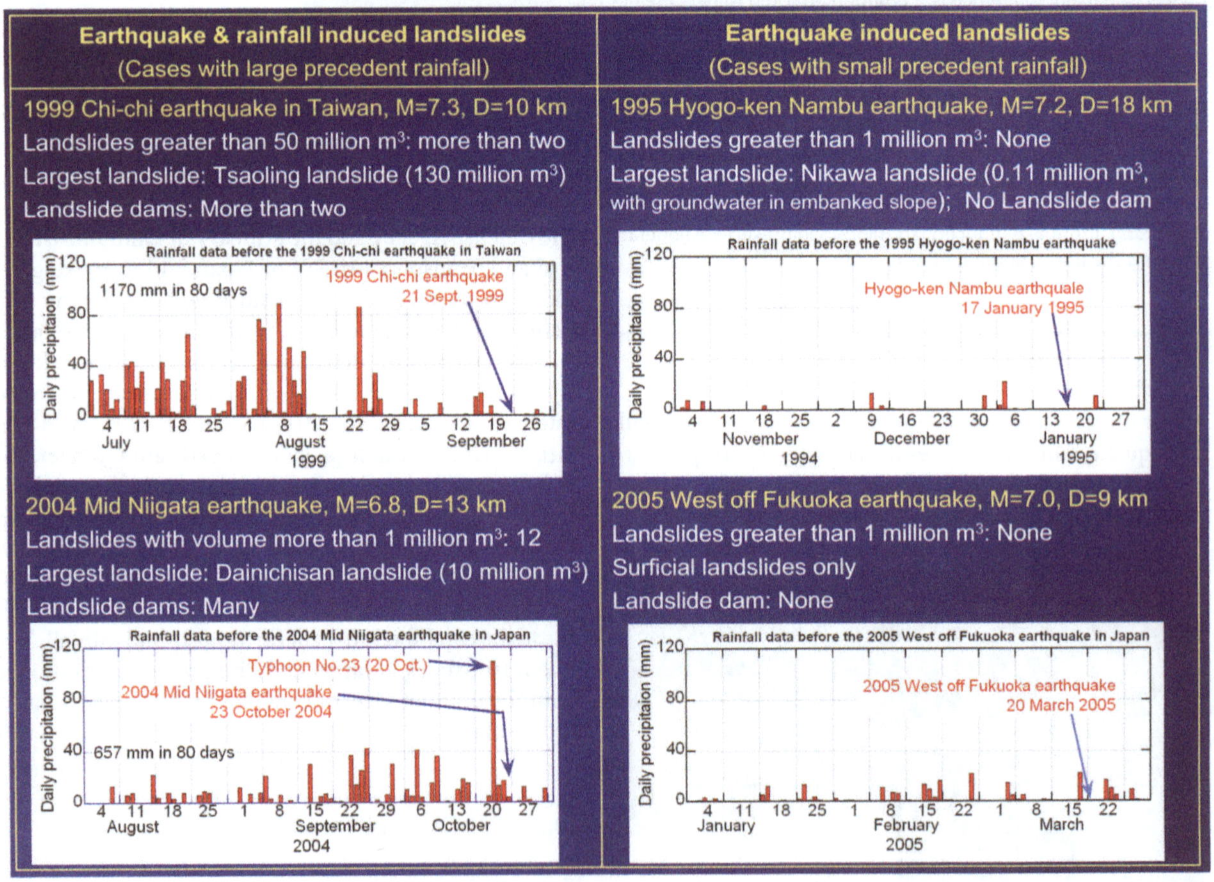

Fig. 14.3. Comparison of landslides triggered by earthquakes. *Left*: Cases with large precedent rainfall; *right*: cases with small precedent rainfall

in Japan and in Taiwan. In the left photo of Fig. 14.2, landslides triggered by the west off Fukuoka Prefecture earthquake on 20 March 2005 are shown. Surficial landslides occurred. The earthquake magnitude is M 7.0, Depth is 9 km. It was the dry season in southern Japan, therefore a small amount of rain was recorded. The Nikawa landslide was triggered by the 1995 Hyogo-ken Nambu earthquake ($M = 7.2$, Depth 18 km) on 17 January 1995 which was also in very dry season. Almost no rain fall was recorded before the earthquake as shown in Fig. 14.3. No major landslide occurred except the Nikawa landslide. The Nikawa landslide occurred in a slope filled in an original concave slope. Soils of the Osaka-group layer (limnic and marine deposits of granitic sands and clays in Pliocene to Middle Pleistocene) in the slope retained the ground water during winter. The ground water flowed out from the slope after landsliding (Sassa 1996; Sassa et al. 1996). A landslide with 0.11 million m^3 volume occurred in the filled soils and it killed 34 residents because of rapid and long travel motion. However, no greater landslide occurred and no landslide dam was made. On the contrary, the 2004 Mid-Niigata Prefecture earthquake occurred only three days after the typhoon No. 23. In Figs. 14.2 and 14.3, examples of landslides triggered by the 1999 Chi-Chi earthquake were introduced. The authors did not have statistics of landslides induced by this earthquake in Taiwan, however, at least two big landslides: the Tsaoling landslide (130 million m^3) and the Chiu-fen-erh-shan landslide (50 million m^3) were triggered, and both made large landslide dams. The rainfall just before the earthquake was not great, but the total rainfall during 80 days before the earthquake was more than 1 000 mm. Possibly such a long continued rainfall maintained the ground water in deep soil layers where sliding-surfaces of large-scale landslides were formed. The photo in the right side of Fig. 14.2 presents a slide-debris flow which was caused by a heavy rain (maximum rate was 91 mm h^{-1}, and total rainfall was 314 mm before landslide) in the rainy season of Japan and killed 15 people. The social impact of earthquake and rainfall induced landslides is great because of its sudden occurrence and rapid motion which do not allow people evacuate. Often earthquake and rainfall induced large-scale landslides form landslide dams because of its large-mass to fill the river and its long travel distance reaching to rivers. Failure of unstable landslide dams is very dangerous because they may trigger large-scale debris flows. The combined effect of earthquake and rainfall on initiation of rapid landslide should be investigated. Two cases are introduced from recent investigation in 2004 and 2006.

14.3 The 2006 Leyte Landslide Triggered by a Small Near-by Earthquake (M 2.6) Five Days after a Heavy Rainfall

14.3.1 Outline

A rapid and long-traveling landslide occurred on 17 February 2006 in the southern part of Leyte Island, Philippines.

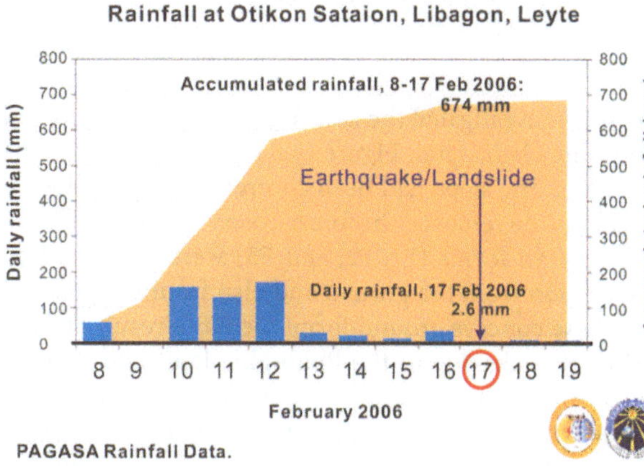

Fig. 14.4. Rainfall record at the nearest station from the landslide (PAGASA 2006)

The landslide resulted in 154 fatalities and 990 people disappeared in the debris. A Japanese and Philippine joint team investigated the site and took samples from the landslide. From field investigation and dynamic-loading ring-shear tests on a sample taken from the landslide site, we report here:

1. A small ($M = 2.6$), near-by earthquake was strong enough to trigger the initiation of the landslide which occurred in five days after a heavy rainfall. Figure 14.4 presents rainfall monitored by PAGASA (Philippine Atmospheric, Geophysical and Astronomical Services Agency). The peak of ground water level in the slope likely passed before the earthquake. Thus, this landslide was rainfall and earthquake induced.
2. The subsequent rapid motion of the landslide was a consequence of the "sliding-surface liquefaction phenomenon". It was the result of the generation of high pore-water pressure within the shear zone, which was caused by crushing of grains of the volcanoclastic debris at the site subjected to shearing under a high normal stress.

This landslide on Leyte Island was the greatest single landslide disaster in the world since the 2001 Las Colinas landslide (about 200 000 m^3) triggered by the El-Salvador earthquake ($M_s = 7.8$), which killed 747 people (Konagai et al. 2002; Fukuoka et al. 2002). A joint Japanese and Philippine team of 22 scientists and engineers investigated the landslide on 19–26 March 2006 with support from Japanese and international organizations through the International Consortium on Landslides (ICL). The sample taken from the site was tested using a new dynamic-loading ring-shear apparatus. The test result provided key information of the role of a small earthquake on the initiation of landslide and the mechanism of rapid and long-traveling motion which caused the catastrophic disaster.

14.3.2 Field Investigation

The team investigated the landslide from the ground and also from a helicopter. Figure 14.5 is a frontal view of the landslide. The slope had been covered by tropical forest. The landslide removed the forest-cover. Three linear concave ground forms (F1, F2, and F3) were identified by observation from the ground and also from the helicopter. These concave ground forms were the result of fracturing of rocks due to the Philippine fault and its subfaults. Along one of these concave ground forms (F3), a secondary landslide, which occurred after the main landslide, was found at point A in Fig. 14.1. The occurrence of the landslide suggests lower shear strength along these fractured zones. The central line of the landslide seemed to bend at B and C; it is likely a combination of two parallel lines (Fig. 14.5).

The section of the central line of the landslide was surveyed by a non-mirror total station and a ground-based laser scanner in the field and compared to a SRTM (Shuttle Radar Topography Mission) map before the landslide (Fig. 14.6). The red-color part shows the initial landslide mass while the blue-color part presents the displaced landslide debris after deposition. The initial landslide mass shown in red seems to consist of two blocks (Block 1 and Block 2) corresponding to the two straight lines above/below the bending part in Fig. 14.5. Though exact verification is difficult, we can speculate a landslide development process from the photo and section as well as the field observation: (*1*) The ground-water level rose upslope from point B in Fig. 14.5 because the ground-water flow probably was blocked at this bending point and likely dammed up upslope of this point; (*2*) this blockage necessarily lowered the stability of the lower part of Block 1 in Fig. 14.6; (*3*) the loss of stability in the lower part of Block 1 should

Fig. 14.5.
Location of the 2006 Leyte landslide (*bottom-left*) and a frontal view from a helicopter. *F1* was identified as the Philippine fault in the field. It was also reported by others (Catane et al. 2006; Lagmay et al. 2006). *A*: Secondary landslide, *S*: sampling point. Photo by Kyoji Sassa

Fig. 14.6. Central section (**a**) of the landslide and a flow mound (**b**) at the sampling point. Many flow mounds that maintained their original structure from the source area were found in the deposition area. The photo shows one of the big flow mounds from which samples were taken at its base (*S* in Fig. 14.5)

have almost simultaneously caused the motion of the rest of Block 1, including the head scarp, due to loss of support at its bottom, and, at the same time, the movement of Block 1 provided additional forces onto the top of Block 2 initiating the movement of Block 2; and (*4*) both landslide masses traveled onto the flat residential and farming areas. The inclination connecting the top of the initial landslide and the toe of the displaced landslide deposit is approximately 10 degrees, which indicates the average apparent friction angle mobilized during the whole travel distance. The value is much smaller than the usual friction angle of debris (sandy gravel) of 30–40 degrees. Therefore, it suggests that high excess pore-water pressure was generated during motion. Fig. 14.6b shows a flow mound that traveled from the initial slope to this flat area without much disturbance. Movement without much disturbance is possible when the shear resistance on the sliding surface is very low; thus, movement of the material is like that of a sled.

14.3.3 Sampling

The material of the flow mound is volcanoclastic debris, including sand and gravel. Though the sizes of flow mounds on this landslide were diverse, most of them consisted of volcanoclastic debris. We also observed from the surface and excavation that the valley-side slope in the source area consisted of volcanoclastic debris or strongly weathered volcanoclastic rocks. Therefore, we took a sample of about 100 kg from the base of the flow mound shown in Fig. 14.6 (S in Fig. 14.5). We subjected this sample to dynamic-loading ring-shear tests assuming the sample to be the soil in which the sliding surface was formed.

14.3.4 Testing Apparatus

The initial dynamic-loading ring-shear apparatus (DPRI-3) was developed by K. Sassa in 1992 (Sassa1992). Then it was improved and four versions with different specifications (DPRI-4, 5, 6, 7) were developed one by one from 1994-2003 (Sassa et al. 2004). The detailed testing procedure using DPRI-6 was explained in 2003 (Sassa et al. 2003). These apparatuses have been used to investigate the mechanisms of earthquake-induced landslides (Sassa et al. 2004, 2005) and rainfall-induced landslides (Sassa et al. 2004, 2005), debris flows transformed from slides (Sassa 2000; Sassa et al. 2004), and a landslide causing a tsunami (Boldini et al. 2005). The basic concept of this apparatus is illustrated in Fig. 14.7. The major characteristics of this apparatus that differ from those of conventional apparatuses (Bishop et al. 1971; Hungr and Morgenstern 1984; Garge and Sendano 2002) are: (*1*) pore-water pressure can be monitored during shearing by preventing any leakage of water from the contact between two rings during shearing; and (*2*) dynamic loading, up to 5 Hz, and high-speed shearing, up to 224 cm s^{-1} (DPRI-6), are possible.

The testing procedure of this apparatus is as follows: (*1*) a sample is taken from the soil layer in which the sliding surface was or will be formed; (*2*) the sample is set in the ring-shaped shear box, saturated, and consolidated under the stress due to the self-weight of the soil layer; (*3*) seismic stress or pore-pressure change due to rainfall will be provided; (*4*) if failure occurs, pore-water pressure and mobilized shear resistance with progress of shear displacement are monitored.

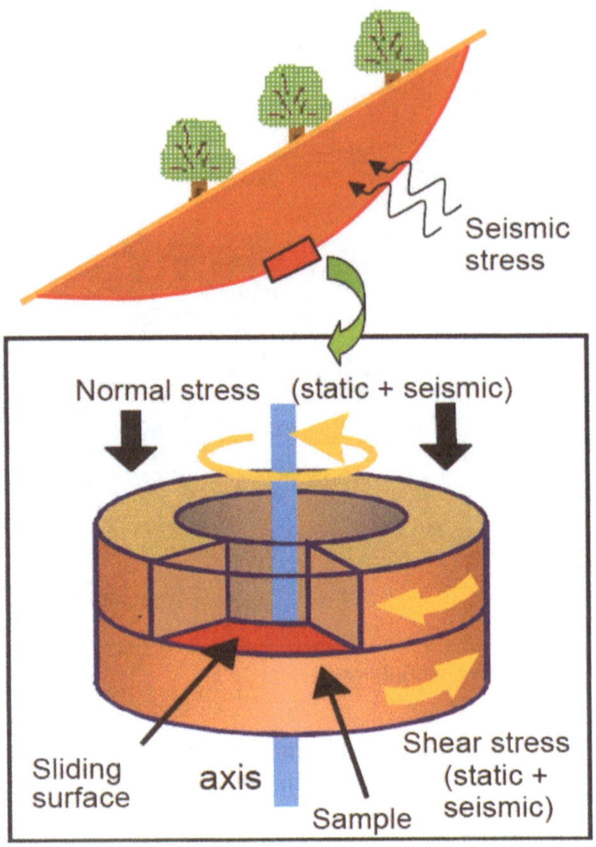

Fig. 14.7. Concept of the stress-controlled dynamic-loading ring-shear apparatus. This device aims to geotechnically simulate the formation of a sliding surface and the resulting post-failure motion by reproducing the stresses acting on the potential sliding surface in the slope: the static stress due to the self-weight of the soil layer, seismic stress due to an earthquake, and pore-pressure increase due to rainfall

14.3.5 Triggers of Landslide

We examined the triggering factors of the Leyte landslide. A small earthquake occurred near the site at the time of occurrence of the landslide, and the location of the hypocenter of this earthquake was estimated by the U.S. Geological Survey (USGS) and the Philippine Institute of Volcanology and Seismology (PHIVOLCS). According to PHIVOLCS, the earthquake occurred at a location (10.30° N, 124.90° E) 22 km west of the source area of the landslide, 6 km deep, with magnitude M_s 2.6, at 10:36 o'clock on 17 February 2006. Using the standard attenuation function between peak ground acceleration and hypocentral distance (Fukushima and Tanaka 1990), the peak ground acceleration at the landslide site was estimated at 10 gal for this magnitude. We then estimated the expected peak acceleration at the bottom of the landslide mass as about 60 gal. This was based on about three times amplification of ground accelerations at the sliding surface due to the difference in the compressional (P) wave speeds between soft volcanoclastic debris ($V_p = 0.5–1.5$ km s^{-1}) and hard volcanic bed rock ($V_p = 2.5–5$ km s^{-1}) that outcropped in the head scarp, because the amplification level is proportional to the velocity contrast between two layers. Though the shear (S) wave speeds of the volcanoclastic debris and the bed rock are unclear, similar level of velocity contrast to the P wave is expecting at the sediment/bed rock interface. An additional magnification of two times is also expecting in the landslide site due to the focusing of seismic waves on the steep mountain topography; this resulted in a total magnification of six times.

Heavy rainfall (459.2 mm for 3 days on 10–12 February and 571.2 mm for 5 days on 8–12 February 2006) occurred in this area before the day of the landslide as shown in Fig. 14.4. This rainfall should have increased the ground-water level and pore-water pressure inside the slope. However, the peak ground-water level had likely passed before the occurrence of the landslide on 17 February because the rainfall on 13–17 February was small. We simulated the ground-water level using a tank model that had been developed to simulate the ground-water level in the Zentoku landslide, Japan (Hong et al. 2005), which had a depth and inclination similar to that of the Leyte landslide. When inputting 10 days' precipitation records at the nearest monitoring station in Otikon (about 7 km west of the landslide) on 8–17 February, the peak ground-water level occurred on 13 February 2006. Because the peak ground-water level had already passed when the landslide occurred on 17 February, we deduced that a small earthquake was the trigger of the landslide.

14.3.6 Dynamic-loading Ring-shear Test

Based on this consideration, a dynamic-loading ring-shear test on the sample from the landslide was conducted as follows: The sample was set in the shear box (250 mm inside diameter, 350 mm outside diameter) of DPRI-6, was fully saturated, and was consolidated under the stresses acting on the sliding surface in the lower part of Block 1. From Fig. 14.6 it can be seen that the depth of the sliding surface at the lower part of Block 1 is 50–70 m. However, a lesser soil depth was assumed because of the capacity of this apparatus: the sliding surface was assumed for the test to be 35 m deep and at an inclination of 25 degrees. The unit weight of the soil was assumed to be 20 kN m^{-3}. In the preliminary test to increase pore-water pressure until failure, the critical ground-water level causing a landslide without earthquake loading in this slope was obtained. In the simulation test of a rain- and earthquake-induced landslide, the normal stress corresponding to that of 5 m lower than the critical ground-water level (i.e., further 5 m of ground-water level elevation will trigger the landslide) was first loaded on the sample in the ring-shear apparatus and consolidated. Then, the shear stress due to the self-weight of the soil layer was loaded. Finally, a peak seismic stress corresponding to a seismic acceleration of 60 gal using the ground acceleration observed in Maasin (PHIVOLCS, Code number: MSLP, Latitude: 10.1340, Longitude: 124.8590, Elevation: 50.0) was loaded on the sample.

14.3.7 Test Result and Mechanism

The test result is presented in Fig. 14.8. The black line in Fig. 14.8a shows the normal stress loaded on the sample, the red line shows the shear resistance mobilized at the sliding surface; the blue line shows the monitored pore-pressure change inside the sample; and the shear displacement at the center (300 mm diameter) of the ring-shape sample is shown in green. The static normal stress and shear stress working on the sliding surface before seismic loading are shown at time = 0 of Fig. 14.8a. Then, the estimated seismic stress acting on the potential sliding surface (a peak seismic stress corresponding to a seismic acceleration of 60 gal using the ground acceleration observed in Maasin) was loaded on the sample. The shear displacement started after 19 seconds of loading when the peak of shaking had passed. With progress of shearing, the pore-water pressure increased while the shear resistance decreased. Then, the pore pressure reached close to the normal stress and a steady-state of high-speed shearing proceeded at a very low shear resistance of only 7 kPa which corresponds 0.9 degrees in the apparent friction angle. This phenomenon is called "sliding-surface liquefaction" (Sassa 1996, 2000; Sassa et al. 1996).

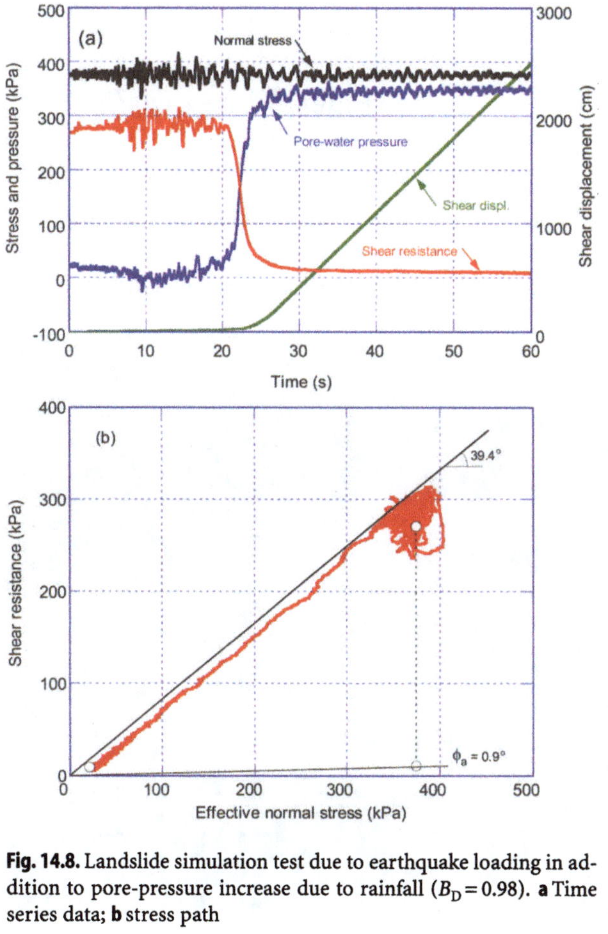

Fig. 14.8. Landslide simulation test due to earthquake loading in addition to pore-pressure increase due to rainfall ($B_D = 0.98$). **a** Time series data; **b** stress path

The volcanoclastic debris tends to dilate during seismic loading before the initiation of failure; so a negative pore-water pressure was monitored before the failure as shown in Fig. 14.8a. The shear displacement started after 19 seconds loading. Thereafter, the shear displacement increased and caused breaking and crushing of grains of volcanoclastic debris under the normal stress (Sassa et al. 2004). The grain crushing necessarily caused volume reduction, leading to a high pore-pressure generation. Pore-pressure continued to increase until a certain value under which no further grain crushing occurred. Every material has a certain value of effective normal stress (normal stress – pore-water pressure) under which grains can be sheared without any crushing. Thereafter, shear displacement continued to increase with a steady state of stress much smaller than the static shear stress due to self-weight of soil mass. The state of grain crushing during shearing was observed through the transparent shear box (DPRI-7). The photos were presented in (Sassa et al. 2004). The change of grain-size distribution after shearing was presented in (Sassa 1996, 2000).

This ring-shear simulation test of the Leyte landslide using the undrained stress-controlled dynamic loading ring shear apparatus clearly showed that even a very small earthquake can be the critical trigger of a landslide when the stability of the slope has already been reduced due to rainfall. Earthquake-induced landslides have been caused by heavy earthquakes, and they are convinced that strong earthquake shaking caused landslides. When the Leyte landslide occurred at the same time of a small earthquake, it is not easily convinced by both landslide researchers and earthquake researchers that such small landslide can trigger a landslide. Suwa (2006) mentioned the possibility that earthquake did not trigger the landslide, but the landslide caused ground shaking and it was monitored as an earthquake.

The combined effect of rainfall and earthquake should be studied in landslide risk analysis. A small earthquake can trigger a large-scale landslide when the ground water table is near a critical level to cause a landslide. A strong earthquake does not always trigger a large scale landslide when the ground water level is lower than the possible sliding surface. Another implication of this important case study is that a rapid and long-traveling landslide like the Leyte landslide can be predicted if the depth of possible landslide is estimated in advance and the shear behavior of the soil at the site is investigated by undrained stress-controlled ring shear test.

14.4 The 2004 Higashi-Takezawa Triggered by the Magnitude (*M* 6.8) Earthquake Three Days after the Typhoon No. 23

14.4.1 Introduction

The 2004 Mid-Niigata Prefecture earthquake (*M* 6.8) triggered 362 landslides more than 50 m wide, and 12 large-scale landslides of more than one million cubic meters, and the total volume of all landslides was about one hundred million cubic meters (Ministry of Land, Infrastructure and Transport 2005). It was the greatest earthquake disaster in Japan after the 1995 Hyogoken-Nambu earthquake (*M* 7.2). Mostly, the landslides were reactivation of past large-scale landslide masses. Forty five landslide dams were formed by this earthquake. Within those, two major landslides, the Higashi-Takezawa landslide and the Terano landslide which formed landslide dams more than 25 m high (Fig. 14.9) were investigated, This research presents the dynamic properties of these landslides and the mechanism of large-scale rapid landslides induced by earthquake within past landslide masses. In this section, the case of Higashi-Takezawa landslide is introduced as another example of landslide by the rainfall and earthquake combined effect.

14.4.2 The Higashi-Takezawa Landslide

Figure 14.10 presents the view of the Higashi-Takezawa landslide. It created the largest landslide dam that has posed the risk of debris flow disaster by dam failure. The head scarp of the current landslide is shown on the figure by red arrows, and the head scarp of the previous landslide is shown by a curved line with red arrows. The building in the foreground was an elementary school.

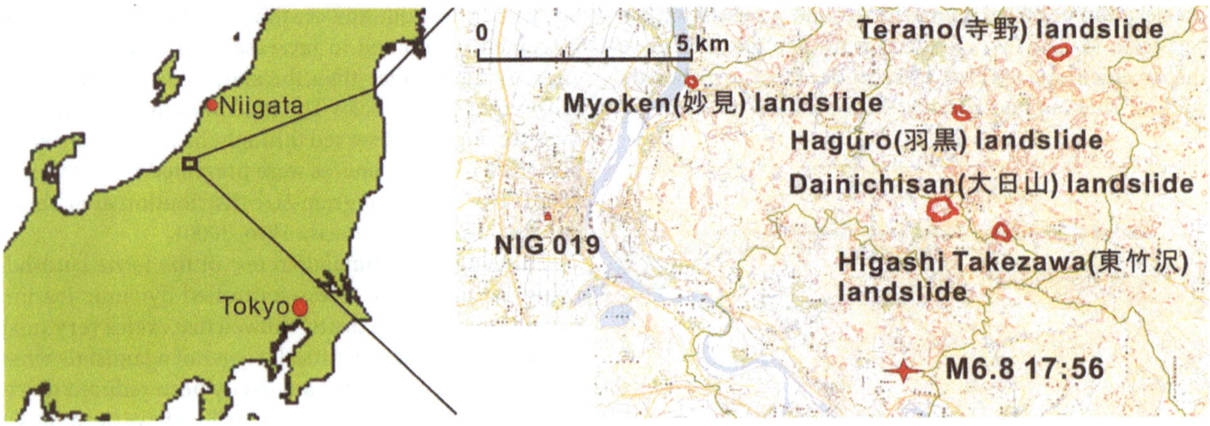

Fig. 14.9. Site map of the Higashi Takezawa landslide and Terano landslide with the epicenter and the seismograph gauge NIG019 of K-net

Fig. 14.10. Higashi Takezawa landslide and the head scarp of past landslides (taken by Sassa on 6 November 2004)

Fig. 14.11. Sampling from the sand layer (*left*) behind the sliding surface and over the stiff silt layer (*right*)

This landslide mass rapidly moved around 100 m, and hit the opposite bank of Imokawa River. A part of landslide mass spread cross the road and hit the school. We investigated the head scarp of this landslide. A very straight, gently-dipping (around 20 degrees) stiff silt (stone) layer outcropped as shown in Fig. 14.11 (right). This layer seemed to be a part of sliding surface of this landslide. It is relatively impermeable and groundwater flowed over this stiff silt layer. The sand layer over this stiff silt layer was soft. It was probably a part of previously moved landslide mass. A sample was taken from this sand layer behind the current sliding surface as seen in Fig. 14.11 (left).

The first question arose on the position of sliding surface from this observation. The outcropped silt (stone) layer was very stiff and apparently too strong to slide. However, the silt near the border to the sand layer could be soft due to weathering because an adequate groundwater flow existed. Accordingly there are two options for the location of sliding surface of this outcropped head scarp, (*1*) on the weathered top part of lower silt layer, (*2*) on the bottom of upper sand layer. This will be examined later in this paper based on the dynamic loading tests and a detailed section of this landslide.

Three days after (26 October 2004) the earthquake, an air borne laser scanning survey was conducted, and Fig. 14.12 presents its result. The gentle slope found in the contours clearly presented the previous landslide topography above the head scarp of this landslide. The central section A-A of the landslide is presented in Fig. 14.13 that includes an interpretation of landslide process based on the topographical map made by the air borne laser scanner after the landslide, the 1 : 25 000 topographical map before the landslide, and field observations. In Fig. 14.13, the black line presents the ground surface before the landslide, and the green line shows the ground surface after the landslide. The mass A (red dots) and the mass B (black dots) are the unstable mass before the current landslide movement.

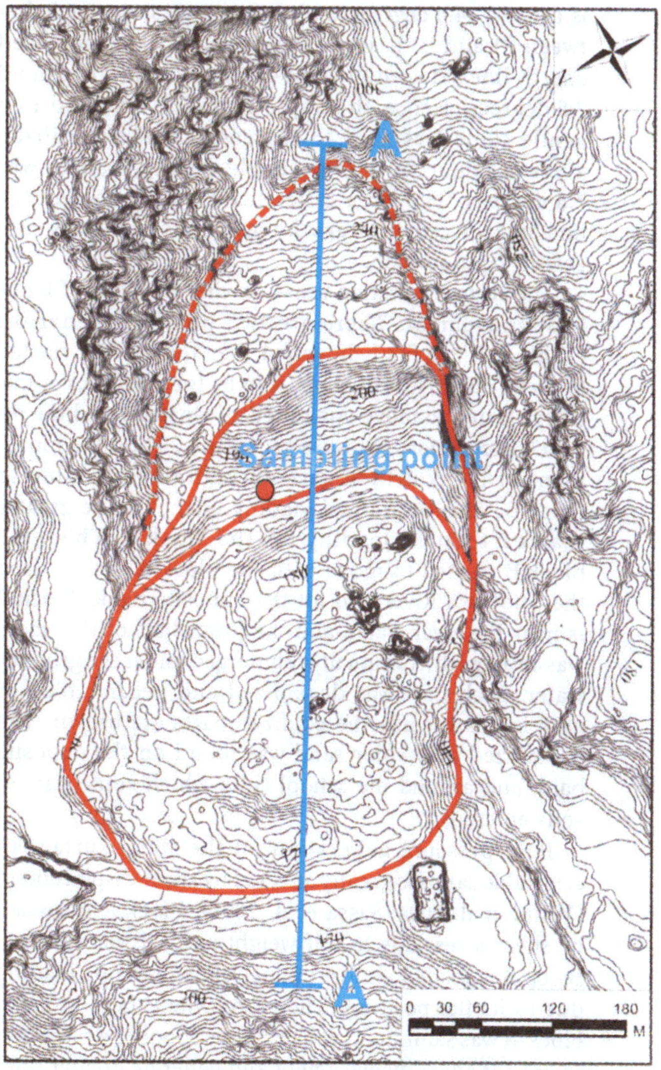

Fig. 14.12. Plan of the landslide (The map was made by the air borne laser scanning in the courtesy of Aero Asahi Corporation. Contour lines are 2 m spacing)

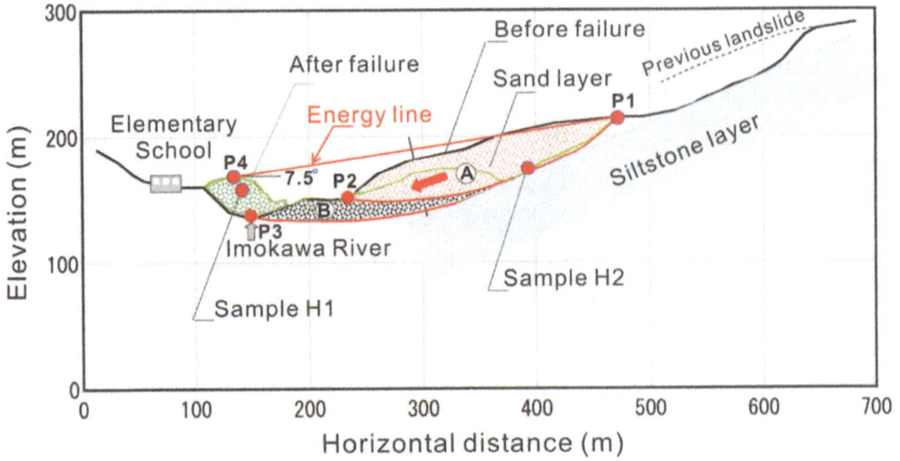

Fig. 14.13. Section of the landslide (the location of section is shown in Fig.12 as A-A line)

The initial average slope angle between P1 and P2 is around 14.6 degrees, and the average slope angle between P1 and P3 is around 13.5 degrees. The mobilized energy line between P1 and P4 (highest point) is around 7.5 degrees. Hence, this rapid landslide occurred in a gentle slope less than 15 degrees, and the mobilized average apparent friction angle during motion was 7.5 degrees. The large difference between the energy line and the center of gravity of the moving mass suggests a rapid motion.

The second question arose on the sequence of landslide movement. There are two possible scenarios: (*1*) block B was firstly liquefied, then the block A started to move due to loss of support in its toe; (*2*) block A initially started to move due to seismic loading and the resulted excess pore pressure generation. Thus, the movement of block A then gave an undrained loading to block B. This undrained loading effect can be greater than earthquake loading for the block B. Then, the block B started to move with the block A and crossed the Imokawa River and hit the opposite bank. Block B is near the river water, and the lower half of this mass was apparently continually submerged under the groundwater. Such part was deoxidized and became blue or gray in color, while the oxidized parts are brown. The deoxidized gray color sand was found on the opposite bank and around the school (into which some part of soils entered).

Pore pressure generation on the shear surface in crushable sands is greatly affected by shear displacement (Wafid et al. 2004; Sassa et al. 2004). Block A is loaded by shear stress due to self weight and seismic loading. Block B is loaded almost exclusively by seismic loading due to its flat position. Therefore, when the bottom of block A was saturated, the shear zone should be subjected to greater shear displacement and generate greater pore pressure compared to block B. There were many terraces along the river, but no liquefaction failure in such terraces near the river was noticed. Another reason is the hardness of sand in this area. The sand is a marine deposit from the Tertiary period. It is much stronger than tuff, pumice or weathered granitic sands, because those sands experienced a long downriver transportation and weaker parts was necessarily removed already through this flow with water. The depth of block B (around 13 m above the river bed) is much smaller than block A (around 40 m). Shaking or shearing under a greater overburden pressure could facilitate grain crushing and resulting pore pressure generation. The rapid movement landslides were the cases for more than 25 m in depth such as the Higashi-Takezawa landslide, the Terano landslide and the Dainichisan landslide (the biggest landslide in this earthquake). Accordingly, and based on this consideration, it has been concluded regarding the second question that block A moved firstly due to strong seismic shaking plus the static shear stress due to self-weight of the around 40 m thick soil layer. Therefore, dynamic loading ring shear tests were conducted to simulate the initial movement of block A.

14.4.3 Landslide Dynamics Tests

To investigate the dynamic properties of two earthquake-induced rapid landslides, dynamic (seismic and cyclic) loading ring shear tests were conducted by DPRI-5 apparatus, one of five dynamic-loading ring-shear apparatuses (DPRI-3,4,5,6,7) developed in DPRI, Kyoto University (Sassa et al. 2003 and 2004). The size of sample box in donuts shape is 120 mm in inside diameter and 180 mm in outside diameter, the maximum shear speed in the center of sample is 10 cm s^{-1}, maximum frequency of cyclic loading is 5 Hz, and the maximum rate of data recording is 200 readings per second. Used samples are the sand taken from the Higashi-Takezawa landslide, and the silt taken from a nearby landslide (Terano). The grain size distribution curves are presented in Fig. 14.14.

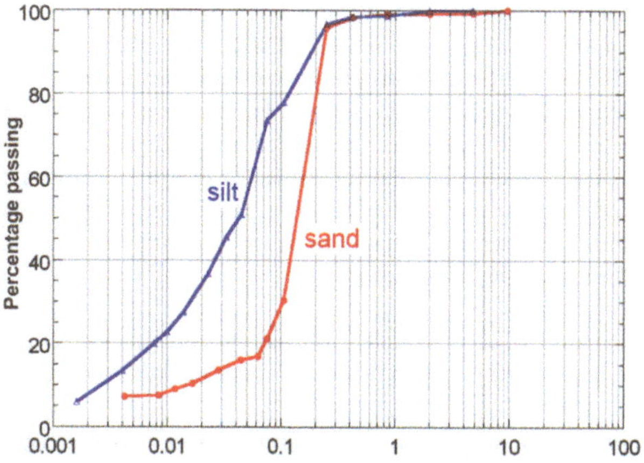

Fig. 14.14. Grain size distribution of sand and silt taken from the landslide sites

14.4.4 Real Earthquake Wave Loading Test for the Higashi-Takezawa Sand

Seismic loading test was performed to simulate the initiation of the Higashi-Takezawa landslide using the real monitoring record of the Mid-Niigata earthquake in the closest site. The monitored record is from one observation station of K-NET (Kyoshin Net), which is the strong earthquake motion monitoring network by the National Research Institute for Earth Science and Disaster Prevention (NIED). K-NET is a system which sends strong-motion data on the Internet. The data are obtained from 1000 observatories (25 km mesh) deployed all over Japan. The nearest monitoring site to the Higashi-Takezawa landslide is the observation station of NIG019 at Ojiya, around 10 km west of the Higashi-Takezawa landslide, and WNW 7 km from the epicenter of the main shock. The Higashi-Takezawa landslide is ENE 3.6 km from the Epicenter. The location of the epicenter is between the NIG019 and the Higashi-Takezawa landslides, but closer to the landslide.

The real acceleration and its wave form that affected the Higashi-Takezawa landslide can not be known in detail as it is complex due to inter alia topography, geology, underground structure, and distance from the fault and the epicenter. Therefore, the monitored earthquake record in NIG019 was applied as the first approximation of input acceleration to the sliding surface. Using three components of acceleration records, normal stress component and shear stress component on the bottom of landslide were calculated based on the section of the Higashi-Takezawa landslide. The estimated seismic stress acting on the potential sliding surface is shown in Fig. 14.15.

The test procedure is as follows:

The sliding surface is approximated as the one dimensional slope (20 degrees and 40 m deep, and a half of the layer was saturated, namely groundwater level was 20 m deep). The total density of soil layer was assumed to be 1.8 Mg m^{-3}. Sample was placed dry into the shear box and saturated by firstly replacing pore air by CO_2 gas and then replacing CO_2 by de-aired water. The pore pressure parameter ($B_D = \Delta u/\Delta\sigma$) was measured as the confirmation of full saturation ($B_D = 0.98$). Then, the shear stress before the earthquake on the sliding surface of 230 kPa was loaded under drained conditions. Thereafter, the shear box was shifted to the undrained state by closing the drainage valve, and then the calculated seismic stresses were applied. The data is shown in Fig. 14.15. The result of seismic loading of the sample was presented in the time series monitoring result of loaded normal stress, mobilized shear resistance (not the same with the applied shear stress because failure occurred in the sliding surface at the failure line), generated pore pressure and the resulting shear displacement as seen in Fig. 14.16a. Due to seismic shaking, pore water pressure was generated, then shear failure occurred and shear displacement started. As shear displacement progressed, a typical sliding surface liquefaction phenomenon was produced (Sassa 1996, 2000; Sassa et al. 2004), namely grain crushing along the shear zone proceeded and a higher pore pressure was generated in progress with shear displacement. The sand was originally a marine deposit and much stronger than volcanic deposits such as pyroclastic flow deposits and pumice, and also weathered granitic sands. However, the shearing stress under 40 m overburden pressure is enough to cause grain crushing and resulting volume reduction. It caused sliding surface liquefaction. The mobilized appar-

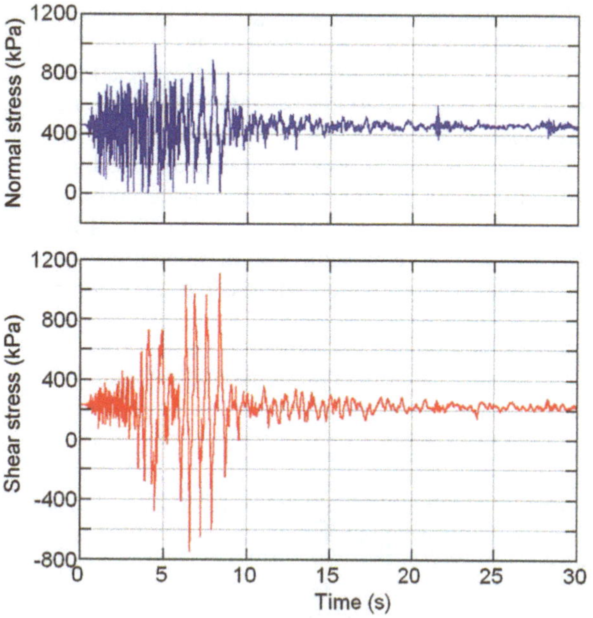

Fig. 14.15. Normal stress and shear stress during the Mid-Niigata earthquake working on the sliding surface of the Higashi-Takezawa landslide, which was calculated from the monitored earthquake record at the monitoring site NIG019 in K-NET (NIED)

ent friction angle is only 2.5 degrees at the steady state. The stress path of the test is shown in Fig. 14.16b (in which a red color stress path expresses the effective stress path and a green color stress path presents the total stress path). The apparent friction angle is obtained from the ratio of mobilized steady state shear resistance divided by initial normal stress. The effective friction angle mobilized in the post failure process is not so clear in Fig. 14.16b. So after this test, pore pressure was dissipated by opening the drainage valves. After full dissipation, sample was sheared under the speed control condition at 0.2 mm s^{-1}. Normal stress was reduced at the unloading rate of 0.5 kPa s^{-1} after reaching the peak shear resistance. The failure line mobilized in the post failure process (residual state) gave 36.9 degrees for the friction angle (Fig. 14.17).

A series of cyclic loading tests to compare the dynamic properties of the sand taken from the Higashi-Takezawa landslide and the silt taken from a nearby landslide (Terano) were conducted.

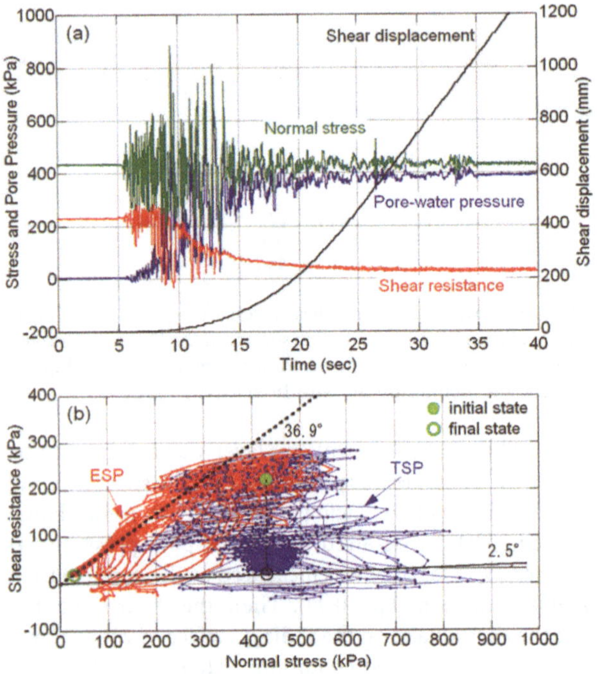

Fig. 14.16. Undrained ring shear test using the monitored earthquake-wave to simulate the Higashi-Takezawa landslide ($B_D = 0.98$). **a** Time series data; **b** stress path

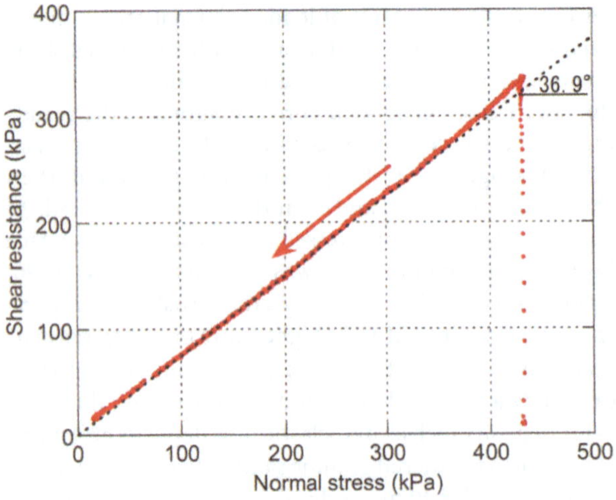

Fig. 14.17. Stress path obtained in the drained constant speed test after the real wave loading test (Fig. 14.16)

14.4.5 Cyclic Loading Test for the Higashi-Takezawa Sand

The sample preparation and test condition were the same with the above mentioned test. The main differences were the loading stress, frequency and wave form. During the test, normal stress was kept constant, because the effective normal stress practically would keep constant in the fully saturated undrained condition when subjected to any change in the normal stress. Shear stress with the wave form of sine curve at 1 Hz frequency was applied. To ensure the failure, the shear stress was increased step by step until 15 cycles. The test result for the Higashi-Takezawa sand is presented in Fig. 14.18. The sample was subjected to the sliding surface liquefaction as that shown in the real wave seismic loading test. The mobilized apparent friction angle at the steady state is 3.3 degrees, almost the same level with that shown in Fig. 14.16b. The failure line estimated from the effective stress path in the post failure stress path is 36.9 degrees, see Fig. 14.18b.

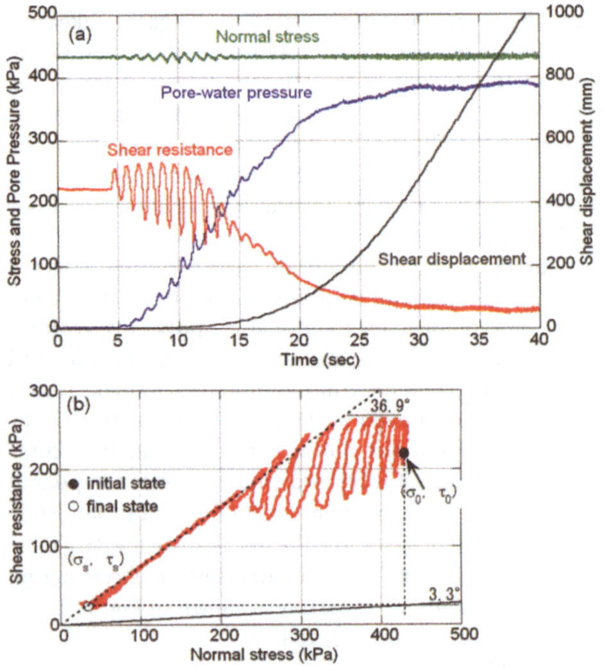

Fig. 14.18. Undrained cyclic loading test of the Higashi Takezawa sand ($B_D = 0.98$). **a** Time series data; **b** stress path

14.4.6 Cyclic Loading Tests for the Silt Taken from a Nearby Landslide (Terano)

The slope angle of the ground surface of the Terano landslide is 17.5 degrees at average from the toe of the slope at the Imokawa River to the top of head scarp. The slope condition for the cyclic loading test was approximated as a 17 degree slope with the depth of landslide mass above the sliding surface being 20 m. Around half of the landslide mass was assumed to be saturated, namely the groundwater table was assumed to be 10 m deep from the ground surface. During the test, normal stress was kept constant, and shear stress of sine curve of 1 Hz loading frequency was applied. The shear stress was increased step by step until 15 cycles. All test condition of cyclic loading was the same for the sand in Fig. 14.18.

The results of cyclic loading test on the silt is presented in Fig. 14.19. The dynamic properties of the Terano silt are quite different to the sands. The sliding surface liquefaction and its resulting rapid landslide motion were not produced in this test. Shear displacement was increased only during the cyclic loading, precisely say, only in the time span when the loaded stress reached or exceeded the peak shear strength. It was the repeated movements

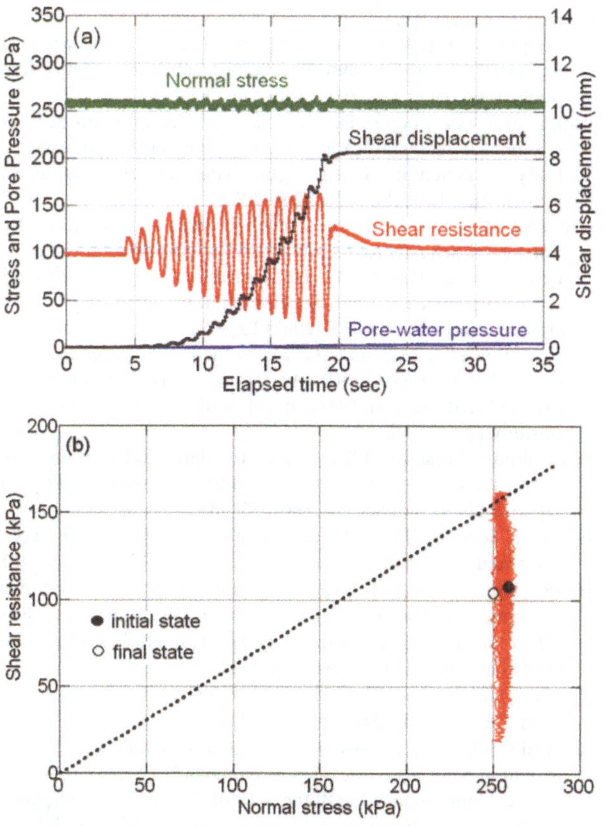

Fig. 14.19. Undrained cyclic loading test on the silt ($B_D = 0.98$). **a** Time series data; **b** stress path

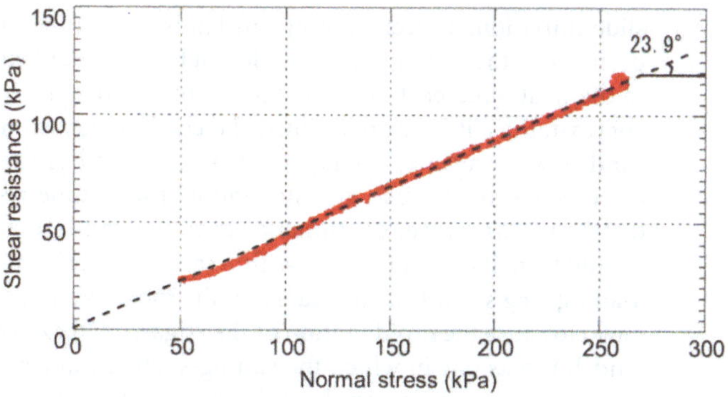

Fig. 14.20. The residual state of friction angle for the silt

of stop and move, and the shear movement stopped at the same time with the final cycle of loading. High pore pressure was probably not generated or at least not monitored as seen in Fig. 14.19a.

To obtain the effective friction angle in the residual state or the post failure condition, the sample was sheared at a constant shear speed of 0.05 mm s^{-1} by using the shear-speed-controlled method, while the normal stress was increased from 50 kPa to more than 250 kPa at a loading speed of 0.1 kPa s^{-1} in the drained condition. The residual friction angle was 23.9 degrees for the friction angle (Fig. 14.20). The friction angle was 13 degrees less than that of the sand (36.9 degrees in Fig. 14.17). The silts and sands display quite different shear characteristics.

14.4.7 Solution for the First Question

In the Higashi-Takezawa landslide, a stiff silt (stone) layer outcropped in the head scarp. And a very similar stiff silt (stone) layer outcropped in the Dainichisan landslide (Sassa et al. 2005). Therefore, whether the location of sliding surface is formed on the weathered part of silt or on the bottom of sand layer is of great importance for understanding the triggering mechanisms of these landslides. The weathered part of silt can not be found in this site of Higashi-Takezawa landslide (Fig. 14.11), probably it is very thin there. However, a thick weathered silt layer was found in the nearby landslide (Terano). The effective friction angle of the silt was much smaller than that of the sand. It means a landslide should occur in the silt layer when rain or melting snow is the triggering factor of landslides. This area is known as a Tertiary mud stone landslide area. Many landslides cover this area (Shimizu et al. 2004 or NIED Web page for landslide map). The area of Niigata Prefecture is known as a heavy snow area and snow melt is the usual main source of water supply for the ground. Slow and limited shear displacement is the common mode of the movement of landslides in this area. In the case of water supply due to snow melting or long rain, it is certain that the silt is much weaker than the sand for landslide initiation. Therefore, previous landslides probably occurred at the sliding surface in the silt layer, most likely in the weathered part of silt just below the sand layer. In contrast, the silt is strong against the earthquake, while sand is weak against earthquake loading. Sand grains are much easy to crush and susceptible for volume reduction. Therefore, earthquake-induced landslides should form its sliding surface within sand. The dynamic loading ring shear tests indicate that the most likely scenario to cause the rapid motion of the Higashi-Takezawa landslide was one in which the sliding-surface liquefaction phenomenon occurred within the sand layer under a high overburden pressure.

14.5 Conclusions

1. Two landslides triggered by the combined effect of rainfall and earthquake were investigated. One is the 2004 Higashi-Takezawa landslide triggered by a strong earthquake (M 6.8) in three days after a typhoon, and another is the 2006 Leyte landslide triggered by a nearby small earthquake (M 2.6) in five days after rainfall.
2. The earthquake triggering the Leyte landslide was very small. The peak ground acceleration inputting to the bedrock at the landslide site was estimated at 10 gal. The peak acceleration at the bottom of the landslide mass may be magnified due to the velocity contrast between the volcanoclastic debris and the bed rock, and also due to the focusing of seismic waves on the steep mountain topography. Then, 60 gal of the peak seismic acceleration in the real monitored wave was assumed and loaded after the ground water rise in the undrained stress-controlled ring shear test (Fig. 14.8). It could produce a rapid landslide motion with high pore pressure generation and the resulting very low mobilized shear resistance. Namely the sliding-surface liquefaction occurred in the shear zone in this test.
3. The Mid-Niigata Prefecture earthquake triggered twelve large-scale landslides more that one million m^3. The size of landslides were much greater than those triggered by previous similar scale of earthquakes in the dry season. A typhoon in three days before the earthquake increased ground water level in the sand layer in the slope. The saturated sand layer was subjected to a pore-pressure generation and the sliding-surface liquefaction during the seismic shaking and the post-failure shear displacement. The silt layer is always or almost saturated because of small permeability without typhoon or strong rainfall in the same area. However, the silt, even saturated, was proved to be strong against seismic loading by the dynamic loading ring shear test, on the contrary, it was weaker in the static loading than that of the sand. Thus, in the area with alternation of layers of sand and silt, the sliding surface may be formed in the silt layer in the case of rainfall or snowmelt, while it will be formed in the sand layer during earthquake.
4. A small earthquake can trigger a large-scale landslide when the ground water table is near a critical level to cause a landslide. A strong earthquake does not always trigger a large-scale rapid landslide when the ground water level is lower than the possible sliding surface. No excess pore pressure is generated during shaking and post-failure displacement in the shear zone. Thus, even though a landslide occurs, it does not exhibit catastrophic rapid movement.

References

NIED (National Research Institute for Earth Science and Disaster Prevention) (nd) Web for Landslide map: http://lsweb1.ess.bosai.go.jp/jisuberi/jisuberi_mini/jisuberi_top.html

Bishop AW, Green GE, Garga VK, Andersen A, Brown JD (1971) A new ring shear apparatus and its application to the measurement of residual strength. Géotechnique 21(1):273–328

Boldini D, Wang FW, Sassa K, Tommasi P (2005) Mechanism of landslide causing the December 2002 tsunami at Stromboli Volcano. In: Sassa K, Fukuoka H, Wang FW, Wang G (eds) Landslides-risk analysis and sustainable disaster management. Springer-Verlag, New York Heidelberg Berlin, pp 173–180

Catane SG, Cabria HB, Jr Tomarong CP, Jr Saturay RM, Zarco MAH, Pioquinto WC (2006) Catastrophic rockslide-debris avalanche at St. Bernard, Southern Leyte, Philippines. Landslides, Online first publication, DOI: 10.1007/s10346-006-0050-3

Chigira M, Duan F, Yagi H, Furuya T (2004) Using an airborne laser scanner for the identification of shallow landslides and susceptibility assessment in an area of ignimbrite overlain by permeable pyroclastics. Landslides 1(3):203–209

Cruden DM, Varnes DJ (1996) Landslide types and processes. In: Turner AK, Schuster RL (eds) Landslides: investigation and mitigation. Special Report 247, Transportation research board, US National research council, Washington, D.C., pp 36–75

Fukuoka H, Sassa K, Okada Y (2002) Experiment to reproduce the rapid flow phenomenon triggered by the January 2004 El Salvador. In: Proceedings APERIF (Areal Prediction of earthquake and rain induced rapid and long-traveling flow phenomena) Symposium, pp 301–312

Fukushima Y, Tanaka T (1990) A new attenuation relation for peak horizontal acceleration of strong earthquake ground motion in Japan. Bull Seismol Soc Am 80(4):757–83

Garga VK, Sendano JI (2002) Steady state strength of sands in a constant volume ring shear apparatus. Geotech Test J 25(4):414–421

Hong Y, Hiura H, Shino K, Sassa K, Fukuoka H (2005) Quantitative assessment on the influence of heavy rainfall on the crystalline schist landslide by monitoring system-case study on Zentoku landslide, Japan. Landslides 2(1):31–41

Hungr O, Morgenstern NR (1984) High velocity ring shear tests on sand. Géotechnique 34(3):415–421

Konagai K, Johansson J, Mayorca P, Yamamoto T, Miyajima M, Uzuoka R, Pulido NE, Duran FC, Sassa K, Fukuoka H (2002) Landslides caused by the January 23, Off the coast of El Salvador Earthquake. In: Proceedings International Symposium "Landslide Risk Mitigation and Protection of Cultural and Natural Heritage", Kyoto, pp 139–154

Lagmay AA, Ong JBT, Fernandez DF, Lapus MR, Rodolfo RS, Tengonciang AP, Soria JLT, Baliatan EG, Quimba ZP, Uichanco CL, Paguican ER, Remedio ARC, Lorenzo GRH, Avila FB, Valdibia W (2006) Scientists investigate recent Philippine landslide, EOS, Trans AGU 87(12):121

Ministry of Land, Infrastructure and Transport (2004) Quick report for Landslides triggered by 2004 Niigata-Chuetsu earthquake. http://www.mlit.go.jp/kisha/kisha04/05/051101_2_.html

National Research Institute for Earth Science and Disaster Prevention (NIED) (nd) Web of K-Net ttp://www.k-net.bosai.go.jp/k-net/data/

PAGASA (Philippine Atmospheric, Geophysical and Astronomical Services Agency) (2006) Precipitation record at Otikon. Libagon

Sassa K (1988) Geotechnical model for the motion of landslides. In: Special Lecture of 5th International Symposium on Landslides, "Landslides", 10–15 July, 1, pp 37–55

Sassa K (1992) Access to the dynamics of landslides during earthquakes by a new dynamic loading high-speed ring shear apparatus. In: Proc. 6th International Symposium on landslides, pp 1919–1937

Sassa K (1996) Prediction of earthquake induced landslides. In: Proceedings of 7th International Symposium on Landslides, A.A. Balkema, Trondheim, 17–21 June, 1, 115–132

Sassa K (2000) Mechanism of flows in granular soils. In: Proceedings of the International Conference of Geotechnical and Geological Engineering, GEOENG2000, Melbourne, 1, pp 1671–1702

Sassa K (2005) Landslide Disasters in the 2004 Mid-Niigata earthquake in Japan. Landslides 2(2):135–142

Sassa K, Fukuoka H, Scarascia-Mugnozza G, Evans G (1996) Earthquake-induced-landslides: distribution, motion and mechanisms. Special issue for the great Hanshin Earthquake Disasters, Soils and Foundations 53–64

Sassa K, Wang G, Fukuoka H (2003) Performing undrained shear tests on saturated sands in a new intelligent type of ring shear apparatus. Geotech Test J 26(3):257–265

Sassa K, Fukuoka H, Wang G, Ishikawa N (2004a) Undrained dynamic-loading ring-shear apparatus and its application to landslide dynamics. Landslides 1:7–19

Sassa K, Wang G, Fukuoka H, Wang FW, Ochiai T, Sugiyama M, Sekiguchi T (2004b) Landslide risk evaluation and hazard mapping for rapid and long-travel landslides in urban development area. Landslides 1(3):221–235

Sassa K, Fukuoka H, Wang FW, Wang G (2005) Dynamic properties of earthquake-induced large-scale rapid landslides within past landslide masses. Landslides 2(2):125–134

Sekiguchi T, Sato H (2004) Mapping of micro topography using airborne laser scanning. Landslides 1(3):195–202

Shimizu F, Oyagi N, Miyagi T, Inoguchi T (2004) Landslide Topography Map, vol. 17 (Nagaoka & Takada), 1/50,000 and 1/25,000

Suwa H (2006) Catastrophe caused by the 17 February 2006 Southern Leyte landslide in Philippine. Natural Disaster Science 25(1):83–97 (in Japanese)

Wafid MA, Sassa K, Fukuoka H, Wang GH (2004) Evolution of shear-zone structure in undrained ring shear tests. Landslides 1(2): 101–112

Chapter 15

Landslide Experiments on Artificial and Natural Slopes

Hirotaka Ochiai* · Toshiaki Sammori · Yasuhiko Okada

Abstract. An almost real-size slope model was used to study the initiation process of landslide fluidization during torrential rain. Experiments were conducted by filling an inclined flume with loose sand under the rainfall simulator to induce the sand to collapse. Both the movement, volumetric strain and the pore water pressure of the sand were monitored throughout the experiments, from the start of spraying to the cessation of the landslide. These experiments showed: (*1*) Landslide fluidization caused by undrained rapid loading undergoes three stages: compaction of the sand layer by the sliding mass from upper slope, generation of excess pore water pressure in saturated zone, and induction of fast shearing; (*2*) Fluidization at the collapse source area undergoes also three stages: destruction and compaction of sand layer skeleton by outbreak of shearing, increase of pore water pressure in saturated zone, and shift to hi-speed shearing, these three stages take place almost simultaneously.

An experiment to induce a fluidized landslide by artificial rainfall was conducted on a natural slope at Mt. Kaba-san in Yamato village, Ibaraki Prefecture, Japan. The experimental slope was 30 m long, 5 m wide, and the average slope gradient was 33 degrees. A landslide initiated 24627.5 s (410 min 27.5 s) after the start of sprinkling at a rainfall intensity of 78 mm hr^{-1}. The landslide mass was 14 m long and 1.2 m deep (at maximum). It first slid, then fluidized, and changed into a debris flow. The travel distance was up to 50 m in 17 seconds. The apparent friction angle of the fluidized landslide was 16.7 degrees. Formation of the sliding surface was detected by soil-strain probes. Motion of the surface of the failed landslide mass was determined by stereo photogrammetry.

Keywords. Landslide, fluidization, flume experiment, pore-water pressure, natural slope, sliding surface formation, stereo photogrammetry, soil surface movement

15.1 Introduction

Fluidized landslides, which travel long distances at high speed, are one of the most dangerous types of landslides (Sassa 2000). Fluidized slope movement takes place both in artificial cut slopes and natural slopes and often results in extensive property damage and significant loss of life (Sassa 1984, 1998). Many fluidized landslides have been observed in Japan, some of them have caused great disasters.

A debris flow that occurred in 1996 in Gamahara-zawa, Nagano, Japan, is an example of a fluidized landslide. The debris flow was triggered by the collapse at around 1 300 m in altitude, and moved approximately 39 000 m^3 of soil, of which 8 000 m^3 was deposited on the landslide section and 31 000 m^3 traveled down a slope 16.4° in mean inclination along a river for approximately 3 km. The debris flow eroded another 37 000 m^3 of soil, of which 15 000 m^3 was captured by an check dam. As a result, approximately 53 000 m^3 of soil reached the alluvial fan, which is 300 m in altitude, and 14 people were in the death by the debris flow (Committee Investigation for Gamahara-zawa Debris Flow Damage in Dec. 6, 1997). As this soil balance shows, the debris flow expanded the volume while traveling down the slope by gathering up river-bed sediment, and almost doubled in the amount of soil. Sassa (1998) called this phenomenon a landslide-induced debris flow. The fluidization of the river-bed sediment was caused by the rapid loading of the landslide and expanded moving mass led to the further fluidization of the lower part which increased the volume.

Liquefaction is an important mechanism in causing the fluidized motion of some landslides, where fluidization occurs along the sliding surface, or within the sliding zone during a rise in pore-water pressure which reduces shear resistance by decreasing the effective normal stress. Bishop (1973) noted that fluidization can be distinguished from general sliding, which usually has an intact soil mass above the sliding surface. Hutchinson (1986) noted that flow-like motion subsequent to fluidization is a neglected and little-understood group of movements with confusing terminology. Liquefaction phenomena as a result of cyclic loading have received much attention from many researchers since the drastic effects of liquefaction were noted after the 1964 Niigata earthquake Japan. Seed and Lee (1966), Yoshimi et al. (1977), Seed (1979), Ishihara et al. (1990) and (1993) discuss extensive laboratory soil tests attempting to reveal the liquefaction mechanism. Effects such as rainfall, as well as motion effects, can trigger fluidized landslides (Eckersley 1985, 1990; Sassa et al. 2004).

In order to reproduce a fluidized landslide at almost field scale, and to investigate its fluidization mechanisms, the National Research Center for Disaster Prevention, Tsukuba Japan, constructed an indoor rainfall simulator and large-scale flume model 10 m long, 4 m wide, and 1.2 m deep. By using such facilities, Fukuzono (1985) proposed that the initiation time of rainfall-induced slope failure was a function of inverse of shear velocity. Iverson and LaHusen (1989) pointed out that pore-water pressures within the soil packed in the slope model were dynamically fluctuating during the rapid shearing at failure. These findings were obtained from indoor flume tests in which the soil sample was quasi-uniformly packed. Iverson et al. (2000) verified the effects of void ratio on initial landslide speed by ring shear tests and flume experiments. They showed that shearing of a soil layer in large void ratio caused a sudden rise in pore water pressure and flow failure. Whereas, in a case of smaller void ratio, pore water pressure was reduced, soil mass was separated into blocks, some of which slipped discontinuously, did not flow in flume experiments.

These studies reviewed above were concerning to the mechanism and conditions for flow failure generation. Several theories have been proposed in them, such as the undrained shear in loose soil layer, rearrangement of soil particle blocks near the shear surfaces and particle breakage causing soil contraction and excess pore water pressure. However, the contraction of soil layers has not been observed during actual landslides and is merely a conjecture supported by soil tests and model experiments. Our study aimed to reproduce the contraction of soil layers and the generation of excess pore water pressure, both of which have only been reproduced inside soil testers, in a real-size landslide model, and to prove the fluidization mechanisms of landslide masses.

Landslide experiments on natural slopes by sprinkling or water supplying from trenches have been reported four times in Japan and United States of America. The respective conditions for each experiment are shown in Table 15.1. In our study, we conducted a rainfall-induced landslide experiment on a natural slope which had more complex and heterogeneous characteristics than the indoor models, in an attempt to investigate the dynamic movements of the soil surface, the formation of the sliding surface, and hydrological characteristics, based on the results of the indoor flume testing. The experimental slope was 30 m long and 5 m wide, and mainly covered by weathered disintegrated granite sand. Soil-surface movements were monitored by using stereo photogrammetry. Hence white-coloured targets were placed on the experimental slopes and the movements of these targets were traced by image analysis. To detect the formation of the sliding surface, soil-strain probes were inserted into the soil to 2 m depth at deepest. Tensiometers were used to measure changes in pore-water pressures within the soil.

Table 15.1. Soil properties of the sand for the experiments

Dry density (surface)	γ_d	1.33 g cm^{-3}
Void ratio (surface)	e	0.73
Saturated water content	w_{sat}	32%
Internal friction angle	φ'	30.6°
Cohesion	c'	0.75 kPa
50% diameter of soil particle	D_{50}	0.51 mm
Silt + Clay content		0.6%
Uniformity coefficient	U_c	3.5
Coefficient of permiability	k	1.2×10^{-2} cm s^{-1}
Approximate mobility index	AMI	1.14

15.2 Landslide Fluidization Process by Flume Experiment

15.2.1 Experimental Flume

A schematic diagram of the experimental flume is shown in Fig. 15.1. Figure 15.2 shows the model flume before collapse. The sizes are shown in the Fig. 15.1. One side of the flume was covered by reinforced glass to observe

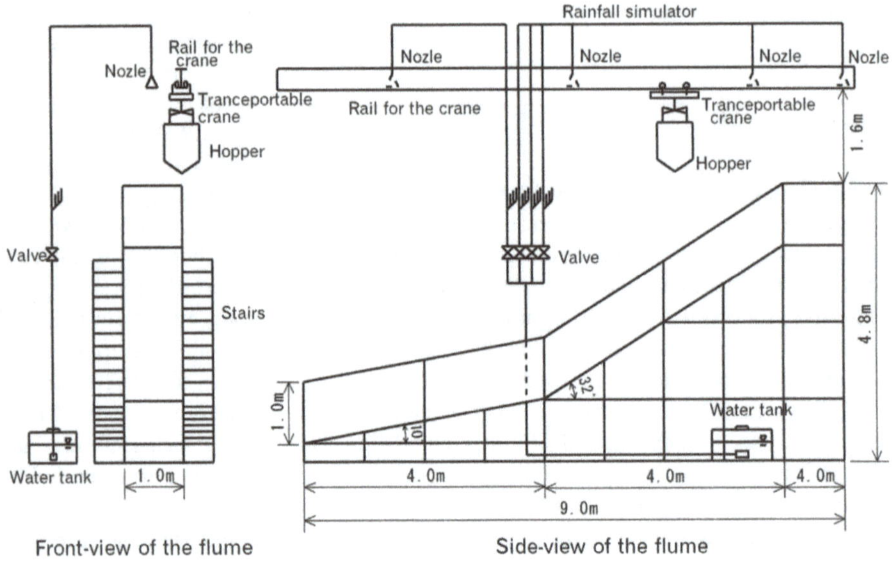

Fig. 15.1.
The front-view and side-view of the flume. Rainfall simulator and the rail for transportable crane are equipped over the flume

sand movement. The inclination of the slope was different between the upper and lower sections to induce collapse at the upper part of the slope and observe how the collapsed soil induces new slide on the lower. Sand was loosely filled into the flume by gently dropping the sand from a mobile hopper.

The rain simulator was equipped with nozzles at four points all 6.4 m above the floor along the flume. For all experiments, rain intensity of 100 mm hr^{-1} was used.

15.2.2 Observation System

The concept of the observation system is illustrated in Fig. 15.3. The system was developed to simultaneously monitor the change in pore water pressure and sand movement. Synchrony of data was attempted by generating time codes from a time-code generator, recording the codes together with pore water pressure measurements, and displaying them in video camera images of the sand movement.

Fig. 15.2. The flume before collapse

To measure dynamically fluctuating pore water pressure, a small cylindrical pressure transducer (8.3 cm long and 1.8 cm in diameter) was embedded within the sand. The transducer moved together with the sand but did not hinder the movement.

The movement of the sand was filmed with three video camera sets located along the glass side of the flume, which filmed the 2–4 m, 4–6 m, and 6–8 m lengthwise sections. The movements of several points within the sand layer were traced by burying cylinders of 3.0 cm in length and 1.6 cm in diameter within the sand so that one bottom side of each cylinder was adjacent and visible through the glass. The display of the time-code generator was placed along the flume so that the elapsed time codes were also filmed in the video camera images. The codes were also recorded in the data files of pore water pressure. Pore water pressure measurements were conducted at 100 Hz, and 30 video camera images were recorded per second.

15.2.3 Sand Layer Shapes and Configuration of the Instruments

Four experiments were conducted by changing the thickness of the sand layer at the upper and lower slope sections. The thickness of the sand layer at the upper and lower slope sections was 0.5 m and 0.5 m in Experiment 1, 0.5 m and 0.7 m in Experiment 2, 0.7 m and 0.5 m in Experiment 3, and 0.7 m and 0.7 m in Experiment 4, respectively. Figure 15.4 shows the initial sand layer cross-sections and the positions of the instruments in Experiment 1. The properties of the sand used for the experiments were tested in advance.

Fig. 15.3. The measurement system

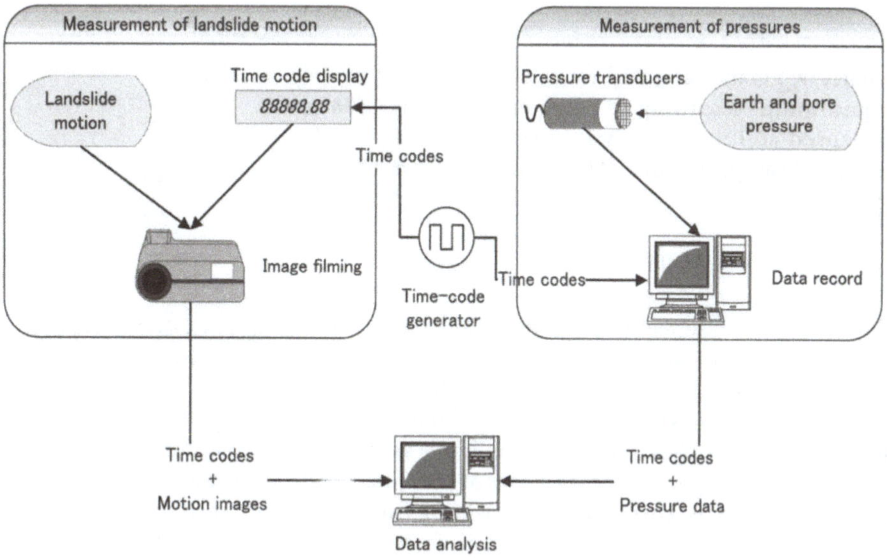

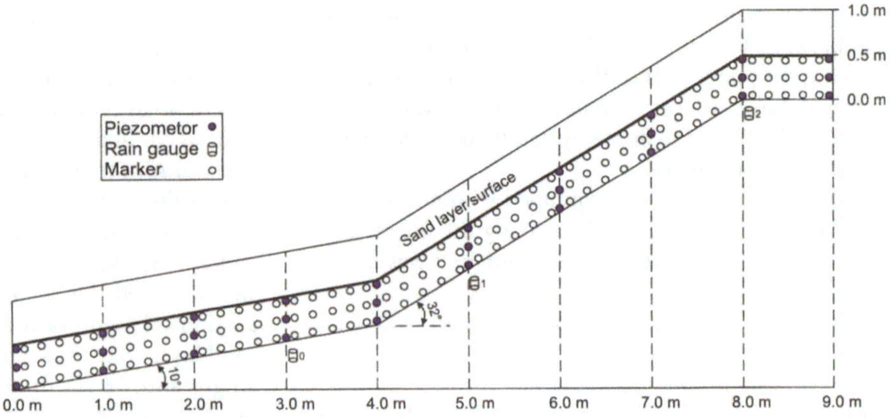

Fig. 15.4. The side-view of sand layer and configuration of instruments in Experiment 1

Table 15.2. The thickness of sand layer, time of collapse initiation and mean water content in upper slope right before collapse in each example of experiment

Exp. No.	Upper slope thickness (m)	Lower slope thickness (m)	Accumulation amount of rainfall (mm)	Water content in upper slope (%)
1	0.5	0.5	81.5	27.1
2	0.5	0.7	72.2	26.1
3	0.7	0.5	108.2	27.3
4	0.7	0.7	111.2	27.3

15.2.4 Results and Discussion

15.2.4.1 Properties of the Sand Used for the Experiments

The same sand was used for all experiments. The results of soil tests are listed in Table 15.1. The internal friction angle φ' and cohesion c' were derived by direct shear tests. AMI values were determined with the following equation:

$$\text{AMI} = \frac{w_{\text{sat}}}{w_{\text{L}}} \tag{15.1}$$

where, w_{sat} is the saturated water content and w_{L} is the liquid limit. According to Ellen and Fleming (1987), collapsed soil mass easily fluidizes when AMI > 1.0. Our sand sample satisfied AMI conditions of easily fluidizing soil.

15.2.4.2 Cumulative Rainfall until the Collapse

The thickness of the sand layers, the cumulative amount of rain, and the mean water content of the upper slope section are listed in Table 15.2 for each experiment. In all experiments, the sand layer collapsed from the upper section. The table shows that the cumulative amount of rain until the collapse and the thickness of the upper sand layer have a positive correlation. The wetting front travels vertically downward from the sand surface at a constant speed. When the wetting front reaches the bottom of the flume, the water forms a water table and destabilizes the slope. Therefore, there is a positive correlation between the thickness of the sand layer and the cumulative rainfall until the formation of the water table.

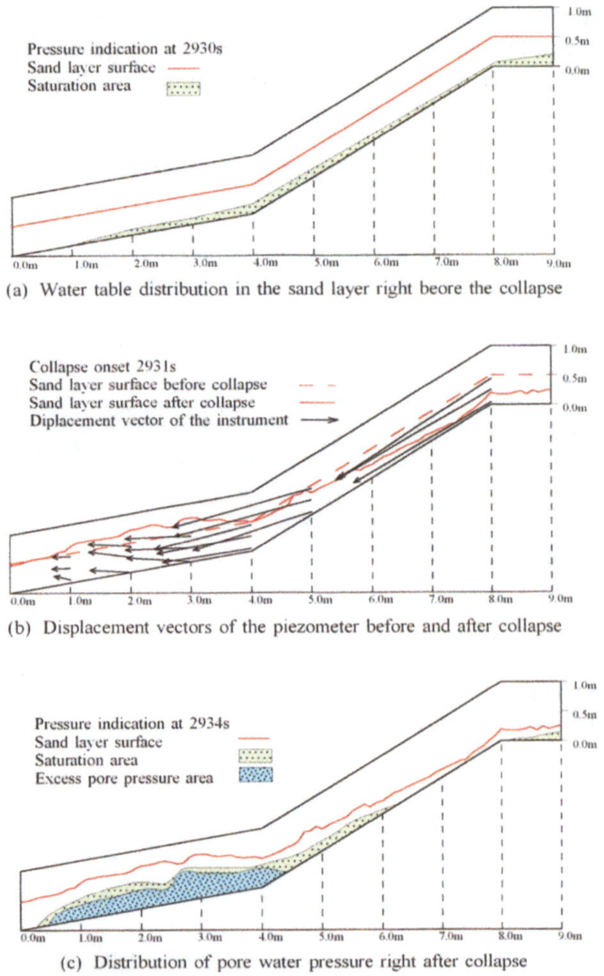

(a) Water table distribution in the sand layer right beore the collapse

(b) Displacement vectors of the piezometer before and after collapse

(c) Distribution of pore water pressure right after collapse

Fig. 15.5. The side view of the sand layer right before and after collapse in Experiment 1

15.2.4.3 Landslide Motion and Pore Water Pressure Distribution

Figures 15.5 to 15.8 show: (*a*) the water table within the sand layer immediately before the collapse; (*b*) the form of the sand layer surface and the displacement vector of the pressure transducer embedded within the sand before and after the collapse; and (*c*) the pore water pressure distribution within the sand layer immediately after the collapse and sedimentation, for Experiments 1 to 4.

Figure 15.5 shows the results of Experiment 1, which used uniform sand layer thickness of 0.5 m. As shown in Fi. 15.5a, the water table was uniformly distributed over the flume floor at both the upper and lower slope sections. Figure 15.5b shows that the slip surface was generated along the water table. As shown in Fig. 15.5c, the excess pore water pressure after the collapse was generated in a compressed sand section. Here, the section of excess pore water pressure is a section below the water table and showed water pressure values larger than estimated hydrostatic pressure.

Figure 15.6 is for Experiment 2, which was 0.5 m in sand thickness at the upper section and 0.7 m at the lower section. The water table was found only in the upper slope (Fig. 15.6a), and the wetting front did not reach the bottom of the flume at the lower section (Fig. 15.6b). Therefore, the slip surface was observed only in the upper slope section, and the excess pore water pressure was recorded in a compressed section near the border of the lower and upper sections (Fig. 15.6c).

On the other hand, Experiment 3 (Fig. 15.7) was 0.7 m at the upper section and 0.5 m at the lower section. The water table was far from the bottom at the lower section (Fig. 15.7a). Since the upper slope section was thicker than the lower section, a relatively heavy mass acted on the lower sand layer, and the traveling soil flowed over the wall at the lowest side of the flume. The excess pore water pressure was observed in the compressed section near the bottom side.

In Experiment 4 (Fig. 15.8), the thickness of the sand was 0.7 m throughout the flume. The water table was ob-

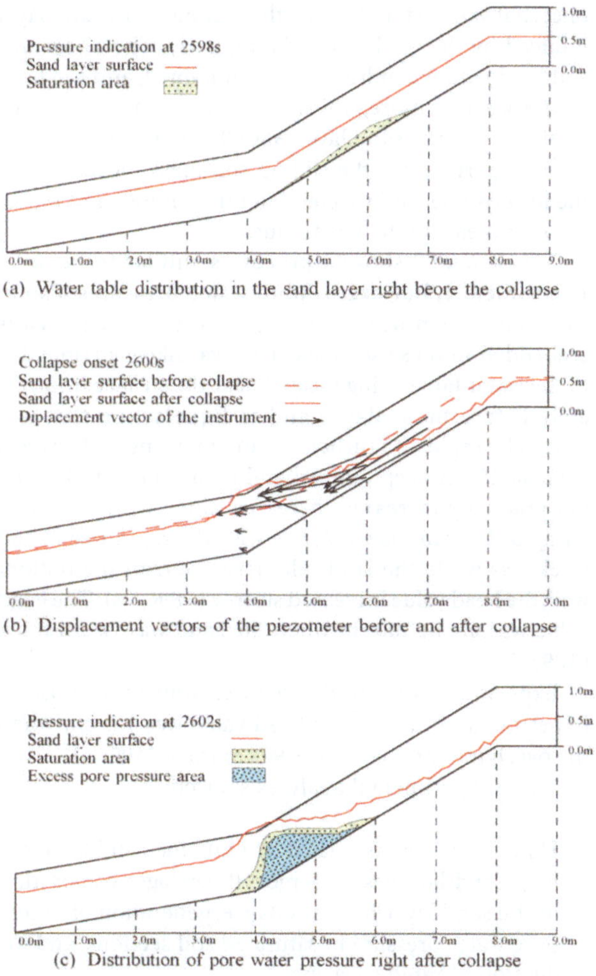

Fig. 15.6. The side view of the sand layer right before and after collapse in Experiment 2

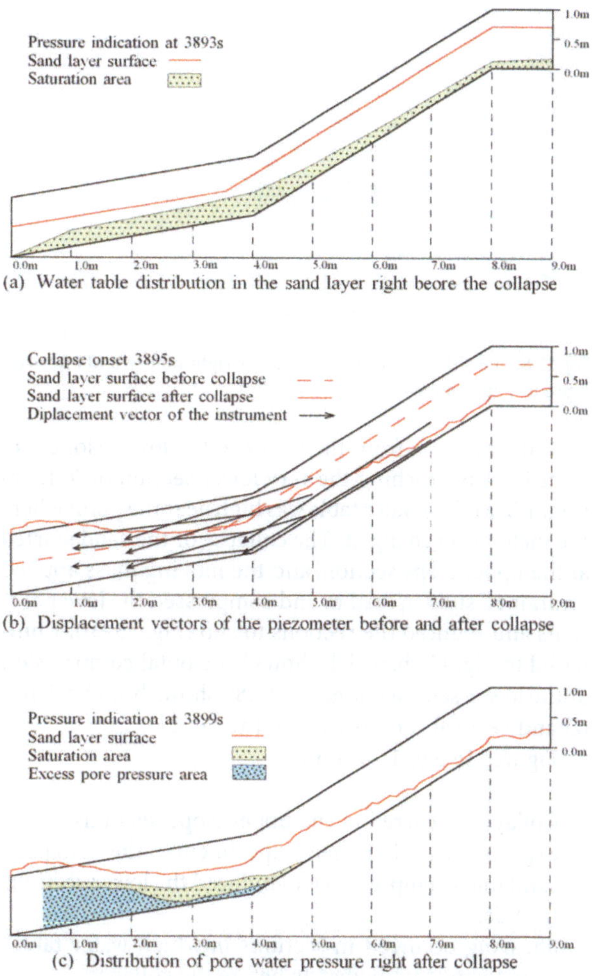

Fig. 15.7. The side view of the sand layer right before and after collapse in Experiment 3

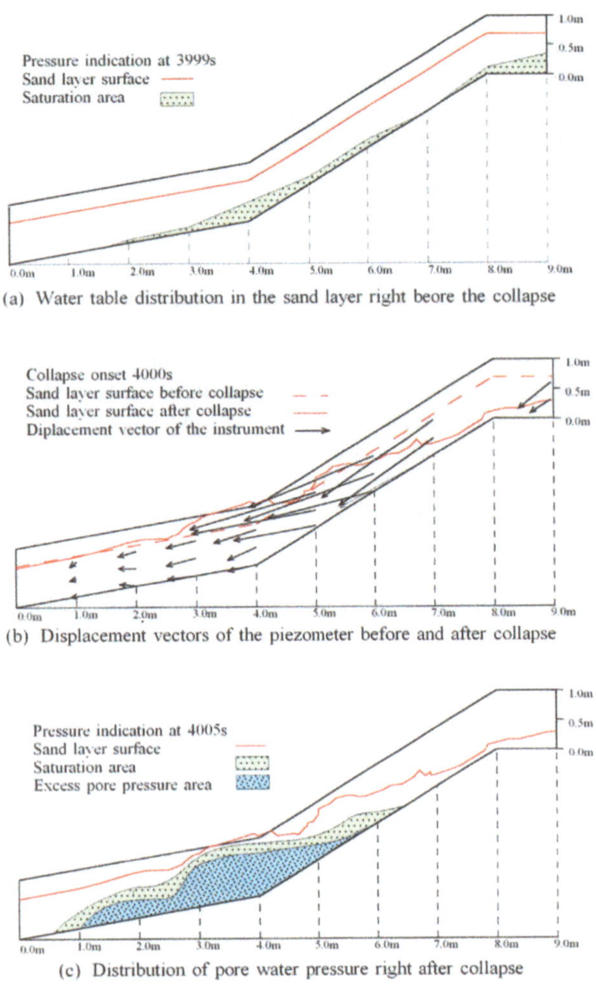

Fig. 15.8. The side view of the sand layer right before and after collapse in Experiment 4

served extending from the upper to the lower slope sections but not reaching the very lower section of 0–1.5 m (Fig. 15.8a). The water table was high near the point where the inclination changed. The collapse of the sand started at the upper slope section, and the moving mass applied undrained sudden load to and compressed the lower sections and induced the sections to slip (Fig. 15.8b). Compared to Fig. 15.5b, which shows horizontal compression of the lower sand section, Fig. 15.8b shows both horizontal and vertical compression of the sand.

Figures 15.5 to 15.8 show:

1. Collapse occurred at the upper slope sections. In all experiments other than Experiment 2, the collapsed sand mass compressed and induced the lower sections to slide.
2. Shearing occurred in sections in which water tables were formed immediately before the collapse.
3. Excess pore water pressure was generated in compressed sand sections.

15.2.4.4 Volumetric Strain, Velocity, and Pore Water Pressure Fluctuation

Figure 15.9 shows the speed of soil movement, volumetric strain, pore water pressure, and change in sand thickness all at the compressed soil section 3 m from the lowest end of the flume for Experiments 1 to 4. Volumetric strain and the speed of movement were determined from video camera images of the markers embedded in the sand. In all experiments, it was difficult to trace the markers as the landslide progressed, or to determine the volumetric strain and traveling speed.

As shown in the figure, in Experiment 1, the lower slope section received horizontal compression by the collapse of the upper slope. As soon as the semi-consolidated top section started to move, both the negative volumetric strain and head value increased at the 45 cm-deep section, or near the floor of the flume (2931.5 s). The head value exceeded the thickness of the sand layer, causing excess pore water, and induced rapid shearing even in the 45 cm-deep section (2932 s). The head value dropped once, but was restored by the thickening of the sand layer induced by the load of the falling mass. The undrained sudden loading condition was maintained, and the head value rose once more, which was 78 cm at the 45 cm-deep section when the sand layer was 60 cm thick (2932.8 s).

In Experiment 2, the sliding soil mass did not reach the lower slope, and we could not determine sand movement or measure the head values.

Experiment 3 showed processes similar to those in Experiment 1, but excess pore water pressures also at 5 cm- and 25 cm-deep sections, according to the sensors embedded in these sections. This was likely attributable to the water table being located near the surface immediately before the collapse and to liquefaction that was caused by rapid and turbulent sand movement. The head value at 45 cm deep jumped and then dropped (3895 s), but gradually increased thereafter along with the thickening of the sand layer. Along with the rapid rise in soil thickness while the landslide mass overrun the bottom wall, the head value increased steeply (3896.5 s). The head value was about 100 cm when the sand was 70 cm thick (3896.7 s).

Experiment 4 showed processes similar to those in Experiments 1 and 3. The head value was 125 cm when the overlying sand layer was 90 cm thick (4005 s).

These experimental analyses showed:

1. Fluidization that is caused by undrained sudden loading of landslide mass undergoes three stages: compaction of the sand layer by the collapse, generation of excess pore water pressure in saturated soil sections, and induction of rapid shearing.
2. Pore water pressure increases more rapidly in saturated area than in unsaturated through rapid compaction.

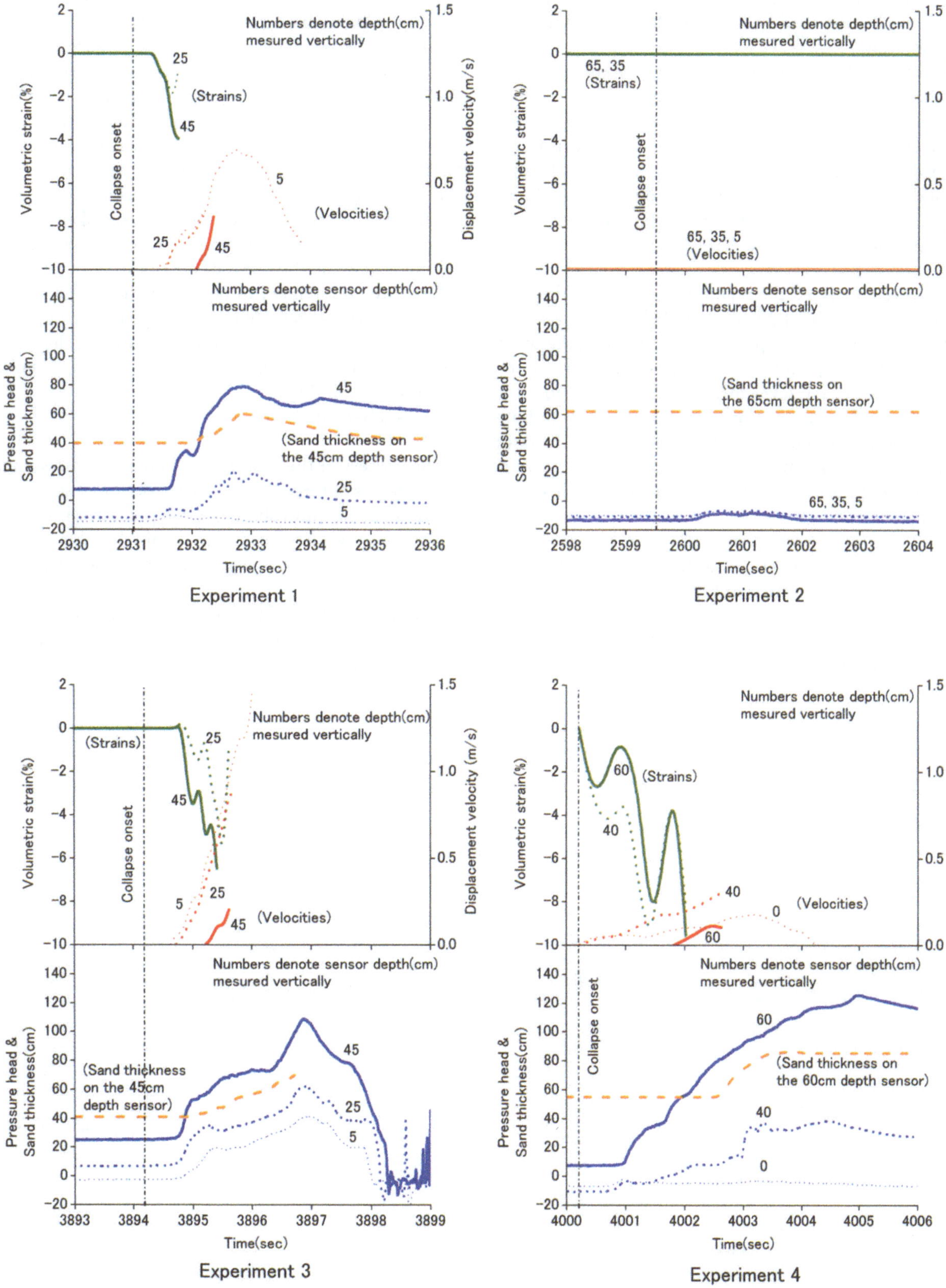

Fig. 15.9. The volumetric strain, thickness, pressure head and displacement velocity of sand layer at 3 m point from the lowest end of the flume in the case of Experiment 1 to 4

This mainly because that the void between sand particles absorbs a volumetric shrinkage in unsaturated condition.

3. Sections that suffered excess pore water pressure showed a positive correlation between the thickness of the overlying sand and pore water pressure. This is likely attributable to water pressure supporting the overlying load. This also means that a large-scale landslide reduces effective stress ratio on the shear plane, thus shows a gentle slope of compensation and long travel distance.

We then analyzed the relationship between the sand movement near the slip surface and change in pore water pressure in upper slope. Figure 15.10 shows the speed of soil movement, volumetric strain, pore water pressure, and change in sand layer thickness, all at 7 m from the lowest end of the flume, for Experiment 1 to 2. In both experiments, the slip surface was formed near the floor of the flume at 45 cm deep. Experiments 3 and 4 were not analyzed since there was no piezometer installed near the slip surface.

In Experiment 1, shearing and volumetric shrinkage around slip surface generated increase of pore water pressure, and the slope lost stability and collapsed entirely since the coefficient of friction was smaller than the inclination of the slope (2 931 s). The rapid changes in head value at the 45 cm-deep section were probably attributable to the noises caused by the roughness on the flume floor. But it maintained constant value in average (2 932 s). Besides, sand layer thickness decreased because of velocity inclination from slip surface to sand layer surface (2 931~2 932 s). There was no correlation between pressure head and sand layer thickness, and pressure head did not exceeded sand layer thickness. It is considered that the generation of the excess pore water pressure was locally and temporary. The gradual rise in pore water pressure during the late landslide movement was likely caused by the landslide mass colliding against the lower slope and increasing the thickness of the sand.

Experiment 2 tended to be similar to 1. The rise of water pressure was retarded in a volumetric shrinkage. It seems to be because measurement position of the water pressure has deviated with calculation area of the volumetric strain a little.

Our analyses showed:

1. The process of the fluidization in the collapse source progresses in three steps: the destruction of sand particle skeleton and volumetric compaction with shear, rise in pore water pressure, and rapid shearing around the slip surface, these steps occurred almost simultaneously.
2. With conversion to high speed shearing, the velocity gradient from slip surface to sand layer surface arises, and the sand layer thickness decreases.

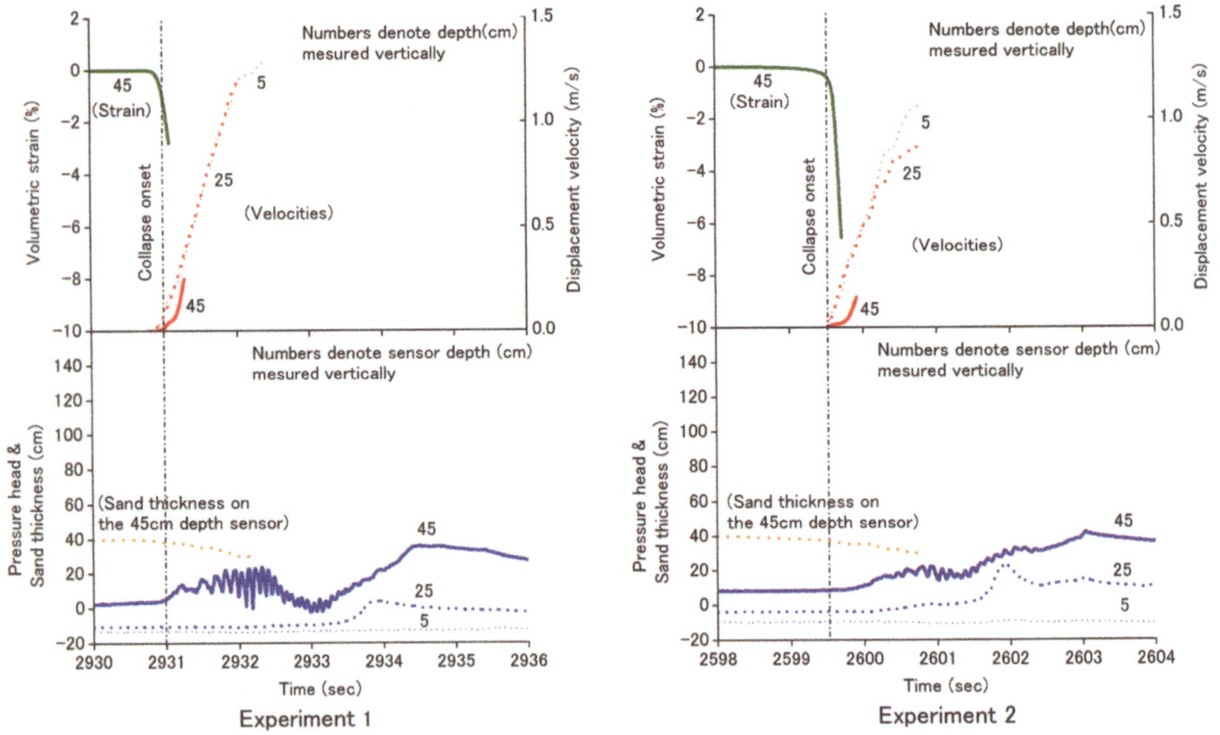

Fig. 15.10. The volumetric strain, thickness, pressure head and displacement velocity of sand layer at 7 m point from the lowest end of the flume in the case of Experiment 1 and 2

3. In high-speed shearing area at the collapse source, pressure head greatly did not rise, as it exceeded the sand layer thickness. This seems to be because that decreasing of thickness and disturbing of sand layer by high-speed shearring do not maintain the undrainage condition.

15.2.4.5 Landslide Fluidization Process by Flume Experiment

To clarify the fluidization process of heavy rain collapse demonstratively, we conducted landslide experiments using an almost real-size slope model equipped with a rainfall simulator. Both the movement and the pore water pressure of the sand layer were monitored throughout the experiment, from the start of spraying rain to the cessation of the landslide. Our experiments showed:

1. Landslide fluidization caused by undrained sudden loading of landslide mass undergoes three stages: compaction of the sand layer caused by the collapsed mass from upper stream, generation of excess pore water pressure in saturated soil sections, and induction of fast shearing. Besides, sections that suffered excess pore water pressure showed a positive correlation between the thickness of the overlying sand and pore water pressure. This is likely attributable to water pressure supporting the overlying load. This also means that a large-scale landslide reduces effective stress ratio on the shear plane, thus shows a gentle slope of compensation and long travel distance.
2. Landslide fluidization at the collapse source area undergoes also three stages: the volumetric compaction with shear, rise in pore water pressure in saturated zone, and rapid shearing around the slip surface, these steps occurred almost simultaneously. With conversion to high speed shearing, the velocity gradient from shear plane to sand layer surface arises, and the sand layer thickness decreases. The rise in pore water pressure is limited near the slip surface, and it does not rise as exceeding the sand layer thickness. This seems to be because that decreasing of thickness and disturbing of sand layer by high-speed shearing did not maintain the undrainage condition.

15.3 A Fluidized Landslide on Natural Slope Experiment

15.3.1 Experiment Site and Testing Procedure

The purpose of our experiment is to produce hopefully a fluidized landside on a natural slope by artificial rain fall. The test site was selected by two conditions: (*1*) to secure complete safety during the experiment, (*2*) to have a possibility producing a fluidized landslide. Then, a natural slope in the Koido National Forest at Mt. Kaba-san, Yamato village, 25 km north of Tsukuba-city, Ibaraki Prefecture, Japan was selected for the controlled experiment on landslide and possible fluidization in cooperation with the Forestry Agency of Japan.

The selected portion of hillslope (Fig. 15.11) was 30 meters long, with an average gradient of 33 degrees (maximum 35 degrees). The soil was 1 to 3 meters deep. A 5 m wide experimental slope was isolated from its surroundings by driving thin steel plates about 1 m deep into the soil. These plates prevented lateral diffusion of infiltrated rain water and cut the lateral tree root network that imparts resistance within the soil layer. The surface of the slope was covered by straw matting to prevent surface erosion and promote rainfall infiltration. Surface material on the slope consisted of fine weathered disintegrated granite sand, called "Masa" in Japan. Loamy soil blanketed the upper portion of the regolith to a depth of about 1 m; this soil mainly originated from tephra of Mt. Fuji, Mt. Akagi, and other volcanoes located west of Mt. Kaba-san.

Table 15.3. Examples of landslide experiments on natural slopes

References	Location	Volume $W \times L$	Excavation etc.	Water supply	Sensors	Fluidization
Oka (1972)	Ikuta, Kawasaki, Japan	500 m^3		Sprayed from fire hose		Yes
Yagi et al. (1985)	Matsuyama, Japan	10 × 25 m	Trench cut at upper and both sides	Rainfall simulator (sprayed on upper half slope)	Extensometer, Piezometer, Strain meter	Yes
Yamaguchi et al. (1989)	Yui, Shizuoka, Japan	10 × 30 m	Upper trench cut, Lower open cut	Supplied from upper trench	Extensometer, Piezometer, Inclinometer	No
Harp et al. (1990)	Utah, USA	1.6 × 1.3 m 2.0 × 2.2 m 3.2 × 4.0 m	Upper trench cut, Lower open cut	Supplied from upper trench	Extensometer, Piezometer	No

Artificial rain at the rate of 78 mm hr^{-1} was applied to the slope segment during the experiment by way of a rainfall simulator. The simulator consisted of a framework of steel pipes with 24 sprinkling nozzles arranged 2 m above the soil surface. Water for sprinkling was pumped from a dam constructed in a small creek at the base of the slope into 80 water storage tanks (1 m^3 for each) on the neighboring hillslope prior to the experiment.

Soil-surface movement was measured by means of stereo photogrammetry. Five CCD video cameras (camera 1, 2, 3, 4, and 5) were placed on A, and another five (6, 7, 8, 9, and 10) on B in Fig. 15.11. The five stereo pairs of CCD video cameras were adjusted to film five parts of the experimental area from bottom to top. For image analysis, 40 white flat targets, 20 cm long × 20 cm wide were distributed over the slope and nailed 20 cm deep. For three-dimensional stereo photogrammetry of the targets, *X*, *Y* and *Z* coordinates of eight control points (at least six for each camera) were obtained before photogrammetric analysis. The control points were 24 pieces of 15 cm diameter domed foam-polystyrene base points, attached to the steel pipes forming the framework of the sprinkling system. A schematic diagram of this measuring system is presented in Fig. 15.12. *Y*- and *Z*-axes were in the longitudinal plane of the experimental slope. The *Z*-axis was vertical and *Y*-axis horizontal. The *X*-axis was horizontal and normal to the *Y*-axis.

Film images were transmitted from the ten CCD video cameras to a 10-channel video time generator under the control of a video-synchronized distributor. Through the 10-channel video time generator, film images of the landslide movement with time-codes were recorded on hard-disk 1 to hard-disk 10, respectively. The 10-channel video time generator counted in time-steps of 0.01 s. The frequency of film imaging was 30 Hz. Calculation of target coordinates was carried out based on the DLT (Direct linear transformation) method. Each image was 480 pixels long × 640 pixels wide, and one pixel represented about 1 cm × 1 cm on the experimental slope. Trajectories of

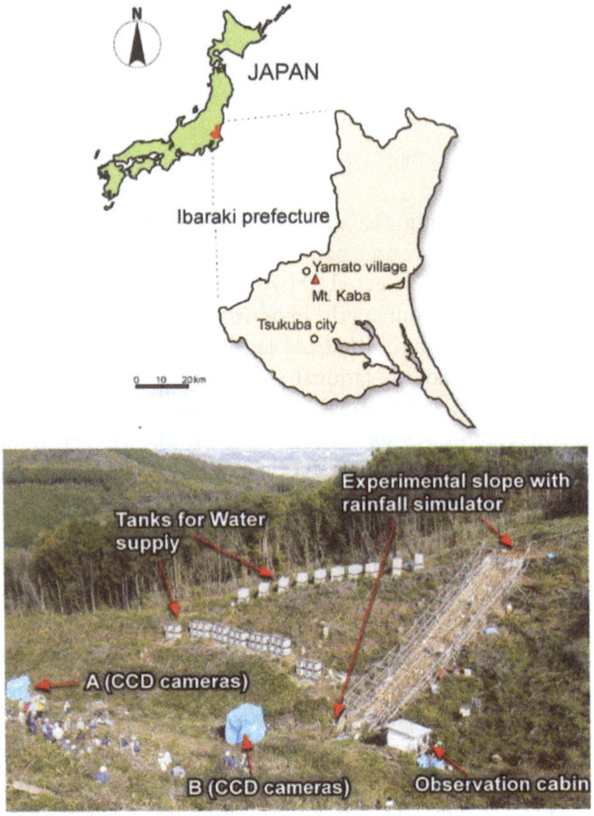

Fig. 15.11. View of the experiment site at Mt. Kaba-san

Fig. 15.12.
Schematic of measuring system for stereo photogrammetry

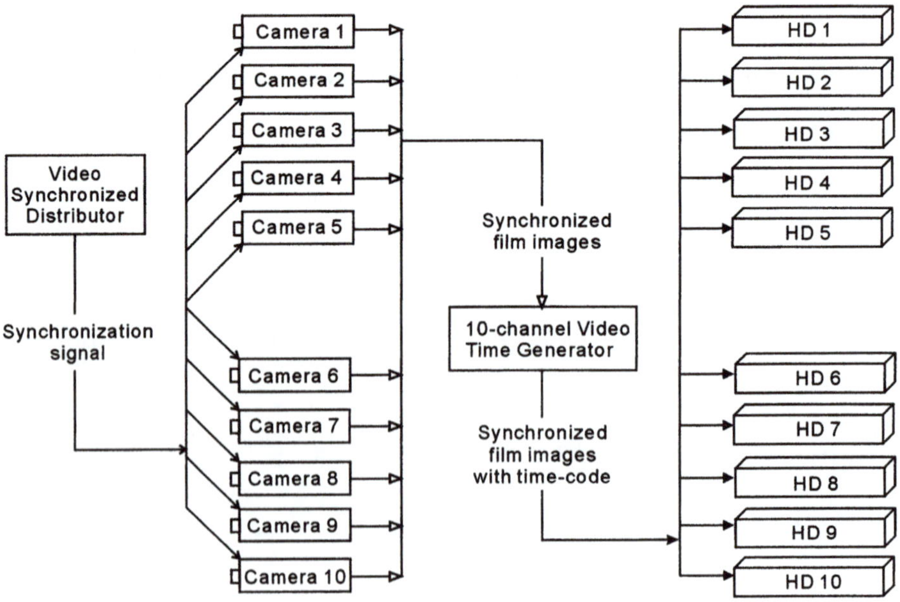

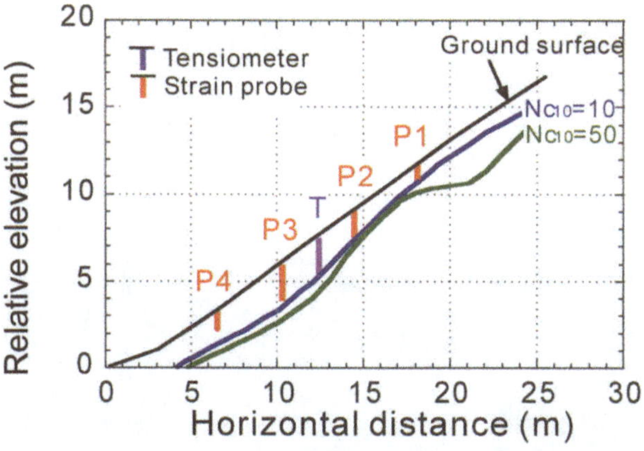

Fig. 15.13. Location of soil-strain probes and tensiometers on the slope. Soil depths are indicated by cone-penetrometer N_{c10} values

targets were detected using differences in brightness in image analysis. The white-coloured targets had a greater brightness than the other parts of the slope covered by straw matting. Coordinates of the targets were taken at the centres of pixel clusters with certain ranges of brightness.

To obtain information on the formation of the sliding surface, four soil-strain probes were inserted into the soil at P1, P2, P3, and P4, as shown in Fig. 15.13. Note that in Fig. 15.13, lines are drawn at the depths where values of $N_{c10} = 10$ and 50 were obtained by light weight cone-penetrometer testing (Noguchi et al. 1997). The N_{c10} values are the number of times it took to drop a weight (5 kg) from height of 50 cm to drive the cone 10 cm into the soil. These values defined the surface soil layer ($5 \geq N_{c10} \geq 0$) and total soil depth ($50 \geq N_{c10} \geq 0$), respectively. The strain probes were made by attaching strain gauges to both the front and back sides of cylindrical polyvinylchloride rods (diameter = 10 mm, 100 cm in length). The distance between each strain gauge was 10 cm. Each rod was coated by a waterproofing tube (Fig. 15.14). To measure the parts in which the sliding surface was expected to be deeper than 1 meter, two strain probes were connected to make a probe 2 m in length. The data acquisition rate was at 1/60 Hz and data were recorded on a data logger under PC control.

To measure saturation conditions within the soil, tensiometers with porous ceramic cups were set into the slope at T as shown in Fig. 15.13. Tenisiometer can measure negative pore water pressure in unsaturated soils and positive pore water pressure in saturated conditions. The data from respective sensors were transmitted through cables to the observation cabin, where they were monitored and recorded. The data acquisition interval for tensiometers was one minute

Several digital video cameras were placed in positions of safety on a hillslope facing the experimental slope to film the entire scene of the fluidized landslide.

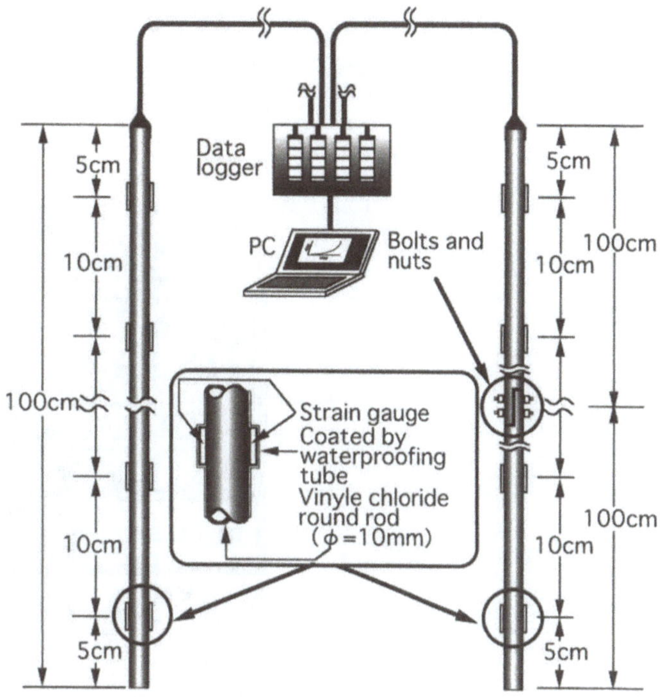

Fig. 15.14. Schematic of the strain probes

15.3.2 General Landslide Movement

On the 12 November 2003, artificial rainfall was given to the slope at a rainfall intensity of 78 mm hr^{-1} for four hours and a half until sunset. No slope movement was observed. The second experiment was conducted on 14 November 2003. Artificial rainfall was started from 9:13 at a rainfall intensity of 78 mm hr^{-1}, the slope deformation was detected from around 15:00, then a clear movement was observed to start at 16:03. The initiated landslide was a type of an expected fluidized landslide, the landslide mass rapidly moved and traveled long.

The cover of the tensiometer started to incline downslope at 24 627.5 s (410 min 27.5 s) after sprinkling commenced. We interpret this as indicating that slope failure initiated at 24 627.5 s. Some images from the digital video camera are presented in Fig. 15.15a–d. Figure 15.15a is at 24 628.5 s (one second after failure initiation), Fig. 15.15b at 24 629.5 s (two seconds), Fig. 15.15c at 24 630.5 s (three seconds), and Fig. 15.15d at 24 631.5 s (four seconds). As soil surface movement increased, a tension crack became visible at the head (Fig. 15.15a), and a compressive bulge resulting from downslope movement was observed 5 m above the base of the slope (Fig. 15.15b). The bulge enlarged (Fig. 15.15c) before the main landslide mass began to undulate and rapidly enter the stream (Fig. 15.15d). The compressive bulge was observed only in the left part of the landslide.

Fig. 15.15. Views of the landslide fluidization between one second and four seconds after failure

Fig. 15.16. Views of the debris flow resulting from the landslide fluidization between five seconds and seven seconds after failure

Fig. 15.17. The landslide deposit one day after the experiment

Images at 24632.5 s (five seconds after failure) and 24634.5 s (seven seconds) are shown in Fig. 15.16a,b. Figure 15.16 presents the movement of liquefied landslide mass. The failed landslide mass had entered the stream and was about to collide with the confronting slope (Fig. 15.16a). After collision, the fluidized landslide turned to the right (Fig. 15.16b), changed into a debris flow, and traveled downstream for 10 seconds on a ~10 degree gradient, as much as 30 m. It took 17 seconds from the initiation of the landslide to the end of deposition. Figure 15.17 shows the landslide deposit one day after the experiment. The straw matting, the cover of the tensiometers, and the white-coloured targets were conveyed to the toe of the fluidized landslide.

15.3.3 Dynamic Movement of the Failed Landslide Mass

For stereo photogrammetry, imaging continued from the start of sprinkling until sunset, a duration of more than seven hours. However, the images in bright sunshine could not be analyzed, and motion of the targets was obtained from stereo photogrammetry only during failure initiation, and after slope failure had taken place. Data from 24620 s (410 min 20 s) through 24632 s (410 min 32 s) are shown. The arrangement of the targets on the failed landslide mass is shown in Fig. 15.18.

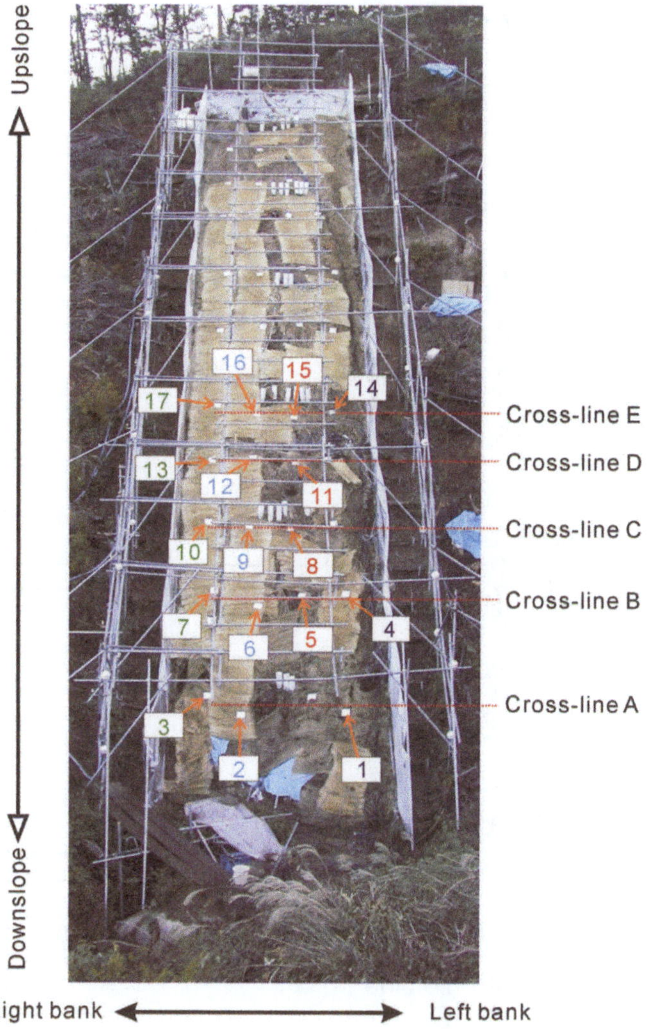

Fig. 15.18. Arrangement of targets for image analysis

Displacements of the targets in the *YZ*-plane are presented in Fig. 15.19. Targets on the cross-lines of C, D, and E, showed similar dynamic movement. From about 24627.5 s (410 min 27.5 s), displacements were observable, and greatly increased in value from about 24628.5 s. Although the available duration times for stereo photogrammetry differed for different targets, differences in the shapes of their motion curves were negligible. For targets on the cross-line B (targets 4 to 7), motions were very different: targets 6 and 7 which were nailed in the right hand bank of the experimental slope (left part of the experimental slope), showed rather similar behavior to that of targets 8 through 17. Target 5 however showed very small movement and target 4 showed no movement. Targets 1–3 on cross-line A did not move. This showed that parts in left hand bank (right side in the experimental slope) at cross-line B and parts around cross-line A neither failed nor slid, but remained stable.

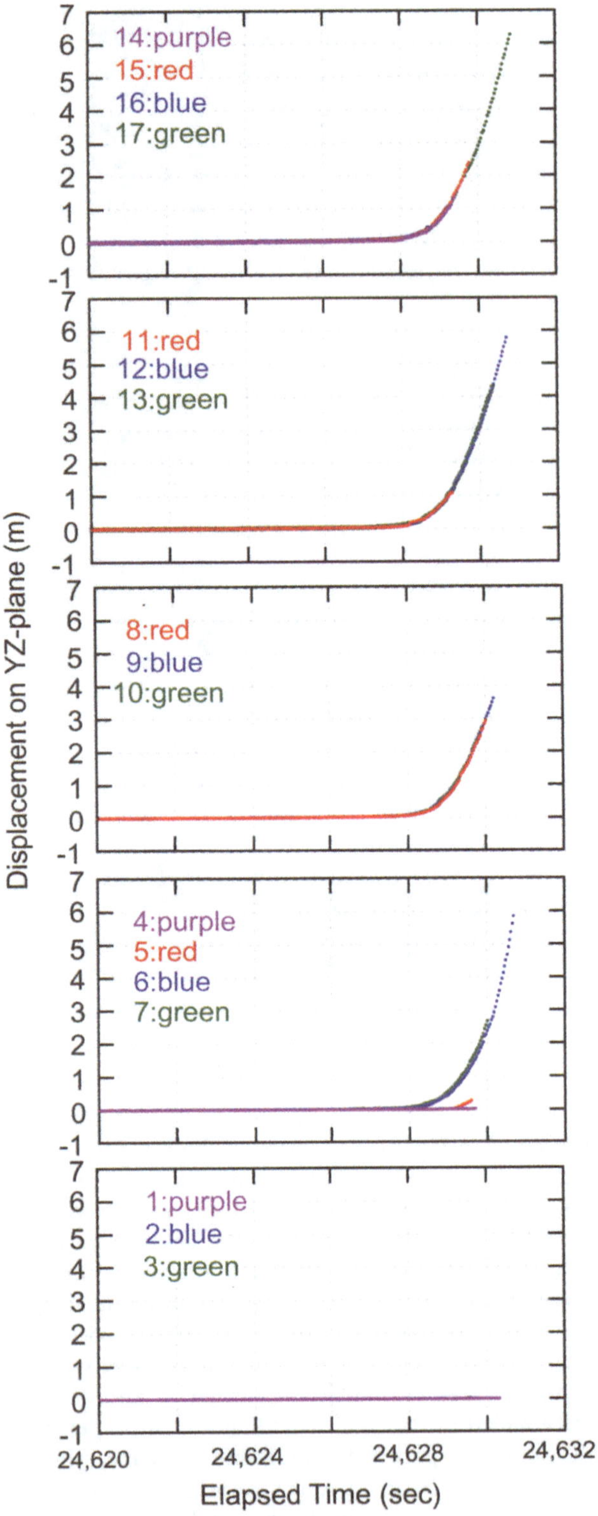

Fig. 15.19. Displacements of targets in the *YZ*-plane

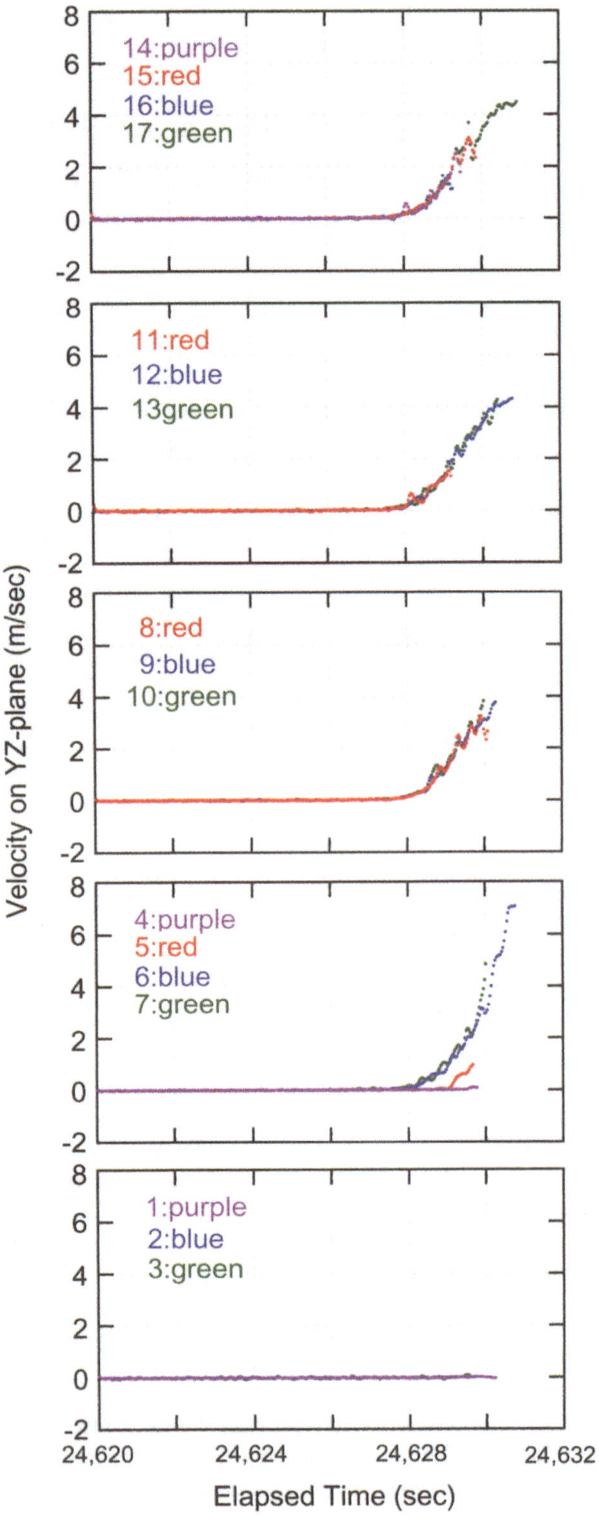

Fig. 15.20. Velocities of targets in *YZ*-plane

Velocities of targets in the *YZ*-plane are presented in Fig. 15.20. Since targets 1 through 4 did not move, their velocities were zero. Other targets increased their velocities after about 24 627.5 s, the time of failure initiation based on film images showing inclination of the cover of the tensiometer at this time. The target-velocity curves in the YZ-plane confirmed the failure initiation time. Although the velocities of targets 5, 6, and 7 were different, the velocities of targets 8 through 17 during 24628.5 to 24630 s increased in proportion to time, and the shapes were similar to one another. Calculated average accelerations during this period were about 2 m s^{-2}. This value of acceleration in a sled model (Sassa 1988) on a 33 degree slope gives an apparent mobilized friction angle of approximately 22 degrees. The internal friction angle of weathered disintegrated granite sand is generally about 33 degrees, indicating a reduction of 11 degrees in apparent friction. This probably was due to excess pore pressure generated within the shear zone, followed by the landslide fluidization. The reason why targets 5 through 7 initially increased their velocities more slowly than targets 8 through 17 was that the former targets were located in the neighborhood of the break-out of the sliding surface (the foot). Hence they were in a zone of initial compression where the surface was observed to bulge upwards at the time of failure - see Fig. 15.22b.

Distances between longitudinally adjacent targets normalized by the initial distances between them before failure in the YZ-plane indicate longitudinal strain in the slope (Fig. 15.21, normalized distances along the south side are not presented). Longitudinal distances between cross-lines D & E (green color in Fig. 15.21) increase in value over time, whereas those between cross-lines A & B (purple), B & C (red), and C & D (blue) decrease. This indicates that tensional stresses were mobilized in the slope between cross-line D and E, whereas, compressive stresses occurred between cross-lines C and D, B and C, A and B. The normalized distance between 13–17 increased in value faster than those of between 11–15 and 12–16 (all of them were between cross-line D & E). We assume that expansive displacement was first mobilized around targets 13 and 17. While longitudinal surface strains between cross-lines D & E and C & D were small, strains between cross-lines A& B and B & C were large. Hence we assume that the slope between cross-line C and E likely failed and slid largely as a block. This interpretation is in general agreement with the velocity results (Fig. 15.20), in which targets 8 through 17 moved with similar velocities at any instant. The break-out of the sliding surface (the foot) appears to have been located above cross-line A. Note that the displacement behaviours of the targets on cross-line B (target 4, 5, 6, and 7) differ from all of the others (in Fig. 15.19). Targets 6 and 7 moved greatly, whereas target 4 and 5 showed little or no movement. Thus, the break-out of the sliding surface in the right hand bank (left part) of the experimental slope was likely between cross-line A and B, whereas, the one in the left hand bank (right part) was likely between cross-line B and C. Only one landslide was generated in this experiment, but the movement characteristics and stress conditions in the right and left hand banks were very different. The differences may be attributed to lateral heterogeneity in the soil on the slope.

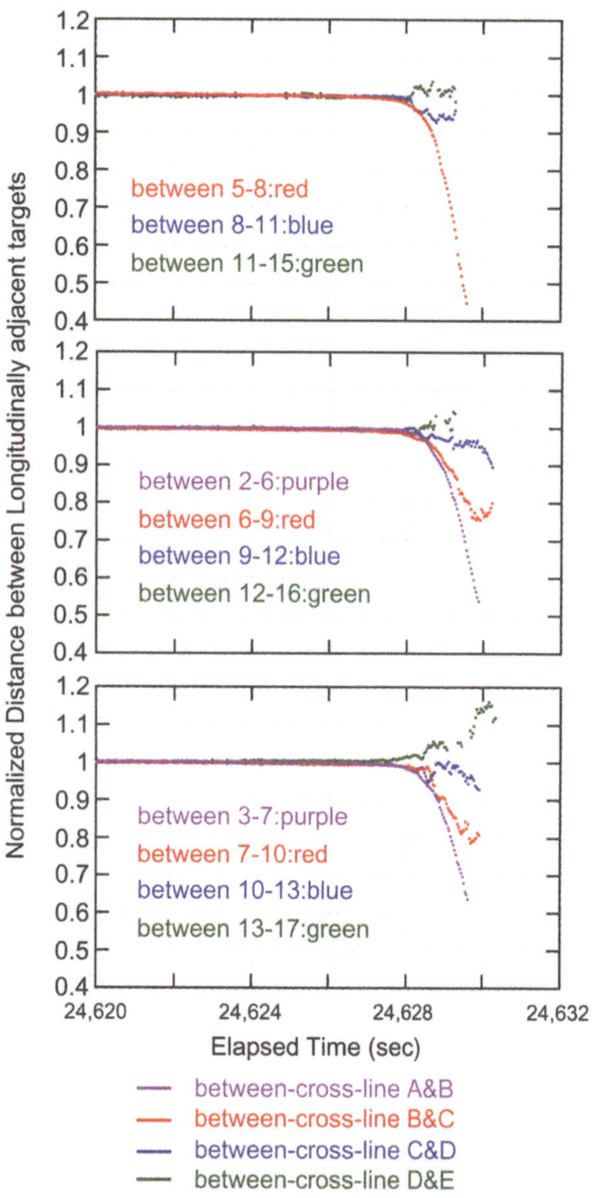

Fig. 15.21. Normalized distances between longitudinally adjacent targets in the YZ-plane

Snapshots of the three-dimensional movements of the targets are shown in Fig. 15.22. Figure 15.22a is 0.5 second before failure (24627 s), Fig. 15.22b is two seconds after failure (24629.5 s), and Fig. 15.22c is 3.5 seconds after failure (24631 s). Targets 6 and 7, located just above the foot of the landslide, followed curved paths, which could not have been determined by conventional displacement measurement using extensometers. Figure 15.22 shows that the landslide mass overrode the slope around targets 2 and 3 and then moved rapidly into the stream at the foot of the experimental slope.

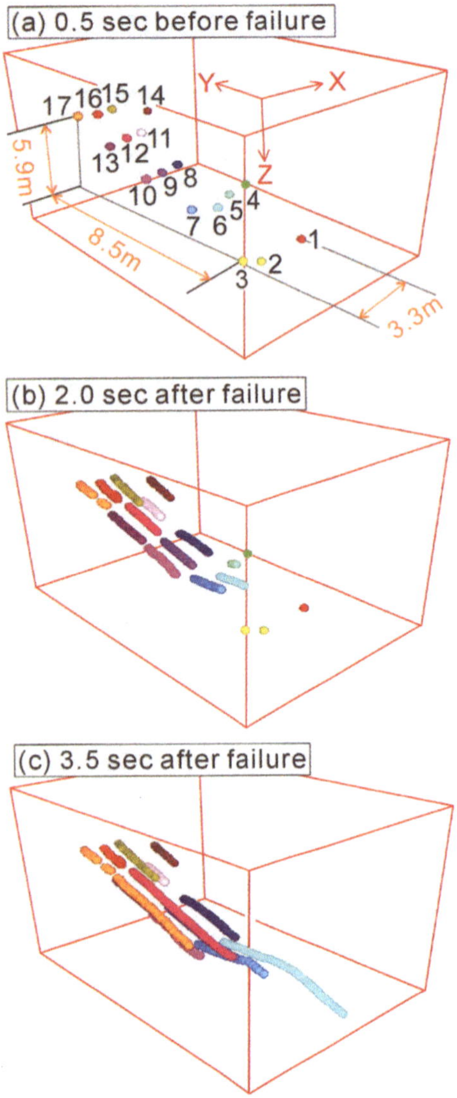

Fig. 15.22. Snapshots of the three-dimensional movements of the targets. **a** 0.5 second before failure (24627 seconds); **b** two seconds after failure (24629.5 seconds); **c** 3.5 seconds after failure (24631 seconds)

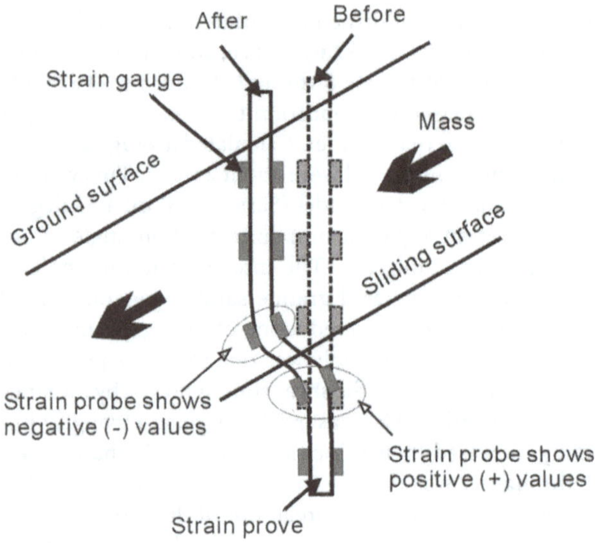

Fig. 15.23. Detection of soil strain by means of strain probes

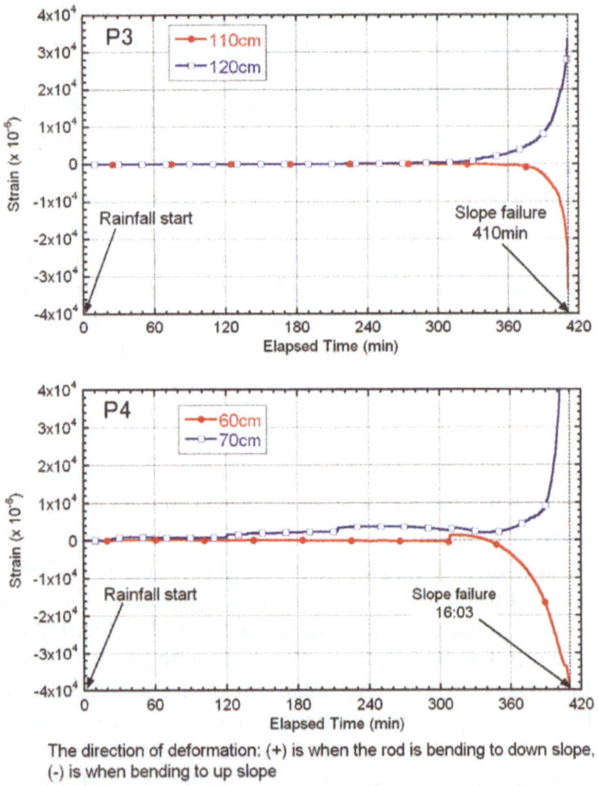

Fig. 15.24. Soil-strain accumulation curves (**a**) at P3 and (**b**) at P4

15.3.4 Sliding-surface Formation

When a sliding surface forms, soil above the sliding surface tends to move downslope, whereas soil below it remains stable. In this situation, a soil-strain gauge located above the sliding surface shows negative values, and the gauge below the sliding surface shows positive values (Fig. 15.23). Hence, if adjacent strain gauges show paired positive and negative values, we assume that the sliding surface has formed between them. Soil-strain probes at P1 and P2 were outside of the slope failure. Figure 15.24 shows the accumulation of soil strain during the experiment. The results from soil-strain probe P3 is shown in Fig. 15.24a, and for P4 in Fig. 15.24b. For the strain probe at P4 (Fig. 15.24b), positive values of strain were observed at a depth of 70 cm from about 20 minutes after sprinkling began, and negative strains at a depth of 60 cm. We interpret from the paired positive and negative values in Fig. 15.23, that the sliding surface formed between 60 cm and 70 cm depth. At failure (410 minutes), the accumulated strain at a depth of 70-cm was larger than 0.04, which

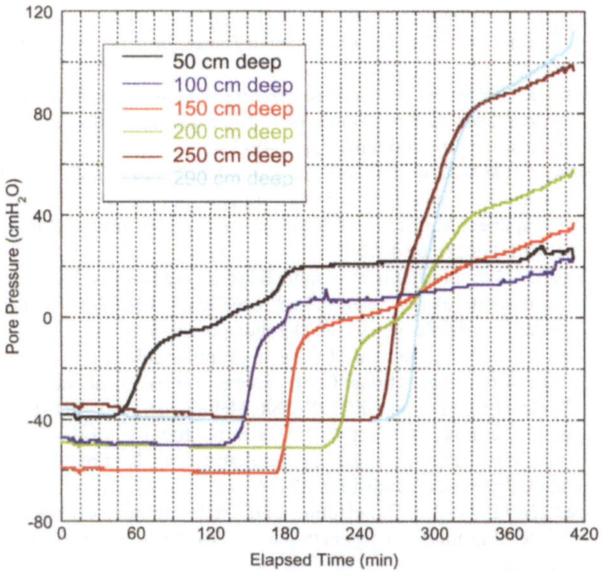

Fig. 15.25. Changes in soil-water pressure in tensiometers at *T*

was the capacity of the strain gauge. Similar results were obtained at P3 (Fig. 15.24a). Positive values of strain were observed at 120 cm depth from about 300 minutes and negative values at 110 cm depth, indicating a sliding surface between 110 cm and 120 cm depth. The depth of sliding at P3 was twice as deep as at P4.

Changes in pore water pressure including suction monitored by the tensiometers at T are presented in Fig. 15.25. The tensiometers at T were placed at depths of 50, 100, 150, 200, 250, and 290 cm. All tensiometers showed negative pore pressures at the start of sprinkling, indicating that the soils at all depths were unsaturated or partly saturated. When the wetting front passed, the tensiometers showed increases in pore pressure in sequence of the depths. At 410 min, when the failure took place, all of the tensiometers showed positive pore pressures. The pore pressure of the deepest tensiometer (290 cm) rapidly increased its values from about 290 min. This almost coincided with the time when the strain gauge at 110 cm depth in the strain probe at P3 started to show strain (Fig. 15.24a). Hence, it can be deduced that general slope instability increased from 290 min, before final failure at 410 min.

15.3.5 A Fluidized Landslide on Natural Slope by Artificial Rainfall

An experiment to induce a fluidized landslide by artificial rainfall was conducted on a natural slope at Mt. Kabasan in Yamato village, Ibaraki Prefecture, Japan. A landslide initiated 24 627.5 s (410 min 27.5 s) after the start of sprinkling at a rainfall intensity of 78 mm hr^{-1}. The landslide mass was 14 m long and 1.2 m deep (at maximum). It first slid, then fluidized, and changed into a debris flow. The travel distance was up to 50 m in 17 seconds. The apparent friction angle of the fluidized landslide was 16.7 degrees.

1. The entire landslide movement and subsequent debris-flow mobilization were successfully recorded by the digital video cameras, allowing qualitative interpretation of the shape of the landslide and definition of failure initiation and deposition times.
2. The soil-surface movements were measured by tracing the three-dimensional displacement of targets on the experimental slope using image analysis. Soil-surface movement of the failed landslide mass on the right and left banks were different. This may be attributed to heterogeneity in the composite soil on the slope.
3. Formation of the sliding surface was detected by soil-strain probes. Strain probes inserted in the middle or lower parts of the mass that failed showed a sliding surface at depths of between 110–120 cm and 60–70 cm, respectively. The tensiometer showed a rapid increase in pore-water pressure after about 290 min from the start of sprinkling. This almost coincided with the time when the strain was first observed on the sliding surface.

Acknowledgment

We acknowledge the help of Hiroshi Fukuoka, Ryo Sasaki, Sergio Lourenco, and Ivan Gratchev of Disaster Prevention Research Institute, Kyoto University, and Kotaro Makihara, Seishi Aihara and Yoshinori Yatabe of Chiba University. Also we wish to thank Hiromu Moriwaki, National Research Institute for Earth Science and Disaster Prevention for the advice for planning the field experiment. And also we appreciate the support of a number of the staff at Forestry and Forest Products Research Institute, especially to Yoshitsugu Takeuchi, Yukio Mashima, and Shozo Nakamura. We are indebted to Ibaraki Prefectural Government and Kanto Regional Forest Office for the permission to conduct the experiment in the Suigo-Tsukuba Quasi-National Park and Koido National Forest.

Flume experiments were conducted as part of the joint research for investigating the conditions of fluidization of landslide, by the Chubu Regional Forestry Office of Forestry Agency and Forestry and Forest Products Research Institute.

Landslide experiment on natural slope was a part of a project called APERIF (Areal Prediction of Earthquake and Rainfall Induced Rapid and Long-traveling Flow Phenomena), launched by the Special Coordinating Fund for Science and Technology of the Ministry of Education, Cultures, Sports, Science and Technology (MEXT) of Japan. In 2002, this project was approved as part of the International Programme on Landslides (IPL M101-APERITIF Project to the International Consortium on Landslides).

References

Bishop AW (1973) The stability of tips and spoil heaps. Q J Eng Geol 6:335–376

Committee of investigation for Gamahara-zawa debris flow damage in Dec. 6 (1997) The research report of investigation for Gamahara-zawa debris flow damage in Dec.6. J Jpn Soc Erosion Control Engng 50(3):89–94 (in Japanese)

Eckersley JD (1985) Flowslides in stockpiled coal. Eng Geol 22:13–22

Eckersley JD (1990) Instrumented laboratory flowslides. Géotechnique 40(3):489–502

Ellen SD, Fleming RW (1987) Mobilization of debris flows from soil slips, San Francisco Bay region, California. Geol Soc Am Rev Engng Geol 7:31–40

Fukuzono T (1985) A new method for predicting the failure time of a slope. In: Proceedings of IVth International Conference and Field Workshop on Landslides, Tokyo, 23–31 August, Japan Landslide Society, 1, pp 145–150

Harp EL, Wells WG, Sarmiento JG (1990) Pore pressure response during failure in soils. Geol Soc Am Bull 102:428–438

Hutchinson JN (1986) A sliding-consolidation model for flow slides. Can Geotech J 23:115–126

Ishihara K (1993) Liquefaction and flow failure during earthquakes. Géotechnique 43(3):349–451

Ishihara K, Okusa S, Oyagi N, Ischuk A (1990) Liquefaction-induced flowslide in the collapsible loess deposit in Soviet Tajik. Soils and Foundations 30(4):73–89

Iverson RM, LaHusen RG (1989) Dynamic pore-pressure fluctuations in rapidly shearing granular materials. Science 246:769–799

Iverson, RM, Reid ME, Iverson NR, LaHusen RG., Logan M, Mann JE, Brien DL (2000) Acute sensitivity of landslide rates to initial soil porosity. Science 290:513–516

Noguchi S, Abdul Rahim N, Zulkifli Y, Tani M, Sammori T (1997) Soil physical properties and preferential flow pathways in tropical rain forest, Bukit Tarek, Peninsular Malaysia. J Forest Res 2:125–132

Oka H (1972) Impacts by the "artificial landslide": re-examine the rage of nature. Kagaku Asahi, January issue, pp 152–153 (in Japanese)

Sassa K (1984) The mechanism starting liquefied landslides and debris flows. In: Proceedings of 4th International Symposium on Landslides, Toronto, June, 2, pp 349–354

Sassa K (1988) Motion of landslides and debris flows-prediction of hazard area. In: Report for Grant-in-Aid for Scientific Research by Japanese Ministry on Education, Science and Culture (Project No. 61480062), pp 4–52

Sassa K (1998) Mechanisms of landslide triggered debris flows. Environmental Forest Science. In: Proceedings of IUFRO Division 8 Conference, Kyoto, 19 – 23 October, pp 499–518

Sassa K (2000) Mechanism of flows in granular soils. In: Proceedings of the International Conference of Geotechnical and Geological Engineering, GEOENG2000, Melbourne, 1, pp 1671–1702

Sassa K, Fukuoka H, Wang G, Ishikawa N (2004) Undrained dynamic-loading ring-shear apparatus and its application to landslide dynamics. Landslides 1:7–19

Seed HB (1979) Soil liquefaction and cyclic mobility evaluation for level ground during earthquakes. J Geotech Eng-ASCE 105: 201–255

Seed HB, Lee KL (1966) Liquefaction of saturated sand during cyclic loading. J Geotech Eng-ASCE 92(CM6):105–134

Yagi N, Yatabe R, Enoki A (1985) Laboratory and field experiments on prediction method of occurring time of slope failure due to rainfall. Landslide (J Jpn Landslide Soc) 22(2):1–7 (in Japanese)

Yamaguchi I, Nishio K, Kawabe H, Shibano H, Iida C (1989) Initiation and fluidization of an artificial landslide. -Field experiment in Yui, Shizuoka Prefecture, Japan. Shinrin Kosoku (Areal Survey) 158:3–9 (in Japanese)

Yoshimi Y, Richart FE, Prakash S, Balkan DD, Ilyichev, VA (1977) Soil dynamics and its application to foundation engineering. In: Proceedings of the 9th International Conference on Soil Mechanics and Foundation Engineering, Tokyo, July, 2, pp 605–650

Part III Landslide Monitoring

Chapter 16

Enlargement of a Failed Area along a Sliding Surface

Kiminori Araiba* · Akira Suemine

Abstract. Progressive failure in short-term slope stability is examined in this study. The relative displacement between a sliding mass and a basement in a landslide was monitored precisely. The onset time of relative displacement at each borehole was determined and the speed of propagation of local relative displacement was calculated. The lowest speed of propagation was 4.6 m hr^{-1}. Stresses inside the specimen were measured in the direct shear apparatus with a 400 × 400 mm shear plane. Results show that the horizontal sliding surface determined by the apparatus does not coincide with the plane in which Mohr's failure criteria are first fulfilled, but with the plane of the maximum shear stress. Shear stress reductions were observed and the drop position migrated with the shear.

Based on those results, we can explain the kinematic processes of formation of a sliding surface during landslides: First local rupture occurs at a certain position where stress conditions fulfill the rupture criteria. The rupture surface that is mobilized by this rupture, in general, does not coincide with the potential sliding surface. Plastic deformation resulting from local failure increases in terms of its area and amount until plastic deformation along the potential sliding surface is eventually mobilized throughout the potential sliding surface. The enlargement process is dependent on material characteristics, variations in external forces, and boundary conditions. The discrepancy between the orientation of local rupture surface is determined mainly by stress and the orientation of potential sliding surface, which are determined mainly by boundary conditions and heterogeneity of materials, which are essential factors for progressive failure during landslides.

Keywords. Progressive failure, field monitoring, direct shear test

16.1 Introduction

Progressive failure during landslides has been investigated by many researchers. Softening of material, non-uniform stress distribution, and stress re-orientation during its progress are considered to be the main factors of progressive failure (Bjerrum 1967; Chowdry 1978; Potts et al. 1997). Most studies have specifically examined the long-term stability problem of slopes. Recently, slope deformation preceding a collapse has been analyzed from the viewpoint of progressive failure. Ohmura and Tsuchiya (1988) suggested a model in which the enlargement rate of a sliding surface is inversely proportional to the power of length of the not-yet-failed potential sliding surface; their model can explain the so-called "creep curve" that precedes the slope collapse. Petley (2005) suggested a model in which accumulation of micro-fractures caused by slight fluctuation of external forces generates and expands the shear surface. Consequently, the slope stability decreases and the shear surface is eventually enlarged dynamically. Thus, generation of local failure and enlargement of the failed area seem to be acceptable to many researchers. However, direct observation of that phenomenon has been only scarcely reported. Burland et al. (1977) and Cooper et al. (1988) observed a deep-seated shear zone, which was considered to be the result of the release of horizontal stress, but a temporal process of formation of such zone has not yet been observed. In laboratory tests, Kamai (1993) observed non-uniform strain inside a specimen of a ring shear test and recorded its enlargement during creep deformation. Kishimoto et al. (1997) produced a long box shear apparatus and investigated soil structure using X-ray photographs of specimens after tests in which shear displacement was as great as 300 mm. They identified three stages of deformation: the initial simple shear, local failure, and relative movement with a shear zone. Those results show the progressive failure of soil, even in a laboratory test. However, stress conditions, which are considered a major factor of progressive failure have not yet been well investigated.

To investigate the mechanism of formation of the sliding surface, we carried out field monitoring of landslide movements using a high-frequency data acquisition system and measurement of stress field around the sliding surface in a direct-shear apparatus.

16.2 Field Monitoring of Deformation Area Enlargement in a Landslide

In this study, we monitored detailed deformation at the sliding surface during a landslide. Figure 16.1 shows a model of this study. The relative displacement between stable ground and the sliding mass is monitored using boreholes for observation (BVs). The onset time of the relative displacement at each borehole might differ among them because of progressive failure. In an elastic condition, redistribution of stress propagates with elastic wave velocity. For a landslide, sliding might cause plastic strain

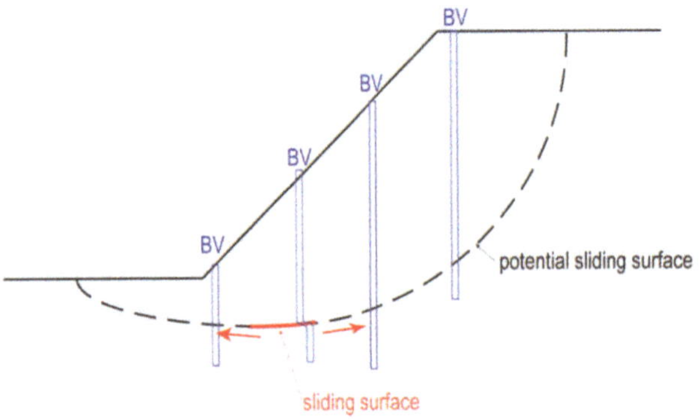

Fig. 16.1. Model for observation of the process of enlargement of sliding surface in landslide. The sliding surface is generated in certain part(s) of the potential sliding surface; it then spreads. Relative displacement caused by formation of the sliding surface is measured the using bend strain of the borehole. The onset time of bend strain is determined for each borehole. Differences in the onset time between boreholes indicate the speed of enlargement of the sliding surface

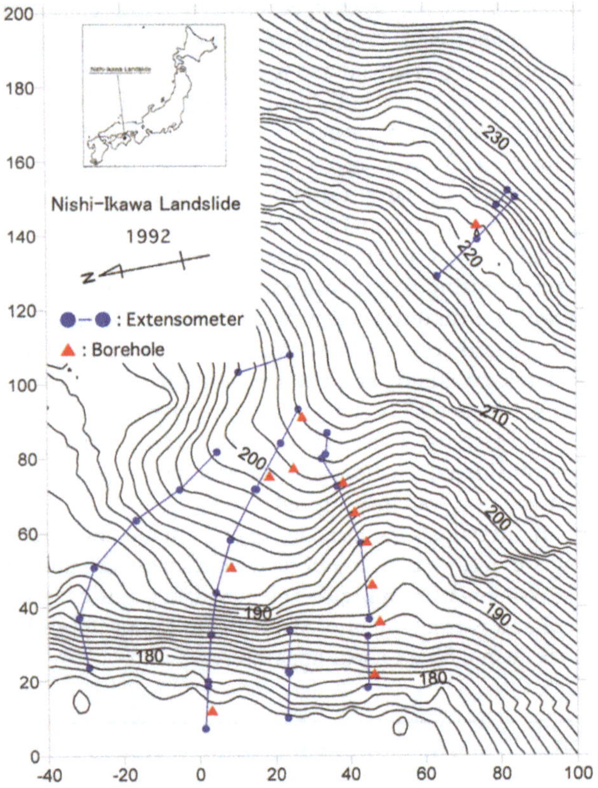

Fig. 16.2. Topography of Nishiikawa landslide and arrangement of the borehole for observation

and the propagation speed might be slower. Although the spreading rate of local movement differs from that of stress re-orientation, it will indicate the order of the role of progressive failure in landslide movement.

The test field was the Nishiikawa landslide (Fig. 16.2) in the Sanbagawa metamorphic rock region of Shikoku Island in southwestern Japan. No sign of landslide activ-

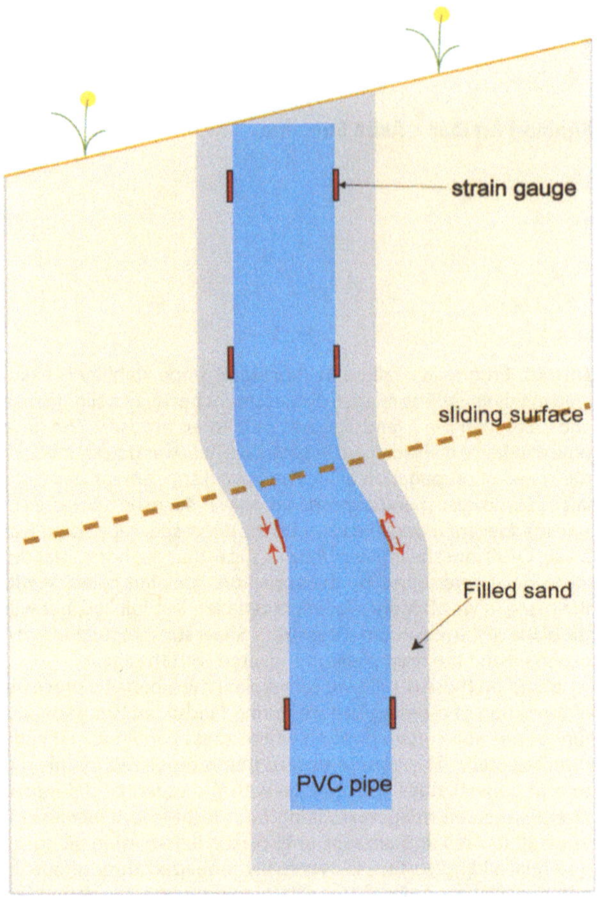

Fig. 16.3. Schematic structure of an observation borehole. A PVC pipe is inserted in a borehole and filled with fine sand. The bending strain of the PVC pipe is measured using strain gauges bonded on both sides of the pipe surface. The interval of pairs of strain gauges is 1 m in this research

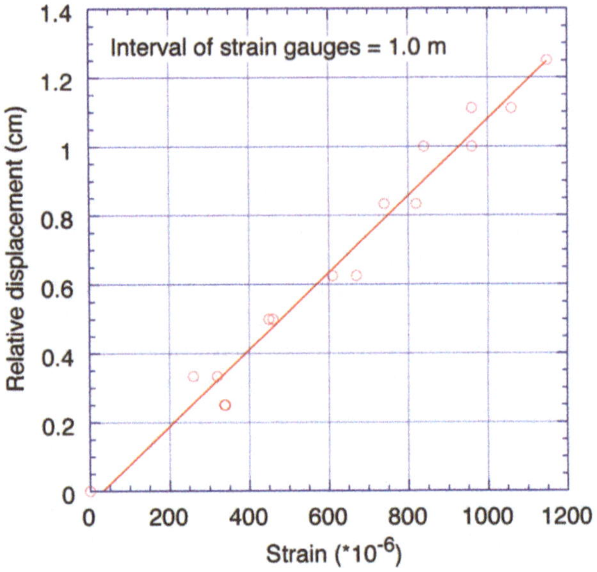

Fig. 16.4. Result of calibration of pipe strain gauges

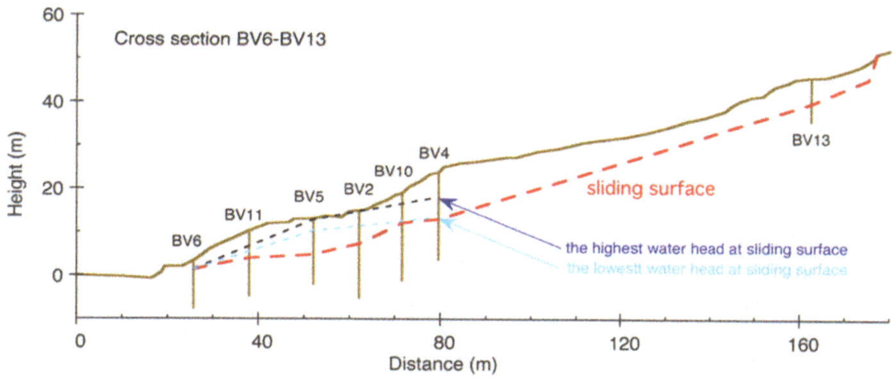

Fig. 16.5. Cross section of Nishiikawa landslide along the monitoring line. Depth of the sliding surface at each borehole is determined by measuring the pipe bending at each borehole

ity was apparent before 1973, when an artificial cutting at the slope toe made the landslide active. Since 1973, sliding has occurred repeatedly during heavy rainfall periods. Suemine (1983, 1992) and Araiba (1998) monitored the behavior of the landslide using wire extensometers, pipe-strain-meters, and pressure transducers. In this document, results obtained using pipe-strain-meters will be presented.

Figure 16.3 shows a pipe-strain-meter structural schematic. A PVC pipe with a 48-mm outer diameter and 3.6-mm wall thickness was embedded in a borehole. After insertion, the clearance between the pipe and the ground was filled carefully with sand to obtain uniform back-filling. The pairs of strain gauges were bonded on the surface of the PVC pipe at many depths before installation. The PVC pipe is bent and surface strain is measured using strain gauges as the landslide material moves. At each depth, two gauges were used to compensate for the effect of temperature: one on the up-slope side, the other on the down-slope side. The distance between the neighboring pairs was 1.0 m in this study. The interval of data acquisition was 1 min. Results of calibration tests are shown in Fig. 16.4.

Figure 16.5 shows a cross-section of monitoring line. The sliding surface depth was identified through detailed inspections of the pipe bend. We measured the pipe bending only at this sliding surface in past heavy rainfall events; for that reason, we infer that the sliding surface was well developed. Therefore, reduction of shear resistance force can result only from a decrease in the effective stress. The sliding surface geometry is neither simply linear nor circular, and the distribution of pore water pressure is irregular. These characteristics are usual for landslides in Japan. Without plastic deformation, the soil mass cannot be displaced long on such a sliding surface. The variation of pore water pressure can cause re-orientation of stress and plastic deformation can delay the propagation of stress re-orientation.

Figure 16.6 shows data of pipe-strain-meters obtained during three heavy rainfall periods. The respective names of boring holes and depths of gauge-pairs are written at the left of the diagrams. Ishikawa and Miyatake (1978)

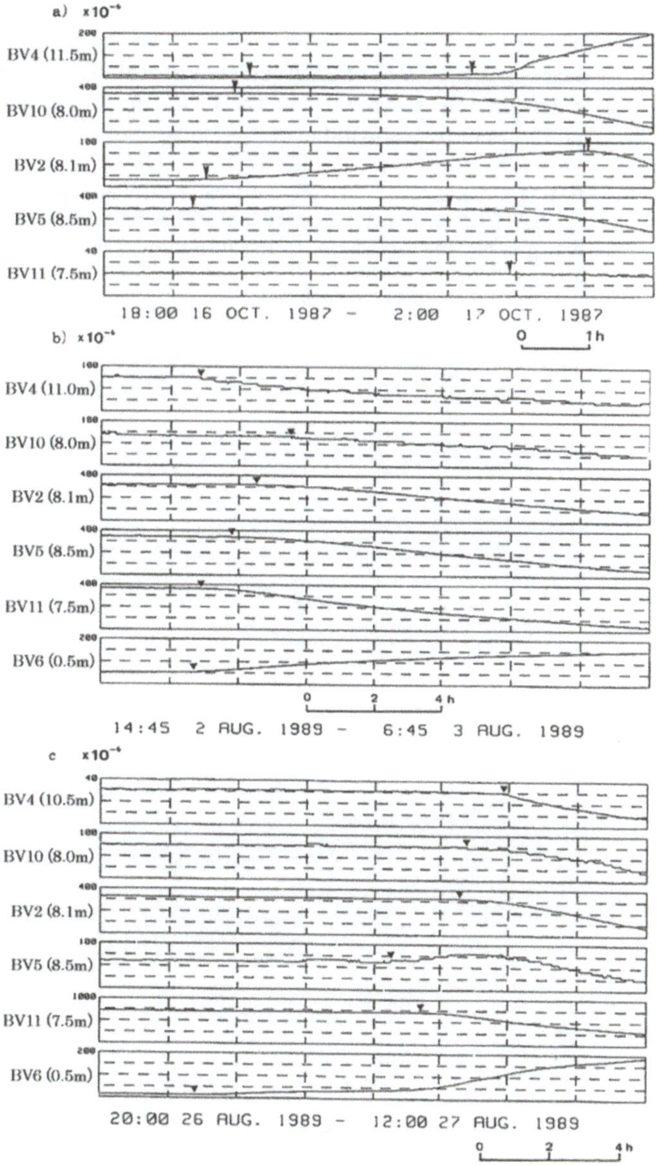

Fig. 16.6. Result of measurement of pipe strain gauges in three periods (**a**), (**b**) and (**c**) when great precipitation was observed. The borehole name and gauge depth are written at the left of each diagram. Black triangles indicate onset times determined using Wiener's prediction filter

showed that the onset times of long-period phenomena can be determined using Wiener's prediction filter. The onset time determined by the filter is shown as a black triangle in each diagram. A typical event (Fig. 16.6b) shows that local deformation starts at the lower end of slope and propagates upward. Early onset is also visible in BV 4. That result shows that the initial position of local sliding is not unique. The speed of enlargement of local sliding is calculated by lags in the onset time (Table 16.1). In the monitoring part of this landslide, local displacement started from the lower end in many cases and the slowest speed was almost 4.6 m h^{-1} (1.3×10^{-3} m s^{-1}). This very slow displacement indicates that stress re-orientation is not elastic. The speed varies among positions and rainfall events. These findings indicate that the speed of enlargement of local movement is determined strongly by the variation speed of external forces in this landslide.

Table 16.1. Determined onset time of local deformation at each borehole for four activities observed in three heavy rainfall events shown in Fig. 16.6. Velocity of enlargement of local deformation between boreholes is calculated from the time lag in the onset time and distance between boreholes

B.V.		Onset time		Spreading speed of local deformation (m h^{-1})
4	11.5 m	20:07	16/10/87	51.3
10	8.0 m	19:54	16/10/87	18.6
2	8.1 m	19:29	16/10/87	50.2
5	8.5 m	19:18	16/10/87	
4	11.5 m	23:22	16/10/87	
10	8.0 m			
2	8.1 m	01:04	17/10/87	4.7
5	8.5 m	23:02	16/10/87	15.9
11	7.5 m	23:56	16/10/87	
4	11.0 m	17:38	02/08/89	
10	8.0 m	20:16	02/08/89	7.9
2	8.1 m	19:17	02/08/89	13.0
5	8.5 m	18:33	02/08/89	16.2
11	7.5 m	17:40	02/08/89	54.0
6	0.5 m	17:25	02/08/89	
4	10.5 m	7:48	27/08/89	10.8
10	8.0 m	6:46	27/08/89	42.2
2	8.1 m	6:35	27/08/89	4.6
5	8.5 m	4:29	27/08/89	15.6
11	7.5 m	5:25	27/08/89	
6	0.5 m	22:45	27/08/89	

16.3 Study of the Formation of Sliding Surface in Laboratory Tests

The continuous sliding surface on which soil mass can be displaced for long distances is considered to be determined strongly by a boundary condition, variation in the external force and heterogeneity of material during an actual landslide. Using the result of Kihimoto et al. (1997), a microscopic rupture surface which is mobilized by the stress condition does not always agree with the macroscopic continuous sliding surface determined by other conditions. In this study, we accessed stress variation during formation of a continuous sliding surface in a laboratory test scale.

A series of direct shear tests was performed. The direct shear apparatus reproduces a clear sliding surface at a pre-determined position. Therefore, it is suitable for

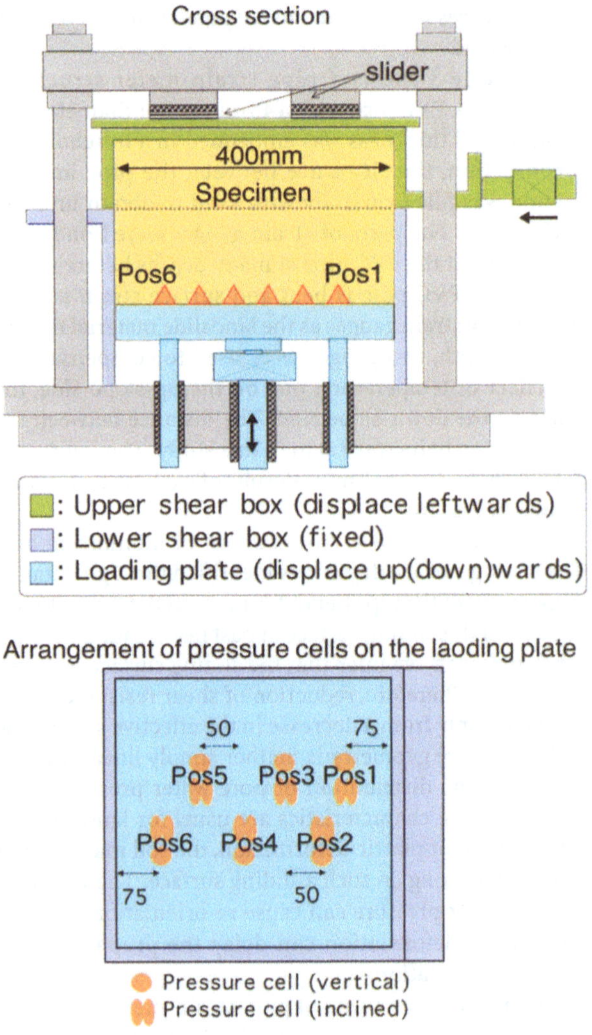

Fig. 16.7. Schematic structure of the direct shear apparatus used for this study

investigation of soil behavior with large relative displacement at the sliding surface. On the other hand, it has a defect by which a stress tensor cannot be determined because the inclination of principal stress is unknown. We measured stress acting on the basement of the shear box to obtain all stress components. Wood et al. (1980) measured boundary stress in a simple shear apparatus to observe non-uniformity of stresses inside the specimen. Non-uniformity is undesirable from the viewpoint of an "element test", but it is a fundamental aspect of geo-materials and it might affect the formation process of a sliding surface.

Figure. 16.7 shows the scheme for the direct shear apparatus used in this study. It has a 400 × 400 mm shear plane. The upper shear box and loading plate are constrained respectively by low-friction sliders and assistant pistons. Earth pressure cells are installed on the loading plate in six positions. Three cells are equipped in each position (Fig. 16.8): one faces vertical and two face slanting directions (45 degrees to the vertical). For the vertical one, a cell is simply buried in the specimen. For inclined ones, an iron attachment of an isosceles triangle shape is affixed at the loading plate and two cells are attached, one to either side of it. We can obtain Mohr's stress circle for each position from those three observed stresses. In the figure, the stress point to the plane of the maximum shear stress (*T*) and the plane of the maximum stress ratio (τ/σ; *M*) are shown. For an angle of zero dilatancy, the *T* plane coincides with the Roscoe's shear band orientation. The *M* plane is that on which Mohr-Coulomb-type rupture criterion is first fulfilled. Vermeer (1990) stated that the shear band in a granular material is formed in the orientation between *M* and the Roscoe angle.

Toyoura sand, a clean seashore sand ($D_{50} = 0.164$, $U_c = 1.46$), was used. Air-dried specimens were set in the shear box by air-pluviation method and consolidated. Vertical force and shear speed were kept constant during shear. Table 16.2 shows the test conditions. Because of technical problems, data of Pos. 1 were not obtained. Figure 16.9 shows typical test results in terms of macroscopic behavior of the specimen.

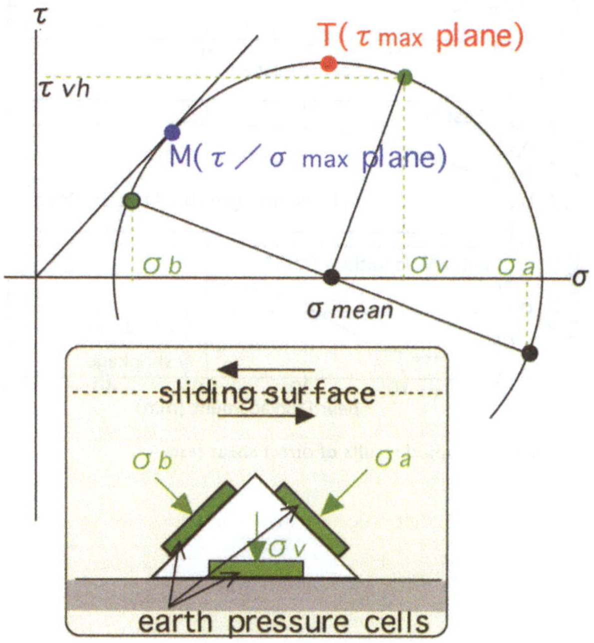

Fig. 16.8. Arrangement of pressure cells (inside the rectangle) and Mohr's stress circle determined by measured pressures. The stress point *T* indicates the point of maximum shear stress and *M* indicates the point of the maximum stress ratio

Table 16.2. Test conditions

Test No.	Material	Depth of lower box (mm)	Shear speed (mm min^{-1})	Consolidation puressure (kPa)	Normal stress during shear (kPa)	Void ratio after consolidation
1	Toyoura sand	50.6	2.0	98	98	0.853
2		50.1	2.0	98	98	0.854
3		49.3	2.0	196	98	0.845
4		49.4	2.0	294	98	0.843
5		49.2	2.0	392	98	0.839
6		48.9	2.0	490	98	0.834
7		46.5	2.0	196	196	0.826
8		47.3	2.0	196	196	0.824
9		48.0	2.0	196	196	0.841
10		48.4	2.0	392	196	0.835
11		19.5	2.0	392	196	0.832
12		48.7	2.0	490	196	0.826

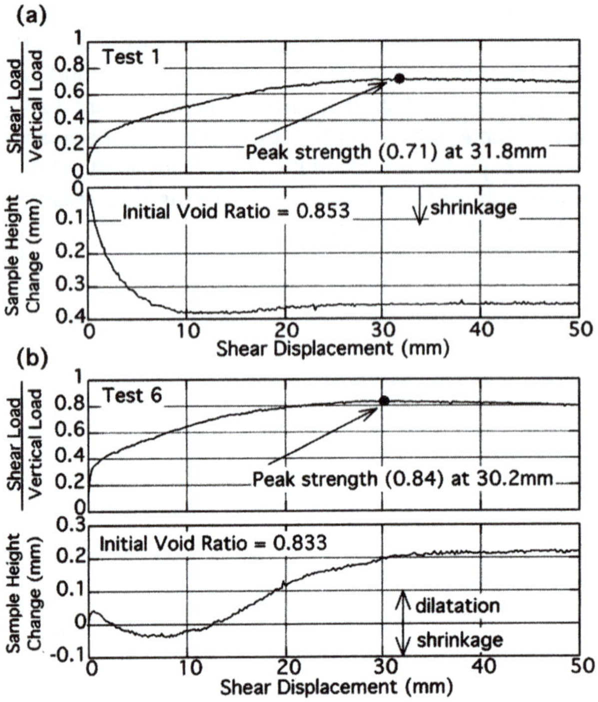

Fig. 16.9. Two typical results of direct shear tests

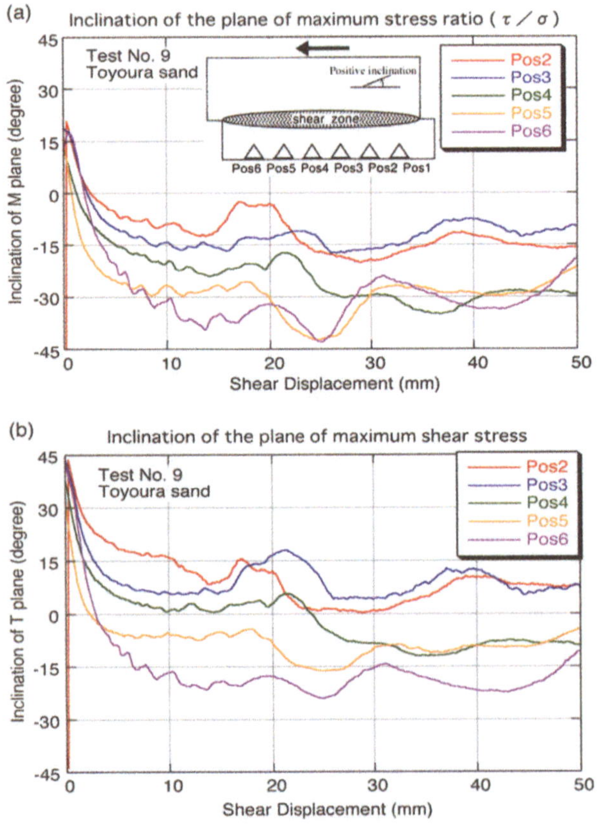

Fig. 16.10. Inclination of the *T* plane (**a**) and inclination of the *M* plane (**b**). The origin of inclination is the horizontal plane. Positive values indicate a counterclockwise direction

Variations in inclination of the *M* plane (a) and *T* plane (b) to the horizontal plane are shown in Fig. 16.10. In most positions, inclination of the *M* plane is negative (upper left side in the cross section of Fig. 16.7), which indicates that local rupture occurs on the inclined plane on which soil particles cannot displace for a long distance because of boundary of the apparatus. Inclinations of the *T* plane concentrate around zero; consequently, the macroscopic sliding surface is coincident with the *T* plane.

Figure 16.11 shows variations in the shear stress acting on the horizontal plane (τ_{vh}). They increase uniformly between shear displacement smaller than 8 mm except

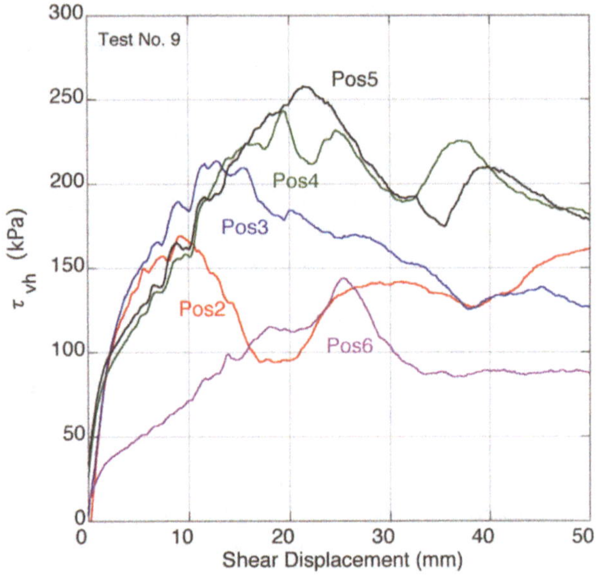

Fig. 16.11. Shear stress variations on the horizontal plane. Peaks and stress drops appeared from Pos. 2 to 6

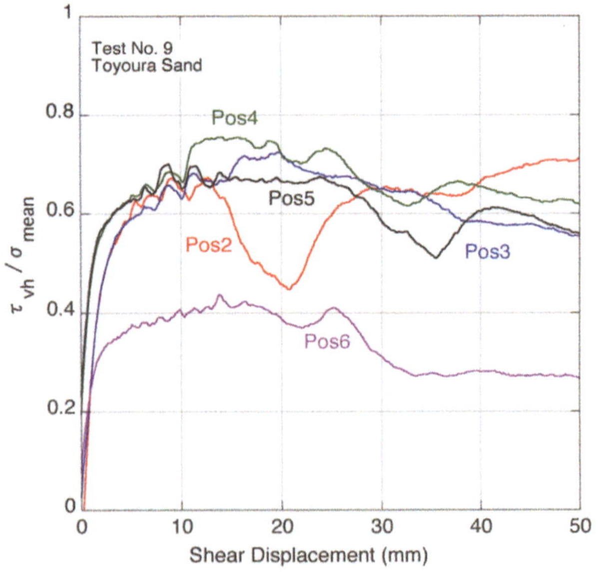

Fig. 16.12. Variations in the stress ratio (τ_{vh}/σ_{mean})

for Pos. 6, which might be affected by the sidewall. Then, they show peaks and drop one by one from Pos. 2 to Pos. 6. Apparently, the yielding zone moves from Pos. 2 to Pos. 6. Between shear displacement larger than 40 mm, differences in τ_{vh} become smaller, indicating that the sliding surface has developed completely and that the relative movement of lower and upper shear boxes occurs without causing a large strain inside.

Peak values of τ_{vh} in Fig. 16.11 differ among positions. Therefore, they are not a material constant. Although the sliding surface is coincident with the *T* plane, stress conditions on the surface are not governed solely by shear stress. Figure 16.12 shows variations in τ_{vh}/σ_{mean}, where σ_{mean} is considered to be related to local volumetric strain. Most concentrate in a narrow range, except for Pos. 6. Figure 16.13 shows four (Pos. 2–5) maximum values of τ_{vh}/σ_{mean}. Solid circles indicate average values. Average values converge on a certain range irrespective of the initial void ratio. Based on these results, we infer that a relationship exists between local volumetric strain and formation of continuous sliding surface.

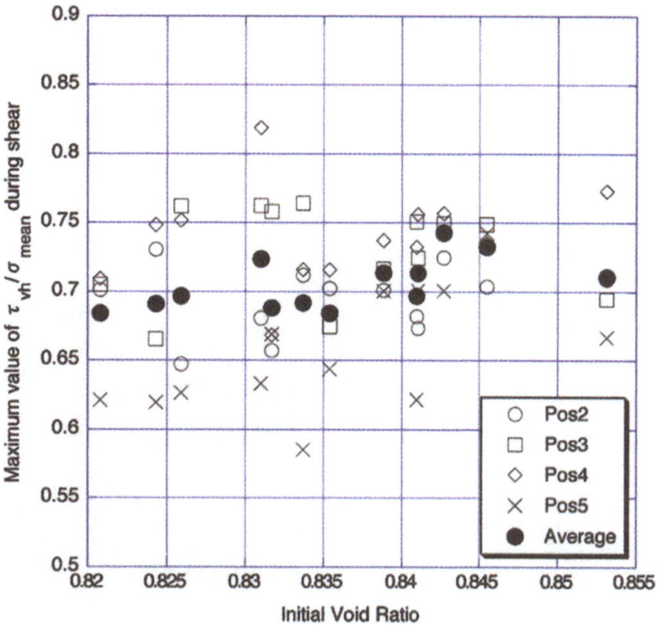

Fig. 16.13. Relation between the initial void ratio and the maximum values of τ_{vh}/σ_{mean}. Average means average over Pos. 2 to Pos. 5

16.4 Discussion

The processes of formation of a continuous sliding surface in a direct shear apparatus are explainable as follows. Microscopic rupture first occurs on the *M* plane, as determined by stress. That plane does not coincide with the potential sliding surface determined mainly by boundary conditions. Local plastic deformation caused by those local ruptures continues until sliding on the potential sliding surface is mobilized. Then the continual sliding surface is formed locally at a certain condition, which might be related to local volumetric strain. The role of volumetric strain remains unclear. We presume that local volumetric strain changes dramatically at the moment of breakdown of the gap between two microscopic rupture surfaces oriented in the *M* plane; this process might affect the appearance of volumetric strain in the condition of a continuous sliding surface. As shear displacement increases, the clear continuous sliding surface increases its area.

This explanation is applicable to an actual landslide. External forces might cause local ruptures in some parts of the potential sliding surface. The rupture plane orientation is not always coincident with the potential sliding surface. Local plastic deformations resulting from those local failures continue until sliding on the potential sliding surface is mobilized; then a clear continual sliding surface begins to form, which causes stress re-orientation and variation in the external force changes stress field. The stress propagates with the elastic wave speed but the plastic strain necessary for formation of a continuous sliding surface delays expansion of the local deformation area.

Two points of difference between laboratory tests and the Nishiikawa landslide must be emphasized. The first is the difference of material: sand was used in the laboratory test, but the sliding surface was formed in residual state soils in the Nishiikawa landslide. The second difference is size. The difference of material is not thought to be important. Sandy soil has a large stress drop (Fig. 16.11), which causes large stress re-orientation and promotes enlargement of the sliding surface. On the other hand, soil in a residual state can cause a stress drop only by the decrease of effective stress. Moreover, it is generally considered to be small. Consequently, enlargement of the sliding surface is slow or not self-excited. In such a material, formation of the sliding surface might be controlled strongly by variation of external forces. However, we can say that failure is progressive because generation of local ruptures and their enlargement still occur. The second point of difference is difficult to conclude at this moment. Phenomena observed in a small scale sliding surface are not always dominant in a large surface. The size effect is a topic for further research.

Enlargement of the deformation area is visible even in a shallow slope with granular material. Figure 16.14 (Araiba 2006) shows pre-failure deformation of a slope in a model slope failure experiment. The model slope was 5 m high, 4 m wide, 1 m deep, with an inclination of 30 degrees. The material is river sand and failure was induced by artificial rainfall. In the figure, the deformation amount is multiplied by 20. In the upper slope, subsidence occurred at the top of the slope before 11:00:36

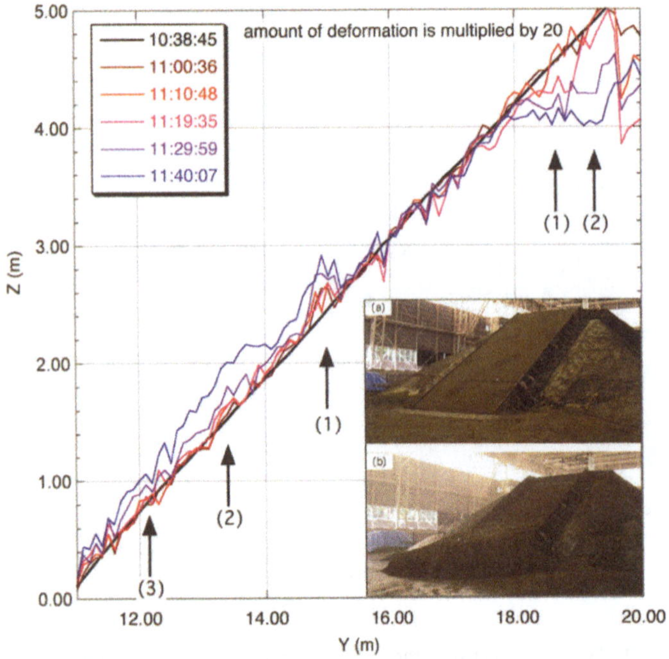

Fig. 16.14. Change of longitudinal section of the models slope before the failure. The amount of deformation is multiplied by 20. Local strain appeared in two parts of the slope shown by arrows (*1*); then deformation spreads upward in the upper part and downward in the lower part

and extended its area over time. In the lower slope, uplifting deformation first appeared around $Y = 15$ then at $Y = 14.0$ at 11:19:35 and the deformation spread between 12.0 to 15.0 at 11:29:59. This example indicates that progressive failure might occur, even in a shallow and sandy slope. Consequently, progressive failure in the short-term slope stability problem remains a topic for further research.

References

Araiba K (2006) Study on the method for detecting and monitoring of pre-failure deformation in slope. In: Proceedings INTERPRAEVENT 2006, 2, pp 581–589

Araiba K, Suemine A (1998) In-situ measurement of internal earth pressure during landslide movement. Soils and Foundations 38(3):97–107

Araiba K, Suemine A (2002) On the process of formation of sliding surface in landslides. In: Proceedings INTERPRAEVENT 2002, 2, pp 495–504

Bjerrum L (1967) Progressive failure in slopes of overconsolidated plastic clay and clay shales. ASCE SM5:3–11

Burland J, Longworth T, Moore J (1977) A study of ground movement and progressive failure caused by a deep excavation in Oxford Clay. Géotechnique 27(4):557–591

Chowdry R (1978) Slope analysis. Elsevier, pp 69–77

Cooper M, Bromhead E, Petley D, Grant D (1998) The Selborn cutting stability experiment. Géotechnique 48(1):83–101

Ishikawa Y, Miyatake T (1978) An application of the Wiener's prediction filter to the records of crustal deformations and seismicity. Jishin 31(1):73–86

Kamai T (1993) Failure propagation process in landslide clay. In: Proceedings 7th ICFL, pp 243–248

Kishimoto R, Mukada Y, Sato K (1997) The changes of soil structure in shear process. J Jpn Landslide Society 34(1):8–14

Ohmura H, Tsuchiya S (1988) Analysis of creep phenomenon by the slip surface enlarging model. J Jpn Landslide Society 25(1):1–6

Petley D, Higuchi T, Petley D, Bulmer M, Carey J (2005) Development of progressive landslide failure in cohesive materials. Geology 33(3):201–204

Potts D, Kovacevic N, Vaughan P (1997) Delayed collapse of cut slopes in stiff clay. Géotechnique 47(5):953–982

Suemine A (1983) Observational study on landslide mechanism in the area of crystalline schist (Part 1)-An example of propagation of Rankine state. Bull Disaster Prev Res Inst Kyoto Univ 33(3):105–127

Suemine A (1992) On the mechanism of landslide in the area of crystalline schist. Tikyuu 14(2):105–111

Vermeer PA (1990) The orientation of shear bands in biaxial tests. Géotechnique 40(2):223–236

Wood DM, Drescher A, Budhu M (1979) On the determination of stress state in the simple shear apparatus. Geotech Test J 2(4):211–222

Airborne LIDAR Data Measurement and Landform Classification Mapping in Tomari-no-tai Landslide Area, Shirakami Mountains, Japan

Hiroshi P. Sato* · Hiroshi Yagi · Mamoru Koarai · Junko Iwahashi · Tatsuo Sekiguchi

Abstract. Detailed landform classification is important if effective measures against landslides are to be taken. Conventional techniques can only measure the detailed terrain in vegetated areas with difficulty. Airborne light detection and ranging (LIDAR) is a promising tool to precisely and directly measure a digital elevation model (DEM). Using a two-meter-grid DEM we attempted to understand landslide characteristics, namely, we produced manual and automated landform classification maps in Tomari-no-tai area in Shirakami Mountains, Japan. In advance, 1 : 2500-scale two-meter-interval contour map was newly printed using the LIDAR-DEM. It was found that valleys and other geomorphological features could be seen in better detail in the airborne LIDAR contour map than in the existing photogrammetric contour map. The map and 1 : 8000-scale aerial photographs were interpreted, and manual landform classification map was produced. As a result, 17 classifications were identified in the map.

In producing the automated landform classification map, in advance, three variables such as slope, surface texture (feature frequency, or spacing), and local convexity were calculated from the DEM. The three variables were subdivided into three, two, and two classes, respectively, and 12 classifications, which mean the combination of $3 \times 2 \times 2$ classes, were identified in the map. The manual landform classification map can give useful information and ideas about landform evolution of the study area, but it may not fully extract geomorphological features. The automated landform classification map can objectively describe the surface morphology, but it of itself does not give information about landform evolution. Interpreting and extracting geomorphological features from the automated landform classification map will help us to revise the manual landform classification map and to comprehensively understand landform and landslide processes.

Keywords. LIDAR, landslide, landform, classification, DEM, Shirakami Mountains

17.1 Introduction

Japan is subject to tectonic movements which render its topographical features easily eroded. Furthermore, since Japan is located in a humid or snowy climatic region, the high precipitation or heavy snow promotes active mass movements. Earthquake-induced and rain-induced landslides may cause disastrous damage to houses and buildings.

On July 20, 2003, debris flow took place in Minamata City, Kumamoto Prefecture, Japan, triggered by the heavy rainfall (Sassa et al. 2004). In the 2004 mid Niigata Prefecture earthquake of Japan, large landslides occurred in and around Yamakoshi Village in hilly areas (Chigira and Yagi 2006). To avoid such disasters in the future, it is necessary to scientifically assess the areas susceptible to landslides (Lee et al. 2003). Until now, photogrammetry has been used to produce contour maps and the landslides which caused these disasters, have been investigated using these maps. It is important to understand the fundamental mechanism of landsliding by using contour maps which show the internal deformation features in landslides in detail. It has recently become possible to use a new technology called airborne light detection and ranging (LIDER) for direct and detailed measurement of the digital elevation model (DEM). Investigators have begun to extract airborne LIDAR to identify geomorphological and topographical characteristics in landslide areas. For example, Sekiguchi and Sato (2004) reported that contour maps from airborne LIDAR-DEM had been useful for producing a landslide inventory and micro-topographic classification map in the Tama Hills, near Tokyo, Japan. McKean and Roering (2004) used DEM derived from airborne LIDAR to quantify the local surface roughness of a landslide complex near Christchurch, New Zealand. It is expected that LIDAR will accelerate high-precision topographic mapping and detailed landform analysis. In this study we describe the principle of this new technology and we show examples of landform classification mapping in landslide area of Tomari-no-tai area in Shirakami Mountains, Japan.

17.2 Study Area

17.2.1 Location

Shirakami Mountains, which are the largest virgin remnant of the cool-temperate beech forest, are located in northeast Japan (Fig. 17.1). From November to next May, the area is annually covered with heavy snow, whose maximum depth is about 4 m. The forest of beech (*Fagus crenata*), which is broadleaf deciduous tree, is virtually entirely undisturbed, and the area is a wilderness without access trails or man-made facilities.

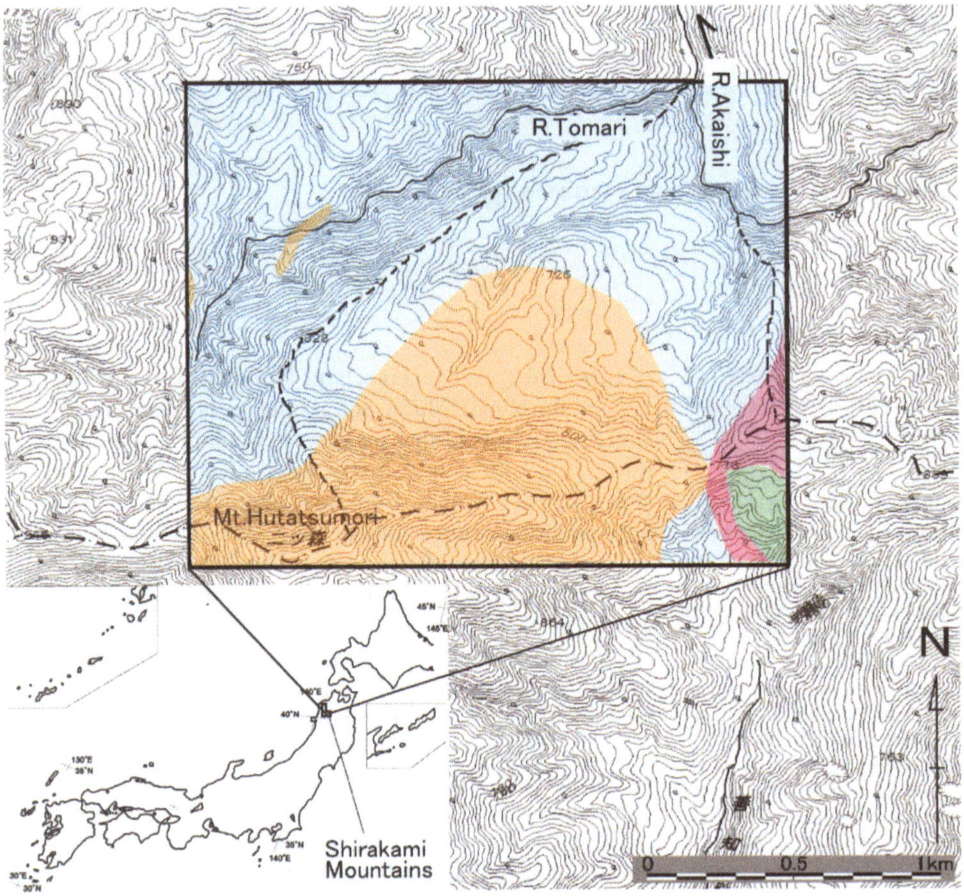

Fig. 17.1. Topographic map and overlaid geological map in the study area. *Pale orange*, late Miocene granite; *pale blue*, middle Miocene mudstone member in Hayaguchigawa Formation; *pale pink*, middle Miocene dolerite and basalt; *pale purple*, middle Miocene basalt member in Hayaguchigawa Formation; *pale green*, middle Miocene acid pyroclastic rock (main part of Hayaguchigawa Formation). The area enclosed by the *black-dotted line*, main ridge, and Akaishi River shows landform classification area in Fig. 17.9

The forest is significant in the traditional Japanese view of nature, therefore, in 1993, United Nations Educational, Scientific, and Cultural Organization (UNESCO) registered the core zone 101.39 km^2 and the buffer zone 6 800 ha in the Mountains as World Heritage Natural Resource based on Convention Concerning the Protection of the World Cultural and Natural Heritage.

17.2.2 Topography and Geology

Study area of Tomari-no-tai is 3.8 km^2 in area, and Fig. 17.1 also shows the area of 1 : 25 000-scale topographic map published by Geographical Survey Institute of Japan, and this map is the existing detail topographic map in the study area. As shown in the geological map (Ozawa et al. 1983) overlaid on Fig. 17.1, the peak of Mt. Hutatsumori (1 086 m) consists of late Miocene granite (pale orange), remaining almost all of the study area occupies the area of mudstone (pale blue) which was deposited in middle Miocene.

Though north face of Mt. Hutatsumori forms steep granite slope, gentle slope surface extends from 600 m to 780 m at the north foot of Mt. Hutatsumori. Yagi (1995) indicated that these geomorphological features were generated by the catastrophic rock slide and the steep slope corresponds to the contact part between the granite and granite-intruded mudstone. Furthermore, 30 cm beneath the gentle slope surface, Yagi (1995) found falling volcanic ash layer and identified it as "To-a" tephra, which was ejected from Volcano Towada 1 000 years ago. Judging from the chronological material, he finally concluded that this Tomari-no-tai landslide area was generated at least 1 000 years ago.

The forest of beech is mainly distributed over the gentle slope surface (Fig. 17.2), and to keep this precious forest, it is important to know landslide topography in detail. In this study 3-D topography data were measured using airborne LIDAR.

17.3 Airborne LIDAR Data Measurement

17.3.1 Airborne LIDAR

In the study area airborne LIDAR data were measured at the beginning of winter, on 30 October, 2004. More information about Airborne LIDAR is available in Ackermann (1999), Wehr and Lohr (1999) and other articles in the same issue of the ISPRS Journal of Photogrammetry and Remote Sensing.

Fig. 17.2.
The beech forest in the study area

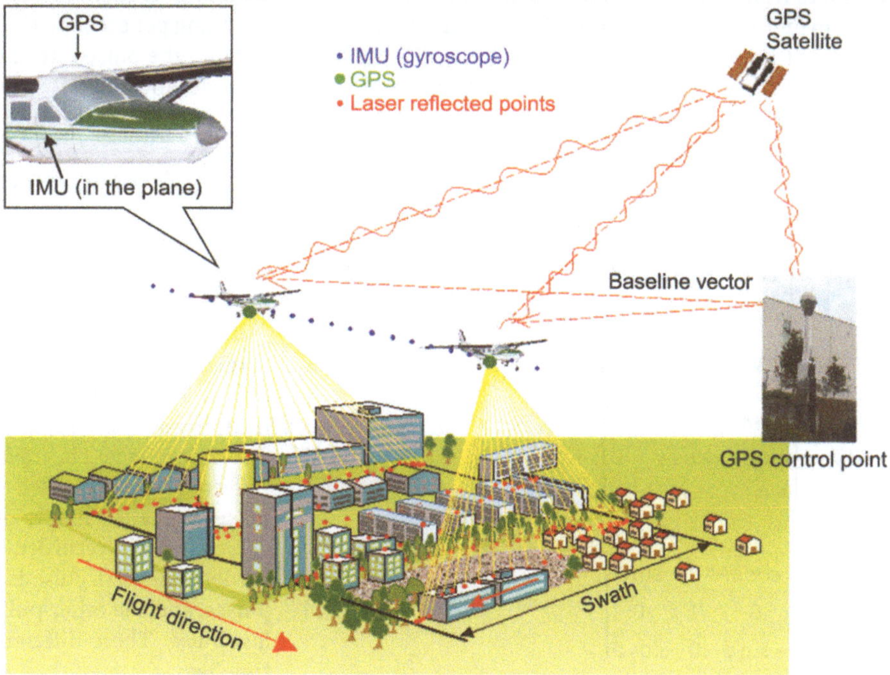

Fig. 17.3.
Concept of the airborne LIDAR

Described simply, airborne LIDAR is an aircraft-mounted electro-optical distance meter that measures the distance between an aircraft and the ground. Figure 17.3 shows the concept of DEM measurement by airborne LIDAR. Its development began in the 1970s and 1980s with an early U.S. National Aeronautics and Space Administration (NASA) system and various other attempts in the United States and Canada. It first became possible by using a pulse laser operating in the near-infrared, which provided clearly recordable return signals after diffusion and reflection from the ground (Ackermann 1999).

Airborne LIDAR measurements offer the following advantages: (*1*) to obtain detailed DEM in digital form rapidly in comparison to digital photogrammetry; (*2*) to measure the ground under vegetative cover provided it is not too dense (Masaharu et al. 2001); (*3*) to obtain dense measurement points, for example, several points per square meter; and (*4*) to measure at night under conditions in which aircraft can fly safely. Daytime flying is undoubtedly safer but when a night-time landslide occurs and it is urgent to map the landform deformation, this advantage will be useful for rapid deployment of disaster rehabilitation activities.

If the laser beams were transmitted in only one direction, the reflected points on the ground would all lie on a line along the flight path. Hence, the aircraft might have to flight many different courses to cover the whole study area. In practice, using a swinging mirror mounted in front of the laser transmitter and receiver, the laser beams are reflected to the cross-flight directions (Fig. 17.3) and the sensor can measure many points not only in the flight direction but also in a cross-flight direction; as a result the sensor can cover the whole study area much more efficiently.

Describing the principle more closely, the airborne LIDAR sensor is an active sensor that measures the distance from the sensor to the target from which the laser beam is reflected. The onboard Global Positioning System (GPS) and Inertial Measurement Unit (IMU) measure position and attitude of the sensor, respectively. The scanning angle of the mirror is also simultaneously measured on the aircraft. The three-dimensional positions of ground targets are obtained by combining these data with measured distances from the sensor (Masaharu et al. 2001).

As in Fig. 17.4, when the aircraft position is (X_0, Y_0, Z_0), and the attitude of the aircraft is $(\kappa, \varphi, \omega)$, the distance separating the aircraft and the target is D, scanning angle of the mirror is θ (the direction directly below the aircraft is zero), and the target position on the ground (X, Y, Z) is calculated as

$$\begin{pmatrix} X \\ Y \\ Z \end{pmatrix} = \begin{pmatrix} X_0 \\ Y_0 \\ Z_0 \end{pmatrix} + \mathbf{R}_Z \mathbf{R}_Y \mathbf{R}_X \mathbf{R}_m D \begin{pmatrix} 0 \\ 0 \\ -1 \end{pmatrix} \quad (17.1)$$

Here, $\mathbf{R}_Z$, $\mathbf{R}_Y$, $\mathbf{R}_X$, and $\mathbf{R}_m$ are

$$\mathbf{R}_Z = \begin{pmatrix} \cos\kappa & -\sin\kappa & 0 \\ \sin\kappa & \cos\kappa & 0 \\ 0 & 0 & 1 \end{pmatrix} \quad (17.2)$$

$$\mathbf{R}_Y = \begin{pmatrix} \cos\phi & 0 & \sin\phi \\ 0 & 1 & 0 \\ -\sin\phi & 0 & \cos\phi \end{pmatrix} \quad (17.3)$$

$$\mathbf{R}_X = \begin{pmatrix} 1 & 0 & 0 \\ 0 & \cos\omega & -\sin\omega \\ 0 & \sin\omega & \cos\omega \end{pmatrix} \quad (17.4)$$

$$\mathbf{R}_m = \begin{pmatrix} 1 & 0 & 0 \\ 0 & \cos\theta & -\sin\theta \\ 0 & \sin\theta & \cos\theta \end{pmatrix} \quad (17.5)$$

An RAMS system made by EnerQuest company was used in this study. The measurement resolution was at least one point per square meter. In the study area, the LIDAR measurement parameters were applied in consideration of an eye-safe factor (Hasegawa and Okamatsu 2001). In the table, the pulse rate (Hz) indicates the number of times the laser pulses are transmitted per second; scanning frequency (Hz) indicates the rate of the mirror's swinging per second. The measurement density is dependent on the pulse rate, the scanning frequency, the flight altitude, and aircraft speed.

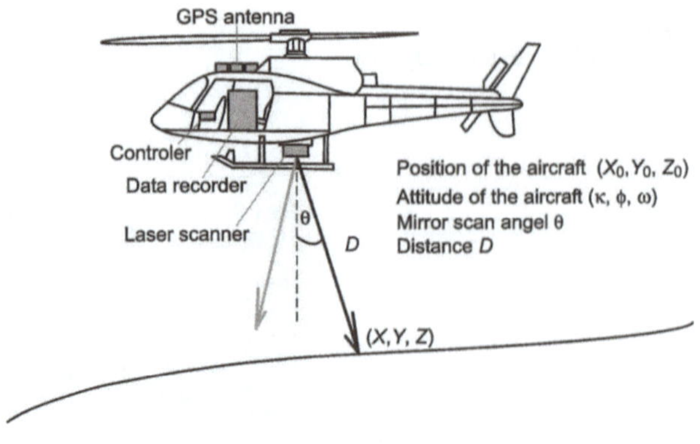

Fig. 17.4. Relation between the positions of an aircraft and a laser reflected point (Masaharu et al. 2001, modified)

17.3.2 First and Last Pulses

When the laser beam is transmitted directly below the instrument and reflected on the terrain surface, it leaves a round "footprint" on the ground (Fig. 17.5).

In Table 17.1, a flying altitude of 1 200 m and beam divergence of 0.3 mrad produces an 36-cm-diameter "footprint". The beam is reflected first from the tree canopy, next, inside the tree, lastly, from the ground (Fig. 17.5). The first reflected beam is called the 'first pulse'; the final reflected pulse is called the 'last pulse'. Other reflected pulses are called simply 'other pulses'. The reflected pulse is received and recorded by the onboard receiver and recorder. The first pulse is received earlier than the other pulses, and the other pulses are received earlier than the last pulse. These difference pulses are identified by the time lag.

In this study return signals from the first, the second, the third, and the last pulses were received (Table 17.1). When the laser beam can penetrate fully into the canopy, the last pulse may be reflected on the ground. But when it cannot penetrate into the dense canopy, the last pulse may be reflected on trees. The points on the trees were filtered out to select only the point data on the ground, namely, DEM. Initially, this filtering was done automatically. Briefly explaining, the study area was subdivided into the search areas, and the points in each search area are approximated by quadratic polynomials, which gave a threshold value from 2 m to 9 m. Those measurement points over the threshold value were eliminated from the original data. Later this filtering was done by manual inspection (Raber et al. 2002). The obtained DEM comprises *x*, *y*, and *z* coordinates of randomly distributed points. Two-meter-grid DEM was generated by interpolating the available original points, using the Triangulation Irregular Network (TIN) method.

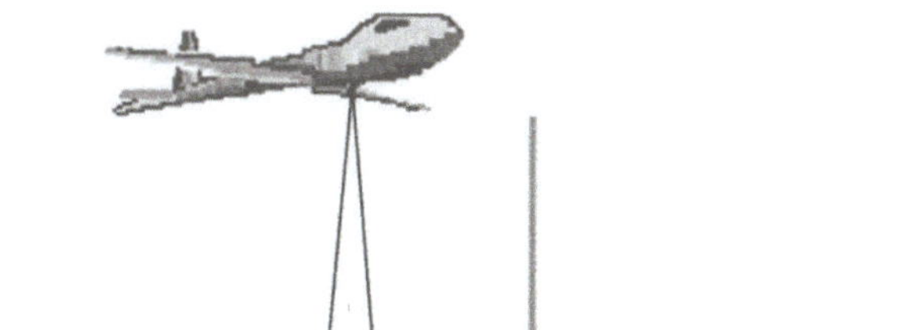

Fig. 17.5. Relation between vegetation and return signal: *1* first pulse, *2* other pulses; *3* last pulse

17.3.3 Measurement Precision

In this study, the measurement precision was not checked by referred ground control points (GCPs). Here, GCPs means the points whose three-dimensional coordinates were surveyed and precisely known. However, even if GCPs were settled, it is difficult to evaluate plane measurement precision, because we cannot precisely indicate where the actual measurement point exists in the footprint.

In terms of the elevation measurement precision, it is able to evaluate it using GCPs. Sato et al. (2004) settled 36 GCPs in the 2 km^2 test site of southwest Japan, measured the topography by airborne LIDAR (flight height: 1200 m) in winter, and compared the elevations between GCP and measurement point of LIDAR-DEM. As a result, root mean square errors were 0.15 m on the flat and bare area, 0.41 m and 1.02 m on the slope less than 20° of broad leaf forest and evergreen needle leaf forest, respectively. This means winter-measured LIDAR-DEM may contain 0.4 m elevation measurement errors in the mountain, broad leaf forest.

17.4 Result

17.4.1 LIDAR-DEM Contour Map

The contour map obtained from LIDAR-DEM is shown in Fig. 17.6. Figure 17.7 shows both (a) photogrammetric and (b) airborne LIDAR-contour maps in the northern portion of the study area. The photogrammetric map, which is same as Fig. 17.1, reveals a smoother topography than the airborne LIDAR map. Using photogrammetric techniques, the contour lines can be obtained directly. Since the two-meter-grid DEM was obtained through interpolation, it may contain more noise than the photogrammetric map. However, as far as LIDAR foot prints reflected on the ground were densely recorded, narrower-interval contour map can be

Table 17.1. Measurement parameters applied in this study

Pulse rate	69000 Hz
Scanning frequency	45 Hz
Mirror scan angle	25° (12.5° in one side)
Measurement mode	First, second, third, and last pulse observation mode
Measurement point interval	1.26 m along flight direction 0.72 m cross flight direction
Footprint	36 cm in diameter
The number of flight course	9 courses (east and west direction)
Overlap ratio between two courses	53.1%
Flying altitude	1200 m above the ground
Flight speed	203.7 km h^{-1} (110 knots)
Beam divergence	0.3 mrad

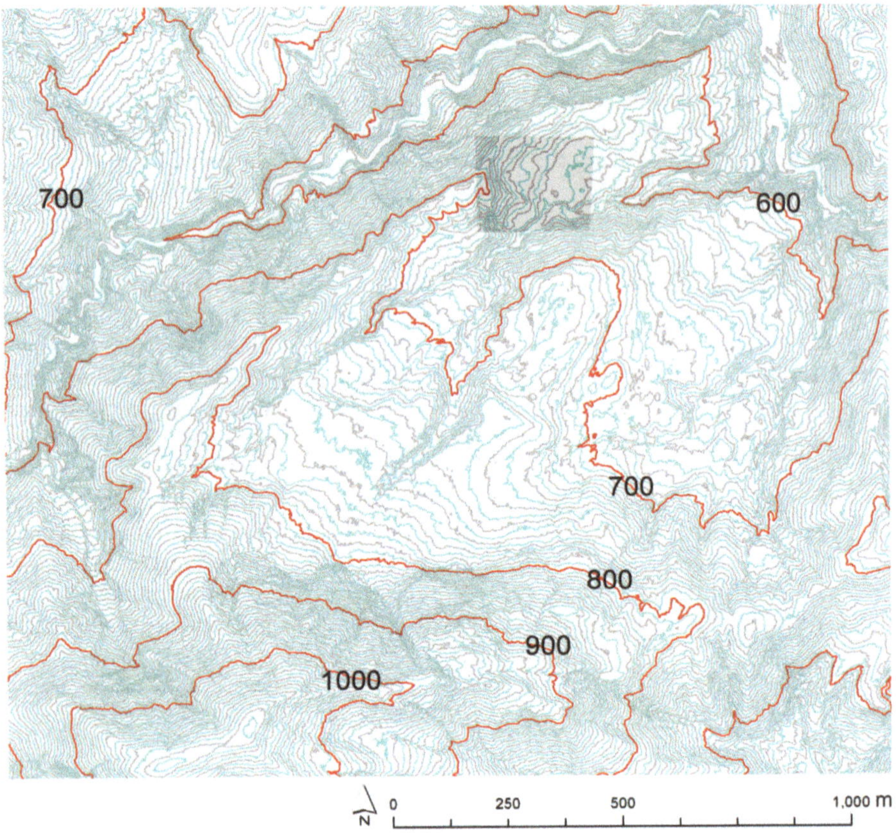

Fig. 17.6.
Airborne LIDAR contour map in the study area. Interval of the contour is 5 m. *Shade area* illustrates the area shown in Fig. 17.7

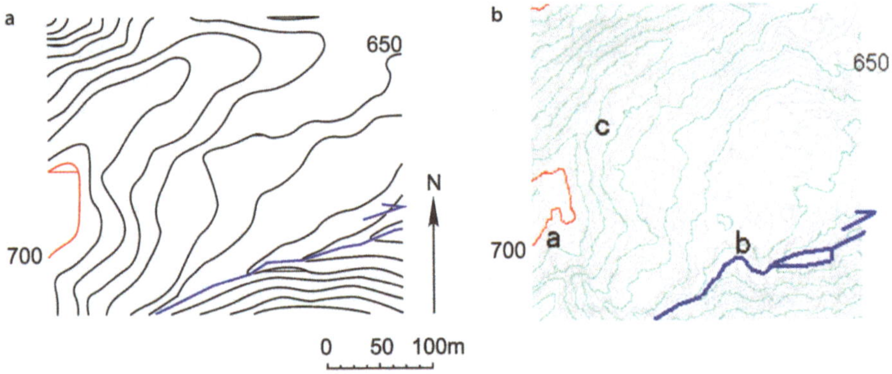

Fig. 17.7.
Comparison of contour maps by (**a**) photogrammetry and (**b**) airborne LIDAR. Contour interval is 10 m and 2 m, respectively

printed out such as Fig. 17.7b. Furthermore, based on the interpretation of aerial photographs and the field survey, it was found that valleys and other geomorphological features could be seen in better detail in the airborne LIDAR contour map than in the photogrammetric contour map.

For example, the deeply incising valley is identified in "a" of Fig. 17.7b, but Fig. 17.7a illustrates it as shallow valley. The meander of the channel in "b" of Fig. 17.7b is straight channel in Fig. 17.7a. Main scarp of large landslide in "c" of Fig. 17.7b is illustrated more clearly than in Fig. 17.7a. Furthermore, detail landslide characteristics such as subordinate scarp or landslide sub-block could be seen in better detail in Fig. 17.7b than in Fig. 17.7a.

17.5 Discussion

17.5.1 Manual Topography Classification Map

LIDAR-contour map at the scale of 1 : 2500 whose contour interval is 2 m was printed, and it was interpreted using 1 : 8000 aerial photographs taken in 1975 by Geographical Survey Institute. In interpreting the map, detailed topographic characteristics were paid attention to. Tamura (1981) studied hill-landform characteristics in six areas, which are located at geologically and climatically different region in Japan. He generalized micro topographic classification as follows: crest slope, side slope,

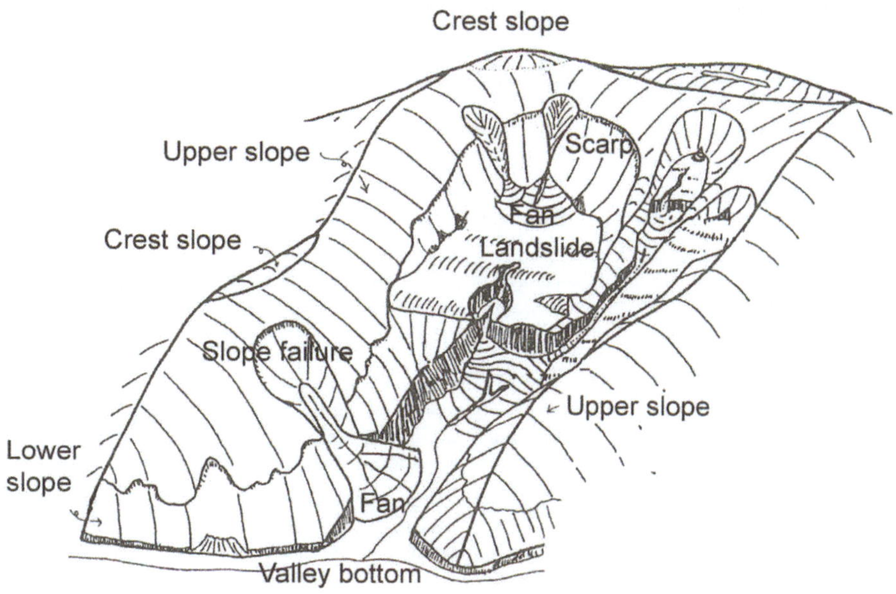

Fig. 17.8. Concept of landform classification (Yagi 1995, simplified)

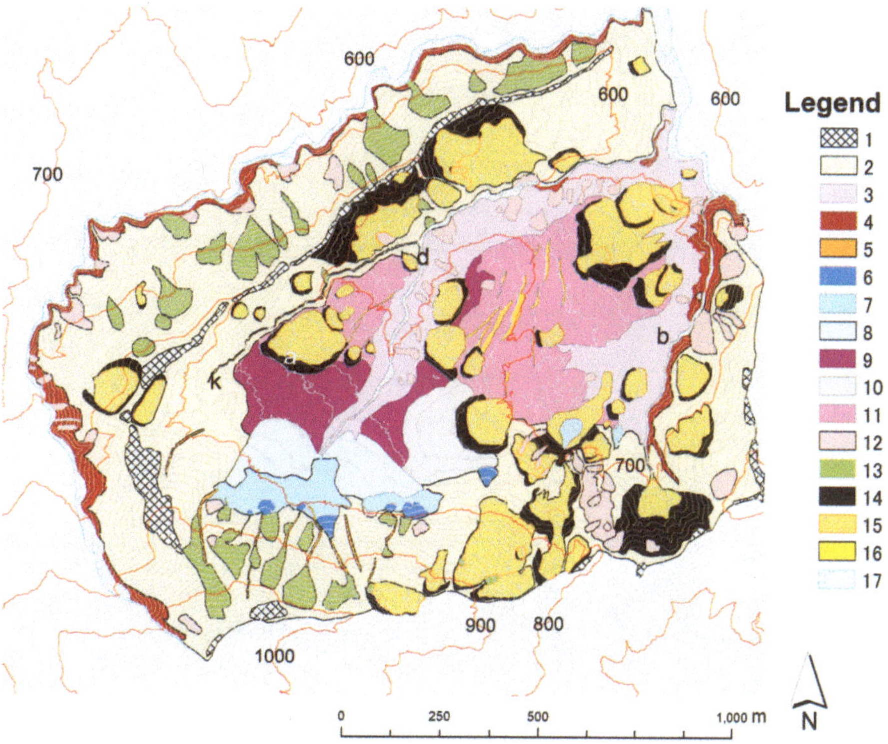

Fig. 17.9. Landform classification map in the study area, interpreted on LIDAR-contour map. Refer to Fig. 17.10 about the explanation of "a", "b", and "d" in the figure

head hollows, head floors, and channel. Yagi (1995) applied this generalized classification to the landslide area in Shirakami Mountains and illustrated concept of the landform classification as shown in Fig. 17.8. This study's topography classification is based on Yagi (1995), however, the classification is simplified and revised to describe the landform characteristics of the study area as shown in Table 17.2.

By interpreting the LIDAR-contour map, landform was classified into 17 categories as shown in Fig. 17.9. The classifications related to the catastrophic rock slide is thought to be "upper slope", "landslide side slope", "high-level surface", "high-level gentle surface", "low-level surface", and "enchelon cliff". The classifications related to the mass movement after the catastrophic rock slide is "lower slope", "slope failure", "scarp", and "landslide". The classifications related to the landform change by snow is "avalanche furrows", "steep slope alluvial fan", "medium slope alluvial fan", "gentle slope alluvial fan", and "slope failures incised by avalanche furrows". The remains "crest

Table 17.2. Explanation of landform classification in Fig. 17.9

No.	Legend	Explanation
1	Crest slope	A little gentle slope around crest flat surface. It is located along ridge, and it includes flat surface on the summit. Relatively thick soil covers both crest flat surface and crest slope
2	Upper slope	It includes steep slope on the north face of Mt. Hutatsumori, namely, main scarp formed by the catastrophic rock slide occurred at least 1 000 years ago
3	Landslide side slope	It consists of mainly side slope of the low-level surface. Upper part more than 660 m in elevation mainly consists of weathered granitoid flow-deposits, and lower part less than 620 m in elevation consists of weathered and fractured mudstone
4	Lower slope	It is located at the lowest part of the lower slope next to the valley bottom, and it forms remarkable steep slope
5	Avalanche furrows	Snow avalanche furrows (Sekiguchi and Sugiyama 2003), which are U-shaped shallow in cross section, incise not only granite but also mudstone steep slopes. Occasionally steep alluvial fan exists at the foot of avalanche furrow. There are many furrows which do not reach foot of slopes
6	Steep slope alluvial fan	Debris whose origin is weathered granite forms steep fan (22–25°) at the foot of avalanche furrows and slope failures. According to the field survey, grain size is not uniform, from medium sand to cobble. And granite cobble is sub-angular. It is thought that deposited debris was given by snow avalanche
7	Medium slope alluvial fan	Debris whose origin is weathered granite forms medium slope fan (9–14°)
8	Gentle slope alluvial fan	Debris whose origin is weathered granite forms gentle fan around the gentle alluvial fans (5–8°). According to the field survey, grain size is not uniform, from fine sand to pebble. Fine deposits may have been flown and deposited by temporal streams of the spring snow-melted water or the summer heavy rain
9	High-level surface	The surface whose strata are flow deposition of weathered granite. According to the field survey, outcrop along the surface-incising valley shows thick massive sand layer includes angular granite cobble layer. Total thickness of the deposition is 50 m, according to the observation of the outcrops along the central valley incising the high-level surface
10	High-level gentle slope	Sediments are same as high-level surface, but its surface indicates concave gentle slope
11	Low-level surface	The surface is thought to have been produced by the catastrophic rock slide. According to the field survey, outcrop along the surface-incising valley shows flow-deposits of weathered granite, whose layer is 30 m at thickest
12	Slope failure	Shallow disrupted landslide and rock slide avalanche
13	Slope failure incised by avalanche furrows	Avalanche furrows are clearly identified on the slope failure
14	Scarp	Landslide scarp
15	Landslide	A coherent landslide corresponds to the Keefer's (2000) Category II slide (*Jisuberi* in Japanese). This classification means landslide-moved soil and debris mass. Large landslides are mainly identified on the high- and low-level surface. Channels and gullies usually erode the landslide mass at the lateral sides and/or center
16	Enchelon cliff	It may indicate pressure ridges formed by the catastrophic rock slide
17	Valley bottom	It includes not only small channel but also main streams such as Akaishi and Tomari Rivers

slope" keeps original landform before the catastrophic rock slide, "valley bottom" have always changed its topography before, in process of, and after the catastrophic rock slide.

1. *The classifications related to the catastrophic rock slide.* The upper slope of the granite slope on the north face Mt. Hutatsumori is correspondent to the main scarp, and the higher- and lower-level surfaces are correspondent to the landslide mass of the catastrophic rock slide (Yagi 1995). The landform process seems to be different between the high- and low-level surfaces.

Figure 17.10a was taken in the spot "a" in Fig. 17.9, where the high-level surface was incised by the gully and eroded by the scarp whose relative height is ca. 20 m. As shown in Fig. 17.10a, the sand from medium to coarse size were massively piled up, and interbedded granite angular cobble and boulder layers exists below Fig. 17.10a. Judging from the facies, these sequences are thought to be yielded by debris flow. Surely snow avalanche, the stream from melted snow and heavy rain may transport debris from the granite mountain, however, they cannot give such the thick fine-medium deposi-

Fig. 17.10. Photographs taken in the field. The positions where (**a**), (**b**), and (**d**) are taken is shown as "a", "b", and "d" in Fig. 17.9

tion at a time. Angular gravels overlain on the sand is the deposition along the gully. These strata-sequence was observed from top to bottom of the scarp.

These strata-sequence was also observed on the outcrops of landslide side slope next to the low-level surface, however, the lower the elevation is, the coarser the granite angular sediments are, and a few weathered-granite big boulders were observed near the spot "d" and on the spot "b" in Fig. 17.9. Figure 17.10b shows such the boulder, which was heavily weathered and turned into clay. Figure 17.10c shows all outcrop and at the bottom of the outcrop, mudstone as the basement rock was observed. The basement mudstone was also observed in the spots "k" and "d" in Fig. 17.9, Fig. 17.10d shows the outcrop of mudstone at the bottom of the landslide side slope in the spot "d". Weathered mudstone were observed on the downstream landslide side slope from the spot "b", neither granite debris-flow deposition nor weathered-granite boulders were observed along the downstream landslide side slope.

Judging from the above observation, spots "k", "b", and "d" are thought to indicate slip surface of the catastrophic rock slide. And the following catastrophic rock slide process is proposed. Weathered-granite was moved downward and big boulders and coarse materials formed pressure zone of the slide and formed low-level surface. The enchelon cliff on the low-level surface may indicates the lateral pressure ridge or subordinate scarp formed according to the movement. In the depression zone formed by the slide, liquefied fine granite-weathered materials were moved downward at the high speed, filled the depression part and partially covered low-level surface.

2. *The classifications related to the mass movement after the catastrophic rock slide.* Large and small landslides are identified not only near the boundary between high-level surface and low-level surface but also on the upper slope. The landslides located at the end of the low-level surface are thought to have been triggered by the tensions according to down cutting of the valley. Almost all landslides are thought to have been mainly formed after the catastrophic rock slide. A few landslides may have been formed secondarily in the catastrophic rock slide, but it is difficult to decide the forming age because chronological materials seen in the stratum are scarce.

Slope failures means disrupted shallow landslides and rock slide avalanche (Houkai in Japanese) correspond to the Category I slide of Keefer (2000). They are mainly identified on the upper slope, and they are thought to have been formed after the catastrophic rock slide. The lower slope is located next to the valley bottom, and it is identified as the eroded belt below the upper slope. It is thought to have been formed according to the down cutting of valley.

3. *The classifications related to the landform change by snow.* Avalanche furrows incise the upper slope, annually occurring snow avalanche shave surfaces of the upper slope and feed the granitoid debris to the foot of the upper slope. The transported debris forms fans at steep, medium, and gentle slopes, and cover the high-level surface, however, total thickness of the fan deposition is thought to be thinner than the thickness of the debris-flow deposition which comprises high-level surface.

   Avalanche furrows also incise mudstone slope at the past-occurred slope failure sites. Such type of "Slope failure incised by avalanche furrows", which is thought to have been formed in the older time than the "slope failure", was shown in Fig. 17.9.

4. *"Crest slope" and "valley bottom".* Crest slope is located at the main and derivative ridges, whose width is narrow. It is thought that granite and mudstone are weathered and thick soil layer covers the crest slope, however, the crest slope on the main ridge also includes flat surface around the summit of the main ridge and the depression on the main ridge, formed by nivation (snow erosion).

   Valley bottom in this study includes debris-flow deposit terraces along the channel. For example, such the large terraces are identified along Akaishi River, but they are not classified in Fig. 17.9. At the west side of the study area debris deposition fed by snow avalanche forms large flat surface along the channel, the valley bottom also includes such the deposition surface.

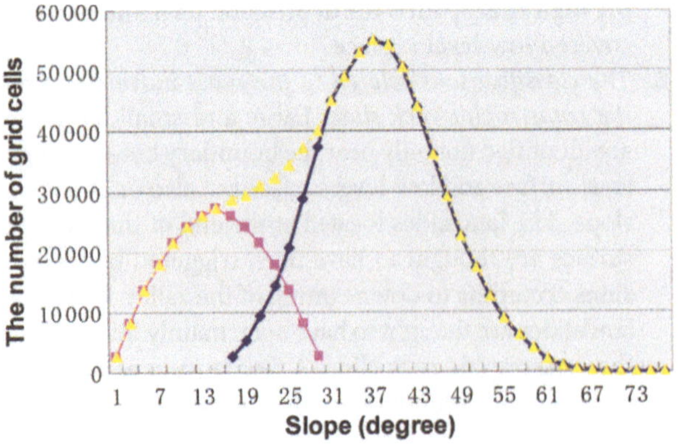

Fig. 17.11. Frequency of slope calculated from LIDAR-DEM data

17.5.2 Automated Landform Classification Map

To classify the landform automatically, Iwahashi and Pike (2007) proposed unsupervised method using just three variables such as slope, surface convexity, and local texture (feature frequency, or spacing). According to Iwahashi and Kamiya (1995), the three variables are sufficient to for a geometric signature designed to create terrain-unit maps that approximated aerial photographs interpretation maps in Japan.

17.5.2.1 Slope

Steepness of slope is so basic property of the land surface. To deal fine-resolution slope quantitatively and investigate topographic characteristics statistically, DEM data are more useful than contour map. In this study slope was calculated from two-meter-grid LIDAR-DEM, the calculation was operated on conventional 3×3 matrix of neighboring grid-cell elevations by the algorithm as follows: when the elevation data, whose interval is "d", were arranged as H_{ij} $(i, j = 1, 2, 3)$, $S_x = \{(H_{11} + H_{21} + H_{31}) - (H_{13} + H_{23} + H_{33})\}/6d$, $S_y = \{(H_{11} + H_{12} + H_{13}) - (H_{31} + H_{32} + H_{33})\}/6d$, then, slope is square root of individual S_x and S_y products.

As shown in Fig. 17.6, the study area is essentially all in slopes without plain areas. The frequency of slope is shown as yellow curve in Fig. 17.11, peaks at 37° with a "shoulder" at about 15°; little low-sloping terrain is evident. Because these two values coincide with mean slope for the area's dominant slope processes, respectively rapid slope failure and slower landsliding, the peak and "shoulder" on the yellow curve in Fig. 17.11 indicate how the graph might be divided into two process-related domains, as pink and dark-blue curves in Fig. 17.11. Slope for the entire study area, however, averages 31°, twice that of the 15° characteristics of high- and low-level surface in Fig. 17.9.

Figure 17.12a shows slope mapping, furthermore, the slope was divided into three parts according to the peak and the "shoulder", namely, "1" (gentle), "2" (medium), and "3" (steep) in Fig. 17.12d, respectively.

17.5.2.2 Surface Texture

The variable "surface texture" is used to express the surface's "fine" or "coarse". Texture is calculated by extracting grid cells that outline the distribution of valleys and ridges in the DEM. These cells are identified from differences between the original grid-DEM and an obtained grid-DEM derived by passing the original through the 3×3 median filter; the filter was originally used to remove high-frequency spatial "noise" from a digital scene by replacing original values with value of central tendency (Iwahashi and Pike 2007). After the filtering the original grid-DEM, the obtained grid-DEM shows smooth surface where ridges are shaven

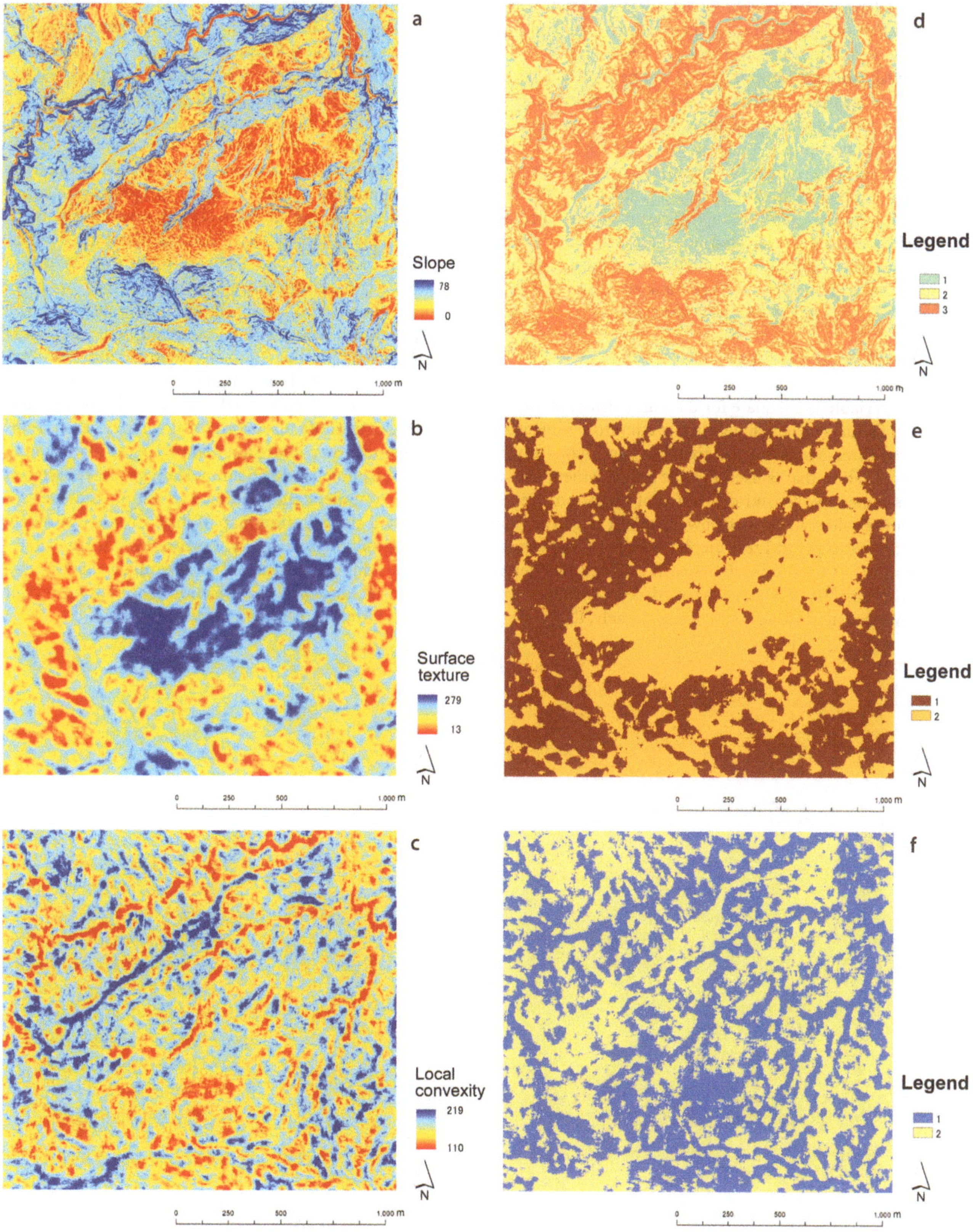

Fig. 17.12. Distribution map of (**a**) slope, (**b**) roughness, (**c**) local convexity. Classification map of (**d**) slope: *1* gentle, *2* medium, *3* steep; (**e**) surface texture: *1* coarse, *2* fine, (**f**) local convexity: *1* concave, *2* convex

and valleys are reclaimed. So the differences between the two kinds of grid-DEM shows grid cells of "peak" and "pit" (coincident with ridge and valley, respectively).

Iwahashi and Kamiya (1995) defined surface texture at each grid cells as the sum number of peaks and pits within a circle whose constant radius is ten cells, namely, the circle occupies 314 cells in area. The higher the local convexity is, the higher the magnitude of the convexity is. The Frequency of the surface texture indicates symmetrical one-modal graph. Mapping result of the calculated surface texture is indicated in Fig. 17.12b, furthermore, the surface texture was divided into low and high value types according to the average value, namely, "1" (coarse) and "2"(fine) in Fig. 17.12e, respectively.

17.5.2.3 Local Convexity

This variable is adequate for discriminating among low-relief features on landslide. Surface curvature is measured by the three-by-three Laplacian filter, an image-processing operation that is used in edge enhancement and approximates the second derivative of elevation, yielding positive values in convex-upward areas, negative values in concave areas, and zero on planar slopes (Iwahashi and Pike 2007). Local convexity at each grid cell was defined as the number of convex-upward cells within a circle whose constant radius is ten cells, namely, the circle occupies 314 cells in area (Iwahashi and Kamiya 1995). The higher the local convexity is, the higher the magnitude of the convexity is. The Frequency of the local convexity indicates symmetrical one-modal graph. Mapping result of the calculated local convexity is indicated in Fig. 17.12c, furthermore, the local convexity was divided into low and high value types according to the average value, namely, "1" (concave) and "2"(convex) in Fig. 17.12f, respectively.

As mentioned above, the three variables were subdivided into three, two, and two classes, respectively, and as shown in Fig. 17.13, 12 classifications, which are the combination of 3×2×2 classes, were identified. When we compare the automated classification map with the manual classification map, the automated map describe morphological features in better detail than the manual map. For example, the automated map can indicates shallow depressions and low mounds on the high-level surface in the manual map, and the automated map describes narrow north-south direction cliff along the enchelon cliff, but these narrow cliff are not mapped in the manual map. These morphology may be important geomorphological features to understand landform evolution, however, the automated classification map of itself does not show the meaning of the feature. By interpreting the automated map and adding the extracted features on the manual map, the manual landform classification map will be revised to understand landslide processes.

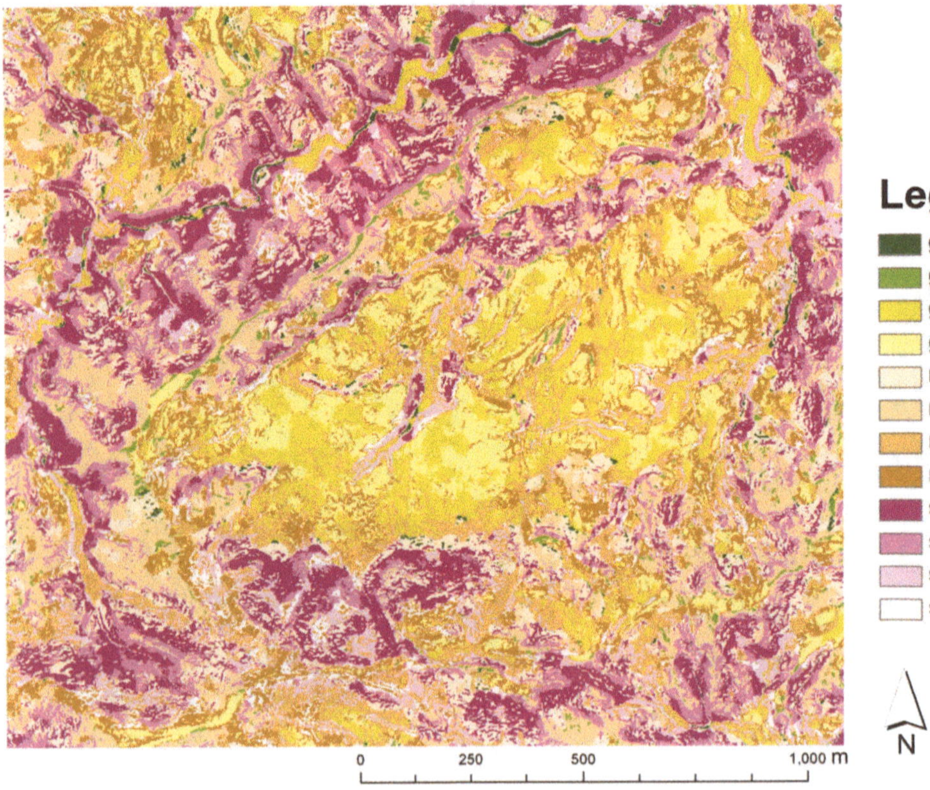

Fig. 17.13. Automated landform classification map using slope, surface texture, and local convexity

17.6 Conclusion

It was found that the contour map using LIDAR-DEM illustrates geomophological features in better detail than the photogrammetric contour map does. Different from photogrammetric technique, LIDAR techniques directly measure 3-D coordinates of the ground. However, even if there are many measurement points on the trees and on the grasses and they are scarcely plotted on the ground, LIDAR-DEM may not correctly illustrate geomorphological features in the study area. If the study area is covered with deciduous trees, the non-leaf season should be selected to measure LIDAR-DEM.

Using LIDER-contour map the manual landform classification map was produced, and considering the field survey result, the catastrophic rock slide process were discussed. Furthermore, the study produced the automated landform classification map, and revealed it as the promising tool to understand landform and landslide processes.

References

Ackermann F (1999) Airborne laser scanning – present status and future expectations. Isprs J Photogramm 54:64–67

Chigira M, Yagi H (2006) Distribution of landslides triggered by the 2004 Mid Niigata Prefecture earthquake. Eng Geol 45:83–90

Hasegawa H, Okamatsu K (2001) Detailed landform feature and characteristics extraction with high dense DTM data. In: Proceedings of the autumn Conference of the Japan society of photogrammetry and remote sensing, pp 189–192 (in Japanese)

Iwahashi J, Kamiya I (1995) Landform classification using digital elevation model by the skills of image processingainly – mainly using the digital national land information. Geoinformatics 6(2):97–108 (in Japanese with English abstract)

Iwahashi J, Pike RJ (2007) Automated classifications of topography from DEMs by an unsupervised nested-means algorithm and a three-part geometric signature. Geomorphology (in press)

Keefer DV (2000) Statistical analysis of an earthquake-induced landslide distribution – the 1989 Loma Prieta, California event. Eng Geol 58:231–249

Lee S, Ryu JH, Min K, Wo JS (2003) Landslide susceptibility analysis using GIS and artificial neural network. Earth Surf Proc Land 28:1361–1376

Masaharu H, Hasegawa H, Ohtsubo K (2001) Three-dimensional city modeling from airborne laser scanning. In: Proceedings of the 20th International Cartographic Conference, International Cartographic Association, Beijing, 2, pp 1337–1343

McKean J, Roering J (2004) Objective landslide detection and surface morphology mapping using high-resolution airborne laser altimetry. Geomorphology 57:331–351

Ozawa A, Tsuchiya N, Sumi K (1983) Geology of the Nakahama district. Geological Survey of Japan, 62 p (in Japanese with English abstract)

Raber GT, Jensen JR, Schill SR, Schuckman K (2002) Creation of digital terrain models using an adaptive lidar vegetation point removal process. Photogramm Eng Remote Sens 68: 1307–1315

Sassa K, Fukuoka H, Wang G, Ishikawa N (2004) Undrained dynamic-loading ring-shear apparatus and its application to landslide dynamics. Landslides 1:7–19

Sato HP, Sekiguchi T, Orimo K, Nakajima T (2004) Accuracy validation of airborne laser scanning DTM using the ground control points. J Jpn Soc Photogrammetry Remote Sensing 43(4):13–21 (in Japanese with English abstract)

Sekiguchi T, Sato, HP (2004) Mapping of micro topography using airborne laser scanning. Landslides 1(3):195–202

Sekiguchi T, Sugiyama M (2003) Geomorphological features and distribution of avalanche furrows in heavy snowfall regions in Japan. Z Geomorphol NF 130:117–128

Tamura T (1981) Multiscale landform classification study in the hills of Japan: Part II, application of the multiscale landform classification system to pure geomorphological studies of the hills of Japan. Science Report of Tohoku University 7th Series (Geography) 31:85–154

Wehr A, Lohr U (1999) Airborne laser scanning – an introduction and overview. Isprs J Photogramm 54:68–82

Yagi H (1995) Landform and its evolution in Shirakami Mountains. In: National Parks Association of Japan (ed) Comprehensive research report on Shirakami Mountains Natural environment conservation region. pp 45–75 (in Japanese)

Chapter 18

Integration of Remote Sensing Techniques in Different Stages of Landslide Response

Paolo Canuti · Nicola Casagli* · Filippo Catani · Giacomo Falorni · Paolo Farina

Abstract. Recent advances in remote sensing techniques have yielded numerous new applications which provide benefits for all stages of landslide management. This paper describes how the data produced from different types of sensors and platforms have been used to estimate, model and mitigate landslide risk in sites in Italy and in other parts of the world. EO data have been utilized to update landslide inventories, with the identification of new landslides and the ratification or modification of landslide boundaries and states of activity, and to improve landslide hazard and risk assessment procedures at regional scales. Radar data provided a global topographic dataset that was used to model lahar inundation hazard while an example of the millimetric resolutions attainable from repeat-pass satellite radar data, with its beneficial implications for landslide monitoring, is also illustrated. Ground-based systems are shown to be innovative early warning systems for slow-moving landslides while coupled techniques involving both optical and radar images can provide support for the management of emergencies. The illustration of these case histories demonstrates the increasing importance of remote sensing in all facets of landslide management, with significant advantages for both policy makers and society.

Keywords. Remote sensing, landslides, landslide risk, landslide management

18.1 Introduction

In recent years the development of remote sensing techniques has produced significant benefits for landslide risk management practices. The data produced from the numerous different types of sensors available today have practical applications in all stages of landslide response: from prediction and early warning to post-crisis management.

The strategic importance of remote sensing in the risk assessment related to all natural hazards is underscored by the proliferation of international initiatives and programmes sponsored by global agencies such as the United Nations (UN), the European Space Agency (ESA), the World Meteorological Organization (WMO), the European Union (EU) and many others. Notable programmes include the Global Landslides Observation Strategy of the International Consortium on Landslides (ICL), the International Global Observing Strategy – Geohazards (IGOS-G) and Global Monitoring for Environment and Security (GMES).

In this paper, we illustrate examples of applications and contributions of several earth observation techniques to different phases of landslide management. The reported cases regard various types of sensors and processing techniques, ranging from radar interferometry (InSAR) to Very High Resolution (VHR) optical images. The sites chosen for the applications are mainly located in Italy which, due to its varied topographic, geologic and climatic settings, in combination with its high density of elements at risk and the fact that these often have a high specific worth (i.e. cultural heritage), make it an ideal site.

The paper is divided into seven sections, each containing at least one relevant case history, that describe the contributions of remote sensing to different aspects and stages of landslide management.

18.2 Contributions to Landslide Inventories

Landslide inventories report the spatial distribution of existing slope movements, frequently including details about the landslide typologies and their state of activity (Wieczorek 1984). Today the most common methodologies for their compilation are based on aerial-photo interpretation, on field surveys and on the collection of local databases (Crozier 1984; Soeters and Van Westen 1996). The relevant advantage gained from the coupled application of remote sensing systems regards the rapid and easily updatable acquisitions of data over wide areas, such as large river basins. The ESA SLAM (Service for Landslide Monitoring) project is an initiative with the specific aim of defining a service based on the integration of earth observation (EO) data within current practices of landslide mapping and monitoring. The implemented methodological approach combines InSAR data with the interpretation of optical images at basin scale (Colombo et al. 2003; Farina et al. 2006).

In Italy, the SLAM project study site is the Arno River Basin in Tuscany. The catchment has an area of 9 131 km^2 and a mean elevation of 353 m a.s.l. while the Arno River is 241 km long. The basin is representative of the typical Mediterranean conditions, both in terms of landslide type and environmental setting. Within the Arno Basin the landslide inventory contains about 28 000 individual landslides, which were identified and mapped through the integration of conventional geomorphologic tools and different data sources. Most of the landslides are earth slides/earth flows (75%), solifluctions and other slow shallow movements (17%) and flows (5%) while soil slips, and in general shallow landslides, seem to be of limited importance within the basin.

The methodology proposed within the SLAM project to update the Arno Landslide Inventory relies on the integration of ground displacement measurements over a sparse grid of points provided by a Permanent Scatterers (PS) analysis within the Arno Basin area. The PS technique is based on the identification of stable natural reflectors within long temporal sequences of InSAR images (Ferretti et al. 2001) which allow millimetric terrain motion detection to be achieved on image pixels containing PS.

About 350 ERS images were interferometrically processed with the PS technique, leading to the identification of approximately 600 000 PS within the study site. Satellite and airborne optical images were also used for performing a geomorphological analysis and for landslide identification (Fig. 18.1). This led to the mapping of over 1 500 landslides with the relative assessment of their boundaries and state of activity.

In general, the integration of PS data requires cartographic and optical supports, such as VHR satellite imagery and aerial-photos, for identifying possible diagnostic morphologies and terrain features related to landslides in order to spatially expand the point-wise PS information. The main benefits obtained from the application of the PS technique to landslides regard:

a better defined boundaries of already detected mass movements and/or;
b better defined states of activity and/or;
c the detection of previously unmapped unstable areas.

The main benefits reported by the local authorities involved in the project as end-users are: (*1*) the capability of the methodology to perform ground displacement measurements over urbanized areas, where risk conditions are usually higher, and (*2*) the possibility of utilizing the PS dataset for fieldwork over areas

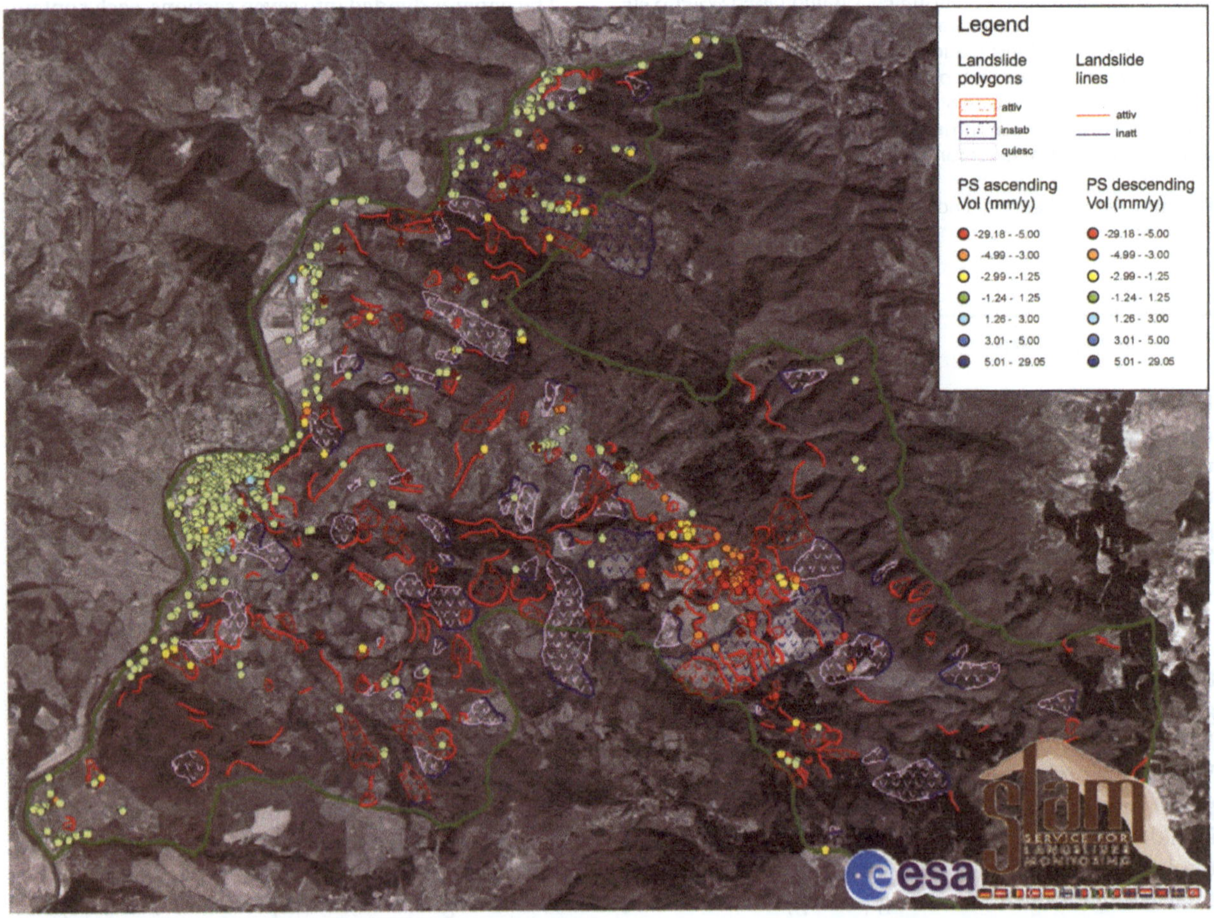

Fig. 18.1. Portion of the updated Arno Basin landslide inventory map (near the city of Florence) obtained through the combined analysis of satellite HR images (SPOT5), aerial photos and Permanent Scatterers

where investigations with traditional instrumentation do not exist.

The positive feedback implies that within the next years, due to the imminent launch of space-borne SAR missions with acquisition parameters better tuned for landslide investigations, the methodology has the potential to become a fully operational tool for landslide mapping.

Although this methodology already represents a promising tool for landslide investigations some challenges still need to be overcome to be able to fully exploit its potential. One limitation is the required presence of buildings or other man-made artefacts within the unstable area; the vast majority of PS, in fact, are anthropogenic structures. Another constraint derives from the limited displacement rates (i.e. up to 4–5 cm yr^{-1} with C-band data) that can be measured with this technique.

18.3 Improvements to Landslide Hazard Assessment

The ESA – SLAM project also illustrated a method for the creation of a landslide hazard map of the Arno River Basin by means of remote sensing techniques, Artificial Neural Networks (ANN) and Geographic Information Systems (GIS).

Varnes and IAEG (1984) defined hazard as the expected probability of a mass movement of a given intensity which takes place in a certain area within a given time span, which implies that landslide hazard assessments should include both spatial and temporal occurrence predictions. However, as the spatial probability of occurrence is much simpler to perform, temporal forecasts are often neglected; in these cases the term "landslide susceptibility" is more appropriate. In the case of the Arno River Basin, where the SLAM project produced a high quality landslide inventory with updated information on state of activity, it is possible to perform temporal hazard predictions in areas where reactivations predominate over first-time failures. Reactivation dates, obtained by combining traditional and remote sensing techniques, allow the expected return period for each of the mapped landslides to be assessed. To extend the assessment of landslide hazard to those areas currently free of landslides, statistical methods implemented within an artificial neural network were used (Catani et al. 2005a).

In the Arno Basin the size of the territory and the variability of the environments involved require the application of a statistical method flexible enough to allow for a spatially-variable calibration method. Landslide hazard characterization was based on the definition of suitable statistical estimators by means of a set of artificial neural networks. Processing was performed through a discrete pixel analysis followed by the definition of unique condition units (Bonham-Carter 1994; Chung et al. 1995; Ermini et al. 2005) for statistical treatment within a GIS environment. For application of the method, due to the ample variability of physiographic settings, the basin was divided into five relatively homogeneous domains regarding lithology and climate, for each of which different ANN model parameterizations were applied.

A preliminary monovariate statistical analysis of the parameters within mapped landslides in the basin produced five mass movement preparatory factors: slope angle, lithology, profile curvature, land cover and upslope contributing area. These parameters were used to construct a series of model vectors for the training and testing of the ANN. Following model training a reclassification scheme was applied to convert ANN output values into four levels of susceptibility which were then transformed into the five classes shown in the landslide hazard map (Fig. 18.2). Validation was performed by comparing predicted and observed mass movements, using two different methods: (*1*) the inventory map and dedicated field surveys and; (*2*) information from the PS dataset.

The final map was adopted by the Arno River Basin Authority and represents one of the first examples in Europe in which remote sensing data is utilized to update land use regulations regarding landslide hazard.

18.4 Basin Scale Risk Assessment

The landslide risk map of the Arno River Basin represents the end-product of the risk assessment procedure implemented within the SLAM project, which comprises the phases described in the previous sections. To determine total risk for the study area a quantitative risk assessment (QRA) procedure was carried out, which included, in addition to an assessment of landslide hazard, a detailed evaluation of the elements at risk, with particular importance assigned to the estimation of their worth in terms of economic value, and an estimate of the vulnerability of the elements at risk (Catani et al. 2005a; Farina et al. 2006). In practical terms, the QRA was calculated on a pixel- by-pixel basis, by computing the product of pixel hazard by potential worth of losses for five different time spans. Remote sensing data was used for determining land use over the entire area of the basin with the implementation of automatic image classification procedures.

The resulting risk map of the Arno River (Fig. 18.3) indicates that more than 20% of the basin is exposed to a relatively high level of risk. Hilly areas in the north-eastern and southern parts of the catchment are especially susceptible, due to the distribution of scattered but densely populated towns and of unstable terrains.

Although a pixel-based approach was used in the SLAM project, as it provided a more detailed level of information, a UCU-based method is also possible; this is generally the best solution for regional land use planning and

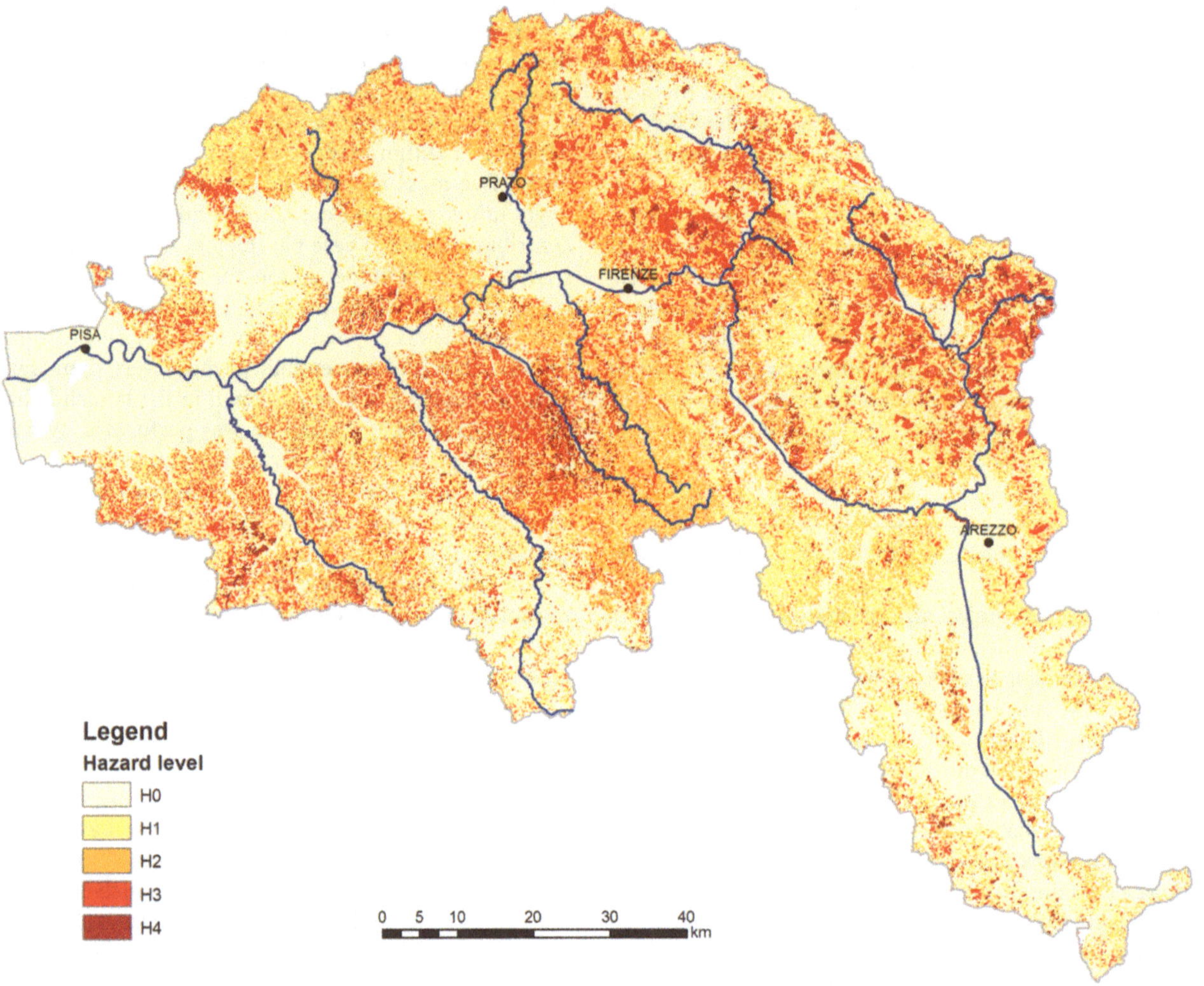

Fig. 18.2. Landslide hazard map of the Arno River Basin. Hazard is divided into 5 classes ranging from H0 (no hazard) to H4 (very high hazard)

regulation design, since it offers a synoptic view of large areas and is easy to implement.

The final results of the research are now undergoing a process of integration and implementation within land planning and risk prevention policies and practices at local and national level.

18.5 Landslide Modeling Support

Topographic data plays a fundamental role in the modeling of many geomorphic processes. However, until recently accurate digital topographic data were unavailable for many areas of the world, particularly developing countries. The recent Shuttle Radar Topography Mission has filled this significant void by providing a freely available near-global topographic dataset.

In 2000 the SRTM mission acquired high resolution topographic data spanning 80% of the Earth's landmass using a dual frequency single-pass InSAR system (Bamler 1999; van Zyl 2001). Much of the instrumentation employed by SRTM derived from experience gained during several previous satellite and shuttle missions that also implemented SAR technology. To achieve the goal of global mapping with no gaps in the data, the design target was a swath width of 225 km, which required operation of the radar antenna in ScanSAR mode. The implementation of this method led to the quasi-simultaneous imaging of four subswaths by steering the beam from small look angles (17°) to large ones (65°) and back.

The native resolution of the acquired topographic dataset is one arc-second, which approximately corresponds to a horizontal spacing of 30 m at the equator. However, the full resolution data was released only for the US while for the rest of the world the elevation postings were downgraded to 3 arc-seconds (~90 m). The mission accuracy specifications state that absolute vertical accuracy is less than 16 m while the relative vertical accuracy is less than 10 m (Duren et al. 1998; Bamler 1999; Kretsch 2000). Investigations have highlighted that the accuracy

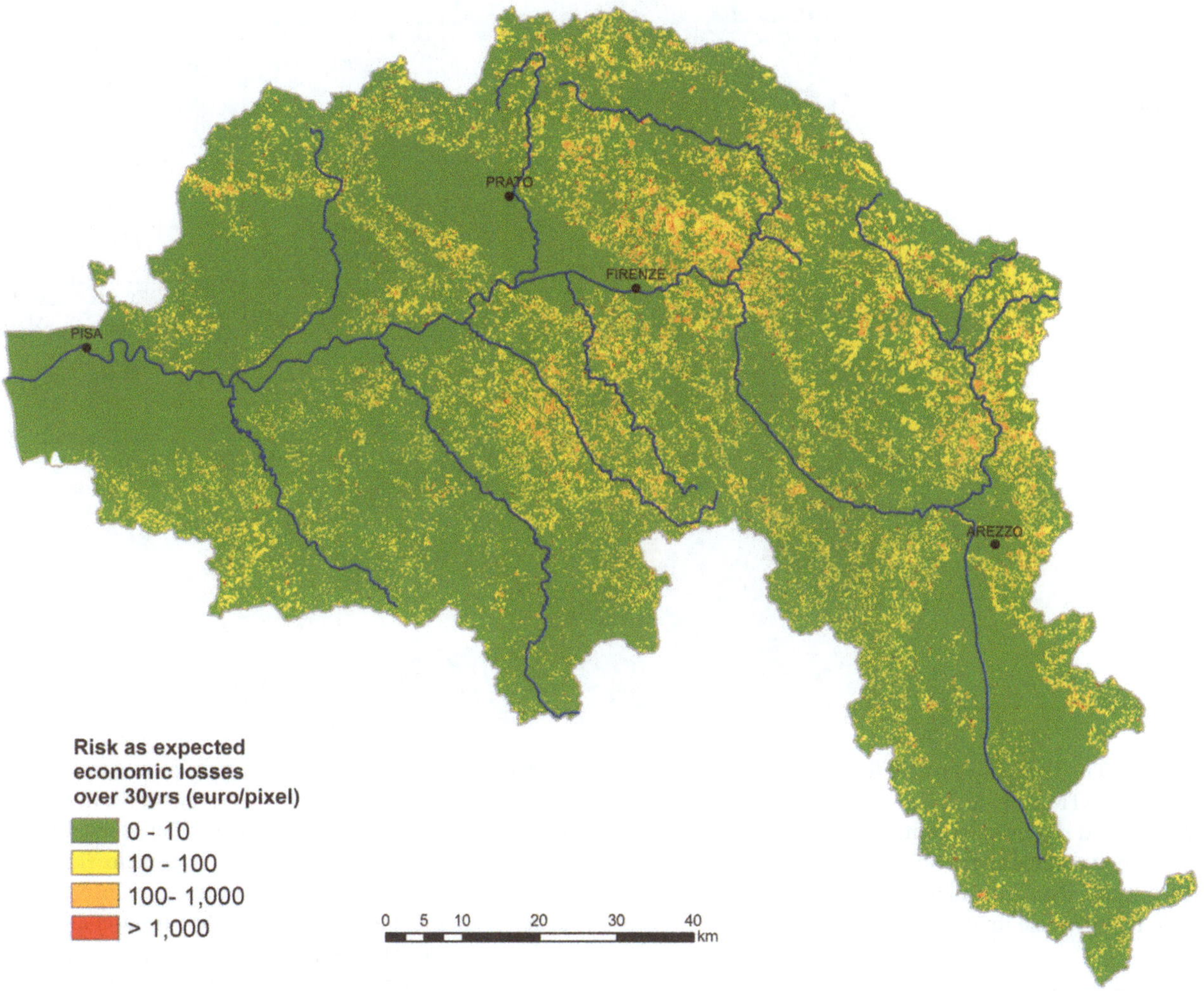

Fig. 18.3. Landslide risk map of the Arno Basin. Risk is expressed as expected economic loss in Euro per pixel

of the dataset decreases with increasing topographic relief (e.g., Falorni et al. 2005; Hall et al. 2005).

SRTM data has been used as the topographic base for geomorphic analyses and modeling at several volcanoes (e.g., Mouginis-Mark et al. 2001; Lagmay and Valdivia 2006). Volcanic debris flows, or lahars, represent one of the deadliest volcanic hazards. They can travel tens of kilometers from volcanoes and give little or no warning of their impending arrival.

The Cotopaxi volcano (5 897 m, 0° 38' S, 78° 26' W) has long been recognized as one of the most dangerous volcanoes in the world. The combination of numerous factors such as high relief, elevation, the presence of an ice cap, the eruptive frequency and the distribution of the elements at risk all make this volcano a natural disaster waiting to happen. Since 1534, eruption cycles at Cotopaxi volcano have averaged roughly one per century. These eruptions have spawned pyroclastic flows, regional tephra and ash falls, short lava flows and long-distance, highly destructive debris flows.

Many streams head on the volcano and are fed by glacial melt water. These include the Tambo and Tamboyacu Rivers to the southeast, the Rio Pita to the north and the Rio Cutuchi, which drains the western and southern flanks of the volcano and travels through the cities of Latacunga and Salcedo. Latacunga, the capital of the Cotopaxi province, is the largest city in the area: in recent decades the urban population of Latacunga has increased steadily, rising from ca. 10 000 inhabitants in 1950 to over 51 000 in 2001.

The city is located ca. 45 km downstream from the volcano and has been inundated by lahars several times in its history. Unregulated urban expansion, military infrastructure, the Panamerican highway and the presence of many other elements at risk are all factors that contribute to the high risk conditions of the Latacunga area.

Lahar modeling of the potential lahars in the Rio Cutuchi was performed to delineate inundation areas and provide estimates of flow depths and velocities (Fig. 18.4). The results will be used to provide information to civil protection authorities and update hazard maps of the

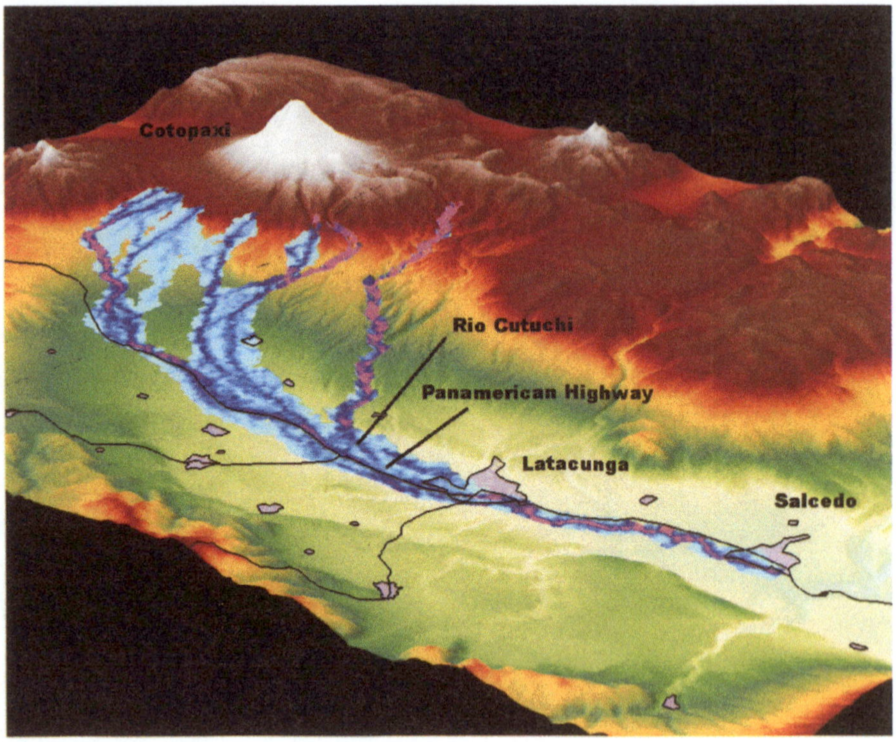

Fig. 18.4.
Prospective view of the Cotopaxi volcano and surrounding area showing the results of the lahar modeling. Flow depths and inundation area are shown. Lahar volume = 1 500 000 m^3

volcano. The topographic base for the modeling was provided by SRTM data. Although the downgraded 90 m data was utilized it proved to contain fewer errors and to have more topographic detail in the flat alluvial plain of the Rio Cutuchi than an available DEM derived from 1 : 50 000 scale topographic maps.

18.6 Landslide Monitoring by Remote Sensing

Landslide monitoring is very often based on traditional geotechnical techniques and, especially at local scales, usually represents the most effective way for assessing superficial displacements, for observing the response to triggering factors and for gauging the effectiveness of corrective measures. However, the inclusion, within a landslide monitoring framework, of space-borne and ground-based InSAR techniques can provide significant additional advantages as these technologies have the capability of providing precise, detailed information regarding ground movements. The results obtained from InSAR instruments are suitable for accurate analyses of temporal and spatial displacement fields, for producing activity maps and, combined with additional information from in-situ instrumentation, for interpreting movement geometry. Moreover, the availability of satellite images since 1992 allows a non-invasive assessment of the effectiveness of remedial works in the monitored area, which represents a fundamental step for planning and managing mitigation activities.

Several InSAR techniques, ranging from conventional DInSAR in C-band and L-band, Permanent Scatterers and the ground-based LISA system, have been applied to several landslides located in Italy (e.g., Antonello et al. 2004; Berardino et al. 2003; Catani et al. 2005b; Corsini et al. 2006; Farina et al. 2006; Strozzi et al. 2005; Tarchi et al. 2003a; Tarchi et al. 2003b). The analysed cases represent a wide range of landslides characterized by different movement mechanisms, displacement rates and land-uses and are usually equipped with in-situ instrumentation. In a majority of these cases remedial measures had been implemented in the past to mitigate the risk, but the persistence of instability conditions led to continuing investigation. The experience gained from these case histories has allowed the tailoring and fine tuning of ad hoc monitoring services which take advantage of the different available InSAR configurations for landslide investigation.

A successful application of InSAR-based techniques regards a small town in Tuscany. Cutigliano is located in the Italian Apennines, close to Pistoia, and has been affected by slope instability problems since the early 1900s. The movements, mainly related to the location of the town on an ancient landslide, have seriously endangered the town and its residents, damaging buildings, infrastructure and agricultural areas.

The PS analysis of 75 ERS1/2 images has resulted in a dense network of radar benchmarks (more than 200 in the central part of the town) that allowed the spatial distribution of the movements to be mapped and the creation of a slope activity map. The satellite data have

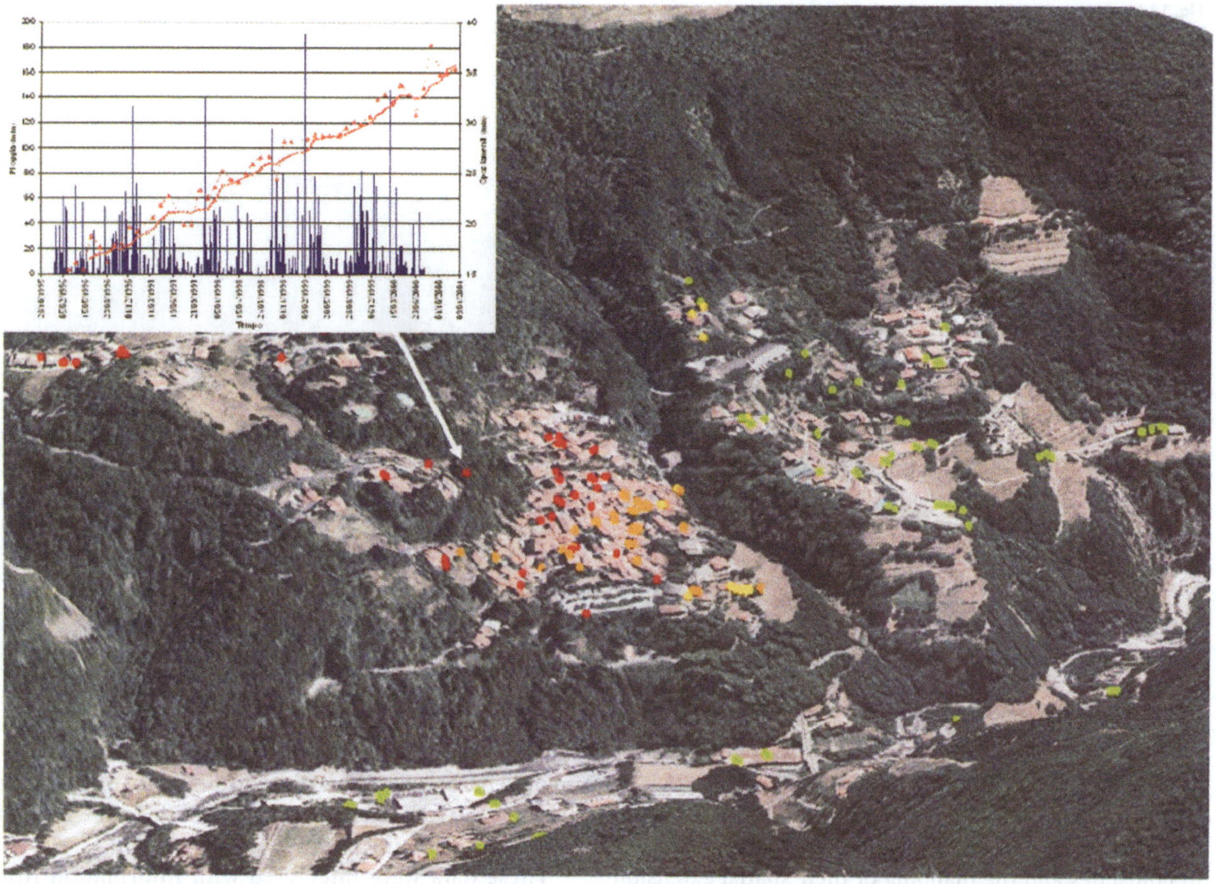

Fig. 18.5. 3D view of Cutigliano showing PS distribution on an aerial-photo. The *inset graph* reports cumulative rainfall and cumulative ground displacements

been integrated with in-situ measurements and geomorphologic interpretation of stereoscopic aerialphotos. In addition, the time series of displacement have been compared to precipitation data to attempt a possible correlation between rainfall and ground movements (Fig. 18.5).

18.7 Innovative Early Warning Systems

In urban areas exposed to landslide hazard, warning systems are often the only viable solution for mitigating risk, especially during emergencies. Civil protection officials and local authorities faced with the task of managing crises produced by slope movements require operative landslide monitoring systems able to predict landslide evolution and draw up emergency plans. Monitoring and analysis of ground displacements has been demonstrated as an effective means to predict the moment of landslide failure (Fukuzono 1985). Ground-based SAR interferometry, coupled with near real-time processing, is an instrument with the proven capability to provide effective early detection of ground movements with the potential of leading to catastrophic slope failures.

The use of GBInSAR as an early warning tool has been tested in several cases in Italy by the Engineering Geology group at the University of Firenze. The most interesting example of a GBInSAR application, however, is undoubtedly represented by the Stromboli Volcano. At the end of December 2002, Stromboli Volcano, located in the Mediterranean sea, erupted, triggering a massive landslide followed by a locally destructive tsunami. No deaths occurred only because the town of Stromboli, a resort town, is nearly uninhabited in the winter. However, concern over the possibility of a further failure of the precarious volcano slope led to the installation of a ground-based InSAR system on the flank of the Sciara of Fuoco, a collapse depression formed 5 000 yr BP. Since installation the instrument has functioned continuously and is employed as an early warning system by the Italian National Civil Protection Department (Fig. 18.6).

The last two years of measurements have revealed a deceleration of residual movements affecting the unstable slope and provide key insight into the instability problems. This application is an excellent demonstration of the unique capabilities of ground-based InSAR for volcano monitoring and, in particular, for the early detection of slope movements potentially leading to flank failures.

Fig. 18.6.
Stromboli volcano. *Upper left*: the LISA ground-based InSAR system; *lower left*: time series of displacements obtained from LISA

18.8 Support for Emergency Management and Scenario Analysis

The detection of precursory movements in areas subject to landslides and the mapping of their spatial extension are fundamental for establishing landslide volumes and runout characteristics. InSAR analyses performed on extensive archives of satellite SAR images acquired since 1992 are ideal for these types of applications. This differed-time monitoring provides invaluable support for predicting landslide evolution, for drawing up the relative emergency plans and for implementing opportune mitigation strategies.

As reported by Casagli et al. (2005), these techniques were demonstrated in the emergency induced by the activation of the Cerzeto landslide (southern Italy) in March 2005. Following initial landslide movements, the Italian Civil Protection asked our research group to monitor the area using remote sensing data from both very high resolution (VHR) optical satellites (IKONOS, QUICKBIRD) and radar sensors (PS analysis) (Fig. 18.7).

New generation VHR images are considered to have the potential to be a powerful tool for Civil Protection activities in risk assessment, emergencies and disaster management, particularly since data acquisition can be programed on demand, making new images available in a matter of days. An archive Quickbird image from before the landslide was obtained and a new IKONOS satellite acquisition was requested that resulted in a very good, clear image for 17 March 2005, shortly after the events occurred. Following orthorectification and radiometric and spectral enhancement, image photo interpretation was performed to highlight the changes between the two satellite images, leading to the identification and mapping of the landslides.

These data were integrated with information from a Standard Permanent Scatterers Analysis (SPSA), performed using an ascending and descending image data set collected by ERS satellites between 1992–2001. The PS interpretation within a GIS environment, integrated with ancillary data, provided superficial deformation rates in the study area from 1992 to 2001 and produced evidence of other older unstable areas, in addition to those affected by the March landslide. The entire set of analyses was performed in only 5 working days, due to the rapid availability and processing of optical and radar remote sensing data integrated with ancillary data. This timeframe was deemed as effective by civil protection authorities for emergency management and response. In the end, the costs involved in stabilizing Cerzeto led the Italian Civil Protection to decide to relocate the entire town to a more stable area.

18.9 Conclusions

The case histories illustrated above demonstrate the increasingly important role played by remote sensing techniques in all facets of landslide management. From early warning to recovery the information gained from these

Fig. 18.7. 3D view of the Cerzeto area with the overlay of PS on a Quickbird image

innovative approaches produce benefits for both scientists and policy makers that range from increased lead times for catastrophic collapses to tools for assessing the feasibility or effectiveness of remedial measures. The advent of the next generation of sensors, some of which are specifically designed for the observation of natural hazards, will further increase the capabilities of the instruments and provide more useful data for the management of natural hazards.

Acknowledgment

The many cases described in this manuscript were made possible by the contributions of many people and agencies. The authors would particularly like to thank all members of the Engineering Geology group at the Earth Sciences Dept. of the University of Firenze, The Italian Civil Protection Agency, the Joint Research Center of the EU, LISALAB, the Arno River Basin Authority, and the European Space Agency.

References

Antonello G, Casagli N, Farina P, Leva D, Nico G, Sieber AJ, Tarchi D (2004) Ground-based SAR interferometry for monitoring mass movements. Landslides 1(1):21–28

Bamler R (1999) The SRTM Mission: A World-wide 30 m Resolution DEM from SAR interferometry in 11 days. Photogrammetric Week 1999:145–154

Berardino P, Costantini M, Franceschetti G, Iodice A, Pietranera L, RizzoV (2003) Use of differential SAR interferometry in monitorino and modelling large slope instability at Maratea (Basilicata, Italy). Eng Geol 68 (1–2):31–51

Bonham-Carter GF (1994) Geographic information systems for geoscientists: modeling with GIS. Pergamon, Ottawa, 198 p

Casagli N, Guerri L, Righini G, Ferretti A, Colombo D, Prati C (2005) Integrated use of PS and very high resolution optical images for supporting landslide risk management. In: URSI, Commission F (ed) Symposium on Microwave Remote Sensing of the Earth, Oceans, Ice and Athmosphere, 20–21 April 2005, Ispra, Italy

Catani F, Casagli N, Ermini L, Righini G, Menduni G (2005a) Landslide hazard and risk mapping at catchment scale in the Arno River Basin. Landslides 2(4):329

Catani F, Farina P, Moretti S, Nico G, Strozzi T (2005b) On the application of SAR interferometry to geomorphological studies: estimation of landform attributes and mass movements. Geomorphology 66:119–131

Chung CF, Fabbri AG, van Western CJ (1995) Multivariate regression analysis for landslide hazard zonation. In: Carrara A, Guzzetti F (eds) Geographical information system in assessing natural hazards. Kluwer, Dordrecht, pp 107–142

Colombo D, Farina P, Gontier E, Fumagalli A, Moretti S (2003) Integration of Permanent Scatterers analysis and high resolution optical images within landslide risk analysis. In: Proceedings of FRINGE 2003 Workshop, Advances in SAR interferometry from ERS and ENVISAT missions, ESA-ESRIN, Frascati, Italy

Corsina A, Farina P, Antonello G, Barbieri M., Casagli N, Coren F, Guerri L, Ronchetti F, Sterzai P, Tarchi D (2006) Space-borne and ground-based SAR interferometry as tools for landslide hazard management in civil protection. Int J Rem Sens 27(12):2351–2369

Crozier MJ (1984) Field assessment of slope instability. In: Brundsen D, Prior D (eds) Slope instability. John Wiley & Sons, Chichester, pp 103–140

Duren R, Wong E, Breckenridge B, Shaffer S, Duncan C, Tubbs E, Salomon P (1998) Metrology, attitude, and orbit determination for spaceborne interferometric synthetic Aperture Radar, in SPIE AeroSense Conference on Acquisition, Tracking and Pointing XII, Orlando, Florida, pp 51–60

Ermini L, Catani F, Casagli N (2005) Artificial neural networks applied to landslide susceptibility assessment. Geomorphology 66:327–343

Falorni G, Teles V, Vivoni ER, Bras RL, Amaratunga K (2005) Analysis and characterization of the vertical accuracy of digital elevation models from the Shuttle Radar Topography Mission. J Geophys Res-Earth Surface 110(F2)

Farina P, Colombo D, Fumagalli A, Marks F, Moretti S (2006) Remote Sensing techniques for landslide risk analysis: outcomes from the ESA-SLAM project. Eng Geol 88:200–217

Ferretti A, Prati C, Rocca F (2001) Permanent scatterers in SAR interferometry. IEEE T Geosci Remote 39(1):8

Fukuzono T (1985) A new method for predicting the failure time of a slope. In: Proceedings of IVth International Conference and Field Workshop on Landslides, Tokyo, 23–31 August, Japan Landslide Society, 1, pp 145–150

Hall O, Falorni G, Bras RL (2005) Characterization and quantification of data voids in Shuttle Radar Topography Mission data. IEEE Geosci Remote Lett 2(2):177–181

Kretsch JL (2000) Shuttle radar topography mission overview. Applied imagery pattern recognition workshop, Washington, D.C., IEEE

Lagmay AMF, Valdivia W (2006) Regional stress influence on the opening direction of crater amphitheaters in Southeast Asian volcanoes. J Volcanol Geotherm Res 158(1–2):139

Mouginis-Mark PJ, Rowland SK, Garbeil H, Amelung F (2001) Topographic change on volcanoes from SRTM and other interferometric radars. 757 p

Soeters R, Van Westen CJ (1996) Slope instability recognition, analysis and zonation. Turner AK, Schuster RL (eds) Landslides: investigation an mitigation. Sp. Rep. 247, Transportation Research Board, National research Council, National Academy Press, Washington, DC, pp 129–177

Strozzi T, Farina P, Corsini A, Ambrosi C, Thuring M, Zilger J, Wiesmann A, Wegmüller U, Werner C (2005) Survey and monitoring of landslide displacements by means of L-band satellite SAR interferometry. Landslides. Journal of International Consortium on Landslides 2(3):193–201

Tarchi D, Casagli N, Fanti R, Leva D, Luzi G, Pasuto A, Pieraccini M, Silvano S (2003a) Landslide monitoring by using ground-based SAR interferometry: an example of application to the Tessina landslide in Italy. Eng Geol 68:15–30

Tarchi D, Casagli N, Leva D, Moretti S, Sieber AJ (2003b) Monitoring landslide displacements by using ground-based SAR interferometry: application to the Ruinon landslide in the Italian Alps. J Geophys Res 108:2387–2401

van Zyl JJ (2001) The Shuttle Radar Topography Mission (SRTM): a breakthrough in remote sensing of topography. Acta Astronautica 48(5–12):559–565

Varnes DJ, IAEG Commission on Landslides (1984) Landslide hazard zonation – a review of principles and practice. UNESCO, Paris, p 63

Wieczorek GF (1984) Preparing a detailed landslide-inventory map for hazard evaluation and reduction. IAEG Bull 21(3):337–342

Chapter 19

Rock Deformation Monitoring at Cultural Heritage Sites in Slovakia

Ján Vlčko · Vladimir Greif* · Rudolf Holzer · Lucia Hencelova · Michal Jezny

Abstract. The article deals with a socially important topic – preservation of cultural heritage in Slovakia. Current results of an engineering geological investigation of five medieval castles, one of them under the patrimony of UNESCO (Spis Castle) located in different geotechnical and engineering geological conditions are discussed. The obtained results are based upon an investigation program sponsored by the Ministry of Environment in which about 40 most prominent medieval castles or castle ruins have been investigated with the aim to recommend the remedial measures and thus to prevent their gradual disintegration. A following project focused on monitoring of the selected castle rocks showing instability problems was later carried out.

The results of engineering geological study and following monitoring showed, that the castles suffer from extremely slow displacement of creep character. This movement is caused mainly by the geologic structure and it is impossible to stop it or even reduce it. Since the monitoring results are influenced by seasonal periodic temerature changes a numerical modeling was used in order to understand better the landslide failure mechanism at Spis Castle site. Further a study of the thermal expansion as one of the potential triggering factors of the displacements observed in the castle's subgrade was carried out.

The results gained during the study are of great value and are helpful in design of stabilization and preservation works.

Keywords. Medieval castles, cultural heritage, monitoring, thermal expansion, modeling

19.1 Introduction

A great number of medieval castle ruins (dating back mostly to the 13th century) located at steep ranging cliffs belongs to the typical landscape of Slovakia. Since the beginning of the 18th century the castles were loosing their primary defense function, many of them were destroyed by fire or during military actions and later abandoned. At present the majority of the castles have the character of ruins. Some of them endanger the residents in nearby settlements or the traffic routes; others are located in dense forests, in places difficult of access. The latter being parts of nature reserves present a hazard mostly to individuals.

The majority of castle ruins is located in areas of flysh and neovolcanic uplands and core middle mountains at a height of 350 to 700 m a.s.l. (Fig. 19.1). Due to intensive neotectonic movement of the West Carpathians in the periods of late Neogene and in the Quaternary a complicated geologic-tectonic structure developed. This development resulted in the formation of a highly dissected relief and a climatic diversity of the territory. The effect of the climatic conditions is quite evident, manifesting sudden daily and seasonal temperature fluctuations ranging from −20 to +20 °C, the daily temperature fluctuations range from 10 to 15 °C. The mean annual amount of precipitation varies between 600 to 800 mm. In this respect, the prolonged snow cover (100 to 140 days a year) is of great importance. The vertical climatic diversity and the variety in lithology control not only the exogenous processes like weathering, erosion, slope movements, etc. but also superficial sediment formation and groundwater conditions.

The presence of overthrusted tectonic units is a typical feature of the West Carpathians and has a substantial influence on the character and the rock mass heterogeneity of the castle rocks. A very frequent case is when soft rocks mostly claystones and marlstones are overlain by rigid limestone-dolomitic units. Overthrust lines act, when affected by shear forces, as an important plain of weakness and potential shear plains. Such geological structure forms favorable conditions for deep-seated creep deformations inhibiting ridge loosening, block failures (block rifts and block field), lateral spreading followed by rock toppling and rock falls.

An investigation program sponsored by the Ministry of Environment in which about 40 most prominent castles or castle ruins were going to be investigated with the aim to recommend the remedial measures and thus to prevent their gradual disintegration both due to natural and man-made influence was set up and later on a successive project focusing on monitoring of the selected castle rocks showing instability problems was established. The paper presents some results of both projects.

19.2 Works and Techniques Employed

The preliminary stage of investigation carried out on a large regional basis, within a limited period of time, with low financial costs requires a cost-effective approach. The

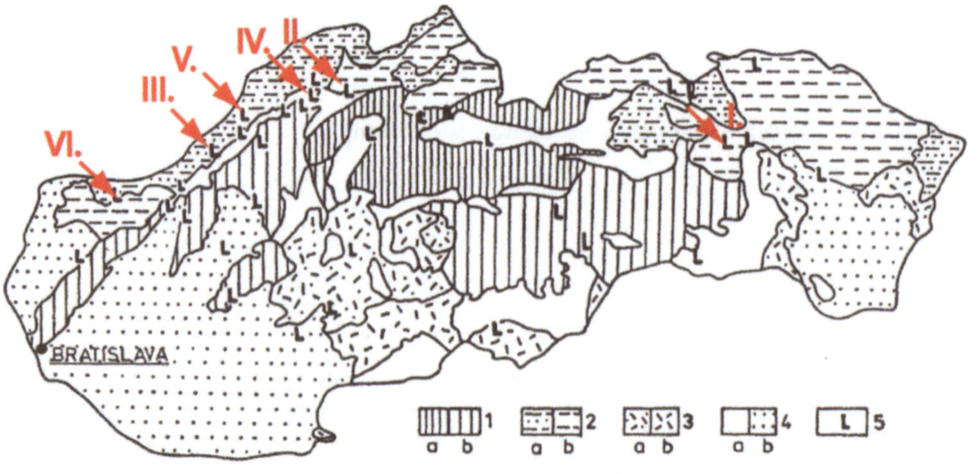

Fig. 19.1. Location of investigated sites in the map of engineering geological zoning (after Matula 1969). *1* Region of the core mountains, *1a* high core mountains, *1b* middle core mountains; *2* region of Carpathian flysh, *2a* flysh highlands, *2b* flysh upplands; *3* region of the Neogene volcanites, *3a* volcanic highlands, *3b* volcanic upplands; *4* region of the Neogene tectonic depressions, *4a* intra-mountain basins, *4b* Intra-Carpathian lowlands; *5* medieval castles: *I.* Spis Castle, *II.* Strecno Castle, *III.* Skalka, *IV.* Lietava, *V.* Lednica, *VI.* Čachticky hrad Castle

experience gained from previous studies and practical works aimed toward the protection and stabilization of historic sites proved that each historic site represents a completely different site of interest from geological, geotechnical and historical point of view. This was the real reason why no standardized methods and techniques were applied and the survey was strictly focused at these priorities:

- to undertake an engineering geological study of all rock slopes and to determine in the rock mass and in the castle walls those sections which represent a potential source of instability;
- to recommend the appropriate stabilization measures in designated sections;
- to prepare graphic outputs easily understood by nongeologists, mostly experts dealing with historic building conservation, landscape architecture, etc.

The engineering geological survey was conducted largely by visual techniques mostly without drilling, applying traditional and special laboratory tests. The main accent was laid on engineering geological mapping and consequently engineering geological maps at a scale of 1 : 500 were prepared from each site. Further activities were focused on measuring the joint orientation, on location of past and active slope failures, rock falls, etc., on study of weathering processes in rock mass and rock material and on karst processes. Special attention was paid to the character of anthropogeneous sediments and their thickness, which is quite different from the thickness found there at the time of construction. These sediments act now as a source of active pressure on the castle ruins and when accompanied by other natural and man-made factors they may cause serious damage.

Use was made of terrestrial photogrammetry for surveying and calculating joint orientation patterns in places difficult to access, as well as in places which have been identified as potential sources of instability. Finally typical profiles from stereo pairs were drawn.

In the case of UNESCO World Heritage Site Spis Castle distinct element modeling and thermal dilation studies were employed in order to determine the failure mechanism and thus potential risk to the castle.

19.3 Spis Castle

The Spis Castle is a Natural cultural Monument under the Patrimony of UNESCO represents the largest medieval fortification system in Central Europe (Fig. 19.1). The castle was founded in 1120. The historic development of it was rather complicated showing traces of many historic epochs up to the Baroque. In 1780 the castle burned out and since that time it was abandoned and the process of destruction caused both by the natural and man-made factors was going on.

From a geological point of view the studied area is located in a zone referred to as Hornadska kotlina Basin (Eastern Slovakia). Spis Castle is built on a travertine mound, which is underlain by Paleogene soft rocks formed by claystone and sandstone strata (flysh-like formation). The travertine body reaching more than 52 meters in thickness reflects several features of destruction and is disturbed by a series of faults, cracks and joint systems. Two prevailing joint sets can be found, sub-vertical joints striking approximately NW to SE with a general dip to SW (dip direction/dip 220°–250°/80°–90°) and joints striking approximately N-S dipping to the W (250° to 270°/85°). The origin of the Temna jaskyna Cave is strongly bound to this system, as well. The destruction is the result of gravitational slope failures due to spreading of rigid travertines blocks on relatively plastic claystone strata. The central part of the travertine rock is formed by a block rift (travertine cliffs separated by persistent tensional joints and cracks), the marginal parts of the castle rock are formed by a block field consisting of displaced and tilted cliffs reaching the height from 25 to 30 m, sloping at an inclination of 70° to 80°, in some places up to 90° with a number of overhangs. The absence of a block field in the SW part of the castle rock is due to the uplift of Palaeogene claystones along the fault line (220°/80°), which inhibited the total disintegration of the block field, followed by rock falls, toppling and tilting of huge cliffs of travertines (Vlcko 2004).

19.3.1 Monitoring

In 1980 three TM-71 type crack gauges (TM-71-P1–P3) and in 1992 six additional were installed (TM-71-1–6). It should be noted that from a number of 9 crack gauges only four (TM-71-1, TM-71-2, TM-71-P1 and TM-71-P2) operate; the unexpected visitors demolished the rest (Fig. 19.2).

Measurement data for the period 1980–1984 show that the greatest crack displacement occurred in the eastern part of the lower courtyard between the transversal and the outer defense walls. Total extension of this crack for the four years was 14 mm. Following repair of the masonry and a three year hiatus in monitoring a new and similar crack gauge was installed in 1997. Since then no further displacement has been recorded. The reason for this maybe:

- The restoration works stabilized the movement effectively – if there is any creep movement of the travertine block in the bedrock potential failures in the wall are likely to develop;
- The observed displacements did not record the movements in the bedrock but there maybe tilting of the wall in horizontal direction (along axis y).

The total displacement in axis y (horizontal shear) recorded by TM-71-P2 reached 14 mm, extension (axis x) 4.3 mm and vertical displacement (axis z) 4.0 mm (Fig. 19.3a). After the repair of masonry (in 1994) and a three years break in monitoring a new crack gauge of the same type was installed in 1997. Since that time the displacements were reduced.

Further monitoring of this location is required in order to explain the direct link between the rock body and the superstructure.

Another significant displacement was observed in the southeast part of the Lower Courtyard near the castle entrance at a high travertine cliff called Perun's rock. Given past failures here, three monitoring locations were selected (Fig. 19.2): one (TM 71-2) in a 74-cm-wide tension crack outside the castle near the entrance to Podhradska jaskyna Cave and two (TM 71-1, TM 71-P1) on the opposite side at Courtyard II.

Data obtained from crack gauge TM-71-2 (Fig. 19.3b) indicate progressive crack closing of 2.3 mm for the period 1992–1994. Since then, and for the next six years, displacements oscillated about 2.5 mm as a function of the influence of seasonal temperature and precipitation. Since December 2000 there was a marked change in the rate of displacement. This is likely due to a large boulder which fell from the upper slope into the entrance of the Podhradska jaskyna and thus formed a temporary barrier to further

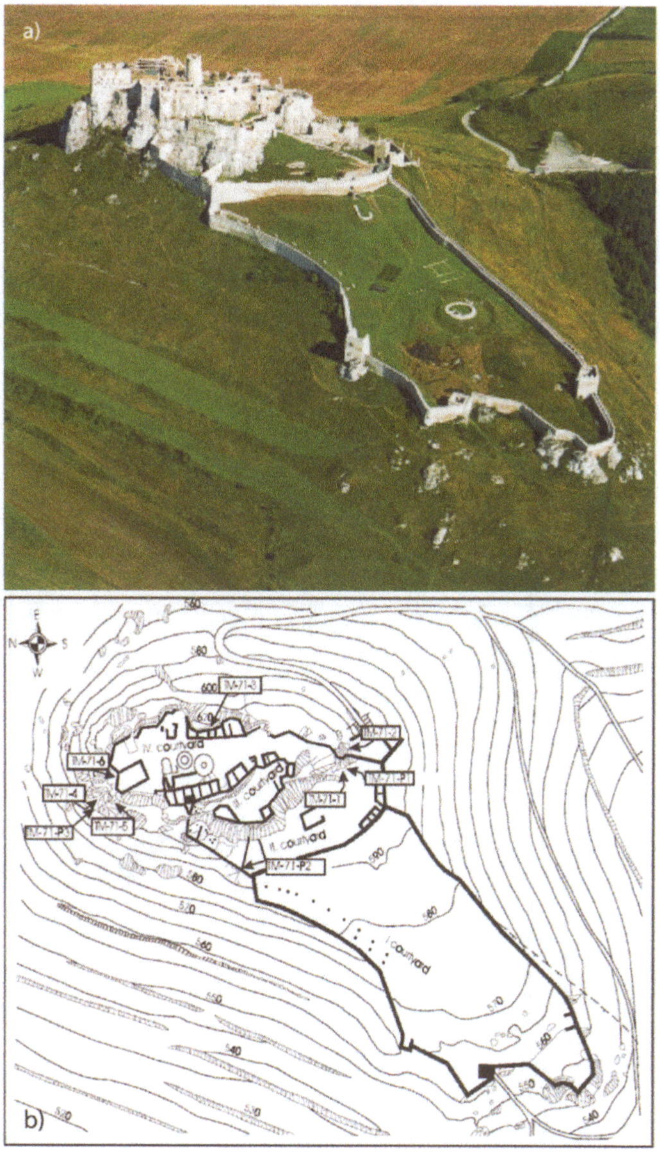

Fig. 19.2. Location of TM-71 crack gauges at the Spis Castle

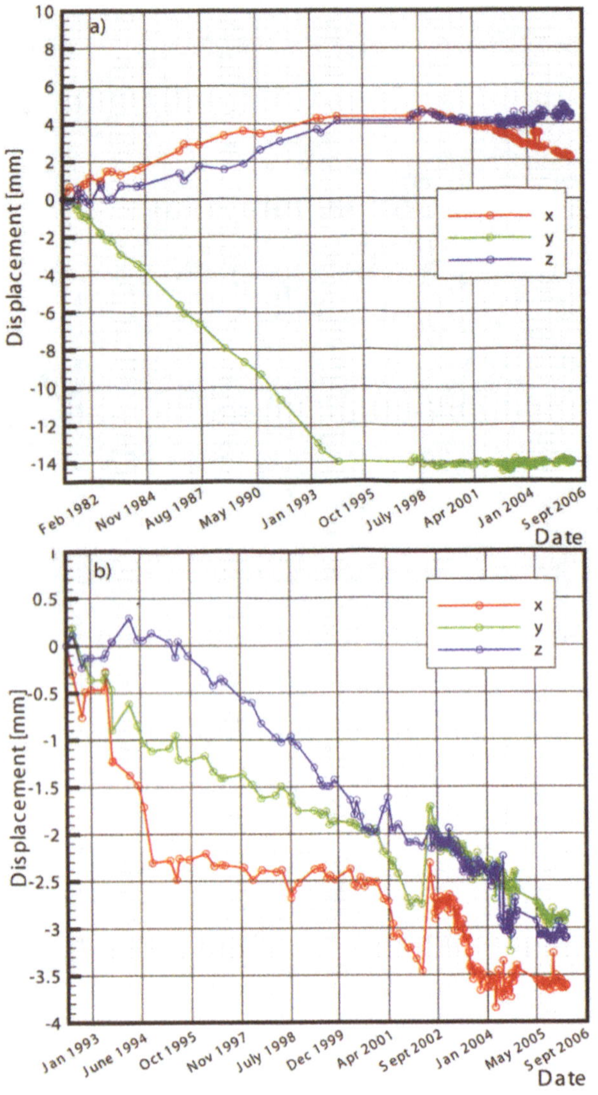

Fig. 19.3. Records from crack-gauges TM-71-2 (**a**) and TM-71-P2 (**b**) (x-direction = opening-closing of the crack; y-direction = horizontal shearing; z-direction = vertical shearing)

compression. The direct evidence for lateral pressure of Peruns rock was the progressive cracking of the boulder and the fissuring that began to occur in the repaired masonry walls. The latter effects required the eventual removal of the boulder from the open crack with explosives. Subsequent measurements confirmed an acceleration of the movements (1.0 mm) until the summer of 2003 when compression reached a maximum value of 3.4 mm. The only exception to this trend occurred in spring 2002 when abrupt extension was observed in response to relaxation of the opposite cliff. The recorded data confirm that boulder removal clearly accelerated the rate of displacement (approximately 1 mm yr^{-1}). Data recorded at Peruns rock indicates movement and tilting in a southeast direction.

19.3.2 Numerical Modeling

In order to represent the slow movements of the blocs a numerical modeling approach was selected to determine the failure mechanism. A 2D model of the travertine mound (castle subgrade) divided by faults and other types of discontinuities were created. The model has been analysed using the Universal Distinct Element Code (UDEC, ITASCA 1993). This program simulates the mechanical behavior of the discontinuous medium represented as an assemblage of discrete blocks subjected to either static or dynamic loading. The main features of the code can by summarised as follows:

- The discontinuities are treated as boundary conditions between blocks; finite displacements along discontinuities and rotation of block are allowed;
- Blocks may be rigid or deformable; contacts are always deformable;
- The program recognizes new contact as the calculation proceeds;
- Several constitutive behavior models following linear or non-linear laws are available for both joints and blocks;
- The program can simulate steady or transient fluid flow through the discontinuities.

As introduced in the UDEC code, vertical sides of the model have been assumed to move vertically only and the horizontal boundaries were free to move horizontally. The rock material and the discontinuities are assumed elastic-perfectly plastic when the Mohr-Coulomb failure envelope or tensile failures are reached. Travertine and bedrock have been assumed as fully deformable blocks and then discretized into finite difference triangular elements. The model ran over a number of iterations until the initial equilibrium conditions were attained.

In geotechnical terms, the stratigraphical sequence can be (schematically) as described above represented as the superimposition of a rigid travertine body over claystone strata, in the upper parts highly weathered. Numerical presentation of a representative geological profile is in Fig. 19.4.

A stepwise modeling procedure we adopted was based upon the back analysis comprising:

a Simulation of travertine sequences over the Paleogene rocks until the state of equilibrium (initial state of stress) was determined. The travertines were assumed as an ideal homogeneous rock body.
b Introduction of gradual decrease of bedrock material properties (weathering, softening) as well as gradual decrease of tensile strength along the joints in travertines (mainly joint normal stiffness and shear stiffness) were considered.

Fig. 19.4.
Representative geological profile (**a**) and its numerical representation (**b**)

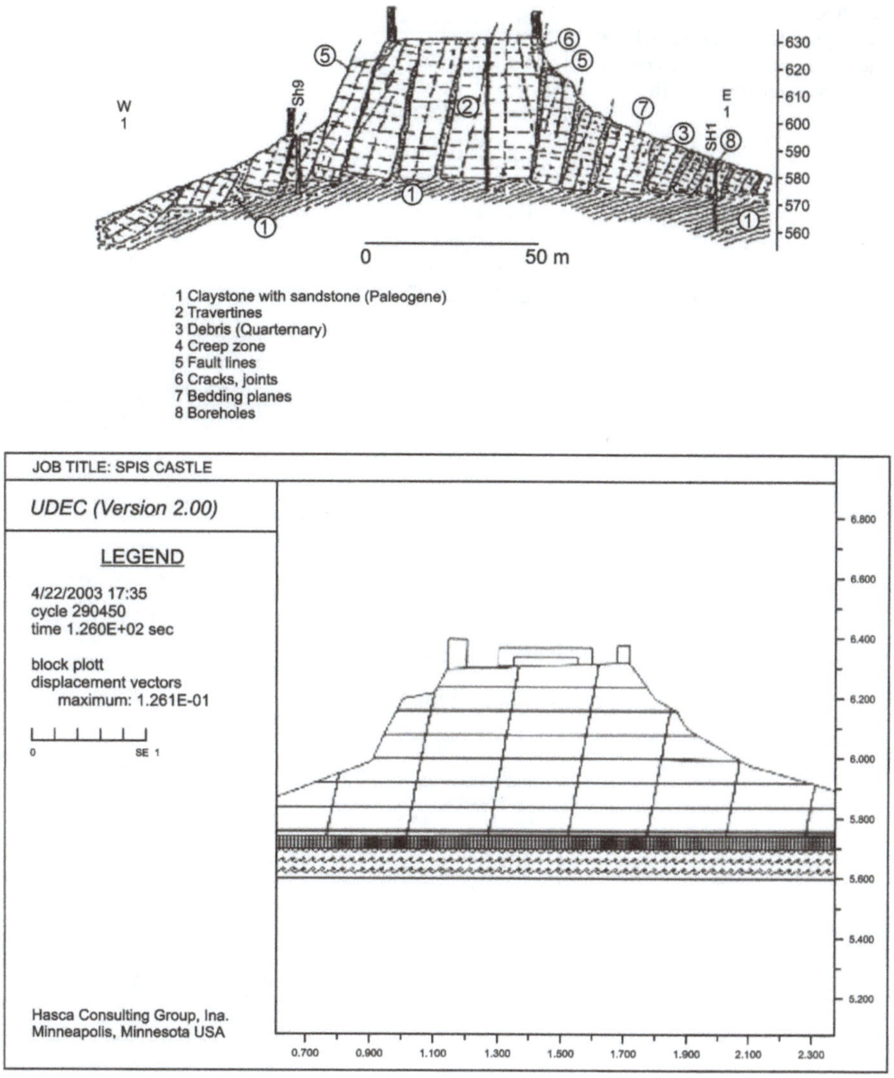

The physical and mechanical parameters as input data for modeling are summarized in Table 19.1.

Numerical analysis confirmed that the instability of the travertine castle rock is related to significant shear strength reduction of the subgrade formation. The failure mechanism involves differential subsiding of the travertine cliffs into soft bedrock and formation of persistent tension cracks in the central part of castle rock (Baskova and Vlcko 2003). The development of tension cracks transmitted the strain to the castle walls and inhibited their rupture, in some places even collapse (Fig. 19.5a). On the marginal parts of the castle rock, the cliffs are toppled up-slope (western part) while in the eastern part these are toppled down-slope. Displacement vectors (Fig. 19.5b) indicate block spreading as a representative mode of failure, which is in coincidence with that observed by the field study and the monitoring of travertine cliffs displacements.

Table 19.1. Physical and mechanical parameters

Litological type	ρ (kg m^{-3})	σ_c (Mpa)	E (Gpa)	ν (–)
Travertine	2500	63	56.6	0.19
Silty clay	1850	10	17.0	0.35
Clay stone	2310	33	20.0	0.25

19.3.3 Thermal Expansion Study

The reliability and accuracy of monitored data recorded by the help of any crack gauge is on one hand based on the material temperature correction of the used device, thermal properties of the monitored rock body are generally not considered. For that reason we included into our work the study on thermal characteristics of rocks and thus to bring the realistic knowledge concerning

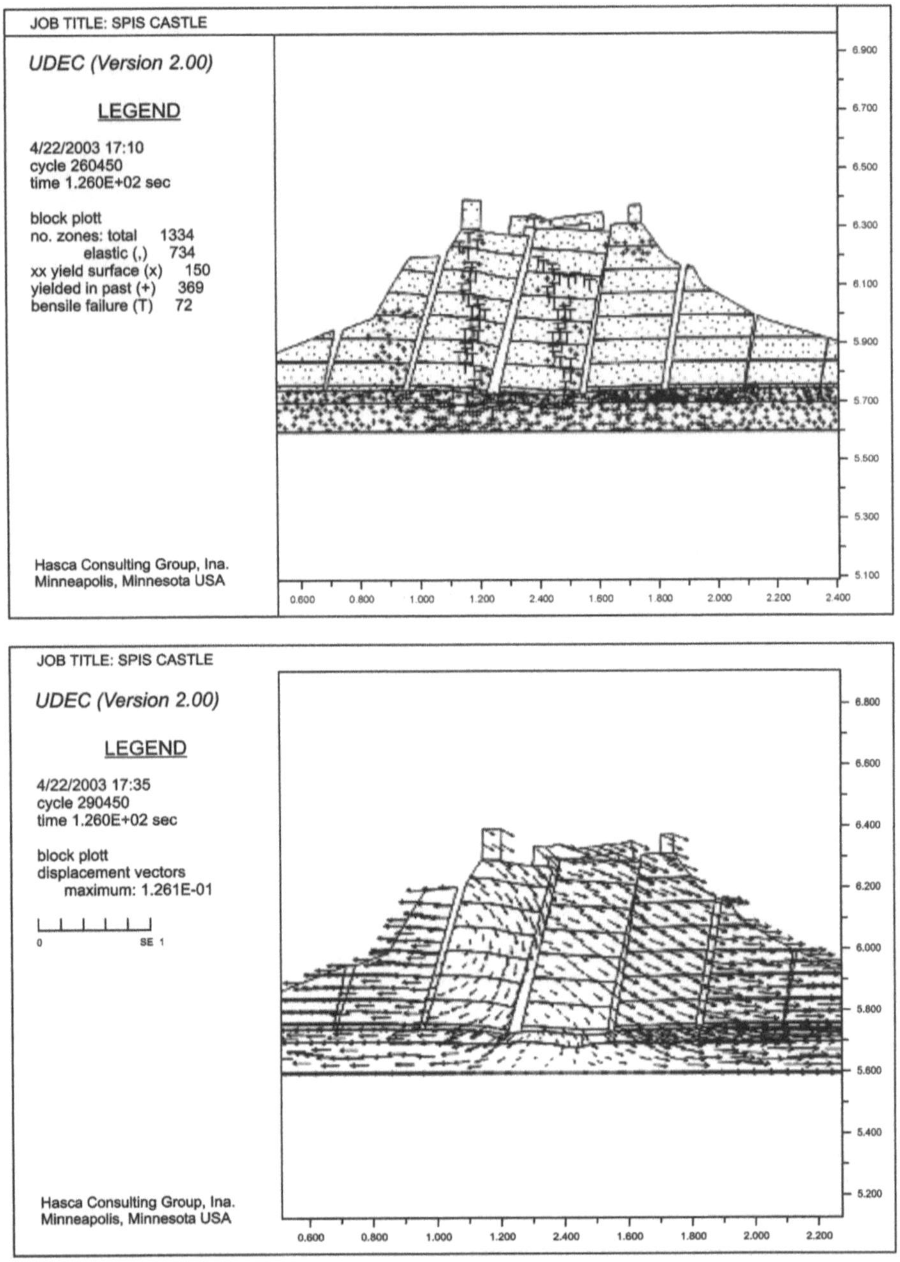

Fig. 19.5. Developments of tension cracks (**a**) and block spreading indicated by the displacement vectors (**b**)

monitoring data sets as well as the information if the data presented are essentially correct.

As far as the laboratory methodology concerns we followed the European Standards Proposal "Natural stone test methods: Determination of the thermal expansion coefficient" which specifies methods to determine the linear thermal expansion coefficient of natural stone based on mechanical length-change measurements.

The linear thermal characteristics are:

- unitary expansion – relative dilation: ε
- linear thermal expansion coefficient: α
- thermal residual strain: L_r

Unitary expansion – relative dilation (ε) expresses relative length change within certain temperature range and is expressed by the formula:

$$\varepsilon = \Delta L / L_i \ [\text{mm m}^{-1}]$$

- ΔL = change of length of the sample between initial temperatures T_i and final experimental temperature T_{fin}
- L_i = initial sample length at the temperature $T_i = 20$ °C).

Linear thermal expansion coefficient (α) of rocks can be defined as unitary expansion of the sample – relative dilation (ε) related to the temperature interval $\Delta T = T_{fin} - T_i$,

$$\alpha = \frac{\varepsilon}{\Delta T}$$

and can also be expressed as:

$$\alpha = \frac{\Delta L}{L_i \Delta T}$$

Temperature change ΔT corresponds to the temperature range $\Delta T = T_{fin} - T_i$.

Apart from the temperature the thermal rock expansion properties depend on mineralogical composition, texture, structure and physical properties.

Thermal residual strain ΔL_r represents a measurable quantity corresponding to the destruction of the sample after heating/cooling treatment (Siegesmund et al. 2000, 2001; Kirschner et al. 2003). Thermal residual strain corresponds to the length increment of the sample length before (L_i: initial sample length) and after experiment until in the heater/cooler chamber initial temperature (T_i) is fixed. Actually it represents a non-reversible sample length.

Thermal expansion tests were performed on travertine using a thermo dilatometer VLAP 01 especially constructed for these purposes with heater/cooler chamber capable of raising the temperature from (−20 ± 0.2) °C to (+80 ± 0.2) °C at a rate of 0.1 °C min^{-1} and maintaining temperatures within that range at least two hours with an accuracy of at least ±0.5 °C (Fig. 19.6). The apparatus is linked with data acquisition system in order to record the temperature and length change of tested sample and to calculate linear thermal characteristics. Thermo dilatometer VLAP 01 is capable to measure thermal expansion on two samples simultaneously. The rock samples were of cylindrical shape 25×50 mm in size cut in three *x*, *y*, and *z*-axis orientations. A dummy sample corresponding to the same rock type with a thermocouple in its center regulates the temperature and secures the identical experimental conditions for both rock samples.

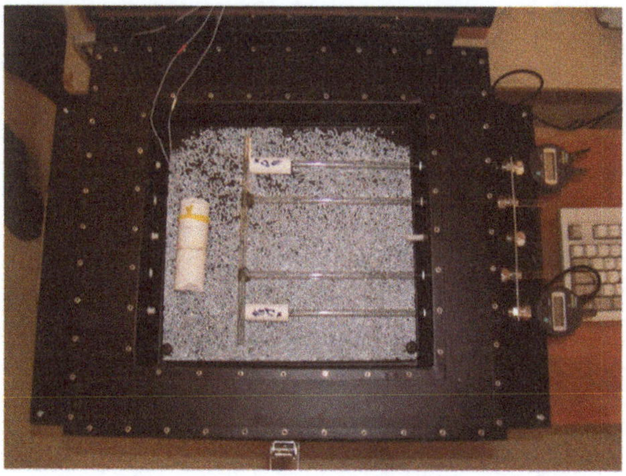

Fig. 19.6. Photo of the thermal dilatometric apparatus VLAP 01

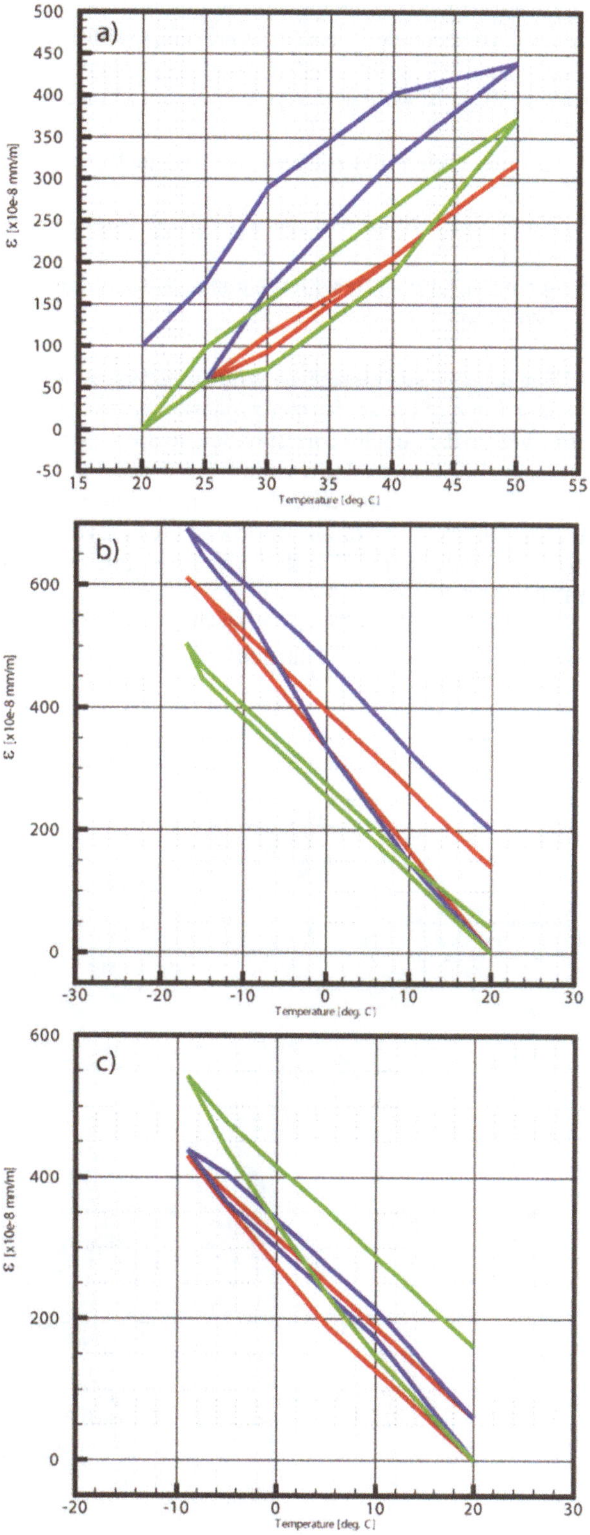

Fig. 19.7. Temperature dilations for summer (**a**), winter (**b**) and spring/autumn (**c**) season temperature ranges. *Red line* = sample prepared in the direction of *x* axis (paralell to bedding), *blue* = sample prepared in the direction of *y* axis (paralell to bedding), *green* = sample prepared in the direction of *z* axis (perpendicular to bedding)

Based on the outside temperature recordings three thermo cycles corresponding to real seasonal temperature conditions were distinguished and in laboratory tests using apparatus VLAP 01 were performed. These are as follows:

- summer cycle with the temperature range from +20 °C to +50 °C,
- winter cycle with the temperature range from –5 °C to –20 °C,
- spring/autumn cycle with the temperature range from +20 °C to –5 °C.

The results we gained maybe summarised as follows: the highest value of linear thermal expansion coefficient was observed on the samples corresponding to the y-axis direction with the mean α value reaching $15.060 \times 10^{-6}\,°C^{-1}$, then in the x-axis direction with the mean α value reaching $13.310 \times 10^{-6}\,°C^{-1}$ and finally the lowest mean α value was determined on samples with z-axis direction reaching $13.299 \times 10^{-6}\,°C^{-1}$ (Fig. 19.7a–c).

The highest value of thermal residual strain (in some papers assigned as residual dilation) was observed in direction of y-axis with the mean value $\Delta L_r = 3.9 \times 10^{-6}$ m, following direction in z-axis with mean value $\Delta L_r = 1.5 \times 10^{-6}$ m and the lowest one was determined in direction of x-axis with mean value $\Delta L_r = 1.23 \times 10^{-6}$ m (Vlcko et al. 2005). The observed differences in thermal residual strain can be explained by means of the thermal anisotropy of the calcite grains in connection with the natural preferred orientation in travertine rocks.

Since the maximum temperature within rock mass reached the value of 35 °C it is possible to conclude that, the dilation effects triggered by temperature change at recent climatic conditions have small influence on the monitoring records but the thermal expansion is causing physical rock deterioration and acts as a potential trigger in time cumulative pre-cursory stage of slope movements at Spis Castle. The thermal expansion results support the idea that the lateral spreading is the representative mode of failure confirmed by long term monitoring as well as by the orientation of displacement vectors gained by numerical modeling.

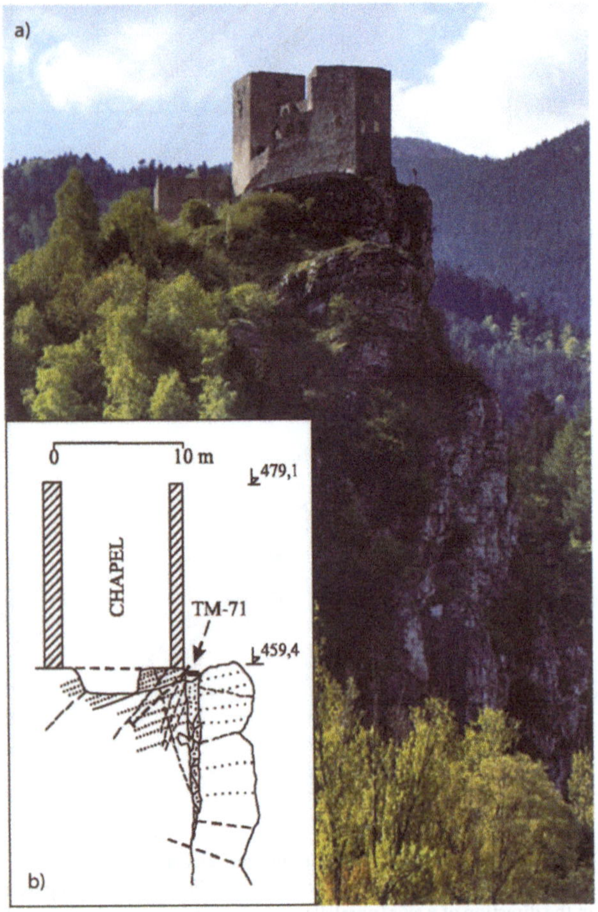

Fig. 19.8. Strecno castle with typical crossection showing the location of the crack gauge TM71

19.4 Strecno Castle

The original structure dates back to the thirteenth–fourteenth century, when the troops of King Leopold I pulled down the outer defense walls and destroyed several buildings. Since that time the castle was abandoned and became a ruin. The restoration works started in 1978 and today the castle represents one of the best-preserved monuments in Slovakia.

The Strecno Castle is located in Malá Fatra Mts. at a cliff reaching 103 m above the flood plain of Váh River (Fig. 19.8). The castle rock is formed by overthrusted lithological formations of the Choc nappe consisting of various carbonate rocks mainly limestones of Gutenstein

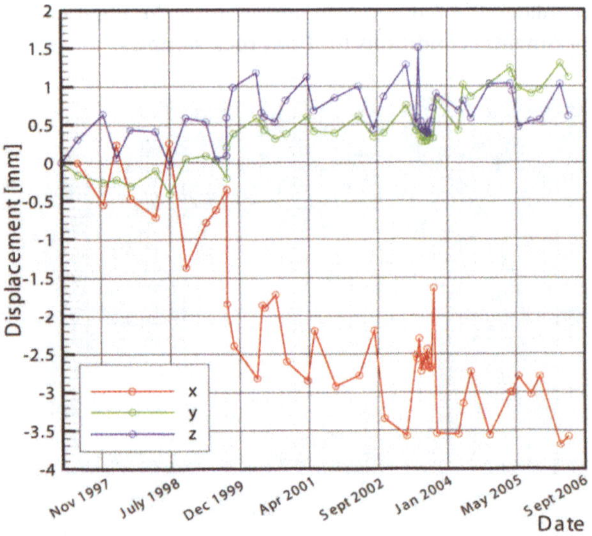

Fig. 19.9. Records from crack-gauge at Strecno Castle

type as well as gray limestones with intercalations of dolomitic limestones and dolomites. The castle rock has a character of a fault zone with intense manifestation of cambering (gravitational loosening). The potential danger of instability generally occurs at places where tectonically disintegrated and weathered dolomite rocks prevail and where a great number of overhangs gradually loosing their stability developed. From the structural point of view the Eastern part of the castle rock where several castle objects were built upon unstable overhangs represents potential danger. The monitoring site has been placed beneath the chapel in open tension crack reaching the width 148 cm, which separates the castle rock-body and a gravitationally separated huge cliff. Since 1996 10-year measurements have confirmed the tendency to extension of the tension crack along axis *x* resulting in slight cliff tilting towards ESE. The total crack opening reached the value of 3.5 mm (Fig. 19.9). The sudden change in the displacements in 1999 is probably response to the activity in a nearby quarry.

19.5 Skalka Monastery

The Skalka Monastery, a well-known place of religious interest in Slovakia consists of six different buildings. In the most important building of the Monastery Baroque chapel, serious failures and cracks in the chapel walls, which might be an evidence of deep-seated deformations were detected (Fig. 19.10).

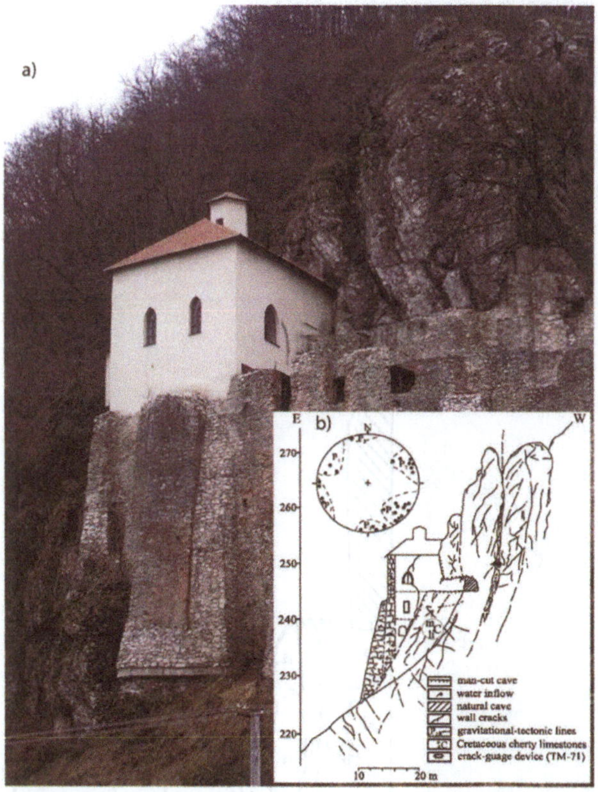

Fig. 19.10. Skalka Monastery, cross section with location TM-71 crack gauge

The Skalka Monastery is located near the city Trencín in the Biele Karpaty Mts. at a height of 250 m above sea level. The broader study area belongs to the contact zone of the Central West Carpathians formed predominantly by a granite core overthrusted by Mesosoic carbonate rocks and the Pieniny Klippen Belt (outer zone) characterized by intensive folding and extreme tangential compression. Due to intensive neotectonic movements associated with an uplift of the Biele Karpaty Mts. as well as with a differentiated subsidence of the River Vah valley a complicated geological-tectonic structure with a highly dissected relief developed. The Monastery is situated on a steep ranging cliff built by Cretaceous marly limestones with cherts. The limestones are rather massive, at places with bedding, gray in color. The discontinuities (gravitational-tectonic lines, cracks and joints) are in general widely spaced, persistent, open, karstified and unfilled, with a rough surface.

Along the prevailing gravitational-tectonic lines a cave developed, reaching 35 m in length and 4 m in width. At some places the cave is man-cut and presents the only way of access to the chapel. Important gravitational-tectonic lines of N-S to NNE-SSW direction cut the bedrock beneath the northern part of the chapel and caused open cracks in the walls. Since November 1995 a crack-gauge of the type TM-71 was installed in an open discontinuity (Fig. 19.11). The initial setting was carried out in November 1995. Since that time no distinct displacement was recorded, the crack-width proved to be constant and a rock mass stable. A small displacement along the axis *y* could indicate some tendency to horizontal slip. The total recorded displacement up to year 2006 is 0.20 mm.

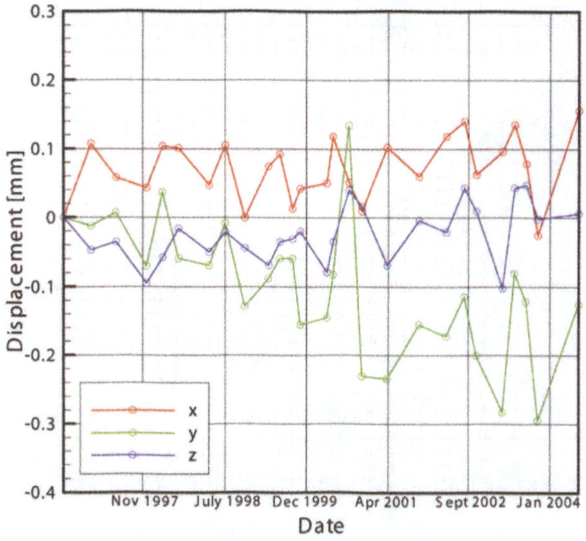

Fig. 19.11. Records from crack-gauge Skalka Monastery

19.6 Lietava Castle

The castle is located (Fig. 19.12a). on a massive carbonate conglomerates castle rock, which form a rather bizarre landscape with a number of cliffs. The basal conglomerates found in the bedrock are underlain by Cretaceous claystones and marlstones. This geological structure led to deep-seated creep movement and a subsequently to a differentiated displacement of conglomerate cliffs along subvertical tectonic planes running deep into rock mass took place. The castle ruin shows a high degree of destruction and the displacement of several cliffs is quite evident from open cracks separating individual castle objects (Fig. 19.12b). On the basis of terrestrial photogrammetry the orientation of faults and joints (dip direction/dip) exhibited on places difficult of access was calculated and profiles showing places of potential instability were drawn by computer (Fig. 19.12c).

19.7 Lednica Castle

A very complicated geological structure developed in the Klippen-Belt zone of the West-Carpathians. It is characterized by intensive folding and extreme tangential compression and it reminds of a "tectonic megabreccia" (Matula 1969). Due to disharmonic folding rigid Jurassic limestones-sandstone strata are sunken into plastic Cretaceous marlstones in the form of blocks and lenses which in the present relief rise actively and form landscape dominants, on the top of which castles were built in the past.

The Lednica castle in Biele Karpaty Mts. represents a typical example of the fact that the physical state of the object is substantially influenced by the complex geotectonic structure of the bedrock. The geology of the Lednica Castle Klippe (Fig. 19.13a,b) represented on a sketch drawn on the basis of photogrammetric survey showing that important tectonic lines found in the bedrock continue in the castle

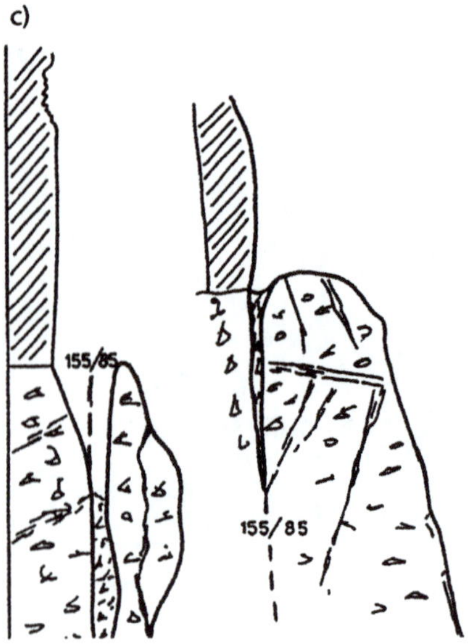

Fig. 19.12. Lietava Castle (**a**) with the open cracks in the walls the castle as the result of a creep movement (**b**) and the typical profile drawn from the photogrammetric survey (**c**)

walls in the form of open cracks and visible damage. It may be assumed that the process of crack formation is the result of a creep movement which is necessary to monitor so that appropriate remedial works may be designed.

19.8 Čachticky hrad Castle

Failures of quite different origin brought about the destruction of the Čachticky hrad Castle in Male Karpaty Mts.(Fig. 19.14a). The castle rock is formed by Triassic dolomites (Hauptdolomite) in which the main discontinuities are the bedding plains. A constant dynamic process of loosening and damages associated with the collapse can be observed on the castle ruins. The main reason for this process is an active earth pressure of anthropogeneous fills of great thickness occurring at critical places. Calculation proved that the coefficient of earth pressure at rest reaches 80% of limiting state of equilibrium.

The western part of the castle ruin manifests signs of destruction illustrated in Fig. 19.14b. For restoration purposes we have delineated in the western castle walls zoning units in terms of their physical state, where the highly endangered parts (1) are affected by active earth pressure of anthropogeneous fills and by the presence of bedding planes (7) which have resulted in the formation of open cracks found in the castle walls (6). The stability of the castle walls (1) has been temporarily secured by means of supporting wooden beams. The zoning of the physical state of the castle objects has been carried out on all investigated castles in a similar way as in the case of Čachtice Castle – either with the use of photogrammetric photography or in a ground plan.

Fig. 19.13.
Lednica Castle (**a**). A typical sketch (**b**) drawn from the photogrammetric survey. Composition material: *1* sandstone, travertine; *2* limestones, claystones; *3* sandstones; *4* limestones, marlstones, travertines, sandstones; *5* tension cracks; *6* faults; *7* block slide; *8* boundaries between different construction materials; *9* Jurassic limestones; *10* Cretaceous marls and marlstones; *11* slope sediments

Fig. 19.14. Zoning sketch of western walls at Čachtice Castle. Failures in the walls: *1* major; *2* moderate; *3* minor; *4* walls reconstructed in 1992; *5* fallen down walls; *6* open cracks; *7* faults; *8* Triassic dolomites; *9* supporting wooden beams

19.9 Conclusions

The safeguarding of the cultural and natural heritage is the priority of every nation. As shown in this paper, the casues and triggering factors of the damage to the medieval castles in Slovakia can be identified using engineering geological methods, with minimal or virtualy no effect on castles or other historical structures itself. The Spis Castle site has shown, that in order to understand better landslide failure mechanism a numerical modeling can significantly contribute to the hazard/risk assessment at various sites. In case of Spis Castle, a monument under patrimony of UNESCO, the results gained during the study are of great value and are helpful in design of stabilization and preservation works. Further methods end techniques were used to investigate the possibility of the temperature changes as the triggering factor of the movements in the castle's subgrade as was the case for instance in the slope of Rocher de Valabres in France (Gunzburger et al. 2005).

In order to understand the mechanism of failure and ongoing processes in the subsurface of the structures it is vital to employ the rock monitoring techniques, which can be of substantial help in proper selection of mitigation works in unstable slope cases.

Acknowledgment

This paper was prepared as part of the framework of Partial monitoring system of geological factors of the environment in Slovakia, VEGA grant project No. 1/1028/04, IPL Project No. C 101-2. The authors are thankful to the Ministry of the Environment of the Slovak Republic, Ministry of Education of the Slovak Republic, grant agency VEGA and the International Consortium on Landslides for their kind support.

References

Baskova L, Vlcko J (2003) Numerical analysis of block movements at the Spis Castle. In: Natau O, Fecker E, Pimentel E (eds) Geotechnical measurements and modelling. Swets & Zeitlinger, Karlsruhe

Gunzburger Y, Merrien-Soukatchoff V, Senfaute G, Piquet JP (2005) Influence of daily surface temperature fluctuations on rock slope stability: case of the Rochers de Valabres slope (France). Int Jour Of Rock Mech and Mining Sciences 42(3):331–349

Itasca (1993) UDEC version 3.10–user manual. Itasca Consulting Group Inc., Minneapolis, Minnesota

Matula M (1969) Regional engineering geology of Czechoslovak Carpathians. Publishing House of Slovak Academy of Science, Bratislava, 224 p

Vlcko J (2004) Extremely slow slope movements influencing the stability of Spis Castle, UNESCO site. Landslides 1(1):67–71

Vlcko J, Jezny M, Pagacova Z (2005) Influence of thermal expansion on slope displacements. Landslides – Risk analysis and Sustainable Disaster Management. pp 71–74

Part IV Landslide Risk Assessment

Extracting Necessary Pparameters from Real Landslide Mass for Mitigating Landslide Disaster

Kazuo Konagai* · Jörgen Johansson · Muneyoshi Numada

Abstract. This paper introduces ideas for extracting important parameters for landslide risk assessment from real landslide masses. The parameters included velocities estimated from mud spatters and therefore equivalent viscosities of fluidized soil masses. Mud spatters remaining on dwelling walls etc., which often show parabolic patterns, are used to estimate flow velocities. One example shows that the viscosity of a liquefied pyroclastic material is about two to three digits smaller than those obtained from torsional shear tests for sandy soil, indicating high potential for the fluidized volcanic products to travel fast a long distance even on this extremely gentle slope.

Geometric parameters describing a deformed shape of a coherent landslide mass indicate ultimate load capacity that the landslide mass can bear. This capacity is less sensitive to the entire length of the landslide mass, and determines the distal reach of the soil mass.

Keywords. Landslide, risk assessment, velocity, volcanic product, motion

20.1 Introduction

Geological and geomorphologic features of surface soils have been greatly responsible for serious destructions in massive earthquakes. Kobayashi (1981) found that more than half of all deaths in large ($M > 6.9$) earthquakes in Japan between 1964 and 1980 were caused by landslides. Geomorphologic surveys, however, have seemingly been viewed as just a reference for more advanced stages of detailed geotechnical investigations probably because they have provided only origin, evolution and/or process that shaped landforms, and less effort has been made to extract measurable parameters from landforms. Eventually, only a few percents of entire pages have been gone for describing geomorphologic features in many reports.

On the other hand, geotechnical studies have been pointwise taking samples from boreholes and blocks cut out from real landslide masses. Though many parameters for describing complicated nature of a single soil sample were obtained, analyzing behavior of an entire soil mass is yet based on interpolation and/or extrapolation of these soil samples' nature. Performing numerical simulations thus facees difficulty even in these days of highly manipulative numerical solutions to problems of increasing complexity.

The authors have been making attempts in field surveys so that important parameters for landslide risk assessment would be extracted from real landslide masses in massive earthquakes. The parameters included velocities and therefore equivalent viscosities of fluidized soil masses. For coherent mass movements, numerical simulations using Material Point Method (MPM), allowing for large soil deformation analyses, have been conducted. The results hint an idea for extracting parameters that can describe possible run-out-distances of similar landslide masses. Below, the authors describe these attempts in more detail.

20.2 Estimation of Velocity of Fluidized Soil Flows

Landslides can range in size from small movements of loose debris to massive collapses of entire summits. For short to medium-length slopes, some measures will be effective for assessing and mitigating landslide hazards. Extremely large slope failures, however, are very difficult to mitigate, and the importance of run-out analysis emerges. Velocities of fluidized soil flows depend largely upon water contents and materials. When the soil includes noticeable amount of porous volcanic products (pyroclastic material) and water, its flow is called "lahar", and can be extremely dangerous because of its energy and speed. Estimating run-out speed from real landslide mass will be thus very important for future preparedness. Examples follow.

20.2.1 Las Colinas Landslide in the El Salvador Earthquake of 2001 (Konagai et al. 2002, 2004a)

Landslides caused by the January 13, 2001, El Salvador Earthquake ($M = 7.6$) were found nationwide in El Salvador. Among all, Las Colinas landslide was the largest (Fig. 20.1). A considerable amount of soil (about 200 000 m^3) slipped down the slope rising behind a new residential district of Neuva San Salvador, west of the capital city, San Salvador. It destroyed hundreds of houses and killed more than 700 people.

Fig. 20.1. Bird's eyes view of the Las Colinas landslide

Fig. 20.2. House wall spotted with splashes in deposition zone

There were many splash marks of mud remaining on house walls, trees, etc. at the time of the authors' visit. In general, the splash marks were higher near the toe of the slope than in the middle or distal reaches of the landslide deposition zone. The highest splash was found about 8 m above ground level on a tree trunk near the toe of the head slope. This indicates that the landslide mass slowed on the deposition zone, while the following mass pushed it from behind. Figure 20.2 shows the walls of a dwelling on the eastern perimeter of the landslide deposit. Mud spatters on the wall follow a parabolic shape. The parabola, having a peak height of 4.5 m, drops downward and reaches the ground after about a 5 m horizontal run. This geometry suggests that the time, t, needed for the splashes to reach the ground from their peak height Δh (4.5 m) was about 1 s ($\Delta h = gt^2/2$). During this time the splashes ran about 5 m horizontally, indicating a horizontal speed of 5 m s^{-1}. Because the estimated velocity is for the eastern edge of the landslide mass, the main stream of debris might have flowed faster despite moving over, around, and through dwellings standing close together.

20.2.2 Tsukidate Landslide in the 2003 Off Miyagi Earthquake

An intense earthquake, with a moment magnitude of 7.0 (Japan Meteorological Agency, JMA) took place at 18:24 JST on June 24, 2003. The epicenter was located at latitude 38.8° N and longitude 141.8° E. Its intense shake was responsible for a landslide at Tsukidate, Miyagi (Fig. 20.3). The horizontal distances from the top end of the scar to the toe of the slope and to the farthest reach of the soil mass are 100 m and 180 m respectively. The landslide descended 27 m over a horizontal distance of 180 m. Thus, the average inclination from the top of the source area to the toe of the deposit is about 6–7 degrees. Figure 20.4, a pair of aerial photographs taken in 1962, is perceived as a single image in terms of depth, and a valley is seen cutting into a hillside. This valley had been filled in for cultivation since 1962, and the landslide took place exactly along this valley.

Mud spatters were found on a mortar block wall of a dwelling on the western perimeter of the landslide mass. From the parabolic geometry of the mud spatters, a horizontal speed of about 3 m s^{-1} was estimated for this western edge. At this point, the maximum depth of the landslide mass of 1.2 m was reached. Assuming that the soil mass exhibited nature of incompressible Newtonian fluid and its flow reached steady state when its maximum depth of about 1.2 m was reached, the velocity v of the soil mass flow satisfies the following equation:

$$\mu \frac{\partial^2 v}{\partial z^2} + \rho g \sin\theta = 0 \tag{20.1}$$

where, μ = equivalent viscosity of fluidized soil, ρg = specific gravity of soil, θ = inclination of base line (see Fig. 20.5). Given the following parameters and boundary conditions,

Fig. 20.3.
Tsukidate landslide (Yahoo news)

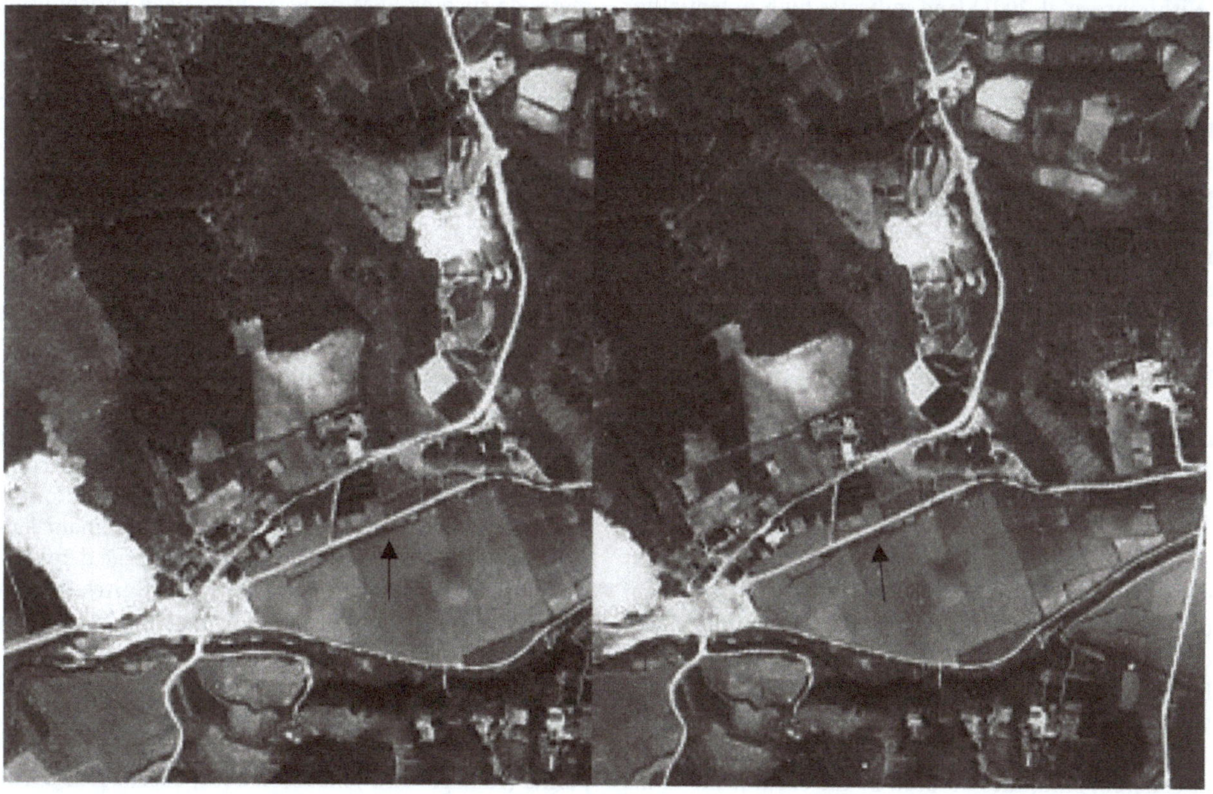

Fig. 20.4. Aerial photos of the landslide area (see *black arrow*) (photographed in June, 1962, provided by Active Fault Research Center, AIST)

$$\rho g = 1.7 \times 10^4 \text{ N m}^{-3}, \quad \sin\theta = 0.04 \tag{20.2a,b}$$

$$v = 3.2 \text{ m s}^{-1} \text{ at } z = H = 1.2 \text{ m}, \quad v = 0 \text{ m s}^{-1} \text{ at } z = 0 \text{ m} \tag{20.2c,d}$$

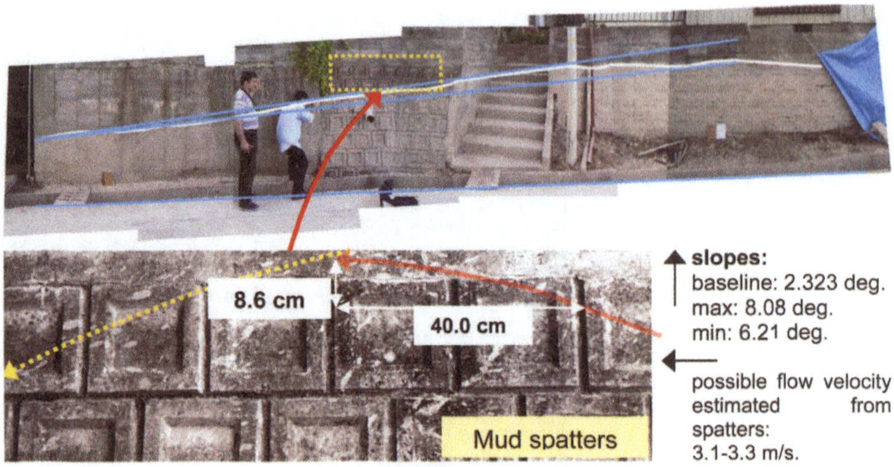

Fig. 20.5.
Mud spatters remaining on retaining wall, Tsukidate (2003 Off Miyagi Earthquake)

Equation 20.1 can be solved for the equivalent viscosity μ as:

$$\mu = \frac{1}{2v}\rho g H^2 \sin\theta = 1.5 \times 10^2 \text{ Pa s} \tag{20.3}$$

This viscosity is about two-to-three digits smaller than those obtained from torsional shear tests for sandy soils (Nishimura et al. 2001), and indicates high potential for the fluidized pyroclastic material to travel fast a long distance even on this extremely gentle slope.

The estimated velocity is for the western edge of the landslide mass. If the same viscosity was reached in the main stream of the soil flow, the flow velocity would be proportional to the square of H. Judging from depths of the landslide mass measured point-wise, the mainstream may have flowed about two times as fast as the edge stream, namely, about 6 to 7 m s^{-1} speed was reached. This estimate is seemingly consistent to the distorted pattern of vegetation (bamboos and other plants carried by the landslide mass: see Fig. 20.1).

20.3 Extracting Parameters for Estimating Travel Distances for Coherent Mass Movement

Though a coherent mass movement may not be as serious as rapid soil flows, it surely affects greatly the lives of people. The strong earthquake (M 6.8) that hit the inland district of Niigata prefecture on October 23, 2004 was followed by strong aftershocks in rapid succession. The district suffered four jolts with a magnitude of 6 or greater within 38 minutes after the main shock. A large number of coherent landslides induced by the earthquake in the Higashiyama mountain district resulted in the closure of 233 segments of several prefecture routes and the national route 249, and 61 localities were completely isolated. The number of closed roads segments increased even more in some intense aftershocks that included the one occurred at 10:40 AM, October 27 ($M = 6.1$), in which the maximum intensity of 6-weak on JMA scale was recorded. The threat of additional aftershocks and heavy rains, at the beginning of the snow season, had made it very difficult to restore the damaged segments and further analyze them for implementing better preventive measures for the upcoming snow-melting season.

For studying large deformations of soils, numerical methods such as FEM or FDM have been widely used. For example, the finite difference based FLAC (Fast Lagrangian Analysis of Continua) calculates large strains by using low-order strain elements (Cundall 1988). However, when dealing with large strains, highly distorted elements often account for inaccurate results. In the field of computational fluid dynamics, where history-dependent materials are less common, purely Eulerian methods are often used. Sulsky et al. (1994) extended one of these methods to solid mechanics. Their method evolved from a particle-in-cell (PIC) method is referred to as the Material Point Method (MPM) and is categorized as a mesh-less methods. In MPM, a body to be analyzed is described as a cluster of material points. The material points, which carry all Lagrangian parameters, can move freely across cell boundaries of a stationary Eulerian computational mesh, which should cover the position of the analyzed body. The computational mesh can remain constant for the entire computation, thus the large deformation disadvantage of the conventional finite element method related to the problem of mesh distortions is eliminated. The authors have developed two dimensional LPFDM (Lagrangian Particle Finite Difference Method) based on simple finite difference scheme of calculation, but with the inclusion of Lagrangian particles/material points (Konagai and Johansson 2001, 2002, 2004b). The method is intended to be a combination of the schemes of FLAC and MPM so that the present method allows for extremely large deformations of soils retaining the simplicity of FLAC. The method has been further extended to model a rapid and long-traveling soil flow keeping its

Table 20.1. Lagrangian parameters

Young's modulus	5×10^7 N m^{-2}
Poisson's ratio	0.30
Density	1 700 kg m^{-3}
Internal friction angle	0.5 rad
Cohesion	9 800 N m^{-2}
Strength reduction	Both cohesion and internal friction angle are reduced by 50%
Initial friction angle on the slip surface	0.5 rad
U_{ref} in Eq. 20.9	0.1 m
α for local non-viscous damping	0.8
L: Cell size on x–y plane	1 m

planar geometry. The landslide mass is represented by a plane assembly of soil columns (material points) in contact with each other, free to deform and retaining fixed volumes in their descent down a curving path.

Examples shown in authors' previous paper (Table 20.1 and Fig. 20.6, for example) provided a perspective on the capability of the present method for describing long-traveling soil movements. The method however leaves much to improve by comparing these numerical simulations with real landslides, and it is extremely difficult to do it, because all Lagrangian parameters for the entire landslide mass are hardly obtained. For example, it is often that plants growing on a landslide mass shoot their roots all through the soil mass in such a way that overall characteristics of the soil mass is largely different from those obtained through tests of soil samples taken point-wise from the landslide mass.

One possible breakthrough will be to consider a real landslide as a huge "simple soil test". The following pseudo 3D-LPFDM simulation has been performed for some hints. A rectangular-planner soil mass of one meter thick is assumed to be resting on a flat slope dipping to a level ground, which spreads in immediately front of the toe of the landslide mass. Constant frictional coefficients between the landslide mass and level/inclined sliding surfaces were respectively set at μ_1 and μ_2 as shown in Table 20.2. Necessary mechanical parameters for the landslide mass include internal friction angle, ϕ, Young's modulus, E, Poisson's ratio, ν, as shown in Table 20.2. Cohesion of the soil mass was set in such a way that it would allow a 1 m-high column of the soil to stand upright in the gravity.

For different lengths L of the landslide mass, the mass movement was examined in the time-marching calculation of MPM and lengths of the landslide mass segments L_1 and L_2 on the level and inclined sliding surfaces, respectively, were obtained at every time step as shown in

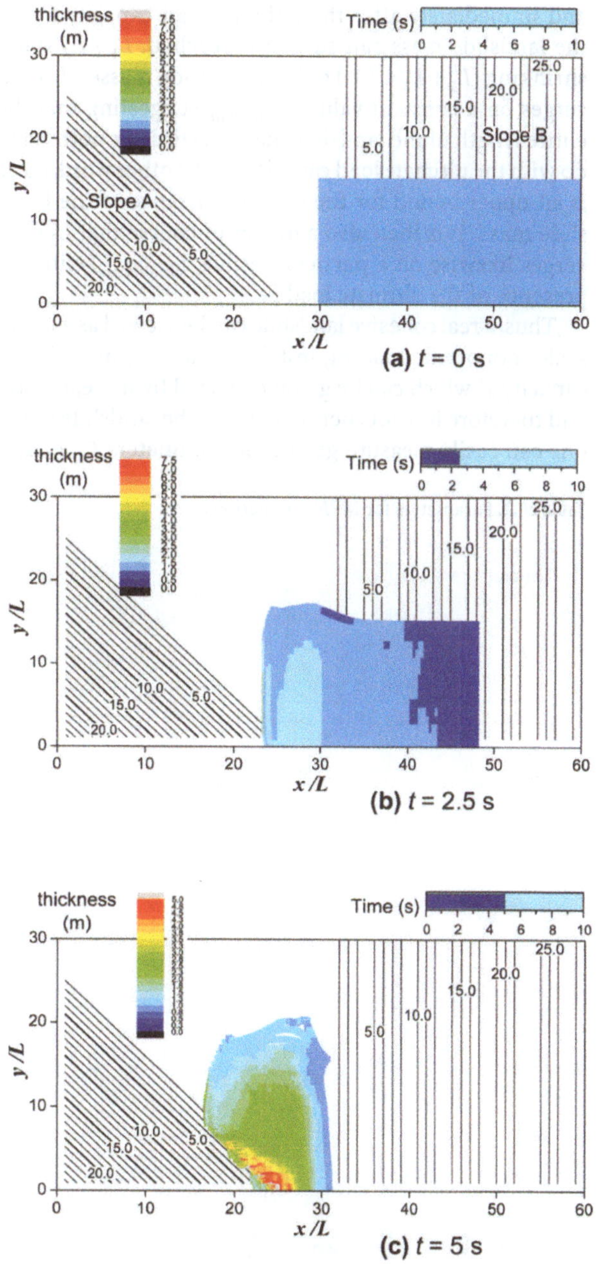

Fig. 20.6. Long traveling soil flow: Contour lines in this figure show that there are two slopes A and B making up the configuration (see **a**). The soil mass from slope B spreads wide as it surged across the horizontal plane, and after hitting slope A, the direction of the mass flow turned avoiding slope A (after Konagai et al. 2004b)

Fig. 20.7. Initially the landslide mass keeps $L_1 + L_2$ to be completely identical to the entire length L of the landslide mass, namely, the entire length is kept unchanged. However, as the landslide mass runs further forward over the level ground (deposition zone), its movement slows down due to larger frictional coefficient and with no driving force induced on the level ground. Eventually the following segment of the landslide mass pushes the soil from behind, and immediately after the ultimate load capacity F that the landslide mass can sustain is reached, $L_1 + L_2$ starts shrinking ($L_1 + L_2 < L$). For long landslide masses, L_1 converges on a constant value, $L_{1,\mathrm{ultimate}}$ suggesting that the entire length of the landslide mass has little or nothing to do with the ultimate load capacity. In the other word, there is an upper bound for the run-out-distance of the landslide mass. It is then also a matter of course that L_2 converges likewise on a particular value, $L_{2,\mathrm{ultimate}}$, with the presence of the ultimate load F.

Thus, a real cohesive landslide can be viewed as a large-scale monotonic loading test for obtaining its ultimate capacity F, which can be greatly affected by its vegetation and therefore its roots network. From the landslide mass, one can easily measure geometric parameters L_1, L_2 and its initial length L. When $L_1 + L_2$ is identical to L, the ultimate force is not reached in the landslide mass yet. On the other hand, when $L_1 + L_2$ is noticeably smaller than L, the ultimate capacity F is considered to have been reached, and L_1 and L_2 will be very close to $L_{1,\mathrm{ultimate}}$ and $L_{2,\mathrm{ultimate}}$, respectively.

Width of a landslide mass can also affect F and therefore $L_{1,\mathrm{ultimate}}$ and $L_{2,\mathrm{ultimate}}$, as illustrated in Figs. 20.8 and 20.9. Figure 20.8 shows plans (top views) of landslide masses with different widths. Figure 20.9 shows both L_1 and L_2 for the entire sliding sequences for all landslide masses examined in Fig. 20.8. As the width increases, the curve converges on the plane-strain solution, while it is rather closer to plane-stress solution for thinner soil cases ($W = 2$ m for example). In this figure, dL_2/dL_1, after the landslide mass starts yielding, becomes less steep for wider soil masses. This can be explained as follows: this dL_2/dL_1, when multiplied by the cross-sectional area of the landslide mass, is identical to $-dV_2/dV_1$, where dV_2 and dV_1 are soil volumes which pass through the cross-section per unit time for segments L_2 and L_1, respectively. The soil volume dV_1 will be pushed forward over the deposition zone by the following soil volume dV_1, while the volume of $dV_2 - dV_1$ is pushed up and/or sideways making a bulge at around the toe of the slope. It is thus clear that less steep dL_2/dL_1 indicates larger confinement of $dV_2 - dV_1$ causing slight increase of F even after the initial ultimate capacity F was reached. In other words, the effect of confinement can be quantitatively estimated from geometry of the deformed landslide mass.

Table 20.2. Parameters for MPM simulations

MPM parameters	Cell size	0.2 × 0.2 m
	Initial arrangement of particles in a cell	2 × 2
	Time increment	5×10^{-4} s
Sliding surface	Slope angle	20 degrees
	Frictional coefficients μ_1 and μ_2	0.8 and 0.2
Landslide mass	Young's modulus	2×10^6 Pa
	Specific density	1 600 kg m^{-3}
	Possson's ratio	0.3
	Internal friction angle	30 degrees

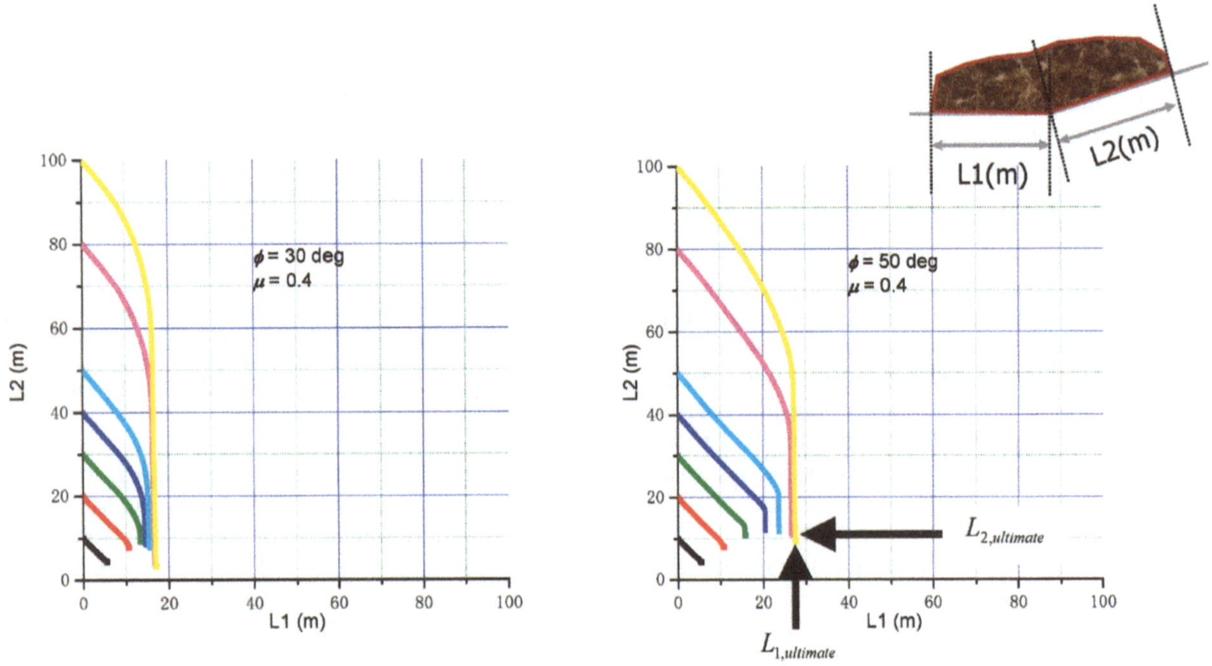

Fig. 20.7. Lengths of the landslide mass segments L_1 and L_2 on the level and inclined sliding surfaces, respectively

Fig. 20.8.
Top view of landslide masses with different widths: Colors show thickness of the mass in meters

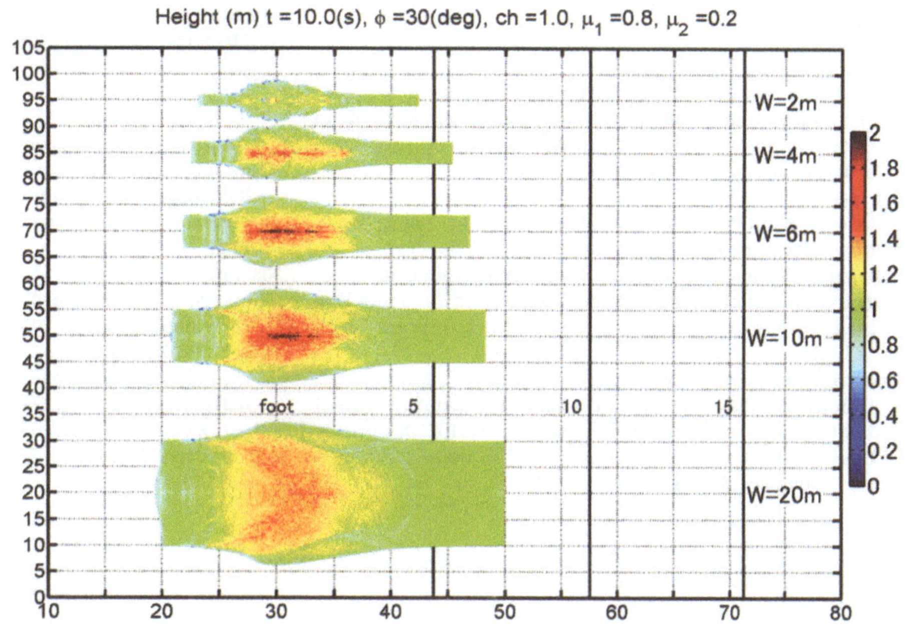

If either μ_1 or μ_2 is obtained in addition to the above-mentioned geometric parameters for the landslide mass, one will be able to obtain the ultimate bearing load F. For this, the ring-shear apparatus (Sassa et al. 1966, for example) will be one of powerful tools. Sassa has demonstrated in a number of his tests that un-drained samples, when started shearing, can exhibit sudden decrease of apparent friction angle suggesting that grain crushing caused sudden increase in pore-water pressure (Sassa 1996). Therefore, with drastically decreased value of μ_2, even a coherent mass can have some large run-out distance L_1, and the ultimate load capacity F would be more crucial in determining its run-out distance.

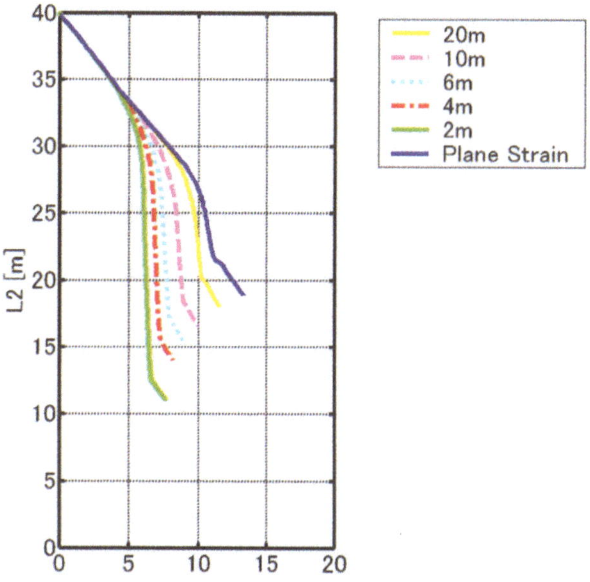

Fig. 20.9. Lengths of the landslide mass segments L_1 and L_2 on the level and inclined sliding surfaces, respectively, for different widths

20.4 Summary

This paper introduced some ideas for extracting important pieces of information for landslide risk assessment from real landslide masses. The parameters included velocities estimated from mud spatters and therefore equivalent viscosities of fluidized soil masses. One example from the 2003 Off Miyagi Earthquake showed that the viscosity of the liquefied pyroclastic material was about one to two digits smaller than those obtained from torsional shear tests for sandy soils (Nishimura et al. 2001), and indicates high potential for the fluidized volcanic products to travel fast a long distance even on this extremely gentle slope.

For coherent mass movements, numerical simulation of landslide mass movements using Material Point Method (MPM) did hint an idea for extracting parameters from real landslide masses. A real cohesive landslide can be viewed as a large-scale specimen of a monotonic loading test for obtaining its ultimate load capacity F, which can be greatly responsible for distal reach of the landslide mass. Geometric parameters such as lengths of the landslide mass segments L_1 and L_2 on the level and inclined sliding surfaces, respectively, can be compared with the initial entire length of the landslide mass. When $L_1 + L_2$ is noticeably smaller than L, the ultimate capacity F is considered to have been reached, and L_1 and L_2 will converge on $L_{1,\text{ultimate}}$ and $L_{2,\text{ultimate}}$, respectively, which ultimate lengths are less sensitive to the initial entire length of the landslide mass. Both $L_{1,\text{ultimate}}$ and $L_{2,\text{ultimate}}$ will provide a good perspective for landslide risk assessments.

References

Cundall PA, Board M (1988) A microcomputer program for modeling large-strain plasticity problems. Numerical Methods in Geotechnics (Proceedings, 6th International Conference, Innsbruck, Austria, April 1988), pp 2101–2108

Konagai K, Johansson J (2001) Two dimensional Lagrangian particle finite difference method for modeling large soil deformations. Structural Eng/ Earthquake Eng. 18(2):91s–95s

Konagai K, Numada M (2002) Pseudo-three dimensional Lagrangian particle finite difference method for modeling long-traveling soil flows. J Jpn Soc Dam Engineers 12(2): 123–128

Konagai K, Johannson J, Mayorcap P, Yamamoto T, Miyajima M, Uzuoka R, Pulido NE, Duran FC, Sassa K, Fukuoka H (2002) Las Colinas landslide caused by the January 13, 2001 off the Coast of El Salvador Earthquake. Journal of Japan Association for Earthquake Engineering 2(1):1–15

Konagai K, Johansson J, Mayorca P, Uzuoka R, Yamamoto T, Miyajima M, Pulido N, Sassa K, Fukuoka H, Duran F (2004a) Las Colinas landslide: Rapid and long-traveling soil flow caused by the January 13, 2001, El Salvador earthquake. GSA Paper 375 – Natural Hazards in El Salvador, William I. Rose eds, Geological Society of America, pp 39–53

Konagai K, Johansson J, Itoh H (2004b) Pseudo-three dimensional lagrangian particle finite difference method for modeling earthquake induced soil flows. 13th World Conference on Earthquake Engineering, No 547

Nishimura S, Honda T, Towhata I (2001) Hollow torsion shear tests on viscosity of liquefied sand under great deformation, Proc., 36th Annual Meeting, Japan Geotechnical Society, Tokushima

Sassa K (1996) Prediction of earthquake induced landslides. In: Proceedings of 7th International Symposium on Landslides, A.A. Balkema, Trondheim, 17–21 June, 1, 115–132

Sulsky D, Chen Z, Schreyer HL (1994) A particle method for history dependent materials. Comput Methods Appl Mech Engrg 118:179–196

Chapter 21

Landslide Dams Formed by the 2004 Mid-Niigata Prefecture Earthquake in Japan

Hideaki Marui* · Hiroyuki Yoshimatsu

Abstract. It was a remarkable aspect that a tremendous number of landslides were triggered by the Mid-Niigata Prefecture Earthqauke. Because the epicenter was located at a depth of 13 km just in the landslide prone area of central part of the Niigata Prefecture, severe damage was caused by the earthqauke-induced landslides. Furtehermore many landslide dams were formed mainly in the watershed of the Imogawa-River by the displaced soil mass of the earthquake-induced landlslides. Some large landslide dams should pose a great threat of flood and debris flow in case of dam collapse to the settlemnt of the Ryuko-District in the downstream area of the watershed. It was urgently needed to arrange the emergency operations to avoid the destructive collapse of the major landslide dams. This contribution illustrates overview of landslide dams and emergency operations against dam collapse.

Keywords. Landslide dam, earthquake, emergency operation

21.1 Introduction

On 23 October 2004, an earthquake with magnitude 6.8 took place in central part of the Niigata Prefecture; the epicentral location was approximately 70 km south of Niigata-City. As a consequence of the earthquake, 46 people were killed, about 4700 people were injured, and 2800 houses were completely and 10000 houses were partially damaged by structural failures, landslides and so on. Furthermore important public infrastructures like railways and highways as well as other major roads were also heavily damaged. About 100000 people were to be evacuated to 600 refuges soon after the earthquake. The total amount of the material damage was estimated to be about thirty billion dollars. The earthquake triggered about 3800 landslides in various types and dimensions, causing extensive damage to settlements, farmlands and infrastructures. It was general understanding until now that because of strong ground shaking during earthquakes a lot of slope failures occur on steep slopes with concave shape but only few reactivated landslides occur on relatively gentle slopes. However, this time also many reactivated landslides occurred in consequence of the Mid Niigata Earthquake in the neighborhood of the epicenters including strong aftershocks especially on hillslopes in the watershed of the Imogawa-River. As an effect, river channels of main stream and tributaries of the Imogawa-River were blocked by landslide dams at more than 50 places in the watershed. For the mitigation of subsequent disasters, it was a matter of great urgency to implement emergency countermeasures against overtopping and successive failure of the landslide dams. In this paper, the characteristics of the earthquake, the outline of landslide disasters and landslide dams and concepts of emergency operations undertaken as well as some practical remarks for mitigation of future disasters are reported.

21.2 Characteristics of the 2004 Mid-Niigata Prefecture Earthquake

The hypocenter of the 2004 Mid-Niigata Prefecture Earthquake was located at shallow depth (13 km) in an active fault-and fold sytem overlain by thick sediments of geologically young formations. An earthquake intensity of 7.0 (by Japan Meteorological Agency) was recorded in Kawaguchi Town close to the hypocenter for the first time since biggining of its recording using seismometers. The event was followed by strong aftershocks in rapid succession. The area suffered much succesively from four major aftershoks with a magnitude of 6 or greater within 38 minutes after the main shock.

The aftershoks were distributed 35 km along the NNE-SSW strike of the geological structure within a 20 km wide zone between the Yukyuzan fault and the Shibata-Koide tectonic line (Fig. 21.1). In the figure the red circles show epicenters of individual aftershoks and the black rectangle shows a projection of the fault model used for analysis by the National Research Institute for Earth Science and Disaster Prevention (NIED). NIED has anlaysed the distribution of asperities by using seismic waveform data observed by strong motion seismographs (Fig. 21.2). High asperitiy zones with yellow color are recognized at three places, namely at the location of the epicenter of the main shock, at eastern side and southwestern side of the epicenter. The area with the highest asperity at eastern side of the epicenter corresponds to the watershed of the Imogawa-River, where especially many landslides occurred.

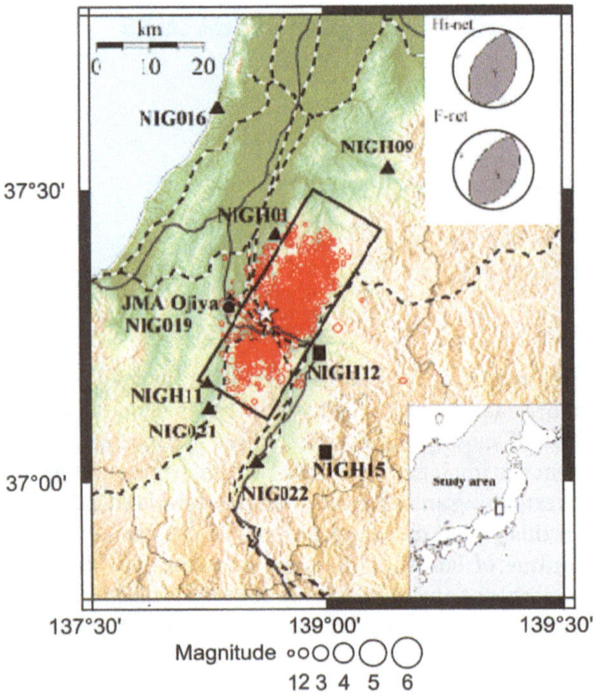

Fig. 21.1. Epicentral distribution of aftershocks (by NIED)

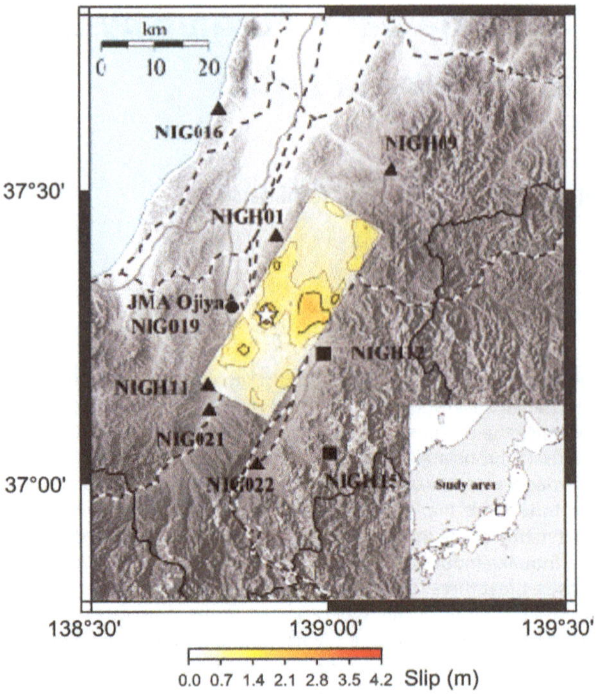

Fig. 21.2. Distribution of asperities (by NIED)

21.3 Distribution of Landslides

According to the report of the Ministry of Land, Transport and Infrastructure, altogether about 3800 landslides and slope failures have occurred and the total amount of the displaced soil mass was estimated to be 100 million m^3. Slope displacements induced by the earthquake can be classified mainly into the following categories: (*1*) shallow slope failures on steep slopes near ridges; (*2*) shallow slope failures on steep slopes along river channels, (*3*) reactivated landslides on relatively gentle hillslopes; (*4*) landslide dams formed by the displaced soils mass by the previous three categories of slope movements. The heavily affected areas were famous for their typical hillslope landscape and most of slopes were used mainly for terraced paddy rice fields and farm ponds for irrigation or carp fish cultivation. Such terraced fields and farm ponds were severely damaged and connecting roads were blocked at many places by the landslides and slope failures. Many houses were destroyed directly by the landslides. Not a few houses were also inundated with the impounded water of reservoir in the upstream area of landslide dams. For the mitigation of subsequent disasters, it was a matter of great urgency to implement emergency countermeasures against overtopping and successive failure of the landslide dams. Furthermore, abnormally heavy snowfall since 1986 struck the heavily damaged area by the earthquake. It was also necessary to pay attention to the successive landslides caused by snowmelt.

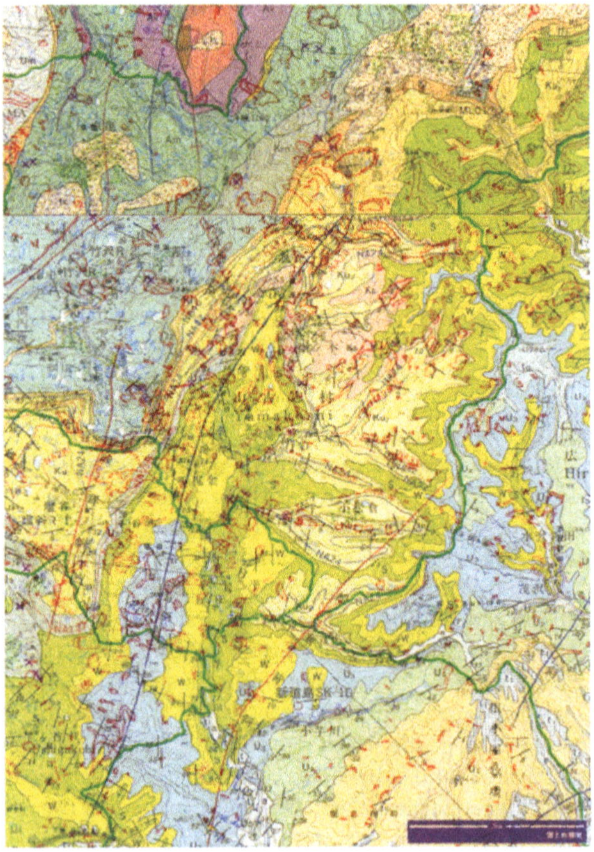

Fig. 21.3. Landslide distribution on the geological map (landslide distribution by Geographical Survey Institute). *Red marking*: landslides

Landslides and slope failures occurred mostly at heights of 300–700 m on the Higashiyama Hill which stretches from north-northeast to south-southwest and especially densely in the watershed of the Imogawa-River along the Kajigane Syncline. The area is characterized as an active folding area. The interval between fold axes is about 1km. Overlay of landslide distribution on the geological map is shown in Fig. 21.3. Detailed distribution of landslides and slope failures in the watershed of the Imogawa-River is shown in Fig. 21.4. The geology of the area consists of thick sequence of Pliocene to lower Pleistocene sediments. Many landslides occurred along dip direction of geological formations. Originally, heavily landslide prone areas are widely distributed in the Tertiary mudstone areas in Niigata Prefecture. Usually reactivated landslides occur frequently in the Mushigame-District which consists of mudstone and locates northwestern side of the watershed of the Imgawa-River. This time, however, most of landslides occurred in the watershed of the Imogawa-River which consists of sandstone and sandy siltstone.

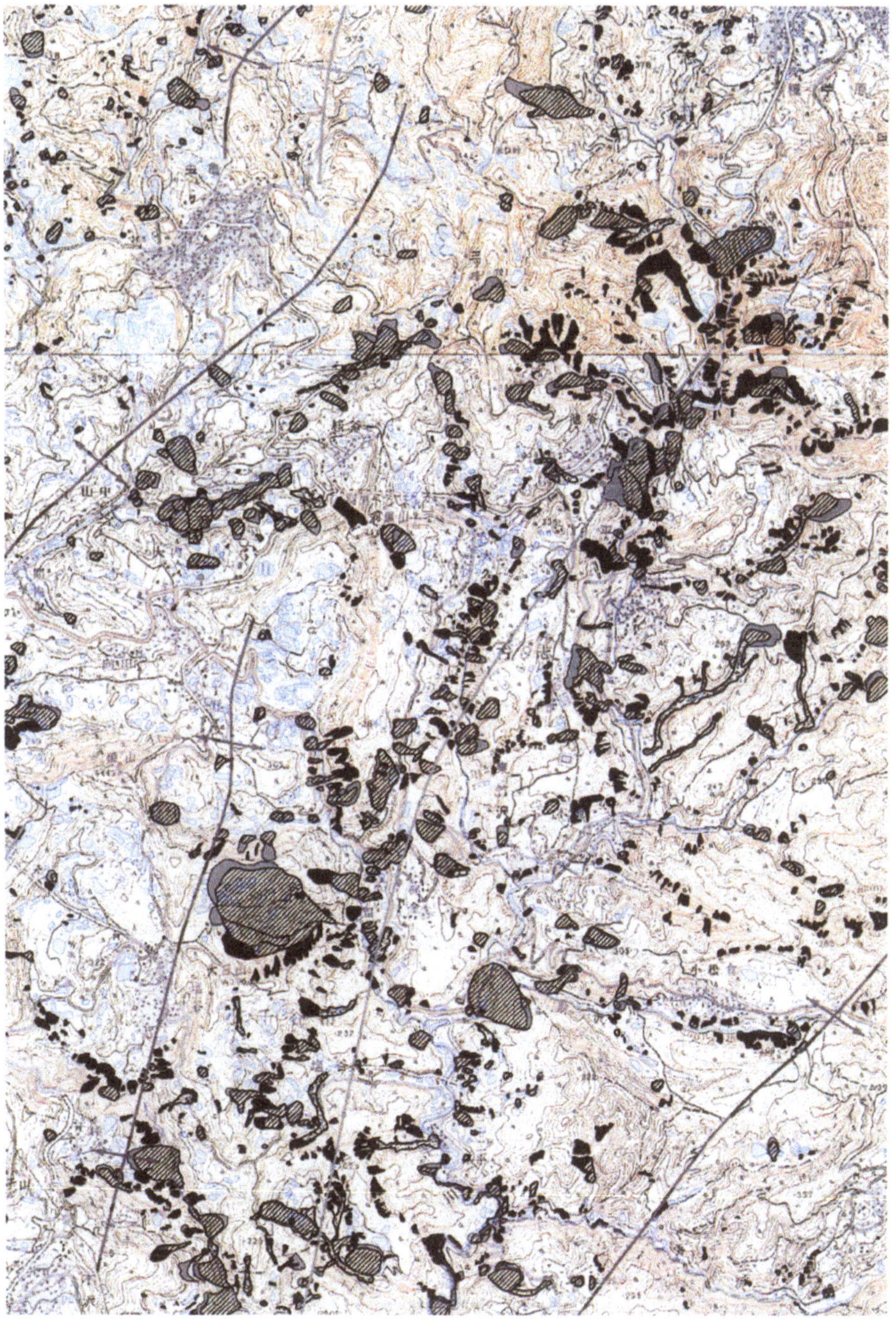

Fig. 21.4. Landslide distribution in the watershed of the *black marking*: landslides (landslide distribution was interpreted by Prof. Yagi) (geological map by Geological Survey)

21.4 River Blockage by Landslide Dam

The watershed of the Imogawa-River has a drainage area of 38.4 km^2, a channel length of 17.2 km and an average bed inclination of 1/70. It is a right tributary of the Uonogawa-River of the Shinanogawa-River system. It originates from Mt. Sarukradake (679 m a.s.l.) and joins the Uonogawa-River at Ryuko-District.

More than 50 landslide dams were formed along the main stream of the Imogawa-River and its tributaries by the landslides triggered by the earthquake (Fig. 21.5). Many of them were naturally overtopped without catastrophic collapse soon after their formation. However, five prominent landslide dams with certain large volumes were formed along the main stream of the Imogawa-River (at the districts of Terano, Nanpei, Naranoki, Higashi-Takezawa and Junidaira). Two of them were critical landslide dams to which emergency operations were to be urgently managed because of their dimensions, namely Higashi-Takezawa landslide dam (Fig. 21.6–21.9) and Terano landslide dam. In both cases, each landslide has a length of about 300 m and a volume of more than 1 million m^3. It was important to evaluate the stability of the individual landslide dam.

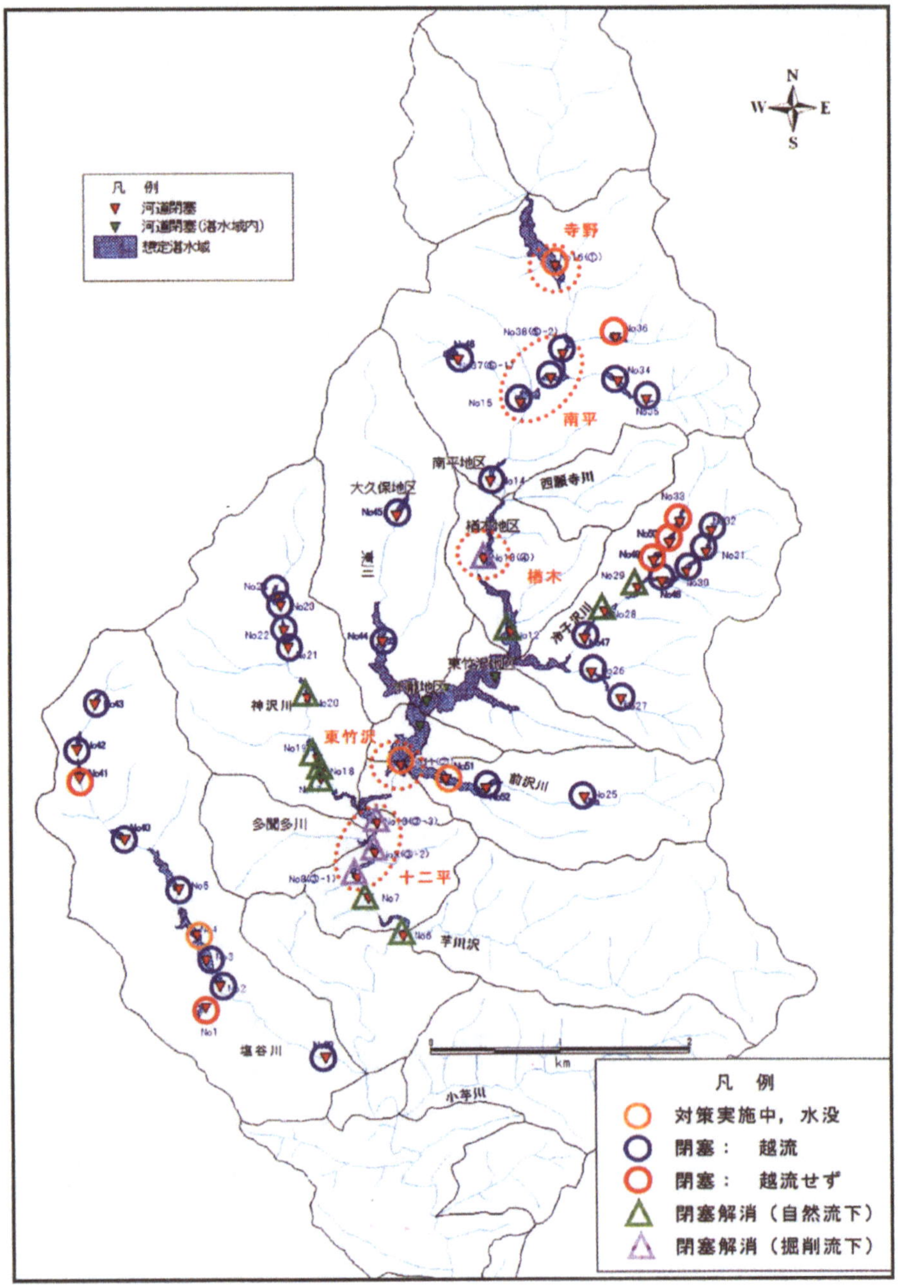

Fig. 21.5. Distribution of landslide dams (Hokuriku Regional Development Bureau) (situation on March 2005). Legends: *blue color* along river channels: reservoir; landslide dams are shown by *circles* and *triangles*: blockages are still existing; *orange circle*: countermeasures installed; *blue circle*: overtopping; *red circle*: without overtopping, blockages are already solved; *green triangle*: naturally washed away; *violet triangle*: artificially

Fig. 21.6. River blockage by the Higashitakezawa landslide

Fig. 21.7. Main scarp of the Higashi-Takezawa landslide

Fig. 21.8. Plane view of the Higashi-Takezawa Landslide

Fig. 21.9. Front view of the Higashi-Takezawa Landslide

21.5 Emergency Operations against Collapse by Landslide Dam

Generally, the stability of landslide dams against collapse should be evaluated on three mechanisms, namely on collapse by water pressure, by overtopping and by piping. In both cases, the length of the blockage of the river channel is about ten times of the maximum water depth of the reservoir. Therefore, the possibility of the destructive collapse of either dam by water pressure and/or piping was estimated to be low. However, the water levels of the both reservoirs were quickly raised up soon after the blockage of the river channel by continuous rainfall. The settlement of Kogomo-District in the upward area was inundated and many houses were damaged by inundation with the impounded water. Because water level of the reservoir has significantly risen, there was an apparent danger of the collapse of the landslide dams by overtopping. In the case of overtopping, collapse of the landslide dam could cause outburst floods or debris flows which would endanger downstream residential areas at the Ryuko-District. Therefore, the inhabitants of the downstream area had to evacuate.

It was urgently necessary to lower the water level of the reservoir. Emergency operations to prevent collapse of the landslide dams were first started by the local government of the Niigata Prefecture and shifted to the responsibility of the central government from November 5. Yuzawa Sabo Office of the Ministry of Land, Infrastructure and Transportation was assigned to carry out the concrete emergency operations. At that time, the water level of the Higashi-Takezawa reservoir has risen up to 6 m below the level of overtopping (161 m a.s.l.).

The reservoir at Higashi Takezawa, which was formed by the largest landslide dam along the main channel of the Imogawa-River, has a critical significance. In order to reduce the danger of overtopping, the following emergency operations were undertaken. At first, the water level was tried to lower by means of pumps and siphons. At the beginning 6 pumps and after then additional 6 pumps were installed. It was not suitable to use drainage pumps for the long term; thus, they should be used for only emergency purposes in the initial stages. Temporal variation of the water level of the Higashi-Takezawa reservoir is shown in Fig. 21.10. In the middle of the November the water level was significantly raised up to the level near to 158 m a.s.l. just 3 m below the overtopping level. Also very critical erosion occurred at the outlet area of the temporal channels because of water leakage form drainage hoses. Therefore, the outlet area of the temporal channel was to be shifted to the more stable area namely through the building of the abandoned elementary school (Fig. 21.11). The heavily eroded area was also to be stabilized by using protection measures against erosion. Because of maintenance problems for the pumps, diversion pipelines were installed as an additional alternative measure to prevent overtopping. These alternative diversion pipelines were quite effective (Fig. 21.12). As a result, the water table was kept lower than the overflow elevation and lowered to a safer level.

Finally, an open channel with a sufficient cross-sectional area for water discharge including snowmelt dur-

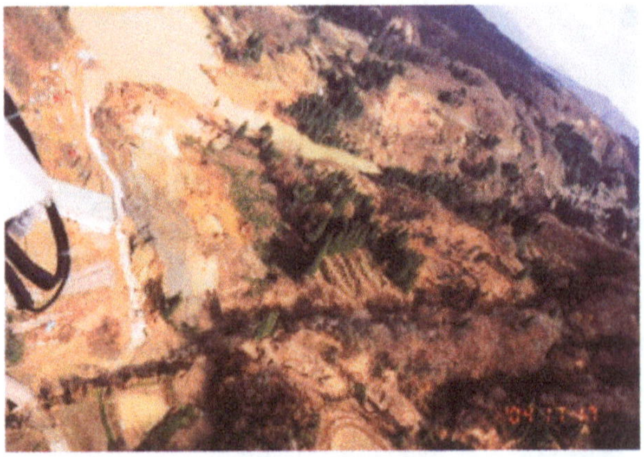

Fig. 21.11. Critical erosion during drainage pumping

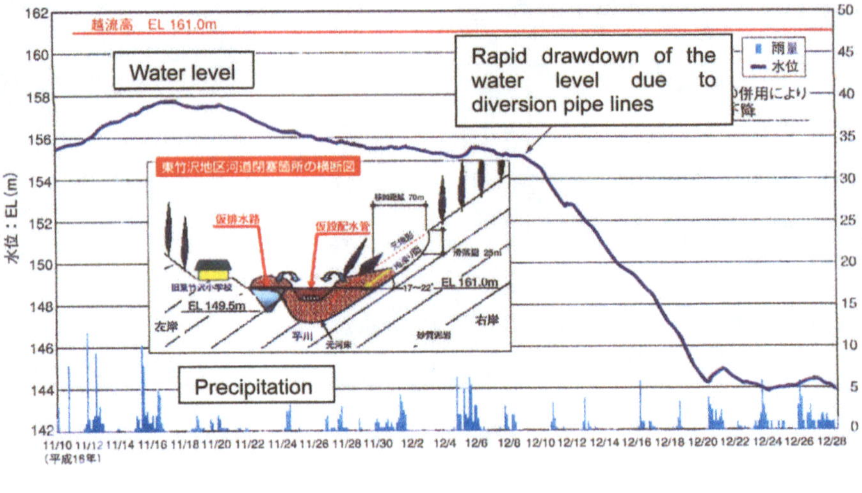

Fig. 21.10. Variation of the water level of the Higashi-Takezawa reservoir (Hokuriku Regional Development Bureau)

ing early spring was constructed (Fig. 21.13). It was absolutely necessary to retain the stability of the displaced soil mass by the earthquake-induced landslide against secondary motion during construction of the open channel at the foot of the landslide mass. Therefore, excavation of the upper part of the displaced soil mass (Fig. 21.14) was immediately carried out as an appropriate countermeasure to compensate the reduction of safety due to construction of the open channel at the foot of the displaced soil mass.

Fig. 21.12. Alternative diversion pipe-lines

Fig. 21.13. Open channel with sufficient cross-sectional area

Fig. 21.14. Excavation of the displaced soil mass

21.6 Monitoring and Observation System

Parallel to the countermeasures to prevent the collapse of the landslide dams, monitoring and observation systems were installed for the case of emergency (Fig. 21.15). For the security of the inhabitants at Ryuko-District, monitoring system including water level gauge, debris flow sensor and monitoring camera was installed around the Higashi-Takezawa landslide dam and the Terano landslide dam respectively. A necessary information transfer system was arranged for evacuation of the inhabitants in case of debris flow occurrence. Monitoring on secondary displacement of soil mass was also carried out for the security during the construction works for the Higashi-Takezawa landslide and the Terano landslide.

21.7 Successive Landslides Caused by Snowmelt

In addition to actually slid slopes, there remained a lot of potentially unstable slopes with cracks and gaps induced by the earthquake in especially heavily damaged areas like the watershed of the Imogawa-River. These areas were struck the heaviest snowfall in 19 years. As a result, in the early spring successive landslides occur on various slopes in heavily damaged area by snowmelt. It is also necessary

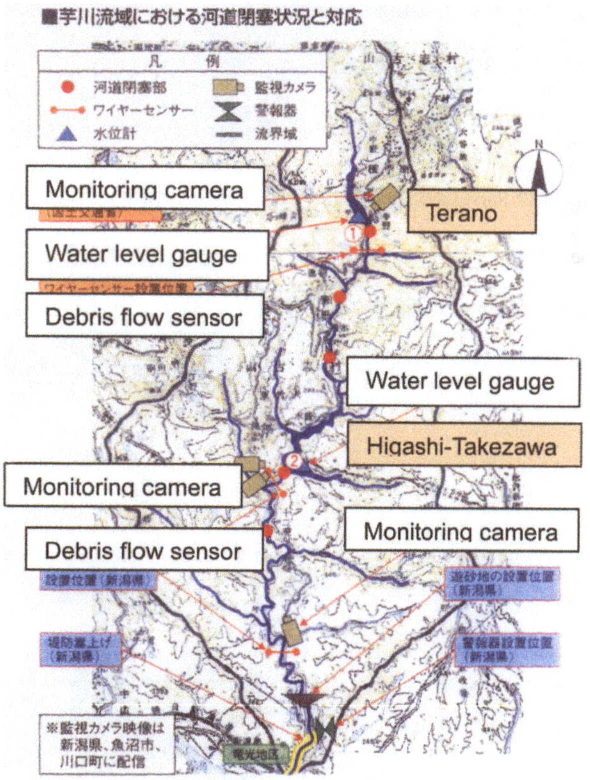

Fig. 21.15. Monitoring system in the watershed of the Imogawa-River

to consider the danger of successive landslide occurrence by snowmelt in recovery planning of settlements in damaged areas.

Fig. 21.16. Early stage of snowmelt at Higashi-Takezawa (March 2005)

Overview of the surroundings of the Higashi-Takezawa landslide dam at early stage of snowmelt is shown in Fig. 21.16. No significant displacement can be seen. Discharged water including snowmelt is flowing to downstream through the open channel without problem. Examples of successive slope failures caused by snowmelt were shown in Figs. 21.17 and 21.18. A slope failure with a volume of about 30 000 m^3 occurred at the main scarp of the Higashi-Takezawa landslide at the last stage of snowmelt. However, this slope failure gives no significant effect to the safety factor of the main landslide at Higashi-Takezawa. Also in Kogomo-District, a few slope failures successively occurred by snowmelt. Displaced soil mass was reached to the houses of the settlement. However, in general no catastrophic slope failure or landslide was caused by snowmelt in the whole watershed of the Imogawa-River, although continuous monitoring and watching out for slope failures and landslides are to be recommended at least several years.

Fig. 21.17. Successive slope failure caused by snowmelt. At the main scarp (Higashi-Takezawa)

Fig. 21.18. Successive slope failures by snowmelt at Kogomo-District

21.8 Remarks for Mitigation of Future Disasters

In relation to hazard mitigation on successive landslides and slope failures it is urgently necessary to carry out reconnaissance of slid slopes and evaluation of slope instability immediately after the earthquake as a first step. Furthermore, it is again necessary to carry out the same reconnaissance of slid slopes and evaluation of slope instability after snowmelt.

In recovery planning of heavily damaged roads by earthquake-induced landslide, it is necessary to consider the alteration of the original route plan in case that the road is extremely heavily damaged at many places. In some cases stabilization of landslides is a prerequisite condition for reconstruction works of damaged roads.

Besides evaluation of instability of individual slopes, it is also necessary to evaluate the danger degree on possible future disasters by transport of unstable debris materials along river channels in various torrent watersheds after the earthquake. The potential of the movement of unstable materials in torrent watershed near to the epicenter should be completely changed because of the strong influence of the earthquake.

Further, it is also an important task to develop appropriate methods of hazard zoning for prediction of landslides induced by future earthquakes.

References

Geographical Survey Institute (2004) Disaster map at 1:30,000 (in Japanese)

Hokuriku Regional Development Bureau (2005) Damages and recovery situation after the Mid-Niigata prefecture earthquake (the second report)

Honda R et al. (2004) Inversion of the epicenter analyzed by the seismic waveform record of the Mid-Niigata Prefecture Earthquake. http:www.k-net.bosai.go.jp/k-net

Yagi H, et al. (2005) Distribution map of the landslides induced by the Mid-Niigata Prefecture Earthquake

Yanagisawa Y, Kobayashi I, Takeuchi K, Tateishi M, Kayahara K, Kato T (1986) Geology of the Ojiya district, with geological sheet map at 1:50,000. Geological Survey of Japan (Japanese with English abstract)

Chapter 22

Shear Behavior of Clay in Slope for Pore Water Pressure Increase

Satoru Ohtsuka

Abstract. Shear behavior of clay in slope for pore water pressure increase was investigated with both triaxial compression and ring shear apparatuses. Not only shear strength but also shear deformation was examined. Pore water pressure loading test was conducted with tiaxial compression apparatus. Yield and failure points for pore water pressure increase were obtained. It was shown that yield point was on the critical state line described by internal friction angle of ϕ'. It means the design method with ϕ' indicates the design to prevent slope from instability and deformation occurrence. Shear displacement in pore water pressure loading test revealed time dependent behavior as creep. It was made clear to be caused by water migration and successive progressive failure. Creep behavior of landslide was discussed from the viewpoint of coupling theory in continuum.

Stress controlled ring shear test was implemented for both virgin and re-sliding type landslide conditions. In virgin type landslide test, shear displacement increased with the decrease in normal stress. Yield and failure points were obtained. It was clarified that yield points were on the internal friction angle line obtained by strain controlled ring shear test. On the contrary, shear displacement was found to be fragile and to increase sharply at yield point in case of re-sliding type landslide test. Yield points were found on the residual internal friction angle line. Discussion on consolidation effect on yield and failure points was carried out in detail.

Keywords. Ring shear test, triaxial compression test, strength, deformation, stress history

22.1 Introduction

Landslide commonly occurs by pore water pressure increase in slope due to rainfall and melted snow. Shear strength of clay in slope is generally determined by strain controlled ring shear test under fully drained condition. The test gives peak and residual strength parameters, but no information on shear deformation property. This reports empirical results of both triaxial compression and ring shear tests to clarify a shear behavior of clay for pore water pressure increase.

Clay in slope is applied shear stress caused by slope inclination. With the increase in pore water pressure, the stress attains to a failure criteria and clay deforms significantly. This phenomenon was investigated by both pore water pressure loading test and stress controlled ring shear test. It is a feature of this study to implement stress controlled tests contrary to strain controlled tests which are commonly employed in practice. A series of tests were conducted to investigate a shear behavior of clay by simulating in-situ conditions in landslide slope.

22.2 Pore Water Pressure Loading Test

Pore water pressure loading test is conducted with a triaxial compression apparatus to investigate a soil behavior of slope for rising underground water level. Feature of triaxial compression apparatus is to measure various strains of soil specimen such as axial and volumetric strains. It is interesting to examine soil deformation with the constitutive equation of original Cam clay model. It will help to understand a soil behavior for pore water pressure increase.

22.2.1 Testing Method

The testing soil was sampled from slope at Nagaoka University of Technology. The liquid and plastic limits of soil were $w_l = 49.8\%$ and $w_p = 35.7\%$, respectively. The soil was sifted with 425 μm sieve and preconsolidated at 47 kPa. The soil specimen was consolidated at prescribed pressure to handle an overconsolidation ratio in triaxial compression apparatus. It was loaded with deviator stress q_i under undrained condition. Keeping deviator stress q_i as constant, pore water pressure was applied at the bottom of specimen. The pore water pressure was measured at the top of specimen. Pore water pressure was increased step by step after confirming transmission of pore water pressure from the bottom to the top of soil specimen. The increase in pore water pressure for each step was 5 kPa. Deviator stress q_i expresses shear stress of slope caused by slope inclination which indicates failure potential of slope.

22.2.2 Deformation for Pore Water Pressure

Figure 22.1 expresses the measured relationship between axial strain and mean effective stress in pore water pressure loading test for $q_i = 75$ kPa. It shows the relationship from initial state of isotropic consolidation to failure. Deviator stress q_i was applied under undrained condition

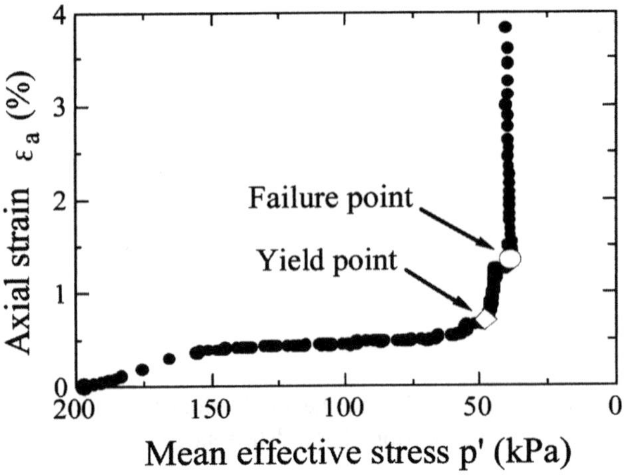

Fig. 22.1. Axial strain for pore water pressure loading

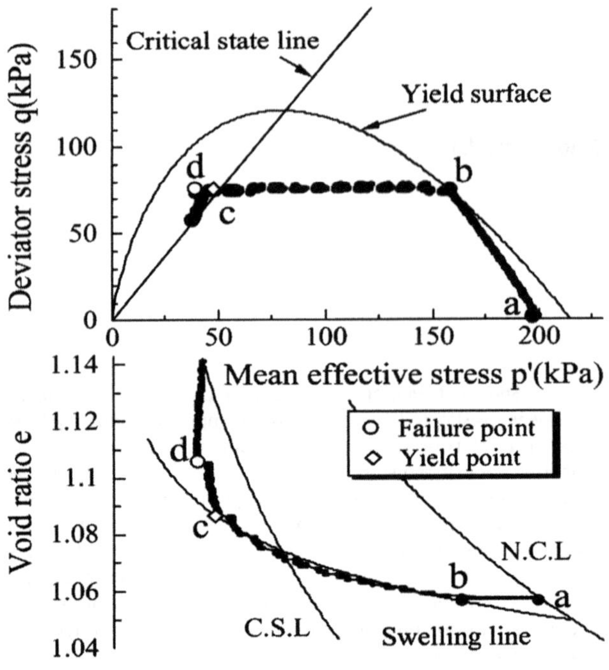

Fig. 22.2. Stress and void ratio paths for pore water pressure loading

before pore water pressure loading. The test started from about $p' = 160$ kPa. The axial strain was small in the former stage of pore water pressure loading, but it increased at about $p' = 50$ kPa sharply. Ogawa (1986) pointed out the existence of threshold magnitudes in normal stress for soil to generate deformation with ring shear apparatus by reducing normal stress under constant shear stress condition. He named the threshold magnitudes as the lower and the upper yield limits which correspond to yield and failure states, respectively.

Figure 22.2 indicates effective stress and void ratio paths for pore water pressure increase. Point a expresses the initial state and path a-b, undrained loading path for deviator stress $q_i = 75$ kPa. Pore water pressure loading test started at Point b. With the increase in pore water pressure, mean effective stress moved leftward on path b-c keeping deviator stress as constant. Void ratio traced the swelling line as path b-c and increased rapidly at Point c. The increase in void ratio from Point c was caused by plastic deformation. In this report Point c was defined as yield point based on the relationship between void ratio and mean effective stress. In pore water pressure loading test, soil specimen was enforced to be overconsolidated and sheared due to loading. Dilation property of overconsolidated clay caused the increase in void ratio as path c-d. For the further increase in pore water pressure, deviator stress could not keep constant at Point d. Passing Point d, deviator stress decreased and the soil specimen rapidly failed. It corresponds to peak strength of overconsolidated clay in fully drained test. Point d was defined as failure point. In Fig. 22.2 the yield function of original Cam clay model is drawn for pore water pressure increase. It is clear that the soil specimen generated plastic deformation even though effective stress of soil was still in elastic domain of original Cam clay model. This tendency was commonly observed in other tests. It is difficult for original Cam clay model to present the soil behavior for pore water pressure increase.

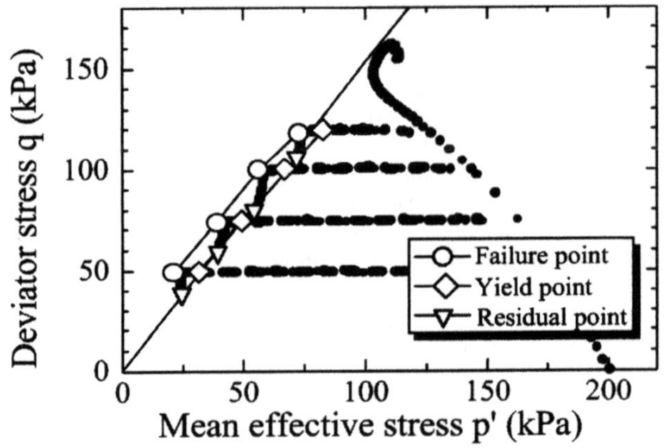

Fig. 22.3. Tests for various deviator stresses

Figure 22.3 shows test results for various deviator stresses, q_i. The critical state line is exhibited in the figure, which corresponds to internal friction angle ϕ' of soil. It is readily seen that yield points were on the critical state line and failure points located leftward. In the figure effective stress moved toward the critical state line after passing failure point. It is interesting that both yield and residual points located almost on the critical state line independent of deviator stress. It is therefore very important to catch the failure and the critical state lines in practice. Internal friction angle (critical state line) generally expresses the strength parameter of soil, but it also exhibits the threshold for soil to generate deformation for pore water pressure increase. It is noted that design methods based on internal frictional angle ϕ' might take into account not only stability but also slope movement.

22.2.3 Simulation with Constitutive Equation

Hashiguchi(1989) developed the sub-loading surface model to simulate a behavior of overconsolidated clay. It is employed to simulate yield and failure points for pore water pressure increase. Figure 22.4 shows the simulation result for pore water pressure loading test with Hashiguchi model employing the original Cam clay yield function. Table 22.1 indicates employed parameters in the model. Figure 22.4 exhibits that the model properly simulated deformation behavior in comparison with experiment, especially on yield and failure points. Figure 22.5 expresses computed yield and failure points of overconsolidated clays where overconsolidation ratio was widely varied. It is possible to compare the predicted values for OCR = 1 with those of experiment in Fig. 22.3. From the comparison between experiment and simulation, it was found that yield and failure points were fairly simulated by the model for various deviator stresses. It is clear heavily overconsolidated clay could resist higher pore water pressure.

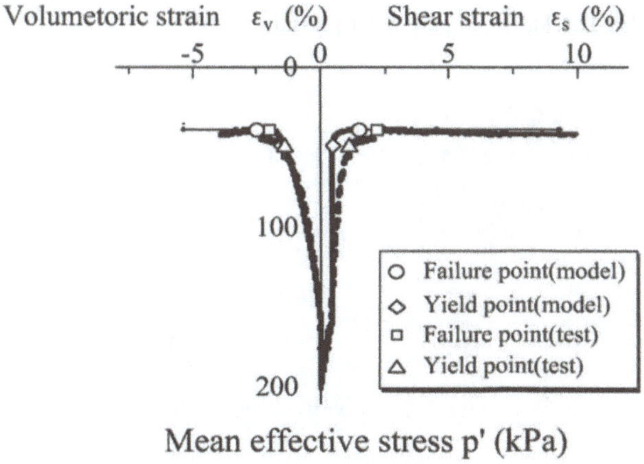

Fig. 22.4. Performance of constitutive equation

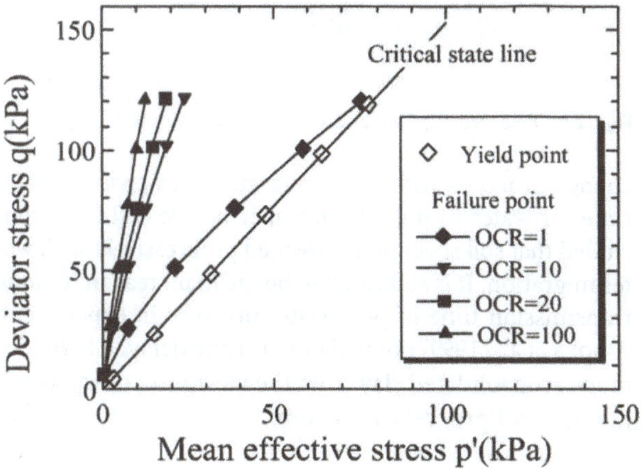

Fig. 22.5. Computed yield and failure points

Table 22.1. Constants for constitutive equation

M	1.53	ν	0.2
μ	20	e_0	1.057
κ	0.025	λ	0.111

22.3 Consideration on Landslide Mechanism

22.3.1 Repetitive Pore Water Pressure Loading

Figure 22.6 indicates the computed relationship between void ratio and mean effective stress for repetitive pore water pressure loading. It simulated repetitive change in ground water level which caused slope to failure. In the figure heavily overconsolidated clay was repeatedly applied pore water pressure under constant deviator stress of $q_i = 75$ kPa. Yield and failure points were exhibited for repetitive pore water pressure increase. Effective mean stresses at yield points were obtained as constant although void ratios changed for repetitive increase. It is because deviator stress was hold constant as $q_i = 75$ kPa. On the contrary, mean effective stresses and void ratios at failure points were obtained to change. They are readily seen to form an envelopment line for repetitive pore water pressure increases. It indicates (*1*) both mean effective stress and void ratio at failure changed following this line and (*2*) overconsolidated clay approached to normally consolidated state by repetitive loading. They finally attain to the steady states after repetitive increases. The figure indicates the existence of specific failure envelopment line for prescribed deviator stress. In the figure three failure lines were exhibited for deviator stresses of $q_i = 50$, 75 and 100 kPa. They show soil could resist higher pore water pressure in case of lower deviator stress.

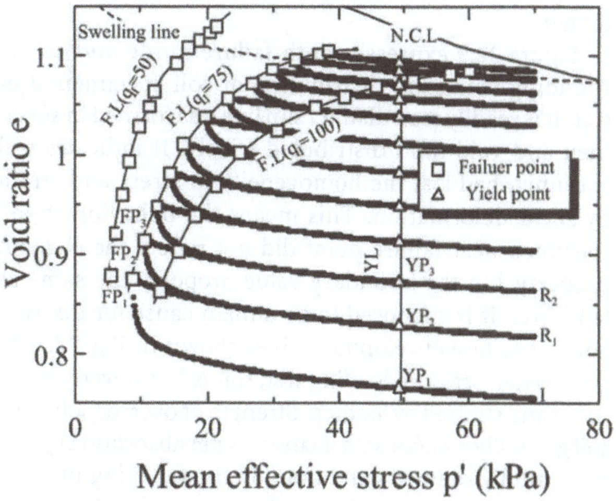

Fig. 22.6. Failure lines for repetitive loading

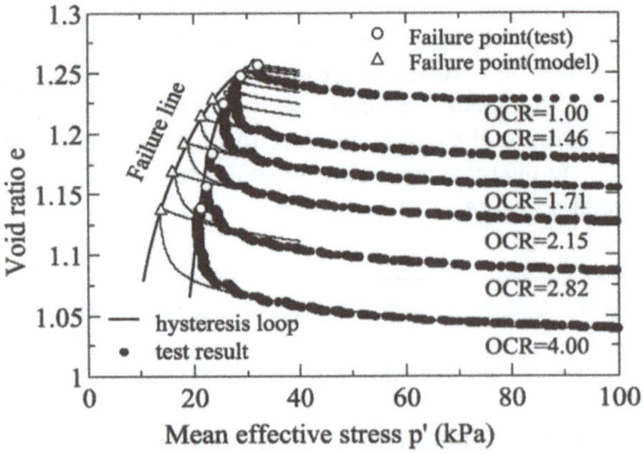

Fig. 22.7. Tests for failure line

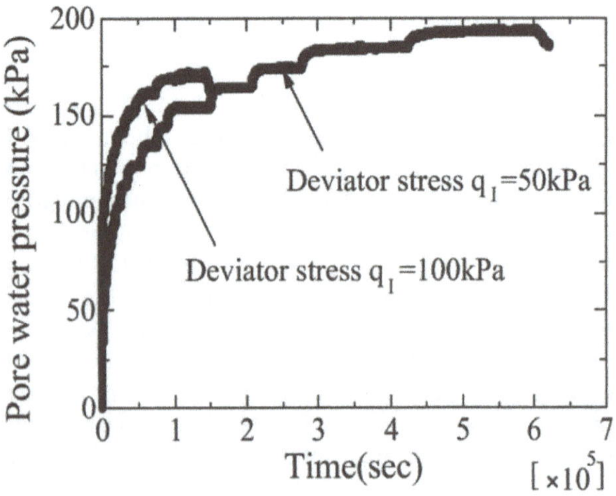

Fig. 22.8. Measured pore water pressure

Figure 22.7 shows the empirical results of pore water pressure loading tests for $q_\mathrm{i} = 50$ kPa. Failure envelopment line was obtained for empirical data as that predicted by simulation. However, large discrepancy between two was observed for higher overconsolidation ratio. The reason for this discrepancy will be discussed in the following section.

22.3.2 Progressive Failure with Water Migration

Figure 22.8 shows measured pore water pressures at the top of soil specimen for deviator stresses, 50 and 100 kPa. The measured pore water pressure formed a convex function of time in each stage of increase as commonly observed in consolidation behavior. It is because a certain time is necessary for pore water pressure to transmit from the bottom to the top of soil specimen. The transmission time seems to be almost same in case of lower pore water pressure. It is, however, the transmittance time becomes longer with the increase in pore water pressure beyond the yield point. At the last stage of pore water pressure increase the transmission time is observed significantly longer.

Figure 22.9 expresses both failure mode and corresponding void ratio distribution of soil specimen after test. It is readily seen distinct slip line was formed in specimen and void ratio distributed widely. It indicates soil specimen had lost the homogeneity in stress and strain by shear deformation. This means the behavior of soil specimen near failure point did not reflect the element property, but the boundary value property the same as landslide. It is supposed to be a main cause for discrepancy of failure envelopment lines shown in Fig. 22.7. In pore water pressure loading test, soil is both overconsolidated and sheared by loading. Strength of overconsolidated soil generally reduces with dilation (water absorption) property in fully drained condition. Strain softening of overconsolidated soil caused the soil specimen to deform progressively. In saturated soil, deformation with volumetric

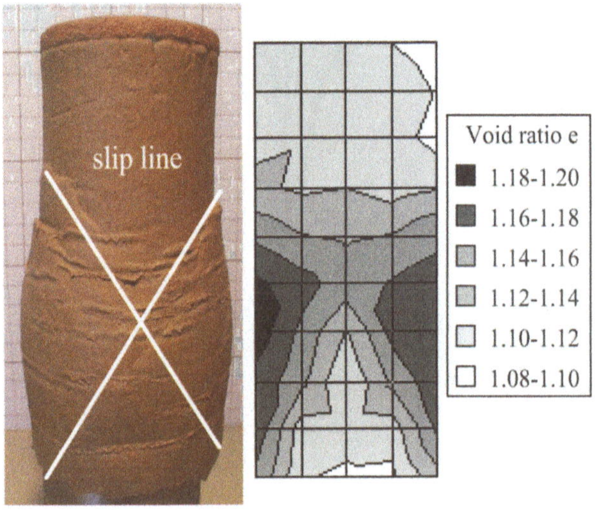

Fig. 22.9. Observed slip line and void ratio distribution

change proceeds with water migration. It needs a certain time for water to migrate in impermeable soil. It is concluded that soil specimen deformed progressively with water migration. It is supposed to be the main reason for long transmission time of pore water pressure in experiment. Asaoka et al. (1999) pointed out the time delayed behavior of overconsolidated clay from the viewpoint of water migration and progressive failure.

22.3.3 Observed Time Dependent Behavior

Figure 22.10 shows the measured pore water pressure at the top of soil specimen in case of $q_\mathrm{i} = 50$ kPa and OCR = 4. The transmission time of pore water pressure gotten longer at the final step of increase although pore water pressure was almost constant. Figure 22.11 reveals axial strain of soil specimen in time at final increase. It is ar-

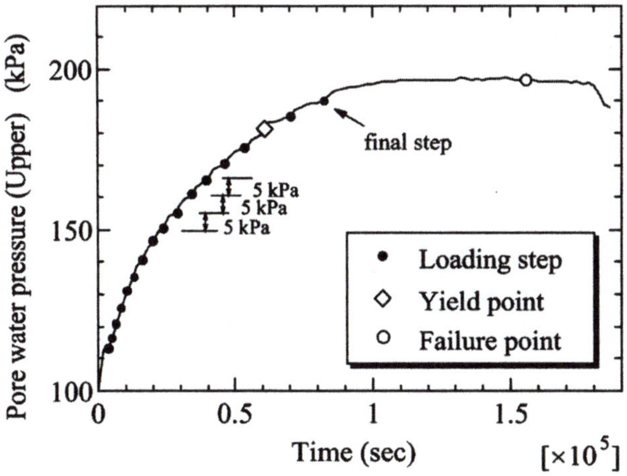

Fig. 22.10. Measured pore water pressure at top end

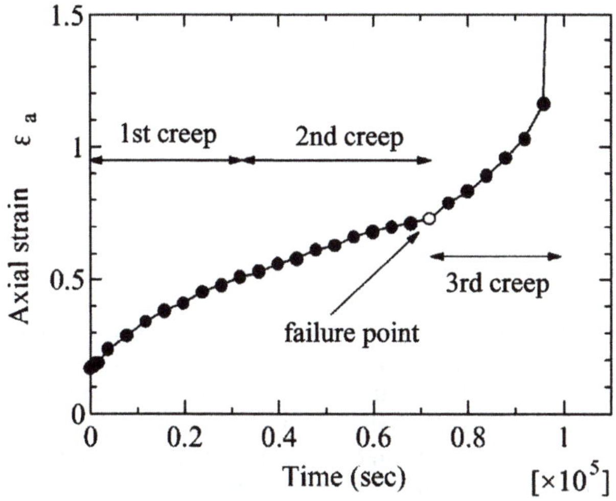

Fig. 22.11. Creep like time dependent behavior

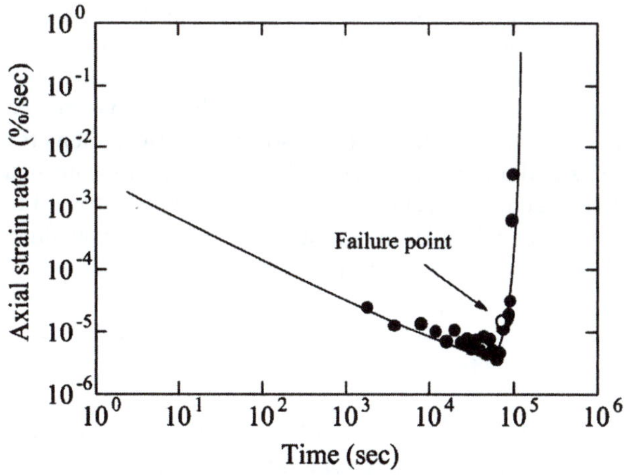

Fig. 22.12. Axial strain rate change with time

ranged in terms of strain rate as shown in Fig. 22.12. These figures express creep like behaviors consisting of first, second and third creep behaviors. The failure point defined in this study almost coincided with the threshold in pore water pressure between second and third creep behaviors as shown in Fig. 22.11. As mentioned above, the specimen had lost the element property at final step of increase. It is possible to explain this behavior by not creep theory but progressive failure of soil specimen. The specimen was not an element, but constituted a boundary value problem the mechanism of which is similar with natural slope even though the dimension of specimen is small. It indicates creep behavior observed in landslide can be explained by combination of progressive failure and water migration the same as pore water pressure loading test. This result proposes a possible interpretation on creep behavior based on soil water coupling theory.

22.4 Stress Controlled Ring Shear Test

Ring shear test is generally applied to investigate residual strength of soil where soil deforms largely along slip line. Strain controlled testing condition is widely employed, but it is different from in-situ deformation condition of soil. Stress controlled ring shear test is implemented to investigate a soil behavior of slope due to pore water pressure increase. By applying constant shear stress supposed to slope inclination, normal stress is gradually reduced to simulate the increase in pore water pressure. In landslide there are two types of sliding modes, where one is a virgin type landslide and the other is a re-sliding type landslide. Mobilized strength of soil is different between two landslide types. This reports empirical results of stress controlled ring shear tests for both conditions.

22.4.1 Testing Method

The testing soil was sampled from Okimi landslide in Niigata prefecture. Physical property of soil is exhibited in Table 22.2. It was remolded by mixing with water and preconsolidated for 24 hours before setting of soil specimen. Mechanical property of soil was examined by strain controlled ring shear test. The results are exhibited in Fig. 22.13. Internal friction angle was obtained as $\phi' = 24.7°$ at peak strength and $\phi'_r = 6.7°$ at residual strength. For peak strength, cohesion was obtained as 12.1 kPa in appearance. In stress controlled ring shear test, two testing soils were prepared to investigate the behav-

Table 22.2. Physical property of soil

Liquid limit W_L (%)	101
Plastic limit P_L (%)	29
Plasticity index I_p (%)	72
Clay fraction content CF (%)	55

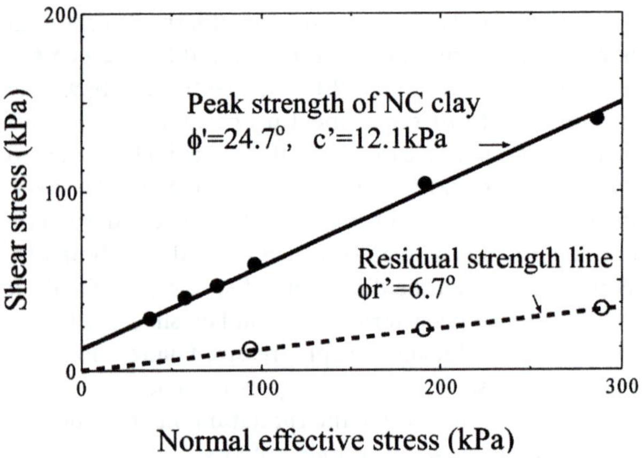

Fig. 22.13. Strain controlled ring shear tests

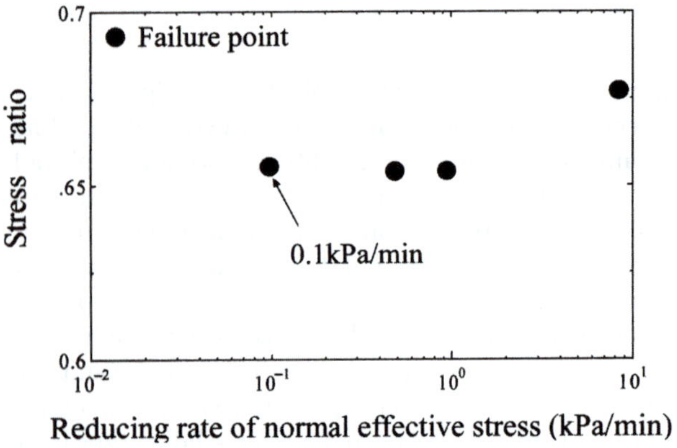

Fig. 22.14. Effect of reduction rate in normal stress

iors of both virgin and re-sliding type landslides. Testing soils were consolidated with prescribed pressure and then applied shear stress to. For re-sliding type landslide, testing soil was especially sheared by strain controlled ring shear test to attain to the residual state before consolidation process.

In stress controlled ring shear test, normal stress was reduced with constant rate by keeping shear stress as constant. Fully drained condition was assumed to estimate the effective normal stress. Rate of reduction in normal stress was determined prior to experiments. Figure 22.14 reveals the empirical result of reduction rate effect on stress ratio at failure in stress controlled ring shear test. To achieve fully drained test, the reduction rate of normal stress was determined as 0.1 kPa min^{-1}. The figure illustrated the stress ratio at failure settled to a unique magnitude around 0.1 kPa min^{-1}.

22.4.2 Test for Virgin Type Landslide

Figure 22.15 shows the results of stress controlled ring shear test for virgin type landslide. Testing soil was consolidated with normal stress of 294 kPa. Applied shear stress was varied from 20 to 80 kPa. In the figure internal

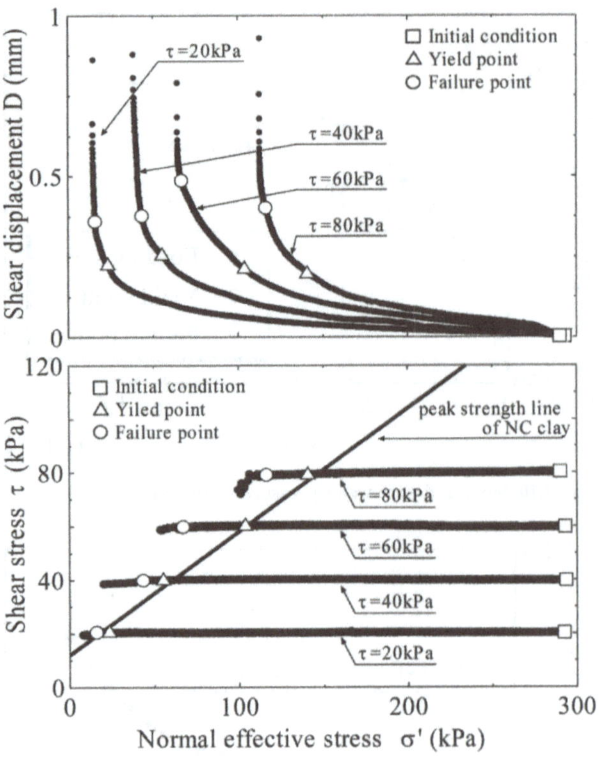

Fig. 22.15. Stress controlled ring shear test for virgin type landslide

friction angle ϕ' is illustrated. Yield point was defined for the condition where displacement rate increased sharply. On the other hand, failure point was also defined for the condition that shear stress reduced due to large deformation. According to the reduction in normal stress, soil specimen behaved elastically at the begging of test. Displacements at yield points were almost same as 0.2 mm in a series of tests although the reduction in normal stress to yield point was varied. In stress paths figure, yield points were on the line of internal friction angle $\phi' = 24.7°$ and it indicates the same trend with the result of pore water pressure loading test. This fact expresses the internal friction angle is strength parameter of soil, but it also exhibits the threshold for soil to generate the deformation for normal stress reduction. Failure points were obtained leftward to yield points. Displacements at failure point were obtained almost constant as 0.4 mm independent of shear stress.

22.4.3 Test for Re-sliding Type Landslide

Figure 22.16 reveals the results of stress controlled ring shear test for re-sliding type landslide. Testing soil was completely sheared before test and slip plane formed a complete mirror plane. Testing soil was consolidated with normal stress of 294 kPa. Shear stress was applied from 20 to 80 kPa. In the figure the residual internal friction angle ϕ'_r is illustrated. It is readily seen that the soil

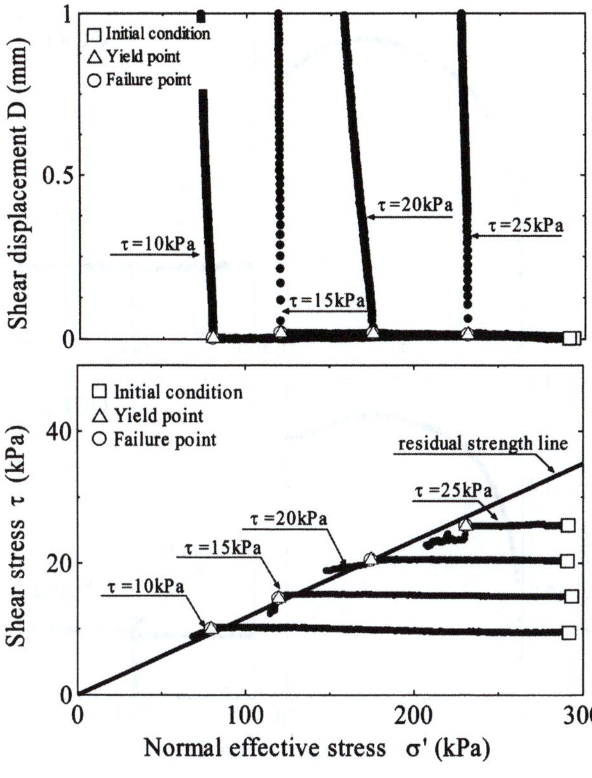

Fig. 22.16. Stress controlled ring shear test for re-sliding type landslide

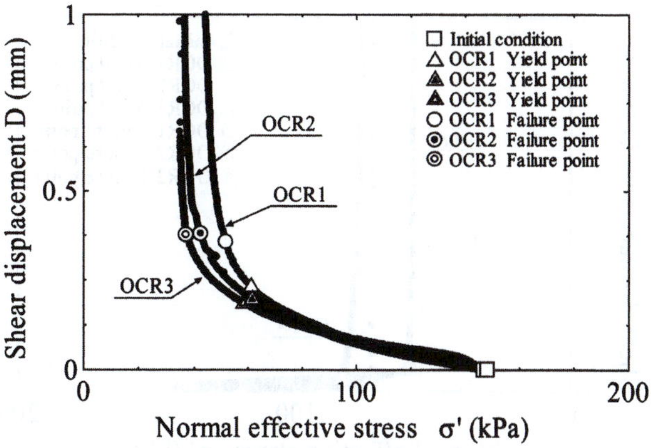

Fig. 22.17. Consolidation effect on shear deformation for virgin type landslide

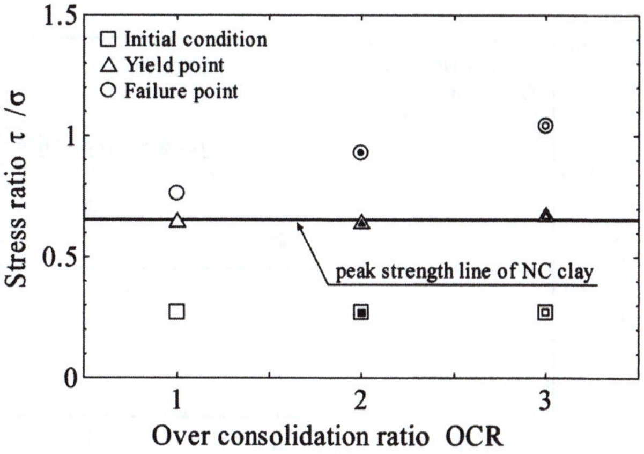

Fig. 22.18. Consolidation effect on yield and failure points for virgin type landslide

behaved fragile contrary to virgin type landslide in Fig. 22.15. In the figure there was little displacement before yield point and the soil specimen rapidly failed at yield point. It is difficult to distinguish failure point from yield point. It is clearly seen that yield and failure points were on the line of residual internal friction angle $\phi_r' = 6.7°$ and it reveals the same tendency with the test for virgin type landslide. It is interesting the threshold line is correlated with the residual internal friction angle.

22.4.4 Consolidation Effect on Yield and Failure Points

It is well known that stress history such as over consolidation ratio (OCR) affects shear behavior of soil. In landslide slope, the change in underground water level causes consolidation and swelling to ground. It is interesting to investigate the effect of stress history on both yield and failure points of soil.

Figure 22.17 expresses the empirical results of stress controlled ring shear tests for OCR from 1 to 3 in case of virgin type landslide tests. Tendency of shear displacement was similar with that of OCR = 1. Shear displacements at yield points were obtained about 0.2 mm and those at failure points, 0.35 mm. It is arranged in terms of stress ratio as shown in Fig. 22.18. It is readily seen that stress ratios at yield points were almost identical with internal friction angle (ϕ') independent of OCR. On the contrary, stress ratios at failure points increased with OCR to resist more reduction in normal stress. It is clearly observed that the stress history affected on failure point in case of virgin type landslide test.

Figure 22.19 expresses the empirical results of stress controlled ring shear tests for OCR from 1 to 3 in case of re-sliding type landslide tests. Stress history (consolidation and swelling) was applied to testing soil after setting a slip line. Since a slip line is supposed to be a contact plane and different from a continuum body, it is interesting to investigate the effect of stress history on shear behavior. Tendency of shear displacements was similar with that of OCR = 1. It is arranged in terms of stress ratio as shown in Fig. 22.20. It is remarkable that stress ratios at yield and failure points increased with OCR to resist more reduction in normal stress. Stress history clearly affected on yield and failure points even in case of re-sliding type landslide test.

Strain controlled ring shear test was conducted to examine the consolidation effect on failure point in re-sliding type landslide test. It was implemented on testing soil which was applied stress history of both loading and unloading after setting a slip line. Figure 22.21 illustrates

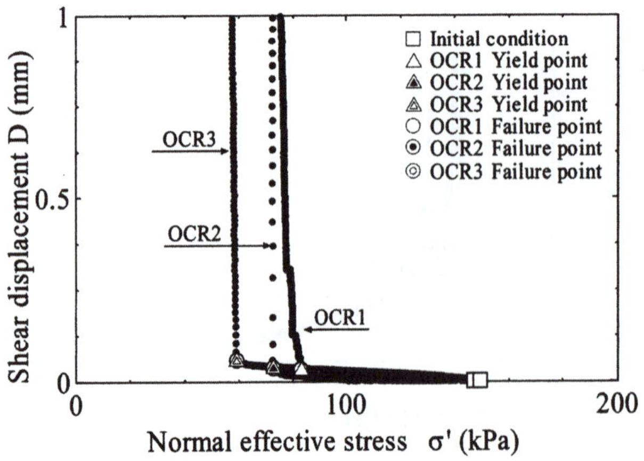

Fig. 22.19. Consolidation effect on shear deformation for re-sliding type landslide

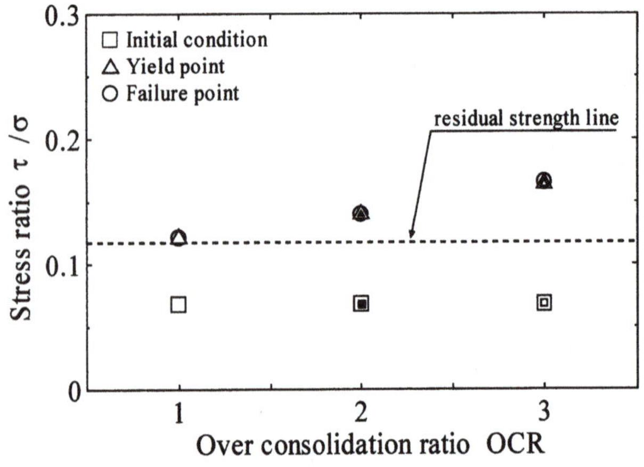

Fig. 22.20. Consolidation effect on yield and failure points for re-sliding type landslide

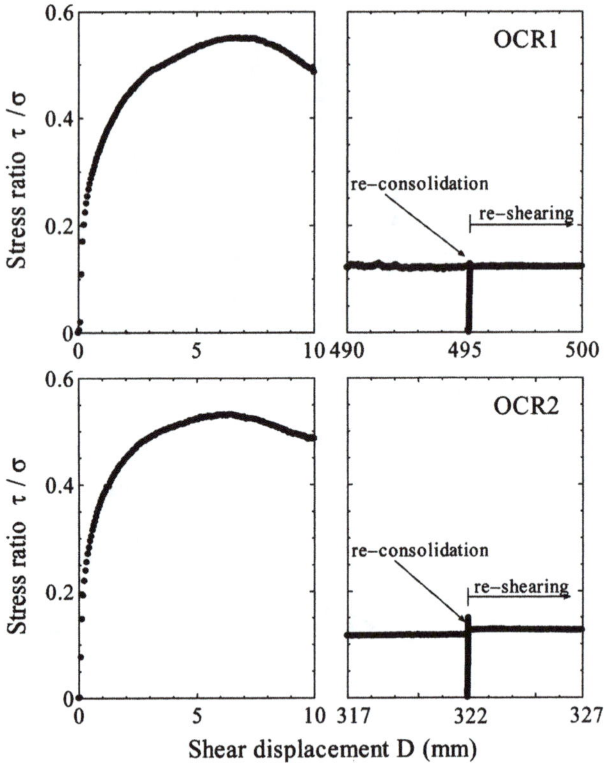

Fig. 22.21. Strain controlled ring shear test of reconsolidated soil

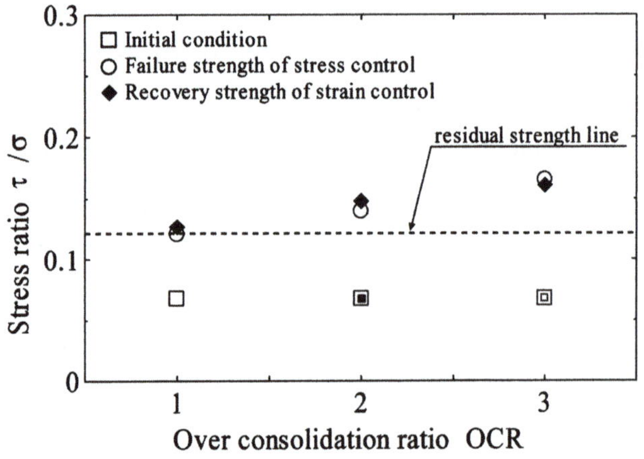

Fig. 22.22. Consolidation effect on yield and failure points by strain controlled ring shear test

the obtained empirical result of stress ratio and shear displacement relationship. The figure shows two kinds of ring shear tests in which re-consolidation pressure was different. OCR expresses the stress history after setting a slip line. In the figure it is clearly seen the recovery in stress ratio at failure point. Test results were arranged in terms of stress ratio–OCR relationship as shown in Fig. 22.22. In spite of strain controlled ring shear tests, stress history as consolidation clearly effected on yield and failure points. The increase in stress ratio for OCR matched with that of stress controlled tests well.

22.5 Conclusions

The followings were concluded in this study.

1. In pore water pressure loading tests, yield points were found on the critical state line described by internal friction angle. It reveals the design method with internal friction angle corresponds to two conditions. One is the limit state of slope and the other is the occurrence of deformation in practice.
2. In pore water pressure loading tests, time dependent shear displacement was observed. It is a creep behavior in appearance. This behavior was shown to be caused by water migration and progressive failure due to strain softening of clay.
3. In stress controlled ring shear tests for virgin type landslide, yield point for reduction in normal stress was found on internal friction angle line obtained by strain controlled ring shear tests.

4. In stress controlled ring shear tests for re-sliding type landslide, shear displacement for reduction in normal stress was obtained fragile and increased rapidly at yield point. Yield points were found on residual internal friction angle line.
5. Effect of loading history as ground water level change on shear behavior was investigated by stress controlled ring shear tests. Although re-sliding type landslide tests, the effect of loading history on failure points were clearly observed. It was examined by strain controlled ring shear tests.

References

Asaoka A, Nakano M, Noda T, Takaine T, Kaneda K, Constantinescu DT (1999) Progressive failure of heavily overconsolidated clays. Soils and Foundations 39(2):105–118

Hashiguchi K (1989) Advanced elasto-plasticity. Asakura-shoten, pp 95–139 (in Japanese)

Ogawa S (1986) Behavior of underground water and mechanical property of soil at Yomogihira and Nigorisawa landslides. In: 14th report of Niigata branch of Japanese Landslide Society. pp 27–38 (in Japanese)

Ohtsuka S (2002) Time dependent behavior of clay in pore water pressure loading test. In: Proceedings of International Congress of Interpraevent 2002 in the Pacific Rim, 2, pp 531–536

Ohtsuka S, Miyata Y (2001) Consideration on landslide mechanism based on pore water pressure loading test. In: Proceedings of 15th International Conference on Soil Mechanics and Geotechnical Engineering, ISSMGE, pp 1233–1236

Ohtsuka S, Yoshida K, Kawashima K (2004) Normal stress decrease test on clay with ring shear apparatus. In: Proceedings 49th Geotechnical Engineering Symposium of JGS, pp 21–26 (in Japanese)

Ohtsuka S, Miyata Y, Iwabe T (2005) Shear strength of overconsolidated clay for repeats of pore water pressure loading. J Jpn Landslide Soc 41(5):29–36 (in Japanese)

Chapter 23

Static and Dynamic Analyses of Slopes by the FEM

Keizo Ugai · Akihiko Wakai* · Fei Cai

Abstract. The slope stability is commonly assessed using limit equilibrium methods. However, the limit equilibrium methods cannot calculate displacements of slopes, thus they cannot evaluate correctly the soil-slope stabilization interaction. Here we discuss how to use the finite element analysis to evaluate the stability of slopes reinforced piles, anchors, or drainage pipes under static condition, and to calculate the behavior of slopes under dynamic loads of earthquakes.

In the first half, the finite element method with shear strength reduction technique (SSRFEM) is explained. This method is effective for the prediction of the stability of slope improved with piles, anchors, or drainage pipes in order to consider the three-dimensional effects in slopes. Based on numerical comparisons between the finite element method and limit equilibrium methods, it is suggested that the finite element method with shear strength reduction technique is a reliable, robust, and maybe unique approach to evaluate the slope stability and slope stabilization. In the second half, the dynamic response analysis of a slope based on the elasto-plastic finite element method is explained. A simple 3-D cyclic loading model that can be applied to the seismic design of slope is introduced, and it is suggested that the dynamic elasto-plastic FEM makes it possible to evaluate the residual deformation of a slope induced by the earthquake precisely.

Keywords. Numerical analysis, Finite element method (FEM), Elasto-plasticity

23.1 Introduction

The slope stability is commonly assessed using limit equilibrium methods. However, the limit equilibrium methods cannot calculate displacements of slopes, thus they cannot evaluate correctly the soil-slope stabilization interaction. Here we discuss how to use the finite element analysis to evaluate the stability of slopes reinforced piles, anchors, or drainage pipes under static condition, and to calculate the behavior of slopes under dynamic loads of earthquakes.

23.2 Finite Element Analysis of Slopes and Slope Stabilization

The finite element method with shear strength reduction technique (SSRFEM) has been gradually used to analyze the slope stability in two-dimensional situations (e.g., Zienkiewicz et al. 1975; Ugai 1989), and three-dimensional situations (Ugai and Leshchinsky 1995). Numerical comparisons have shown that SSRFEM is a reliable and robust method for assessing the safety factor of slope and corresponding critical slip surface. One of the main advantages of SSRFEM is that the safety factor emerges naturally from the analysis without the user having to commit to any particular form of the mechanism *a priori*. Furthermore, it is possible for SSRFEM to evaluate the slope stability under a general framework even for slopes stabilized with piles or anchors if the soil-structure interaction is reliably considered.

The stabilizing measures, introduced and analyzed here, include (*1*) piles, (*2*) anchors, and (*3*) drainage pipes, which are usually used to stabilize slopes in the practical situations. Numerical comparisons between SSRFEM and limit equilibrium methods are conducted, and it is suggested that SSRFEM is a reliable, robust, and maybe unique approach to evaluate the slope stability and slope stabilization under a general framework.

23.2.1 Finite Element Method with Shear Strength Reduction Technique (SSRFEM)

In SSRFEM, a non-associated elasto-plastic constitutive model is adopted, where the Mohr-Coulomb yield criterion is used to define the yield function:

$$f = -c'\cos\phi' - \frac{1}{3}I_1\sin\phi' + \sqrt{J_2}\ \cos\Theta - \frac{1}{3}\sin\Theta\sin\phi' \quad (23.1a)$$

where

$$\Theta = \frac{1}{3}\sin^{-1}\left(-\frac{3\sqrt{3}}{2}\frac{J_3}{J_2^{\,3/2}}\right), \quad \left(-\frac{\pi}{6} \le \Theta \le \frac{\pi}{6}\right) \quad (23.1b)$$

and the Drucker-Prager criterion to define the plastic potential function

$$g = -\alpha I_1 + \sqrt{J_2} - \kappa \quad (23.2a)$$

where

$$\alpha = \frac{\tan\Psi}{\sqrt{9 + 12\tan^2\Psi}}, \quad \kappa = \frac{3c'}{\sqrt{9 + 12\tan^2\Psi}} \quad (23.2b)$$

In the above equations c', ϕ', and Ψ are the effective cohesion, friction angle, and dilatancy angle, respectively. I_1, J_2, and J_3 are the first invariant of the effective stress, and the second and third invariants of the deviatoric stress, respectively.

The global safety factor of slopes in SSRFEM is identical to the one in limit equilibrium methods. The reduced strength parameters c'_F and ϕ'_F are defined by

$$c'_F = \frac{c'}{F}, \quad \phi'_F = \tan^{-1}\left(\frac{\tan\phi'}{F}\right) \tag{23.3}$$

The reduced shear strength parameters c'_F and ϕ'_F then replace the shear strength parameters c' and ϕ' of the Mohr-Coulomb yield criterion in the elasto-plastic finite element analysis. Firstly, a gravity turn-on is implemented under elastic state to determine the initial stress distribution inside the slope. Then, stresses and strains are calculated by the elasto-plastic finite element method. The shear strength reduction factor, F, is then increased incrementally until the global failure of the slope reaches, which means that the finite element calculation diverges under a physically real convergence criterion. The global safety factor at failure lies between the shear strength reduction factor, F, at which the iteration limit is reached, and the immediately previous value.

One of the main advantages of SSRFEM is that the safety factor emerges naturally from the analysis without the user having to commit to any particular form of the mechanism *a priori*. When the slope stability is evaluated with the effective stress method, the pore water pressure is usually predicted with the finite element seepage analysis or Biot's consolidation theory. If the same finite element mesh is used for the seepage or consolidation analyses and SSRFEM, the water pressure, predicted in the seepage or consolidation analysis, can be directly used in SSRFEM. This can simplify the slope stability analysis, and consider accurately the influence of the seepage force. Furthermore, it is possible for SSRFEM to evaluate the slope stability under a general framework even for slopes stabilized with piles and anchors if the soil-structure interaction is reliably considered in the finite element analysis.

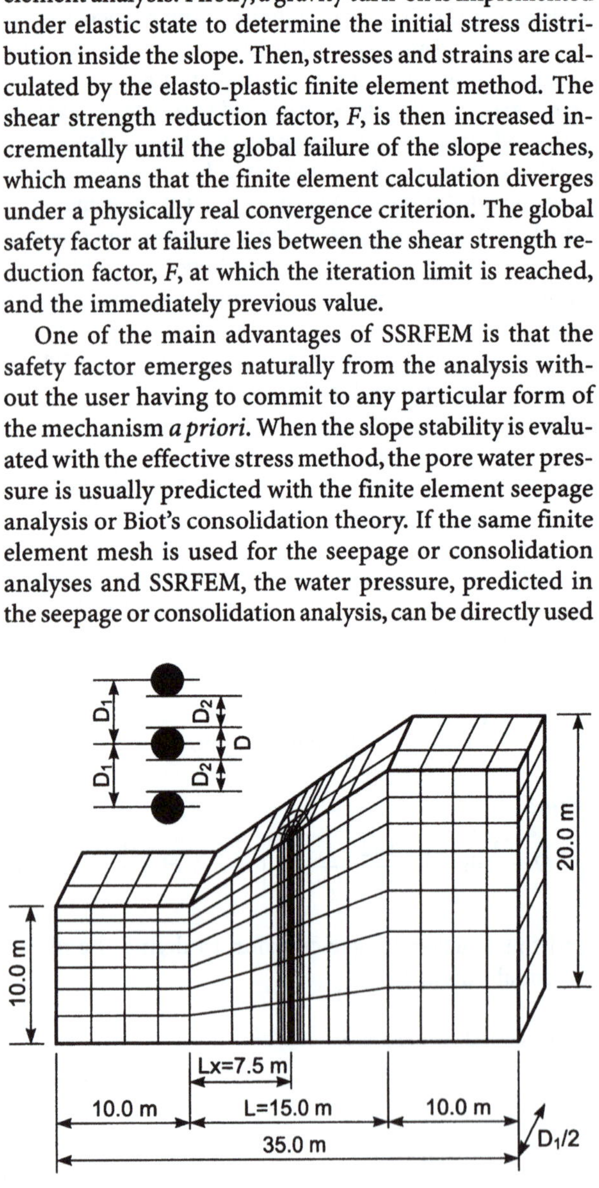

Fig. 23.1. Model slope and finite element mesh

23.2.2 Stability of a Slope with Piles

An idealized slope with a height of 10 m and a gradient of 1 : 1.5 and a ground thickness of 10 m is analyzed with a three-dimensional finite element mesh, as shown in Fig. 23.1. The isoparametric element with 20 nodes is used for the soil, the prism element with 15 nodes for the pile, and the interface element with 16 nodes for the soil-pile interaction. Two symmetric boundaries are used, so that the problem analyzed really consists of a row of piles with planes of symmetry through the pile centerline and through the soil midway between the piles.

A steel tube pile with an outer diameter of 0.8 m is modeled by 7 elements in the vertical direction and 6 wedge elements in the circumferential direction that form a semicircular shape. The piles are treated as the linear elastic solid material. The piles are installed with the center-to-center spacing $D_1 = 3D$. The piles are embedded and fixed into the bedrock or the stable layer. The material parameters of the soil, the soil-pile interface, and the pile are shown in Table 23.1.

Bishop's simplified method of slip circle analysis is employed to determine the safety factor of a slope stabilized with piles. Based on the resisting moment, M_R, of the soil, and the driving moment, M_D, of the sliding body, the safety factor, F_s, is given by

$$F_s = (M_R + M_P)/M_D \tag{23.4}$$

The resisting moment by the pile row, M_P, can be determined by the lateral force per unit thickness acting on the pile, which is evaluated by the equation of Ito and Matsui (1975).

Table 23.1. Material parameters

Parameter	Soil	Interface	Pile
Young's modulus, E (MPa)	200	200	60 000
Poisson's ratio, ν	0.25	0.25	0.20
Unit weight, γ (kN m^{-3})	20.0		
Cohesion, c (kPa)	10.0	10.0	
Friction angle, ϕ (deg)	20.0	20.0	
Dilatancy angle, ψ (deg)	0.0	0.0	

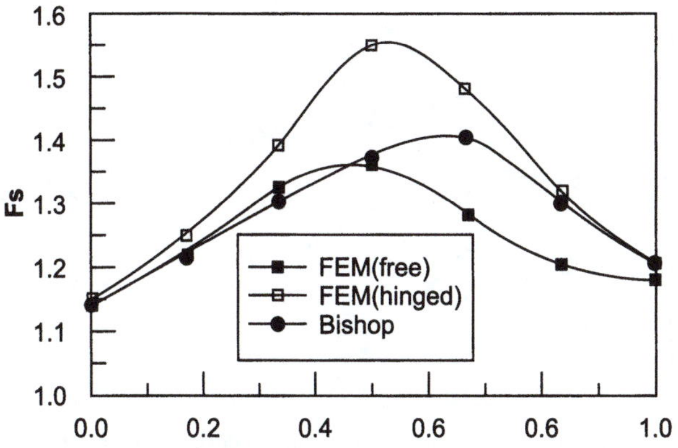

Fig. 23.2. Effect of pile position on safety factor

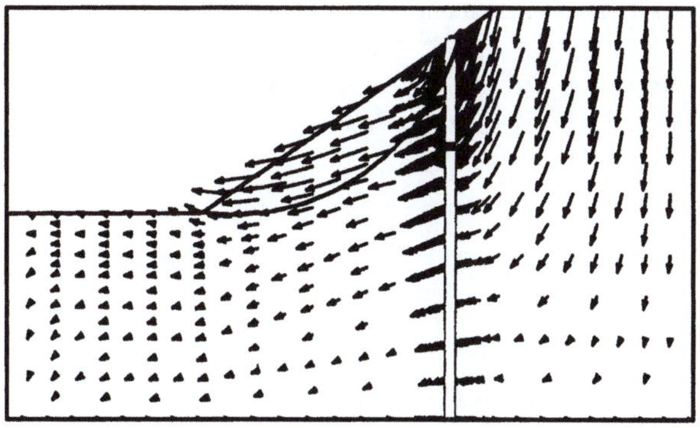

Fig. 23.3. Nodal displacement increments induced by shear strength reduction and critical slip circle located by Bishop's simplified method

The pile positions in the slope are indicated with the ratio of the horizontal distance between the slope toe and the pile position, *Lx*, to the horizontal distance between the slope toe and slope shoulder, *L*, as shown in Fig. 23.1. The influence of the pile positions on the safety factor of a slope stabilized with piles is shown in Fig. 23.2. The numerical results of SSRFEM show that the safety factor of the slopes reinforced with the hinged head piles is larger than that with free head piles. The difference is more significant when the piles are installed in the middle portion of the slopes. Therefore, the restrained pile head (hinged or fixed) is recommended. The improvement of the safety factor of the slopes reinforced with piles is largest when the piles are installed in the middle of the slopes, regardless of the pile head conditions. However, Bishop's simplified method shows that the safety factor increases with the horizontal distance between the slope toe and the pile position, *Lx*, until the piles are placed considerably closer to the top of the slopes. This is the same as the results of limit equilibrium methods reported by Ito et al. and Hassiotis et al. When the piles are installed in the lower portion of the slopes, the pressure on the piles is positive even for the free head condition due to the shallow sliding soil mass. Figure 23.2 indicates that the difference in the safety factors obtained by these two methods is small when *Lx*/*L* is less than 0.3. When the hinged head piles are installed in the middle portion of the slopes, the safety factor of SSRFEM is significantly larger than that predicted by Bishop's simplified method.

Numerical comparisons have shown that the critical slip circle of Bishop's simplified method is nearly the envelope curve of soil body with large displacement calculated by the finite element method when there are no piles in the slope. However, the critical slip circle is clearly not the envelope curve of soil body with large displacement when the slope is reinforced with piles (Fig. 23.3). This simplifies that the critical slip circle and thus the safety factor of slope of Bishop's simplified method are not reliable when the slope is reinforced with piles.

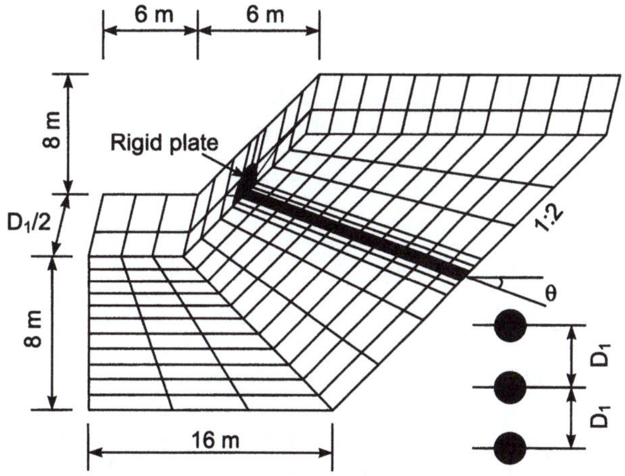

Fig. 23.4. Model slope and finite element mesh

23.2.3 Stability of a Slope with Anchors

An idealized slope with a height of 8 m, a gradient of 1 : 1, and a ground thickness of 10 m is analyzed with a three-dimensional finite element mesh, as shown in Fig. 23.4. Two symmetric boundaries are used, so that the problem analyzed really consists of a row of anchors with planes of symmetry through the centerline and through the soil midway between the anchors. A small thickness mesh is used around the anchor to reduce the computational ill-conditioning of the interface elements. The diameters of the tendon and the grouted body are 32 mm and 90 mm, respectively. The total length of the anchor is 12 m, in which the lengths of the tendon and grouted body are 6 m and 6 m, respectively.

The anchor is modeled by 9 elements in the axial direction and 6 wedge elements in the circumferential direction that form a semicircular shape. The tendon is assumed to smoothly contact with the soil, so that the shear strength of the soil-tendon interface is equal to zero, as shown in Table 23.2. The shear strength of the soil-grout interface is assumed to be the same as the soil. Material

Table 23.2. Material parameters

Parameters	Soil	Soil-tendon interface	Soil-grout interface
Young's modulus, E (MPa)	200	200	200
Poisson's ratio, ν	0.25	0.25	0.25
Unit weight, γ(kN m^{-3})	20.0		
Cohesion, c (kPa)	12.0	0.01	12.0
Friction angle, ϕ(deg)	20.0	0.0	20.0
Dilatancy angle, ψ(deg)	0.0	0.0	0.0

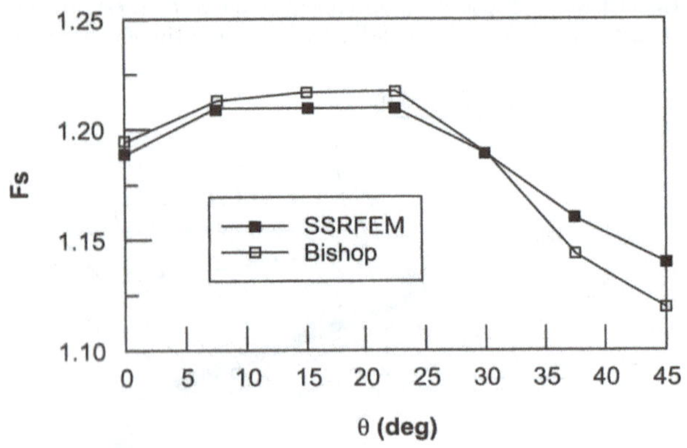

Fig. 23.5. Effect of direction angle on safety factor

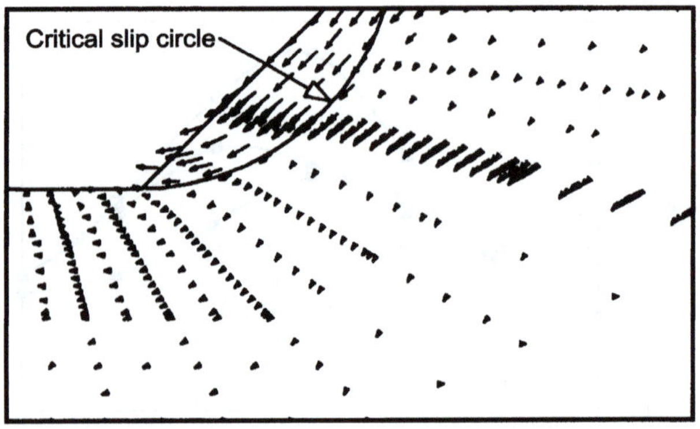

Fig. 23.6. Nodal displacement increments induced by shear strength reduction and critical slip circle located by Bishop's simplified method

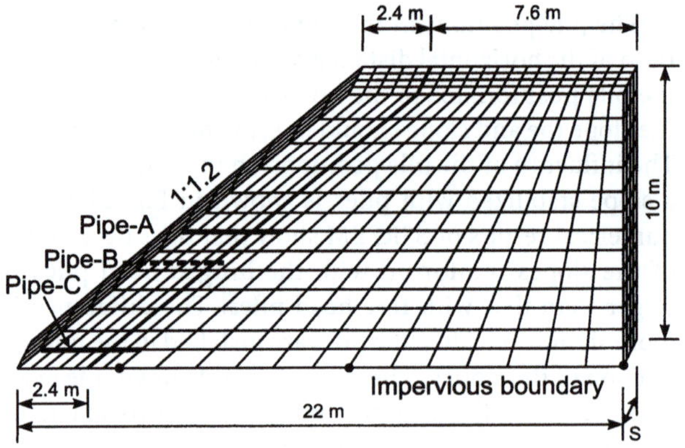

Fig. 23.7. Model slope and finite element mesh

parameters of the tendon and the grouted body are assumed to be the same as those of the steel and the reinforced concrete, respectively. The rigid circular plate on the slope surface is with a diameter of 30 cm. The anchors are installed in the middle of the slope, i.e. the horizontal distance between the slope toe and the anchor head, $Lx = 4$ m, and the center-to-center spacing, $D_1 = 1.5$ m. The material parameters of the soil, soil-tendon interface, and soil-grout interface are shown in Table 23.2.

The direction angle of the anchor, θ, as shown in Fig. 23.4, has significant influence on the safety factor of the slope reinforced with anchors, as shown in Fig. 23.5. It shows that there is an optimized direction angle of the anchor for the safety factor of the slope reinforced with anchors. SSRFEM and Bishop's simplified method predict the same optimized direction angle of anchors. The safety factor of the two methods is consistent with each other. Figure 23.6 indicates that the slip surface predicted by the two methods compares also well with each other.

23.2.4 Stability of a Slope with Drainage Pipes

An idealized slope with a height of 10 m and a gradient 1 : 1.2 is analyzed with a three-dimensional finite element mesh, as shown in Fig. 23.7. The slope is underlain with an impervious layer. Three rows of the perforated drainage pipes are inserted in the slope, of which the length is 3.5 m. The position of the perforated drainage pipes is shown in Fig. 23.7, which is a practically designed layout of them in a slope along a highway. The lowest row is 0.7 m higher than the ground surface, and it is the smallest height for the installation machine to insert the perforated drainage pipes in the slope. The slope is divided into two inclined layers: shallow layer and deep layer, and the shear strength of the shallow layer is lower than that of the deep layer. The mechanical properties of the layers are shown in Table 23.3.

It is for simplicity that the rainfall with a uniform intensity is assumed to last for enough time to reach the steady-state ground water level in the slope under rainfall. The water pressure in the slope, predicted by the finite element analysis of water flow through unsaturated-saturated soils, is used into SSRFEM. The details can be found in Cai et al. (1998). The safety factor of the slope improved with the perforated drainage pipes, evaluated with Fellenius' method, simplified Bishop's method, and SSRFEM, is shown in Fig. 23.8. The safety factor of Fellenius' method is slightly smaller than that of simplified Bishop's method, and the difference of the safety factor between the two methods is only about 0.02. However, the safety factor of SSRFEM

Table 23.3. Mechanical properties

Parameters	Shallow layer	Deep layer
Young's modulus, E (MPa)	9.81	9.81
Poisson's ratio, ν	0.3	0.3
Unit weight, γ(kN m^{-3})	17.66	17.66
Cohesion, c' (kPa)	8.83	24.53
Friction angle, ϕ' (deg)	27.0	27.0
Dilatancy angle, ψ(deg)	27.0	27.0

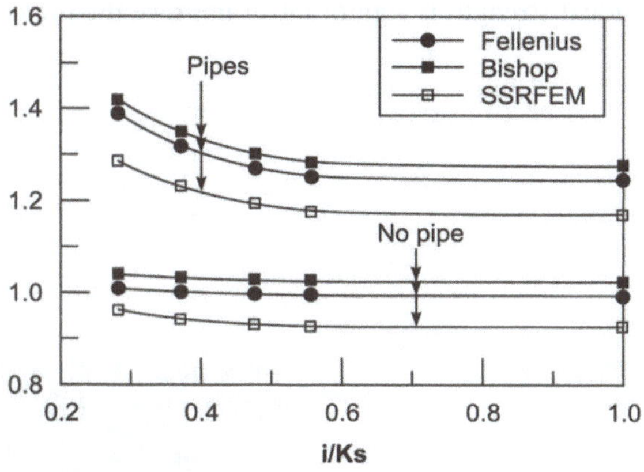

Fig. 23.8. Safety factor versus rainfall intensity

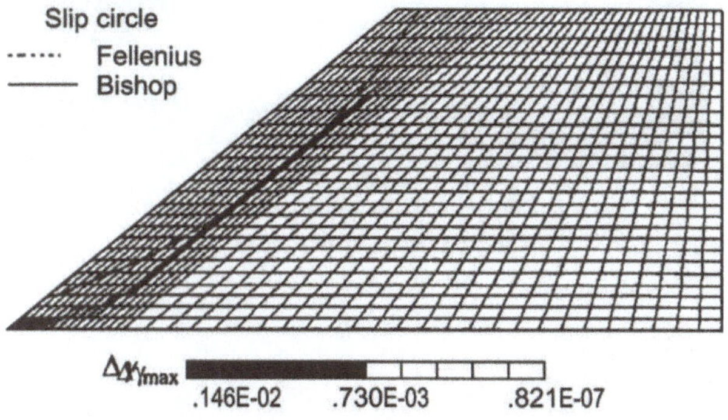

Fig. 23.9. Critical slip surface of Fellenius' method and Bishop's simplified method, and of SSRFEM indicated by maximum incremental shear strain $\Delta\gamma_{max}$

is significantly smaller than that of limit equilibrium methods, and the difference is about 0.1. One of the main reasons is that they predict different failure mechanism for the slope under rainfall. SSRFEM predicts a non-circle translational slip surface, which is significantly different from the circle slip surface, used in limit equilibrium methods. When the normalized rainfall intensity, $i/Ks = 0.278$, the predicted failure surfaces are shown in Fig. 23.9 for the slope with drainage pipes. The slip surface of SSRFEM is hereby indicated with the maximum incremental shear strain, $\Delta\gamma_{max}$, just before the slope failure.

23.3 Dynamic Elasto-plastic Finite Element Method

23.3.1 Analytical Model

23.3.1.1 Equation of Motion for the Total System

The dynamic elasto-plastic finite element method is often used for the simulation of the slope failure induced by the earthquake. In this chapter, the outline and an example of the analytical method is introduced briefly.

The equation of motion is written with

$$[M]\{\ddot{u}\} + [C]\{\dot{u}\} + \{P\} = -[M]\{\ddot{U}\} \tag{23.5}$$

where $\{P\}$is the nodal force vector equivalent to total stress acted inside each element. If the system behaves as elastic body, $\{P\} = [K]\{\dot{u}\}$. $[M]$, $[C]$, and $[K]$ are the mass, damping and stiffness matrices, respectively. $\{\ddot{u}\}$ and $\{\ddot{U}\}$ are the relative displacement vector at each node and the absolute displacement vector at the base. Each matrix and vector are turned separation on the basis of finite element method. Newmark's β method is used to achieve time integration of Eq. 23.5. The effect of viscous damping based on Rayleigh damping is adopted. That is,

$$[C] = \alpha[M] + \beta[K] \tag{23.6}$$

23.3.1.2 Elasto-plastic Constitutive Model

Several constitutive models of varying complexity have been proposed for use in estimating the seismic response of slopes. However, it seems more important to propose and demonstrate a 3-D simple cyclic loading model that can be applied to slope design. In order to achieve it, Wakai and Ugai (2004) proposed an effective elasto-plastic constitutive model, which is often called as the UW model. Both the $G - \gamma$, $h - \gamma$ relationships and the shear strength parameters c and ϕ can be appropriately accommodated by the model. This is useful because while the $G - \gamma$ and $h - \gamma$ relationships of soil have been frequently used for the prediction of seismic response, c and ϕ are more commonly used in conventional calculations of slope stability based on global factors of safety. Incorporating both parameterizations in a seismic design strategy is therefore likely to prove beneficial. In many previous studies, it has been shown that the UW model is very effective in simulating the non-linear dynamic behavior of slopes during strong earthquake.

Detailed explanation of the UW model is omitted here, because of the limitation of the space.

23.3.1.3 Example of Analysis (Wakai and Ugai 2004)

In this section, the appropriateness of the aforementioned model (the UW model) is examined through the FE simulations of the simulation of centrifuge test of a simple slope. Sato et al., which is summarized by Wakai et al. (2001), performed a shaking table test of a slope under centrifugal acceleration of 20 g. In the experiment, the slope was made of homogeneous unsaturated silty clay (DL clay). The prototype height and inclination of the slope were 7.0 m and 45°, respectively. The input acceleration is close to regular sinusoidal waves whose amplitude gradually increases. In the experiment, a failure with a clear slip surface can be seen.

It is clear that the consideration of the shear strength of soil is necessary for the simulation, because most of the residual displacement is produced by the movement of the sliding body. In addition, the strain-softening characteristics of soil and the shear band effect may be also significant to improve the prediction accuracy.

Table 23.4 shows the soil parameters used in the analysis. E was assumed to be $2E_{50}$ on the basis of a hyperbolic stress-strain curve that is adopted in the proposed model.

Table 23.4. Material constants for the simulations of a centrifuge test of a simple slope

Material		Silt
E [G_0] (kPa)		17 000 [6 071]
ν		0.40
c (kPa)		11.1
ϕ (deg)		33.3
K_{cv}		3.24
$b\gamma_{G0}$		8.0
n		1.5
γ_t (kN m^{-3})		15.7
Rayleigh damping	α	0.172
	β	0.00174

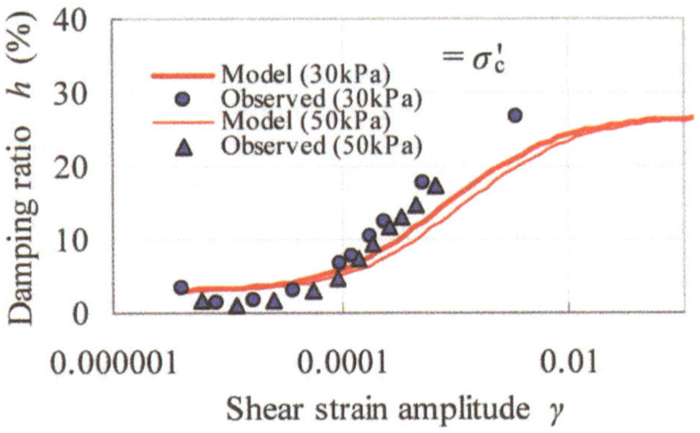

Fig. 23.10. Comparison of $h-\gamma$ relationships under each confining pressure

E_{50} was determined by triaxial tests. The stress dependency of E_{50} is not so obvious in this time, and G_0 is assumed to be constant during and after earthquake. The above assumption is used even in the remaining exercises. Strictly speaking, of course, the stress dependency of G_0 during earthquake exists. The dependency expressed with Eq. 23.2 should be considered in the cases with sufficient soil data according to this matter. ν was assumed to be 0.4. The strength parameters c and ϕ were determined based on the peak strength of triaxial tests. The strain-softening characteristics of soil observed in the triaxial tests actually are not considered in the analysis. The residual strength is significant to improve the prediction

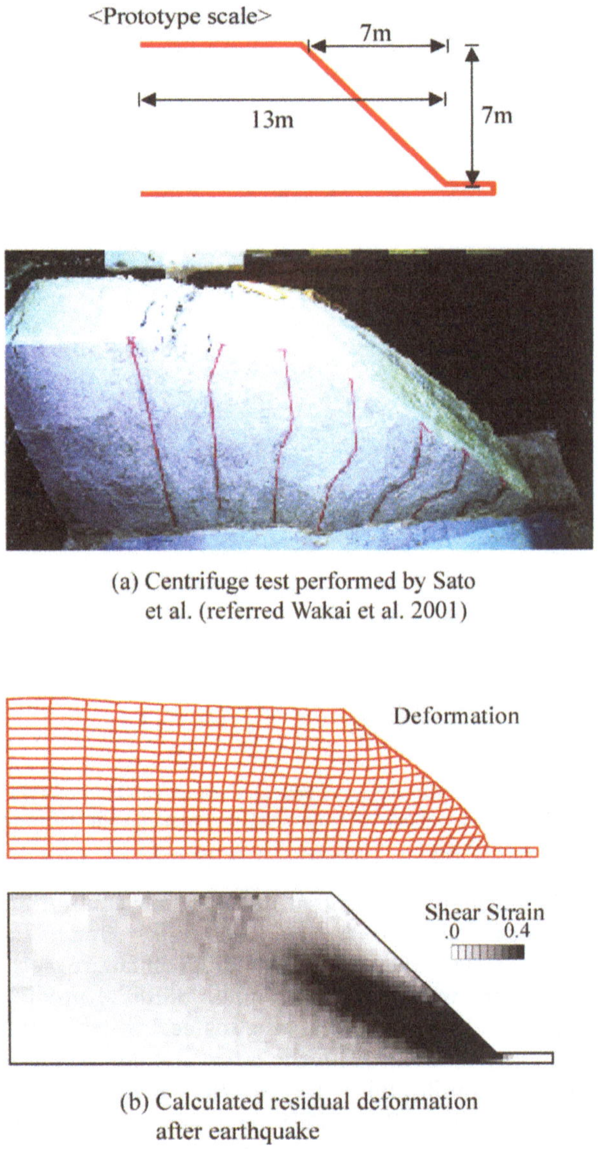

Fig. 23.11. Simulation for centrifuge test of a simple slope. **a** Centrifuge test performed by Sato et al. (referred Wakai et al. 2001); **b** calculated residual deformation after earthquake

accuracy and this may be a future subject. The value of K_{cv} was measured as the ratio of (σ_1/σ_3) and $(-2\dot{\varepsilon}_3/\dot{\varepsilon}_1)$ in triaxial compression tests directly. The parameters for Rayleigh damping α and β are chosen so that h due to viscous damping is maintained at approximately 3%, although this is an assumption. $b\gamma_{G_0}$ and n were determined as the analytical $h-\gamma$ curves fit with the measurements in undrained cyclic triaxial compression tests shown in Fig. 23.10, on the basis of trial and error. It is seen that the two curves agree well with each other. Because the soil is silty, the influence of the rate effect is thought to be very small.

In the analysis, the calculation of the initial stress distribution was based on the shear strength reduction method. The shaking table acceleration history measured at the time of the experiment was used as the input waves in the analysis. The observed and calculated residual deformations after shaking are shown in Fig. 23.11. The overall pattern of deformation of the observed results is simulated extremely well by the numerical analysis. The slip surface, which is represented in the model as the area of highest shear strain, is clearly observed in the slope. Figure 23.12 shows the time histories of the input acceleration, and resulting vertical and horizontal displacements near the top of the slope. A good agreement between the analysis and the test can be seen in the figure. The agreement of the final displacement values is very important from a design point of view. It is concluded that the proposed model is effective in the simulation of failure processes that include strain localization and total slope failure

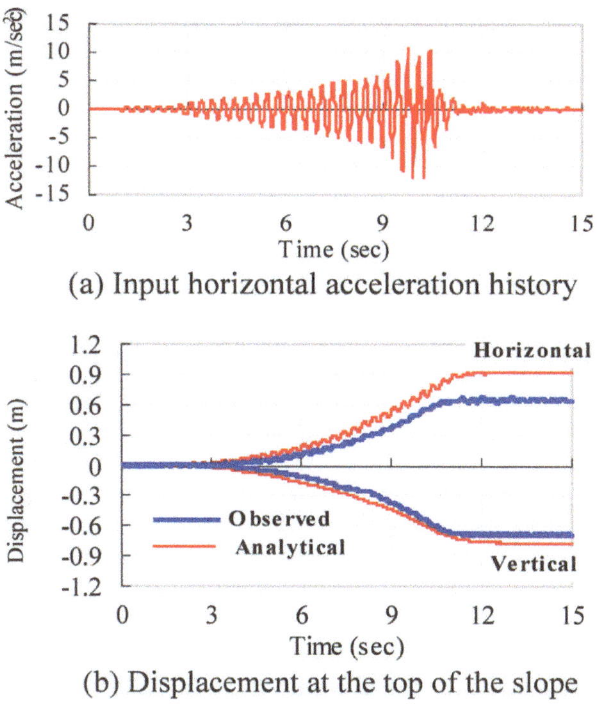

Fig. 23.12. Comparison of time histories of displacement. **a** Input horizontal acceleration history; **b** displacement at the top of the slope

23.4 Conclusions

The slope stability is commonly assessed using limit equilibrium methods. However, the limit equilibrium methods cannot calculate displacements of slopes, thus they cannot The finite element method with shear strength reduction technique (SSRFEM) is proposed to predict the stability of slope improved with piles, anchors, or drainage pipes in order to consider the three-dimensional effects in slopes. Based on numerical comparisons between the finite element method and limit equilibrium methods, it is suggested that the finite element method with shear strength reduction technique is a reliable, robust, and maybe unique approach to evaluate the slope stability and slope stabilization.

On the other hand, the dynamic elasto-plastic FEM makes it possible to evaluate the dynamic response of slopes by considering appropriate stress-strain relationships. If a reasonable constitutive model is used with FEM, the residual displacement of slopes induced by an earthquake can be predicted precisely. In this report, a simple 3-D cyclic loading model that can be applied to the seismic design of slope was shown. Only seven input parameters are necessary for the model. The model incorporates both $h-\gamma$ relationships and shear strength parameters c, ϕ, which are indispensable in the design of slope. Based on such a simple concept that is easy to understand, a dramatic change in the existing design system is avoided, and it brings about the improvement of prediction accuracy in design.

References

Cai F, Ugai K (2000) Numerical analysis of the stability of a slope reinforced with piles. Soils and Foundations 40(1):73–84

Cai F, Ugai K, Wakai A, Li Q (1998) Effects of horizontal drains under rainfall by three-dimensional finite element analysis. Comput Geotech 23:255–275

Ito T, Matsui T (1975) Methods to estimate lateral force acting on stabilizing piles. Soils and Foundations 15(4):43–59.

Ugai K (1989) A method of calculation of global safety factor of slopes by elasto-plastic FEM. Soils and Foundations 29(2):190–195 (in Japanese)

Ugai K, Leshchinsky D (1995) Three-dimensional limit equilibrium and finite element analyses: a comparison of results. Soils and Foundations 35(4):1–7

Wakai A, Ugai K (2004) A simple constitutive model for the seismic analysis of slopes and its applications. Soils and Foundations 44(4):83–97

Wakai A, Ugai K, Sato M, Tazo T (2001) Numerical analysis for seismic behavior of a slope based on a simple cyclic loading model. In: Proceedings 4th International Conference on Recent Advances in Geotechnical Earthquake Engineering and Soil Dynamics, Paper No. 5.06 in CD-ROM, San Diego.

Zienkiewicz OC, Humphson C, Lewis RW (1975) Associated and non-associated visco-plasticity and plasticity in soil mechanics. Géotechnique 25(4):671–689

Chapter 24

Debris Flows in the Vicinity of the Machu Picchu Village, Peru

Jan Klimeš · Vít Vilímek* · Ján Vlčko

Abstract. The work summarizes information about spatial and temporal occurrence and damage caused by debris flows within last 50 years in the vicinity of the Machu Picchu village. Size and dynamic characteristics of limited number of events reveal, that two main classes of debris flows occurred in the study area. These classes are distinguished by specific triggering conditions, amount of delivered material to the main river valley (Urubamba River), duration time and speed. Incomplete weather information and debris flows allowed only preliminary comparison of relationship between rainfalls and debris flow occurrence. The results suggest coupling between 6-days preceding precipitation index and debris flow occurrence. Several low cost mitigation measures and risk management practices were suggested to reduce risk of the economically important area of the Machu Picchu village.

Keywords. Debris flows, Machu Picchu, natural hazards, Peru

24.1 Introduction

Several debris flows in the close vicinity of the archaeological area of Machu Picchu occurred in the last years. The archaeological site itself was built on the mountain ridge at the elevation about 2400 m a.s.l. and has not been endangered by these types of phenomena. Nevertheless, the archaeological site is at present under study from the point of view of possible influence of other types of slope movements (Sassa et al. 2000; Vilímek and Zvelebil 2002).

The Urubamba River (Vilcanota) created around the Machu Picchu deeply incised valley with steep slopes (up to 70°) and with short but steeply inclined tributary valleys. According to huge well rounded rock blocks deposited in the Urubamba Valley floor fare above the lower water level limit, there are big differences in the discharge during the dry (April–October) and wet (November–March) seasons. The largest blocks could be moved or replaced only occasionally during floods. Rock falls contribute periodically to the formation of large block accumulations. Beside this, input of sediments through tributary valleys is very important.

The debris flows occurred periodically within last decades. It could be documented by large alluvial accumulation fans at the mouth of tributary valleys to Urubamba Valley. On one hand the glory of the archaeological site brought the attention to this area, on the other hand an increase of population (due to the largely enhanced tourism) created a significant pressure to urbanize also sites not very favourable for housing. And if we take into account global climate changes, which are usually followed by an increase of frequency and intensity of hazardous processes one can speak about an urgent need to study debris flows in this area. Investigations of slope movements in the area of Machu Picchu were incorporated in the Tokyo Action Plan dealing with strategy in landslides research (*http://ICL.dpri.kyoto-u.ac.jp/*).

The list of debris flows which in the last fifty years affected the territory in the close vicinity of Machu Picchu is shown in Table 24.1. Carreño and Kalafatovich (2006) studied debris flow in the Alcamayo River Valley, which occurred on April 10th 2004. A year later on January 8th debris flow ran down through the valley of Lucumayo – a watershed northward from Aquas Calientes Valley. Another event occurred few weeks later on February 12th 2005 in the valley of Mandormayo. Both Mandormayo and Lucumayo are right-hand tributaries of the Urubamba Valley.

24.2 Preliminary Field Survey

The detailed field survey and interpretation of archive pictures, revealed the relation between the recent catastrophic processes and general geomorphologic evolution of the alluvial fan. Due to the relatively high frequency of these events in this area, the former accumulations are still present. According to the variety in size of the boulders and their spatial distribution were significant differences both in the volume of transported material and in the fluidity of the mass of the past events, which also included rock falls.

Debris flows usually contribute to the formation of the alluvial fans on the riverbanks of Urubamba Valley. But in the cases of February 12th 2005 event, part of the transported material was deposited below waterfall (Fig. 24.1), where the speed of flow was reduced. The height of the inundation is indicated by stones and sands, which were pressed into the bark of the trees 0.45–1.2 m above present ground.

Rock blocks with slickenside were observed in the deposition area. This might be the evidence that the valley is formed within a fault zone. Two important fault lines could be described – one following the course of the Mandormayo River

Table 24.1. List of debris flows occurred between 1946 and 2005 in the vicinity of the Machu Picchu Village (based on information of INRENA Cuzco, Perú and Carlotto et al. 2001)

Date of origin	Basin	Causes	Damages	6-days preceding (ppt mm^{-1})	Daily totals (ppt mm^{-1})
??.02. 1946	Two events 20 days apart, Aguas Calientes	Temporary stream damming[a]	Railroad bridge, local electric power plant	–	–
01.02. 1950	Aguas Calientes		Thermal baths and train bridge destruction	–	–
16.05. 1950	Aguas Calientes		Thermal baths destruction	–	–
1961	Aguas Calientes		Without major damages	–	–
01.01. 1962	Santa Teresa		Human lives lost	–	–
01.02. 1970	Aguas Calientes		Residential houses	–	–
01.02. 1970	Alcamayo			–	–
05.03. 1995	Aguas Calientes	Temporary stream damming[a]	2 houses, hanging bridge and the thermal baths destruction	–	–
05.03. 1995	Alcamayo			–	–
12.07. 1996	Aobamba River (originated in Sisaypampa Lake bellow Nev. Salcantay, Orcospampa stream)	Ice fall to the glacial lake, which outburst	5 people killed, 13 houses destroyed	–	–
01.12. 1997	Aguas Calientes		Without major damages	–	–
22.12. 1997	Aguas Calientes		Thermal baths destruction	–	–
13.01. 1998	Santa Teresa (Sacsara Valley)	High precipitation and glacier melting, minor earthquake occurred on 10.1.1998	Village of Santa Teresa and Yanatile and railroad destroyed	–	–
27.01. 1998	Santa Teresa (Sacsara Valley)	Subsequent event involved material of the 13.1.1998 debris flow	Without major damages	–	–
27.02. 1998	Aobamba River (originated in the Rayancancha stream bellow Nev. Salcantay)	High precipitation and glacier melting	Railroad and Machu Picchu hydro-power plant damaged	–	–
12.03. 1998	Aobamba River (originated Sisaypampa Lake bellow Nev. Salcantay, Orcospampa stream)	Ice avalanche to glacial lake		–	–
??.03. 1998	Aguas Calientes		Without major damages	–	–
??.03. 1998	Aguas Calientes		Without major damages	–	–
22.11. 1998	Aobamba River (originated in Sisaypampa Lake bellow Nev. Salcantay, Orcospampa stream)	Ice fall to the glacial lake, which outburst		–	–
17.01. 2004	Cedrobamba	High precipitation	Railroad track and electric power line pole damaged	87	24
23.03. 2004	Cedrobamba	High precipitation		43	12
10.04. 2004	Cedrobamba	High precipitation	100 m of railways and electric power line pole destroyed	74	31
10.04. 2004	Leonchayoq (close to Cedrobamba River)	High precipitation	Heavy machinery destroyed	74	31
10.04. 2004	Alcamayo	High precipitation	11 people killed, 10 residential houses and 2 railroad bridges destroyed, restaurant Incaterra, part of the railway station and 300 m of railroad track destroyed	74	31
08.01. 2005	Lucumayo	High precipitation	Railroad track destroyed	25	2.4
12.02. 2005	Mandormayo	High precipitation	Railroad track destroyed	116	1.5

[a] Observed temporary drop in the water flow of the streams for two or three hours before the event.

Fig. 24.1.
Debris flows accumulation on the Mandormayo and Lucumayo alluvial fan (*1* – 8.1.2005 debris flow accumulation, *2* – 12.2.2005 debris flow accumulation, *3* – older debris flow accumulations mapped from the 2003 photograph, *4* – waterfall, altitude is indicated at selected contour lines)

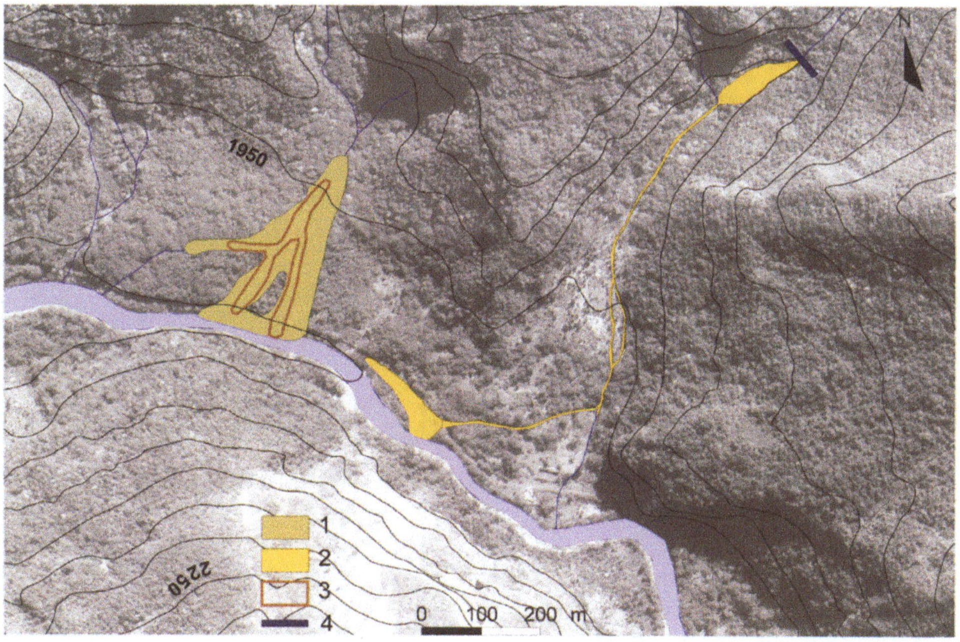

Table 24.2. Size and dynamic characteristics of selected debris flows in the vicinity of the Machu Picchu Village

Characteristics	Alcamayo, 10.04.2004	Lucumayo, 08.01.2005	Mandormayo, 12.02.2005	Aobamba, 27.02.1998	Santa Teresa, 13.01.1998
Area (m^2)	6800	37546	10576	720000[a]	no data
Volume (m^3)	6000	81000	12000	28000000[a]	no data
Size class according to Jakob (in print)	3	4	5	6	7–8 (estimation)
Horizontal distance between initiation point and river mouth (km)	2.1	4.5	3.7	20[b]	30[b]
Vertical distance between initiation point and river mouth (km)	1	1.5	1.45	3[b]	3[b]
Gradient between initiation point and river mouth (deg)	25.4	18	21	8.5	6
Duration of event (h)	0.5	no data	no data	ca. 14[b]	9.5[b]

[a] According to Kuroiwa (2002).
[b] According to Carlotto et al. (1999).
Other information are based on calculations according Jakob (2005).

and the second one runs perpendicularly to the previous one. The crossing of both faults can be observed few meters bellow waterfall. The intersection of both fault lines is accompanied by ferrite springs, which are also found in the Alcamayo River Valley (Vilímek et al. 2006).

The Lucumayo debris flow deposited (compared to the last 3 events around the Machu Picchu Village) the largest amount of debris (Table 24.2). The detailed study of photos (Fig. 24.2) taken from the top of Mt. Huaynapicchu revealed that the 2005 event followed an older one in a similar way. The width of the accumulation from 2005 was 175 m (observed and measured along the rail track) and estimated thickness of the accumulation was about 2 m.

24.3 Triggering Factors of the Debris Flows in the Machu Picchu Village and Surrounding Area

Basic assumptions about triggering factors of the debris flows in the vicinity of the Machu Picchu village (Aquas Calientes) may be made based on the inventory of the historic events presented in Table 24.1 as well as on the field inspections. Even though, the inventory is not complete mainly for older and smaller events occurring in places without any infrastructure. According to Carreño and Kalafatovich (2006) and Vilímek et al. (2006) following two main debris flow triggering conditions are identified:

Fig. 24.2. Comparison of 2003 and 2005 photographs of the Mandor alluvial fan taken from the Mt. Huaynapicchu (photo by J. Klimeš)

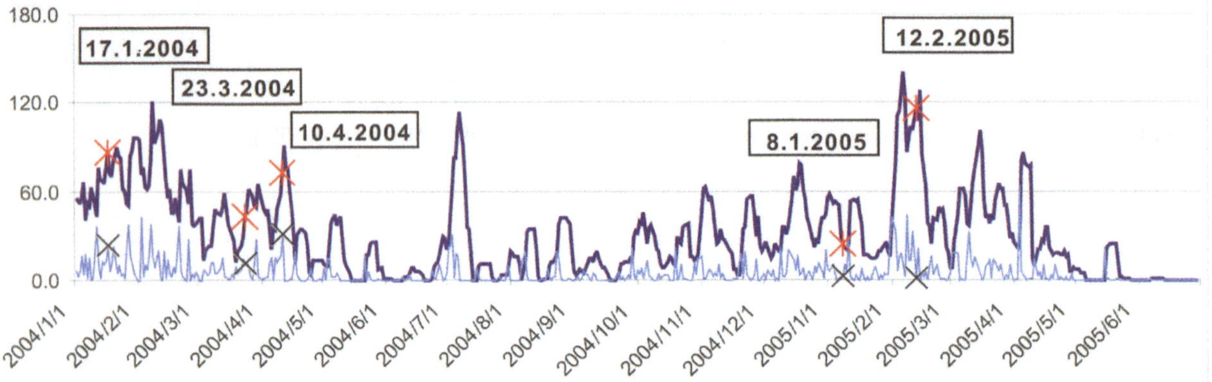

Fig. 24.3. The total 6-day preceding precipitation (*dark blue curve*) and daily precipitation totals (*light blue curve*) of the period between 1.1.2004 and 1.6.2005 (precipitation data SENAMHI Cuzco)

ice falls or ice avalanches caused probably by warm and rainy weather and high precipitation rate initiating landslides or rock falls which further transform into high saturated debris flows while traveling to the main river bottom. The ice falls cause overflow of glacial lakes producing floods, which mobilized moraine and river sediments.

All the debris flows listed in Table 24.1, except the event from 12.7.1996, 12.3.1998 and 22.11.1998, were triggered by high precipitation and occurred in rainy season (between November and March) with majority events between January and March (17 events from 26 recorded). The two exceptions occurred in May 1950 and July 1996, with the later caused by glacial processes at the altitude above 4800 m a. s. l. Distinct is also clustering of the events. In most of the cases (except the years of 1946, 1961, 1962 and 1996) more than one event occurred within the same rainy season. In some cases (e.g., Santa Teresa 13.1. and 27.1.1998), it can be probably linked with repeated saturation of older debris flow accumulations. Table 24.2 summarizes basic size characteristic of selected events in the vicinity of the Machu Picchu village.

The authors did not get the possibility to perform detailed statistical analyses of daily precipitation records in relation to debris flows occurrences, since only limited number of precisely dated debris flows were recorded for the time period covered by daily rainfall data, which are available from January 1999 to June 2005 (Machu Picchu weather station – 2519 m a.s.l., SENAMHI Cuzco). Despite this, some qualitative observations based on 7 precisely dated debris flows events were performed. Three of these events (Table 24.1) occurred on the same date – 10.4.2004, but at three different locations: one debris flow destroyed the part of the Machu Picchu village and two other damaged railroad 3.5 km E of the Machu Picchu village along the Cedrobamba River.

The precipitation record was evaluated using variety of methods (e.g., preceding precipitation index, moving average, cumulative line of differences of daily and monthly precipitation totals from their long-time averages, Guzzetti et al. 2004; Kopecký 2002; Rybár et al. 1995) assessing relationship between short-term precipitations and landslide or debris flows occurrence. The best results were gained adopting 6-days preceding precipitation curves, which in four dates (17.1.2004, 23.3.2004, 10.4.2004 and 12.2.2005), placed six events on the rising part of the curve shortly before the local 6-days preceding precipitation maximums were achieved (Fig. 24.3). In all these cases the 6-days preceding precipitation exceeded the mean value for the evaluated period, which is 34 mm. Great scatter may be observed in the daily precipitation totals of the days when debris flow occurred (Table 24.1). These results suggest that the short term preceding precipitation is more important for triggering the studied debris flows than the daily totals.

The precipitation characteristics of the debris flow from 8.1.2005 were in a great contrast with the others. The 6-day preceding precipitation was only 25 mm and the daily precipitation reached only 2.4 mm. These ambiguous results may be explained by the hypothesis, that the actual precipitation triggered the debris flow were not recorded on the Machu Picchu weather station. This may be caused by high spatial variability of the precipitation (supported also by witnesses of local inhabitants) due to elevation differences between the highest point and the river mouth of the basins surrounding Aquas Calientes village. The calculated gradients of the selected basins are in Table 24.2. Formation of temporary dam caused by landslide or rock fall in the narrow stream valley may block the river for more than one day and therefore delay the debris flow arrival to the Urubamba Valley. That is another possible explanation of the mismatch between precipitation and debris flow record.

The analyses of monthly totals for period 1965–1976 and 1999–June 2005 suggest some relation between short and long-time (e.g., seasons, years) rainfalls patterns and debris flows occurrence. Figure 24.4 shows curve of cumulative monthly differences between monthly precipitation sums and long-time monthly precipitation averages, which gains high positive values since the turn of March and April 2002 and which steeply descends towards the very dry winter months of 2005. All of the 7 precisely dated debris flow events occurred in the time period of high positive values shown on the curve.

It can be summarised, that the events involving smaller amount of material (e.g., 10^3–10^4 m^3, 10.4.2004 Alcamayo River debris flow) occurring in smaller catchments implying for short travel distances may be more dangerous for local inhabitants (in terms of fatalities) than much larger events (e.g., 10^6–10^7 m^3, 27.2.1998, Aobamba River debris flow) traveling on long distances, which on other hand may damage infrastructure in much larger areas due to blocking of the Urubamba River stream. This implies that the hazard connected with the Mandormayo and Lucumayo debris flows was very high and only absence of objects at risk (e.g., developed and inhabited areas) on the alluvial fan prevented from serious consequences. The Lucumayo debris flow damaged about 200 m of railroad and the traffic was interrupted for several days. In the case of Mandormayo debris flow around 60 m of railroad were damaged (Fig. 24.5) and some crop was also lost. The other part of the fan have not been cultivated before, which along with Fig. 24.2 may indicates more frequent occurrence of debris flows there. On the other hand, more elements at risk can be identified on the Mandormayo part of the alluvial fan where a tourist attraction (waterfalls) and a framer's house are located. To evaluate their risk, detailed debris flow hazard map (similar to one prepared for the Aquas Calientes village, Carlotto et al. 2001) needs to be prepared. The map should pay attention to the local morphological changes of the alluvial fan surface and river beds due to previous debris flow events, e.g., the rise of the river bed by 1 m and thus alternating the surface drainage conditions, was observed on the Lucumayo part of the alluvial fan after the 8.1.2005 event.

In the case of higher speed and/or larger volume of debris flows, their accumulation may reach 30–40 m wide riverbed of the Urubamba River (Fig. 24.6) and cause its temporal damming. Up stream flooding and possible damage caused by the dam outburst may cause huge infrastructure damage. The 10.4.2004 Alcamayo debris flow blocked most of the 30 m wide Urubamba River bed.

Since the area damaged by the Mandormayo and Lucumayo debris flows is not inhabited (Fig. 24.7), there is only very low possibility that the local authorities will decide to construct any technical protective measures there. Therefore the most successful solution to mitigate the debris flow hazard is creating a debris flow warning system. It should be based upon relationship

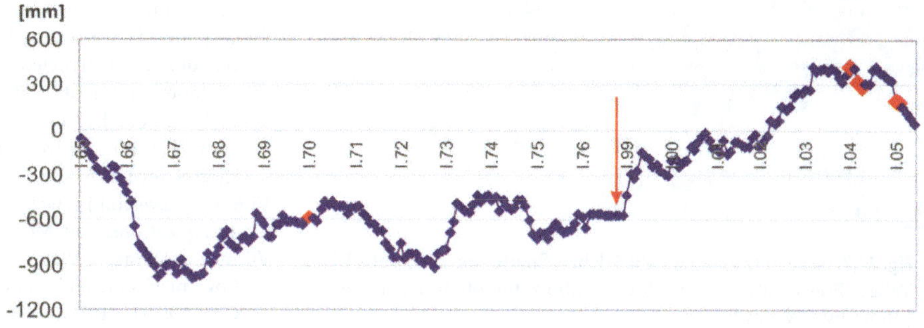

Fig. 24.4.
Cumulative monthly differences between measured and long-time total monthly precipitation averages (long-time monthly average calculated for years 1965–1976, precipitation data SENAMHI Cuzco). *Red diamonds* show months with debris flow occurrence, *red arrow* shows time gap in the data between 1976 and 1999

Fig. 24.5. The railroad after reparation – situation in September 2005 (photo by V. Vilímek)

Fig. 24.6 The riverbed of Urubamba River. All the material from the debris flow was already removed from this area (photo by V. Vilímek)

Fig. 24.7. Issue of the January 2005 debris flow to the Urubamba River Valley. Fresh sediments are located above the older accumulation (photo by V. Vilímek)

between rainfall data and debris flow occurrence, precipitation monitoring and preparation of debris flow risk map. To establish such system requires installing automatic rain and hydrologic gauge stations in river basins in the vicinity of the Machu Picchu village, and the installation of an automatic weather station with a real time data transfer. The warning system, especially during rainy seasons, can be supplemented by special guards who may be very helpful especially in case, when the event occurs during night.

All the technical measures and precautions mentioned in the paper will help to predict the debris flow hazard and to reduce or to absolutely diminish fatalities for the one of the most beautiful site in the world.

Acknowledgment

The authors would like to thank the Ministry of Education, Youth and Sports of the Czech Republic (Projects MSM 0021620831 and INGO, LA 157) for their financial support.

References

Carlotto V, Cárdenas J, Romero D, Valdivia W, Tintaya D (1999) Geología de los cuadrángulos de Quillabamba y Machupicchu. Boletín No. 127, Serie A Carta Geológica Nacional, INGEMET, Setiembre 1999, Lima, 318 p

Carlotto V, Galdos B, Solís M, Arias V, Machicao O, Mar P (2001) Plan para la mitigacion de desastres del poblado de Machu Picchu – Aguas Calientes. PROFONAPE-Programa Machu Picchu, Cuzco, Peru (in Spanish)

Carreño R, Kalafatovich S (2006) The Alcamayo and Cedrobamba catastrophic debris flow (January, March and April 2004) in Machupicchu area – Peru. Landslides 3:79–83

Guzzetti F, Cardinali M, Raichenbach P, Cippola F, Sebastiani C, Galli M, Salvati P (2004) Landslides triggered by 23 November 2000 rainfall event in the Imperia Province, Liguria, Italy. Eng Geol 73:229–245

Jakob M (2005) A size classification for debris flows. Eng Geol 79: 150–161

Kopecký M (2002) Influence of climatic and hydrogeologic conditions on the origin of landslides in Slovakia. In: Rybar J, Stemberk J, Wagner P (eds) Landslides – Proceedings of the 1st European Conference on Landslides, Prague, 24–26 June 2002. Swets & Zeitlinger, Lisse, pp 367–380

Kuroiwa J (2002) Reducción de desastress – Viviendo en armonía con naturaleza. Quebecor World Perú S.A., Lima, 429 p

Rybár J, Košťák V, Novotný J, Zvelebil J (1995) Examples of evaluation of individual meteorological factors affecting different types of slope movements. CEC Project MeFISSt. Final Report

Sassa K, Fukuoka H, Shuzui H (2000) Field investigation of the slope instability at Inca's world heritage, in Machupicchu, Peru. Landslide News 13:37–41

Vilímek V, Zvelebil J (2002) Slope instability at Machu Picchu: ideas and questions. Acta Montana 19:75–89

Vilímek V, Klimeš J, Vlèko J, Carreño R (2006) Catastrophic debris flows near Machu Picchu village (Aguas Calientes), Peru. Environ Geol 50:1041–1052

Engineering Geology and Cultural Heritage: the Conservation of Remaining Bamiyan Buddhas (Central Afghanistan)

Claudio Margottini

Abstract. The present paper describes deformation processes and geomorphological hazards affecting the historical site of Bamiyan (Central Afghanistan). The major cultural heritage of the site were the two standing Buddhas that, carved in rock since the 2nd–3rd century A.D., were destroyed by Taliban in March 2001. Slope instablity from both the consequences of the explosion as well as from natural processes are quite evident: rock slides and rock falls have already occurred in the recent past and most areas are prone to collapse.

Under the UNESCO coordination, a global feasibility project for the needed restoration work was developed; field data were collected and mechanisms for potential cliff and niche evolution were provided. In the meantime a first practical consolidation work for the most critical rockfall-prone area, was implemented to avoid any further collapse in the period 2003–2005. This last is also considered essential to allow archaeologists the safe cataloguing and recovering of the Buddha Statue remains, still on the floor of the niches.

Keywords. Buddha statues, explosion, rock fall, toppling, consolidation, Afghanistan, Bamiyan

25.1 Introduction

In the great valley of Bamiyan, 200 km north-west of Kabul, Central Afghanistan, two big standing Buddha statues were carved out of the sedimentary rock of the region, at 2 500 meters of altitude. The Emperor Kanishka ordered their construction around the 2nd century A.D. for the smaller one, while the initiation of the Western Giant Buddha seems to have been in the 3rd or 4th century A.D. (Dupree 2002). Recent ^{14}C measurements moved the construction to the 6th century A.D. (ICOMOS, personal communication). Some descendants of Greek artists who went to Afghanistan with Alexander the Great started the construction that lasted till the 4th century A.D. (Gruen et al. 2002).

The Buddhist art of the Hindu Kush mountain region, of which the Bamiyan Valley is a part, represents the final age of Buddhism in Afghanistan. The kingdom of Bamiyan was a Buddhist state positioned at a strategic location along the trade routes that for centuries linked China and Central Asia with India and the West. Bamiyan served as an important monastic and spiritual center, as well as a hub of intense commercial activity. The site was constructed between approximately the 5th and 9th centuries A.D. during a distinctive phase in the history of Buddhist art, a period of intense cultural and religious exchanges between East and West, and a time of great cultural change within Buddhism itself. Bamiyan served as a ceremonial and spiritual center that attracted and comforted crowds of pilgrims and merchants traveling between Central and South Asia.

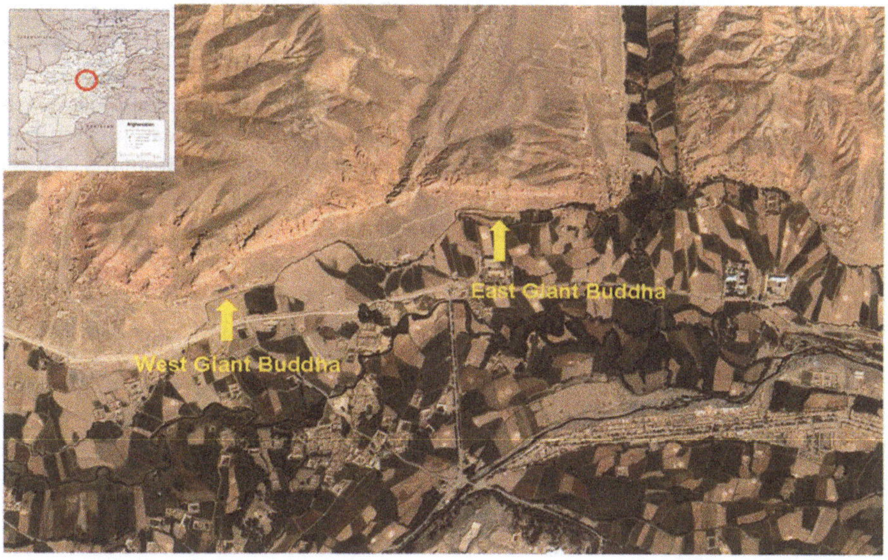

Fig. 25.1. Geographical settlement of the area

Fig. 25.2. The valley of Bamiyan, Central Afghanistan

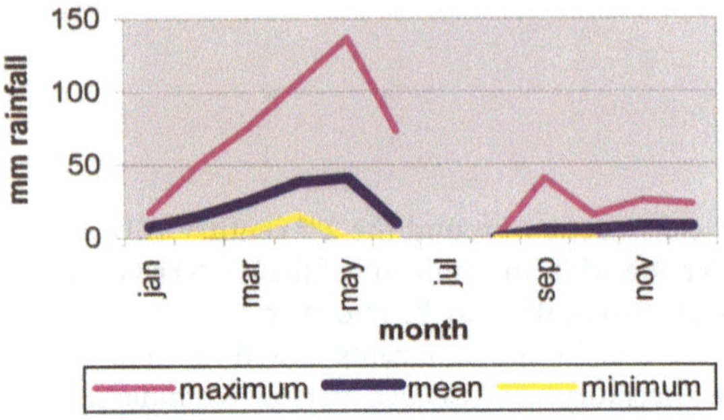

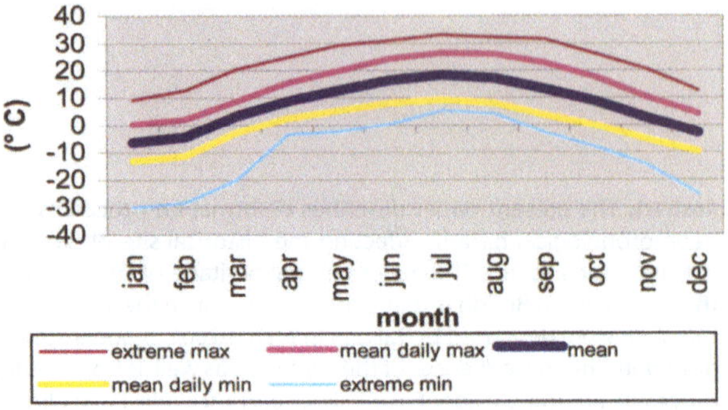

Fig. 25.3. Meteorological data for the Bamiyan Station for the period 1958–1977 (added October 2001) (source Operational Climatic Data Summary 2002)

During this extended period of Bamiyan's Buddhist florescence two massive Buddha images were carved out of a high stretch of cliff facing the widest part of the valley. These colossal images are the largest Buddhist sculptures in the world. The greater of the two images stands 53 m in height at the western end of the cliff-face; the second massive Buddha, at the eastern end of the cliff, is some 35 m tall. The niches containing the statues were respectively 58 and 38 m high. All along the cliff face between these monolithic images are carved hundreds of caves of varying size used as chapels for both private and communal worship (Fig. 25.2). Ambulatories, off of which are further rock-cut chapels and image-niches, surround the larger Buddha at the level of his feet and again at the level of his head high up the cliff face. Most of the rock-cut chapels and ambulatories at Bamiyan are covered with paintings over plastered walls that display an incredibly rich, varied, and important body of early Buddhist painting.

The two statues were demolished in March 2001 by the Taliban, using mortars, dynamite, anti-aircraft weapons and rockets. The Buddhists, the world community, UN and UNESCO failed to convince the Taliban to not harm such works of cultural heritage. The fundamentalist Islamic militia, which governed most of Afghanistan from 1996 to December 2001, followed an edict of its supreme leader who ordered a campaign of destruction to rid the land of all un-Islamic graven images (Grun and Remondino 2002).

The result of demolition was immense: the two large statues were destroyed, the niches where the Buddhas were carved are almost to collapse, many mural paintings have been covered by asphalt.

Since 2001, UNESCO was involved both before and after the destruction of the statues. Presently, one of the most important conservation aspect is the consolidation of the two niches, prone to collapse partially. The present paper is the result of two UNESCO missions on the site and laboratory investigations, aimed at understanding the ongoing processes on the cliff and niches and providing the most appropriate stabilization measures, according to the high environmental and cultural value of the site.

The potential reconstruction of the statues or conservation in a local museum, is not the topic ofthe present paper and will involve the Government of Afghanistan and UNESCO, with archaeologist and expert of restoration, mainly from the International Council of Monuments and Site (ICOMOS).

25.2 Meteo-climatic Setting

Afghanistan is a mountainous country in a dry part of the world which experiences extremes of climate and weather. Winters are cold and snowy, and summers hot and dry. The wet season generally runs from winter through early spring, but the country on the whole is dry, falling within the desert or desert-steppe category of climatic classification. Very little snow falls in the lowland deserts of the southwest, but the snowy season averages roughly from October–April in the mountains and varies considerably with elevation.

The site of Bamiyan exhibits a mean annual precipitation of 162.56 mm and mean annual temperature of 7.40 °C (Afghan Air Authority Meteorology Department, Climate Section, operating for 8–9 years in the period 1958–1977, plus October 2001) (data from Operational Climatic Data Summary 2002).

Figure 25.3 synthesizes available meteorological data for the site of Bamiyan.

25.3 Geological, Mineralogical and Petro-geophysical Setting

In the area of both the Eastern and Western Giant Buddha are outcropping materials, likely coming from dismounting of surrounding morphological peaks and deposition in a flood plain and a small lagoon. The subsequent river erosion produced the present morphology.

The almost vertical cliff where the Buddha statues are carved, probably experienced also human excavation, to obtain a vertical cliff suitable to host the monastic civilization. As confirmation of human influence on present morphology it can be pointed out that such a step morphology is not recognizable in other parts of the valley.

The cliff and niches are composed of an alternance of conglomerate and siltstone (yellow at the bottom and red in the middle of the cliff).

The conglomerate is composed of quartz, calcite, mica, feldspar group minerals, clay minerals and heavy minerals, in order of importance (X-ray diffraction); quartz is associated to individual grains whereby calcite is present in grains as well as secondary deposition in terms of carbonatic cement. Matrix is composed mainly of clay minerals.

Microscopic thin sections of the conglomerate (Fig. 25.4) reveal mono-mineral grains and lithic fragments, moderately classed, cemented by carbonates. Mono-mineral grains comprise: mono-crystalline quartz scarcely rounded and polycrystalline aggregation of quartz moderately rounded. Less frequent are polycrystalline aggregations of carbonates moderately rounded and rarely mono-crystals or aggregation of feldspar slightly rounded. Dimensions of mono-crystals may vary from tens of ~μm to about 1.5 mm; the dimension of aggregations can reach 3 mm.

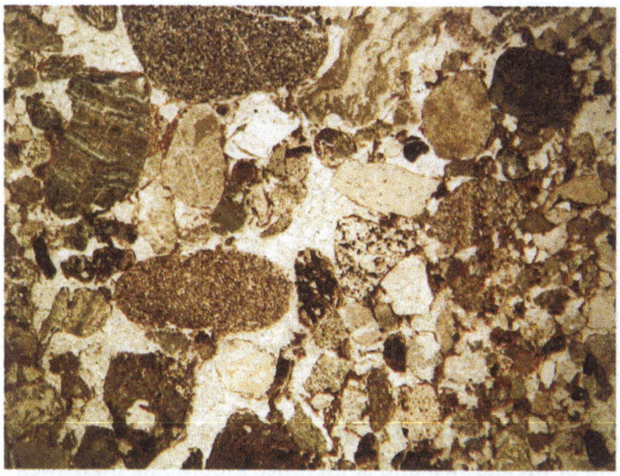

Fig. 25.4. Thin section view of conglomerate, with the presence of white carbonatic cement strengthening the lithotype

Lithic fragments exibit dimensions from 1 mm up to 15 mm. The majority of these can be grouped into three main categories:

- fragments of rocks constisting of small grains (sandstone?), composed mainly ofquartz, calcite and clay minerals;
- fragments moderately rounded of claystone-siltstone; this family seems to be similar to the siltstone formation surveyed on the cliff;
- fragments of most likelyvolcanic/subvolcanic origin highly altered in clay minerals, oxides and/or hydroxides and, rarely epidot. Some fragments can still display feldspar fenocrystals scarcely altered and femic minerals altered in oxides and hydroxides, in a microcrystalline background paste. Rarely can be detected rounded quartz crystals; more often they are fragments of altered background paste.

According to the above consideration the recorded field uniaxial compression strength of about 24–35 MPa is mainly determined by carbonatic cement, developed secondarily during the diagenetic phase.

The siltstone has been surveyed on the lower part of the cliff (yellow component) and in some strata in the middle-upper part of the cliff. It is composed ofquartz, calcite, mica, clay minerals and heavy minerals, in order of importance (X-ray diffraction); roughly speaking, quartz can be estimated as 70%, calcite 20%, and all the others about 10%. According to Brown (1972) the following clay minerals have been detected: illite, chlorite, smectite.

X-ray diffraction of yellow (Fig. 25.5) and red material does not reveal any mineralogic difference.

In microscopic thin sections the rock can be recognized as an aggregation of clay minerals lacedwith small irregular veins of carbonates. Clay minerals seems to be associated with iron hydroxides and/or sulphates. Within the aggregate are evident quartz grains, scarcely rounded with dimensions ranging between 0.01–1 mm as well as fragments of moderately rounded crypto-crystalline carbonatic rocks, with dimensions of lower than 2 mm.

No cement as been identified in microscopic thin sections, as confirmed by the total loss of mechanical cohesion when the sample is weathered; only the red siltstone flocculates in water which testifys to the presence of salts, residual of an evaporitic depositional environment. In conclusion, the siltstone can be considered as a desiccated mud, with an apparent cohesion depending, at least, in part from dryness.

In both conglomerate and siltstone (yellow) X-ray diffractometry does not reveal soluble salts (gypsum, alite, etc.). This is in contrast with the flocculation of red siltstone but it testifys to an alternance of environmental deposition in time.

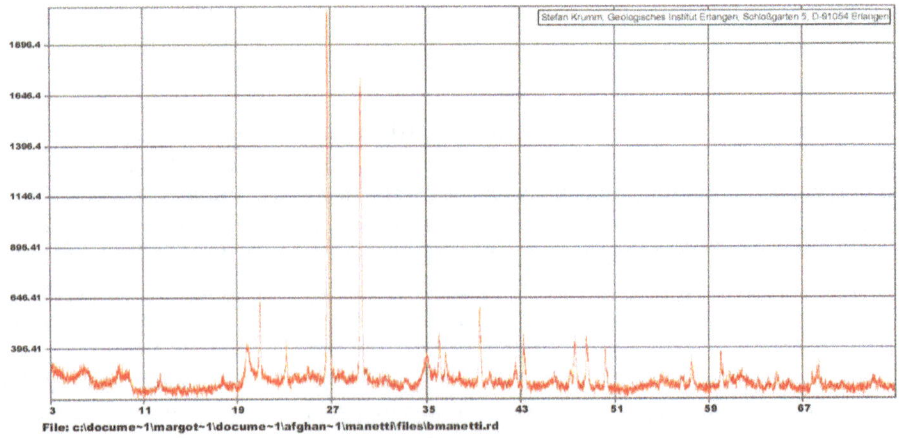

Fig. 25.5. X-ray diffractometry of yellow siltstone, for tout venant sample

The depositional environment and diagenetic conditions probably affected the present-day characters of lithotypes since the conglomerate exhibits a permanent cohesion, while the siltstone easily looses the apparent cohesion when saturated, as depicted in Fig. 25.6. The mechanism for this behavior, probably a slaking process (Franklin and Chandra 1972), has been investigated by means of scanning electron microscopy, as well as the determination of physical and mechanical properties in soil and rock mechanical laboratory tests.

Values of Vp have been measured in the laboratory for both conglomerate and siltstone (Fig. 25.7). Conglomerate exhibits values of about 2.2–2.5 km s^{-1} and Poisson ratio of about 0.300–0.400 (samples perpendicular and transverse to deposition flow); the value of Vp of about 1.5–1.7 km s^{-1} and Poisson ratio of about 0.100–0.200 has been measured in the direction of depositional flow.

The siltstone exibits a Vp of about 1.7–2.0 km s^{-1} and Poisson ration is about 0.150–0.200, without any geometrical anisotropy.

The lower seismic velocity in conglomerate is in the direction parallel to sedimentary flux (stratification plane). This implies that, in this direction, likely, seismic waves exhibit lower values since they are crossing mainly the matrix composed ofcemented silt material; when the same seismic waves are crossing the samples perpendicular and transverse to sedimentary flux/stratification plane the velocity is higher probably as a consequence of greater thickness of rocks to the intersect. Finally, the values obtained in the sedimentary flux/stratification plane of conglomerate, when matrix is predominant, are quite closer to the values measured in the siltstone, indirectly confirming a similar origin and composition.

Fig. 25.6. Stability of conglomerate (*right*) and degradation of siltstone (*left*) when weathered, after only few minutes of submersion in water

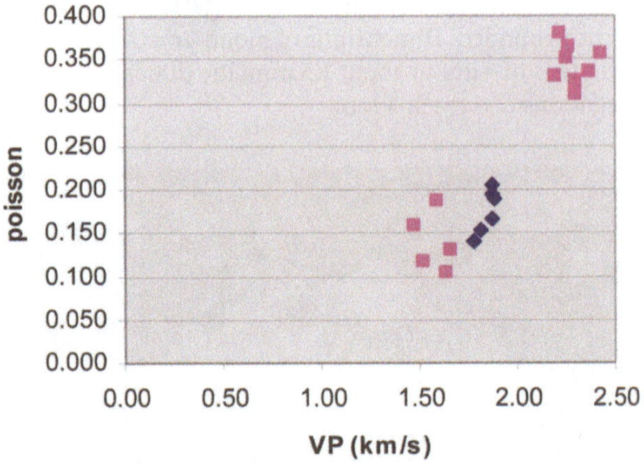

Fig. 25.7. Vp velocity and Poisson ratio in conglomerate (*red*) and siltstone (*blue*). The lower value for conglomerate is parallel to sedimentary flux

25.4 Scanning Electron Microscopy (SEM) of Siltstone

The swelling behavior of fine-grained sedimentary rocks (claystones, mudstones, shales, etc.) is a complex phenomenon. When in contact with water, these materials can present considerable volume change resulting in swelling or breakdown and sometimes mobilizing high swelling pressures. The behavior of swelling rock is controlled by numerous factors acting jointly and resulting in an effect whose main agent is frequently hard to identify. The last two decades have been marked by a growing concern about weak rock behavior and by an increased number of studies. This type of rock generally presents a behavior that is intermediate between rock and soil, which complicates the sampling and testing required to forecast its geotechnical behavior. Some weak rocks are more sensitive to wetting and drying cycles than other types resulting in an almost total breakdown after a number of cycles or, like in the siltstone from the Bamyian cliff, just after weathering. Identification of this behavior is crucial in planning works in areas where weak rocks occur (Pejon and Zuquette 2002).

Sample preparation and microscopic observations were carried out, following the recommendations of Goldstein et al. (1992) and Le Roux (1971). Obtained by simple rupture, all the observed samples were about 15 to 20 mm long, 5 to 10 mm wide, and 10 mm high. They were dried by a cryosublimation process (Le Roux 1971; Shi et al. 1999) before being viewed under the microscope. This procedure is intended to preserve rock microtexture and original structure and to allow observation of modifications caused by the swelling process. The surfaces perpendicular to the rock lamination were preferred for observation with SEM although some surfaces parallel to the bedding plane were also examined, mainly when generated during the swelling tests. All the samples were observed with the electron microscope both before and after the swelling tests, to analyze textural, structural or mineralogical modifications caused by the swelling process. Figure 25.8 shows SEM images for 1 kx, undisturbed and after weathering and breakdown of the structure.

In spite of the low concentration of swelling clay mineral in the studied materials, breakdown was intense, and occurred in few minutes when the samples were submitted to wetting.

Observation of the rocks under SEM demonstrated that the presence of a compact and laminar structure (left side of Fig. 25.8) no longer existed after the breakdown (right side of Fig. 25.8). Laminar structure could be the consequence of clay minerals orientation, filling completely the voids of the siltstone, and producing a high density material. Wetted and reconstituted material does no longer presents the typical structure of undisturbed samples; it clearly presents an evident grain composition, some of them laminar, probably clay minerals. The number of

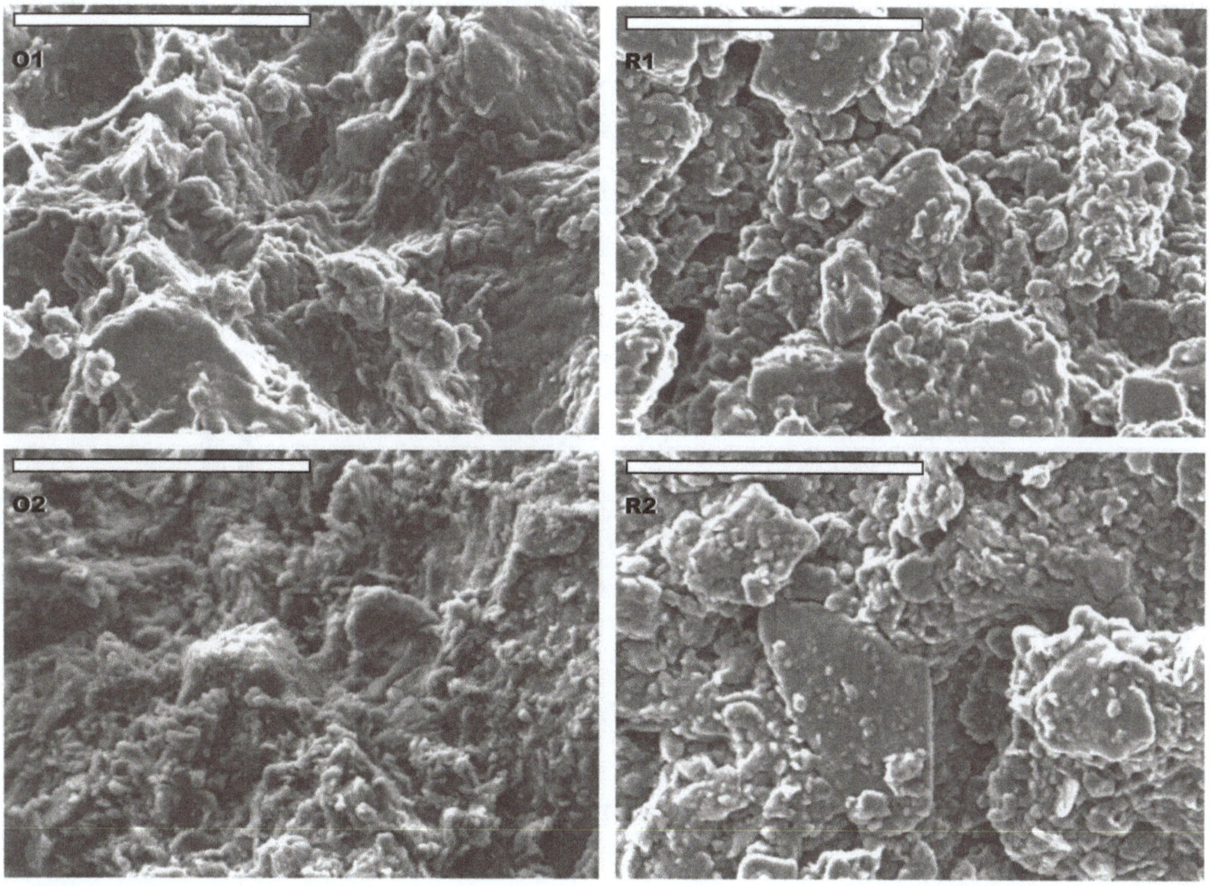

Fig. 25.8. Observation under scanning electron microscope at 1 000×. *O1* and *O2* are undisturbed sample, while *R1* and *R2* are from material weathered, broken down and thean dried to obtain a feature similar to natural conditions

voids is clearly higher with respect to undisturbed siltstone, so that the original high density is not reached after a wetting and drying cycle, on the same material.

Possibly the small amounts of swelling clay (like smectite) in certain positions in the rock texture could promote their swelling and breakdown, as observed by Pejon and Zuquette (2002), in French mud rocks with similar granulometric and mineralogical composition of Bamiyan samples.

The compression of entrapped air was shown to possibly also be an important factor in swelling and breakage of mud rocks, as also observed by Taylor and Spears (1970).

In essence, the resistance to slaking is a function of different controlling parameters often mentioned in the literature; among these are permeability and porosity, adsorption, solution or disruption of the bonds, disruptive forces and initial degree of weathering (Crosta 1998).

25.5 Physical and Mechanical Properties of Materials

A field survey as well as laboratory tests have been conducted for collecting the largest as possible amount of data. Laboratory tests comprise physical properties (e.g., Atterberg limits and grain size distribution) for siltstone and point load test as an indicator of uniaxial compressive strength for both conglomearate and siltstone. Field investigation includes in situ density for conglomerate, rock classification (RMR and Q system), and uniaxial compressive strength by means of Schmidt hammer test for both siltstone and conglomerate.

Soil and rock mechanic tests on both materials was necessary since the two lithotypes show completely different behavior, when weathered. As already described, the conglomerate exhibits a permanent cohesion due to the carbonate cement in the matrix while the siltstone, a desiccated mud, easily looses the apparent cohesion when saturated. From a physical and mechanical point of view it is also extremely important to define the major parameters characterizing the two lithotypes as well as to define their respective behavior to the possible consolidation techniques or, the other way around, to define the consolidation measurements according to the real behavior and property of local materials.

The grain size distribution of siltstone, determined with an electronic sedigraph, determined clay for 7%, sand for 17%, and silt for 76%. This distribution identifies a "silt" in the Shepard (1954) diagram (Fig. 25.9). According to Atkinson (1997) this grain size distribution is typical for a river estuary environment.

The grain size distribution on a second sample of siltstone, but investigated with a standard hydrometer, point out clay for 28%, silt for 66%, and sand for 6%.

The plasticity character reflects the low clay content pointing out a Liquid Limit of 39.80% and a Plastic Limit of 16.96%. The material is characterized as CL, according to the Unified Soil Classification System reported in ASTM (1977), which fits in the category of "inorganic clays of low to medium plasticity, gravelly clays, sandy clays, silty clays, lean clays".

The siltstone exibits also a high natural density (2.28 g cm^{-3}), despite of a dry density of 2.23 and as displayed in the phase diagram of Fig. 25.10. Quite low are the values of porosity and void index, respectively 14.93% and 0.176, because of the high compaction of siltstone. These data are from samples taken from the site and could not reflect exactly the real conditions. Nevertheless the high natural density is a notable characteristic.

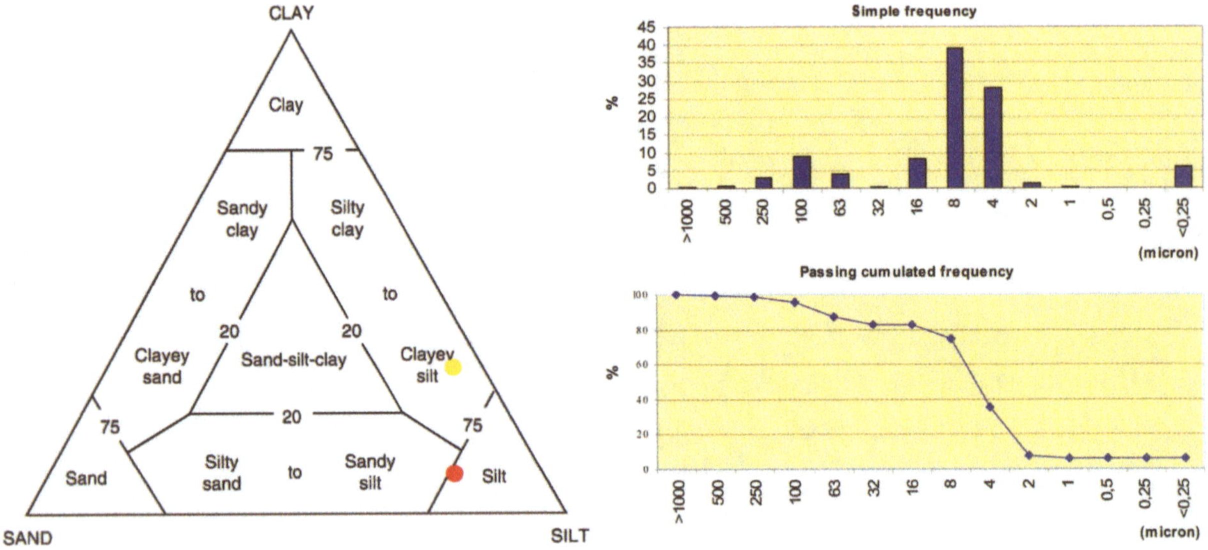

Fig. 25.9. Shepard (1954) diagram for siltstone grain size distribution (*red* is from sedigraph and *yellow* from hydrograph), classifying the material as "silt" (*red*) and "clayey silt" (*yellow*). Curves are from sedigraph

A major consideration from Fig. 25.10 is that the siltstone is almost in a state of critical natural density (maximum allowed); any breakdown and reconsolidation may occur only with a volume increase, as testified from SEM.

The uniaxial compressive strength, from field Schmidt hammer, highlight the general satisfactory condition of materials, at least in dry season (Fig. 25.11). With respect to standards, and in the area of both the Eastern Giant and Big Buddha, the average value of about 30 MPa for conglomerate and of about 34 MPa for sandstone classify these materials respectively as of moderate strength (ISRM 1981). It is interesting to notice that sandstone values are, roughly speaking, higher than for conglomerate, in the dry state. When weathered the situation is completely different but, likely, the presence of siltstone in the cliff is no more than 20–30%.

The results obtained in situ have been confirmed by point load tests of samples from Afghanistan. Both siltstone and conglomerate have been investigated. The results are plotted in Fig. 25.12. From this it is possible to notice a generally higher uniaxial compressive strength of sandstone, with respect to conglomerate. In general terms, the uniaxial compressive strength of siltstone (11.0 MPa) is about 30% of the one exhibited on in situ tests, while the strength of conglomerate in point load (5.6 MPa) is about 20% of the in situ Schmidt hammer tests. This difference is not surprising because of the limitations of both methods in the undirectional definition of uniaxial compressive strength, as well as the different confined conditions of in situ samples with respect to the cubic ones tested in the laboratory. Nevertheless, the field survey may give useful information about the distribution of mechanical parameters along the entire investigated cliff.

Lithotechnical classifications of materials (Barton Q-system, Beniawsky RMR, Geological Strength Index GSI) have been performed on the site. The results show mainly a different behavior between the intact material and the blasted ones. Furthermore, the fractured siltstone exhibits a general low performance, due to the presence of discontinuities spacing about 30 cm to 1 m. Table 25.1 reports the obtained values for RMR (basic value, Beniawsky 1989) and Q system (Barton et al. 1974).

In situ density for conglomerate has been recorded in Bamiyan, with a value of 17.65 kN m^{-3}, corresponding to an standard deviation of 0.03. This value is sensibly lower than the density for siltstone (22.35 kN m^{-3}), obtained in the laboratory.

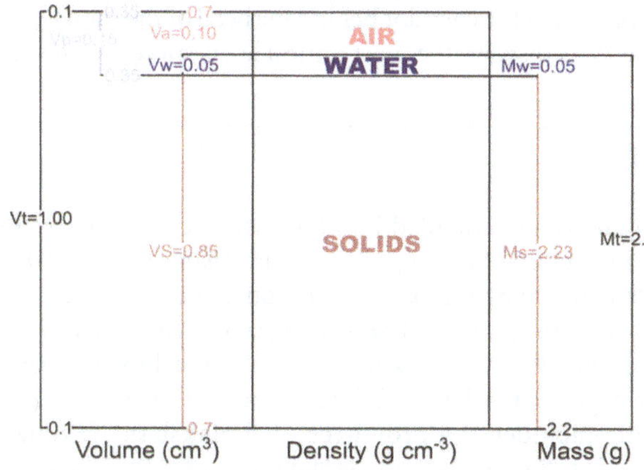

Fig. 25.10. Phase diagram for solids, water and air constituting the siltstone lithotype of Bamiyan

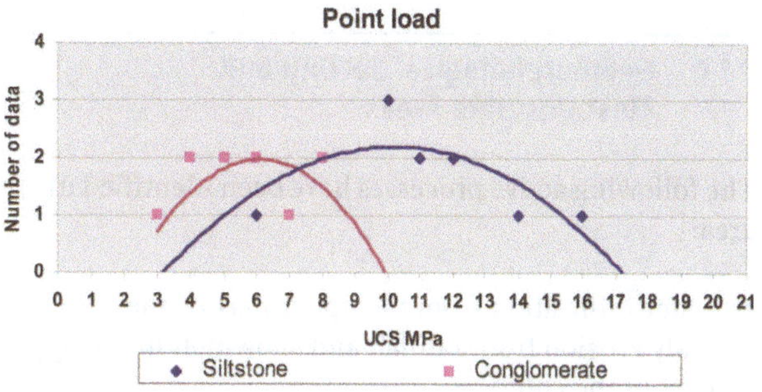

Fig. 25.12. Uniaxial compressive tests from point load in both conglomerate and siltstone

Fig. 25.11. Uniaxial compressive strength from Schmidt hammer field survey

Table 25.1.
Rock mechanic classification in terms of RMR (basic value) and Q system estimation

	Conglomerate				Siltstone			
	Undisturbed		Blasted		Undisturbed		Blasted	
	RMR	Q	RMR	Q	RMR	Q	RMR	Q
Eastern Giant Buddha	74	95	66	13.3	64	21.25	57	8.3
Western Giant Buddha	69	90	64	16	69	22.5		

Shear strength parameters of both rock formations has been considered to exhibit a Hoek and Brown (1980) behavior. On the basis of this standard assumption, c' and ϕ' have been determined simulating the behavior of a sample in a triaxial test, by means of an explicit difference finite code, FLAC (ITASCA Consulting Group 2000). The obtained results, in the Eastern Giant Buddha site, are a friction angle of 32° and cohesion of 1.42 MPa for sandstone, a friction angle of 52° and cohesion of 1.13 MPa for conglomerate.

According to AGI (1977) standards, a field permeability test has been conducted in the site, in order to acquire information on possible hydrogeological characters of rocks. The test, performed on conglomerate rock located not far from the Eastern Giant Buddha, exhibited a value of $K = 2 \times 10^{-4}$ m s^{-1}.

25.6 Geomorphological Setting and Most Unstable Areas

The following active processes have been identified in the area:

- water infiltration from the upper part of the cliff;
- gully erosion from rainfall and snowmelt in the upper part of the cliff;
- accumulation of debris materials at the toe;
- occurrence of mud flow, in the upper part, probably when the sandstone saturates;
- toppling of some isolated blocks;
- sliding in the large portion of the slope, mainly when discontinuities from the top are reaching the lower highly fractured sandstone formation;
- progressive opening of discontinuities in the more external part of the cliff;
- weathering of siltstone levels.

The detonations of March 2001, apart from causing the collapse of the statues, produced a deterioration of the stability conditions, mainly in the shallower part of the niches. In the Eastern Giant Buddha niche, apart from the collapsed statue, three minor rockfalls occurred in higher portions of the site. Also, the blasting produced a degradation of the back side of the niche's highest right part, where stairs are located inside the cliff and the section between the stairs and the niche is quite narrow (about 30–50 cm). This part is presently the most critical for future stability (A3 in Fig. 25.12, *right*). The left side, as a consequence of the existing buttress, did not suffer substantially. Only in the upper part a rockfall occurred and some instabilities are now evident.

In the Great Buddha, the major effect was the collapse of statue and the consequent instability of the back side of the niche. A small rockfall occurred on the top of the niche, to the left side. Probably, the large thickness of section between the stairs going up into the cliff and the niche (about 1 m), did not allow a large propagation of the effects of blasting. A large discontinuity, of about 20–30 cm, is present in the corridor behind the head of the statue. Figure 25.13 displays the most critical areas surveyed on site.

25.7 Seismological Setting

Afghanistan is divided into four seismic zones in terms of likelihood of damages (Amateur Seismic Centre 2003): Main, medium, minor, and safe zones (non-seismic). According to the Global Seismic Hazard Assessment Programme (Peizhen Zang et al. 2003), the northeastern parts of Afghanistan have the highest hazard in the country. This includes the provinces of Badakshan, Takkhar, Kondoz, Balkh, Jowzjan, Samangan, Sar-e-Pol, Baghlan, Parvan, Kabul, Kapisa, Konar, Laghman, Nangarhar, and Lowgar. Here, maximum peak ground acceleration (PGA) ranges from 0.24 g to in excess of 0.48 g in the easternmost regions.

The site of Bamiyan borders this area. Without direct information on historical earthquakes affecting the site, a possible estimation of maximum expected vibratory ground motion can be derived from Peizhen Zang et al. (2003) and according maps. Peak ground acceleration (PGA), given in units of m s^{-2}, with a 10% chance of exceedance in 50 years, for rock site classification ranges between 1.6 and 2.4 m s^{-2}. Historical data and earthquake lists for Afghanistan are reported in Ambraseys and Bilham (2003).

25.8 Geomechanic Characters of Discontinuities

Since discontinuities play a major role in rock stability they have been investigated in terms of roughness and

Fig. 25.13. Most unstable prone areas, according to the field survey, for Great (*left*) and Eastern Giant (*right*) Buddha niches. The A3 block in the Eastern Giant Buddha is certainly the most acute situation

friction angle. JRC obtained from about 30 field tests exhibit a value of about 2 for sandstone and a value of about 15 for conglomerate. Roughness profiles have been used to investigate the first- and second-order asperities (wavelength less than 50 cm and higher), in order to differentiate the contribution of asperities to the total friction angle ($\phi' = \phi_{basic} + i_{1st\text{-}2nd}$)

Reconstruction of the shear strength parameters of fractures (ϕ') was made by means of tiltmeter tests (Barton et al. 1985) on an average of about ten different measurements for any sample. The results for the site of the Eastern Giant Buddha are 42° (s.d. 6°) and 38° (s.d. 4°) for conglomerate, and 32° (s.d. 4°) for siltstone.

The tests have been performed in rough material, so the result should include partially the contribution of asperities. This is more true for conglomerate, where JRC is about 15 and less significant for the sandstone with a JRC of about 2. In fact, the result of the tilt test, for sandstone, is almost coincident with that from simulation of the triaxial test for the Hoek and Brown (1980) approach. The result of ϕ' for conglomerate is relatively smaller than that obtained in the simulation of triaxial tests, possibly demonstrating the importance of asperities in the total behavior of the rock, with respect to the tilt test that uses a small sample of material (about 20×20×20 cm).

As far as it concerns the Great Buddha the properties of discontinuities are not too much different from the site of the small one. Joint Roughness Coefficient (JRC), obtained as an average of eight different sites of sampling, exhibit a value of 16 for conglomerate and 4 for siltstone.

Tiltmeter tests (Barton et al. 1985), aimed at identifying the strength parameters of discontinuities ($\phi' = \phi_{basic} + (\phi' = \phi_{basic} + i_{1st\text{-}2nd}$), were developed forthe site of the Great Buddha, for both conglomerate and siltstone; sandstone has also been investigated for the different behavior of the material with asperity ($\phi' = \phi_{basic} + i_{1st\text{-}2nd}$)

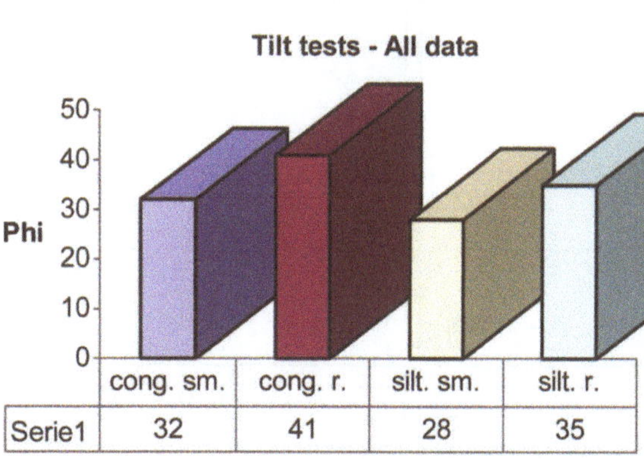

Fig. 25.14. Mean value of tilt tests for conglomerate (smoothed and rough) and siltstone (smoothed and rough)

and without (smoothed with carborundum stone – e.g., $\phi' = \phi_{basic}$). Conglomerate shows a value ranging from 38° to 41°, siltstone has a value of 28° for the smoothed sample and 38° for the case of natural conditions.

All tilt tests are grouped in Fig. 25.14, for conglomerate and siltstone, smoothed and rough (natural), to give a general idea of data distribution.

25.9 Structural Analysis of Discontinuities

Discontinuities play a major role in slope stability of rocks. The material of Bamiyan presents a natural distribution of discontinuities, probably as a consequences of unloading during the excavation of the valley from the river. More recent also the explosion produced some new fractures in the rocks, but mainly enlarged the pre-existing ones.

Distribution of discontinuities for the Eastern Giant Buddha and Western Giant Buddha sites are reported in Fig. 25.15. From this it is possible to notice only a major pattern in autochtonous materials, and some orthogonal

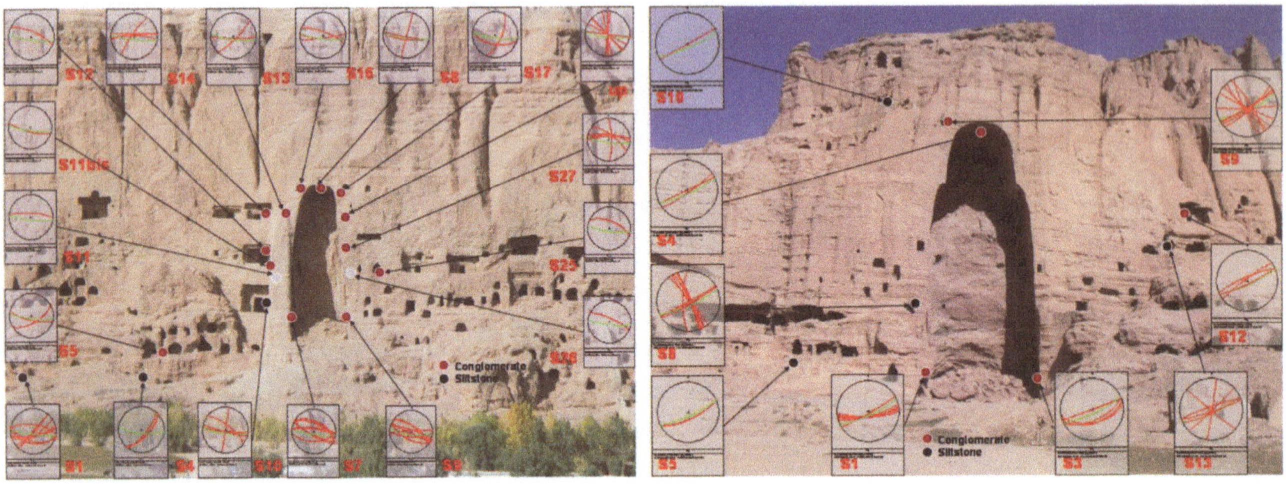

Fig. 25.15. Discontinuity trends for both Eastern and Western Giant Buddha sites

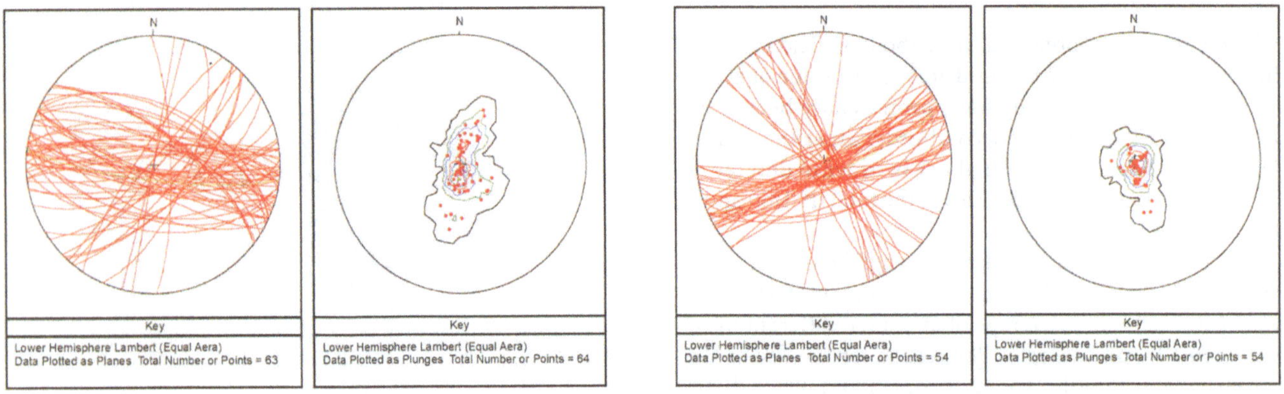

Fig. 25.16. Stereo-plot for kinematic analysis in the Great and Eastern Giant Buddha sites. *Green line* represent slope face

discontinuities. The lower sandstone formation, especially when belonging to the old sliding phenomena laying left of the Eastern Giant Buddha niche, show two major sets and some random pattern. Only in the upper part, on the right side of the Eastern Giant Buddha niche, there is clear evidence of discontinuities generated by the explosion.

In the Western Giant Buddha a major pattern exhibits two sets which are only evident in the cave top of the niche, suggesting a potential long-term instability in this area.

25.10 Kinematic Analysis

The possible collapse mechanisms have been investigated through comparison of slope morphology and discontinuity patterns (wedge failure, planar sliding, etc.). The data are reported in the following figures. From Fig. 25.16 it is possible to notice the near coincidence of slope face and major discontinuity patterns, not satisfying the condition for planar sliding or wedge failure (Turner and Schuster 1998). It has also to be considered that some discontinuities plotted in the figure are representing the same fracture, detected in different vertical levels in the field survey (see Fig. 25.15 for localization of measurement sites).

25.11 Stability Analysis

Investigations of the possible stability conditions of the cliff have been carried out by computing the explicit-difference-finite code, FLAC (ITASCA Consulting Group 2000). For a given element shape function, the set of algebraic equations solved by FLAC is identical to that solved with finite element mode. However, in FLAC, this set of equations is solved using dynamic relaxation, an explicit time-marching procedure, in which the full dynamic equations of motion are integrated step by step. Static solutions are obtained by including dumping terms that gradually remove kinetic energy from the system (Dawson et. al. 1999).

In the area of Bamiyan the following has has been investigated:

- the behavior of present-day situation, according to Hoek and Brown (1980) theory on shear strength of rocks,

- the behavior of modified conditions, to investigate the possible consequences of a deepening of discontinuities today, present at the top of the niche,
- the possible contribution of a lower cohesion of sandstone settled in the lower part of the cliff. The mesh density of the grid has been computed to allow the best resolution in the area of possible movements (Fig. 25.17).

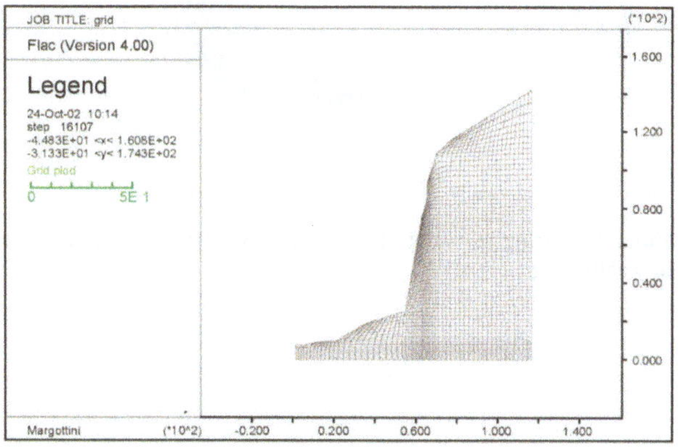

Fig. 25.17. The adopted mesh for the Eastern Giant Buddha stability analysis

The results, for the Eastern Giant Buddha, can be summarized as follows:

- According the Hoek and Brown (1980) shear strength parameters of conglomerate and sandstone, and a major discontinuity ranging from the middle of the cliff to the middle of the niche (only friction value for shear strength), the deformation of the cliff is relatively low and it seems to be nowadays in a condition of stability (Fig. 25.18);
- since we modify the shear strength parameters to the value obtained in the field, we decrease the cohesion of sandstone at 25% due to the level of fracturing, and we consider the fracture in conglomerate reaching the top of sandstone, the cliff is still stable, in similar static conditions as above, with Factor of Safety = 2.04 (Fig. 25.19);
- if we consider now the cohesion of lower sandstone equal to 0 (as probably happened in the past as testified by the two slidings respectively on the left of the Eastern Giant Bhudda and on the right of the Great Buddha, and as possible to happen if a weathering process occurs), the cliff is unstable, with maximum displacement and vector, at the base of the niche (Fig. 25.20).

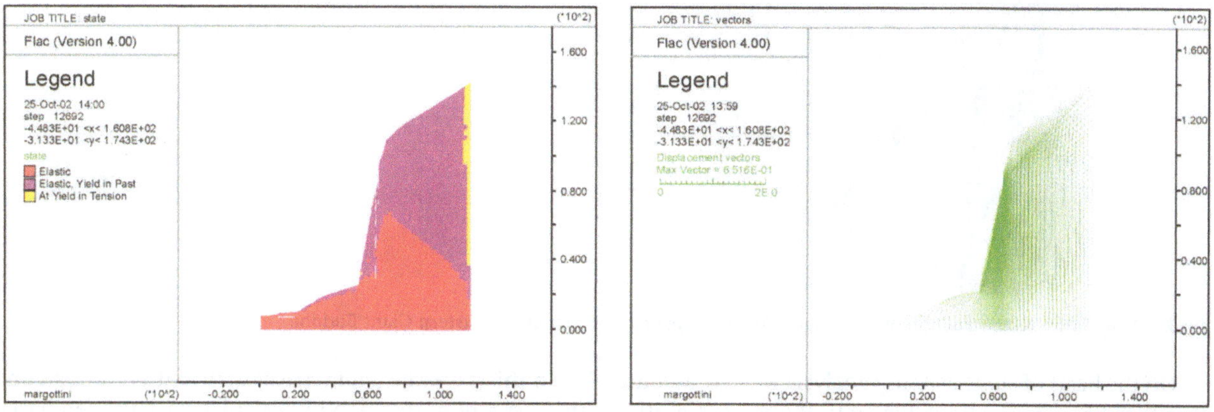

Fig. 25.18. State and displacement for the cliff of the Eastern Giant Buddha under the present conditions

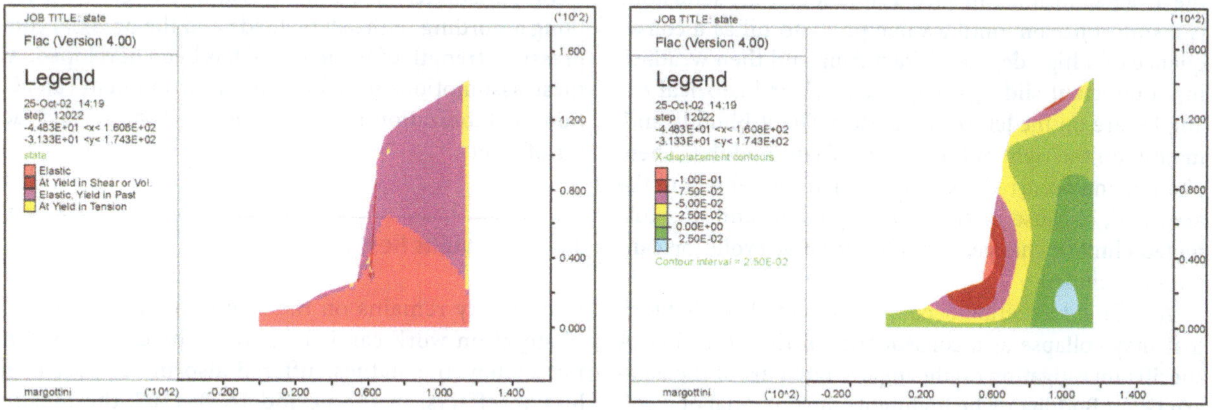

Fig. 25.19. State and displacement for the cliff of the Eastern Giant Buddha with fracturing from top reaching the base of the cliff and 25% reduction of cohesion in the lower siltstone

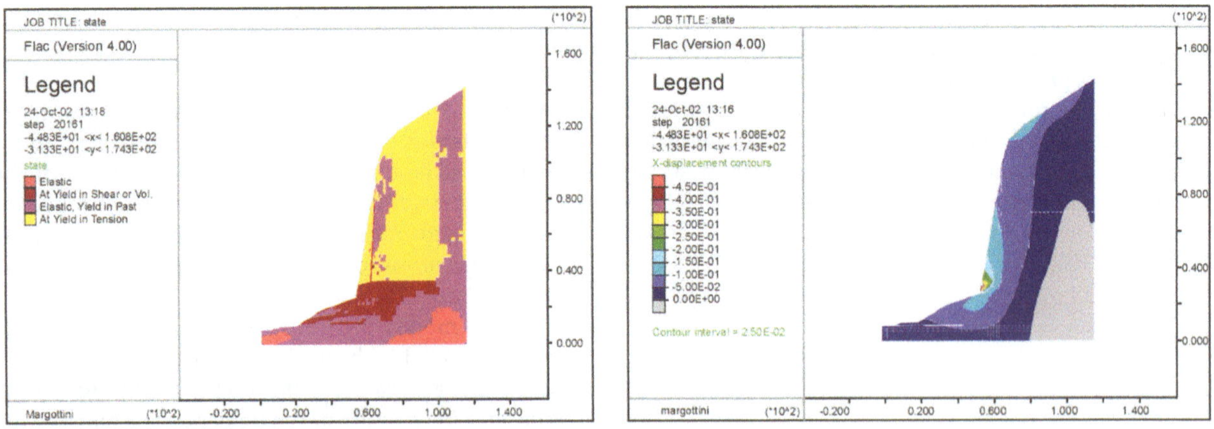

Fig. 25.20. State and displacement for the cliff of the Eastern Giant Buddha, fracturing reaching the lower part of the cliff and the lower siltstone exhibits no cohesion as a consequence of internal fracturing or weathering

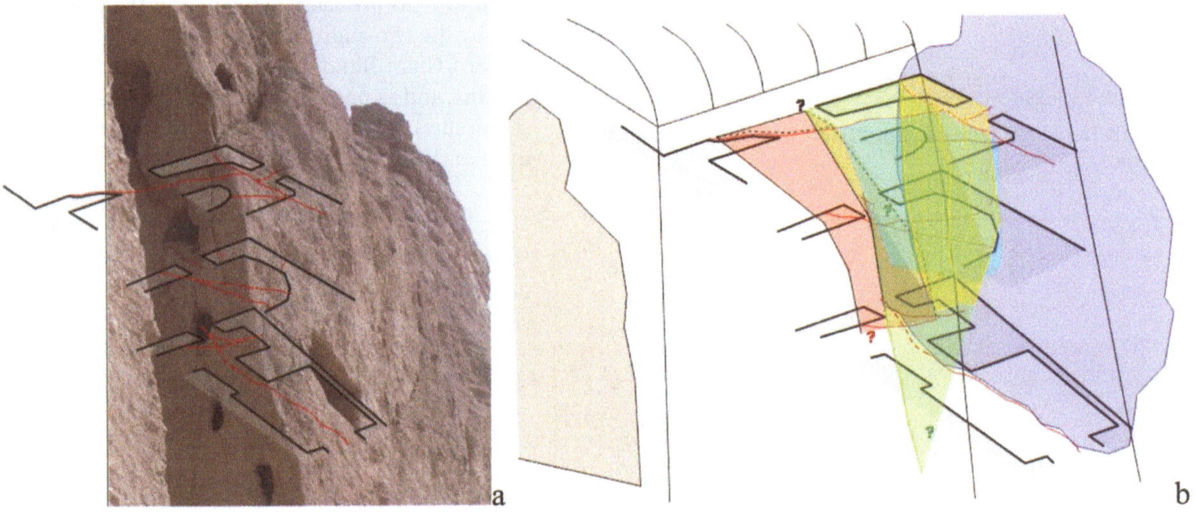

Fig. 25.21. Tentative reconstruction of individual blocks in upper-right part of the Eastern Giant Buddha

On the basis of slope stability analysis we may assume that the cliff is generally stable, until the discontinuities of conglomerate from the top of the cliff reach the top of the lower sandstone. In such a situation, if the sandstone is losing cohesion, until a value proxy to nil as a consequence of a high degree of fracturing and then weathering, important slidings may occur; clear historical examples are on the left of the Eastern Giant Bhudda and, mainly, on the right of the Western Giant Buddha. These phenomena occurred also after human settlement in the area and, because of the 2001 explosion and possible meteo-climatic change, a possible negative evolution cannot be excluded.

A different situation occurs for the small single block that may collapse as a consequence of the explosion. A specific investigation on the upper-right part of the Eastern Giant Buddha niche highlights the potential of a collapse even in short time. Five blocks (Fig. 25.21) have been detected, gaping up to 20–30 cm in the upper part.

In this case, the geometrical distribution of load and the assumption for stability calculation, has been based on mechanics of the rigid mass, and the related moment; a comparison test about the possibility to generate toppling according to the static loads and the uniaxial compressive strength of the material has been developed. All these assumptions have been used for moment calculation of stabilization measurements and the related factor of safety.

25.12 Previous Restoration Work

Uncertainty remains on the sequence of damage and restoration work carried out over the centuries. Unfortunately, the statues suffered also in the past from human effects: in the period 1678–1707, the muslim ruler, Aurangzed, shot with a big gun against the Buddhas.

In recent time some restorative work on the statue had been carried out; reconstructing the nature of these investigations may help in understanding the effects of future restoration work on the local materials and long-term effects. However, these investigation sofar have only been based on surface remains after a visual inspection and comparison with some old photography (some of the photos are from the private collection of Prof. Kurt Lambek, University of Camberra, who visited the Bamyian Valley in August 1967); this approach is certainly not exhaustive and does not cover all the interventions carried out over recent decades. It only attempts to describe the typology of previous work and presently still noticeable effects:

1. Filling of fractures with concrete: a similar intervention was conducted, probably in the 1960s, in the Western Giant Buddha, as testified from K. Lambek's photos (courtesy offered for this work), in August 1967. Nowadays there is just little evidence of these filling activities, suggesting a possible effect of the explosion or, an enlargement of fractures. This second hypothesis seems to be unconfirmed by the evolution pattern of discontinuities.
2. Reconstruction of missing parts: photos from 1967 as well as current ones reveal that a small portion of the border niche of the Western Giant Buddha has been reconstructed, in a perfect manner. Also the contact between the large slide in front of the Western Giant Buddha and the cliff has been filled with concrete. Similar intervention of reconstruction of weak or missing parts also has been done for the Eastern Giant Buddha, with excellent results.
3. Protection from water runoff. This measure has been considered extremely important in the past, as it really is, as testified by the Indian Archaeological Survey's construction of two major protection channels above the two Buddhas. Their perfect maintainance and conservation it is quite essential.
4. Bolts have been used in the past to anchor the statue of the Western Giant Buddha to the cliff. The approach was correct and the result fully satisfactory, until the blast occurred.
5. A buttress was carried out on the left side of the Eastern Giant Buddha, to sustain the cliff probably at the limit equilibrium. The buttress, even if questionable for the modification of the landscape, was perfectly covered by cement producing a camouflage not easy to recognize at first sight. On the other hand, this side of the niche did not suffer too much as a consequence of the explosion of March 2001, confirming the efficacy of this intervention.

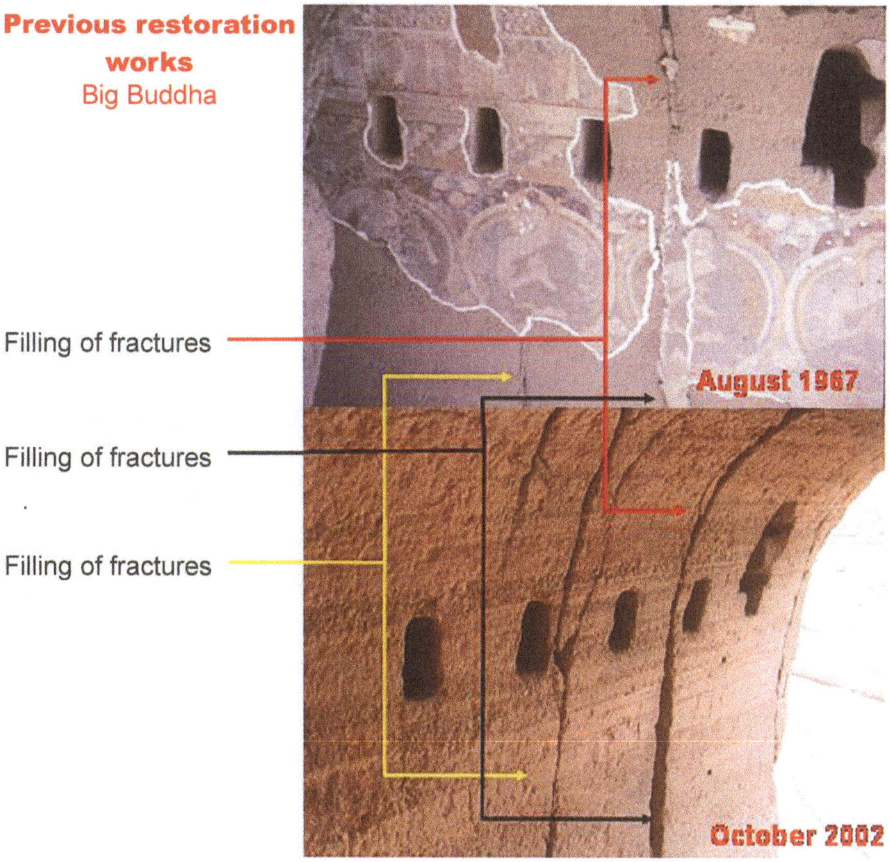

Fig. 25.22.
Filling of fracture with cement in the Western Giant Buddha

Fig. 25.23. Reshaping and consolidation with cement in both Western and Eastern Giant Buddha

Fig. 25.25. Use of bolt for anchoring the statue of the Western Giant Buddha

Fig. 25.26. Construction of a buttress for the protection of the left side of the Eastern Giant Buddha niche

Fig. 25.24.
Protection from water runoff and infiltration

25.13 Long-term Conservation Strategy for Repair, Enhancement, Research, and Risk-preparedness for the Preservation of the Site

Considering the above processes, any stabilization strategy has to solve the causes of active processes and not to focus only on mitigating the effects. Different typology of stabilization measures have been investigated, considering also the need of implementing the work in an area of high cultural value but low, at the moment, technological support. Also the consolidation works should not be restricted to the statues themselves but need to consider the entire area, especially for those processes that are affecting the entire cliff (e.g., rainwater drainage).

The final proposed solution includes the following typologies of stabilization techniques for both the Eastern and Western Giant Buddha statue:

- the protection from water circulation and infiltration from the upper part of the cliff (Fig. 25.27); a preliminary work has considered the restoration of the existing channel constructed by the Indian Archaeological Survey in the 1970s, but a complete revision of the system has to be done. In fact, the Indian channel adequately collects the water in the upper part of the cliff but, in some cases, releases the water directly onto the cliff inducing a considerable gully erosion (e.g., close to the large Buddha); possibly, a more extensive channel has to be considered, in order to contribute also to the preservation of the most important cave areas. This work consists mainly in the canalization for the drainage of rainwater to avoid water infiltration, gully erosion and debris deposition hiding part of the monumental complex (Bruno 1996). Finally, also cracks have to be grouted and filled, to avoid water infiltration from the surface through open discontinuities;
- stabilization of cliff and potential rockfalls by means of passive anchors (stainless steel) and bolts, grouted with cement with low water release (Fig. 25.28). This intervention can be restricted to the Buddhas site and surroundings and has to be planned directly on site, in order to define exactly where to place individual anchors and bolts. Strength parameters for anchors have been calculated by Colombini and Margottini (2003) and tested on-site with pulling out tests on two sample bars (see later in this report). The field tests for both conglomerate and siltstone shows an anchor/ground adherence larger than 50 t ml^{-1}. Composition of grouting was defined to avoid slaking, namely adding superplasticizer to water and cement mortar.
- on the back side of both Buddha niches a very fragile area exists. This is the place where original plaster still exists. These areas should be safeguarded by a restorer as soon as possible. Removal of as few as possible unstable blocks of rock (barring down) using hand tools, and installation of a net is required:
  - to allow the protection and restoration of the few remains of the Buddha Statues (some parts are still in the cave and it would be advisable to preserve these as much as possible). As said before, further restoration must be conducted jointly by experts of art conservation, earth science, and rock mechanics;

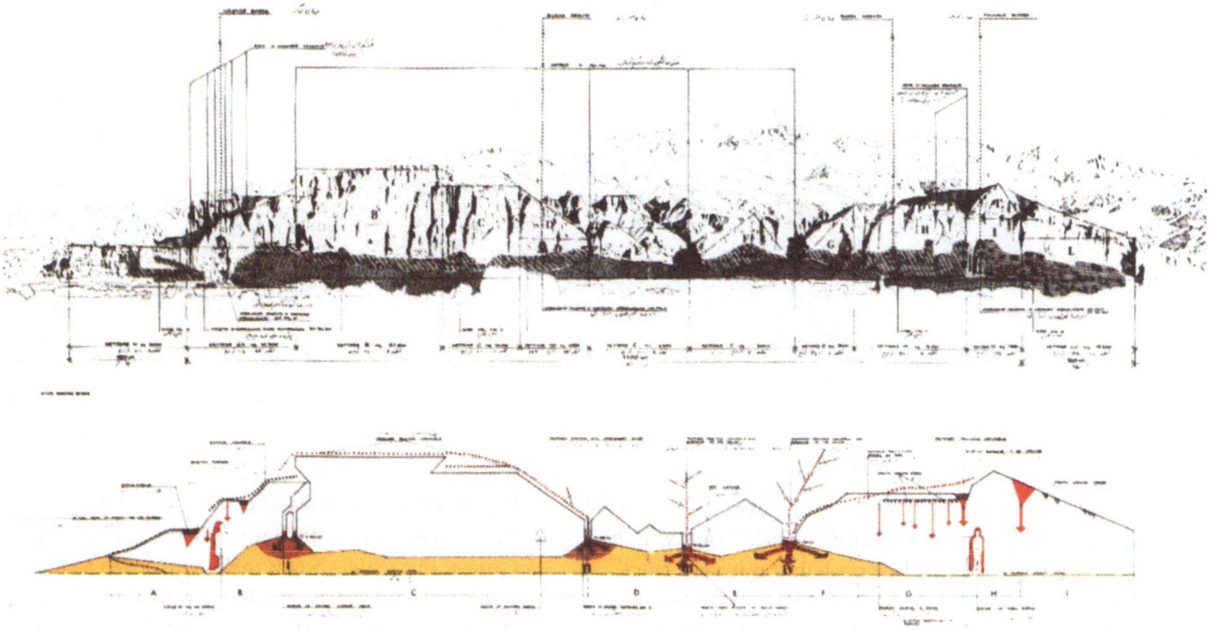

Fig. 25.27. Proposed canalization work form Bruno (1996)

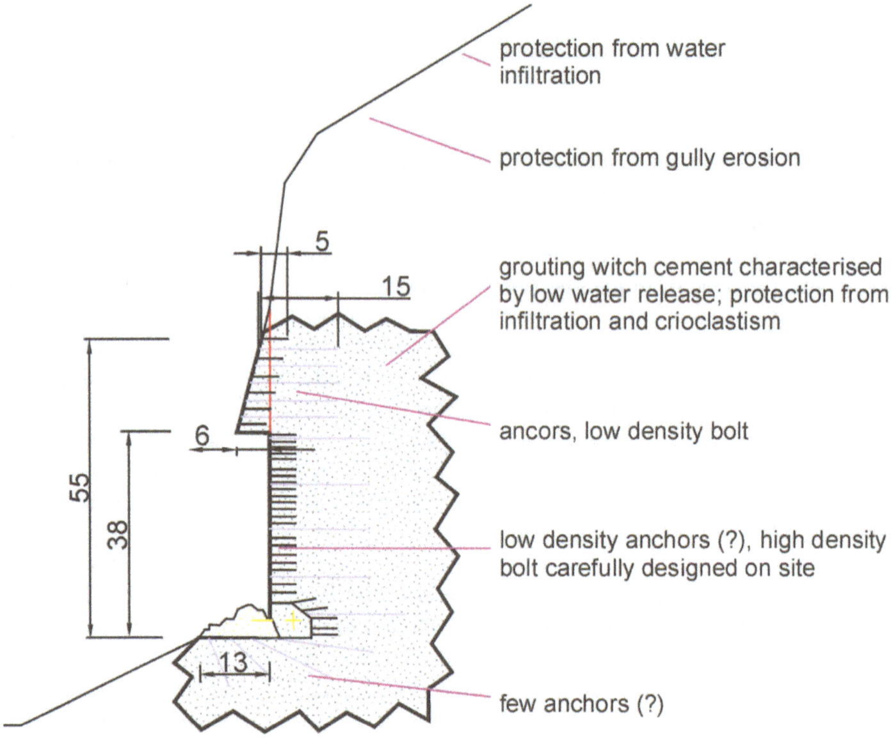

Fig. 25.28.
Identification of stabilization techniques in different areas of the cliff and niche. Anchors and bolt can regularly foreseen in the cliff, outside of the niche, while mainly bolts and a low number of anchors can be proposed inside the niches

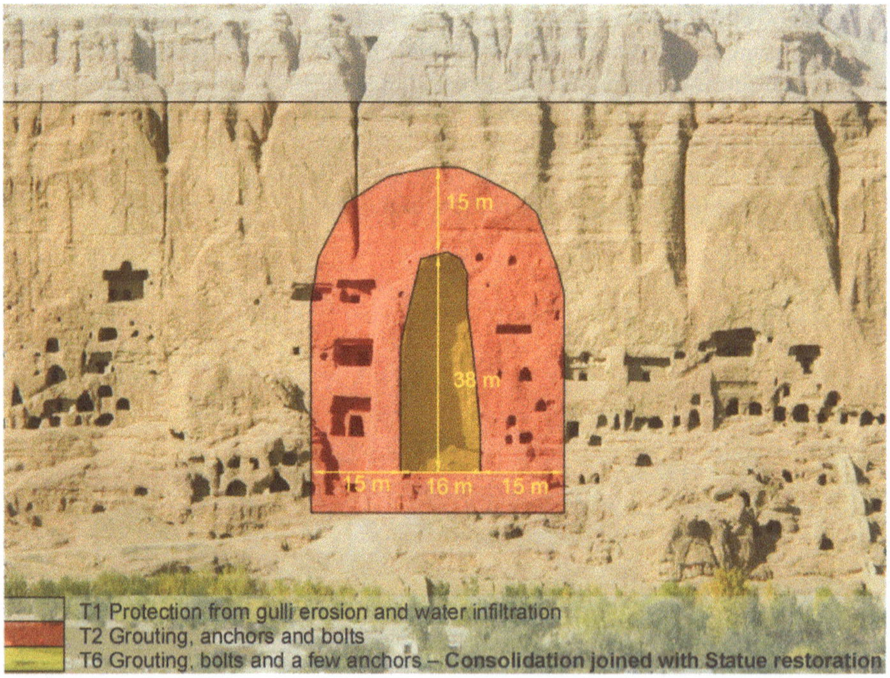

Fig. 25.29.
Identification of major areas of intervention and related typology for the Eastern Giant Buddha

- to allow archaeologist to begin recovering and restoring the fragments; this is in the area of the collapse of the Buddha and within a shear zone, which is extremely unstable but probably not very thick. The work has to be performed by professional climbers (scalers) who will remove the unstable hanging blocks and install the net; in case of need, instead of removing, the scalers will stabilize some bigger blocks by means of anchors/bolts in boreholes executed with hand-portable drilling machines; the scalers work their way down to the face to ensure that there is no loose rock above them. In these areas any solution must be jointly decided by earth science experts, archaeologists, and restores.

The proposed solution considers, in general, the use of nails, anchors, and grouting (also for reconstructing some missing parts of windows). Such measures would be intended for immediate intervention at the most critical rockfall areas. Decisions on the specific locations for such interventions, the direction of drillings and depth, grouting, and typology of injection materials, would need to be made directly on the spot. For this reason a very high profile, in terms of competence and expertise in managing field works is required.

Parallel to the identification of the main typology of stabilization (see also Fig. 25.29), the recognition of the exact location of the most critical rockfall-prone areas was developed (Figs. 25.30 and 25.31). The survey was performed in September 2002 compared with ground deformation and movements inspected in September 2003 and finally calibrated once more in March 2004; the survey of 2004 was required since the stabilization of some very critical sites on which inspection was not possible

Fig. 25.30. Identification of most critical rockfall-prone areas in the Western Giant Buddha niche

Fig. 25.31. Identification of most critical rockfall-prone areas in the Eastern Giant Buddha niche

before, due to the high risk of rockfall. These sites certainly are the ones that most urgently need to be looked at and consolidated before any further intervention.

25.14 Identification of Most Unstable Areas

The detonation of March 2001, as well as demolishing of the statues, reduced the stability of the slope, mainly in the outer parts of the niches.

In the Eastern Giant Buddha niche, through the collapse of the statue, there were three minor rockfalls from the top of the niche. The blasting also degraded the upper-eastern part of the niche where a stairway is located inside the cliff, and the wall between the stairs and the niche is quite thin (about 30–50 cm). This part is presently the most critically unstable site. The western side, as a consequence of an existing buttress, suffered less damage. Nevertheless, a rock fall occurred and some instabilities are now also evident only in the eastern part.

Major effects in the Western Giant Buddha niche were the collapse of the statue and the consequent instability of the rear of the niche. A small rockfall occurred from the left side of the top of the niche. Probably, the strength of the greater thickness of the wall between the stairway going up into the cliff and the niche (about 1 m), reduced the effects of the blasting and resulted in less severe damage. The possible stability conditions of the cliff were computed using the explicit-difference-finite code, FLAC (Itasca Consulting Group 2000). Considering the Hoek and Brown (1980) shear strength criteria for conglomerate and siltstone, and with a major discontinuity ranging from the middle of cliff to the middle of the niche (only friction value for shear strength) the deformation of the cliff is relatively low and it seems presently to be in a condition of stability. Since we consider the fracture in conglomerate reaching the lower sandstone formation and we decrease gradually the cohesion of siltstone due to fracturing/weathering, the cliff has become unstable if cohesion is proxy to nil. In this situation maximum displacement and vector are at the base of the niche (Fig. 25.20).

In general, the niche and the cliff need holistic stabilization work and not episodic and local intervention. Nevertheless, it must be recognized that one cannot propose a specific stabilization plan at the moment because any intervention has to be specified for the local conditions. At the present stage, it is convenient to set up a general master plan, to be locally adapted according to further specific investigations and data. The master plan includes mainly nails, anchors and grouting, that will have a low environmental impact on the site.

Finally, the field data (Colombini and Margottini 2003a; Margottini 2004), kinematic analysis, mathematical modeling, caves and crack distribution, and detail inspection of the effect of the explosion has led to the results of Figs. 25.30 and 25.31 which report the most endangered sites for both niches.

25.15 Emergency Measures in the Upper Eastern Part of the Eastern Giant Buddha Niche

With the general strategy for stabilization in place, a follow up of activities was performed in September 2003, aimed at the identification of potential negative evolution of the cliff and niches during winter 2003–2004. The

Fig. 25.32. Unstable blocks in the upper East side of Eastern Giant Buddha niche and pattern of existing discontinuities

result of a field mission suggested an immediate response to the upper East side of the Eastern Giant Buddha niche where the existing large fissures were widening and the risk of an imminent rockfall was estimated to be very high. This collapse could involve large parts of the upper Eastern section of the cliff and then totally destroy the niche (Fig. 25.32).

Emergency consolidation work was instantly planned and carried out in this most critical rockfall-prone area to avoid any further collapse in the coming winter season, but also to enable archaeologists the safe cataloguing and recovering of the Buddha statues' remains, still laying on the floor of the niches. The stabilization activities started in October 2003 and included four different steps:

1. The installation of a monitoring system, to evaluate in real-time any possible deformation of the cliff. 11 sensors were designed to monitor the entire working area, connected with an alarm system, to provide safe conditions for workers;
2. The realization of temporary protection. These include two temporary beams to laterally support the cliff deformation, and steel rope at 0.6 inches diameter, with light pre-tensioning (Fig. 25.33). Each beam was calculated to offer a resistance of about 40 tons, similar to two designed long anchors. Among the temporary work, a wire net was installed on the back side of both niches to allow archaeologists to safely work on the ground floor, just after the consolidation of the niche's wall;
3. The final stabilization (Fig. 25.34) is regarding the corner of the niche were the situation is most dangerous. In this area the following has been implemented:
   - six short passive anchors (steel nails, dia. 16 mm, FeB 44K, threaded, with couplers anchor plates and nut – single bar length $L = 2.5$ m) with diameter = 36 mm and length about 5 m, placed on the internal side of the niche (diamond head rotary machine);
   - 29 stainless steel passive anchors with diameter = 26 mm and a length 5–10 m (in any case double the distance of the last encountered fissure from the surface), 20 on the internal side and 9 on the external;
   - 17 passive anchors, pre-grouted to avoid oxidation, with diameter = 90 mm and length 15 m, for a total length of 200 m, placed on the external part of the cliff (anchor bars VSL, dia. 26.5 mm, st 835/1030, preinjected, with external corrugated sheating, including plates and nuts).

A detailed map of these interventions is given in Fig. 25.9 (courtesy of TREVI div. RODIO), with the exact typology, position, and length of executed anchors.

Fig. 25.33. The temporary beam (*left*) and steel rope (*right*) for the temporary support of upper Eastern part of Eastern Giant Buddha

Fig. 25.34.
Typology, position and length of executed anchors (*red* is for passive anchors, pre-grouted, with diameter = 26 mm and length 15 m; *violet* is for stainless steel passive anchors with diameter = 26 mm and a length 5–10 m; *green* is for short passive anchors with diameter = 16 mm and length about 5 m, placed on the internal side of the niche) (courtesy of TREVI division RODIO)

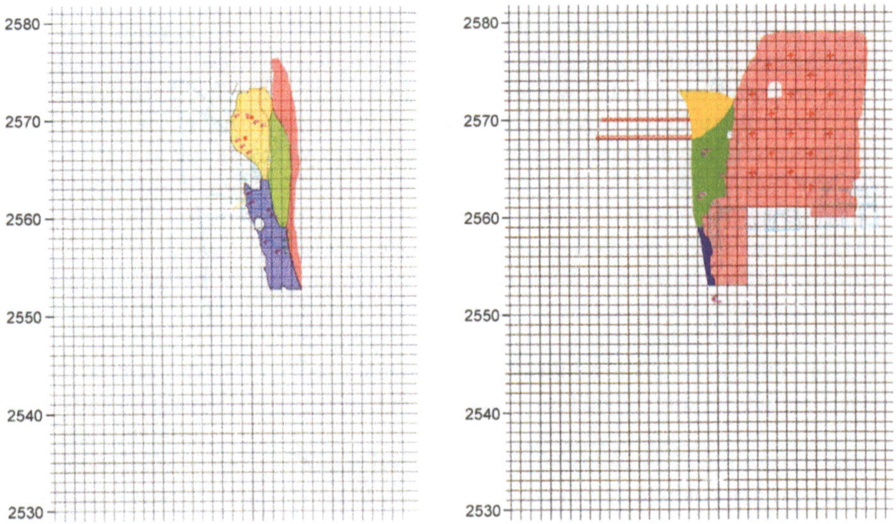

Fig. 25.35. Final consolidation with the use of professional climbers

4. Minimization of intervention (anchor/nail head finishing) has been performed in spring 2004. A number of tests for a better mixture, between cement, local clay/silt, and water also have been carried out in 2003, in cooperation with ICOMOS experts. The results prove the better chromatic stability and robustness of the mixture.

A major concern was the choice of the typology of the approach for executing the work. Certainly, the first idea was the construction of a scaffolding but, due to the very high probability of rockfall and concomittant destruction of the scaffolding with additional risk for the workers staying below hanging rocks, because of the economic cost of the scaffolding itself, and for the reason of the approaching winter season 2003–2004, speediness was given priority and the use of professional climbers was decided to be the way of implementing the job. Climbers, also supported by ground staff, were operating directly on the surface, hanging from the top of the cliff, in a safe area, moving from top to bottom and then in safe conditions with respect to any potential rockfall (Fig. 25.35).

A major difficulty was also the calibration of drillings with respect to existing cavities. In fact a large number of caves (around 800) and tunnels exist on the cliff constituting a unique example of rupestrial settlement. The selection of drilling was then required a detailed investigation in their orientation and inclination to avoid drilling and grouting into the archaeological caves.

As mentioned previously, major attention was devoted to the methodology of consolidation. Short (16 mm) and medium (20 mm) length passive anchors (stainless steel) were installed using rotary drilling machines, with diamond head, to avoid any possible vibration. Cooling fluid led to the occurrence of slaking-prone siltstone in the presence of water: due to this, the usage of water was limited when drilling the conglomerate and a mix of compressed air and water was adopted when discontinuities were detected and when a possible level of siltstone was encountered. Pre-grouted long passive anchors (26 mm) were mounted with roto-percussion machines and use of air as flushing medium. They were drilled only after installing shorter ones and from the further part of the unstable blocks, towards the most critical ones. The purpose of small anchors is to mold together the unstable masses and fix them to the proxy stable geological background. The long anchors are intended to homogenize this part to the most internal and stable geological material. Direction and inclination of anchors has been defined on site but, in any case, direction of deformation and perpendicularity to discontinuities has been taken into consideration. Temporal execution considered the principle to start from the most stable place to the most unstable. This is to commence consolidation from the part where disturbance can better be sustained. In particular, with respect to the internal side of the niche, the lower unstable block must be approached from the bottom to the top. In fact, at the top there is an almost collapsed block that cannot be touched without prior stabilization of the lower part. Grouting was made with cement added with superplasticizer to avoid any water release, capable to interfere with the slaking siltstone as well as to get the best possible adherence between bar and rock, namely composed as in Table 25.2.

Table 25.2. Composition of grouting materials for both anchoring and cracks filling

	Anchoring grout (Kg m^{-3})	Cracks filling mortar (Kg m^{-3})
Water	540	300
Cement	1 360	610
Sand		1 270
Additive superplasticizer	7	7

In total, for grouting and filling, approx. 12 m^3 were injected in the first phase (2003), with 14 000 kg of cement divided into:

1. short anchors grouting: 1 200 kg;
2. anchors grouting: 6 600 kg;
3. crack filling (from top): 6 200 kg.

Proof of the correct implementation of grouting is given from the anchor suitability test, performed to understand the bounding capacity of anchors in both siltstone and conglomerate. The design strength of passive anchors was assumed as 20 tons, for a bounded length, after the major discontinuity, at least of 5 meters (about 4 tons per linear meter); the anchor suitability test was performed for 1 meter length, up to 40 tons, close to the yield capacity of steel. Up to this value no remarkable permanent elongation was detected for the correct bounding effect between siltstone and conglomerate and the anchors (Fig. 25.36).

The solution and the techniques adopted as well as the four steps of improvement activities proved quite satisfactory since the monitoring system did not record any noticeable deformation in the unstable blocks through the working period (Fig. 25.37).

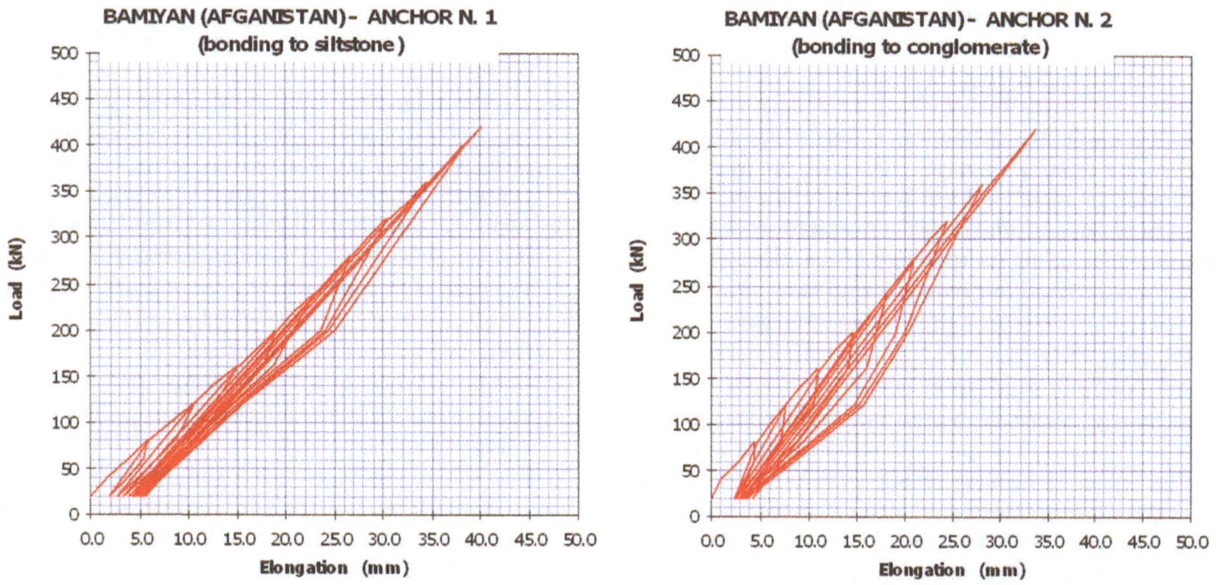

Fig. 25.36. Anchor suitability tests for siltstone and conglomerate, in 1-m-length anchor. The load (kN) and respective elongation (mm) are reported showing, up to 40 tons, the uphold of elastic domain and the missing of any permanent deformation for the tested anchor

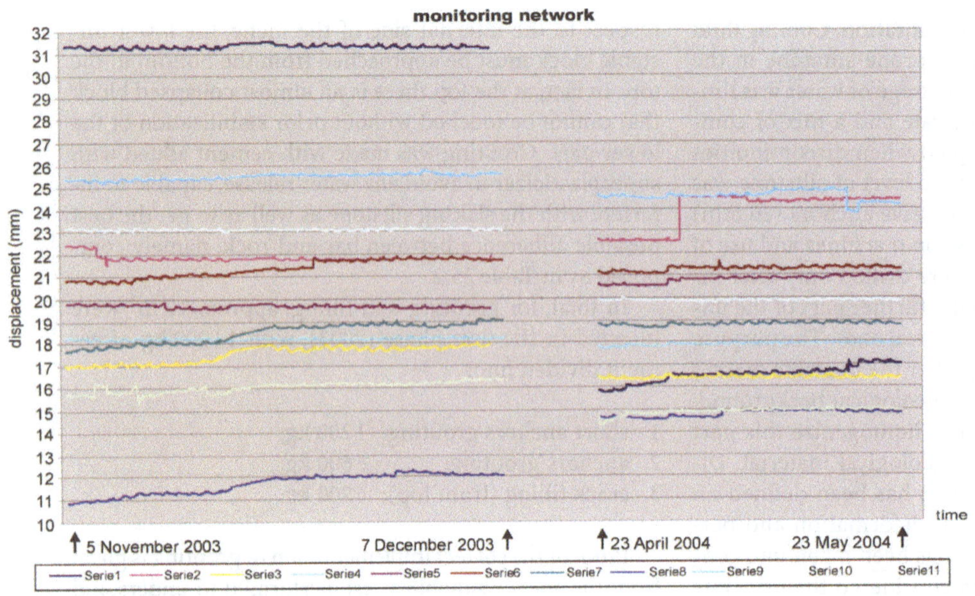

Fig. 25.37. Time evolution of the 11 extensometers operating in the period 5 November–7 December 2003 and 23 April–23 May 2004, and showing no remarkable movements on the cliff. Some minor steps have been caused by workers on the cliff

25.16 Completion of Emergency Intervention in the Eastern Giant Buddha Niche

The western side of the niche is suffering from the effect of the detonation as well as from the sunk existing buttress. The buttress was probably constructed to reduce the risk of collapse of this flank which was considered extremely unstable justifying a very massive intervention by a French archaeological expedition in the late 1950s and early 1960s, which was then finally strengthened and mitigated in impact by the Indian Archaeological Survey in the 1970s. Since the buttress is connected with bolts to the cliff it is possible that its sinking may induce horizontal stress toward the outside, inducing additional instability as testified by the intervention of the Indian Archaeological Survey (Fig. 25.38, courtesy of Prof. Maeda). It is possible today that the buttress is hanging towords the cliff, more than sustaining it. This situation may increase the existing damage. Effects of the March 2001 explosion are mainly evident at the top of the niche, probably in an area of maximum concentration of stress in consequence of the morphology of the niche (arch and pillar, as described in Colombini and Margottini 2003a). In particular, there is a small pillar (Margottini 2004b, 2006) (Fig. 25.39) that needs immediate emergency intervention before it collapses, which in consequence could induce large deformation processes to the entire western part of the niche. This part was also completely restored by the Indian Archaeological Survey in late 1970s. In this area any intervention should include geotechnical investigation on present buttress foundation and, later on, the complete stabilization of the niche. Likely, the manual monitoring system installed in 2003, presently does not exhibit any further deformation of the most severe cracks.

An emergency intervention can be planned in the upper part of the niche, beware of not fixing the buttress to the cliff (e.g., extensive grouting) since its possible evolution has not yet been investigated.

Also, in this situation the general strategy of an emergency intervention has to be developed in four steps:

1. A monitoring system on the most relevant discontinuities.
2. A temporary support (Fig. 25.39), by means of:
   a the two existing iron beams in the upper part of the niche, to provide some lateral support to the niche;
   b two iron/wood beams capable to support any lateral deformation of the small pillar;
   c steel ropes bounding completely the pillar and cliff.
3. Emergency intervention including nails and grouting as follow:
   a a set of small nails (phi 16–20 mm) to connect internally the pillar, in both directions parallel to the face and perpendicular to it to create a robust net (number of nails can vary from 7 to 10);
   b two short passive nails (phi 20–26 mm) located below and above the critical pillar, with depth less than the back-side plane where a large crack has been detected after the removal of fragments in the lower caves behind the feet of the statue;
   c one passive anchor (phi 20–26 mm) parallel to the surface aimed at stabilizing the upper gallery where a large fissure is present;
   d low water release grouting should maintain the composition of water and cement successfully adopted on the eastern side which includes: W/C = 1/2.0 + superplasticizer;
4. Minimization of impact will follow the same criteria established for the eastern side.

Fig. 25.38.
Consolidation works from the Archaeological Survey of India (Courtesy of Prof. Maeda)

Fig. 25.39.
The most unstable element (pillar) in the western side of the Eastern Giant Buddha niche, with the firsts temporary protections (iron beams, iron/wood beams, steel ropes)

Figure 25.40 shows the distribution of the anchors as well as their chronological sequence. This is quite important to avoid disturbance to the most critical part of the cliff without having stabilized the boundary conditions.

Finally, on the external top of the niche five passive anchors, each 15 meters deep, were installed (drill diameter = 90 mm; anchor bars VSL, diameter 26.5 mm, st 835/1030, pre-injected, with external corrugated sheating, including plates and nuts), to provide an additional strength, and also, with the grouting, to provide a further protection against water infiltration potentially slaking the siltstone as well as generating additional pore pressure.

25.17 First Interventions in the Western Giant Buddha Niche

Despite the destruction of the statue, the Western Giant Buddha niche did not suffer extensively as a consequence of the explosion (Margottini 2006). First emergency interventions include (Fig. 25.41):

a grouting of the large fissure placed in the corridor, at the back side of the niche. This operation can be done from inside the niche as well as from the outside (top of the cliff). In the internal part of the niche, the fis-

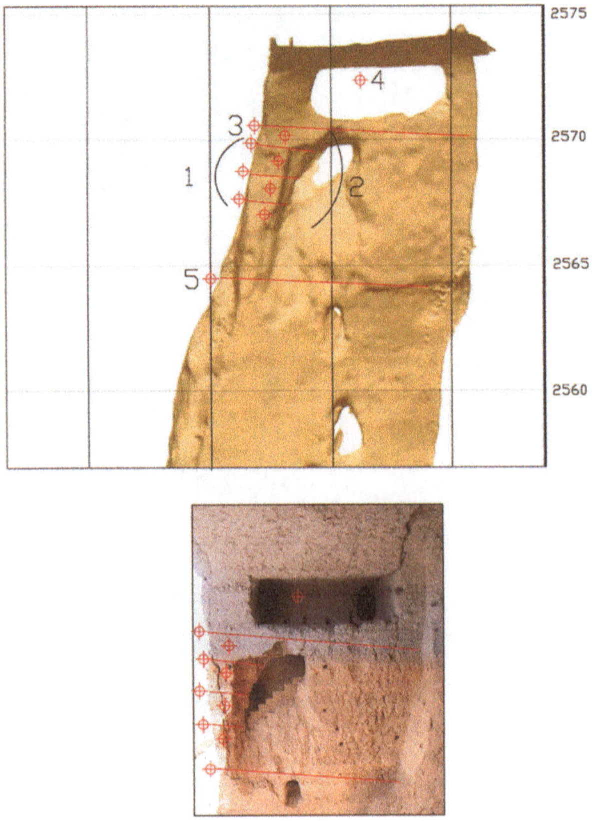

Fig. 25.40. Distribution of anchors and temporal insertion of them (topographic data from PASCO 2003)

sure will be grouted and closed to keep away from cement infiltration from the top. Small pipes are requiring inside the cement to avoid internal overpressure. From the top of the cliff inclined drills will be performed and, when the fissure is encountered in the perforation, grouted with the same mixture of cement and superplasticizer described in Margottini (2003). The area of grouting should considers some tens of meters on both side of the niche. Possibly, a detail identification of the fissure outcrop on surface must be done, and than grouted;

b other minor sites to consolidate;

c any intervention on the top cannot be developed without a complete de-mining of the area.

25.18 The Back Side of Both Buddha Niches

25.18.1 Temporary Support of Endangered Vaults

The removal of fragments from the bottom of the niches, mainly the Eastern Giant Buddha niche, has confirmed the large instability of blocks settled in the lower part of the niche's back side, and the risk of collapse, in the vaults. Additionally, in the Eastern Giant Buddha niche, the dis-

Fig. 25.41. Emergency intervention in the Western Giant Buddha niche (topographic data from PASCO 2003)

covering of a master crack just behind the fragments and parallel to the cliff, is increasing the risk of fall also for a large portion of the back side. A numerical evaluation of the stability conditions of the cliff is not presently available but the overall shear strength of rock mass is relatively low, in consideration of the wide opening (about 10 cm) of the fissure. Nevertheless, this large crack was investigated in the nearby caves, showing a probable limited penetration into the lateral sides. It is interesting to notice that the crack is probably limiting the back side of the niche (Fig. 25.42)

In the vault there is now the risk for: collapse of small pieces of rock (up to 30×30 cm); fall of large blocks of rock (around 1–2 cubic meters); instability of the entire back side due to the large discovered fissure.

This situation warrants the installation of a temporary support by means of wooden pillars. They should be arranged perpendicular to the vault, and stably fixed to the ground. In case the pillars can not be mounted perpendicular to the ground surface, an appropriate small excavation would be needed to host it.

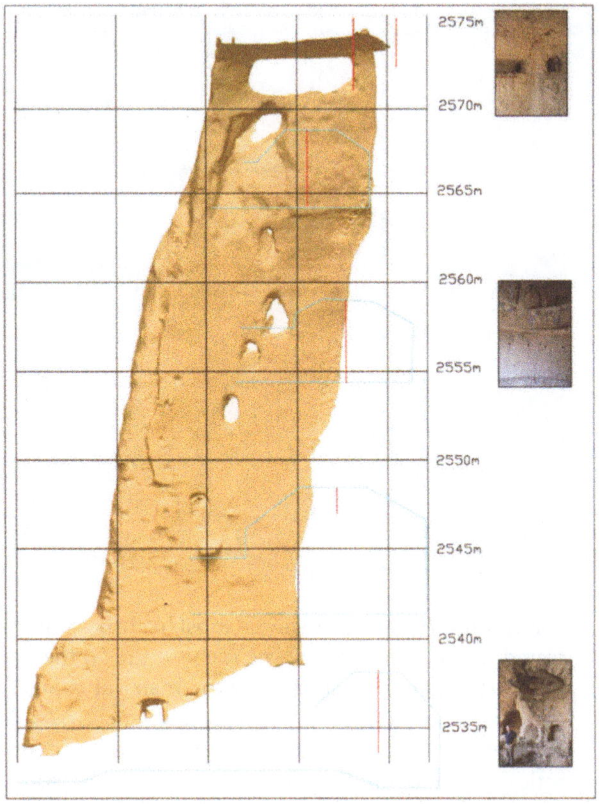

Fig. 25.42. Reconstruction of upward and lateral propagation of the large crack, in *red*, discovered at the bottom of the vault, after the removal of statue's fragments (topographic data from PASCO 2003). In *green* is the projection of lateral caves that were investigated to evaluate the propagation of the crack

25.18.2 Preliminary Consideration for the Consolidation of the Back Side of Both Niches

During this field work, a first strategy for the consolidation of the back side of the niches was developed. The preliminary ideas need to be further confirmed by mathematical modeling and compared with the experience of restorers. This, in fact, is the connection point between earth science specialists and experts of restoration of cultural heritage. The objects is certainly a rock cliff modeled by human intervention to carve a statue: any solution has to fit consider variables from both domains – in this paper we refer mainly to the geological aspects.

Also, any strategy has to start with the identification of the most appropriate conservation approach for the remaining fragments. In other words, whether the cliff should be preserved as it is. In any case there is the need of an accurate preservation, maintaining as much as possible of present state. If, for instance, the final solution considered the reconstruction of a supporting substructure for the relocation of original fragments, such a support may play the most important part in the stabilization of the cliff. The present work recommends maintaining the present state in terms of archaeological remaining structures also in reference to the stability of the niches.

The general strategy should include:

1. Temporary support for the unstable blocks including the vault in the bottom; the existing net is certainly operating properly (Fig. 25.43), even if a better fixing at the bottom of the niche is required. In the execution of a possible consolidation of the back side, by means of a scaffolding, the installation of more temporary steel ropes, horizontally placed, is required.
2. Damage assessment (surficial degradation and crack pattern distribution), to be assessed with a detailed local survey, may be carried out with the support of proper geophysical investigation (e.g., georadar is presently considered the most appropriate technique, but its application is not suitable with the on-site metallic net).
3. Establishing a proper consolidation strategy which, in this first hypothesis, should include namely three phases. The first, to be conducted with extremely low vibration and low water cooling equipment, for a depth of about 50 cm and a diameter lower than 20 mm (nail 10 mm), providing the opportunity to grout and nail, the most surficial crust of the back side. In such a way a first resistant surface can be provided. The grouting material to be inject should be identified in coordination between restorers and consolidation experts, likely through field tests. In any case, the low water release in the mixture, as a consequence of slaking for some siltstone strata, is a fundamental requisite. The proposed technique is similar to the one in use for the consolidation of masonry with micro-injection. The spacing of such small grouting/injection can depend on the surface conditions of the materials and, as a first approximation, within the order of 50 cm. The second phase is aimed at stabilizing the remaining shape of the figure along the master crack identified in 2006, in the underneath caves behind the feet of the Eastern Giant Buddha. In this phase it is expected to consolidate the material with grouting and nails, in order to obtain an homogeneous and stable rock mass, down to a depth of no more than 3 meters. In this case the drills holes will be conducted with low vibration equipment, with a diameters of about 32 mm (including the stainless steel nail of 16 mm and grouting). The spacing is here relatively constant at about 1.5–2 m. The third phase is aimed at fixing the consolidated mass to the stable rock of the cliff, and the grouting of the master crack identified in 2006. The stabilization will be conducted with passive anchors and grouting to a depth of about 10–15 meters. The depth of anchors should be, in this case, twice the distance from the last detected large fissure, encountered during the drilling. The diameters can be, as a first estimate, about 90 mm,

Fig. 25.43. The metallic net protecting from the detachment and fall of small stones from the shear band of destroyed statue

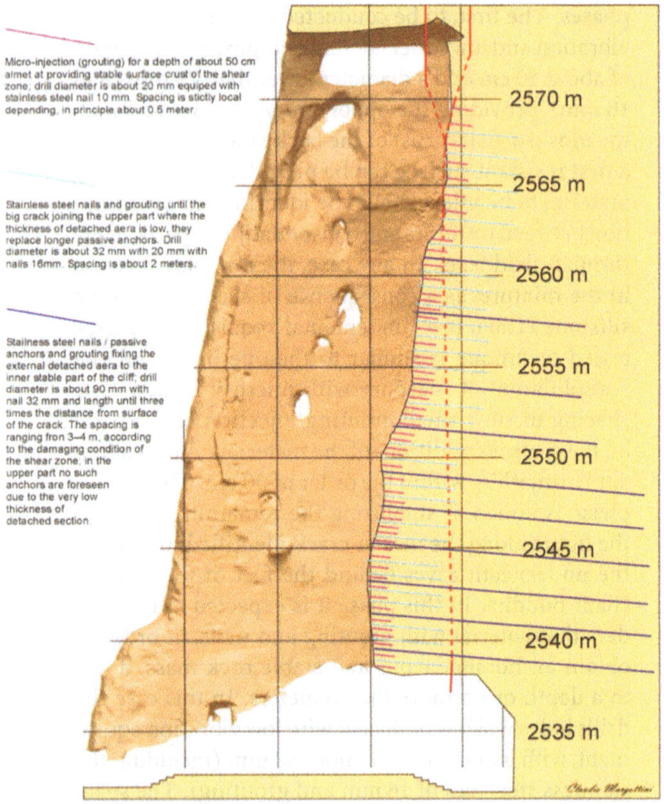

Fig. 25.44. Tentative sketch for the stabilization of the back side of both niches (case study developed for the Eastern Giant Buddha). *Pink*, *green* and *blue* stand for shallow, intermediate and deep anchors; *dashed red line* is the probable reconstruction of a large crack detected at the inner back side of the niche

in order to allow a correct placing of anchors. The spacing of such anchors will be evaluated through stability analysis of the back side of the niche under present conditions. The anchors should most likely be concentrated in the lower part of the niche while, in the upper part where the thickness of the material is lower, the extension could be 4–5 m of passive anchors/nails as foreseen in phase 2. Figure 25.44 shows the theoretical distribution of such anchors and the relationship with the large crack detected at the inner back side of the niche (dashed red line).

25.19 Manual Crack Gauge Monitoring System: Current Results

A crack gauge monitoring network was established in September 2003 (Colombini and Margottini 2003b) to evaluate possible long-term deformation of the cliff and efficacy of consolidation measurements. The site of investigation included the most hazardous discontinuities of both niches. Figures 25.45 and 25.46 report the data on the monitored sites. The measurements, with a theoretical accuracy of 0.001 mm, reveal that:

1. no deformation occurred until now for the site of the Western Giant Buddha (Fig. 25.47);
2. minor enlargement was observed in Stations 3 and 4, having occurred during the execution of emergency consolidation work of the Eastern side of the Eastern Giant Buddha niche in spring 2004. This minor deformation was confirmed also by automatic sensors employed during the implementation of the work (Fig. 25.45). Station 3 revealed about 1 mm of enlargement on November 2003 but this was not confirmed by the electronic sensor Nr. 5 that was positioned just close to this station. Probably the small basement (5 mm) for the crack gauge monitoring received a shot from one of the workers. After the

completion of the work in spring 2004 no deformation was recorded. On the western side of the Eastern Giant Buddha niche, the monitoring of the buttress (Station 1) and the monitoring of a large crack on the top (Station 2), probably the same nowadays recovered in the vault of the lower cave behind the feet of the statue, do not reveal any movement.

25.20 Conclusion

The present paper describes all the emergency intervention measures performed in Bamiyan (Central Afghanistan) for the consolidation of niches and unstable blocks resulting from the detonations, executed by Talibans in March 2001, aimed at destroying the 5th–6th century A.D. giant Buddha statues.

The effect of the explosions was quite drastic: the two statues totally collapsed but also some small parts of the niches were destroyed – a large part of the Eastern Giant Buddha niche was close to the collapse. UNESCO immediately prompted an emergency intervention for securing the remainings of this unique cultural heritage and, thanks to the generous financial support of the Government of Japan,the work started in November 2003.

The activities were developed according to the following general scheme:

1. engineering geological study of the site, including laboratory testing and filed work (the first were conducted in Europe with few samples and the latter conditioned in their execution from the presence of land mines);
2. installation of a high-precision monitoring system;
3. realization of temporary support infrastructure, to maintain stability of the blocks at limit equilibrium, also during the actual implementation of the recovery measures;
4. performing the consolidation work, with professional climbers to avoid any activities below the hanging and unstable blocks, with a system of small and long passive anchors and grouting;
5. minimization of impact of anchor heads, with a mixture of special mortar, investigated in detail with the support of ICOMOS expert.

The obtained results are encouraging, especially taking into account thatthe area is slowly recovering from decades of war, where it was necessary to adopt the maximum of professional judgment in identifying deficiencies and knowledge gaps, and, in the mean time, to adopt technologies capable of solving the problems in a very short time and under safe conditions.

After the investigations of September 2002 and the practical intervention of October–December 2003, March 2004, and October–December 2006, the cliff and niche of the East-

Fig. 25.45. Identification of monitored discontinuities in Giant Eastern Buddha niche

Fig. 25.46. Identification of monitored discontinuities in Giant Western Buddha niche

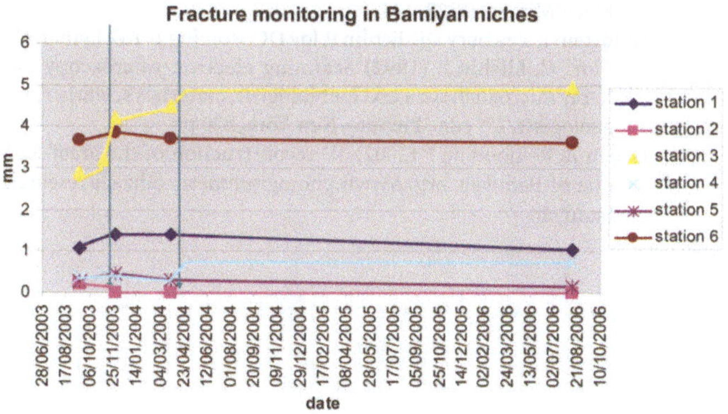

Fig. 25.47. Observed deformation on selected discontinuities on cliff and niches (station 3 and 4 were modified in data development in spring 2004, due to the contribution of overlapping automatic sensors – see also Fig. 25.37)

ern Giant Buddha (the most critical part) is now more stable and the risk of collapse seems to be avoided. Also, the niche of the Western Giant Buddha has been protected from water infiltration. Further work will be needed in the future, especially on the back side of both niches but, at least, the major risk of a collapse involving the few remains not destroyed by the Talibans now seems to be eliminated.

References

AGI (1977) Raccomandazioni sulla programmazione ed esecuzione delle indagini geognostiche

Amateur Seismic Centre (2003) Seismicity of Afghanistan. http://www.asc-india.org/seismic/afghanseis.htm#history

Ambrasey N, Bilham R (2003) Earthquakes in Afghanistan. Seismological Research Letters 74:107–123

ASTM (1977) Annual book of ASTM standards. Natural building stones: soil and rock; peats, mosses, and humus. Part 19. American Society for Testing Materials, W. Conshohochen, Pennsylvania

Atkinson J (1997) Geotecnica. Meccanica delle terre e fondazioni. Mc Graw-Hill, Milano

Barton N, Lien R, Lunde J (1974) Engineering classification of rock masses for the design of tunnel support. Rock Mech 6:183–286

Barton N, Bandis S, Bakhtar K (1985) Strength deformation and conductivity coupling of rock joints. Int J Rock Mech Min Sci 22(3):121–140

Beniawsky ZT (1989) Engineering rock mass classification. John Wiley & Sons, New York

Brown G (1972) The X-ray identification and crystal structures of clay minerals, 2nd ed. Mineralogical Society (Clay Minerals Group), London

Bruno A (1966) Restoration and behind. Tourin

Colombini V, Margottini C (2003a) Reducing the most critical rock fall-prone areas in the Buddhas niches in Bamiyan (Central Afghanistan). UNESCO Report, July 2003

Colombini V, Margottini C (2003b) Buddhas niches in Bamiyan Valley: Emergency measures in autumn 2003, to permit future consolidation. UNESCO Report, September 2003

Crosta G (1998) Slake durability vs. ultrasonic treatment for rock durability determination. Int J Rock Mech Min Sci 35(6):815–824

Dawson EM, Roth WH, Drescher A (1999) Slope stability analysis by strength reduction. Géotechnique 49(6):835–840

Dupree N (2002) Bamiyan. University of Peshwar, Peshawar

Franklin JA, Chandra R (1972) The slake durability test. Int J Rock Mech Min Sci 9:325–341

Goldstein J, Newbury DE, Echlin P, Joy DC, Roming Jr AD, Lyman CE, Fiori C, Lifshin E (1992) Scanning electron microscopy and X-ray microanalysis: a text for biologists, materials scientists, and geologists, 2nd edn., Plenum, New York, 820 pp

Gruen A, Remondino F (2002) 3D reconstruction of the great Buddha of Bamiyan. http://www.photogrammetry.ethz.ch/research/bamiyan

Gruen A, Remondino F, Zhang L (2002) Reconstruction of the great Buddha of Bamiyan, Afghanistan. http://www.photogrammetry.ethz.ch/research/bamiyan

Hoek E, Brown JW (1980) Empirical strength criterion for rock masses. J Geotech Eng-ASCE, 106(GT9):1013–1035

ISRM, International Society for Rock Mechanics (1981) Rock characterisation, testing and monitoring ISRM suggested methods. Pergamon, London

ITASCA Consulting Group (2000) Fast lagrangian analysis of continua, vs. 4.0. Minneapolis, Minnesota, USA

Le Roux A (1971) La lyophilisation, technique d'étude des textures des marnes et argiles. Bull Liaison Lab Ponts Chaussées 55:27–28

Margottini C (2003a) The Buddha niches in the Bamiyan Valley (Central Afghanistan): instability problems and restoration plans in the UNESCO intervention. J Jpn Landslide Soc 450(3):246–249 (in Japanese)

Margottini C (2003b) The geomorphological instability of the Buddha niches and surrounding cliff in Bamiyan Valley (Central Afghanistan). UNESCO Report, June 2003

Margottini C (2004a) Instability and geotechnical problems of the Buddha niches and surrounding cliff in Bamiyan Valley, Central Afghanistan. Landslides 1(1):41–52

Margottini C (2004b) Definition of 2004 activities for implementing the emergency consolidation of Eastern Giant Buddha and for beginning the preservation of Western Giant Buddha from geological hazards. UNESCO report, March 2004

Margottini C (2006) UNESCO: Definition of 2006 activities for implementing the emergency consolidation of Large Eastern Buddha and for beginning the preservation of Large Western Buddha. Report, August 2006

Margottini C, Colombini V, Crippa C, Tonoli G (2005) Emergency intervention for the geo-mechanical conservation of the niches of Bamiyan Buddhas (Northern Afghanistan). In: Sassa K, Fukoka H, Wang F, Wang G (eds) Landslides – risk analysis and sustainable disaster management. Springer-Verlag, Berlin Heidelberg, pp 75–79

Operational Climatic Data Summary (2002) https://www2.afccc.af.mil/ocds_mil/products

PASCO (2003) Bamiyan mapping project. Restricted

Pejon O, Zuquette L (2002) Analysis of cyclic swelling of mudrocks. Eng Geol 67:97–108

Shepard FP (1954) Nomenclature based on sand-silt-clay ratios. J Sediment Petrol 24:151–158

Shi B, Wu Z, Inyang H, Chen J, Wang B (1999) Preparation of soil specimens for SEM analysis using freeze-cut-drying. Bull Eng Geol Environ 58:1–7

Taylor RK, Spears DA (1970) The breakdown of British Coal Measures rocks. Int J Rock Mech Min Sci 7:481–501

Turner K, Schuster R (1996) Landslides, investigation and mitigation. Special report, Transportation research board, National Research Council, 247, Washington

Zang PZ, Yang ZX, Gupta HK, Bhatia SC, Shedlock KM (2002) Global Seismic Hazard Assessment Program (GSHAP) in Continental Asia. http://seismo.ethz.ch/gshap/eastasia/eastasia.html

Chapter 26

Debris Flow Hazard Defense Magnitude Assessment with Numerical Simulation

Ko-Fei Liu* · Hsin-Chi Li · Yu-Charn Hsu

Abstract. Debris flow disasters are usually accompanied by serious loss of lives and properties. However, debris flows are also part of Earth's natural phenomenon, what is the reasonable budget to be spent on mitigation measures becomes an important issue on the budget allocation processes. This paper utilizes economic concept to propose a reasonable estimation of the hazard damage and the cost of proposed mitigation measures. The proposed method is composed of 4 steps, namely, delineating the area of the disaster with different return periods, itemizing the land use within those area, calculating the hazard lose using official values and computing the expected probable maximum lose with a probability distribution. The comparison between the assessment of hazard and the economic gains of any proposed mitigation measures can be used as a reference for future decision making process.

Keywords. Debris flow, Hazard assessment, Risk analysis, Numerical Simulation, GIS, econometric model

26.1 Introduction

Taiwan is located in the typhoon district of the Pacific Ocean and is attacked by typhoon average 3.6 times annually. Many large scale hillside development projects were undertaken in the past decades during the peak of economic growth. These areas become very vulnerable for natural hazard and usually are subject to heavy damage under slope land disasters (Fig. 26.1). The losses of the people's lives and properties are severe, but the social economic problems, such as the livelihood of victims, reconstruction and economic recovery of the disaster area, etc., consumed even more social economic resources. Therefore, how to effectively use limited government resources to reduce the damage as much as possible in the micro economy sense is a new research direction.

The fundamental of risk management is finding the proper mitigation methods such as ecological method, engineering method, evacuation or migration of village base on the possible hazard loss. More detailed assessment gives more practical policy making results. The assessment is also very useful during emergency situation. Therefore, a quantifiable hazard assessment approach is critical to risk management and policy implication for debris flow hazard mitigation.

However, policy making needs considering not only hazard loss itself but also all the costs incurred during policy enforcement. This is called the transaction cost which includes negotiation between agencies, man power adjustment, negotiation with local people and execution quality control etc. From past experience in Taiwan's slope land problems, the most expensive transaction cost occurred during negotiation with local people. The cost is brought about because people consider their loss is not properly compensated. This usually leads to long time negotiation and spending much more budget. This kind of cost is also called external effect or externality problem in economics. Any kind of externality problem will greatly increase the difficulty in resolving the present situation. Also external effects will strongly affect the effectiveness of policies. Therefore, externalities should be one of the important considerations during policy making. However, externalities can be roughly assessed by economic methods (Liu and Li 2004), and transformed into internal cost.

After obtaining the hazards loss and all transaction cost of policies, policy maker can choose the best options based on the available resources and the assessment of all possible mitigation methods. This is the ideal and com-

Fig. 26.1. Debris flow disaster of Song-Her village in central Taiwan. The village is partially buried by debris flows and original small stream outbursts into debris flow river of more than 30 m wide. This scenario is typical in Taiwan (Photo provided by Water and Soil Conservation Bureau, Taiwan)

plete approach for debris flow hazard assessment. However, there is no established assessment method covers all the steps above. This paper will discuss only hazard assessment method and leave risk management and policy making for future research.

But in order to link with future risk management research, we shall define "defense magnitude" for debris flow hazard as the budget one should spend on mitigation the statistically most expected hazard loss. Under this concept, we have to first complete the assessment of the debris flow hazard under different return frequencies, and then use statistical average to obtain the most expected loss. Then the minimum budget required to mitigate the most expected loss is considered as the "defense magnitude". As a result, different counter measurements can be compared through quantitative evaluations. This process becomes even more important for severe disasters because the budget involved can be very large.

26.2 Literature Review on Flood Hazard Assessment

According to the definition in economics, disaster loss means the amount of budget needed for recovery (Grigg and Heiweg 1974). It means that the loss assessment should include not only people and property loss but also money for recovering to pre-disaster situation. However, this definition can easily leads to moral arguments. Therefore, we shall adopt an engineering point of view to deal with the value-wise discussion.

In economics, if a model can monetizes the assessment results, it is called an econometric model. There are many researches on debris flow risk assessment or hazard assessment such as Laigle and Marchi (2000). But all the literatures discuss the range of affected area and the degree of danger for that area. No existing reference demonstrating systematic evaluation of debris flow hazard loss with monetized method can be found. Since our intention is to combine engineering knowledge and economic concept to establish a preliminary econometric model, so we turn to flood hazard assessment research for references.

Breaden (1973), Grigg and Heiweg (1975), Grigg (1976) etc. divided the flood loss into direct damages, indirect damages, secondary damages, intangible damages and uncertain damages. Freeman (1993) used Utility function and Expenditure function to establish his flood damage assessment model. Tsai (1994, 1995) used the flood depth vs. loss curve to perform the flood damage assessment. The value of loss is separated into four categories – agriculture, fishery, salt industry and real properties. They applied the method to Chiayi County, Taiwan with the help of GIS.

For simplicity, we separate the loss into two categories according to period the disaster occurs. All the damages occurred during the disaster belong to direct damages, including human casualties, private and public property damages. Indirect damages include all damages occurred after the disaster. A complete list for indirect loss is case dependent. Extra living cost for victims; shrinking of economy activities; the psychological impact are all examples of indirect damages.

In reality, it is quite difficult to assess all the losses, especially the indirect losses. Since indirect losses can be weighted differently by different people at different time and location, indirect losses are usually assessed by questionnaires. The analysis of the questionnaires takes a long time and explanation can be varied subject to different point of views. However, the direct damages are the most important during the emergency responding stage and are also the immediate remediation targets. In Taiwan, the direct damage is used to weight how serious a hazard is and can be used to evaluate defense magnitude. Therefore, as the first step, this paper will focus on a systematic method of evaluating the direct damages.

26.3 Assessment of Debris Flow Hazard

The assessment procedure we proposed can be separated into four steps (Fig. 26.2) including data collection, delimiting the influence area, land usage analysis and the calculation of expected loss.

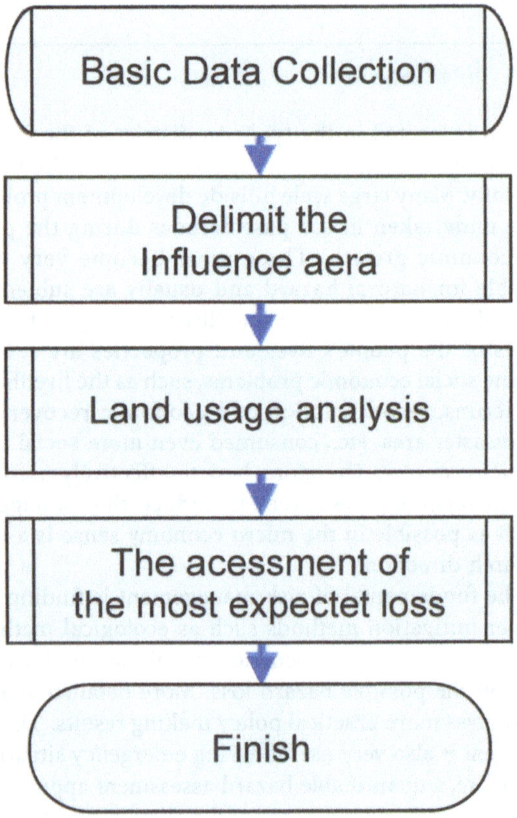

Fig. 26.2. The procedure for debris flow hazard assessment

26.3.1 Data Collection

The data needed include 1 : 25 000 numerical topographic map, 1 : 5000 numerical topographic map or even more detailed maps, rainfall data, past detailed disaster history, existing mitigation constructions, on-site investigation of debris flows, on-site investigation of residents, social and economic activities, land usage, etc. With more completed data, one can obtain more accurate assessment results.

26.3.2 Delimiting the Area Affected by Debris Flows

Many methods can be used to determine the influenced area of debris flows. In general, numerical simulations are considered more accurate providing the input is not faulty. But before applying the numerical program in the present hazard assessment approach, one must make sure it has the ability to perform two dimensional simulations as well as simulations with counter measurements. The after process for useful engineering maps can be performed manually as long as the output can be used in a GIS system. The program Debris-2D (Liu and Huang 2006) was used in delimiting the influenced area of debris flow in our study. The program uses Julien and Lan (1991) model and can be applied to granular flows as well as mud flows. A typical output can be seen from Fig. 26.3.

This numerical program needs digital topographical map, the rheological properties of debris flow, rainfall data and the distribution of debris sources. Numerical topographical map (digital terrain map) can be found easily for most countries. But the resolution has to be as fine as 5 m by 5 m for an accurate and meaningful result. If the resolution is 40 m by 40 m, then field investigation is crucial in order to adjust the interpolated grids to reflect the real situation. Rheological properties can be measured with laboratory test with field samples. The flow rate calculation is described as follows.

We assume the concentration of a debris flow can be approximated by the equilibrium concentration (Takahashi 1980)

$$C_{d\infty} = \frac{\rho \tan\theta}{(\sigma-\rho)(\tan\phi - \tan\theta)} \tag{26.1}$$

where $C_{d\infty}$ is the equilibrium concentration of debris flow; ρ is the density of fluid in debris flow, (for clayey water,

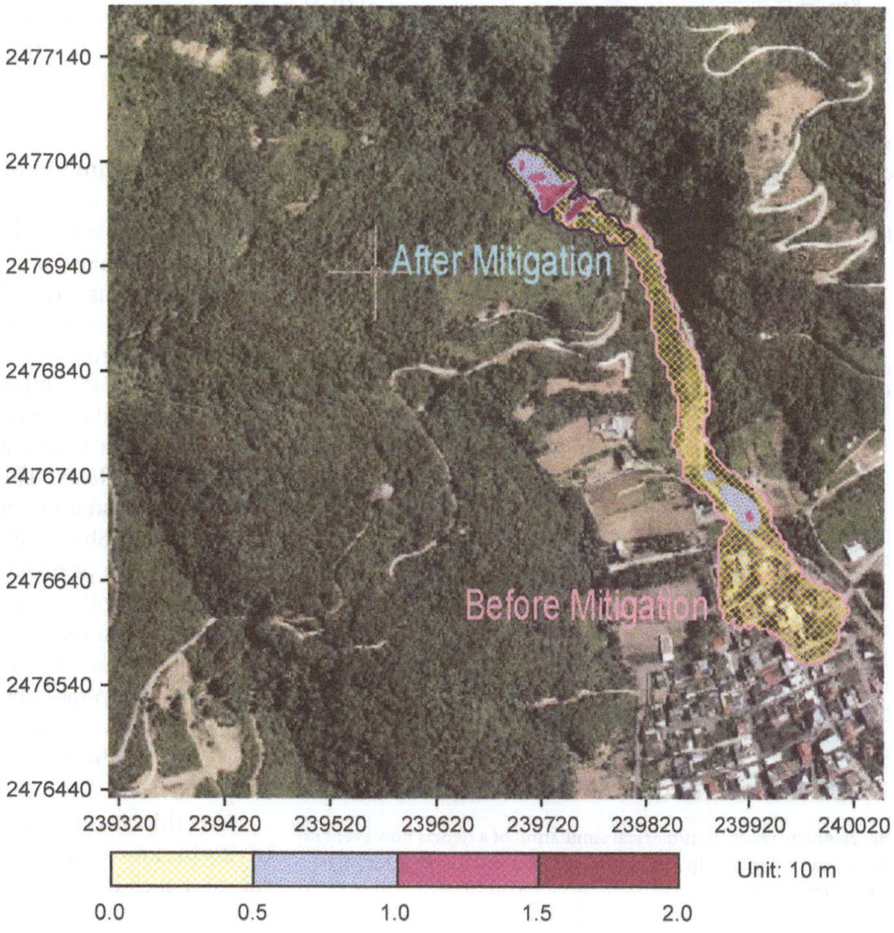

Fig. 26.3. Lower right conner is the village to be protected. *Pink area* indicates the area affected by debris flow originally. *Blue area* is the area affected by debris flow after construction of mitigation measures. *Different colors* indicate the depth of the final deposition of debris flow disaster. The area and depth contour are the necessary features for a numerical program in order to be used for hazard assessment.

the density can be close to $1.2\,\mathrm{g\,cm^{-3}}$); θ is the average slope of riverbed; σ is the density of gravel, ϕ is the interior angle of friction of gravel.

Then the available amount of debris sources in the field and the accumulated rainfall can be used to estimate the flow rate. If accumulated rainfall is small, only part of the material in riverbed or hill slope can be eroded to form debris flows. If the accumulated rainfall is large, all material available can be mobilized, the max flow rate is then determined by the amount of available soil and gravel in the filed. There are many methods for estimating the filed debris volume. But field investigation with aero photos is considered the most accurate. In our research, we also use latter approach.

In summary, if the available field debris source volume is V_S and the accumulated rainfall amount is large, the total volume of debris flow V_{D1} is

$$V_{D1} = \frac{V_S}{C_{d\infty}} \tag{26.2}$$

If the accumulated rainfall amount V_W is small then the total volume is

$$V_{D2} = \frac{V_W}{1 - C_{d\infty}} \tag{26.3}$$

We choose the smaller one from V_{D1} and V_{D2}.

With the above volume and the corresponding debris source distribution from field investigation, we can proceed with the numerical simulation. A typical result is shown in Fig. 26.4. The result is calculated for 5-year return period.

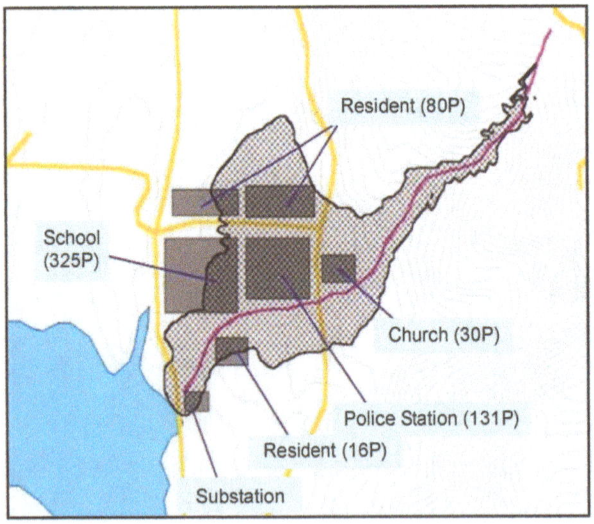

Fig. 26.4. The result of numerical simulation of a debris flow event at Hsin-Dian County, Taipei. The affected area will increase for higher return frequency

After obtaining the debris flow influenced area, we can proceed to calculate the total loss within that area. But the type of land use and degree of damages also affects the loss value. The degree of damages is related to the depth of final accumulation of debris flows. Therefore, the numerical program must also provide this information.

26.4 Land Utilization within the Influenced Area

In real disasters, losses can come from damage of land, construction, agriculture, forest or others. But the most often used first level classification (Liu and Li 2006) is to first classify the loss into two categories, human loss and property loss. Then the property loss was further divided into land loss and the ground surface property losses. By consulting land law (The Land Act, Article 2, Taiwan), land loss is again divided base on land utilization into seven different types: construction use, agricultural use, forest use, tourism use, other production use, transportation and hydraulic use and other uses (Table 26.1).

Different values can be found for different types of land use according to government bulletins. This value is to be used later in the loss assessment.

26.5 The Expected Disaster Loss in the Influenced Area

26.5.1 The Loss Assessment for a Single Event

Bass on the definition of disaster loss in economy, and the work of the empirical flood depth vs damage curve used in flood loss (Smith 1994), Liu and Li (2006) established a preliminary model for assessment of direct loss for debris flow hazard.

The ground surface property is separated into construction and non-construction. For non-construction properties such as plants or transportation facilities, they will be destroyed completely by a debris flow. But for construction, it is a recoverable loss for small scale debris flows or mud flows. Therefore, a simple standard for evaluation is listed in Table 26.2. This standard adopts the result form Shih et al. (1997) and is slightly modified considering the destructive power of debris flows. For final buried depth greater than 3 m, it is impossible or useless trying to recover the construction. If the buried depth is between 1 m and 3 m, the construction itself is subject to recoverable but costly damage. We consider the loss as half of total destroyed loss. For buried depth less than 1 m, the damage is usually minor and can be fully recovered.

After this classification, we can use formulae for quantifying the loss of land and ground surface properties. The value for each item is separately discussed in the follows.

Table 26.1.
The seven different types for land utilization

Land law	Type of land use		Content
Type 1	Construction use		Such as house, industry, commerce, etc.
Type 2	Agricultural use		Such as paddy field, dry land, fruit land, betel field, etc.
	Forest use	Developed	Bamboo field, the cutting area, etc.
		Original	Such as broadleaf wood, coniferous wood, bush woods, collapse place, wasteland, etc.
	Tourism use		Such as holiday villages, forest attraction area, etc.
	Other production use		Such as ore land, herd ground, salt industry, etc.
Type 3	Transportation and hydraulic use		Such as road, bridge, irrigation canals, dyke weir, watercourse, etc.
Type 4	Other use		Not included in the above-mentioned areas, for instance desert or snow-covered mountain, etc.

Table 26.2.
The standard for ground surface property loss

Final buried depth	Objects on the ground	
	Construction	Non-construction (such as agriculture)
More than 3 m	Destroyed completely	Destroyed completely
1~3 m	Half destroyed (loss reduce by half)	Destroyed completely
Under 1 m	Slightly damaged (loss is down to legal fee)	Destroyed completely

26.5.1.1 Human Life Loss

In the general assessments, the loss of human life usually will not be assessed, because it is difficult to be calculated. But according to the definition of loss, human life loss is also one part of disaster damages. In many cases, this loss is valued most important and facilitates the mitigation measures. Thus human life loss can be assessed as follows.

$$HL = HLV \times N \tag{26.4}$$

- HL = human life loss (unit: dollar)
- HLV = human life value (unit: dollar per person)
- N = number of death toll

Admittedly, this is the most arguable item. A lot of people consider people's life is priceless and can not be monetized. However, once the tragedy occurs, there has to be a value associated in order to discuss compensation. Therefore, this paper wouldn't discuss the moral part, and only will focus discuss on the economics point of view.

Base on the researches in economics, Fisher et al. (1989) assessed samples from U.S.A. and showed that the life value of a human was between 56 million NTD (1.6 million US dollars) and 140 million NTD (4 million US dollars). Xue and Wang (1987) also assessed samples from Taiwan from the reports of the labor force survey in 1984. They showed that the life value of a human was between 12 million NT dollars (0.34 million US dollars) and 34 million NT dollars (0.97 million US dollars). Liu, Hammitt and Liu (1997), also assessed samples from the labor force survey reports in 1982–1986. They found that the life value of a human was between 14 million NTD (0.4 million US dollars) and 16 million NTD (0.46 million US dollars). This result is similar to Xue and Wang's, but differs greatly from that of Fisher's. For the application in Taiwan, this paper adopts the average result of both papers and uses the average value, 15 million NTD (0.43 million US dollars), for the value of a human life loss.

Since the value was calculated in 1982–1986, it must be converted to the present value (divided by the price-index of 1986). If we take the index at 2005 as 100%, the index is 148.2% (Monthly Bulletin of Statistics 2005) at 1986. So the present value for a human life loss is 20.74 million NT dollars (0.59 million US dollars).

The value above is for human death. If severe injury occurs instead of death, the corresponding loss is considered less. This is again controversy. An injured people may carry psychological damage which can last for years and can affect other people. The related loss may be humongous and hard to evaluate. However, this belongs to the indirect damage which is not included in the present discussion. Then it comes with the problem that death toll associated with an event is actually not predictable. For simplicity, we assume the situation that all residents are inside their household. Then according to Table 26.2, if the depth of debris flow is greater than 3 m, house is completely destroyed and the residents inside are assumed to be dead. For debris flow depth between 1 m and 3 m, the house is only half destroyed and there is a good chance that people can survive but injured. Therefore, we consider the residents are serious injured and the loss equals half of the human value. For debris flow depth less than 1 m, residents are assumed safe and sound. No countable loss occurs.

26.5.1.2 Property Loss

The calculations of property loss were divided into land loss and the loss above the ground. There are other kinds of losses such as the interior furniture and decoration loss. But these losses are much smaller than the land and construction loss (Lee 2006). As a result, we shall ignore them at the present discussion.

A Land Loss

$$LL = \sum_{i=1}^{N} LV_i \times LLA_i \tag{26.5}$$

- LL = total land loss (unit: NT dollar)
- LV_i = land value (unit: NT dollar per m^2)
- LLA_i = land loss area (unit: m^2)
- i = land no. or parcel no.

There are two approaches for evaluating the value of land. The first one is direct evaluation with market price. The advantage of this method is the result can reflect real market demand. But it needs effort and budget for investigating. One of the characteristic of market price is that it fluctuates rapidly during unusual period. So it's hard to be used for emergent response. The other approach is adopting the value of government's bulletin. Because the price is announced once per year, the value may not truthfully reflect market price, but it represents the lowest price guaranteed by government. Therefore, this paper uses the second approach, adopting government's bulletin, to evaluate the current price of land.

B The Loss above the Ground

After finishing the discussion about the land loss, next were the discussions about the loss above the ground. They were discussed bass on the classifications of Table 26.2.

a Construction Use

$$BL = \sum_{i=1}^{N} \sum_{j=1}^{M} BC_{ij} \times BA_{ij} \tag{26.6}$$

- BL = total builing loss (unit: NT dollar)
- BC_{ij} = unit area cost of floor j in builing i (unit: NT dollar per m^2)
- BA_{ij} = area of floor j in building i (unit: m^2)
- i = index of different building
- j = index of different floor

In the formula, the values for BC can be found from "Standard unit price form for building under different structure", which was published by every city or county government every year (Table 26.3)

Table 26.3. Sample of "Standard unit price form for building under different structure" cited from Taichung City

Structure		Floor 4	Floor 3	Floor 2	Floor 1
Steel (P)	Type 1	4 040	3 940	3 840	3 740
Steel concrete (A)	Type 2	3 740	3 640	3 540	3 440
Steel reinforced concrete (S)	Type 3	3 050	2 950	2 850	2 750
	Type 4	2 390	2 290	2 190	2 090
Ferroconcrete (B)	Type 1	3 260	3 160	3 060	2 960
Pre-stressed concrete (T)	Type 2	2 960	2 860	2 760	2 660
	Type 3	2 700	2 600	2 500	2 400
	Type 4	2 140	2 040	1 940	1 840
Strength brick (C)	Type 1	2 740	2 640	2 540	2 440
	Type 2	2 440	2 340	2 240	2 140
	Type 3	2 300	2 200	2 100	2 000
Iron	Over 200 m^2	–	–	2 300	2 200	2 100
	Under 200 m^2	–	–	900	800	700
Stone and brick		–	–	1 200	1 100	1 000
Wood		–	–	–	600	500

Resource: Revenue Service Office, Taichung County.

Based on Table 26.3, the price per square meters of every kind of construction can be found. Multiplying the area of every floor and adding the cost of every building, the total values of construction can be calculated. This is the value used for the construction destroyed completely.

b Agriculture Use

The production from agriculture land is mainly crops. The actual market price for crops can be used to assess crop loss directly.

$$\mathrm{CL} = \sum_{i}^{N} \mathrm{CO}_i \times \mathrm{CP}_i \times \mathrm{CLA}_i \qquad (26.7)$$

- CL = total crop lost (unit: NT dollar)
- CO_i = crop output (unit: kg ha^{-1})
- CP_i = crop price (unit: NT dollar per kg)
- CLA_i = crop loss (unit: ha)
- i = index for different crops within hazard area

In the formula, the values of crops can be found from Agriculture and Food Agency, Council of Agriculture, Executive Yuan, Taiwan.

c Forest Use

Forest use loss can be evaluated as follows

$$\mathrm{FL} = \mathrm{FAL} \times \mathrm{DLA} \qquad (26.8)$$

- FL = total forstry loss (unit: NT dollar)
- FAL = average unit area forestry loss from previous year (unit: NT dollar per ha)
- DLA = disaster area (unit: ha)

In the formula, the values of FAL are cited from the reports of forestry statistics of previous year, which was announced by Forestry Bureau. In the report, the statistics were separated into five categories, fire alarm, fire hazard, unlawfully chopped trees, unlawfully cultivated and others such as wind hazard, rain hazard. Only the last category accounts for natural hazard, therefore, this paper adopts the data of the last category. Using the report of forestry statistics in 2003 as an example, we can find the value of FAL to be 10 224 NTD (292.1 US dollars) per hectare.

d Tourism Use

The actual tourism loss usually takes the form of indirect loss such as fewer tourists and related income. Since this paper accounts for direct loss only, so only the tourism related building, land and forest should be calculated. Thus the assessment of direct damages in tourism use can follow the calculation of the corresponding loss formulas. Following is the formula

$$\text{direct tourism loss} = \mathrm{BL} + \mathrm{CL} + \mathrm{FL} \qquad (26.9)$$

e Other Production

According to the analysis of the land usage in the potentially dangerous debris flow, the only possible land usage in this category is mining industry in Taiwan. But the mining industry almost ceases to exist in Taiwan. Therefore, we do not have to consider this category in Taiwan. Different situation exists in different countries, it should be considered differently if this is applied elsewhere.

f Transportation and Hydraulic Use

The losses within this category are mainly public facility damages. The method of assessment uses the recovery cost as the loss value. Following is the formula:

$$\mathrm{THL} = \mathrm{SUO}_i + \mathrm{SLN}_i \qquad (26.10)$$

- THL = total traffic and hydraulic structure loss (unit: NT dollar)
- SUO_i = structure unit cost (unit: NT dollar per m^2 or NT dollar per m)
- SLN_i = number of structure loss (unit: m^2 or m)
- i = index of different structures

In the formula, the losses of structures were calculated by adopting the actual repairing or reconstruction costs. But different facilities require different works. For example, road repairing works include cleaning, side slope reinforcement, foundation protection etc. which are different from other constructions (such as river bank and dam construction). So the cost is quite different.

g Other Use

This category includes items in a very wide range. The consideration is really case by case. It needs real examples to explain how to calculate the total loss and how to obtain the real value.

Finally, all types of losses are summarized in Table 26.4. The final total loss is simply the sum of all the losses in Table 26.4

$$\text{Total loss} = \mathrm{HL} + \mathrm{LL} + \mathrm{BL} + \mathrm{CL} + \mathrm{FL} + \mathrm{THL} + \mathrm{TL} \qquad (26.11)$$

This is the total direct loss for a debris flow event under specified return frequency.

Table 26.4. Loss formula for different types of loss

Types		Formula	Value for assessment
▪ Human life loss		$HL = HLV \times N$	20.74 million NT dollars per person
▪ Property loss			
Land loss		$LL = \sum_{i=1}^{N} LV_i \times LLA_i$	Government's bulletin
The loss of thing on the ground	Construction use	$BL = \sum_{i=1}^{N}\sum_{j=1}^{M} BC_{ij} \times BA_{ij}$	Government's announcement
	Agriculture use	$CL = \sum_{i}^{N} CO_i \times CP_i \times CLA_i$	Agriculture product price from Government
	Forest use	$FL = FAL \times DLA$	Forest loss from the Forestry Bureau
	Transportation and hydraulic use	$THL = \sum_{i=1}^{N} SUO_i \times SLN_i$	The fabrication cost of actual item
	Tourism use	$TL = BL + CL + FL$	
	Other production use	Not considered in Taiwan	

26.5.2 The Probability Factor for Different Debris Flow Events

Different definitions and explanations for risk can be seen in different research fields (e.g., insurance, industry, educational, etc.). In engineering, one usually adopts "the probability or chance" of loss as the definition. And it is usually expressed in expectation value, which means the average value of loss. In the case of natural hazard, it can be expressed as follows:

Expected loss of disaster

$$= \sum_{t=1}^{\infty} C_t[\beta_t p(x_t)] \quad t = 1,2,3,\ldots\infty \qquad (26.12)$$

- $p(x_t)$ = probability distribution function of rainfall
- β_t = probability coefficient of debris flows (The probability for debris flow to occur under the given return period)
- C_t = total loss under a specified return period, calculated from Eq. 26.11
- t = recurrence interval of rainfall

Generally speaking, the probability distributions function of different recurrence interval $p(x_t)$ can be calculated from the past rainfall records. The loss C_t can be calculated by Eq. 26.11 with a numerical simulation results. But, the probability coefficient β_t for debris flow

Table 26.5. The calculated value of β_t

Rainfall recurrence interval	Statistical frequency	Number of debris flow	Debris flow recurrence probability
1	27310	9	0.0003
2	22	3	0.1364
3	15	3	0.2
4	5	1	0.2
5	8	1	0.125
10	16	3	0.188
15	10	3	0.3
20	4	1	0.25
25	4	1	0.25
50	6	3	0.5
100	6	3	0.5
150	0	0	–
200	0	0	–
250	0	0	–
300	1	0	0
350	0	0	–
400	2	1	0.5

must be computed from the actually debris flow records. Values of β_t are obtained for Taiwan in the follows.

This research uses the hydrologic data recorded from 1990 to 2004 by Central Weather Bureau and the debris flow records from Soil and Water Conservation Bureau to compute the probability coefficient of debris flows (β_t) in Taiwan. We collected all recorded debris flow events and the associated rainfall records. The collected data is listed in Table 26.5 and plotted in Fig. 26.5. The third row in Table 26.5 is the number of debris flow events recorded under the given return frequency of rainfall. The last row in Table 26.5 is the corresponding probability of debris flow occurrence under that specified rainfall return frequency. An approximated curve is obtained through regression in Fig. 26.5, so β_t for different rainfall return period can be found from the fit curve.

26.6 Case Study – Assessment for Taipei 021 Debris Flow Potential Stream

Taipei 021 was one of the high potential debris flow streams marked by Soil and Water Conservation Bureau.

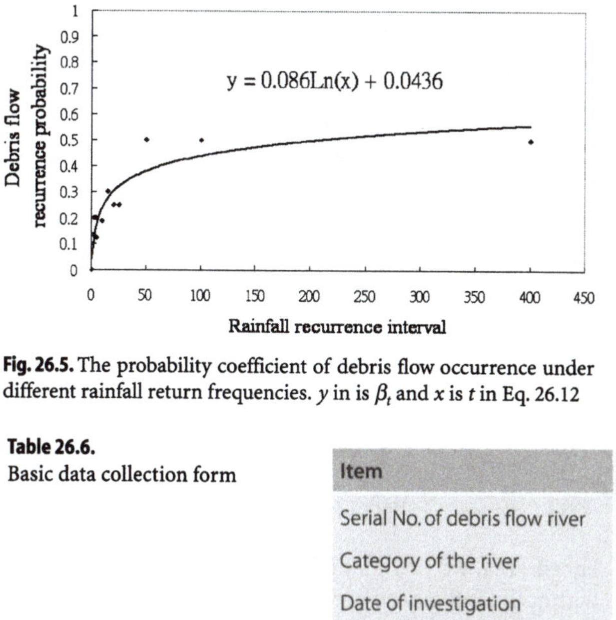

Fig. 26.5. The probability coefficient of debris flow occurrence under different rainfall return frequencies. y in is β_t and x is t in Eq. 26.12

There was a small scale debris flow in 2001 due to heavy rainfall from Typhoon Nari. The debris flow buried two households and some other buildings, but caused no casualty. We shall proceed according to the steps described above.

26.6.1 Data Collection

The typical form used in field investigation is shown in Table 26.6. 1 : 25 000 map and numerical map are shown in Figs. 26.6 and 26.7, respectively.

26.6.2 Delimiting the Influence Area

Field investigation shows the volume of debris source is roughly 9000 cubic meters.

With Eq. 26.1, the equilibrium concentration can be calculated. Using 1.2 g cm^{-3} for ρ, 2.65 g cm^{-3} for σ, 37° for the average slope, 13.5° for θ, we get

$$C_{d\infty} = \frac{\rho \tan\theta}{(\sigma-\rho)(\tan\phi - \tan\theta)} = \frac{1.2\times 0.24}{1.45\times 0.51} \cong 0.39$$

Then substituting the volume of debris source into Eq. 26.3, the maximum volume for debris flow is

$$V_{D1} = \frac{V_S}{C_{d\infty}} = \frac{9000\ \text{m}^3}{0.39} \cong 23077\ \text{m}^3$$

This volume of debris flow corresponds to the case when all available materials in the stream are mobilized. This volume can not be increased even with more accumulated rainfall. The total volume using different return periods is calculated in Table 26.7.

Table 26.6. Basic data collection form

Item	Content	Item	Content
Serial No. of debris flow river	Taipei 021	*X*-coordinates	304 959.6
Category of the river	Slope type	*Y*-coordinates	2 756 267.2
Date of investigation	2001/08/28	Land type	Mountainside and Forest land
County(City) name	Taipei County	Length (m)	1 775
Village name	Sindain City	Average slope (deg)	13.5
Neighborhood name	Chu-Chi	Watershed area (ha)	133
Obvious land mark	Chu-Chi bridge	Lithology	Sedimentary rock
Closing road	Province Route 9	The length of fault (m)	1 064.3
Main watershed name	Dan-Shui River	No. of disaster	1
Secondary watershed	Sindain River	Affected household	10
Third watershed name	Da-Chu-Keng	Danger level	High

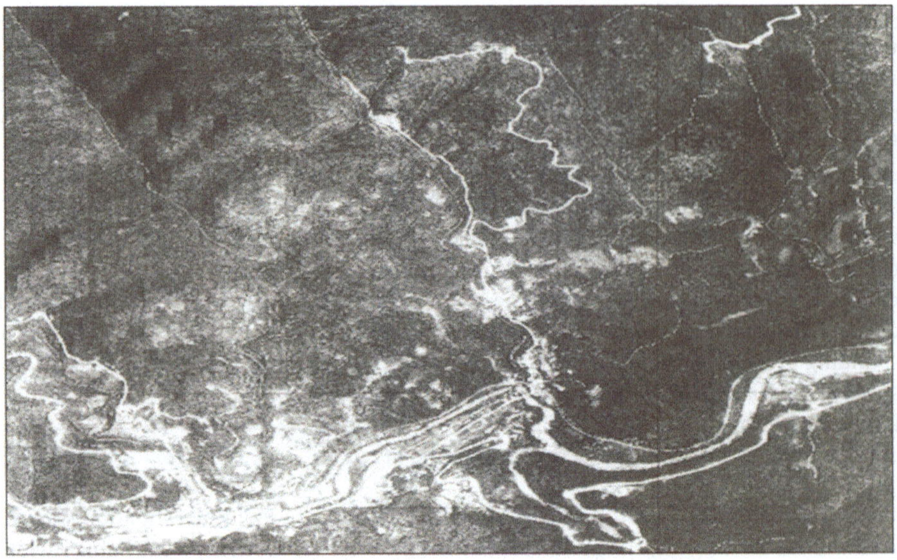

Fig. 26.6.
1 : 25 000 map

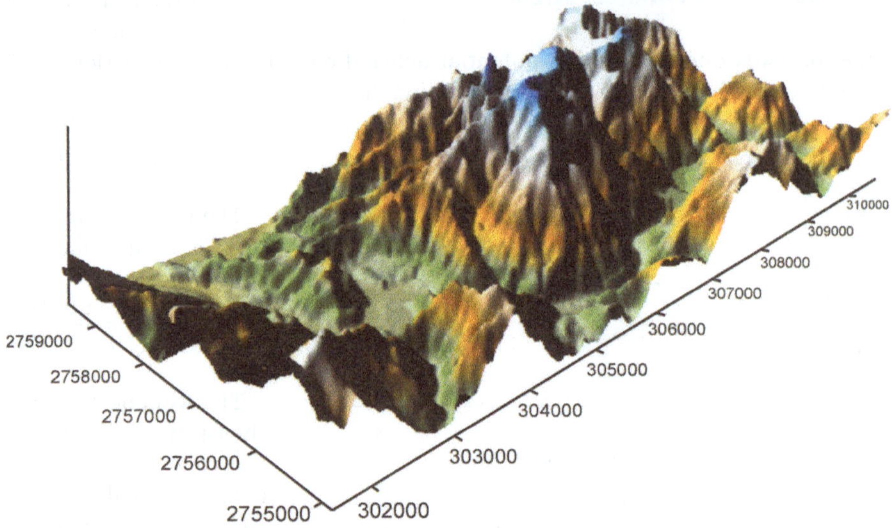

Fig. 26.7.
Numerical map

Table 26.7.
Flow rate for debris flow under different return period

Return period (yr)	1	50	100	150	200	250	300	400
Total volume (m^3)	9844	18983	20603	21550	22222	22743	23169	23529

From Table 26.7, it can be seen that the maximum volume of 23 077 m^3 is exceeded for rainfall of a return period of 260 years. So for return period higher than 250 years, we use the same value 23 077 m^3. The results of numerical simulation under different return periods are shown in Fig. 26.8.

26.6.3 Land Utilization in the Influence Area of Debris Flow

After obtaining the influenced area under different return periods, we have to find the land utilization map from local government. The combined data will be analyzed using GIS. Overlaying the utilization map with the influenced area, we can identify different types of land use within the influenced area as in Fig. 26.9.

26.6.4 The Loss Assessment of Disasters under Different Return Periods

26.6.4.1 50-Year Return Period

Based on the simulated result (Fig. 26.9.a), the loss can be classified in Table 26.8 with the classification listed in Table 26.1. The total direct loss of debris flow hazard, under a 50-year rainfall recurrence interval, can be calculated and listed in Table 26.9.

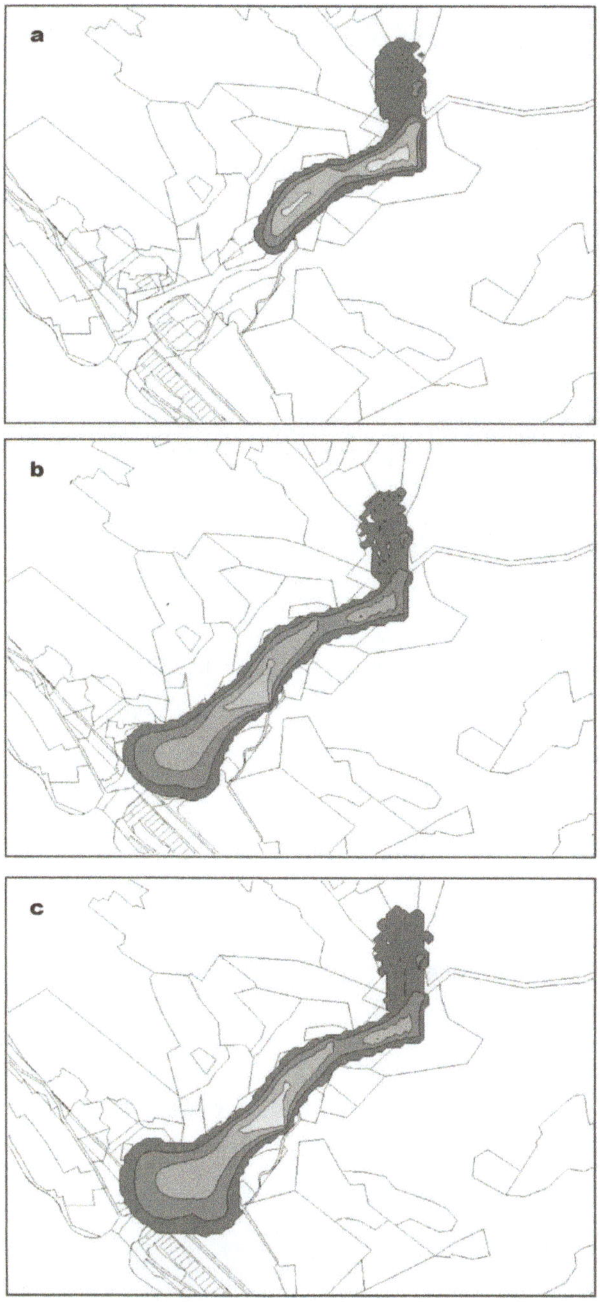

Fig. 26.8. The influenced area by debris flows under rainfall with different period. The influenced area will not increase for return period higher than 260 because there is no more debris sources available. **a** Influence area (return period 50 years); **b** influence area (return period 150 years); **c** influence area (return period 250 years)

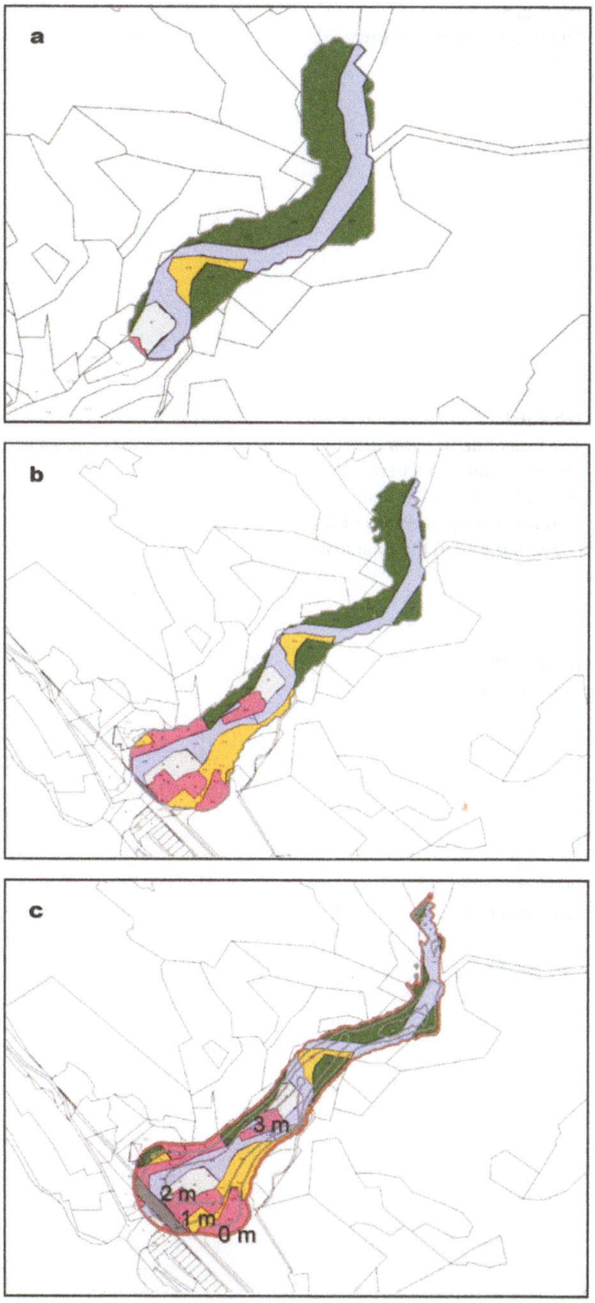

Fig. 26.9. The result of overlaying simulated debris flow area with land utilization information for different return periods. Different colors indicate different land use. *Green color* represent forest use. *Pink color* represent construction use. *Gray color* is other use. *Yellow color* is agriculture use. There is no land for tourism use and transportation use. **a** Land utilizing (return period 50 years); **b** land utilizing (return period 150 years); **c** land utilizing (return period 250 years)

26.6.4.2 150-Year Return Period

From the simulated result of Fig. 26.9b, we found that there is construction loss. Because the construction loss will influence the loss of human life, it needs to be calculated more accurately (Fig. 26.10). Based on the data of on-site investigation and standard in Table 26.2, the loss of human and construction can be counted and listed in Table 26.10.

Besides the loss of human and construction, other losses can be evaluated through Fig. 26.8b also. The results are listed in Table 26.11. Then, the total direct loss

Table 26.8. The loss for 50-year return period

Loss classification		Land use	Content of loss
Human life loss			None
Property loss	Land loss		None
	Loss above the ground	Construction	None
		Agricultural	None
		Forest	The loss area is 0.5 ha
		Other production	None
		Transportation and hydraulic	None

Table 26.9. The loss of debris flow disaster under rainfall of recurrence interval 50 years. The last column is the percentage of the particular loss to the total loss

Loss classification		Loss (NT dollars)	The price index (%)	Present value (2005)	Ratio
Property loss	Forest loss (FL)	5 112 (2003 data)	102.2	5 224	100
	Other production loss	0	0	0	0
Total loss		5 224			100

Table 26.10. The analysis of human and construction loss

The buried depth	Under 1 m	1~3 m	More than 3 m
House situation	Slightly damaged	Half destroy	Destroy completely
No. of houses	4	7	1
Human injury	No injury	Severely injured	Death
No. of people	5	9	2

Table 26.11. The debris flow loss under rainfall of 150 years return period

Loss classification		Land use	Content of loss
Human life loss			2 people die, 9 people are severely injured
Property loss	Land loss		None
	Loss above the ground	Construction	1 house was complete destroyed, 7 houses were half-destroyed
		Agricultural	None
		Forest	The loss area is 0.5 ha
		Other production	None
		Transportation and hydraulic	None

of debris flow hazard, under rainfall of 150-year return period, can be calculated in Table 26.12.

26.6.4.3 250-Year Return Period

From the simulated result in Fig. 26.9c, we can first find that the construction and human life losses (Fig. 26.11 and Table 26.13). Combining with field investigation of residents and the standard (Table 26.2), the loss of human lives and construction can be counted in Table 26.13. Other losses are calculated in Table 26.14 and monetized in Table 26.15.

26.6.4.4 300-Year Return Period

Corresponding to the maximum volume (23 077 m^3) and thus the maximum loss, it is 193 398 744 NT dollars.

26.6.5 The Risk Analysis of Debris Flow

After calculating the losses under different recurrence interval, we can use Eq. 26.12 to calculate the expected loss. The statistical distribution of the local rain gage

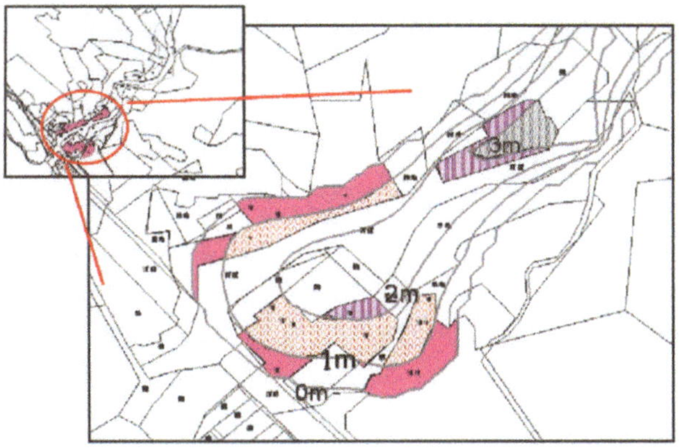

Fig. 26.10. Construction loss according to the depth of debris flow accumulation

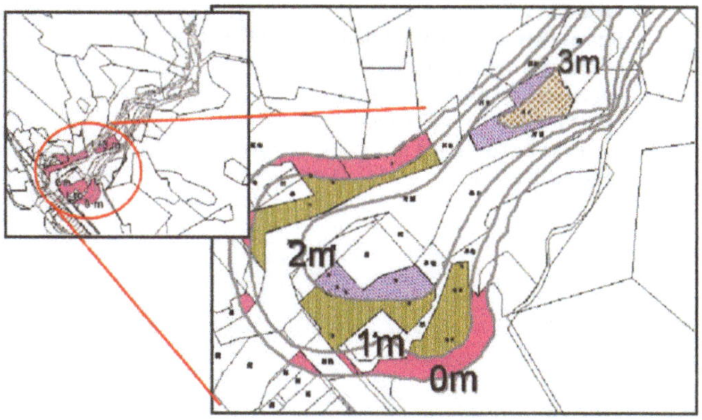

Fig. 26.11. Construction loss according to the depth of debris flow accumulation

Table 26.12. Total loss of debris flow disaster (rainfall return period 150 years). Forest loss data can be found only for 2003. Thus it is necessary to be converted to value of 2005

Loss classification		Loss (NT dollars)	The price index (%)	Present value (2005)	Ratio
Human life loss (HL)		134 862 000	–	134 862 000	98.40
Property loss	Construction loss (BL)	2 182 950	–	2 182 950	1.59
	Forest loss (FL)	5 112[a]	102.2	5 224	0.01
Total loss		137 050 174			100

[a] 2003 data.

Table 26.13. The loss under rainfall of 250 year return period

The buried depth	Under 1 m	1~3 m	More than 3 m
House situation	Slightly damaged	Half destroyed	Destroyed completely
No. of houses	4	10	1
Hunan injury	No injury	Severely injured	Death
No. of people	6	14	2

Table 26.14. The debris flow loss under rainfall of 250 years return period

Loss classification		Land use	Content of loss
Human life loss			2 people die, 14 people are severely injured
Property loss	Land loss		None
	Loss above the ground	Construction	1 house was complete destroyed, 10 houses were half-destroyed
		Agricultural	None
		Forest	The loss area is 0.33 ha
		Other production	None
		Transportation and hydraulic	The loss area of load is 0.06 ha. One bridge was complete destroyed

station can be found from Central Weather Bureau, the accumulated probability of rainfall of different recurrence interval $p(x_t)$ can then be computed. Using the regression formula of β_t (Fig. 26.5), the probability coefficient can be obtained. The corresponding values are listed in Table 26.16.

Finally, substituting the data in Table 26.16 into Eq. 26.12, the expected value of the loss can be calculated.

Table 26.15. Debris flow loss under rainfall of 250-year recurrence period. Forest loss data can be found only for 2003. Thus it is necessary to be converted to value of 2005

Loss classification		Loss (NT dollars)	The price index (%)	Present value (2005)	Ratio
Human life loss (HL)		186732000	–	186732000	96.55
Property loss	Construction loss (BL)	2910600	–	2910600	1.50
	Forest loss (FL)	3374[a]	102.2	3448	0.01
	Transportation and hydraulic loss (THL)	3750000	–	3750000	1.94
Total loss		193396048			100

[a] 2003 data.

Table 26.16. The loss under different recurence intervals

Recurrence interval	Loss	$p(x_t)$	β_t
50 Year C_{50}	5224	0.097	0.38
150 Year C_{150}	137050174	0.394	0.47
250 Year C_{250}	193396048	0.303	0.52
300 Year C_{300}	193398744	0.206	0.53

$$\begin{aligned}\text{expected loss} &\cong \beta_{50}p(x_{50})\,C_{50}+\beta_{150}p(x_{150})\,C_{150}\\&\quad+\beta_{250}p(x_{250})\,C_{250}+\beta_{300}p(x_{300})\,C_{300}\\&=76\,995\,900\text{ NT dollars (2.2 mio \$US)}\end{aligned}$$

This example uses loss value from only 4 different return periods to approximate the expected loss. For more accurate evaluation, more points should be used with the help of GIS system. This loss is close to the loss amount 2 million US dollars declared by the government in 2001. But we did not use any information from the actual event. The surprising accuracy may be coincidence, but this also implies the present statistical average approach is usable even before hazard actual occurs.

This procedure can also be used to evaluate the effect of any mitigation measures without any difficulties.

26.7 Conclusion

This paper describes an assessment method to quantify debris flow hazard loss. The numerically simulated debris flow hazard areas are first obtained for different rainfall return frequencies. Then the land utilization map is overlapped to find loss in different land for all return periods. The statistical average using debris flow occurrence coefficient gives the expected loss. This loss can be used as the defense magnitude for that area. The calculated result in our example is very close to government announced value.

The best benefit of this assessment method is that it can be used before the hazard actually occurs and can be applied to different mitigation measure plans. This feature greatly enhances our ability to finding the most effective mitigation measures.

Acknowledgment

This research is supported by National Science Council, Taiwan.

References

Breaden JP (1973) The generation of flood damage time sequences. University of Kentucky Water Resources Institute Paper, No.32

Fisher A, Violette D, Chestnut L (1989) The value of reducing risks of death: a note on new evidence. J Policy Anal Manag 8(1):88–100

Freeman AM III (1993) The measurement of environmental and resource values, theory and methods. Resources of the Future, Washington, D.C.

Grigg NS, Helweg OJ (1974) Estimating direct residential flood damage in urban areas. Colorado State University

Grigg NS, Helweg OJ (1975) State-of-the art of estimating flood damage in urban areas. Water Resour Bull 11(2):379–390

Grigg NS, Botham LH, Rice L, Shoemaker WJ, Tucker LS (1976) Urban drainage and flood control projects economic, legal and financial aspects, hydrology paper. Colorado State University, Fort Collins, Colorado, February

Julien PY, Lan Y (1991) Rheology of hyperconcentrations. J Hydraul Eng-ASCE 117:346–353

Laigle D, Marchi L (2000) Example of mud/debris-flow hazard assessment, using numerical models, debris-flow hazard mitigation. Mechanics, Prediction, and Assessment Conference, Taipei, pp 417–424

Lee CH (2006) Household debris-flow disaster damage assessment in Taiwan. National Taipei University, Master's thesis in Institute of Natural Resource Management

Liu KF, Huang MC (2006) Engineering planning of debris flow protection measure with numerical simulation. Eng Geosci 10:221–240

Liu KF, Li HC (2004) Application of externality on a hazard potency of debris flow. Bulletin of the College of Engineering National Taiwan University Eng 92:21–44

Liu KF, Li HC (2006) The study of the direct damage estimation of debris flow. Journal of Chinese Soil and Water Conservation 37(2):143–155

Liu JT, Hammitt JK, Liu JL (1997) Estimated hedonic wage function and value of life in a developing country. Econ Lett 57:353–358

Monthly Bulletin of Statistics (2005) Directorate-general of budget, accounting and statistics, executive Yuan, R.O.C, Vol. 420

Shih BJ, Jern KY, Shieh CL, Chen LJ (1997) A study on the hazardous debris flow zoning. In: Proceedings of the 1st National debris flow Conference, Taiwan, pp 141–155

Smith DI (1994) Flood damage estimation- A review of urban stage-damage curves and loss function. Water SA 20(3):231–238

Takahashi T (1980) Debris flow on prismatic open channel. J Hydr Eng Div-Asce 106(HY3):381–396

Tsai CT, Yu PS (1994) GIS system applied in flood disaster II. Department of Hydraulic and Ocean Engineering of National Cheng Kung University

Tsai CT, Yu PS (1995) GIS system applied in flood disaster III. Department of Hydraulic and Ocean Engineering of National Cheng Kung University

Xue LM, Wang SW (1987) Estimation of the value of life for employed workers in Taiwan: a wage-risk approach. Ching chi chuan lun, No.108

The Tokyo Action Plan

"2006 Tokyo Action Plan"

Strengthening Research and Learning on Landslides and Related Earth System Disasters for Global Risk Preparedness

Adopted in the Round Table Discussion on 20 January 2006 in Elizabeth Rose Hall of the United Nations University, Tokyo

The 2006 Tokyo Round Table Discussion "Strengthening Research and Learning on Earth System Risk Analysis and Sustainable Disaster Management within UN-ISDR as Regards Landslides" – towards a dynamic global network of the International Programme on Landslides (IPL) was held at the United Nations University, Tokyo, from 18 to 20 January 2006 to formulate a framework for cooperation and to identify focus areas to reduce landslide risk worldwide. The following action plan was adopted as a summary of the meeting, to be implemented within the scope of the Hyogo Framework for Action 2005–2015, "*Building the Resilience of Nations and Communities to Disasters*", declared at the United Nations World Conference on Disaster Reduction held in Kobe, Japan, in 2005.

A1.1 Preamble

Large and small landslides occur almost every year in nearly all regions of the world. Figure A1.1 shows an example for casualties due to natural disasters in Japan for 1967–2004. Landslide disasters in Japan have occurred every year; the total number of deaths due to landslides is about one half of those caused by earthquakes, including the catastrophic 1995 Kobe earthquake.

Landslides are a complex-disaster phenomenon that can be caused by earthquakes, volcanic eruptions, heavy rainfall (typhoons, hurricanes), sustained rainfall, heavy snowmelt, unregulated anthropogenic developments, mining, and other factors (Fig. A1.2a). Large-scale coastal or marine landslides are known to cause tsunami waves that kill many people; an example was the 1792 Unzen-Mayuyama landslide, which caused a devastating tsunami that resulted in 16 000 fatalities from the landslide and the tsunami in Japan. Also large-scale landslides on volcanoes can deform the mountain tops and trigger volcanic eruptions; such was the case for the 1980 eruption of Mount St. Helens in the USA and presumably for Mt. Bandai in Japan. Landslides also may occur without earthquakes, heavy rains, volcanic eruptions, or human activities due to progress of natural weathering; therefore, they occur almost everywhere in the world.

Landslides are natural phenomena that can only be effectively studied in an integrated, multi-disciplinary fashion, including contributions from natural and engineering sciences (earth and water sciences) and social sciences (Fig. A1.2b). In addition, landslides are strongly related to cultural heritage and the environment. Landslides should be jointly managed by cooperation of different ministries and departments of government, including those representing education, science and technology, construction and transportation, agriculture, forestry, the environment, culture and vulnerable groups (the poor, aged, handicapped, or children). Because landslides are highly localized phenomena, it is crucial to seek the contribution of local governments or autonomous communities (Fig. A1.2c).

The disasters caused by landslides are of very complex nature wherever they occur around the world. Research on landslides should be integrated into a new multi-disciplinary science of landslide study. Landslide risk preparedness is to be managed by multi-ministries.

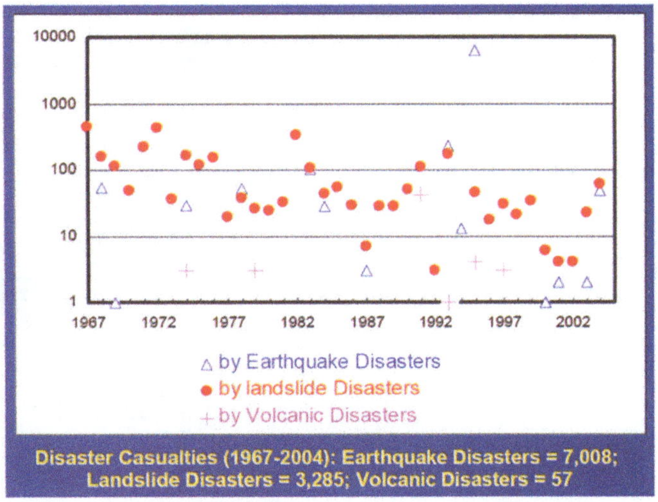

Fig. A1.1. Comparison of the numbers of victims in Japan from 1967–2004 due to landslide disasters, earthquake disasters, including deaths by earthquake-induced-landslides, and volcanic disasters, including deaths due to volcanic gases (This statistic of victims by landslide disasters since 1967 was published by the Sabo Technical Center)

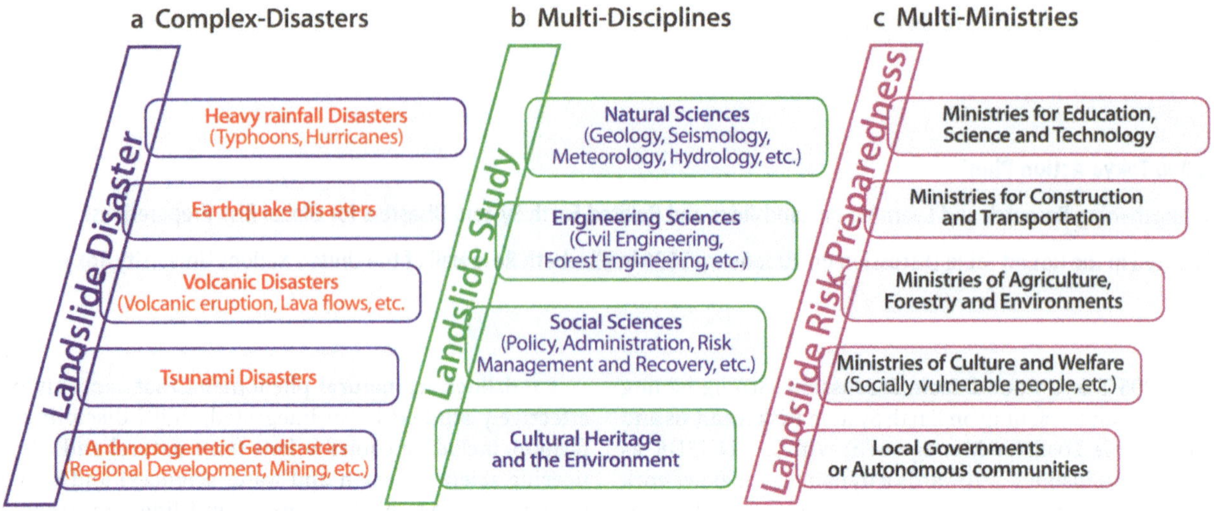

Fig. A1.2. Characteristics of landslide disasters

A1.2 Action Plan

Global cooperation in landslide risk-reduction research and learning will be carried out encompassing related disasters affecting the earth-system, such as heavy rainfall, earthquakes, volcanic eruptions, tsunamis, and disasters of anthropogenic origin. Establishment of a "Dynamic Global Network of the International Programme on Landslides" and its operation will effectively function for landslide and related risk reduction through the implementation of the following Action Items adopting a multi-hazard, multi-sectoral approach.

A1.2.3 Actions

I Establishment of the IPL Framework

1. *Establishment of the IPL Global Promotion Committee.* The IPL Global Promotion Committee shall be established by ICL members and ICL supporting organizations, as illustrated in Fig. A1.3. The committee will meet annually, on the occasion of ICL Board of Representative meetings, or possibly at other occasions and locations. The committee will conceive a strategy to promote the 2006 Tokyo Action Plan, and will

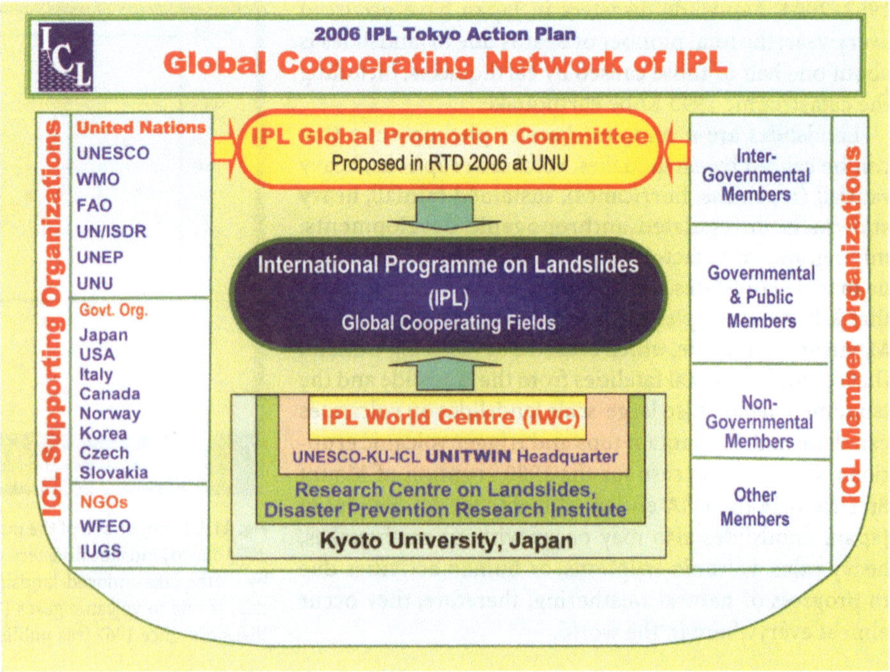

Fig. A1.3. Structure of the IPL global-cooperation framework

discuss the management of IPL global cooperation fields, and their possible modification, selection, and termination.

2. *Establishment of the IPL World Centre.* The IPL World Centre will be established to coordinate and support implementation of the global cooperating fields of the International Programme on Landslides (IPL), which works as the secretariat of the IPL Global Promotion Committee and the International Programme on Landslides (IPL). The Centre will be hosted by the Headquarter of the UNESCO-KU-ICL UNITWIN Cooperation Programme "Landslide Risk Mitigation for Society and the Environment" in the Research Centre on Landslides, Disaster Prevention Research Institute, Kyoto University, where the secretariat of the International Programme on Landslides has been located since its formation in 2002.

II Promotion of the Global Cooperating Fields of the International Programme on Landslides (IPL)

The global cooperating fields of the IPL are identified as follows for the initial phase:

1. *Technology Development*
   a Monitoring and Early Warning
      - Use of various on-site, in-situ technologies, as well as satellite observations, in monitoring landslide effects and contributing factors for early-warning purposes
      - Development of automated monitoring methods covering large spatial extent and real-time data communication, including low-cost monitoring devices
      - Development of early-warning methodologies, in particular for rainfall-induced landslides
      - Applications linking meteorological, hydrological, and landslide models
   b Hazard Mapping, Vulnerability, and Risk Assessment
      - Hazard mapping at local and global scales
      - Vulnerability assessment, considering human life, land resources, structures, infrastructure, and cultural heritage
      - Risk assessment and communicating risk in an easily understood manner
2. *Targeted Landslides: Mechanisms and Impacts*
   a Catastrophic Landslides
      - Catastrophic landslides induced by natural and anthropogenic factors, such as rainfall, earthquakes, volcanic activity, river erosion, and human activities, and their combinations
      - Landslides threatening human lives and societal values
      - Gigantic coastal landslides and marine landslides causing tsunamis
   b Landslides Threatening Heritage Sites
      - Studies for protection of cultural heritage, cultural landscape, and the natural heritage from landslides using non-invasive technologies and appropriate mitigation strategies (e.g., Machu Picchu, Bamiyan, Lishan, Cordillera Blanca)
3. *Capacity Building*
   a Enhancing Human and Institutional Capacities
      - Building human capacities and expertise in landslide management
      - Institution building at national and local levels through Centers of Excellence
      - Enhancing implementation and action at local levels
   b Collecting and Disseminating Information/Knowledge
      - Developing a culture of awareness on landslide risks
      - Developing model policy frameworks, standards, guidelines/checklists, and training modules.
4. *Preparedness, Mitigation, and Recovery*
   a Preparedness
      - Strengthening disaster preparedness of all stakeholders
      - Strengthening capacities of communities and local institutions to cope with landslide hazards
      - Forecasting and providing early warning of adverse conditions likely to lead to landslide activity
      - Preparing contingency recovery plans, including pre-positioning of technical and material resources for likely landslide events
   b Mitigation
      - Development of innovative, low-cost, and ecologically appropriate landslide mitigation techniques
      - Mountain conservation methods, including soil conservation, forest and watershed management, and appropriate land-use techniques
      - Appropriate civil engineering works, including construction and urban and coastal development
      - Restricting inappropriate development in landslide-prone areas
      - Development of appropriate policy and planning mechanisms, such as land-use management (including zoning)
      - Promotion and strengthening of monitoring and warning systems
   c Recovery
      - Post-landslide recovery and rebuilding efforts to integrate landslide mitigation measures
      - Prevention of secondary risks of landslides resulting from inappropriate re-building efforts in response to any disaster (for example, earthquakes, volcanic eruptions, extreme weather events, etc.)
      - Implementation of landslide recovery efforts and programmes (including psycho-social and health

aspects) with the participation of affected communities and local authorities
- Providing long-term support to ensure sustainable recovery

III Promotional Activities

1. *World Landslide Forum.* Capitalizing on the competence, international experience and established organizational network of ICL-IPL, it is proposed to create a global information platform for future joint activities of the worldwide landslide community, named the "World Landslide Forum" that shall be convened every 3 years.

   The first World Landslide Forum – organized by the ICL – can be planned to take place in January 2009, bringing together academics, practitioners, politicians, et al., to a global, multidisciplinary, problem-focused platform. This forum will provide an opportunity for identification of the first World Centre of Excellence on Landslides. Linkages to ISDR activities, as well as other global events, including the World Water Forum, the International Year of Planet Earth, etc., will be established.
2. *Identification and Promotion of World Centres of Excellence on Landslide Risk Reduction.* The IPL Global Promotion Committee will identify and promote World Centres of Excellence (WCoE) every 3 years within eligible organizations, such as universities, institutes, NGOs, government ministries and local governments, contributing to "Risk Reduction for Landslides and Related Earth System Disasters". Linkages to WCoEs at the national level will be used to promote cooperation with the ICL and dissemination of knowledge and information. An independent Panel of Experts, set up by the Global Promotion Committee of IPL, may be appointed to endorse the WCoEs.
3. *Contributions to Global Landslide Issues.* The IPL will mobilize global cooperation for strengthening research and learning on risk reduction for landslides and related Earth-system disasters at sites/areas identified as of great concern to the global community, such as Macchu-Picchu, Kashmir, the Central Asia high mountainous area, and Bamiyan.
4. *Partnerships.* Mutually beneficial partnerships with other global initiatives, such as the International Hydrological Program (IHP), the International Geoscience Program (IGCP), and The Mountain Partnership will be developed.

A1.3 After the 2006 Tokyo Action Plan

The first Memorandum of Understanding between the United Nations Secretariat for Disaster Risk Reduction (UN/ISDR) and the International Consortium on Landslides (ICL) to cooperate for the implementation of the 2006 Tokyo Action Plan until the year 2015 was established on 3 March 2006. Memorandum of Understanding with other global stakeholders are being prepared.

On 17 February 2006, a large landslide occurred on the island of Leyte, Philippines causing approximately 1 000 fatalities. The International Consortium on Landslides, the Disaster Prevention Research Institute, Kyoto University, with support from international organizations involved in the RTD, the Philippine Institute of Volcanology and Seismology, and others organized a mission to investigate the landslide on 19–26 March 2006 as a global landslide issue defined by the 2006 Tokyo Action Plan. Based on the results of this investigation, these organizations plan to propose a project of the International Programme on Landslides to reduce such landslide disasters. The first meeting of the IPL Global Promotion Committee (established in the RTD) and the IPL Symposium 2007– Landslide Risk Analysis and Sustainable Disaster Management– will be held at the United Nations University, Tokyo, on 23–25 January 2007. The IPL symposium will be open for everybody. Information on the first meeting of the IPL Global Promotion Committee and the IPL Symposium can be obtained from the ICL Web: *http://icl.dpri.kyoto-u.ac.jp/.*

Those who are interested in these activities are requested to contact the IPL World Centre:

IPL World Centre
UNESCO-KU-ICL UNITWIN Headquarter
Research Centre on Landslides, Disaster Prevention Research Institute,
Kyoto University, Uji, Kyoto, Japan
jimu@landslide.dpri.kyoto-u.ac.jp

References

Sassa K (2004a) Preface. Landslides 1(1):1–5
Sassa K (2004b) The international consortium on landslides. Landslides 1(1):91–94
Sassa K (2004c) The international programme on landslides. Landslides 1(2):95–99

MoUs between ICL and Global Stakeholders to Promote the 2006 Tokyo Action Plan

ICL has exchanged the Memorandum of Understanding to promote the 2006 Tokyo Action Plan with global stakeholders related to landslides: the United Nations Secretariat for International Strategy for Disaster Risk Reduction (UN/ISDR), the United Nations University (UNU), World Meteorological Organization (WMO), World Federation of Engineering Organisations (WFEO), United Nations Educational, Scientific and Cultural Organization (UNESCO), and International Council for Science (ICSU) in 2006.

A2.1 MoU between ICL and UN/ISDR

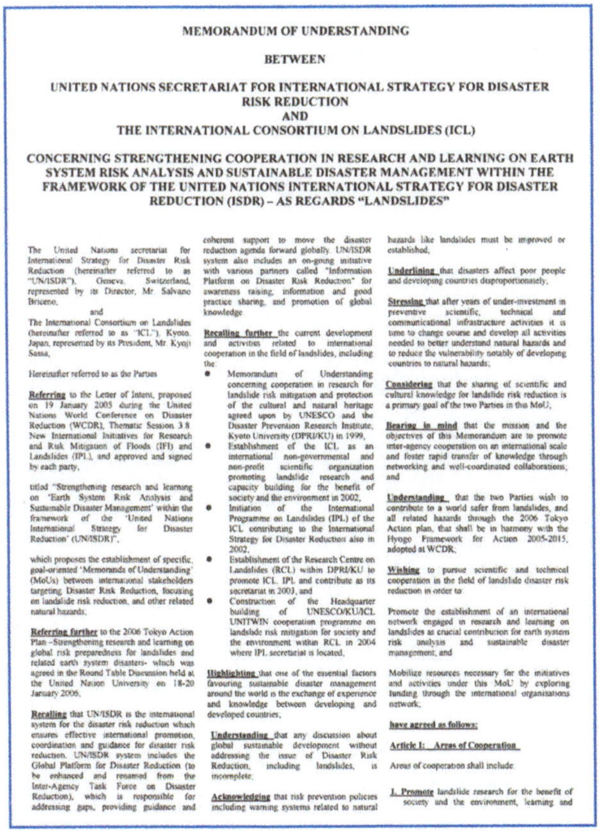

MEMORANDUM OF UNDERSTANDING

BETWEEN

UNITED NATIONS SECRETARIAT FOR INTERNATIONAL STRATEGY FOR DISASTER RISK REDUCTION

AND

THE INTERNATIONAL CONSORTIUM ON LANDSLIDES (ICL)

CONCERNING STRENGTHENING COOPERATION IN RESEARCH AND LEARNING ON EARTH SYSTEM RISK ANALYSIS AND SUSTAINABLE DISASTER MANAGEMENT WITHIN THE FRAMEWORK OF THE UNITED NATIONS INTERNATIONAL STRATEGY FOR DISASTER REDUCTION (ISDR) – AS REGARDS "LANDSLIDES"

The United Nations secretariat for International Strategy for Disaster Risk Reduction (hereinafter referred to as "UN/ISDR"), Geneva, Switzerland, represented by its Director, Mr. Salvano Briceno,

and

The International Consortium on Landslides (hereinafter referred to as "ICL"), Kyoto, Japan, represented by its President, Mr. Kyoji Sassa,

Hereinafter referred to as the Parties

Referring to the Letter of Intent, proposed on 19 January 2005 during the United Nations World Conference on Disaster Reduction (WCDR), Thematic Session 3.8 New International Initiatives for Research and Risk Mitigation of Floods (IFI) and Landslides (IPL), and approved and signed by each party,

titled "Strengthening research and learning on 'Earth System Risk Analysis and Sustainable Disaster Management' within the framework of the 'United Nations International Strategy for Disaster Reduction' (UN/ISDR)",

which proposes the establishment of specific, goal-oriented 'Memoranda of Understanding' (MoUs) between international stakeholders targeting Disaster Risk Reduction, focusing on landslide risk reduction, and other related natural hazards,

Referring further to the 2006 Tokyo Action Plan –Strengthening research and learning on global risk preparedness for landslides and related earth system disasters- which was agreed in the Round Table Discussion held at the United Nation University on 18-20 January 2006,

Recalling that UN/ISDR is the international system for the disaster risk reduction which ensures effective international promotion, coordination and guidance for disaster risk reduction. UN/ISDR system includes the Global Platform for Disaster Reduction (to be enhanced and renamed from the Inter-Agency Task Force on Disaster Reduction), which is responsible for addressing gaps, providing guidance and coherent support to move the disaster reduction agenda forward globally. UN/ISDR system also includes an on-going initiative with various partners called "Information Platform on Disaster Risk Reduction" for awareness raising, information and good practice sharing, and promotion of global knowledge.

Recalling further the current development and activities related to international cooperation in the field of landslides, including the:

- Memorandum of Understanding concerning cooperation in research for landslide risk mitigation and protection of the cultural and natural heritage agreed upon by UNESCO and the Disaster Prevention Research Institute, Kyoto University (DPRI/KU) in 1999,
- Establishment of the ICL as an international non-governmental and non-profit scientific organization promoting landslide research and capacity building for the benefit of society and the environment in 2002,
- Initiation of the International Programme on Landslides (IPL) of the ICL contributing to the International Strategy for Disaster Reduction also in 2002,
- Establishment of the Research Centre on Landslides (RCL) within DPRI/KU to promote ICL, IPL and contribute as its secretariat in 2003, and
- Construction of the Headquarter building of UNESCO/KU/ICL UNITWIN cooperation programme on landslide risk mitigation for society and the environment within RCL in 2004 where IPL secretariat is located,

Highlighting that one of the essential factors favouring sustainable disaster management around the world is the exchange of experience and knowledge between developing and developed countries,

Understanding that any discussion about global sustainable development without addressing the issue of Disaster Risk Reduction, including landslides, is incomplete,

Acknowledging that risk prevention policies including warning systems related to natural hazards like landslides must be improved or established,

Underlining that disasters affect poor people and developing countries disproportionately,

Stressing that after years of under-investment in preventive scientific, technical and communicational infrastructure activities it is time to change course and develop all activities needed to better understand natural hazards and to reduce the vulnerability notably of developing countries to natural hazards,

Considering that the sharing of scientific and cultural knowledge for landslide risk reduction is a primary goal of the two Parties in this MoU,

Bearing in mind that the mission and the objectives of this Memorandum are to promote inter-agency cooperation on an international scale and foster rapid transfer of knowledge through networking and well-coordinated collaborations; and

Understanding that the two Parties wish to contribute to a world safer from landslides, and all related hazards through the 2006 Tokyo Action plan, that shall be in harmony with the Hyogo Framework for Action 2005-2015, adopted at WCDR,

Wishing to pursue scientific and technical cooperation in the field of landslide disaster risk reduction in order to:

Promote the establishment of an international network engaged in research and learning on landslides as crucial contribution for earth system risk analysis and sustainable disaster management; and

Mobilize resources necessary for the initiatives and activities under this MoU by exploring funding through the international organizations network,

have agreed as follows:

Article I: Areas of Cooperation

Areas of cooperation shall include:

1. Promote landslide research for the benefit of society and the environment, learning and

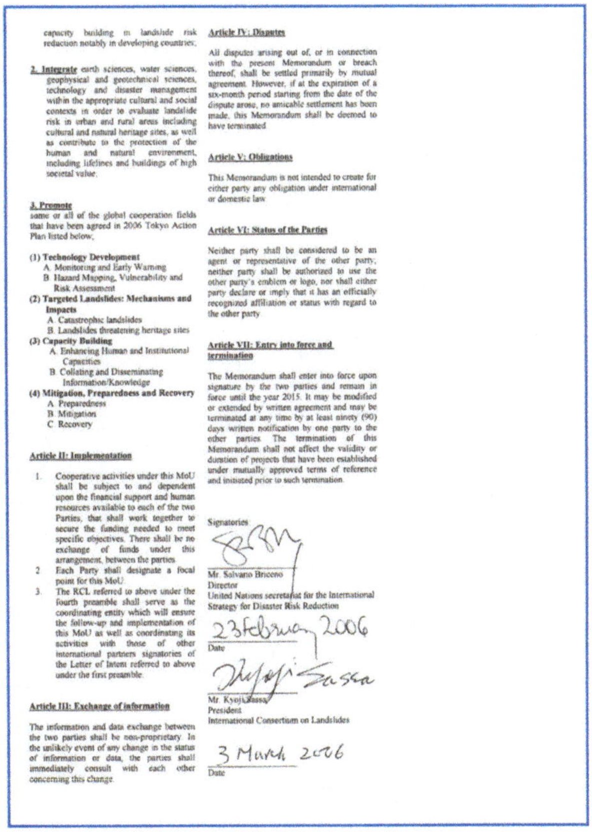

capacity building in landslide risk reduction notably in developing countries;

2. Integrate earth sciences, water sciences, geophysical and geotechnical sciences, technology and disaster management within the appropriate cultural and social contexts in order to evaluate landslide risk in urban and rural areas including cultural and natural heritage sites, as well as contribute to the protection of the human and natural environment, including lifelines and buildings of high societal value.

3. Promote
some or all of the global cooperation fields that have been agreed in 2006 Tokyo Action Plan listed below,

(1) Technology Development
 A. Monitoring and Early Warning
 B. Hazard Mapping, Vulnerability and Risk Assessment
(2) Targeted Landslides: Mechanisms and Impacts
 A. Catastrophic landslides
 B. Landslides threatening heritage sites
(3) Capacity Building
 A. Enhancing Human and Institutional Capacities
 B. Collating and Disseminating Information/Knowledge
(4) Mitigation, Preparedness and Recovery
 A. Preparedness
 B. Mitigation
 C. Recovery

Article II: Implementation

1. Cooperative activities under this MoU shall be subject to and dependent upon the financial support and human resources available to each of the two Parties, that shall work together to secure the funding needed to meet specific objectives. There shall be no exchange of funds under this arrangement, between the parties.
2. Each Party shall designate a focal point for this MoU.
3. The RCL referred to above under the fourth preamble shall serve as the coordinating entity which will ensure the follow-up and implementation of this MoU as well as coordinating its activities with those of other international partners signatories of the Letter of Intent referred to above under the first preamble.

Article III: Exchange of information

The information and data exchange between the two parties shall be non-proprietary. In the unlikely event of any change in the status of information or data, the parties shall immediately consult with each other concerning this change.

Article IV: Disputes

All disputes arising out of, or in connection with the present Memorandum or breach thereof, shall be settled primarily by mutual agreement. However, if at the expiration of a six-month period starting from the date of the dispute arose, no amicable settlement has been made, this Memorandum shall be deemed to have terminated.

Article V: Obligations

This Memorandum is not intended to create for either party any obligation under international or domestic law.

Article VI: Status of the Parties

Neither party shall be considered to be an agent or representative of the other party; neither party shall be authorized to use the other party's emblem or logo, nor shall either party declare or imply that it has an officially recognized affiliation or status with regard to the other party.

Article VII: Entry into force and termination

The Memorandum shall enter into force upon signature by the two parties and remain in force until the year 2015. It may be modified or extended by written agreement and may be terminated at any time by at least ninety (90) days written notification by one party to the other parties. The termination of this Memorandum shall not affect the validity or duration of projects that have been established under mutually approved terms of reference and initiated prior to such termination.

Signatories

Mr. Salvano Briceno
Director
United Nations secretariat for the International Strategy for Disaster Risk Reduction

23 February 2006
Date

Mr. Kyoji Sassa
President
International Consortium on Landslides

3 March 2006
Date

A2.2 MoU between ICL and UNU

UNITED NATIONS UNIVERSITY

MEMORANDUM OF UNDERSTANDING
BETWEEN
THE INTERNATIONAL CONSORTIUM ON LANDSLIDES
AND
THE UNITED NATIONS UNIVERSITY

CONCERNING STRENGTHENING COOPERATION IN RESEARCH AND LEARNING ON EARTH SYSTEM RISK ANALYSIS AND SUSTAINABLE DISASTER MANAGEMENT WITHIN THE FRAMEWORK OF THE UNITED NATIONS INTERNATIONAL STRATEGY FOR DISASTER REDUCTION – AS REGARDS "LANDSLIDES"

The International Consortium on Landslides (hereinafter referred to as "ICL"), Kyoto, Japan, represented by its President, Mr. Kyoji Sassa, and

The United Nations University (hereinafter referred to as "UNU"), Tokyo, Japan, represented by its Rector, Prof. J. A. van Ginkel,

Hereinafter referred to as the "Parties"

Referring to the Letter of Intent proposed on 19 January 2005 during the United Nations World Conference on Disaster Reduction (WCDR), Thematic Session 3.8: New International Initiatives for Research and Risk Mitigation of Floods (IFI) and Landslides (IPL), and approved and signed by each Party,

titled "Strengthening research and learning on 'Earth System Risk Analysis and Sustainable Disaster Management' within the framework of the 'United Nations International Strategy for Disaster Reduction' (UN/ISDR)",

which proposes the establishment of specific, goal-oriented "Memoranda of Understanding" (MoUs) between international stakeholders targeting Disaster Risk Reduction, focusing on landslide risk reduction, and other related natural hazards;

Referring further to the 2006 Tokyo Action Plan – Strengthening research and learning on global risk preparedness for landslides and related earth system disasters – which was agreed upon in the Round Table discussion held at the United Nation University on 18-20 January 2006;

Recalling that UN/ISDR is the international system for the disaster risk reduction which ensures effective international promotion, coordination and guidance for disaster risk reduction. UN/ISDR system includes the Global Platform for Disaster Reduction (enhanced and renamed from the Inter-Agency Task Force on Disaster Reduction), which is responsible for addressing gaps, providing guidance and coherent support to move the disaster reduction agenda forward globally. UN/ISDR system also includes an on-going initiative with various partners called "Information Platform on Disaster Risk Reduction" for awareness raising, information and good practice sharing, and promotion of global knowledge;

- 2 -

Recalling further the current development and activities related to international cooperation in the field of landslides, including the:

- Memorandum of Understanding concerning cooperation in research for landslide risk mitigation and protection of the cultural and natural heritage agreed upon by UNESCO and the Disaster Prevention Research Institute, Kyoto University (DPRI/KU) in 1999,
- Establishment of the ICL as an international non-governmental and non-profit scientific organization promoting landslide research and capacity building for the benefit of society and the environment in 2002,
- Initiation of the International Programme on Landslides (IPL) of the ICL contributing to the International Strategy for Disaster Reduction also in 2002,
- Establishment of the Research Centre on Landslides (RCL) within DPRI/KU to promote ICL, IPL and contribute as its secretariat in 2003, and
- Construction of the Headquarters building of UNESCO/KU/ICL UNITWIN cooperation programme on landslide risk mitigation for society and the environment within RCL in 2004 where IPL secretariat is located;

Highlighting that one of the essential factors favouring sustainable disaster management around the world is the exchange of experience and knowledge between developing and developed countries;

Understanding that any discussion about global sustainable development without addressing the issue of Disaster Risk Reduction, including landslides, is incomplete;

Acknowledging that risk prevention policies including warning systems related to natural hazards, such as landslides must be improved or established;

Underlining that disasters affect poor people and developing countries disproportionately;

Stressing that after years of under-investment in preventive scientific, technical and communicational infrastructure activities, it is time to change course and develop all activities needed to better understand natural hazards and to reduce the vulnerability notably of developing countries to natural hazards;

Considering that the sharing of scientific and cultural knowledge for landslide risk reduction is a primary goal of the two Parties;

Bearing in mind that the mission and the objectives of this Memorandum of Understanding (MoU) are to promote inter-agency cooperation on an international scale and foster rapid transfer of knowledge through networking and well-coordinated collaborations;

Understanding that the two Parties wish to contribute to a world safer from landslides, and all related hazards through the 2006 Tokyo Action plan, that shall be in harmony with the Hyogo Framework for Action 2005-2015, which was adopted at WCDR;

- 3 -

Wishing to pursue scientific and technical cooperation in the field of landslide disaster risk reduction in order to:

- Promote the establishment of an international network engaged in research and learning on landslides as crucial contribution for earth system risk analysis and sustainable disaster management; and
- Mobilize resources necessary for the initiatives and activities under this MoU by exploring funding opportunities through the international organizations network;

HAVE AGREED as follows:

Article I: Areas of Cooperation

Areas of cooperation shall include:

1. Promote landslide research for the benefit of society and the environment, learning and capacity building in landslide risk reduction notably in developing countries;
2. Integrate earth sciences, water sciences, geophysical and geotechnical sciences, technology and disaster management within the appropriate cultural and social contexts in order to evaluate landslide risk in urban and rural areas including cultural and natural heritage sites, as well as contribute to the protection of the human and natural environment, including lifelines and buildings of high societal value;
3. Promote some or all of the global cooperation fields that have been agreed in 2006 Tokyo Action Plan, as listed below:

(1) **Technology Development**
A. Monitoring and Early Warning
B. Hazard Mapping, Vulnerability and Risk Assessment

(2) **Targeted Landslides: Mechanisms and Impacts**
A. Catastrophic landslides
B. Landslides threatening heritage sites

(3) **Capacity Building**
A. Enhancing Human and Institutional Capacities
B. Collating and Disseminating Information/Knowledge

(4) **Mitigation, Preparedness and Recovery**
A. Preparedness
B. Mitigation
C. Recovery

- 4 -

Article II: Implementation

4. Cooperative activities under this MoU shall be subject to and dependent upon the financial support and human resources available to each of the two Parties who shall work together to secure the funding needed to meet specific objectives;
5. Each Party shall designate a focal point for this MoU; and
6. The RCL referred to above under the fourth preambular paragraph shall serve as the coordinating entity which will ensure the follow-up and implementation of this MoU, as well as coordinating its activities with those of other international partners/signatories of the Letter of Intent referred to above under the first preamble.

Article III: Exchange of information

The information and data exchange between the Parties shall be non-proprietary. In the unlikely event of any change in the status of information or data, the Parties shall immediately consult with each other concerning this change.

Article IV: Disputes

All disputes arising out of, or in connection with the present MoU or breach thereof, shall be settled primarily by mutual agreement. However, if at the expiration of a six-month period starting from the date of the dispute arose and no amicable settlement has been made, this MoU shall be deemed to have terminated.

Article V: Obligations

This MoU is not intended to impose on either Party any obligation under international or domestic law.

Article VI: Status of the Parties

Neither Party shall be considered to be an agent or representative of the other Party; nor authorize to use the other Party's emblem or logo, nor declare or imply that it has an officially recognized affiliation or status with regard to the other Party.

Article VII: Waiver

Nothing in this MoU shall be deemed to be a waiver, express or implied, of the privileges or immunities of the United Nations or of the UNU which is an organ thereof.

Article VIII: Entry into force and termination

The MoU shall enter into force upon signature by the two Parties and remain in force until 31 December 2015. It may be modified or extended by written agreement and may be terminated at

- 5 -

any time by at least ninety (90) days written notification by either Party. The termination of this MoU shall not affect the validity or duration of projects that have been established under mutually approved terms of reference and initiated prior to such termination. The Parties shall agree on measures required for the orderly conclusion of ongoing activities.

Kyojji Sassa
President
International Consortium on Landslides

17-04-06
Date

J. A. van Ginkel
Rector
United Nations University

10-04-06
Date

A2.3 MoU between ICL and WMO

MEMORANDUM OF UNDERSTANDING

BETWEEN

THE WORLD METEOROLOGICAL ORGANIZATION
AND
THE INTERNATIONAL CONSORTIUM ON LANDSLIDES

The World Meteorological Organization (hereinafter referred to as "WMO"), Geneva, Switzerland, represented by its Secretary-General, Mr. Michel Jarraud,

and

The International Consortium on Landslides (hereinafter referred to as "ICL"), Kyoto, Japan, represented by its President, Mr. Kyoji Sassa,

Hereinafter referred to as the Parties

Referring to the Letter of Intent, proposed on 19 January 2005 during the United Nations World Conference on Disaster Reduction (WCDR), Thematic Session 3.8: New International Initiatives for Research and Risk Mitigation of Floods (IFI) and Landslides (IPL), and approved and signed by each party,

titled "Strengthening research and learning on 'Earth System Risk Analysis and Sustainable Disaster Management' within the framework of the 'United Nations International Strategy for Disaster Reduction' (UN/ISDR)",

which proposes the establishment of specific, goal-oriented 'Memoranda of Understanding' (MoUs) between international stakeholders targeting Disaster Risk Reduction, focusing on landslide risk reduction, and other related natural hazards;

Referring further to the 2006 Tokyo Action Plan -Strengthening research and learning on global risk preparedness for landslides and related earth system disasters- which was agreed in the Round Table Discussion held at the United Nations University on 18-20 January 2006;

Recalling that UN/ISDR is the international system for the disaster risk reduction which ensures effective international promotion, coordination and guidance for disaster risk reduction. UN/ISDR system includes the Global Platform for Disaster Reduction (to be enhanced and renamed from the Inter-Agency Task Force on Disaster Reduction), which is responsible for addressing gaps, providing guidance and coherent support to move the disaster reduction agenda forward globally. UN/ISDR system also includes an on-going initiative with various partners called "Information Platform on Disaster Risk Reduction" for awareness raising, information and good practice sharing, and promotion of global knowledge.

Recalling further the current development and activities related to international cooperation in the field of landslides, including the:

- Establishment of the ICL as an international non-governmental and non-profit scientific organization promoting landslide research and capacity building for the benefit of society and the environment in 2002,
- Memorandum of Understanding concerning cooperation for the promotion of sciences of meteorology, hydrology and landslide for the safety of human life and property, protection of the environment, sustainable economic and social development, and education and training in relevant fields agreed upon by WMO and ICL in 2002,
- Initiation of the International Programme on Landslides (IPL) of the ICL contributing to the International Strategy for Disaster Reduction also in 2002, and
- Establishment of the Research Centre on Landslides (RCL) within DPRI/KU to promote ICL, IPL and contribute as its secretariat in 2003;

Highlighting that one of the essential factors favouring sustainable disaster management around the world is the exchange of experience and knowledge between developing and developed countries;

Understanding that any discussion about global sustainable development without addressing the issue of Disaster Risk Reduction, including landslides, is incomplete;

Acknowledging that risk prevention policies including warning systems related to natural hazards like landslides must be improved or established;

Underlining that disasters affect poor people and developing countries disproportionately,

Stressing that after years of under-investment in preventive scientific, technical and communicational infrastructure activities it is time to change course and develop all activities needed to better understand natural hazards and to reduce the vulnerability notably of developing countries to natural hazards;

Considering that the sharing of scientific knowledge for landslide risk reduction is a primary goal of the two Parties in this MoU;

Bearing in mind that the mission and the objectives of this Memorandum are to promote inter-agency cooperation on an international scale and foster rapid transfer of knowledge through networking and well-coordinated collaborations; and

Understanding that the two Parties wish to contribute to a world safer from landslides, and all related hazards through the 2006 Tokyo Action plan, that shall be in harmony with the Hyogo Framework for Action 2005-2015, adopted at WCDR.

Wishing to pursue scientific and technical cooperation in the field of landslide disaster risk reduction in order to:

Promote the establishment of an international network engaged in research and learning on landslides as crucial contribution for earth system risk analysis and sustainable disaster management; and

Mobilize resources necessary for the initiatives and activities under this MoU by exploring funding through the international organizations network;

have agreed as follows:

Article I: Objective

Areas of cooperation shall include:

1. **Promote** landslide research for the benefit of society and the environment, learning and capacity building in landslide risk reduction notably in developing countries,

2. **Integrate** atmospheric sciences, earth sciences, water sciences, geophysical and geotechnical sciences, technology and disaster management within the appropriate cultural and social contexts in order to evaluate landslide risk in urban and rural areas including cultural and natural heritage sites, as well as contribute to the protection of the human and natural environment, including lifelines and buildings of high societal value,

Article II: Areas of Cooperation

Promote some or all of the global cooperation fields that have been agreed in 2006 Tokyo Action Plan listed below;

(1) Technology Development
A. Monitoring and Early Warning
B. Hazard Mapping, Vulnerability and Risk Assessment
(2) Targeted Landslides: Mechanisms and Impacts
A. Catastrophic landslides
B. Landslides threatening heritage sites
(3) Capacity Building
A. Enhancing Human and Institutional Capacities
B. Collating and Disseminating Information/Knowledge
(4) Mitigation, Preparedness and Recovery
A. Preparedness
B. Mitigation
C. Recovery

Article III: Implementation

1. Cooperative activities under this MoU shall be subject to and dependent upon the financial support and human resources available to each of the two Parties, that shall work together to secure the funding needed to meet specific objectives.
2. Each Party shall designate a focal point for this MoU.
3. The RCL referred to above under the fourth preamble shall serve as the coordinating entity which will ensure the follow-up and implementation of this MoU as well as coordinating its activities with those of other international partners signatories of the Letter of Intent referred to above under the first preamble.

Article IV: Exchange of information

The information and data exchange between the two parties shall be non-proprietary. In the unlikely event of any change in the status of information or data, the parties shall immediately consult with each other concerning this change.

Article V: Disputes

All disputes arising out of, or in connection with the present Memorandum or breach thereof, shall be settled primarily by mutual agreement. However, if at the expiration of a six-month period starting from the date of the dispute arose, no amicable settlement has been made, this Memorandum shall be deemed to have terminated.

Article VI: Obligations

This Memorandum is not intended to create for either party any obligation under international or domestic law.

Article VII: Status of the Parties

Neither party shall be considered to be an agent or representative of the other party; neither party shall be authorized to use the other party's emblem or logo, nor shall either party declare or imply that it has an officially recognized affiliation or status with regard to the other party.

Article VIII: Entry into force and termination

The Memorandum shall enter into force upon signature by the two parties and remain in force until the year 2015. It may be modified or extended by written agreement and may be terminated at any time by at least ninety (90) days written notification by one party to the other parties. The termination of this Memorandum shall not affect the validity or duration of projects that have been established under mutually approved terms of reference and initiated prior to such termination.

Mr. Michel Jarraud
Secretary-General
World Meteorological Organization

30 June 2006
Date

Mr. Kyoji Sassa
President
International Consortium on Landslides

24 July 2006
Date

A2.4 MoU between ICL and WFEO

MEMORANDUM OF UNDERSTANDING

BETWEEN

THE WORLD FEDERATION OF ENGINEERING ORGANIZATIONS

AND

THE INTERNATIONAL CONSORTIUM ON LANDSLIDES

CONCERNING STRENGTHENING COOPERATION IN RESEARCH AND LEARNING ON EARTH SYSTEM RISK ANALYSIS AND SUSTAINABLE DISASTER MANAGEMENT WITHIN THE FRAMEWORK OF THE UNITED NATIONS INTERNATIONAL STRATEGY FOR DISASTER REDUCTION REGARDING THE IMPLEMENTATION OF THE 2006 TOKYO ACTION PLAN ON LANDSLIDES

The World Federation of Engineering Organizations (hereinafter referred to as "WFEO"), Paris, France, represented by its President, Mr Kamel Ayadi,
and
The International Consortium on Landslides (hereinafter referred to as "ICL"), Kyoto, Japan, represented by its President, Mr Kyoji Sassa,

Hereinafter referred to as "the Parties",

Referring to the Letter of Intent entitled "Strengthening research and learning on 'Earth System Risk Analysis and Sustainable Disaster Management' within the framework of the 'United Nations International Strategy for Disaster Reduction' (UN/ISDR)", which was proposed on 19 January 2005 during the United Nations World Conference on Disaster Reduction (WCDR), Thematic Session 3.8: New International Initiatives for Research and Risk Mitigation of Floods (IFI) and Landslides (IPL), and signed by the Parties, and which proposes the establishment of specific, goal-oriented Memoranda of Understanding (MoUs) between international stakeholders targeting disaster risk reduction, focusing on landslide risk reduction, and other related natural hazards;

Referring further to the 2006 Tokyo Action Plan on "Strengthening Research and Learning on Landslides and Related Earth System Disasters for Global Risk Preparedness", which was agreed in the Round Table Discussion held at the United Nations University from 18 to 20 January 2006;

Recalling that UN/ISDR is the international system for disaster risk reduction which ensures effective international promotion, coordination and guidance; that the UN/ISDR system includes the Global Platform for Disaster Reduction, which is responsible for addressing gaps, providing guidance and coherent support to move the disaster reduction agenda forward globally; that UN/ISDR system also includes an ongoing initiative with various partners called the Information Platform on Disaster Risk Reduction for awareness-raising, information and good practice sharing, and promotion of global knowledge;

Recalling further the current development and activities related to international cooperation in the field of landslides, including:

the Memorandum of Understanding concerning cooperation in research for landslide risk mitigation and protection of the cultural and natural heritage agreed by UNESCO and the Disaster Prevention Research Institute, Kyoto University (DPRI/KU), in 1999;

the establishment of the ICL in 2002 as an international non-governmental and non-profit scientific organization promoting landslide research and capacity-building for the benefit of society and the environment;

the initiation of the International Programme on Landslides (IPL) of ICL contributing to the ISDR, also in 2002;

the establishment of the Research Centre on Landslides (RCL) within DPRI/KU to promote ICL, IPL and contribute as its secretariat in 2003;

and the construction of the headquarters building of the UNESCO/KU/ICL UNITWIN cooperation programme on landslide risk mitigation for society and the environment within RCL in 2004, containing the IPL secretariat;

Highlighting that one of the essential factors favouring sustainable disaster management around the world is the exchange of experience and knowledge between developing and developed countries;

Understanding that any discussion about global sustainable development without addressing the issue of disaster risk reduction, including landslides, is incomplete;

Acknowledging that risk prevention policies, including warning systems related to natural hazards like landslides, must be established or improved;

Underlining that disasters affect poor people and developing countries in particular;

Stressing that, after years of underinvestment in preventive scientific, technical and communication infrastructure activities, it is time to change course and develop all activities needed to understand better natural hazards and reduce the vulnerability of developing countries notably to natural hazards;

Considering that the sharing of scientific and cultural knowledge for landslide risk reduction is a primary goal of the Parties in this MoU;

Bearing in mind that the mission and the objectives of this MoU are to promote inter-agency cooperation on an international scale and foster rapid transfer of knowledge through networking and well-coordinated collaborations; and

Understanding that the Parties wish to contribute to a world safer from landslides and all related hazards through the 2006 Tokyo Action Plan, which shall be in harmony with the Hyogo Framework for Action 2005-2015, adopted at WCDR;

Wishing to pursue scientific and technical cooperation in the field of landslide disaster risk reduction in order to:

2

promote the establishment of an international network engaged in research and learning on landslides as crucial contribution for earth system risk analysis and sustainable disaster management;

and mobilize resources necessary for the initiatives and activities under this MoU by exploring funding through the network of international organizations;

have agreed as follows:

Article I: Areas of Cooperation

WFEO and ICL shall cooperate in the following areas:

(1) the promotion of landslide research for the benefit of society and the environment, learning and capacity-building in landslide risk reduction, notably in developing countries;

(2) the integration of earth sciences, water sciences, geophysical and geotechnical sciences, technology and disaster management within the appropriate cultural and social contexts in order to evaluate landslide risk in urban and rural areas, including cultural and natural heritage sites, as well as to contribute to the protection of the human and natural environment, including lifelines and buildings of high societal value;

(3) the promotion of some or all of the global cooperation fields that have been agreed in 2006 Tokyo Action Plan listed below:

(1) Technology Development
A. Monitoring and Early Warning
B. Hazard Mapping, Vulnerability and Risk Assessment

(2) Targeted Landslides: Mechanisms and Impacts
A. Catastrophic Landslides
B. Landslides threatening Heritage Sites

(3) Capacity-Building
A. Enhancing Human and Institutional Capacities
B. Collating and Disseminating Information/Knowledge

(4) Mitigation, Preparedness and Recovery
A. Preparedness
B. Mitigation
C. Recovery

Article II: Implementation

Cooperative activities under this MoU shall be subject to a separate agreement and are dependent upon the financial support and human resources available to each of the Parties, which shall work together to secure the funding needed to meet specific objectives.
Each Party shall designate a focal point for this MoU.

3

The RCL referred to above under the fourth paragraph of the preamble shall serve as the coordinating entity which will ensure the follow-up and implementation of this MoU as well as coordinating its activities with those of other international partners signatories of the Letter of Intent referred to above under the first paragraph of the preamble.

Article III: Exchange of information

The information and data exchange in the context of this MoU between the Parties shall be non-proprietary. In the unlikely event of any change in the status of information or data, the Parties shall immediately consult with each other concerning this change.

Article IV: Disputes

All disputes arising out of, or in connection with the present MoU or breach thereof, shall be settled primarily by mutual agreement. However, if at the expiration of a six-month period starting from the date of the beginning of the dispute no amicable settlement has been made, this MoU shall be deemed to have terminated.

Article V: Status of the Parties

Neither party shall be considered to be an agent or representative of the other party; neither party shall use the other party's name or logo without its prior written consent, nor shall either party declare or imply that it has an officially recognized affiliation or status with regard to the other party.

Article VI: Entry into force and termination

The MoU shall enter into force upon signature by the Parties and remain in force until 31 December 2015. It may be modified or extended by written agreement. It may be terminated at any time with at least ninety (90) days written notification by one party to the other party. The termination of this MoU shall not affect the validity or duration of projects that have been established under mutually approved terms of reference and initiated prior to such termination.

Mr. Kamel Ayadi President World Federation of Engineering Organizations	Mr. Kyoji Sassa President International Consortium on Landslides
10. July 2006 Date	4 July 2006 Date

4

A2.5 MoU between ICL and UNESCO

MEMORANDUM OF UNDERSTANDING

BETWEEN

THE UNITED NATIONS EDUCATIONAL, SCIENTIFIC AND CULTURAL ORGANIZATION

AND

THE INTERNATIONAL CONSORTIUM ON LANDSLIDES

CONCERNING STRENGTHENING COOPERATION IN RESEARCH AND LEARNING ON EARTH SYSTEM RISK ANALYSIS AND SUSTAINABLE DISASTER MANAGEMENT WITHIN THE FRAMEWORK OF THE UNITED NATIONS INTERNATIONAL STRATEGY FOR DISASTER REDUCTION REGARDING THE IMPLEMENTATION OF THE 2006 TOKYO ACTION PLAN ON LANDSLIDES

The United Nations Educational, Scientific and Cultural Organization (hereinafter referred to as "UNESCO"), Paris, France, represented by its Director-General, Mr Koïchiro Matsuura,
and
The International Consortium on Landslides (hereinafter referred to as "ICL"), Kyoto, Japan, represented by its President, Mr Kyoji Sassa,

Hereinafter referred to as "the Parties",

Referring to the Letter of Intent entitled "Strengthening research and learning on 'Earth System Risk Analysis and Sustainable Disaster Management' within the framework of the 'United Nations International Strategy for Disaster Reduction' (UN/ISDR)", which was proposed on 19 January 2005 during the United Nations World Conference on Disaster Reduction (WCDR), Thematic Session 3.8: New International Initiatives for Research and Risk Mitigation of Floods (IFI) and Landslides (IPL), and signed by the Parties, and which proposes the establishment of specific, goal-oriented Memoranda of Understanding (MoUs) between international stakeholders targeting disaster risk reduction, focusing on landslide risk reduction, and other related natural hazards;

Referring further to the 2006 Tokyo Action Plan on "Strengthening Research and Learning on Landslides and Related Earth System Disasters for Global Risk Preparedness", which was agreed in the Round Table Discussion held at the United Nations University from 18 to 20 January 2006;

Recalling that UN/ISDR is the international system for disaster risk reduction which ensures effective international promotion, coordination and guidance; that the UN/ISDR system includes the Global Platform for Disaster Reduction, which is responsible for addressing gaps, providing guidance and coherent support to move the disaster reduction agenda forward globally; that UN/ISDR system also includes an ongoing initiative with various partners called the Information Platform on Disaster Risk Reduction for awareness-raising, information and good practice sharing, and promotion of global knowledge;

Recalling further the current development and activities related to international cooperation in the field of landslides, including:

the Memorandum of Understanding concerning cooperation in research for landslide risk mitigation and protection of the cultural and natural heritage agreed by UNESCO and the Disaster Prevention Research Institute, Kyoto University (DPRI/KU), in 1999;

the establishment of the ICL in 2002 as an international non-governmental and non-profit scientific organization promoting landslide research and capacity-building for the benefit of society and the environment;

the initiation of the International Programme on Landslides (IPL) of ICL contributing to the ISDR, also in 2002;

the establishment of the Research Centre on Landslides (RCL) within DPRI/KU to promote ICL, IPL and contribute as its secretariat in 2003;

and the construction of the headquarters building of the UNESCO/KU/ICL UNITWIN cooperation programme on landslide risk mitigation for society and the environment within RCL in 2004, containing the IPL secretariat;

Highlighting that one of the essential factors favouring sustainable disaster management around the world is the exchange of experience and knowledge between developing and developed countries;

Understanding that any discussion about global sustainable development without addressing the issue of disaster risk reduction, including landslides, is incomplete;

Acknowledging that risk prevention policies, including warning systems related to natural hazards like landslides, must be established or improved;

Underlining that disasters affect poor people and developing countries in particular;

2

Stressing that, after years of underinvestment in preventive scientific, technical and communication infrastructure activities, it is time to change course and develop all activities needed to understand better natural hazards and reduce the vulnerability of developing countries notably to natural hazards;

Considering that the sharing of scientific and cultural knowledge for landslide risk reduction is a primary goal of the Parties in this MoU;

Bearing in mind that the mission and the objectives of this MoU are to promote inter-agency cooperation on an international scale and foster rapid transfer of knowledge through networking and well-coordinated collaborations; and

Understanding that the Parties wish to contribute to a world safer from landslides and all related hazards through the 2006 Tokyo Action Plan, which shall be in harmony with the Hyogo Framework for Action 2005-2015, adopted at WCDR;

Wishing to pursue scientific and technical cooperation in the field of landslide disaster risk reduction in order to:

promote the establishment of an international network engaged in research and learning on landslides as crucial contribution for earth system risk analysis and sustainable disaster management;

and mobilize resources necessary for the initiatives and activities under this MoU by exploring funding through the network of international organizations;

have agreed as follows:

Article I: Areas of Cooperation

UNESCO and ICL shall cooperate in the following areas:

(1) the promotion of landslide research for the benefit of society and the environment, learning and capacity-building in landslide risk reduction, notably in developing countries;

(2) the integration of earth sciences, water sciences, geophysical and geotechnical sciences, technology and disaster management within the appropriate cultural and social contexts in order to evaluate landslide risk in urban and rural areas, including cultural and natural heritage sites, as well as to contribute to the protection of the human and natural environment, including lifelines and buildings of high societal value;

(3) the promotion of some or all of the global cooperation fields that have been agreed in 2006 Tokyo Action Plan listed below:

3

(1) Technology Development
A. Monitoring and Early Warning
B. Hazard Mapping, Vulnerability and Risk Assessment

(2) Targeted Landslides: Mechanisms and Impacts
A. Catastrophic Landslides
B. Landslides threatening Heritage Sites

(3) Capacity-Building
A. Enhancing Human and Institutional Capacities
B. Collating and Disseminating Information/Knowledge

(4) Mitigation, Preparedness and Recovery
A. Preparedness
B. Mitigation
C. Recovery

Article II: Implementation

Cooperative activities under this MoU shall be subject to a separate agreement and are dependent upon the financial support and human resources available to each of the Parties, which shall work together to secure the funding needed to meet specific objectives.
Each Party shall designate a focal point for this MoU.

The RCL referred to above under the fourth paragraph of the preamble shall serve as the coordinating entity which will ensure the follow-up and implementation of this MoU as well as coordinating its activities with those of other international partners signatories of the Letter of Intent referred to above under the first paragraph of the preamble.

Article III: Exchange of information

The information and data exchange in the context of this MoU between the Parties shall be non-proprietary. In the unlikely event of any change in the status of information or data, the Parties shall immediately consult with each other concerning this change.

Article IV: Disputes

All disputes arising out of, or in connection with the present MoU or breach thereof, shall be settled primarily by mutual agreement. However, if at the expiration of a six-month period starting from the date of the beginning of the dispute no amicable settlement has been made, this MoU shall be deemed to have terminated.

4

Article V: Status of the Parties

Neither party shall be considered to be an agent or representative of the other party; neither party shall use the other party's name or logo without its prior written consent, nor shall either party declare or imply that it has an officially recognized affiliation or status with regard to the other party.

Article VI: Entry into force and termination

The MoU shall enter into force upon signature by the Parties and remain in force until 31 December 2015. It may be modified or extended by written agreement. It may be terminated at any time with at least ninety (90) days written notification by one party to the other party. The termination of this MoU shall not affect the validity or duration of projects that have been established under mutually approved terms of reference and initiated prior to such termination.

Mr. Koïchiro Matsuura	Mr. Kyoji Sassa
Director-General	President
United Nations Educational, Scientific and Cultural Organization	International Consortium on Landslides
22/08/06	22/08/06
Date	Date

5

A2.6 MoU between ICL and ICSU

MEMORANDUM OF UNDERSTANDING

BETWEEN

THE INTERNATIONAL COUNCIL FOR SCIENCE

AND

THE INTERNATIONAL CONSORTIUM ON LANDSLIDES

CONCERNING STRENGTHENING COOPERATION IN RESEARCH AND LEARNING ON EARTH SYSTEM RISK ANALYSIS AND SUSTAINABLE DISASTER MANAGEMENT WITHIN THE FRAMEWORK OF THE UNITED NATIONS INTERNATIONAL STRATEGY FOR DISASTER REDUCTION AS REGARDS LANDSLIDES

The International Council for Science (hereinafter referred to as "ICSU"), Paris, France, represented by its Executive-Director, Prof. Thomas Rosswall,

and

the International Consortium on Landslides (hereinafter referred to as "ICL"), Kyoto, Japan, represented by its President, Prof. Kyoji Sassa,

hereinafter referred to as "the Parties",

Referring to the Letter of Intent signed by ICSU and other partners entitled "Strengthening research and learning on 'Earth System Risk Analysis and Sustainable Disaster Management' within the framework of the United Nations International Strategy for Disaster Reduction' (UN/ISDR)", which was proposed on 19 January 2005 during the United Nations World Conference on Disaster Reduction (WCDR), Thematic Session 3.8: New International Initiatives for Research and Risk Mitigation of Floods (IFI) and Landslides (IPL), and signed by the Parties, and which proposes the establishment of specific, goal-oriented Memoranda of Understanding (MoUs) between international stakeholders targeting disaster risk reduction, focusing on landslide risk reduction, and other related natural hazards;

Referring further to the 2006 Tokyo Action Plan on "Strengthening Research and Learning on Landslides and Related Earth System Disasters for Global Risk Preparedness", which was agreed at the Round Table Discussion held at the United Nations University from 18 to 20 January 2006;

Recalling that UN/ISDR is the international system for disaster risk reduction which ensures effective international promotion, coordination and guidance; that the UN/ISDR system includes the Global Platform for Disaster Reduction, which is responsible for addressing gaps, providing guidance and coherent support to move the disaster reduction agenda forward globally; that the UN/ISDR system also includes an ongoing initiative with various partners called the Information Platform on Disaster Risk Reduction for awareness-raising, information and good practice sharing, and promotion of global knowledge;

Recalling further the current development of, and activities related to, international cooperation in the field of natural and human-induced environmental hazards and disasters, and of landslides, including:

- the establishment of the ICL in 2002 as an international non-governmental and non-profit scientific organization promoting landside research and capacity-building for the benefit of society and the environment;

- the initiation of the International Programme on Landslides (IPL) of ICL contributing to the ISDR, also in 2002; and

- the creation, in 2006, of an ICSU Planning Group charged with preparations for the establishment of an international, interdisciplinary programme on natural and human-induced environmental hazards and disasters;

Highlighting that one of the essential factors favouring sustainable disaster management around the world is the exchange of experience and knowledge between developing and developed countries;

Understanding that any discussion about global sustainable development without addressing the issue of disaster risk reduction, including landslides, is incomplete;

Acknowledging that risk prevention policies, including warning systems related to natural hazards like landslides, must be established or improved;

Underlining that disasters affect poor people and developing countries in particular;

Stressing that, after years of underinvestment in preventive scientific, technical and communication infrastructure activities, it is time to change course and develop all activities needed to understand better natural hazards and reduce the vulnerability of developing countries, notably to natural hazards;

Considering that the sharing of scientific and cultural knowledge for landslide risk reduction is a primary goal of the Parties in this MoU;

Bearing in mind that the mission and the objectives of this MoU are to promote inter-agency cooperation on an international scale and foster the rapid transfer of knowledge through networking and well-coordinated collaborations; and

Understanding that the two Parties wish to contribute to a world safer from landslides and all related hazards through the 2006 Tokyo Action Plan, which shall be in harmony with the Hyogo Framework for Action 2005-2015, adopted at WCDR;

Wishing to pursue scientific and technical cooperation in the field of landslide disaster risk reduction in order to:

- promote the establishment of an international network engaged in research and learning on landslides as a crucial contribution for earth system risk analysis and sustainable disaster management; and

- mobilize the resources necessary for the initiatives and activities under this MoU by exploring funding through the network of international organizations;

have agreed as follows:

2

Article I: Areas of Cooperation

ICSU and ICL shall cooperate in the following areas:

(a) the promotion of landslide research for the benefit of society and the environment, learning and capacity-building in landslide risk reduction, notably in developing countries;

(b) the integration of earth sciences, including water and meteorological sciences, geophysical and geotechnical sciences, as well as biological, technological and social sciences relevant for disaster management, in order to evaluate landslide risk in urban and rural areas, as well as to contribute to the protection of the human and natural environment;

(c) the promotion of some or all of the fields for global cooperation that have been agreed in the 2006 Tokyo Action Plan listed below:

(1) Technology Development
A. Monitoring and Early Warning
B. Hazard Mapping, Vulnerability and Risk Assessment

(2) Targeted Landslides: Mechanisms and Impacts
A. Catastrophic Landslides
B. Landslides threatening Heritage Sites

(3) Capacity-Building
A. Enhancing Human and Institutional Capacities
B. Collating and Disseminating Information/Knowledge

(4) Mitigation, Preparedness and Recovery
A. Preparedness
B. Mitigation
C. Recovery

Article II: Implementation

Cooperative activities under this MoU shall be subject to a separate agreement and are dependent upon the financial support and human resources available to each of the Parties, which shall work together to secure the funding needed to meet specific objectives.

Each Party shall designate a focal point for this MoU.

The Research Centre on Landslides (RCL) established within the Disaster Prevention Research Institute, Kyoto University, Japan in 2003, shall serve as the coordinating entity which will ensure the follow-up and implementation of this MoU, as well as coordinating its activities with those of other international partners signatories of the Letter of Intent referred to above under the first paragraph of the preamble.

Article III: Exchange of information

All information and data exchange in the context of this MoU between the Parties shall be non-proprietary. In the unlikely event of any change in the status of information or data, the Parties shall immediately consult with each other concerning this change.

3

Article IV: Disputes

All disputes arising out of, or in connection with, the present MoU or breach thereof, shall be settled primarily by mutual agreement. However, if at the expiration of a six-month period starting from the date of the beginning of the dispute no amicable settlement has been reached, this MoU shall be deemed to have terminated.

Article V: Status of the Parties

Neither party shall be considered to be an agent or representative of the other party; neither party shall use the other party's name or logo without its prior written consent, nor shall either party declare or imply that it has an officially recognized affiliation or status with regard to the other party.

Article VI: Entry into force and termination

The MoU shall enter into force upon signature by the Parties and remain in force until 31 December 2015. It may be modified or extended by written agreement. It may be terminated at any time with at least ninety (90) days' written notification by one party to the other party. The termination of this MoU shall not affect the validity or duration of projects that have been established under mutually approved terms of reference and initiated prior to such termination.

Prof. Thomas Rosswall Executive-Director International Council for Science	Prof. Kyoji Sassa President International Consortium on Landslides
Date 6 December 2006	Date 01 / 12 / 2006

4

Index

International Consortium on Landslides

An international non-government and non-profit scientific organization promoting landslide research and capacity building for the benefit of society and the environment

President: Kyoji Sassa (Kyoto University, Japan)
Vice Presidents: Peter Bobrowsky (Geological Survey of Canada) / Paolo Canuti (University of Firenze, Italy)
Oddvar Kjekstad (International Centre for Geohazards, Norway) / Peter Lyttle (U.S. Geological Survey, USA)

ICL Supporting Organizations:

The United Nations Educational, Scientific and Cultural Organization (UNESCO) / The World Meteorological Organization (WMO) / The Food and Agriculture Organization of the United Nations (FAO) / The United Nations International Strategy for Disaster Reduction Secretariat (UN/ISDR) / The United Nations University (UNU) / International Union of Geological Sciences (IUGS) / Governments of Japan, USA, Italy, Canada, and Norway

ICL Members:

Geological Survey of Canada / Chinese Academy of Sciences, Chengdu Institute of Mountain Hazards and Environment / Jilin University, Environmental Geological Disaster Research Institute, China / Xi'an Municipal Government, Lishan Landslide Prevention and Control Office, China / Northeast Forestry University, China / Xi'an Jiaotong University, Department of Civil Engineering, China / China Three Gorges University, Institute of Geological Hazard Mitigation / China Geological Survey / Universidad Nacional de Columbia, Columbia / Charles University, Faculty of Science, Czech Republic / Technische Universitat Darmstadt, Institute and Laboratory of Geotechnics, Germany / Mekelle University, Ethiopia / Ministry of Energy and Mineral Resources, Geological Agency, the Republic of Indonesia / Building & Housing Research Center, Iran / International Institute of Earthquake Engineering and Seismology (IIEES), Iran / Soil Conservation and Watershed Management Research Institute (SCWMRI), Iran / ENEA (Italian Agency for New Technologies Energy and Environment), Italy / European Commission's Joint Research Centre, IPSC/HSU, Italy / Instituto Nationale di Oceanografia e di Geofisica Sperimentale - OGS, Italy / University of Firenze, Earth Sciences Department, Italy / Ehime University, Faculty of Engineering, Japan / Forestry and Forest Product Research Institute, Japan / Geographical Survey Institute, Japan / Japan Landslide Society / Kyoto University, Disaster Prevention Research Institute, Research Section for Innovative Disaster Prevention Technology and Policy, Japan / Kyoto University, Disaster Prevention Research Institute, Research Centre on Landslides, Japan / Niigata University, Research Center for Natural Hazards and Disaster Recovery, Japan / University of Tokyo, Department of Civil Engineering, Geotechnical Engineering Group, Japan / University of Tokyo, Institute of Industrial Science, Japan / Korea Institute of Geoscience and Mineral Resources (KIGAM) / Mara University of Technology, Malaysia / International Centre for Integrated Mountain Development (ICIMOD), Nepal / International Centre for Geohazards (ICG) in Oslo, Norway / Instituto Geologico Minero y Metalurgico (INGEMMET), Peru / Grudec Ayar, Peru / Proexrom S.R.L. Technical University, Civil Engineering Faculty, Romania / Federal State Unitary Geological Enterprise Scientific Centre HydGeo, Russia / Russian Academy of Sciences, Institute of Environmental Geoscience (IEG RAS) / Russian Academy of Sciences, Institute of the Geospheres Dynamics / Unified Energy System of Russia, Open Joint-Stock Company Engineering Centre / Comenius University, Faculty of Natural Sciences, Department of Engineering Geology, Slovakia / Swedish Geotechnical Institute / Swiss Federal Institute for Snow and Avalanche Research (SLF) / Ministry of Agriculture and Cooperatives, Land Development Department, Thailand / U.S. Geological Survey, USA / National Aeronautics and Space Administration (NASA), Goddard Space Flight Center, Laboratory for Atmospheres, USA

ICL Supporters:

Marui & Co., Ltd., Osaka, Japan / Eight Consultants Co., Ltd., Okayama, Japan / Geo-Research Institute, Osaka, Japan / GODAI Development Corp., Kanazawa, Japan / Japan Conservation Engineering & Co., Ltd., Tokyo / Kokusai Kogyo Co., Ltd., Tokyo, Japan / Kowa Co., Ltd., Niigata, Japan / MEISAI Co., Suita, Japan / Nippon Koei Co., Ltd., Tokyo, Japan / Ohta Geo-Research Co., Ltd., Nishinomiya, Japan / OSASI Technos Inc., Kochi, Japan / OYO Corporation, Tokyo, Japan / Research Center of Computational Mechanics, Inc., Tokyo, Japan / Sabo Technical Center, Tokyo, Japan / Sakata Denki Co., Ltd., Tokyo, Japan

ICL Secretariat:

Research Centre on Landslides, Disaster Prevention Research Institute, Kyoto University, Uji, Kyoto 611-0011, Japan
Web: http://ICL.dpri.kyoto-u.ac.jp/, E-mail: jimu@landslide.dpri.kyoto-u.ac.jp
Tel: +81-774-38-4110, Fax: +81-774-32-5597

MIX
Papier aus verantwortungsvollen Quellen
Paper from responsible sources
FSC® C105338

If you have any concerns about our products,
you can contact us on
ProductSafety@springernature.com

In case Publisher is established outside the EU,
the EU authorized representative is:
Springer Nature Customer Service Center GmbH
Europaplatz 3, 69115 Heidelberg, Germany

Printed by Libri Plureos GmbH
in Hamburg, Germany